Probability and Statistical Physics in Two and More Dimensions

Clay Mathematics Proceedings
Volume 15

Probability and Statistical Physics in Two and More Dimensions

Proceedings of the Clay Mathematics Institute Summer School and XIV Brazilian School of Probability Búzios, Brazil July 11–August 7, 2010

David Ellwood
Charles Newman
Vladas Sidoravicius
Wendelin Werner
Editors

American Mathematical Society
Clay Mathematics Institute

2000 *Mathematics Subject Classification.* Primary 60–06, 60G60, 60K35, 60K37, 82B20, 82B27, 82B28, 82B41, 82B43, 82B44.

Cover image courtesy of Jason Miller.

Library of Congress Cataloging-in-Publication Data

Clay Mathematics Institute, Summer School (2010 : Armação dos Búzios, Brazil)
Probability and statistical physics in two and more dimensions : Clay Mathematics Institute Summer School and XIV Brazilian School of Probability, Búzios, Brazil, July 11–August 7, 2010 / David Ellwood... [et al.], editors.
p. cm. — (Clay mathematics proceedings ; v. 15)
Includes bibliographical references.
ISBN 978-0-8218-6863-8 (alk. paper)
1. Probabilities—Congresses. I. Ellwood, D. (David), 1966– II. Brazilian School of Probability (14th : 2010 : Armação dos Búzios, Brazil) III. Title.

QC174.85.P76C53 2010
519.2—dc23

2012017930

Published by the American Mathematical Society, Providence, RI,
for the Clay Mathematics Institute, Cambridge, MA.
Printed in the United States of America.

♾ The paper used in this book is acid-free and falls within the guidelines
established to ensure permanence and durability.
Visit the AMS home page at http://www.ams.org/
Visit the Clay Mathematics Institute home page at http://www.claymath.org/

10 9 8 7 6 5 4 3 2 1 17 16 15 14 13 12

Contents

Preface

The Clay Mathematics Institute 2010 Summer School, "Probability and Statistical Physics in Two and More Dimensions" took place in Búzios, Brazil from July 11 to August 7. The final week was a joint event with the XIV Escola Brasileira da Probabilidade.

In the past ten to fifteen years, various areas of probability theory related to statistical physics, disordered systems and combinatorics have undergone intensive development. A number of these developments deal with two-dimensional random structures at their critical points, and provide new tools and ways of coping with at least some of the limitations of Conformal Field Theory (CFT) that had been so successfully developed in the theoretical physics community to understand phase transitions of two-dimensional systems.

One of the new ideas that emerged in the mathematics community just before the new millenium is the Stochastic Loewner Evolution (SLE), introduced by Oded Schramm. This new approach is probabilistic in nature and focuses directly on non-local structures that characterize a given system, such as cluster boundaries in Ising, Potts and percolation models, or loops in the O(n) model. At criticality, these become, in the continuum limit, random curves whose distributions can be uniquely identified thanks to their conformal invariance and a certain Markovian property. There is a one-parameter family of SLE's indexed by a positive real number κ, and they appear to be the only possible candidates for the scaling limits of interfaces of two-dimensional critical systems that are conformally invariant.

A complementary approach has been to understand and control discrete models that exhibit discrete holomorphic features. These now include several important models, such as critical percolation on the triangular lattice, the critical Ising model and its related random cluster model, loop-erased random walks and double-dimer models. Some of these results are very recent–or even ongoing–developments. They make it possible to prove that indeed, these discrete models give rise to Schramm's SLE curves in the large-scale limit, and to provide a detailed description of various aspects of their large-scale behavior.

Different questions correspond to the case where one considers these same models from statistical physics on certain natural planar graphs that are themselves random (they are often called "planar maps") - here conformality is not obvious to formulate, but the combinatorics of the problems turn out to be more tractable. This has led to spectacular recent progress, and the proof of several results in this discrete approach to what is often referred to as "quantum gravity". It is interesting to note that another approach to quantum gravity builds on the Gaussian Free Field, which is another conformally invariant continuous model, that has also been recently shown to be directly related to the SLE processes (and SLE(4) in particular).

We thus believed it was a good time for a school that would provide a complete picture of the current state of the art in these topics and discuss the relations between them as well as other possible future directions.

The School offered three long Foundational Courses: Beffara's course provided an introduction to Schramm's SLE processes and their properties. Garban and Steif's course gave an account of recent results concerning the scaling limit of percolation, and its relation to noise-sensitivity, and the course by Le Gall and Miermont focused on the description of planar maps via tree-like structures and their large-scale limits.

Five advanced mini-courses covered further topics on this theme: Smirnov focused on the conformal invariance of critical percolation and of the critical Ising model, while Kenyon described the conformal invariance of another discrete model, called the double-dimer model. Sheffield described aspects of the relation between the Gaussian Free Field and SLE processes, while Lawler focused on finer studies of the SLE processes themselves. Di Francesco provided an approach to the combinatorial structures related to integrable systems. Courses by Slade (on self-avoiding walks) and by den Hollander (on polymers) were presented during the final week, jointly with the Brazilian School of Probability.

In addition to all these courses, research seminars organized by young participants and evening lectures by prominent senior researchers took place. Given the enormous range of subjects covered during the School and the diversity of scientific topics, it would be pointless to say more about the contents here, but we believe that the high quality of the lectures is reflected in these pages.

Foundational Courses

- SLE and other conformally invariant objects, Vincent Beffara.
- Noise-sensitivity and percolation, Christophe Garban and Jeffrey Steif.
- Large random planar maps and their scaling limits, Jean-François Le Gall and Grégory Miermont.

Mini-Courses

- Random geometry and Gaussian free field, Scott Sheffield.
- Conformal invariance of lattice models, Stanislav Smirnov.
- Integrable combinatorics, Philippe Di Francesco.
- Fractal and multifractal properties of SLE, Gregory Lawler.
- The double dimer model, Rick Kenyon.

Courses joint with XIV Escola Brasileira da Probabilidade

- Random polymers, Frank den Hollander.
- Self-avoiding walks, Gordon Slade.

A School of such scale could not have happened without the generous support of numerous sponsors and the efforts of many individuals. Besides CMI funding, IMPA, CNPq (Brazil) and NSF-PIRE, were major contributors to the budget of the school, we are grateful to all the other foundations who provided financial support.

All sponsors are listed at
http://www.impa.br/opencms/pt/eventos/store_old/evento_1007

We would like to express our gratitude to Professors C. Camacho, C. Aragao and J. Palis for their constant support and also our many thanks to the whole administrative support teams of CMI (Amanda Battese and Katherine Brack) and DAC of IMPA, especially DACs coordinator Suely Lima and Pedro Faro for their personal efforts.

Last but not least, the editors would like to give special recognition to CMI's publications manager Vida Salahi for her work and dedication in managing the editorial process of this volume.

David Alexandre Ellwood, Chuck Newman, Vladas Sidoravicius and Wendelin Werner

April 2012

Clay Mathematics Proceedings
Volume 15, 2012

Schramm-Loewner Evolution and other Conformally Invariant Objects

Vincent BEFFARA

CONTENTS

2010 *Mathematics Subject Classification.* Primary: 60J67; Secondary: 60K35.

FOREWORD

These notes are not meant as a reference manual, but rather as an introduction combined with a kind of "user's guide" to the existing bibliography. I plan to keep them mostly self-contained in the sense that the reader will need no additional information to understand the majority of the statements; but they contain essentially no detailed proofs. In the case of very important results, I give indications about the main ideas of the demonstration, but of course that is hardly sufficient to a motivated student.

In each part, the most important section is therefore the extended bibliography at the end. I chose to gather all bibliographical references there, and to omit them from the main body of the text (in particular, the main results are only attributed to their respective authors, not to a particular publication). The point is to make reading through the text more natural; maybe it failed!

The notes were started while I was giving a graduate course in Lyon during the spring preceding the school. As a result, they cover a certain quantity of material in addition to what will was discussed in Buzios (mostly the parts about random-cluster models and convergence to SLE, which correspond more to Smirnov's course). These can constitute indications towards further reading, or can be ignored completely in a first reading.

For reference, here is a rough outline of the course schedule in Buzios; the contents of the exercise sessions matched these. However, to avoid too much overlap between these notes and the others from the school, and to make them more focused, some of the material is not included here (for instance, the exercise sheet about Brownian intersection exponents was left out).

- Course 1: Percolation and Cardy's formula.
- Course 2: Loop-erased random walks and uniform spanning trees.
- Course 3: Loewner chains in the radial case.
- Course 4: Chordal Loewner chains, and definition of SLE.
- Course 5: First properties of SLE.
- Course 6: The locality property and SLE_6.
- Course 7: The restriction property, $SLE_{8/3}$ and restriction measures.
- Course 8: More exotic objects: CLE, loop soups, Gaussian fields...

Part I . A FEW DISCRETE MODELS

Introduction

The goal of these lectures is to provide a self-contained introduction to SLE and related objects, but some motivation is needed before introducing SLE as such; so it seems natural to start with a quick review of a few two-dimensional discrete models.

The focus of this part will be, for each model, to arrive at the question of scaling limits as quickly as possible, and to justify conformal invariance where it is known to hold in the limit. The proofs of actual convergence to SLE will of course have to be postponed (see Part III) — but providing the key arguments is our main objective here.

I.1. Lattice models

We start with what we want to call *lattice models* — even though that might not exactly be the usual sense of that word. Essentially, given a (two-dimensional) lattice embedded in the plane, a *configuration* is a map from the set of vertices and/or edges of the lattice into a finite alphabet, and a probability measure on the set of configurations is constructed by taking a thermodynamical limit from measures in finite boxes derived from a Hamiltonian.

We choose to limit ourselves to a few representative models, namely percolation, the Ising and Potts models, and the random-cluster model. The uniform spanning tree (UST) is an important case because it was one of the first two-dimensional models for which convergence to SLE was proved; we will briefly come back to it in the next section in association with the loop-erased random-walk.

Besides, we will mostly be interested in models taken at their critical point, and defined on specific lattices for which more is understood about their asymptotic behavior (*e.g.*, we limit our description of percolation to the case of site percolation on the triangular lattice) — even though of course a lot is known in a more general setting.

I.1.1. Percolation. The simplest lattice model to describe is *Bernoulli percolation.* Let $p \in (0,1)$ be a parameter; for each vertex of the triangular lattice $\mathcal{T}$, toss a coin and declare it to be *open* (resp. *closed*) with probability p (resp. $1-p$), independently of the others. Denote by P_p the corresponding probability measure on the set of configurations (it is simply a product measure). One can see a configuration as a random subgraph of the underlying lattice, obtained by keeping the open vertices and all the edges connecting two open vertices.

I.1.1.1. *Basic features of the model.* The question of interest is that of the connectivity structure of this subgraph. Let

$$\theta(p) := P_p\left[0 \leftrightarrow \infty\right]$$

be the probability that the origin belongs to an infinite connected component, or *cluster* (*i.e.*, that it is "connected to infinity"). It is easy to show that the function θ is non-decreasing, and using a simple counting argument (known as a *Peierls argument*), that for p small enough, $\theta(p)$ is equal to 0 and $\theta(1-p)$ is positive; in other words, defining

$$p_c := \inf\{p : \theta(p) > 0\} = \sup\{p : \theta(p) = 0\},$$

one has $0 < p_c < 1$. The value p_c is called the *critical point* of the model. Its value depends on the choice of the underlying lattice; in the case of the triangular lattice, by duality arguments it is equal to $1/2$.

The behavior of the system changes drastically across the critical point:

- If $p < p_c$, then almost surely all connected components are finite; moreover, they have finite expected volume, and the connection probabilities exhibit exponential decay: There exists $L(p) < \infty$ such that, for every x, $y \in \mathbb{Z}^2$,
$$P_p[x \leftrightarrow y] \leqslant C\, e^{-\|y-x\|/L(p)};$$
- If $p > p_c$, then almost surely there exists a *unique* infinite cluster and it has asymptotic density $\theta(p)$; but exponential decay still occurs for connectivity through finite clusters: There exists $L(p) < \infty$ such that for all x, $y \in \mathbb{Z}^2$,
$$P_p[x \leftrightarrow y; x \nleftrightarrow \infty] \leqslant C\, e^{-\|y-x\|/L(p)};$$
- If $p = p_c$, there is no infinite cluster (*i.e.*, $\theta(p_c) = 0$) yet there is no finite characteristic length in the system; the two-point function has a power-law behavior, in the sense that for some $c > 0$ and for all x, $y \in \mathbb{Z}^2$,
$$c\,\|y-x\|^{-1/c} \leqslant P_p[x \leftrightarrow y] \leqslant c^{-1}\,\|y-x\|^{-c}.$$

FIGURE 1. Critical site-percolation on a rectangular region of the triangular lattice (the state of a vertex is represented by the color of the corresponding face of the dual lattice).

The last statement is an instance of what is known as *Russo-Seymour-Welsh theory*, or *RSW* for short: essentially, the largest cluster within a large box of a given size has a diameter of the same order as the size of the box, and it crosses it horizontally with a positive probability, uniformly in the actual size of the box.

To be more specific, if $\mathcal{R}$ is a rectangle aligned with the axes of the lattice, denote by $LR(\mathcal{R})$ the probability that, within (the intersection between $\mathbb{Z}^2$ and) $\mathcal{R}$, there is a path of open edges connecting its two vertical sides. Then, RSW states that for every $\lambda > 0$, there exists $\eta(\lambda) \in (0,1)$ such that, for every n large enough,

$$\eta(\lambda) \leqslant P_p\left[LR([0,\lambda n] \times [0,n])\right] \leqslant 1 - \eta(\lambda). \tag{1}$$

This can easily be used as a black box, but figuring out the proof is a good way to get intuition on the model, so we include it here in the form of an exercise.

EXERCISE I.1.1 (Proof of the RSW bounds). (1) *Since we are working in the triangular lattice, it makes more sense to first prove* (1) *for parallelograms aligned with the lattice; it is easy to see why this is sufficient. In this whole exercise, we will thus use two lattice directions as coordinate axes, so that for instance what is denoted as* $[0,n]^2$ *is in fact a rhombus. What is the probability that there exists a horizontal crossing of* $[0,n]^2$*?*

(2) *Assume* $[0,n]^2$ *is crossed from left to right and set* Γ *to be the lowest horizontal crossing. Let* γ *be a deterministic path from left to right, prove that* $\{\Gamma = \gamma\}$ *is measurable with respect to the* σ*-algebra spanned by the sites* ***below*** γ *and the sites of* γ*. When conditioning on* $\{\Gamma = \gamma\}$*, what can be said about the law of sites above* γ*?*

(3) *Consider the shape in the following figure and assume that the left rectangle* $[0,n]^2$ *is crossed horizontally. Can you bound from below the probability that the lowest crossing* Γ *is connected to the bold part by a black path?* Hint: condition on $\{\Gamma = \gamma\}$ and consider the reflected path $\sigma(\gamma)$ with respect to the line $y = n + \frac{1}{2}$.

(4) *Deduce that the probability of crossing the rectangle* $[0,2n] \times [0,n]$ *horizontally is bounded away from 0 when* n *goes to infinity.*

(5) *a) Let* $\rho > 1$*. Deduce that the probability to cross the rectangle* $[0,\rho n] \times [0,n]$ *horizontally is (uniformly in* n*) bounded away from* 0*;*

b) Prove that the probability of a black circuit surrounding the origin in the annulus $[-2n,2n]^2 \setminus [-n,n]^2$ *remains bounded away from* 0 *when* n *goes to infinity;*

c) Show that almost surely there is no infinite cluster at $p = \frac{1}{2}$*;*

d) What can be said about $\mathbb{P}(0 \leftrightarrow \partial\Lambda_n)$*?*

e) (difficult) *Explain a strategy to prove that* $p_c = \frac{1}{2}$.

A natural question is then the following: Does the crossing probability above actually converge as $n \to \infty$? In fact, that question is still open in the general case, and in particular in the case of bond-percolation on the square lattice $\mathbb{Z}^2$, and it is not quite clear how many new ideas would be needed to prove convergence. But, conjecturally, the limit does exist and does not depend on the choice of the underlying lattice, provided that it has enough symmetry — this is part of what is known as *universality.*

I.1.1.2. *The Cardy-Smirnov formula.* We now turn to the main result which we want to present in this section. In the case of site-percolation on the triangular lattice (which is the one we are considering here), in fact the above probability does converge, and the limit is known explicitly. This was first conjectured by Cardy using mathematically non-rigorous arguments, and later proved by Smirnov.

Before stating the main theorem, we need some additional notation. Let Ω be a smooth, simply connected, bounded domain in the complex plane, and let a, b, c and d be four points on $\partial\Omega$, in this order if the boundary is oriented counterclockwise. Let $\delta > 0$, and consider the triangular lattice scaled by a factor of δ, which we will denote by $\mathcal{H}_\delta$; let $\mathcal{P}_\delta(\Omega, a, b, c, d)$ be the probability that, within percolation on $\Omega \cap \mathcal{H}_\delta$, there is an open path connecting the arc ab to the arc cd of the boundary.[1]

THEOREM I.1.1 (Cardy, Smirnov). *There exists a function f defined on the collection of all 5-tuples formed of a simply connected domain with four marked boundary points, satisfying the following:*

(1) *As $\delta \to 0$, $\mathcal{P}_\delta(\Omega, a, b, c, d)$ converges to $f(\Omega, a, b, c, d)$;*
(2) *f is* conformally invariant, *in the following sense: If Ω and Ω' are two simply connected domains and if Φ maps Ω conformally to Ω', then*
$$f(\Omega, a, b, c, d) = f(\Omega', \Phi(a), \Phi(b), \Phi(c), \Phi(d));$$
(3) *If $\mathcal{T}$ is an equilateral triangle, and if a, b and c are its vertices, then*
$$f(\mathcal{T}, a, b, c, d) = \frac{|cd|}{|ab|}$$
(which, together with the conformal invariance, characterizes f uniquely).

A complete proof of this theorem can be found in several places, so it does not make much sense to produce yet another one here; instead, we briefly describe the main steps of Smirnov's general strategy in some detail. The same overall approach (though obviously with a few modifications) will be applied to other models below; the main point each time will be to find the correct *observable*, *i.e.* a quantity derived from the discrete model and which is computable enough that its asymptotic behavior can be obtained (and is non-trivial).

Step 1: Definition of the observable. Let Ω be as above, and let z be a vertex of the dual lattice $\mathcal{H}_\delta^*$ (or equivalently, a face of the triangular lattice); denote by $E_\delta^a(z)$ the event that there is a simple path of open vertices joining two points on the boundary of Ω and separating a and z on one side and b and c on the other side, and by $H_\delta^a(z)$ the probability of $E_\delta^a(z)$. Define H_δ^b and H_δ^c accordingly. Notice that if we choose $z = d$, we get exactly the crossing probability:
$$\mathcal{P}_\delta(\Omega, a, b, c, d) = H_\delta^a(d).$$
In fact, we will compute the limit of H_δ^a as $\delta \to 0$ in the whole domain; the existence of f will follow directly.

Step 2: Tightness of the observable. Let z and z' be two points within the domain, and let $\mathcal{A}$ be an annulus contained in Ω and surrounding both z and z'. If $\mathcal{A}$ contains an open circuit, then either both of the events $E_\delta^a(z)$ and $E_\delta^a(z')$ occur, or neither of them does. The existence of such circuits in disjoint annuli are independent events, and if one fixes the modulus of the annuli, their probability is bounded below by RSW estimates (1). Besides, the number of such disjoint annuli

[1]Technically, this definition would require constructing a discrete approximation of the domain; we choose to skip over such considerations here, and refer the avid reader to the literature for more detail.

which can be fit around $\{z, z'\}$ is of order $-\log|z'-z|$. This implies a bound of the form

$$|H_\delta^a(z') - H_\delta^a(z)| \leqslant C\,|z'-z|^c$$

for some C, $c > 0$ depending only on the domain Ω, but not on δ. In other words, the functions H_δ^a are uniformly Hölder with the same exponent and the same norm; and this implies, by the Arzelà-Ascoli theorem, that they form a relatively compact family of continuous maps from Ω to $[0,1]$. In particular, one can always choose a sequence (δ_k) going to 0 along which $H_{\delta_k}^a$ (as well as $H_{\delta_k}^b$ and $H_{\delta_k}^c$) converges to some continuous function h^a (resp. h^b, h^c) defined on $\overline{\Omega}$. Proving convergence of (H_δ^a) then amounts to proving the uniqueness of such a sub-sequential limit, *i.e.*, all that remains to be done is to identify the function h^a.

FIGURE 2. The exploration of a critical percolation interface in the upper-half plane.

Step 3: The exploration process. One key tool which we will need in what follows is an algorithmic way to measure features of percolation interfaces. In any simply connected planar domain different from the whole plane, split the boundary into two subsets, and assign "artificial" boundary conditions to be open on the first one and closed on the second one. From each contact point between the two, this creates an *interface* between open and closed vertices of the lattice, and one can follow it by looking at each site one after the other along the interface, turning left or right according to its color. This is easier drawn than formally described; see Figure 2. The outcome of the construction is a lattice path, known as the *exploration process* of the interface. We will use it in the next step, and again when we speak about the scaling limit of geometric objects (the percolation exploration process, as the lattice mesh goes to 0, converges to the trace of SLE_6).

Step 4: Local behavior of the observable. This is essentially the only place in the proof where one uses the fact that the underlying model is site-percolation on a triangulation. Let z be a vertex of $\mathcal{H}_\delta^*$, and let z_1, z_2 and z_3 denote its three neighbors, ordered counterclockwise; define $P_\delta^a(z, z_1)$ to be the probability that $E_\delta^a(z_1)$ occurs, but $E_\delta^a(z)$ does not. This is equivalent to the existence of three disjoint paths in $\mathcal{H}_\delta$, each joining one of the vertices of the triangle around z to one of the three boundary arcs delimited by a, b and c, and of appropriate states (two open, one closed — draw a picture!). The core of Smirnov's proof is then a wonderful relation between these quantities; namely:

$$P_\delta^a(z, z_1) = P_\delta^b(z, z_2) = P_\delta^c(z, z_3).$$

The argument is very simple, but not easy to write down formally; it goes as follows: Assuming the existence of three arms as above, it is possible to discover two of them by exploring the percolation configuration starting from c (say), and always staying on the interface between open and closed vertices. The exploration path reaches z if and only if two of the above arms exist; besides, it gives us no information about the state of the vertices which are not along it, because the underlying measure is a product measure. The key remark is then the *color-swapping argument*: changing the state of each of the vertices in the unexplored portion of Ω does not change the probability of the configuration (because we work at $p = p_c = 1/2$); but it does change the state of the third arm from open to closed. Swapping the colors of all the vertices in Ω (which still does not change probabilities) one then arrives at a configuration with three arms of the appropriate colors, but where the role of a (resp. z_1) is now taken by b (resp. z_2).

Step 5: Holomorphicity in the scaling limit. Now, we need to exhibit a holomorphic function built out of h^a, h^b and h^c; following the symmetry of order 3 in the setup, it is natural to define

$$H_\delta(z) := H_\delta^a(z) + \tau H_\delta^b(z) + \tau^2 H_\delta^c(z)$$

and $h := h^a + \tau h^b + \tau^2 h^c$ accordingly, where $\tau = e^{2\pi i/3}$. To prove that h is holomorphic, it is enough to show that, along every smooth curve γ contained in Ω, one has

$$\oint_\gamma h(z)\,\mathrm{d}z = 0$$

(by Morera's theorem); and to show that, it is enough to pick a sequence of suitable discretizations of γ and estimate the integral using H_δ, and to show that the discrete estimate vanishes as δ goes to 0. It is always possible to approach γ by a discrete path $\gamma_\delta = (z_0^\delta, z_1^\delta, \dots, z_{L_\delta}^\delta = z_0^\delta)$ on $\mathcal{H}_\delta^*$ in such a way that $L_\delta = \mathcal{O}(\delta^{-1})$, and one then has

$$\oint_\gamma h(z)\mathrm{d}z = \sum_{j=0}^{L_\delta - 1} \frac{H_\delta(z_j^\delta) + H_\delta(z_{j+1}^\delta)}{2}(z_{j+1}^\delta - z_j^\delta) + \mathcal{O}(\delta^c)$$

with $c > 0$ by the previous tightness estimate. One can then apply a discrete analog of Green's formula to make discrete derivatives of H_δ appear, and write these in terms of P_δ^a, P_δ^b and P_δ^c: after elementary calculus, one gets

$$\oint_\gamma h(z)\mathrm{d}z = \frac{i\delta\sqrt{3}}{2} \sum_{z\sim z'} \left[P_\delta^a(z, z') + \tau P_\delta^b(z, z') + \tau^2 P_\delta^c(z, z')\right](z' - z) + \mathcal{O}(\delta^c),$$

where the sum extends to all pairs of nearest neighbors in the interior of γ. Applying Smirnov's identity to write everything in terms of P_δ^a only then leads to

$$\oint_\gamma h(z)\mathrm{d}z = \frac{i\sqrt{3}}{2}\sum_{z\sim z'}\left[P_\delta^a(z,z')\sum_{j=0}^{2}\tau^j(z_j-z)\right]+\mathcal{O}(\delta^c)$$

(where the z_j are the neighbors of z, numbered counterclockwise in such a way that $z_0 = z'$). It is then easy to see that the inner sum is identically equal to 0 (because it is always proportional to $1+\tau^2+\tau^4$).

Step 6: Boundary conditions and identification. The same computation as above can be performed starting with $S_\delta := H_\delta^a + H_\delta^b + H_\delta^c$, and the conclusion is the same: The (sub-sequential) limit $s := h^a + h^b + h^c$ is holomorphic as well. But because it is real-valued, this leads to the conclusion that it is constant, equal to 1 by looking at the point $z = a$. This means that the triple (h^a, h^b, h^c) can be seen as the barycentric coordinates of $h(z)$ relative to the points 1, τ and τ^2, respectively, meaning that h maps Ω to the interior of the corresponding equilateral triangle $\mathcal{T}$. Since it sends boundary to boundary in a one-to-one way (the variations of h^a on the boundary are easy to determine), it has to be conformal, and so it has to be the unique conformal map from Ω to $\mathcal{T}$ mapping a (resp. b, c) to 1 (resp. τ, τ^2). Because the sub-sequential limit is thus identified uniquely, one obtains convergence of (H_δ) itself to h, and it is not difficult to conclude the proof.

This concludes the few features of percolation which we will need in the following parts; we will come back to it (and say a little bit more about the exploration process) in Part III. For now, the relevant piece of information to remember is that, at criticality, the scaling limit of percolation (in any reasonable sense) is non-trivial and exhibits conformal invariance.

I.1.2. The random-cluster model. Percolation is very easy to describe, because the states of the vertices are independent of each other; but it is not very physically realistic. We now focus our attention on the *random-cluster model* (sometimes also referred to as *FK-percolation* or simply the FK model, for the names of its inventors, Fortuin and Kasteleyn). It is a dependent variant of *bond* percolation. We choose to keep this section shorter than it could be; the interested reader will find all the details in the notes for Smirnov's course at the same summer school [**38**].

I.1.2.1. *Definitions and first properties.* Let $G = (V, E)$ be a finite graph, and let $q \in [1, +\infty)$ and $p \in (0, 1)$ be two parameters. The random-cluster measure on G is defined on the set of subgraphs of G, seen as subsets of E, by

$$P_{p,q,G}[\{\omega\}] := \frac{p^{o(\omega)}(1-p)^{c(\omega)}q^{k(\omega)}}{Z_{p,q,G}},$$

where $o(\omega)$ is the number of open edges in ω, $c(\omega)$ the number of closed edges, and $k(\omega)$ the number of connected components of the subgraph (counting isolated vertices). The *partition function* $Z_{p,q,G}$ is chosen so as to make the measure a probability measure. Notice that the case $q = 1$ is exactly that of a product measure, in other words it is Bernoulli bond-percolation on G.

The definition in the case of an infinite graph needs a little more care — of course defining the measure as above makes little sense since the terms $o(\omega)$, $c(\omega)$ and $k(\omega)$ would typically be infinite (as well as the partition function). The first step is to define *boundary conditions*, which in this case amounts to introducing additional edges whose state is fixed (either open or closed); if ξ denotes such a choice, then $P^{\xi}_{p,q,G}$ denotes the corresponding measure. Notice that the only effect ξ has is in the counting of connected components within G.

Now consider the square lattice, and a sequence of increasing boxes $\Lambda_n := [-n,n]^2$. We will consider two types of boundary conditions for the random-cluster model on Λ_n: *free* (*i.e.*, ξ is empty) and *wired* (*i.e.*, all the vertices on the boundary of Λ_n are assumed to be connected). We denote these boundary conditions by f and w, respectively. A third boundary condition is known as the *Dobrushin* boundary condition, and consists in wiring the vertices of one boundary arc of the box (with prescribed endpoints) together while leaving the rest of the boundary free.

If $q \geqslant 1$, the model exhibits positive correlations (in the form of the FKG inequality). This implies that, if $n < N$, the restriction of the wired (resp. free) measure on Λ_N to Λ_n is stochastically smaller (resp. larger) than the corresponding measure defined on Λ_n directly. As n goes to infinity, this allows for the definition of infinite-volume measures as monotonic limits of both sequences, which we will denote by $P^{w}_{p,q}$ and $P^{f}_{p,q}$.

For fixed q and either free or wired boundary conditions, these two measure families are stochastically ordered in p; this implies the existence of a critical point $p_c(q)$ (the same in both cases, as it turns out) such that, as in the case of Bernoulli percolation, there is a.s. no infinite cluster (resp. a unique infinite cluster) if $p < p_c$ (resp. $p > p_c$).

It has long been conjectured, and was recently proved [**5**], that for every $q \geqslant 1$,

$$p_c(q) = \frac{\sqrt{q}}{1+\sqrt{q}}. \tag{2}$$

This comes from the following duality construction. Let for now G be a graph embedded into the 2-sphere, and let G^* be its dual graph. To any configuration ω of the random-cluster model on G, one can associate a configuration ω^* on G^* by declaring a dual bond to be open if and only if the corresponding primal bond is closed (see figure 3). As it turns out, if ω is distributed as $P_{p,q,G}$, then ω^* is distributed as P_{p^*,q^*,G^*}, *i.e.* it is a random-cluster model configuration, with

$$q^* = q \quad \text{and} \quad \frac{p^*}{1-p^*} = q\frac{1-p}{p}.$$

It is easy to see that there is a unique value p_{sd} of p satisfying $p_{sd} = (p_{sd})^*$, and it is then natural to expect that $p_c = p_{sd}$, leading to the value above.

EXERCISE I.1.2. *Prove the duality statement.*

Answer: Use Euler's formula to relate the number of open and closed bonds in the primal and the dual configurations with their numbers of faces and clusters. It helps to rewrite the weight of a configuration as

$$\left(\frac{p}{1-p}\right)^{o(\omega)} q^{k(\omega)}$$

and to notice that $o(\omega) + o(\omega^*)$ does not depend on the configuration.

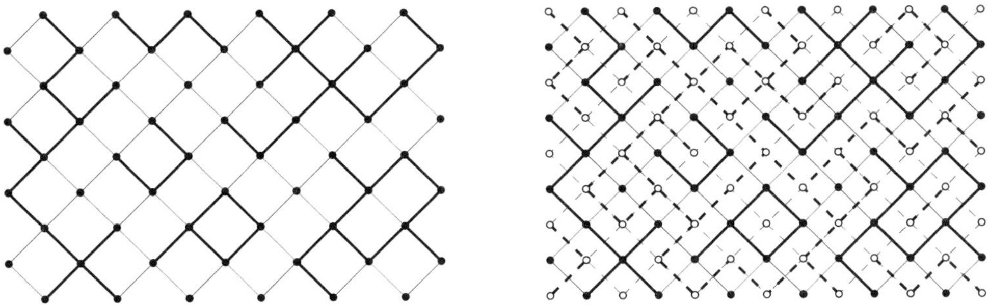

FIGURE 3. Duality between random-cluster configurations.

I.1.2.2. *The para-fermionic observable.* As before, we want to introduce an observable defined from the discrete model and giving us enough information to determine asymptotic properties in the scaling limit. From now on, let Ω be a smooth, simply connected domain in the complex plane, with two marked points a and b on its boundary. Let G_δ be the graph obtained as the intersection of Ω with $\delta\mathbb{Z}^2$ for some mesh $\delta > 0$; we will consider the random-cluster model with parameters $q \geqslant 1$ and $p = p_{sd}(q)$ on G_δ, with Dobrushin boundary conditions, wired on the (positively oriented) arc ab — and we will denote the corresponding measure simply by P (or later by P_δ when we insist on the scaling behavior as $\delta \to 0$).

As in the case of percolation, we briefly describe the main steps in Smirnov's proof of conformal invariance. A big difference is that the statement of convergence needs more notation, so we will have to postpone it a little bit.

Step 1: The loop representation. In addition to the graph G_δ and its dual G_δ^*, we need a third one known as the *medial graph* and denoted by $G_\delta^\diamond$; it is defined as follows. The vertices of the medial graph are in bijection with the bonds of either G_δ or G_δ^* (which are in bijection), and one can think of them as being at the intersection of each primal bond with its dual; there is an edge between two vertices of $G_\delta^\diamond$ if and only if the two corresponding primal edges share an endpoint and the two corresponding dual edges do as well.

One can encode a random-cluster configuration on G_δ using the medial graph, by following the boundary of each of its clusters (or equivalently, each of the clusters of the dual configuration). This leads to a covering of all the bonds of $G_\delta^\diamond$ by a family of edge-disjoint paths, one joining a to b (which we will call the *interface* and denote by γ), and the others being loops. If $l(\omega)$ denotes the number of loops obtained this way, it is possible to rewrite the probability of a configuration ω as

$$P[\{\omega\}] = \frac{x^{o(\omega)}(\sqrt{q})^{l(\omega)}}{Z_{x,q}} \quad \text{with} \quad x := \frac{p}{(1-p)\sqrt{q}}.$$

Since we work at the self-dual point, in fact we have $x = 1$ and the weight of a configuration is written as a function of only the number of loops in its loop representation.

EXERCISE I.1.3. *Prove the equivalence of the random-cluster representation and the loop representation.*

Answer: It works exactly the same way as the previous exercise, use Euler's formula in the natural way and it will work.

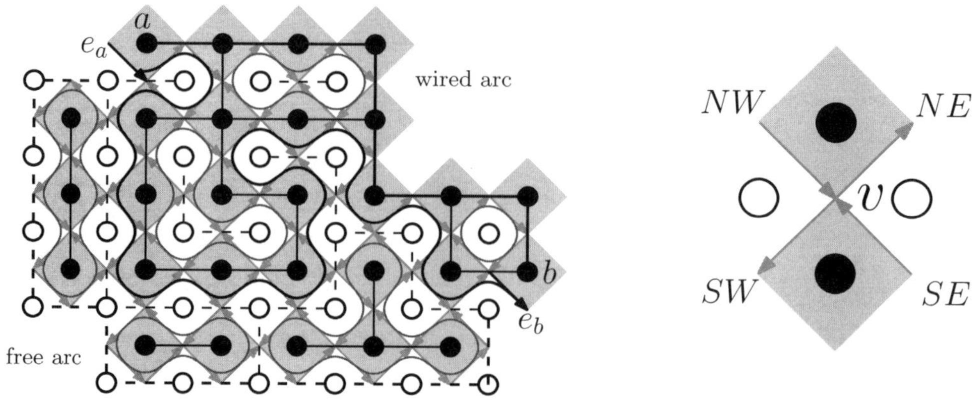

FIGURE 4. The loop representation of a random-cluster configuration, and the possible orientations of the interface γ around a vertex.

Step 2: Definition of the para-fermionic observable. Let e be an edge of the medial graph. We would like to be able to compute the probability that the interface passes through e; unfortunately, this seems to be out of reach of current methods, so we need an alternative. Assume that γ does go through e. Then, it is possible to follow it from a to the midpoint of e, and to follow the variation of the angle of the tangent vector along the way: it increases (resp. decreases) by $\pi/2$ whenever γ turns left (resp. right). The *winding* of the curve at e is the value one gets when reaching e; it is in $(\pi/2)\mathbb{Z}$ and we denote it by $W(e)$. If γ does not pass through e, define $W(e)$ to be an arbitrary value, as it will not be relevant.

The para-fermionic observable is then defined as

$$F_\delta(e) := E\left[e^{-i\sigma W(e)}\mathbb{1}_{e\in\gamma}\right] \quad \text{where } \sigma \text{ satisfies} \quad \sin\frac{\sigma\pi}{2} = \frac{\sqrt{q}}{2}.$$

Notice the difference here between the cases $q \leqslant 4$ (when σ is real) and $q > 4$ (when it is pure imaginary); we will come back to this distinction shortly. The parameter σ is known as the *spin* of the model. Morally, the main convergence result is that, at the self-dual point, $\delta^{-\sigma}F_\delta$ converges (to an explicit limit) as $\delta \to 0$; but giving a precise sense to that statement requires a little more preparation.

Step 3: Local behavior of the observable. Let ω be a configuration, and let e again be an edge of the medial lattice. We will denote by $F_\delta(e,\omega)$ the contribution of ω to the observable, so that $F_\delta(e) = \sum F_\delta(e,\omega)$. Besides, let ℓ be a bond of the primal lattice which is incident to e. There is a natural involution on the set of configurations given by changing the state of the bond ℓ without changing anything else — denote this involution by s_ℓ. It is easy to see how $F_\delta(e,\omega)$ and $F_\delta(e,s_\ell(\omega))$ differ: If both are non-zero, then the winding term is the same and their ratio is therefore either $x\sqrt{q}$ or $x/\sqrt{q}$ according to whether opening ℓ creates or destroys a loop.

Notice that the medial lattice can be oriented in a natural way, by declaring that its faces corresponding to dual vertices are oriented positively. The definitions above ensure that γ always follows the orientation of the bonds it uses. We will use the notation $e \to \ell$ (resp. $\ell \to e$) to mean that the bond e is oriented towards (resp. away from) its intersection with ℓ. The above observations imply that, for

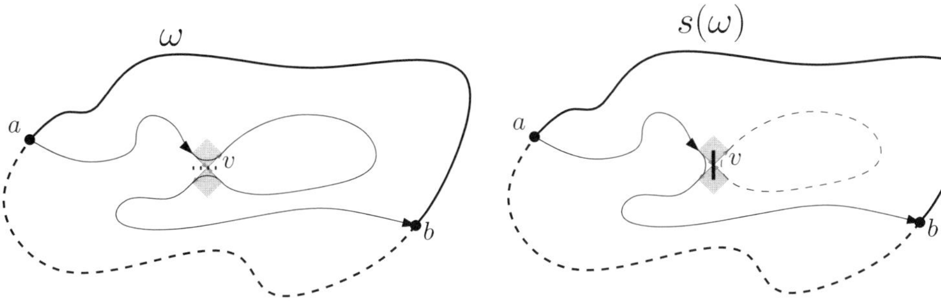

FIGURE 5. One pair of configurations contributing to (3).

every ℓ in the primal lattice and every $x > 0$,

$$(3) \qquad \sum_{e \to \ell} \big[F_\delta(e,\omega) + F_\delta(e, s_\ell(\omega))\big] = \frac{e^{\sigma\pi i/2} + x}{1 + x e^{\sigma \pi i/2}} \sum_{e \to \ell} \big[F_\delta(e,\omega) + F_\delta(e, s_\ell(\omega))\big].$$

The prefactor on the right-hand side is a complex number of modulus 1 if $q \leqslant 4$; it is real and positive if $q > 4$; and in both cases, it is equal to 1 if and only if $x = 1$ — in other words, exactly at the self-dual point. Summing the previous relation over ω, we get

$$\sum_{e \to \ell} F_\delta(e) = \frac{e^{\sigma\pi i/2} + x}{1 + x e^{\sigma \pi i/2}} \sum_{e \to \ell} F_\delta(e)$$

and in particular, at the self-dual point, this relation boils down to the flow condition:

$$\sum_{e \to \ell} F_\delta(e) = \sum_{e \to \ell} F_\delta(e).$$

This is the basis from which all the rest of the proof is built up: we now need to interpret this relation as the vanishing of a divergence, and in turn as the (discrete) holomorphicity of a well-chosen function.

I.1.2.3. *Interlude.* Here we have to stop for a moment. All the preceding reasoning is perfectly general, and the intuition behind it is rather clear. However, a worrisome remark is that we get one linear relation per bond of the primal lattice, but one unknown (the value of F_δ) per bond of the *medial* lattice, of which there are twice as many. That means that these relations cannot possibly characterize F_δ uniquely, and indicates that something more is needed; and in fact, the proof of conformal invariance is indeed not known in all generality.

The first "easy" case is that of $q > 4$. Here the observable is real-valued, and therefore lends itself to more analytic techniques, mostly inequalities. This is quite fruitful if one aims for the value of the critical point, and indeed in that case one can show that $p_c = p_{sd}$ using only elementary calculus. The use of inequalities is not optimal though, as it is not precise enough to derive any information at the critical point — much less to prove conformal invariance.

The second "easy" case is that of $q = 2$, for which $\sigma = 1/2$. Recall that γ always traverses edges in their positive direction; this means that $W(e)$ is known in advance up to a multiple of 2π. In turn, this means that the argument of $e^{i\sigma W(e)}$ is known up to an integer multiple of π; in other words, for every (oriented) edge e, the observable $F_\delta(e)$ takes its value on a line in the complex plane depending only on the direction of e (essentially, viewing e as a complex number, $F_\delta(e)/\sqrt{e}$ is a real number). The previous obstruction can then be bypassed: while we get a (complex) linear relation per primal bond, we only need to determine a *real* unknown per medial bond, which is the same quantity of information. This is morally why a complete derivation of conformal invariance is known only in that particular case, to which we restrict ourselves from now on.

I.1.2.4. *The rest of the proof when $q = 2$ and $x = 1$.*

Step 4: The integrated observable. We would like to prove that $\delta^{-1/2} F_\delta$ converges and be done with it. However, this is not very realistic, in particular because the argument of F_δ oscillates wildly (with a period of one lattice mesh). A solution to that is to consider an integrated version of it, or rather of its square. There is (up to an additive constant) a unique function H_δ, defined on the vertices of both the primal and dual lattices, satisfying the following condition: Whenever W (resp. B) is a vertex of the primal (resp. dual) lattice, and they are adjacent and separated by the bond e_{WB} of the primal lattice, then

$$H_\delta(B) - H_\delta(W) = |F_\delta(e_{WB})|^2 .$$

In some sense, one can think of H_δ as a discrete integral of $|F_\delta|^2$; proving its existence is a matter of checking that the sum of the prescribed increments around a vertex of the medial lattice vanishes; and this in turn is a direct consequence (via the Pythagorean theorem) of the flow condition, noticing that the values of $F_\delta(e)$ for $e \to \ell$ (resp. $\ell \to e$) are always orthogonal. Besides, it is easy to check that H_δ is constant along both boundary arcs of the domain, and that its discontinuity at a (and hence also at b) is exactly equal to 1 — because the interface *has* to pass through a with no winding. From now on, we will thus assume that H_δ is equal to 0 (resp. 1) on the arc ab (resp. ba).

Now, H_δ has two natural restrictions, H_δ^w to the primal vertices and H_δ^b to the dual ones. These two restrictions have nice properties: H_δ^w is superharmonic (its discrete Laplacian is non-positive) while H_δ^b is subharmonic (its Laplacian is non-negative). Besides, assuming that F_δ is small, they differ by very little, so that any sub-sequential scaling limit of H_δ has to be harmonic — in fact, has to be the unique harmonic function h with boundary values 0 on ab and 1 on ba. That is already a non-empty statement, but extracting useful information from H_δ is not easy ...

EXERCISE I.1.4. *Prove that H_δ^w is indeed superharmonic.*

Answer: This actually takes some doing (the last, very tedious 4 pages of Smirnov's article [**37**]), but it is completely elementary. Simply expand the discrete Laplacian in terms of F_δ, and use the flow relation repeatedly to eliminate terms (it allows one to express each value of F_δ in terms of its values at 3 neighboring edges, but projecting on lines brings this down to 2).

Step 5: Tying up the loose ends. The above arguments are rather convincing, but a lot is missing which would not fit comfortably in these notes. Following the scheme of the percolation argument, the main ingredient is relative compactness in the shape of uniform continuity; here, it follows from classical results about the Ising model (essentially, from the fact that the phase transition is of second order, or equivalently that the magnetization of the critical 2D Ising model vanishes) and it does imply the convergence of H_δ, as $\delta \to 0$, to h as defined above.

Then, one needs to come back down from H_δ to F_δ, which involves taking a derivative (the same way H_δ was obtained by integration). Notice in passing that

$|F_\delta|^2$ is an increment of H_δ, so it is expected to be of the same order as the lattice spacing, and thus F_δ itself should be — and, indeed, is — of order $\sqrt{\delta}$, and related to the square root of the gradient of H_δ. The objection we raised before about the argument of F_δ still stands; we need to make the following adjustments.

Each vertex of the medial graph has four *corners* (one per adjacent face); by "rounding up" γ at each of its turns, it is possible to naturally extend F_δ to all such corners, with a winding at a corner defined to be the midpoint between that on the incident edge and that of the exiting one. Then, define $F_\delta^\diamond$ on each medial vertex to be the sum of F_δ over all adjacent corners. $F_\delta^\diamond$ is now a *bona fide* complex-valued function, and in fact F_δ can be recovered from it by appropriate projections.

We are now poised to state the main convergence result. Let Φ be a conformal map from Ω to the horizontal strip $\mathbb{R} \times (0,1)$, mapping a to $-\infty$ and b to $+\infty$. Such a map is unique up to a horizontal translation, which will not matter here since we will look at derivatives anyway; notice that h is simply the imaginary part of Φ.

THEOREM I.1.2 (Smirnov). *As $\delta \to 0$, and uniformly on compact subsets of Ω,*

$$\delta^{-1/2} F_\delta^\diamond \to \sqrt{2\Phi'}.$$

In particular, the scaling limit of $F_\delta^\diamond$ is conformally invariant.

I.1.3. The Ising model. The Ising model might be the best known and most studied model of statistical mechanics; it is amenable to the same kind of study as the random-cluster model, through the use of a similar observable. The two-point function of the Ising model, which encodes the spin-spin correlations, is closely related to the connection probability for the $q = 2$ random-cluster model, through the Edwards-Sokal coupling: starting from a random-cluster configuration with parameter p, color each cluster black or white independently of the others with probability $1/2$; this leads to a dependent coloring of the vertices of the lattice, which is distributed according to the Ising model with inverse temperature $\beta = -\log(1-p)$.

The observable is a little different, because the spin interfaces are not completely well-defined as simple loops (think for instance about the case of a checkerboard configuration); this is the main reason why the construction is more specific. The interested reader will find a detailed description in the notes for the mini-course of Stanislav Smirnov in the same school [**38**].

I.2. Path models

Maybe the simplest model for which conformal invariance is well understood is that of the *simple random walk* on a periodic lattice, say $\mathbb{Z}^2$. Indeed, as the mesh of the lattice goes to 0, the random walk path converges in distribution to that of a Brownian motion, and this in turn is conformally invariant.

More precisely, let Ω be a (bounded, smooth, simply connected) domain of the complex plane, and let $z \in \Omega$; let (B_t) be a standard planar Brownian motion started from z, τ be its hitting time of the boundary of Ω. Besides, let Φ be a conformal map from Ω onto a simply connected domain Ω', and let (W_s) be a Brownian motion started from $\Phi(z)$ and σ its hitting time of $\partial\Omega'$.

It is not true that (W_s) and $(\Phi(B_t))$ have the same distribution in general, because their time parameterizations will be different, but in terms of the path considered as a subset of the plane, they do; the following statement is another instance of conformal invariance:

THEOREM I.2.1 (P. Lévy). *The random compact sets* $\{\Phi(B_t) : t \in [0, \tau]\}$ *and* $\{W_s : s \in [0, \sigma]\}$ *have the same distribution.*

A consequence of this and the study of SLE processes will be (among others) a very detailed description of the *Brownian frontier*, *i.e.* of the boundary of the connected component of infinity in the complement of $B_{[0,\tau]}$. However, the frontier is not visited by the Brownian path in chronological order, and that makes the direct use of planar Brownian motion problematic; it seems that things would be simpler if the random curve had no double point, and correspondingly if the underlying discrete path were self-avoiding.

The most natural way to generate a self-avoiding path in a discretized simply connected domain, from an inside point z to the boundary, would be to notice that there are finitely many such paths and to define a probability measure on the set of paths (morally, uniform given the length of the path). This leads to the definition of the *self-avoiding walk*, but unfortunately not much is known about its scaling limit, so we turn our attention to a different object which is a bit more difficult to define but much easier to study.

I.2.1. Loop-erased random walk. Let again Ω be a simply connected domain in the plane, and let $\delta > 0$; let Ω_δ be an appropriate discretization of Ω by $\delta\mathbb{Z}^2$ (say, the largest component of their intersection), and let z_δ be a vertex in Ω_δ. In addition, let (X_n) be a discrete-time random walk on $\delta\mathbb{Z}^2$, starting from z_δ, and let

$$\tau := \inf\{n : X_n \notin \Omega_\delta\}$$

its exit time from Ω_δ. The loop-erasure $LE(X)$ is defined, as the name indicates, by removing the loops from (X_n) as they are created. Formally, define the (n_i) inductively by letting $n_0 = 0$ and, as long as $n_i < \tau$,

$$n_{i+1} := \max\{n \leqslant \tau : X_n = X_{n_i}\} + 1.$$

Then, $LE(X)_i := X_{n_i}$.

Clearly, the loop-erasure of a discrete path is a self-avoiding path, as the same vertex cannot appear twice in $LE(X)$; when as above X is a simple random walk, $LE(X)$ is known as the *loop-erased random walk* (from z_δ to $\partial\Omega_\delta$ in Ω_δ). If b is a boundary point of Ω_δ, one can condition X to leave Ω_δ at b and the loop-erasure of that conditioned random walk is called the loop-erased random walk from a to b in Ω_δ.

The profound link between the loop-erased random walk and the simple random walk itself will be instrumental in the study of its asymptotic properties as δ goes to 0. For instance, the distribution of the exit point of a simple random walk or an unconditioned loop-erased walk is the same (it is the discrete harmonic measure from z_δ).

The counterpart of RSW for loop-erased walks (in the sense that it is one of the basic building blocks in proofs of convergence) will be a statement that $LE(X)$ does not "almost close a loop" — so that in particular, if it does have a scaling limit, the limit will be supported on simple curves. We defer the exact statement to a later section, but essentially what happens is the following: for $LE(X)$ to form a fjord, without closing it, X itself needs to approach its past path and then proceed to the boundary of the domain without actually closing the loop (as this would

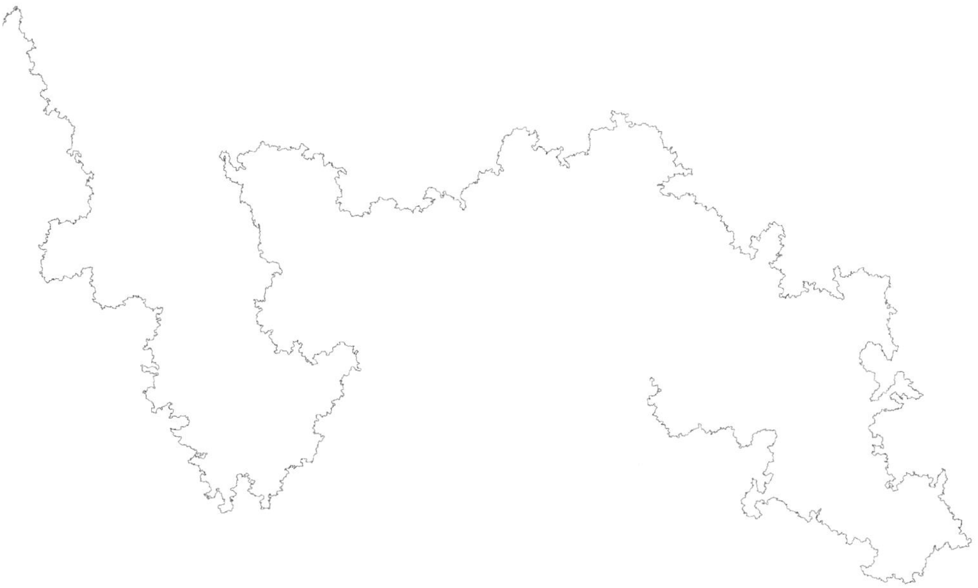

FIGURE 6. A loop-erased random walk.

vanish in $LE(X)$); the escape itself is very unlikely to happen as a consequence of Beurling's estimate.

We finish this definition of the loop-erased random walk with a statement of the *domain Markov property.* Let again Ω be a simply connected domain in the plane, discretized as $\Omega_\delta = \Omega \cap \delta\mathbb{Z}^2$, and let $a \in \delta\mathbb{Z}^2$ be an interior point of Ω and $b \in \delta\mathbb{Z}^2$ be a boundary point (*i.e.*, a point outside Ω with at least one neighbor inside Ω). Consider a loop-erased random walk path γ from a to b in Ω, and label it backwards as $(\gamma_i)_{0\leqslant i\leqslant \ell(\gamma)}$ where $\ell(\gamma)$ is the number of steps in γ, and where $\gamma_0 = b$ and $\gamma_{\ell(\gamma)} = a$.

This defines a sequence of discrete domains $\Omega_i := (\Omega \cap \mathbb{Z}^2) \setminus \{\gamma_j : j < i\}$, and by definition, γ is a loop-erased random walk from a to γ_0 in Ω_0. The statement of the domain Markov property is then the following: for every $k > 0$, the conditional distribution of $(\gamma_i)_{i\geqslant k}$ given $(\gamma_i)_{i\leqslant k}$ is the same as that of a loop-erased random walk from a to γ_k in Ω_k. In other words, the decreasing sequence of domains (Ω_k) can be seen as a Markov chain.

One way of proving the Markov property is, for a given path γ, to write the probability that a loop-erased random walk from a to b follows γ in terms of a product of Green functions and transition probabilities. The outcome of the proof is an alternative description of the backwards loop-erased random walk as a *Laplacian walk*, which is a growth process defined in terms of harmonic functions.

Fix $k \geqslant 0$, and assume that the first k steps of γ do not contain a. Let f_k be the unique function which is harmonic on Ω_k, equal to 0 outside Ω_k, except at γ_k where it is equal to 1. There is a finite, non-empty family $\{z_1, \dots, z_\ell\}$ of neighbors of γ_k at which f_k is positive, which are exactly the possible locations for γ_{k+1}, and one has

$$P\left[\gamma_{k+1} = z_i | \gamma_0, \dots, \gamma_k\right] = \frac{f_k(z_i)}{f_k(z_1) + \cdots + f_k(z_\ell)}.$$

In words, the growth distribution of γ is proportional to the value of f_k at the neighbors of γ_k. Yet another restatement of the same fact is that backwards loop-erased random walk is the same as "DLA conditioned not to branch." [Do not take this statement as a hope that a given branch in the DLA tree would look like a loop-erased walk; this is not true at all, unfortunately.]

I.2.2. Uniform spanning trees and Wilson's algorithm. Let $G = (V, E)$ be a finite graph; let v_∂ be a vertex of G (the "boundary" of the graph). A *spanning tree* of G is a connected subgraph of G containing all its vertices and no loop (a subgraph with all the vertices and no loop is called a *spanning forest*, and a tree is a connected forest). The set of spanning trees of G is finite; a *uniform spanning tree* is a random tree with the uniform distribution on that set.

Given a vertex $v \neq v_\partial$, we now have two ways of constructing a random self-avoiding path from v to v_∂:

- The loop-erased random walk in G from v to v_∂ (defined exactly as in the case of the square lattice above);
- The (unique) branch of a uniform spanning tree joining v to v_∂.

As it turns out, these two random paths have the same distribution. In particular, because in the second definition the roles of v and v_∂ are symmetric, we get an extremely non-obvious feature of loop-erased random walks: the time-reversal of the loop-erased walk from v to v_∂ is exactly the loop-erased walk from v_∂ to v. This is instrumental in the proof of convergence of the loop-erased walk to SLE_2 in the scaling limit.

As an aside, loop-erased walks provide a very efficient method for sampling a uniform spanning tree, which is due to David Wilson. Essentially: pick a point v_1, and run a loop-erased walk γ_1 from it to v_∂; then, pick a vertex v_2 which is not on γ_1 (if there is such a vertex) and run a loop-erased walk γ_2 from v_2 to γ_1; proceed until all the vertices of V are exhausted, each time building a loop-erased walk from a vertex to the union of all the previous walks. When the construction stops, one is facing a random spanning tree of G; and as it happens, the distribution of this tree is that of a uniform spanning tree.

I.2.3. The self-avoiding walk. Another, perhaps more natural probability measure supported on self-avoiding paths in a lattice is simply the uniform measure on paths of a given length. More specifically, let Ω_n be the set of n-step nearest-neighbor, self-avoiding path in $\mathbb{Z}^2$, starting at the origin, and let P_n be the uniform measure on Ω_n: we are interested in the behavior of a path sampled according to P_n, asymptotically as $n \to \infty$. The measure P_n is known as the *self-avoiding walk* of length n.

Obviously the first question coming to mind is that of the cardinality of Ω_n. By a simple sub-multiplicativity argument, there exists a constant $\mu \in [2, 3]$ known as the *connectivity constant* of the lattice such that

$$\frac{1}{n} \log |\Omega_n| \to \log \mu$$

FIGURE 7. A self-avoiding walk.

(which we will denote in short by $|\Omega_n| \approx \mu^n$). In fact, the behavior of $|\Omega_n|$ is conjecturally given by

$$|\Omega_n| \sim C\mu_n n^{\gamma-1}$$

for some exponent γ which, in two dimensions, is expected to be equal to

$$\gamma = \frac{43}{32}.$$

The value of μ depends on the chosen lattice, and is not expected to take a particularly relevant value in most cases; it is only known in the case of the hexagonal lattice, for which it is equal to $(2+\sqrt{2})^{1/2}$ — we refer the reader to the notes for the course of Gordon Slade in this same volume for a proof of this fact. The value of γ however is expected to be *universal* and depend only on the dimension.

Now, let $\omega = (\omega_0, \dots, \omega_n)$ be a self-avoiding path distributed according to P_n. We are still interested in scaling limits as $n \to \infty$, and for that the second relevant piece of information would be the appropriate scaling to apply to ω. One way to determine it is to look at the law of $\|\omega_n\|$; conjecturally, there exists a *scaling exponent* $\nu \in (0,1)$ such that

$$E\left[\|\omega_n\|^2\right] \approx n^{2\nu}.$$

Again, the value of ν is expected to be universal and depend only on the dimension. It is known in high dimension that $\nu = 1/2$, meaning that the self-avoiding walk is diffusive in that case and behaves like the simple random walk; this is the main focus of the course of Gordon Slade in the same summer school, so we simply again refer the reader to the corresponding notes. In the two-dimensional case, it is believed that

$$\nu = \frac{3}{4};$$

we will come back to this after we talk about convergence to SLE.

In the case of the self-avoiding walk (as we will see later also happens for percolation), there is a natural way to bypass the question of the relevant scaling and still be able to define a natural limit as the lattice mesh vanishes. Let U be a bounded, simply connected domain in the complex plane, with smooth (enough) boundary, and let a and b be two points on ∂U. For every $\delta > 0$, let $U_\delta = \delta\mathbb{Z}^2 \cap U$ and let a_δ and b_δ be approximations of a and b in the same connected component of U_δ; let Ω_δ^U be the set of self-avoiding paths from a_δ to b_δ in U_δ.

Since the elements of Ω_δ^U have various lengths, it is not that natural to consider the uniform measure on it. Instead, let $x > 0$ and define a measure $\mu_{x,\delta}^U$ on Ω_δ^U by letting

$$\mu_{x,\delta}^U(\{\omega\}) = \frac{x^{\ell(\omega)}}{Z_{x,\delta}^U}$$

where $\ell(\omega)$ denotes the length of ω and $Z_{x,\delta}^U$ is a normalizing constant (the *partition function* in physical parlance). Then, as $\delta \to 0$, one expects the asymptotic behavior of a walk ω distributed according to $\mu_{x,\delta}^U$ to strongly depend on the value of x. More precisely, letting $x_c := 1/\mu$ where μ is the connective constant of the lattice[2]:

- If $x < x_c$, then ω converges in distribution to a deterministic measure supported on the shortest path joining a to b in $\overline{U}$; its fluctuations around the limiting path are Gaussian and of order $\delta^{1/2}$, and the scaling limit of ω after the corresponding rescaling in the transverse direction is a Brownian bridge.
- If $x > x_c$, then the scaling limit of ω is a random space-filling curve in U, which is conjectured to be be the same as the scaling limit of the exploration path of the uniform spanning tree in U (*i.e.*, SLE_8 for those reading ahead).
- If $x = x_c$, then the scaling limit is believed to be a non-trivial random curve from a to b in $\overline{U}$, and to be conformally invariant. It is known that **if** this is the case, then the scaling limit is $\mathrm{SLE}_{8/3}$, and the previously mentioned conjectures about the values of ν and γ hold.

One key remark about the measures defined above is the following. Let $U' \subset U$ be another simply connected domain of the plane, such that a and b are on $\partial U'$ as

[2]It is unfortunate that every second object in this section seems to be called μ, but each of these notations seems to be classical ... hopefully this is not too confusing for the reader.

well. One can define two probability measures on $\Omega_\delta^{U'}$ in a natural way: the first one is $\mu_{x,\delta}^{U'}$, and the other one is the *restriction* of $\mu_{x,\delta}^{U}$ to $\Omega_\delta^{U'} \subset \Omega_\delta^{U}$, renormalized to be a probability measure. It is very easy to check that those two measures are in fact exactly identical; we will say that the self-avoiding walk has the (discrete) *restriction property*, and we morally expect the scaling limit to exhibit something similar.

We will see in the last part that there is exactly one conformally invariant measure supported on simple curves which has the restriction property, namely $\mathrm{SLE}_{8/3}$; meaning that, **if** the self-avoiding walk converges to a conformally invariant random simple curve, it has to be $\mathrm{SLE}_{8/3}$. In other words, as in the case of percolation, we can *predict* the value of the parameter κ for the scaling limit. Note however that an actual proof of convergence might not use this fact at all (as again is the case for percolation).

I.3. Bibliographical notes

Section I.1.1. A very complete review of percolation theory is Grimmett's book [**13**]; it contains everything mentioned in these notes except for Cardy's formula. Its bibliography section is far more complete than I could hope to gather here, so I will just list a few key papers. An alternative, which is a bit hard to find but well worth reading, is the book of Kesten [**18**]. For more recent progress and conformal invariance (and more exercises), one can *e.g.* consult the lecture notes for Werner's lectures [**41**].

Besides the anecdotal quotation from [**43**], the first proper introduction of percolation as a mathematical model is the article of Broadbent and Hammersley [**10**]. Exponential decay (up to the critical point) was derived in a very general setting by Menshikov [**27**]. The first derivation of the value of a critical parameter was obtained (for bond-percolation on the square lattice) by Kesten [**17**]; RSW estimates were obtained independently by Russo [**31**] and by Seymour and Welsh [**33**].

Cardy's formula was first conjectured by — well, Cardy [**12**], and then proved on the triangular lattice by Smirnov [**35**]. A slightly simplified exposition of the proof (which is the one we followed here) can be found in [**3**], and a very (very!) detailed one in the book of Bollobás and Riordan [**9**].

Section I.1.2. Here again, the reader is advised to refer to the book of Grimmett [**14**] (and references therein) for a general introduction to random-cluster models, including most of the results which are mentioned in this section. The proof of conformal invariance for the $q = 2$ critical random-cluster model was first obtained by Smirnov [**37**]; the approach we follow here is very close to the original, but some notation is borrowed from [**6**] (and technically, the notation $F_\delta^\diamond$ is only used here). All the details can be found in the notes for Smirnov's course in Buzios [**38**].

The equality $p_c = p_{sd}$ is related to the so-called Kramers-Wannier duality [**19**]; while still open in the general case, it is known to hold in the case $q = 1$ (where it is exactly Kesten's result on the percolation critical point in [**17**]); in the case $q = 2$ (where it is related to the derivation of the critical temperature of the two-dimensional Ising model by Onsager [**28**] — see also [**6**]); and in the general case as proved in [**5**].

Section I.2.1. For the contents of this section, and an introduction (possibly the best introduction), one can have a look at Schramm's original paper on LERWs and USTs [**32**]. Wilson's article [**42**] complements it nicely; and for more quantitative results, parts of the paper by Lawler, Schramm and Werner [**25**] can be read without any prior knowledge of SLE.

Section I.2.3. As was apparent in the text, we strongly recommend that the interested reader have a look at the notes for the lectures of Gordon Slade, in this same volume; it contains all we could possibly mention here and more.

Part II . SCHRAMM-LOEWNER EVOLUTION

Introduction

The previous part introduced a few discrete models, and for each of them we saw that, in the scaling limit as the lattice mesh goes to zero, a particular *observable* converges to a conformally invariant limit. It is natural to hope that convergence will actually occur in a much stronger sense, and in particular that the *interfaces* of the discrete model will have a continuous counterpart described as random curves in a planar domain.

Schramm's insight was to realize that, under mild (and reasonable) assumptions in addition to conformal invariance, the limit has to be distributed as one of a one-parameter family of measures on curves, which he named Stochastic Loewner Evolutions. They are now universally known as *Schramm-Loewner Evolutions.* The aim of this part is to define these random curves and give a few of their fundamental properties.

II.1. Definition of SLE

II.1.1. Loewner evolution in the half-plane. Let $\mathbb{H}$ denote the open upper half-plane, seen as a subset of the complex plane, and (for now) let $\gamma : [0, \infty) \to \overline{\mathbb{H}}$ be a continuous, simple curve. In keeping with probabilistic tradition, we will denote the position of γ at time t by γ_t instead of $\gamma(t)$; besides, we will assume that γ satisfies the following conditions:

- $\gamma_0 = 0$;
- For every $t > 0$, $\gamma_t \in \mathbb{H}$ (or in other words, $\gamma_t \notin \mathbb{R}$);
- $|\gamma_t| \to \infty$ as $t \to \infty$.

The results we will state in this section are actually valid in much more generality, but the intuition is not fundamentally different in the general case.

Let $H_t := \mathbb{H} \setminus \gamma_{[0,t]}$ be the complement of the path up to time t. Our assumptions ensure that H_t is a simply connected domain, and therefore Riemann's mapping theorem can be applied to show that there exists a conformal map

$$g_t : H_t \to \mathbb{H}$$

(we refer the reader to Appendix IV for a refresher on complex analysis, if needed). The map g_t is uniquely determined if one imposes the *hydrodynamic normalization*, which amounts to fixing the following asymptotic behavior at infinity:

$$g_t(z) = z + \frac{a(t)}{z} + \mathcal{O}\left(\frac{1}{z^2}\right).$$

With this notation, it is not hard to prove that a is a strictly increasing, continuous function; it need not go to infinity with t, but we will add this as an assumption on the curve. The function a can therefore be used to define a "natural" time parametrization of the curve: up to reparametrization, it is always possible to ensure that $a(t) = 2t$ for all $t > 0$. From now on, we shall assume that γ is indeed parametrized that way.

EXERCISE II.1.1. *Prove the statements made so far in the section, and in particular prove that a is indeed continuous and strictly increasing. Give an example of a curve going to infinity, but for which a is bounded.*

Answer: It is enough to show that $a(t)$ is strictly positive for every $t > 0$ — look at what happens under composition. One can then use Schwarz' Lemma to conclude. a will remain bounded if γ remains close enough to the real line.

The normalizations of g and t are chosen in such a way that the behavior of $g_t(z)$ as a function of t is then easy to describe:

THEOREM II.1.1 (Loewner). *There exists a continuous function $\beta : [0, \infty) \to \mathbb{R}$ such that, for every $t \geqslant 0$ and every $z \in H_t$,*

$$\partial_t g_t(z) = \frac{2}{g_t(z) - \beta_t}.$$

This differential equation is known as *Loewner's equation (in the half-plane)*. The gain is substantial: We were able to encode the whole geometry of γ, up to reparametrization, in terms of a single real-valued function. Indeed, it is not difficult to show that the construction up to now is essentially reversible: Given β, one can solve Loewner's equation to recover (g_t), and hence (H_t) and (γ_t) as well[3].

II.1.2. Chordal SLE. Consider, say, critical site-percolation on the triangular lattice in the upper half-plane, with boundary conditions open to the right of the origin and close to the left — this corresponds to the case in Section I.1.1 with one of the boundary points at infinity. This creates an interface starting from the origin, which is the path of the exploration process and satisfies the previous hypotheses on the curve γ.

Now, assume that, as the lattice mesh goes to 0, the exploration curve converges in distribution to a (still random) curve in the upper half-plane. This scaling limit can then be encoded into a real-valued process β using Loewner's equation; of course, β will be random as well. The question is now whether we can use the results of the previous part to identify β.

Let $R > 0$, and stop the exploration process at the first time τ_R when it reaches the circle of radius R centered at 0. Conditionally on its path so far, the next steps are exactly the exploration process of percolation in a new domain H_{τ_R}, namely the unbounded connected component of the complement of the current path: this is known as the *domain Markov property*.[4]

Cardy's formula being conformally invariant, it is natural to expect that the scaling limit of the exploration process would be as well, or in other words, that

[3]Well, some care is needed here: It is *not* true that one can plug any function β into Loewner's equation and obtain a Jordan curve γ out of it. It is true if β is Hölder with exponent $1/2$ and small enough norm, but a sharp condition is not known. Obviously everything works out fine if β comes from the above construction in the first place!

[4]The counterpart in statistical physics would be the DLR conditions for Gibbsian fields.

the path after time τ_R would be distributed as the conformal image of the path in $\mathbb{H}$ by a map sending $\mathbb{H}$ to the appropriate domain.

We almost have such a map at our disposal, from Loewner's equation: The map g_{τ_R} sends H_{τ_R} to $\mathbb{H}$, so its inverse map looks like what we are looking for. The only difference is that $g_{\tau_R}(\gamma_{\tau_R})$ is equal to β_{τ_R} instead of 0. Taking this into account, we get the following property (assuming of course the existence of the scaling limit): **The image of $(\gamma_t)_{t \geqslant \tau_R}$ by $g_{\tau_R} - \beta_{\tau_R}$ has the same distribution as $(\gamma_t)_{t \geqslant 0}$.**

Besides, all information coming from the path up to time τ_R is forgotten in this map: Only the shape of H_{τ_R} is relevant (because such is the case at the discrete level), and the dependence on that shape vanishes by conformal invariance. **The image of $(\gamma_t)_{t \geqslant \tau_R}$ by $g_{\tau_R} - \beta_{\tau_R}$ is independent of $(\gamma_t)_{0 \leqslant t \leqslant \tau_R}$.**

It remains to investigate what these two properties translate to in terms of the process (β_t). First, notice that the coefficient in $1/z$ in the asymptotic expansion at infinity which we are using is additive under composition. Letting $s, t > 0$, the previous reasoning, applied at time t, leads to the following:

$$g_{t+s} = \beta_t + \tilde{g}_s \circ (g_t - \beta_t),$$

where equality holds in distribution and where $\tilde{g}$ is an independent copy of g_s; the addition of β_t takes care of the normalization at infinity.

Differentiating in s and using Loewner's equation, this leads to

$$\beta_{t+s} = \beta_t + \tilde{\beta}_s,$$

where again equality holds in distribution and $\tilde{\beta}$ is an independent copy of β. **The process $(\beta_t)_{t \geqslant 0}$ has independent and stationary increments.**

Besides, the distribution of γ is certainly invariant under vertical reflection (because this holds at the discrete level), so β and $-\beta$ have the same distribution. So, we arrive at the following characterization: **Under the hypotheses of conformal invariance and domain Markov property, there exist a constant $\kappa \geqslant 0$ and a standard Brownian motion (B_t) such that**

$$(\beta_t)_{t \geqslant 0} = (\sqrt{\kappa} B_t)_{t \geqslant 0}.$$

II.1.3. Radial SLE. We just say a few words here about the case of radial Loewner chains, since not much needs to be changed from the chordal setup. Here, we are given a continuous, Jordan curve γ in the unit disk $\mathbb{D}$, satisfying $\gamma_0 = 1$, $\gamma_t \neq 0$ for all $t > 0$ and $\gamma_t \to 0$ as $t \to 0$. In other words, the reference domain is not the upper half-plane with two marked boundary points, but the unit disk with one marked boundary point and one marked interior point.

Let D_t be the complement of $\gamma_{[0,t]}$ in the unit disk; notice that 0 is in the interior of D_t, so there exists a conformal map g_t from D_t onto $\mathbb{D}$ fixing 0; this map is unique if one requires in addition that $g_t'(0) \in \mathbb{R}_+$, which we will do from now on.

The natural parametrization of the curve still needs to be additive under composition of conformal maps; here, the only choice (up to a multiplicative constant) is the logarithm of $g_t'(0)$: up to reparametrization, we can ensure that for every $t > 0$, $g_t'(0) = e^t$. With this choice, we have the following:

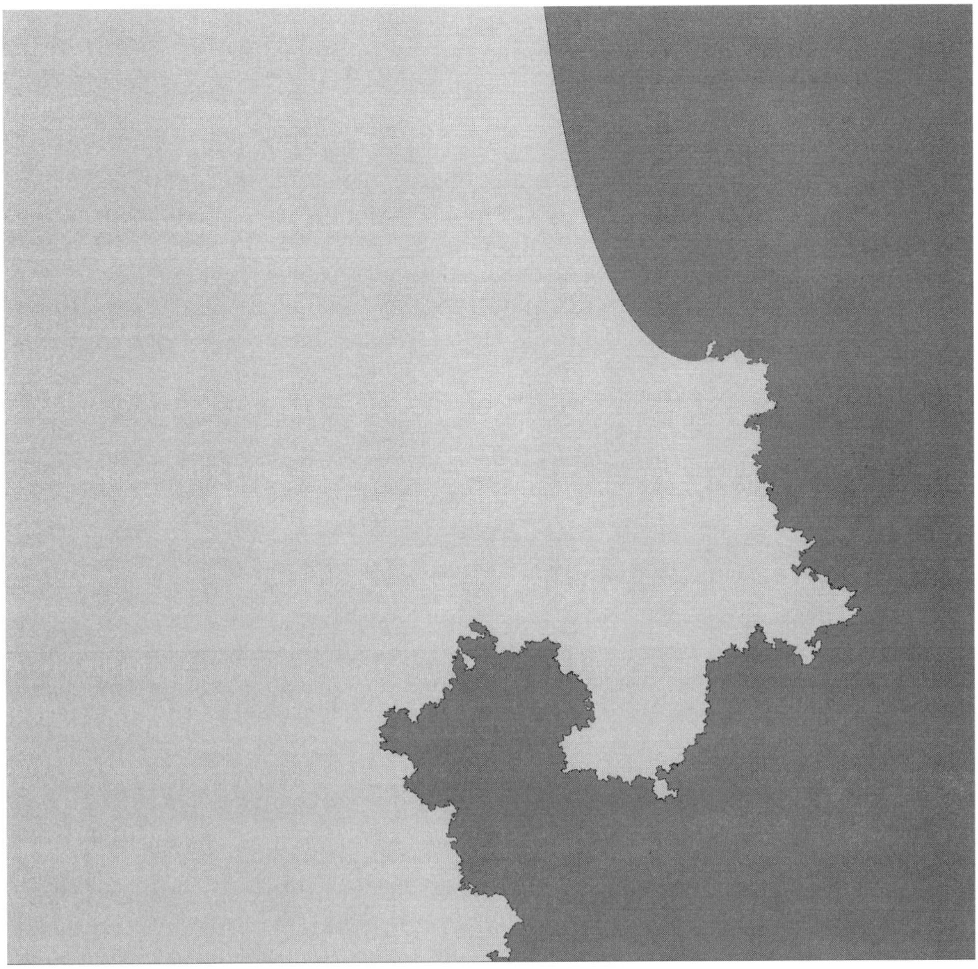

FIGURE 8. A chordal SLE process with parameter $\kappa = 2$. (The driving Brownian motion is stopped at time 1, which explains the smooth "tail" of the curve.)

THEOREM II.1.2 (Loewner). *There exists a continuous function* $\theta : [0, \infty) \to \mathbb{R}$ *such that, for every* $t \geqslant 0$ *and every* $z \in D_t$,

$$\partial_t g_t(z) = \frac{e^{i\theta_t} + g_t(z)}{e^{i\theta_t} - g_t(z)} g_t(z).$$

This is known as Loewner's equation in the disk.

Everything we just saw in the chordal case extends to the radial case. In particular, if the curve is related to a conformally invariant model (say, if it is the scaling limit of the loop-erased random walk), then under the same hypothesis of domain Markov property, one gets that there must exist $\kappa > 0$ such that

$$(\theta_t)_{t \geqslant 0} = (\sqrt{\kappa} B_t)_{t \geqslant 0}$$

(where again (B_t) is a standard real-valued Brownian motion). Solving Loewner's equation in the disk with such a driving function defines *radial* SLE_κ.

Remark. The local behavior of this equation around the singularity at $z = e^{i\theta_t}$ involves a numerator of norm 2; it is the same 2 as in Loewner's equation in the upper half-plane, in the sense that the local behavior of the solution for the same value of κ will then be the same on both sides.

II.1.4. SLE in other domains. We end the definition of the various kinds of SLE processes by a remark on the general case of simply connected domains. If Ω is such a domain and if a and b are two boundary points, then there is a conformal map Φ from Ω to $\mathbb{H}$ sending a to 0 and b to ∞; *chordal* SLE$_\kappa$ *from a to b in Ω* is simply the pullback of chordal SLE$_\kappa$ in $\mathbb{H}$ through Φ. One potential obstruction is that Φ is not uniquely defined; however, this is harmless because all such conformal maps are scalings of each other, and chordal SLE is *scale-invariant.*

The radial case is treated in a similar, and actually easier way: given a boundary point a and an interior point c in Ω, there is a *unique* conformal map Ψ from Ω to the unit disk mapping a to 1 and c to 0. *Radial* SLE$_\kappa$ *from a to c in Ω* is simply defined as the pullback of radial SLE$_\kappa$ in the disk through Ψ.

II.2. First properties of SLE

II.2.1. Geometry. The first, very non-trivial question arising about SLE is whether it actually fits the above derivation, which more specifically means whether the curve γ exists. In the case of regular enough driving functions, namely Hölder with exponent 1/2 and small enough norm, Marshall and Rohde [**26**] proved that it does, but it is possible to construct counterexamples.

It turns out to indeed be the case, up to one notable change. Recall that one can always solve Loewner's equation to obtain $g_t : \mathbb{H} \setminus K_t \to \mathbb{H}$ where K_t is the (relatively compact) set of points in the upper half-plane from which the solution blows up before time t. Then:

THEOREM II.2.1 (Rohde-Schramm, Lawler-Schramm-Werner). *For every $\kappa > 0$, SLE$_\kappa$ is generated by a curve, in the following sense: there exists a (random) continuous curve γ in the closure of the upper half-plane $\overline{\mathbb{H}}$, called the* SLE trace, *such that, for every $t > 0$, $\mathbb{H} \setminus K_t$ and $\mathbb{H} \setminus \gamma_{[0,t]}$ have the same unbounded connected component.*

What this really means is that K_t can be obtained starting up from $\gamma_{[0,t]}$ and then filling up every bounded "bubble" it forms, if there is any; technically, the proof in the article by Rohde and Schramm [**30**] covers all cases but one, namely $\kappa = 8$, for which the existence of the trace is only known as a consequence of the convergence of the UST contour to SLE$_8$.

Given the existence of the trace, it is natural to ask whether K itself is a curve or not. Whether this happens depends on the value of κ:

THEOREM II.2.2 (Rohde-Schramm). *The topology of the* SLE *trace undergoes two transitions:*

- *If $\kappa \leqslant 4$, then γ is almost surely a simple curve, and besides $\gamma_t \in \mathbb{H}$ for every $t > 0$;*
- *If $4 < \kappa < 8$, then γ does have double points, and $\gamma_{[0,t]} \subsetneq K_t$;*
- *If $8 \leqslant \kappa$, then γ is almost surely a space-filling curve,* i.e. $\gamma_{[0,\infty)} = \overline{\mathbb{H}}$.

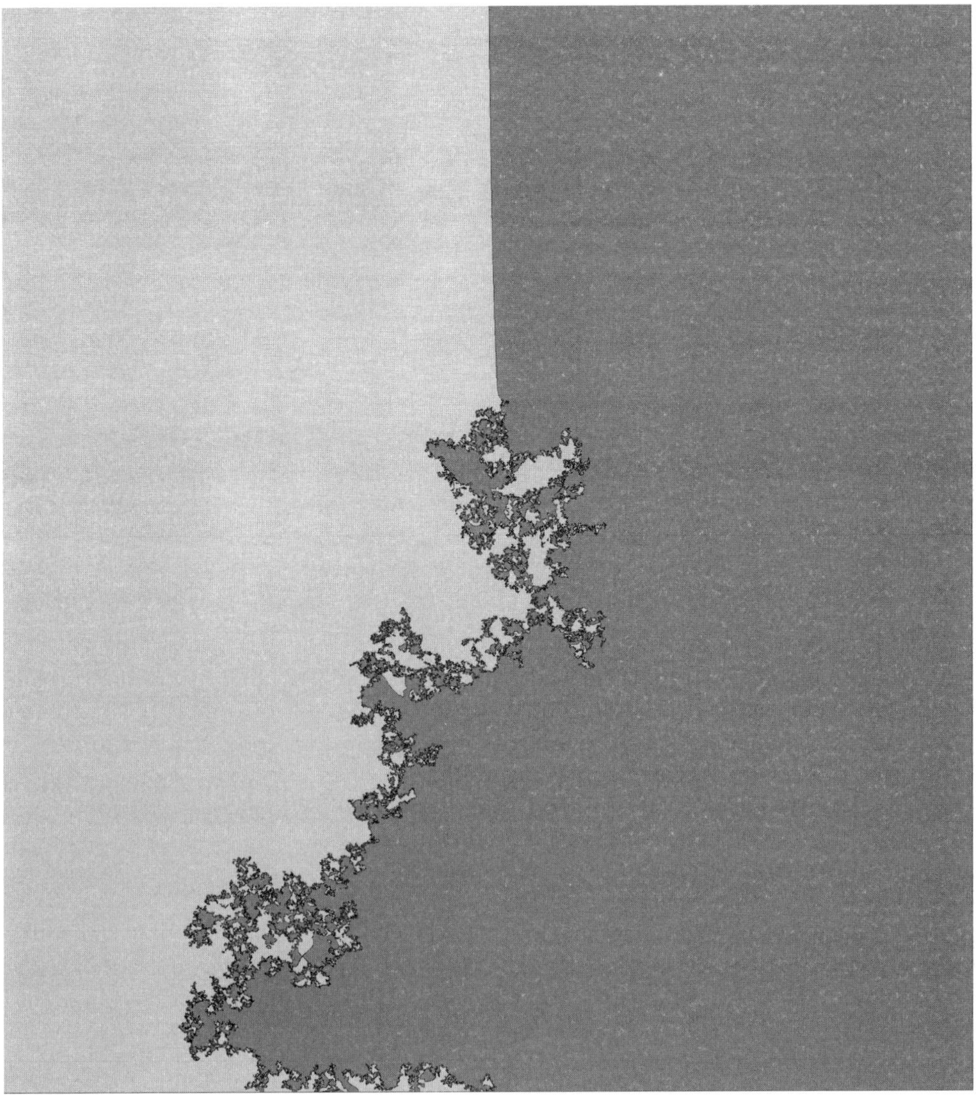

FIGURE 9. A chordal SLE with parameter $\kappa = 6$.

PROOF. The proof of this theorem involves the first use of SLE in computations. Let us start with the transition across $\kappa = 4$. Let $x > 0$, and trace the evolution of x under the (chordal) SLE flow by defining

$$Y_t^x := g_t(x) - \beta_t.$$

From Loewner's equation, one gets

$$\mathrm{d}Y_t^x = \partial_t g_t(x)\mathrm{d}t - \mathrm{d}\beta_t = \frac{2}{g_t(x) - \beta_t}\mathrm{d}t - \mathrm{d}\beta_t = \frac{2\mathrm{d}t}{Y_t^x} - \sqrt{\kappa}\mathrm{d}B_t.$$

Up to a linear time change, this is exactly a *Bessel process of dimension* $1+4/\kappa$. In particular, it will hit the origin (meaning that x is swallowed by the curve in finite time) if and only it the dimension of the process is less than 2, if and only if $\kappa > 4$.

The transition across $\kappa = 8$ is a bit more problematic, and involves estimates for the probability of hitting a ball inside the domain, proving that this probability is equal to 1 if and only if $\kappa \geqslant 8$; we leave that as an exercise, which can be skipped on first reading. □

EXERCISE II.2.1 (Conformal radius, space-filling SLE and Hausdorff dimension). *We consider chordal SLE(κ) in the upper half-plane for $\kappa > 0$. As usual, g_t is the conformal map from H_t to $\mathbb{H}$ with hydrodynamic renormalization. Define*

$$\tilde{g}_t(z) := \frac{g_t(z) - g_t(z_0)}{g_t(z) - \overline{g_t(z_0)}} \quad \textit{for all} \quad z \in \mathbb{H}.$$

(1) *Let $z_0 \in \mathbb{H}$. Prove that there exist constants $0 < c_1, c_2 < \infty$ such that*

$$c_1 d\big[z_0, \gamma[0,t)\big] \leqslant |\tilde{g}_t'(z_0)|^{-1} \leqslant c_2 d\big[z_0, \gamma[0,t)\big]$$

for any $t < \tau(z_0)$.

(2) *a) Assume that $t < \tau(z_0)$. Explain how one could derive the following equality – we do not ask for the (straightforward yet messy) computation.*

$$\partial_t \tilde{g}_t(z) = a_t \times \frac{\tilde{\beta}_t \tilde{g}_t(z)(\tilde{g}_t(z) - 1)}{(1 - \tilde{\beta}_t)(\tilde{g}_t(z) - \tilde{\beta}_t)},$$

where

$$\tilde{\beta}_t = \frac{\beta_t - g_t(z_0)}{\beta_t - \overline{g_t(z_0)}} \qquad \textit{and} \qquad a_t = \frac{2(\tilde{\beta}_t - 1)^4}{(g_t(z_0) - \overline{g_t(z_0)})^2 \tilde{\beta}_t^2}.$$

Check that $\tilde{\beta}_t \in \partial\mathbb{D}$ and $a_t > 0$ for every $t < \tau(z_0)$.

b) We introduce the time-change $s := \int_0^t a_u du$. Show that $h_s = \tilde{g}_{t(s)}$ satisfies a 'radial Loewner-like' equation with 'driving process' α_s where $\exp(i\alpha_s) = \tilde{\beta}_{t(s)}$.

c) By differentiating (with respect to z) the previous equation at z_0, show that

$$\partial_s h_s'(z_0) = \frac{2h_s'(z_0)}{1 - \tilde{\beta}_s}.$$

Deduce that $|h_s'(z_0)| = |h_0'(z_0)|e^{-s}$ when $s < s(\tau(z_0))$.

(3) *Explain how one could prove (we do not ask for the computation)*

$$\alpha_0^x = x \quad \textit{and} \quad d\alpha_t^x = \sqrt{\kappa} dB_t + \frac{\kappa - 4}{2} \cot(\alpha_t^x/2) dt.$$

(4) *To which event for the diffusion α does $\{t(s) = \tau(z_0)\}$ correspond? Show that $d(z_0, \gamma[0,\infty)) \asymp e^{-S}$ where S is the survival time of the diffusion α.*

(5) *Prove that the* SLE(κ) *is dense whenever $\kappa \geqslant 8$. We assume that it is generated by a transient continuous curve. Prove that it is space-filling.*

(6) *What can be said about the Hausdorff dimension of* SLE(κ) *when $\kappa < 8$?*

Much more can be said about the topological and metric properties of γ and the related boundary behavior of the maps g_t; they are the topic of G. Lawler's mini-course in this summer school, so we don't dwell on it much further, simply ending on the following result:

THEOREM II.2.3 (Beffara). *For every $\kappa \in [0, 8]$, the Hausdorff dimension of the* SLE *trace is almost surely equal to $1 + \kappa/8$.*

II.2.2. Probability. We now turn to uses of SLE in computing the probabilities of various events which are of particular interest in the framework of scaling limits of discrete models; we focus on two kinds of estimates, for crossings and for arm events. Before doing that, though, we first mention two special properties which were already mentioned in a discrete setting, for percolation and the self-avoiding walk respectively.

II.2.2.1. *Locality and the restriction property.* Let A be a non-empty relatively compact subset of $\mathbb{H}$, at a positive distance from the origin, and with simply connected complement in $\mathbb{H}$ — such a set is called a *hull* in the SLE literature, though the reason for the choice of this term is not clear. We are in the presence of two simply connected domains, the upper half-plane $\mathbb{H}$ and a subdomain $\mathbb{H} \setminus A$; we wish to compare SLE in these two domains.

Let (K_t) be an SLE_κ in $\mathbb{H}$, and let γ be its trace; let $(\tilde{K}_t)$ be an SLE_κ in $\mathbb{H} \setminus A$, and let $\tilde{\gamma}$ be its trace. Let τ (resp. $\tilde{\tau}$) be the first hitting time of A by (K_t) (resp. by $(\tilde{K}_t)$) — for now, assume that $\kappa > 4$ so that both of these stopping times are a.s. finite.

THEOREM II.2.4 (Locality property of SLE_6). *In the case $\kappa = 6$, with the previous notation, the two random sets*

$$K_{\tau-} := \bigcup_{t<\tau} K_t \quad \text{and} \quad \tilde{K}_{\tilde{\tau}-} := \bigcup_{t<\tilde{\tau}} \tilde{K}_t$$

have the same distribution.

In other words, as long as γ does not touch the boundary of the domain, it does not "know" whether it is growing within $\mathbb{H}$ or $\mathbb{H} \setminus A$, hence the name of this property. Notice that the exploration process of percolation (see section I.1.1 of the previous part) satisfies the same property at the discrete level; this is one way to predict that its scaling limit has to be SLE_6. One interesting corollary of locality is the following:

COROLLARY II.2.1. *Let Ω be a simply connected domain in the plane, and let a, b and c be three points on $\partial\Omega$. Then, until their first hitting time of the boundary arc bc, an SLE_6 in Ω from a to b and an SLE_6 in Ω from a to c have (up to time-change) the same distribution.*

So, not only does SLE_6 not know in which domain it is growing, it does not know where it is going to either. This allows for language shortcuts such as "SLE_6 in Ω from a to the arc bc" which will be useful very soon.

With the same notation as above, assume now that $\kappa < 4$, so that in particular $\tilde{K}$ a.s. never hits A and K avoids it with positive probability. This provides us with two probability distributions on simple curves in the complement of A in the upper half-plane: $\tilde{\gamma}$ on one hand, and on the other hand, γ conditioned not to hit A, which we will (temporarily) denote by $\hat{\gamma}$.

THEOREM II.2.5 (Restriction property of $\mathrm{SLE}_{8/3}$). *In the case $\kappa = 8/3$, with the previous notation, the two random sets*

$$\tilde{K}_\infty := \{\tilde{\gamma}_t : t > 0\} \quad \text{and} \quad \hat{K}_\infty := \{\hat{\gamma}_t : t > 0\}$$

have the same distribution; the curves $\tilde{\gamma}$ and $\hat{\gamma}$ themselves have the same distribution as well, up to appropriate time-change.

We mentioned already that the self-avoiding walk measure has the same property at the discrete level; that is one reason to predict that it converges to SLE(8/3) in the scaling limit. However, this presumes the existence of a scaling limit and its conformal invariance, which in that case remain mostly mysterious.

II.2.2.2. *Interlude: restriction measures.* As an aside to the main text, we now give a short description of another family of measures on random sets which have strong links to SLE. Let K be a closed, connected subset of $\mathbb{H} \cup \{0\}$, containing 0, and having its complement in $\mathbb{H}$ consisting of exactly two (open) connected components, both unbounded, one having $\mathbb{R}_+$ on its boundary and the other, $\mathbb{R}_-$. For the duration of this interlude, let us call such a set a *nice* set.

Let A be a compact subset of $\overline{\mathbb{H}}$; we will say that A is a *hull* if the distance $d(0, A)$ is positive and if $\mathbb{H} \setminus A$ is simply connected. If A is a hull, we will denote by Ψ_A the unique conformal map from $\mathbb{H} \setminus A$ to $\mathbb{H}$ sending 0 to 0, ∞ to ∞ and such that $\Psi(z)/z$ tends to 1 at infinity. (Notice that this is not exactly the same normalization as that of g_t in the case of SLE; they differ by a real constant term.) If A is a hull, K a nice set and if $K \cap A = \varnothing$, then $\Psi_A(K)$ is again a nice set.

Now, let P be a probability measure supported on nice sets. We say that P is a *restriction measure* if the following happens: if K is distributed according to K, then so is λK for every $\lambda > 0$, and moreover, for every hull A, conditionally on the event $K \cap A = \varnothing$, the nice set $\Psi_A(K)$ is also distributed according to P.

We already saw that the whole trace of $\mathrm{SLE}_{8/3}$ is an example of such a restriction measure; in fact, there is a very simple structure theorem:

THEOREM II.2.6. *Let P be a restriction measure: there exists a real $\alpha \geqslant 5/8$ such that, if K is distributed according to P, then for every hull A,*

$$P[K \cap A = \varnothing] = \Psi_A'(0)^\alpha.$$

Moreover, for every real $\alpha \geqslant 5/8$, there is a unique restriction measure P_α satisfying the previous relation for every hull A. The only restriction measure supported on simple curves is $P_{5/8}$, which corresponds to $\mathrm{SLE}_{8/3}$.

There is another one of these restriction measures which is easy to describe and has to do with planar Brownian motion; essentially it is "Brownian motion in the half-plane, conditioned not to touch the boundary", or a variation of a *planar Brownian excursion.*

The easiest description is as follows: Let $\varepsilon > 0$, $R > 0$ and let B be Brownian motion started at $i\varepsilon$, conditioned to reach imaginary part R before hitting the real axis; by the gambler's ruin estimate, the probability of the conditioning event is ε/R. Let $K_{\varepsilon,R}$ be the path of that Brownian motion, up to the hitting time of imaginary part R. It is not difficult to prove that as ε goes to 0 and R to infinity, $K_{\varepsilon,R}$ converges in distribution to a random locally compact set K which intersects the real axis exactly at the origin. For short, while in this interlude we will refer to K as a *Brownian excursion.*

Mapping the picture through Ψ_A as ε goes to 0 and R to infinity sends $i\varepsilon$ to a point close to $i\varepsilon\Psi'(A)$ while it leaves the horizontal line $iR + \mathbb{R}$ close to invariant (because Ψ_A is normalized to be close to a horizontal translation near infinity). From this remark, conformal invariance of the Brownian path, and the gambler's ruin estimate, one directly obtains the fact that K is in fact distributed as the restriction measure P_1.

One final remark is in order. Let α_1 and α_2 both be real numbers at least equal to 5/8; let K_1 and K_2 be independent and distributed as P_{α_1} and P_{α_2}, respectively. Typically, the union $K_1 \cup K_2$ is not a nice set, but there is a well-defined minimal nice set, say $K_1 \oplus K_2$, containing both of them (one can think of it as being obtained from their union by "filling its holes", so we will call it the *filled union* of K_1 and K_2). By independence, it is obvious that $K_1 \oplus K_2$ is distributed as $P_{\alpha_1+\alpha_2}$.

This has a very interesting consequence: since 8 times 5/8 is equal to 5, the filled union of 8 independent realizations of $\mathrm{SLE}_{8/3}$ has the same distribution as the filled union of 5 independent Brownian excursions. In particular, the boundaries of those two sets have the same distribution. On the other hand, the boundary of the first one looks locally like the boundary of one of them, *i.e.* like an $\mathrm{SLE}_{8/3}$, while the boundary of the second one looks like that of a single Brownian excursion, which is th same as the Brownian frontier. This is one possible way to prove Mandelbrot's conjecture that the Brownian frontier has dimension 4/3.

II.2.2.3. *Crossing probabilities and Cardy's formula.* The initial motivation behind the definition of SLE was the conformal invariance of some scaling limits; here we looked particularly at percolation through Cardy's formula. We still have to identify the value of κ though, and crossing probabilities are a natural way to do it: one can compute them in terms of κ and match the result with Smirnov's theorem.[5]

THEOREM II.2.7. *Let $\kappa > 4$, $a < 0 < c$, and let $E_{a,c}$ be the event that the* SLE_κ *trace visits $[c, +\infty)$ before $(-\infty, a]$. Then,*

$$P\left[E_{a,c}\right] = F\left(\frac{-a}{c-a}\right) \quad \textit{where} \quad F(x) = \int_0^x \frac{\mathrm{d}u}{u^{4/\kappa}(1-u)^{4/\kappa}}.$$

This is equal to the Cardy-Smirnov result in the case $\kappa = 6$.

PROOF. The proof follows essentially the same lines as in the exercise on Bessel processes, but it is rather instructive, so we still give a very rough outline here for the benefit of the serious reader willing to do the computation. Let $A_t := g_t(a)$, $C_t = g_t(c)$ and

$$Z_t := \frac{\beta_t - A_t}{C_t - A_t}.$$

From Itô's formula, it is straightforward to obtain

$$\mathrm{d}Z_t = \frac{\sqrt{\kappa}\mathrm{d}B_t}{C_t - A_t} + \frac{2\mathrm{d}t}{(C_t - A_t)^2}\left(\frac{1}{Z_t} - \frac{1}{1 - Z_t}\right),$$

which after the time-change $\mathrm{d}s = \mathrm{d}t/(C_t - A_t)^2$, $\tilde{Z}_s = Z_t$ leads to

$$\mathrm{d}\tilde{Z}_s = \sqrt{\kappa}\mathrm{d}\tilde{B}_s + 2\left(\frac{1}{\tilde{Z}_s} - \frac{1}{1 - \tilde{Z}_s}\right).$$

Finding F now amounts to writing that the drift term of $F(Z_{\tilde{s}})$ should vanish, thus leading to the following differential equation:

$$\frac{\kappa}{4}F''(x) + \left(\frac{1}{x} - \frac{1}{1-x}\right)F'(x) = 0.$$

Proceeding from this is left as an exercise. □

[5]Historically, the first tool used to predict which value of the parameter κ corresponds to which model was to derive estimates for the winding numbers of the SLE curves as a function of κ; this has the advantage of being more general, but if one is only interested in the case of percolation, crossing probabilities give a shorter route.

II.2.2.4. *Arm events and critical exponents.* We now turn to radial SLE in the disk. For the remaining of the section, fix $\kappa > 4$ (as the estimates we are going to consider would be trivial in the case $\kappa \leqslant 4$). We will compute the *one-arm* and *two-arm exponents* of radial SLE_κ; the names should become clear as soon as one sees the SLE trace as an exploration process ...

For $\varepsilon > 0$, let τ_ε be the first time the radial SLE_κ trace visits the circle of radius ε around 0. Besides, let T be the first time t when K_t contains the whole unit circle. In addition, let U be the first time t when $\gamma_{[0,t]}$ contains both a clockwise and a counterclockwise loop separating 0 from the unit circle; note that almost surely $0 < T < U < \infty$.

THEOREM II.2.8. *As* $\varepsilon \to 0$, *the* one-arm probability *scales like*

$$P\left[\tau_\varepsilon < U\right] = \varepsilon^{\lambda_\kappa^{(1)} + o(1)} \quad \textit{with} \quad \lambda_\kappa^{(1)} = \frac{\kappa^2 - 16}{32\kappa};$$

the two-arm probability *(or* non-disconnection probability*) behaves like*

$$P\left[\tau_\varepsilon < T\right] = \varepsilon^{\lambda_k^{(2)} + o(1)} \quad \textit{with} \quad \lambda_k^{(2)} = \frac{\kappa - 4}{8}.$$

The case $\kappa = 6$ is of particular interest for us because of its ties to critical percolation: from these SLE estimates, one gets that the one-arm exponent of 2D percolation is $5/48$ and that the two-arm exponent is equal to $1/4$.

The proof is presented in detail in the form of an exercise (in the case $\kappa = 6$, but this is nothing special here); the overall idea is the same as that of the proof of Cardy's formula, *i.e.* to derive a PDE from Loewner's equation, to identify boundary conditions, and to exhibit a positive eigenfunction.

EXERCISE II.2.2 (Disconnection exponent for SLE_6). *Let $x \in (0, 2\pi)$ and let $\mathcal{H}(x,t)$ be the event that one radial* SLE(6) *starting from 1 does not disconnect e^{ix} from 0 before time t. The goal of this exercise is to show that there exists $c > 0$ universal such that*

$$e^{-\frac{t}{4}}\big(\sin\frac{x}{2}\big)^{\frac{1}{3}} \leqslant \mathbb{P}\big[\mathcal{H}(x,t)\big] \leqslant c e^{-\frac{t}{4}}\big(\sin\frac{x}{2}\big)^{\frac{1}{3}}.$$

(1) *Let ζ_t be the driving process of the* SLE *(it is $\sqrt{6}$ times a standard Brownian motion). Why can one define a real valued process Y_t^x such that $g_t(e^{ix}) = \zeta_t \exp(iY_t^x)$ and $Y_0^x = x$ for every $t < \tau(e^{ix})$ ($\tau(z)$ is the disconnecting time)? Show that*

$$dY_t^x = \sqrt{6} dB_t + cotg(Y_t^x/2)dt.$$

Hint. Recall that the argument is the imaginary part of the logarithm. Moreover, what is $\mathrm{Im}\big[\partial_t \log g_t(e^{ix})\big]$?

(2) *Let $\tau_x := \inf\{t \geqslant 0 : Y_t^x \in \{0, 2\pi\}\}$, prove that $\mathbb{P}[\mathcal{H}(x,t)] = \mathbb{P}[\tau^x > t]$.*

(3) *Assume that $f(x,t) := \mathbb{P}\big[\mathcal{H}(x,t)\big]$ is smooth on $(0, 2\pi) \times [0, \infty)$ (the general theory of diffusion processes guarantees that), show that*

$$3f'' + \cot(x/2)f' = \partial_t f$$

and that $\lim_{x \to 0^+} f(x,t) = \lim_{x \to 2\pi^-} f(x,t) = 0$ *and* $f(x,0) = 1$.

(4) *Define* $F(x,t) := \mathbb{E}\big[\mathbb{1}_{\mathcal{H}(x,t)}(\sin \frac{Y_t^x}{2})^{\frac{1}{3}}\big]$ *and show that*

$$F(x,t) = e^{-\frac{t}{4}}\big(\sin \frac{x}{2}\big)^{\frac{1}{3}}.$$

Hint. We can assume that the solutions of the PDE are determined by boundary conditions.

(5) *Conclude the proof.*

Exercise II.2.3 (Disconnection exponent II). *Let* $x \in (0, 2\pi)$ *and let* $\mathcal{J}(t)$ *be the event that one radial* SLE(6) *starting from 1 does not close any counterclockwise loop before time* t. *Let* ∂_t^1 *be the part of* $\partial K_t \setminus \partial\mathbb{U}$ *lying on the left of the endpoint* γ_t. *We set* Y_t *to be the arc-length of* $g_t(\partial_t^1)$.

(1) *Find an* SDE *which is satisfied by* Y *and express* $\mathbb{P}\big[\mathcal{J}(t)\big]$ *in terms of the survival time of* Y.

(2) *Find a* PDE *which is satisfied by the function* $h(x,t) := \int_0^1 f(x, t+s)ds$ *where* $f(x,t) := \mathbb{P}(2\pi \notin Y[0,t]|Y_0 = x)$. *What are the boundary conditions?* Hint. One can use a relation with discrete models in order to prove that $h(x,t) - h(0,t) = o(x)$; this relation can be assumed to hold.

(3) *Explain how one could prove that there exist* $0 < c_1, c_2 < \infty$ *such that*

$$c_1 e^{-\frac{5t}{48}} \leqslant \mathbb{P}\big[\mathcal{J}(t)\big] \leqslant c_2 e^{-\frac{5t}{48}}.$$

(4) *Show that the probability of a radial* SLE(6) *starting from 1 does not close any counterclockwise loop before touching the circle of radius* ε *is of order* $\varepsilon^{5/48}$.

II.3. Bibliographical notes

It is still difficult to find a self-contained reference on SLE processes. Lawler's book [**20**] is a good start, and contains both the basics of stochastic calculus and complex analysis. Werner's Saint-Flour lecture notes [**40**] assume more preliminary knowledge.

Of course, it is always a good idea to have a look at the articles themselves. The very first paper where SLE was introduced by Schramm, together with the reasoning at the beginning of the part, is about loop-erased walks [**32**]; a reference for the complex-analytic statements on Loewner chains is the book of Pommerenke [**29**].

One can then consult the whole series of articles by Lawler, Schramm and Werner [**21, 22, 23, 24**], as well as the (very technical) article of Rohde and Schramm [**30**] for the existence of the trace. The Hausdorff dimension of the trace is derived in [**2, 4**].

Part III . CONVERGENCE TO SLE

In this part, we gather two things: first, information about convergence of discrete objects to SLE in the scaling limit, *i.e.*, a description of the method of proof and its application in a few cases; and second, uses of convergence to get estimates about the discrete models themselves, mainly in the form of the values of some critical exponents.

III.1. Convergence: the general argument

We already have a few convergence results for *observables* of discrete models; let us focus for now on percolation and Cardy's formula. The question is whether the convergence of crossing probabilities is enough to obtain the convergence of the exploration process to the trace of an SLE (in that case, SLE_6 because of the locality property, or by matching the crossing probabilities). As it turns out, this is not an easy question; it seems that this information in itself is not quite enough to conclude.[6]

Moreover, in other cases such as the random-cluster model, the information contained in the observable is not directly of a geometric nature, and it is not clear at first how to extract geometry from it.

III.1.1. First attempt, percolation. The most natural approach (which is briefly described by Smirnov in [**35, 36**] and can indeed be applied in full rigor in the case of percolation) is the following. Let $\varepsilon > 0$ and let $\delta \in (0, \varepsilon)$; look at critical site-percolation on the triangular lattice of mesh size δ, in the upper-half plane. Fix boundary conditions to be open to the right of the origin and closed to the left, and let γ^δ be the corresponding exploration curve; let τ be its first exit time of the disk of radius ε around 0.

The distribution of $\gamma^\delta(\tau)$, asymptotically as $\delta \to 0$, is precisely given in terms of Cardy's formula in a half-disk. Indeed, for fixed δ, the probability that γ^δ exits the disk to the left of a point $z \in \mathbb{H} \cap \mathcal{C}(0, \varepsilon)$ is exactly the probability that there is an open crossing of the half-disk between the boundary intervals $(0, \varepsilon)$ and $(z, -\varepsilon)$, which we know converges to an explicit limit as $\delta \to 0$ by Theorem I.1.1.

In fact, this distribution is also the same, through the locality property and the computation of SLE crossing probabilities, as that of the exit point of the half-disk by an SLE_6 process. This means that morally, at the scale ε and as $\delta \to 0$, the beginning of an SLE_6 and that of a percolation exploration process look "very similar."

There is a more specific statement about SLE_6 which we will not in fact need later but which is very nice anyway. Let Ω be a simply connected domain in the plane, and let a, b and c be three points (in that order) on $\partial\Omega$. From the locality property, we know that we can define an SLE_6 from a to the boundary arc bc, until it touches that boundary arc; let K be its shape just before it touches. [Technically, recall that this SLE is defined as SLE_6 from a to either b or c, and that these two processes agree until γ disconnects b from c, which happens exactly when the trace hits bc.]

The set K is a relatively compact subset of Ω; its complement in Ω has two connected components, one containing b on its boundary and the other, c. Now, let A be a hull in Ω (*i.e.*, in this case, a relatively compact subset of Ω which does not disconnect a from bc). It is easy to compute the probability that K does not intersect A: indeed, again by the locality property, it is exactly equal to the probability that SLE_6 from a to $bc \setminus \overline{A}$ in $\Omega \setminus A$ touches bc before it touches ∂A.

[6] It might be enough if one can derive relative compactness from it directly; see for instance [**15**], [**16**] and possibly [**34**] (after reading the rest of this section) for current progress in this direction.

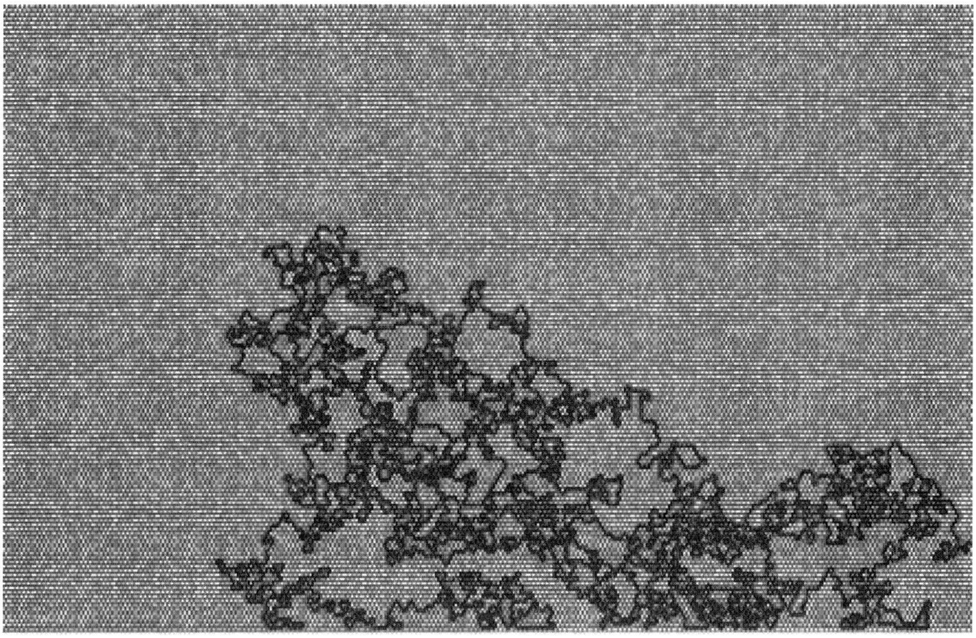

FIGURE 10. The same exploration curve of critical percolation as before, with a smaller lattice mesh.

This last probability can be expressed in terms of crossing probabilities within $\Omega \setminus A$; on the other hand, by a Donsker-class type argument, the data of the non-intersection probabilities $P[K \cap A = \varnothing]$ for all hulls A characterizes the distribution of the set K itself. In other words: morally, the shape of SLE_6 "as seen from the outside" is characterized by its crossing probabilities.

Now, Theorem I.1.1 gives us convergence of percolation crossing probabilities to those of SLE_6, and this implies the convergence of the exploration process itself to the SLE process — still *seen from the outside*. In particular, the outer shape of a large percolation cluster can be described by the boundary of an SLE_6. All this however says very little about the convergence of the exploration process *seen as a curve*.

Coming back to our proof attempt: at the time when they both exit the ball of radius ε around the origin, the exploration process of percolation at mesh δ and the trace of SLE_6 look very similar. In addition to that, they both satisfy the domain Markov property; if z_1 denotes the hitting point of the circle of radius ε by either of them, and K_1 the (filled) shape at the hitting time τ_1, then the distribution of the process after time τ_1 is the same as the initial process started from z_1 in the domain $\mathbb{H} \setminus K_1$.

One can then look at the hitting time τ_2 of the circle of radius ε around z_1: the two processes will "live" in very similar domains, so their outer shape K_2 will still be very similar at time τ_2. Inductively, one can then couple a percolation exploration with an SLE_6 process through a chain of disks of radius ε.

Letting $\varepsilon \to 0$, on the SLE side this gives the whole information about the trace of the process, so it should be possible to leverage the construction into a proof of pathwise convergence. Unfortunately, the main piece missing from the puzzle here is an estimate of the speed of convergence in Theorem I.1.1, meaning that at each step of the process there is an error term which we cannot estimate (and the errors accumulate as the construction proceeds); moreover, even if one manages to produce a fully formal proof, it would rely too strongly on the locality property to be of any more general use. We need another idea.

III.1.2. Proving convergence using an observable. Let us consider the exploration process of percolation a bit more. One way to represent it graphically is to see percolation itself as a random coloring of the faces of the hexagonal lattice, in which case the exploration curve can be seen as a collection of edges of the hexagonal lattice separating hexagons of different colors. In other words, it can be seen as a piecewise linear curve γ^δ in the upper half-plane.

Being a curve in the half-plane, it is amenable to the previous general construction of Loewner chains, which gives it a natural continuous time-parametrization, a family of conformal maps (g_t^δ), and encodes it into a real-valued driving process, say $(\beta_t^\delta)_{t\geqslant 0}$ where again δ is the mesh of the lattice. Since γ^δ is piecewise linear, β^δ is piecewise smooth.

Morally, Loewner chains should depend continuously on their driving functions. This means that a natural notion of convergence of γ^δ to an SLE_6 is the convergence of β^δ to a Brownian motion of appropriate variance as $\delta \to 0$. This is a good plan of attack, for two reasons:

- Proving convergence in distribution of a sequence of real-valued processes is a classical problem, and there are several well-known techniques to choose from;
- Convergence at the level of the driving processes does not seem at first sufficient to obtain pathwise convergence, but what is missing is an *a priori* estimate of interface regularity, similar to the Aizenman-Burchard precompactness criterion [**1**]; and in fact, in most cases such regularity *can* indeed be extracted from convergence (though we won't say more about this here — see [**34**] for details).

So now we need a tool to prove convergence of the "discrete driving process" to the appropriate Brownian motion as $\delta \to 0$. Here is a general framework; the actual implementation will depend on the model. For all $n > 0$, let Γ_n^δ be the discrete exploration up to its n-th step and let $H_n^\delta := \mathbb{H} \setminus \Gamma_n^\delta$; let τ_n^δ be the corresponding time-parameter, so that $g_{\tau_n^\delta}^\delta$ maps H_n^δ conformally back to $\mathbb{H}$.

Let z be a point in the upper half-plane, and let A_z^δ be the event that γ^δ passes to the left of z — to fix ideas; the observable of choice can vary from model to model, but the general argument will be the same in all cases. Let X_n^δ be the conditional probability of A_z^δ, given Γ_n. The key remark is that the sequence (X_n^δ) is a martingale; it converges almost surely, and its limit is either 0 or 1 according to whether γ^δ passes to the left or to the right of z.

The main assumption we will make is that we know how to compute the limit

$$\varphi(z) := \lim_{\delta\to 0} P[A_z^\delta] = \lim_{\delta\to 0} X_0^\delta,$$

that the function φ is smooth, and that we have conformal invariance (in the same sense as in Theorem I.1.1). Notice that, for percolation, this is not exactly the kind of probability that Theorem I.1.1 gives, but it is close enough to present the gist of the argument.

Because of the domain Markov property of the exploration process, X_n^δ is the probability that the exploration curve *defined in the domain* H_n passes to the left of z; by conformal invariance, this is (close to) the probability that the exploration in the initial domain passes to the left of $g^\delta_{\tau^\delta_n}(z) - \beta^\delta_{\tau^\delta_n}$. In other terms, morally

$$X_n^\delta \simeq \varphi\left(g^\delta_{\tau^\delta_n}(z) - \beta^\delta_{\tau^\delta_n}\right).$$

Now let ε be small, and let N be the first time at which either τ_n^δ is larger than ε^2, or $|\beta^\delta_{\tau^\delta_n}|$ is larger than ε; let $\sigma := \tau_N^\delta$. If δ is taken small enough, σ cannot be much larger than ε^2, and $|\beta^\delta_\sigma|$ cannot be much larger than ε (only the N-th step of the exploration process needs to be accounted for). Since σ is small, Loewner's equation gives

$$g^\delta_\sigma(z) \simeq z - \frac{2\sigma}{z}$$

so that, by the previous paragraph,

$$X_N^\delta \simeq \varphi\left(z - \beta^\delta_\sigma - \frac{2\sigma}{z}\right)$$

(the first "small" term β^δ_σ being of order ε, and the second one $2\sigma/z$ of order ε^2). Because X is a martingale, this boils down to

$$E\left[\varphi\left(z - \beta^\delta_\sigma - \frac{2\sigma}{z}\right)\right] \simeq \varphi(z).$$

The point here is that, since we know φ explicitly, we can power-expand it around z inside the expectation and then match the two sides of the relation. This will provide a relation between the powers of β^z_σ, those of σ, and explicit functions of z coming from the appropriate derivatives of φ. This is not enough to identify the driving process, but one can always write the same relation for various values of z, and this typically leads to a pair of equations of the form

$$\begin{cases} E[\beta^\delta_\sigma] & \simeq & 0 \\ E[(\beta^\delta_\sigma)^2 - \kappa\sigma] & \simeq & 0 \end{cases}$$

(where κ is a constant coming out of the computation, which will be the parameter of the SLE in the scaling limit, and where of course the symbol '$\simeq$' means that equality holds up to error terms, which have to be controlled along all the previous steps).

The last step of the proof involves what is known as *Skorokhod embedding.* The basic statement is the following: given a square-integrable random variable Z such that $E[Z] = 0$, there exists a standard Brownian motion (B_t) and a stopping time T such that B_T has the same distribution as Z, and satisfying the equality $E[T] = E[Z^2]$. Applying this to β^δ_σ above, we find that we can write it (still up to error terms) as $B_{\kappa T_1}$ where $E[T_1] = E[\sigma]$.

It remains to iterate the process. Once the discrete interface is explored up to capacity σ, what remains is a random discrete domain H_N^δ in which the exploration can be extended, thus extending β^δ from time $\sigma = \sigma_1$ to some $\sigma_2 > \sigma_1$; by the very same argument as above, we get

$$\begin{cases} E[\beta^\delta_{\sigma_2 - \sigma_1}] & \simeq & 0 \\ E[(\beta_{\sigma_2} - \beta^\delta_{\sigma_1})^2 - \kappa(\sigma_2 - \sigma_1)] & \simeq & 0 \end{cases}$$

and iteratively, we can construct an increasing sequence (σ_k) of stopping times and the corresponding sequence (T_k) through repeated use of Skorokhod embedding.

The increments $(T_{k+1} - T_k)$ are essentially independent and identically distributed, because of the domain Markov property of the underlying discrete model; thus a law of large numbers applies, stating that T_k is approximately equal to its expectation. Combining this with the previous remark, we arrive at the fact that

$$\beta^\delta_{\sigma_k} \simeq B_{\kappa\sigma_k}$$

where again B is a standard Brownian motion.

In other words, provided that δ is small, the discrete driving process is very close to being a Brownian motion with variance parameter κ; and as $\delta \to 0$, the error terms will vanish and we get

$$(\beta^\delta_t) \underset{\delta \to 0}{\to} (B_{\kappa t}).$$

One final remark is in order. What we just did in the case of an observable depending on a point z within the domain does not in fact depend directly on the existence of z; only the martingale (X^δ_n) is relevant. Of course, if (X^δ_n) is not defined in terms of z, the function φ will have to be replaced accordingly, and we will see an example of this below in the case of the UST contour. Nevertheless, the core of the argument is the same in all cases.

Notice how conformal invariance of the scaling limit, and statements of convergence in other domains, come "for free" with the rest of the argument as soon as the scaling limit for φ is itself conformally invariant: since SLE in another domain is defined via conformal mappings anyway, the discrete driving process itself drives an SLE in the upper half-plane, and all that is needed in addition of the above argument is a composition by the conformal map from the domain to $\mathbb{H}$.

III.2. The proof of convergence in a few cases

What remains to be done now is to apply the above strategy to a few actual models. This amounts to two things to do for each model: find an appropriate observable (which is what the reader will find below); and refine driving process convergence into pathwise convergence (which, being of a much more technical nature, will be kept out of these notes and can be found in the literature).

III.2.1. The UST Peano curve. The simplest case to state is that of the uniform spanning tree in the upper half-plane, on a square lattice of mesh δ, with wired boundary conditions to the right of the origin and free boundary conditions to the left. The dual of that tree is a uniform spanning tree with reversed boundary conditions (wired on the left and free on the right); and the curve γ^δ winding between the two is known as the *UST Peano curve*, or UST contour curve. It is a simple exercise to check that the curve γ^δ satisfies the domain Markov property.

Let $z \in \mathbb{H} \cap \delta\mathbb{Z}^2$ and let $x \in \mathbb{R}_+$; let A^δ_z be the event that the branch of the UST containing z lands on the real axis somewhere on the interval $[0, 1]$, and let $\varphi_\delta(z)$ be the probability of this event. The key remark is the following: from Wilson's algorithm, $\varphi_\delta(z)$ is exactly equal to the (discrete) harmonic measure of the interval $[0, 1]$ in $\mathbb{H}$ seen from z, with reflecting boundary conditions on the negative real axis; or equivalently, to the harmonic measure of the interval $[0, 1]$ in the slit plane $(\mathbb{C} \cap \delta\mathbb{Z}^2) \setminus \mathbb{R}_+$, seen from z.

This converges to the continuous counterpart of that harmonic measure, which is easily computable: if $z = re^{i\theta}$,

$$\begin{aligned}\varphi_\delta(z) \to \varphi(z) &:= \omega^z_{\mathbb{C}\setminus\mathbb{R}_+}([0,1]) = \omega^{z^{1/2}}_{\mathbb{H}}([-1,1]) \\ &= \frac{1}{\pi}\cot^{-1}\left(\frac{|z|-1}{2\,\mathrm{Im}\sqrt{z}}\right) = \frac{1}{\pi}\cot^{-1}\left(\frac{r-1}{2\sqrt{r}\sin(\theta/2)}\right)\end{aligned}$$

where the determination of $\cot^{-1}$ is taken in the interval $(0,\pi)$. Because harmonic measure (or equivalently, planar Brownian motion up to time-reparametrization) is conformally invariant, so is φ, in the sense that if we defined the UST in another domain, with 3 boundary intervals as above, the scaling limit of the hitting probability would be conjugated to φ by the appropriate conformal map.

EXERCISE III.2.1. *Show the convergence of φ_δ to φ and prove the formulas giving the value of $\varphi(z)$ as a function of z.*

Answer: The convergence can be obtained by coupling a discrete with a planar Brownian motion and writing the continuous harmonic measure in terms of hitting probabilities. To prove the formula, it is enough to show that the function as given takes values between 0 and 1, and satisfies appropriate boundary conditions: equal to 1 on $[0,1]$, 0 on $[1,+\infty)$ with vanishing normal derivatives along $(-\infty,0)$.

Now, all that remains to do to show convergence of the driving process to a Brownian motion is to apply the strategy described above; one gets $\kappa = 8$ from the computation, so the UST contour curve converges (in the driving-process topology) to SLE_8 in the scaling limit.

III.2.2. The loop-erased random walk. The case of the loop-erased random walk is a little bit more involved. First of all, convergence will be to *radial* SLE rather than chordal as in the case of the UST contour curve — though this is a minor point, as the scheme of the proof is exactly the same in both cases. Hence, we will work primarily with the loop-erased walk from the origin to 1 in the unit disk $\mathbb{U}$; let Γ be the path of a simple random walk from 0 to 1 in $\mathbb{U}\cap\delta\mathbb{Z}^2$ (*i.e.*, conditioned to exit the domain at 1); let γ be the loop-erasure of its time-reversal. As in the first part, this defines a decreasing sequence of domains $\Omega_n := (\mathbb{U}\cap\delta\mathbb{Z}^2)\setminus\{\gamma_i : i < n\}$; besides, γ turns out to have exactly the same distribution as the time-reversal of the loop-erasure of Γ (though of course they differ in general).

Fix $v \in \mathbb{U}\cap\delta\mathbb{Z}^2$, neither too close to the boundary nor to the origin (say for instance $d(0,v)\in[1/3,2/3]$), and let Z be the number of visits of v by Γ before it exits $\mathbb{U}$; this will replace the event defining the observable in the previous section: $X_n^\delta := E[Z|\gamma_0,\dots,\gamma_n]$. As above, the key is to write the fact that (X_n^δ) is a martingale, or in other words that

$$E[X_n^\delta] \quad = \quad E[X_0^\delta] \quad (= E[Z]),$$

and to estimate both sides to the appropriate precision. Notice that $E[Z]$ is nothing but a Green function.

Conditionally on $(\gamma_j)_{j\leqslant\ell}$, let n_j be the first hitting time of γ_j by Γ and let Γ^j be the portion of Γ between times n_j and n_{j-1}; let Z_j be the number of visits of v by Γ^j. The key argument is then that the distribution of Γ up to time n_j is the same as that of a simple random walk conditioned to exit Ω_j through γ_j, so that one can write Z as the sum of the Z_j plus the number of visits to v before time n_ℓ, and this last term has the same distribution as that of the version of Z as defined in Ω_ℓ.

Now, the structure of Γ^1 is simple: it is a finite sequence of random walk excursions in Ω based at γ_1, followed by a jump from γ_1 to γ_0. This enables the computation of the expectation of Z_1 in terms of (discrete) Green functions and hitting probabilities within the domain Ω. Asymptotically as the lattice mesh goes to 0, the behavior of those is well understood (if technical to obtain to the right precision); all the limiting quantities as $\delta \to 0$ are conformally invariant and explicit.

Since we want to focus here on the heuristics of the proof, we refer the more serious readers to the initial article of Lawler, Schramm and Werner [**25**] for the full details of the proof.

III.2.3. Percolation. In the case of percolation, we again have convergence of a discrete observable (the crossing probability of a conformal rectangle) to an explicit scaling limit, so the general framework of the proof is still the same.

Here, the natural way to set up the computation is the following: let $d < 0 < b$, and let $A_{b,d}^\delta$ be the event that the critical percolation exploration curve in the upper-half plane, started at the origin, touches the half-line $[b, +\infty)$ before the half-line $(-\infty, d]$ (so, instead of depending on the location of one point in Ω, it depends on the location of two points on its boundary). The limit of $P[A_{b,d}^\delta]$ as $\delta \to 0$ can be computed from Theorem I.1.1, and is conformally invariant.

III.2.4. The Ising model. The observable in this case is given by the para-fermionic observable introduced in the previous part, and the general scheme of the proof is once again the same; one key difference is that the observable as defined initially goes to 0 with the lattice mesh (as $\delta^{1/2}$), so it has to be normalized accordingly. The reader can find all the details in the notes of Smirnov's course on that very topic at the same school.

III.3. One application of convergence: critical exponents

We saw that the disconnection exponent for radial SLE_6 is equal to 1/4. This has a natural counterpart (and generalization) in terms of critical site-percolation on the triangular lattice:

THEOREM III.3.1 (Arm exponents for critical percolation). *Consider critical site-percolation on the triangular lattice in the plane; recall that Λ_n is the intersection of the lattice with the ball of radius n. As $n \to \infty$,*

$$P\left[0 \leftrightarrow \partial\Lambda_n\right] \approx n^{-5/48}.$$

Besides, let $k > 1$, fix a sequence of k colors $\sigma = (\sigma_i)_{1 \leqslant i \leqslant k}$ and let $A_{k,\sigma}(n)$ be the event that there exist, between the circle of radius k and that of radius n, k disjoint paths (γ_i) (in that order), such that all the vertices along γ_i are of color σ_i. As $n \to \infty$,

$$P[A_{k,\sigma}(n)] \approx n^{-\alpha_k} \quad \textit{where} \quad \alpha_k := \frac{k^2 - 1}{12}.$$

The exponents α_k are called *polychromatic k-arm exponents*; notice that the non-disconnection exponent of SLE_6 corresponds to α_2. The value of α_k does not depend on the precise sequence of colors σ, as long as both colors are present. The main reason for this already appeared in the proof of the Cardy-Smirnov formula (see section I.1.1.2 in Part I): conditioned on the existence of 2 arms of different colors, it is possible to discover those two arms using the exploration process — this actually provides an algorithm to find the leftmost white arm and the rightmost black arm, say. The exploration process depends only on the states of the vertices between those two arms, so one can swap the state of all non-explored vertices, and this gives a correspondence between the events $A_{k,\sigma}(n)$ and $A_{k,\tilde{\sigma}}(n)$ where $k-2$ colors differ between σ and $\tilde{\sigma}$. Variations of the argument allow to relate any two sequences of colors, as long as the exploration process is available, *i.e.* as soon as both colors are present.

In fact, it can be shown that the situation is indeed similar but with different exponents in the monochromatic case (see [**8**]):

THEOREM III.3.2 (Monochromatic arm exponents for percolation). *Consider critical site-percolation on the triangular lattice in the plane; let $k > 1$, and let $B_k(n)$ be the event that there exist, between the circle of radius k and that of radius n, k disjoint open paths. As $n \to \infty$,*

$$P[B_k(n)] \approx n^{-\tilde{\alpha}_k} \quad \text{for some} \quad \tilde{\alpha}_k \in (\alpha_k, \alpha_{k+1}).$$

The values of the exponents $\tilde{\alpha}_k$ are not known for $k > 1$. The case of $\tilde{\alpha}_2$ is of particular interest; that exponent is known as the *backbone exponent* in the physics literature, and its value is very close to 17/48 (see [**7**]). Whether that is its actual value remains open.

As we did before, we present the proof of Theorem III.3.1 in the form of two exercises (one for the one-arm exponent, and one for the others).

EXERCISE III.3.1 (One-arm exponent for percolation). *Consider critical site percolation on the triangular lattice; we want to prove that*

$$\mathbb{P}(0 \leftrightarrow \partial\Lambda_n) = n^{-\frac{5}{48}+o(1)}$$

where $\partial\Lambda_n$ is the boundary of the box of size n. We assume that the result of exercise 3 is known.

(1) *Let $\pi(r, R)$ be the probability that there exists a path between $\partial\Lambda_r$ and $\partial\Lambda_R$. Show that there exists a constant $c > 0$ uniform in r and R such that*

$$c\pi(r_1, r_2)\pi(r_2, r_3) \leqslant \pi(r_1, r_3) \leqslant \pi(r_1, r_2)\pi(r_2, r_3).$$

Hint. For the left side of the inequality, use RSW.

(2) *Consider percolation on a finite subgraph of the triangular lattice with circular shape. Explain how to define a natural exploration process at a discrete level. Towards which process should it converge (we do not ask for a proof!!)?*

(3) *What is the event associated to $\{\partial\Lambda_r \leftrightarrow \partial\Lambda_R\}$ for the exploration process? Show that there exist $0 < c_1, c_2 < \infty$ such that for every R,*

$$c_1 R^{-\frac{5}{48}} \leqslant \pi(n, nR) \leqslant c_2 R^{-\frac{5}{48}}$$

when n is large enough.

(4) *Conclude the proof.*

EXERCISE III.3.2 (Universal exponents). *Let σ be a finite sequence of colors (B for black, W for white). We associate to $n > 0$ and $\sigma = \{\sigma_1, \dots, \sigma_k\}$ the event $A_\sigma(n)$ that there exist paths $\gamma_1, \dots, \gamma_k$ such that:*

- *γ_i has color σ_i*
- *γ_i connects the origin to the boundary of $[-n,n]^2$. When $k>2$, we require only that the paths connect $[-k,k]^2$ to the boundary of $[-n,n]^2$.*
- *$\gamma_1,\dots,\gamma_k$ can be found in counterclockwise order.*

We define the same event in the upper half-plane (which we denote by $A^{\mathbb{H}}_\sigma(n)$). In this case, the paths must be found in counterclockwise order, starting from the right.

(1) *a) Prove that $\mathbb{P}(A^{\mathbb{H}}_{BW}(n)) \geqslant \frac{c}{n}$ for some universal constant c.* Hint. Use the RSW theorem to construct a point in $\{0\}\times[-n/2,n/2]$ which is connected to the boundary of the box by two arms of distinct colors.

b) Assume $A^{\mathbb{H}}_{BW}(n)$ holds. We require that the site on the left of 0 is white and that it is the start of the white path, and the site on the right is black and is the start of the black path. Show that one can explore the interface between the black and the white paths without exploring any other site;

c) Let $B(n)$ be the event that there exist a white path connected to the left side of $[-n,n]\times[0,n]$ and a black path connected to the right side. Show that there exists a universal $c_1>0$ such that

$$\mathbb{P}(A^{\mathbb{H}}_{BW}(n)) \leqslant c_1 \mathbb{P}(B(n));$$

d) Deduce that there exists $c_2>0$ such that

$$\mathbb{P}(A^{\mathbb{H}}_{BW}(n)) \leqslant \frac{c_2}{n}.$$

What was proved?

(2) *Prove that the exponent for BWBWW in the plane is 2.*

(3) (difficult) *Prove that the exponent for BWBW in the plane is smaller than 2.*

III.4. Bibliographical notes

Section III.1. The argument outlined in the beginning of the section is that initially described by Smirnov [**35, 36**], and more details, including the necessary technicalities involved in the proof of convergence of the exploration process to the trace of SLE_6, are in the article of Camia and Newman [**11**]. The general method is fully described in a paper by Lawler, Schramm and Werner [**25**], applied in the cases of the LERW and the UST Peano curve.

Section III.2. In addition to the previously mentioned articles, we simply refer the reader to the notes for Smirnov's Buzios course [**38**] again.

Section III.3. The polychromatic percolation exponents were obtained from SLE by Werner and Smirnov [**39**]; the existence of the monochromatic ones was proved in [**8**].

Part IV . MATHEMATICAL TOOLBOX

We gather in this section a few exercises from the sessions in Buzios, which provide some mathematical background for the main body of these notes. Somehow every result proved here is classical in some mathematical communities, but which community depends on the result — SLE itself being at the interface between probability and complex analysis.

IV.1. Probabilistic tools

IV.1.1. Stochastic calculus and Itô's formula. In this section, B is a standard one-dimensional Brownian motion. Let $(\mathcal{F}_t)$ be the filtration associated to the Brownian motion, *i.e.* $\mathcal{F}_t = \sigma(B_s, s < t)$. The process (M_t) is a *martingale* (with respect to $\mathcal{F}_t$) if for each $s < t$, $\mathbb{E}[|M_t|] < \infty$ and $\mathbb{E}[M_t|\mathcal{F}_s] = M_s$.

EXERCISE IV.1.1 (integration with respect to Brownian motion).

(1) *We call* H *a* simple process *if it is of the form*

$$H_s = \sum_{j=1}^{n} C_j \mathbb{1}_{[t_{j-1},t_j)}(s)$$

where (t_j) *is increasing and* C_j *is* $\mathcal{F}_{t_{j-1}}$*-measurable;*

a) For a (random) process $H = C\mathbb{1}_{[s,t)}$*, where* C *is* $\mathcal{F}_s$*-measurable, find a natural candidate for the integral of* H *against Brownian motion* B*, in other words, what could* $\int_0^\infty H_s dB_s$ *be? How could the notion of integral be extended to any simple process?*

b) We assume that the integral has been constructed as above. For any simple process H*, check that*

$$\mathbb{E}\left[\left(\int_0^\infty H_s dB_s\right)^2\right] = \int_0^\infty \mathbb{E}[H_s^2]ds.$$

(2) *Let* $\mathcal{L}^2$ *the set of square integrable adapted processes (*i.e., processes (H_s) *satisfying* $\int_0^\infty \mathbb{E}[H_s^2]ds < \infty$*). Explain how to extend the definition of integral to* $\mathcal{L}^2$.

(3) *For a bounded adapted process* H*, we define* $\int_0^t H_s dB_s$ *as* $\int_0^\infty H_s \mathbb{1}_{[0,t)} dB_s$. *Show that* $M_t = \int_0^t H_s dB_s$ *is an* $\mathcal{F}_t$*-martingale.* Hint: Check it in the case of simple processes first. ** *Show that it is a continuous process.*

Remark: *Note that for any bounded adapted process* a*,* $\int_0^t a_s ds$ *is straightforward to define. It is also possible to check that* $H_t = \int_0^t a_s dB_s + \int_0^t \sigma_s ds$ *is a martingale if and only if* $\sigma = 0$.

(4) *a) Let* H_s *be a bounded continuous adapted process and* $t > 0$. *Considering subdivisions* $0 = t_1^n < \cdots < t_n^n = t$ *with* $\max(t_{i+1}^n - t_i^n) \to 0$*, show that*

$$\sum_{i=1}^{n-1} H_{t_i^n}(B_{t_{i+1}^n} - B_{t_i^n}) \xrightarrow{\mathcal{L}^2} \int_0^t H_s dB_s.$$

b) Let H_s *be a bounded continuous adapted process and* $t > 0$. *Considering subdivisions* $0 = t_1^n < \cdots < t_n^n = t$ *with* $\max(t_{i+1}^n - t_i^n) \to 0$*, show that*

$$\sum_{i=1}^{n-1} H_{t_i^n}(B_{t_{i+1}^n} - B_{t_i^n})^2 \xrightarrow{\mathcal{L}^2} \int_0^t H_s ds.$$

Hint: Recall that $B_t^2 - t$ is a martingale.

c) Prove Itô's formula: For any a, σ *bounded adapted processes and* $t > 0$*, we set* $Y_t = \int_0^t a_s dB_s + \int_0^t \sigma_s ds$. *Let* $\varphi : \mathbb{R} \to \mathbb{R}$ *be a function twice continuously derivable; then*

$$\varphi(Y_t) = \varphi(Y_0) + \int_0^t \varphi'(Y_s) a_s dB_s + \int_0^t \left[\varphi'(Y_s)\sigma_s + \frac{1}{2}\varphi''(Y_s) a_s^2\right] ds.$$

Remark: *In order to write the equality*

$$H_t = x + \int_0^t a_s dB_s + \int_0^t \sigma_s ds$$

in a concise way, we often write

$$H_0 = x \quad \textit{and} \quad dH_t = a_t dB_t + \sigma_t dt.$$

IV.1.2. An application: Bessel processes.

EXERCISE IV.1.2. *Let $d > 0$. We* ***assume*** *without proof that there exists a unique process, denoted X_t^x, which solves the following stochastic differential equation:*

$$dX_t^x = dB_t + \frac{d-1}{2X_t^x}dt, \quad X_0^x = x$$

*up to time $T_x := \inf\{t : X_t^x = 0\}$. This process is called a d-*dimensional Bessel process. *For integer values of d, this process is the norm of a d-dimensional vector with independent Brownian entries. Let $0 < a < x < b < \infty$, τ the first exit time of the set $[a, b]$, and $\varphi(x) = \mathbb{P}(X_\tau^x = a)$.*

(1) *Show that $\varphi(X_{t\wedge\tau}^x)$ is a martingale with respect to $\mathcal{F}_{t\wedge\tau}$.*
(2) *a) Assume φ is twice continuously differentiable. Using Itô's formula, deduce that*

$$\frac{1}{2}\varphi''(x) + \frac{d-1}{2x}\varphi'(x) = 0, \quad a < x < b,$$

and compute φ when $d \geqslant 2$,

b) When $d > 0$, compute $\mathbb{P}(X_\tau^x = a)$. What can you deduce?

(3) (difficult) *Using Itô's formula, show that $\psi(x, t) = \mathbb{P}_x(\tau > t)$ is the solution of a partial differential equation. Deduce an estimate for $\mathbb{P}_x(\tau > t)$ when t goes to infinity.*

IV.2. Complex analytic tools

IV.2.1. Conformal maps.

EXERCISE IV.2.1 (Around the Riemann mapping theorem). *Recall the statement of the RMT: Let D and D' be two simply connected domains included in $\mathbb{C}$ and different from $\mathbb{C}$, there exists a conformal map (*i.e. *a bijection differentiable in the complex variable) between D and D'.*

(1) *Find a conformal map between the following domains:*
   - *from $\mathbb{R}\times]0, \pi[$ to $\mathbb{H} = \{z, \mathrm{Im}(z) > 0\}$;*
   - *from the disk $\mathbb{D} = \{z, |z| < 1\}$ to $\mathbb{H}$;*
   - *from $\mathbb{H} \setminus [0, ir]$ to $\mathbb{H}$;*
   - *from $\mathbb{D}$ to $\mathbb{C} \setminus (-\infty, -\frac{1}{4}]$;*
   - *from $S_\epsilon = \big(\mathbb{R} \times (0, 2)\big) \setminus \big((i - \infty, i - \varepsilon] \cup [i + \varepsilon, i + \infty)\big)$ to $\mathbb{H}$*
   - *from $\mathbb{H}$ to an equilateral triangle.*
(2) *a) Show that there is no conformal map from $D(0, 1)$ to $\mathbb{C}$.* It confirms that the assumption $D \neq \mathbb{C}$ is necessary.

b) Let D be a simply connected domain and f be a conformal map; why is $f(D)$ simply connected?

(3) *What are the conformal maps from $D(0,1)$ into $D(0,1)$?* Hint. One can guess what they are and make sure none is omitted using Schwarz's Lemma. *Deduce that there are three (real) degrees of freedom in the choice of a conformal map between two domains in the following sense:*
   - *one can fix the image of one point on the boundary and the image of one point inside the domain;*
   - *one can fix the image of one point inside the domain and the direction of the derivative;*
   - *one can fix the image of three points on the boundary (keeping the order).*

EXERCISE IV.2.2 (Estimates for conformal maps).

(1) ***Schwarz' Lemma:*** *Let f be a continuous map from $\overline{\mathbb{D}}$ to $\overline{\mathbb{D}}$ such that $f(0) = 0$ and f is holomorphic inside $\mathbb{D}$. Show that $|f(z)| \leqslant |z|$.* Hint. Think about the maximum principle. *Study the case where $|f'(0)| = 1$.*

(2) ***Koebe's 1/4-theorem:*** *Let*

$$\mathcal{S} := \{f : \mathbb{D} \to \mathbb{C}, \text{ analytic, one-to-one with } f(0) = 0 \text{ and } f'(0) = 1\}$$

a) (Area theorem) Let $f \in \mathcal{S}$ and $K = \mathbb{C} \setminus \{1/z, z \in f(\mathbb{D})\}$, prove that

$$area(K) = \pi \left[1 - \sum_{n=1}^{\infty} n|b_n|^2\right]$$

where $1/f(1/z) = z + b_0 + \sum_{n\geqslant 1} \frac{b_n}{z^n}$. Note that it implies $|b_1| \leqslant 1$.

b) Prove that if $f = z + a_2 z + \cdots$ is in $\mathcal{S}$, then $|a_2| \leqslant 2$. Hint: construct a function $h \in \mathcal{S}$ such that $h(z) = z + \frac{a_2}{2} z^3 + \cdots$ and conclude.

c) Deduce Koebe's 1/4-theorem: if $f \in \mathcal{S}$, then $B(0, \frac{1}{4}) \subset f(\mathbb{D})$.

d) Suppose $f : D \to D'$ is a conformal transformation with $f(z) = z'$. Then

$$\frac{1}{4} \frac{d(z', \partial D')}{d(z, \partial D)} \leqslant |f'(z)| \leqslant 4 \frac{d(z', \partial D')}{d(z, \partial D)}.$$

IV.2.2. Interaction with Brownian motion.

EXERCISE IV.2.3 (Conformal invariance of Brownian motion). *Consider B a Brownian motion in the plane, and for a domain U, $\tau_U := \inf\{t \geqslant 0 : B_t \notin U\}$ the exit time of U. A conformal map is a bijective biholomorphic map. In this exercise, we prove the following theorem:*

THEOREM IV.2.1. *Let $z \in U$, and let $f : U \to V$ be conformal. The law of $\{f(B_t), t \leqslant \tau_U\}$ is the same as the law of the trace of a Brownian motion in V from $f(z)$ to the boundary.*

Let $\tilde{B}$ be an independent Brownian motion in the plane; introduce the time changes

$$\zeta_s := \int_0^s |f'(B_u)|^2 du \quad \text{and} \quad \sigma_t := \inf\{s \geqslant 0 : \zeta_s \geqslant t\}.$$

Define

$$W_t = f(B(\sigma_t \wedge \tau_U)) + \tilde{B}(t) - \tilde{B}(\zeta[\sigma_t \wedge \tau_U]), \quad t \geqslant 0.$$

It is sufficient to prove that W is a Brownian motion.

(1) *What does the previous construction boil down to?*

(2) *Show that W is continuous. Let $\mathcal{G}_s$ be the σ-algebra generated by the set $\{W_u, u \leqslant s\}$. Show that W is a $(\mathcal{G}_s)$-Brownian motion if and only if*

$$\mathbb{E}\left[\, e^{\langle \lambda, W_t\rangle} \mid W_s = f(z)\,\right] = \exp\left(\,\frac{1}{2}|\lambda|^2(t-s) + \langle \lambda, f(z)\rangle\,\right)$$

for every $z \in U$. The quantity $\langle u, v\rangle$ corresponds to the usual scalar product between two complex numbers. In order to simplify, we prove the statement only for $s = 0$. Consider a conformal map f from U to $f(U)$. Assume first that f and f' are uniformly bounded.

(3) *a) Show that for every $z \in U$,*

$$\begin{aligned}\mathbb{E}&\left[\, e^{\langle \lambda, W_t\rangle} \mid W(0) = f(z)\,\right]\\ &= \mathbb{E}_z\left[\exp\left(\,\frac{1}{2}|\lambda|^2(t - \zeta(\sigma_t \wedge \tau_U)) + \langle \lambda, f(B(\sigma_t \wedge \tau_U))\rangle\,\right)\right]\end{aligned}$$

where $\mathbb{P}_z$ is the law of a Brownian motion starting at z.

b) Prove that for every $z \in U$,

$$\Delta e^{\langle \lambda, f(z)\rangle} = |\lambda|^2 |f'(z)|^2 e^{\langle \lambda, f(z)\rangle}.$$

c) Using the two-dimensional Itô formula (see below), show that

$$M_s = \exp\left(\frac{1}{2}|\lambda|^2[t - \zeta(s \wedge \tau_U)] + \langle \lambda, f(B(s \wedge \tau_U))\rangle\right)$$

is a bounded martingale and conclude.

(4) *Where did we use the assumption that f and f' are uniformly bounded? How could one get rid of the assumption on f and f'?*

(5) *What is the probability that a Brownian motion starting at ϵ exits the domain $\mathbb{D} \setminus [-1, 0]$ through $\partial\mathbb{D}$?*

(6) *a) Explain how one could define the Brownian motion in a simply connected domain D between two boundary points a and b. We denote by $\mathbb{P}^{BM}_{(D,a,b)}$ this measure;*

b) Sketch a proof of the following conformal invariance property: let (D, a, b) be a simply connected domain and f a conformal map; then

$$f \circ \mathbb{P}^{BM}_{(D,a,b)} = \mathbb{P}^{BM}_{(f(D),f(a),f(b))};$$

c) Make explicit a construction when $D = \mathbb{H}$, $a = 0$ and $b = \infty$;

d) Let K be a compact set such that $H = \mathbb{H} \setminus K$ is a simply connected domain containing 0 (the set H is called a hull*). Prove that*

$$\mathbb{P}^{BM}_{(\mathbb{H},0,\infty)}(B \textit{ stays inside } H) = \varphi'_H(0)$$

where φ_H is the map from H to $\mathbb{H}$ that maps 0 to 0 and with $\varphi_H(z) \sim z$ when z goes to infinity.

THEOREM IV.2.2 (Itô's formula in dimension 2). *Let $B = B^{(1)} + iB^{(2)}$ be a two-dimensional Brownian motion. Let Σ be an increasing continuous adapted process and*

$$X_t = \int_0^t a_s^{(1)} dB_s^{(1)} + i \int_0^t a_s^{(2)} dB_s^{(2)}.$$

Let $f : \mathbb{C} \times \mathbb{R}_+ \to \mathbb{R}$ be a smooth function; then

$$f(X_t, \Sigma_t) = f(X_0, \Sigma_0) + \sum_{i=1}^{2} \int_0^t \partial_i f(X_s, \Sigma_s) a_s^{(i)} dB_s^{(i)} + \int_0^t \partial_t f(X_s, \Sigma_s) d\Sigma_s + \frac{1}{2} \sum_{i=1}^{2} \int_0^t \partial_{ii}^2 f(X_s, \Sigma_s) \big(a_s^{(i)}\big)^2 ds.$$

References

[1] M. Aizenman and A. Burchard, *Hölder regularity and dimension bounds for random curves*, Duke Mathematical Journal, 99 (1999), pp. 419–453.

[2] V. Beffara, *Hausdorff dimensions for* SLE_6, The Annals of Probability, 32 (2004), pp. 2606–2629.

[3] ———, *Cardy's formula on the triangular lattice, the easy way*, in Universality and Renormalization, I. Binder and D. Kreimer, eds., vol. 50 of Fields Institute Communications, The Fields Institute, 2007, pp. 39–45.

[4] ———, *The dimension of the SLE curves*, The Annals of Probability, 36 (2008), pp. 1421–1452.

[5] V. Beffara and H. Duminil-Copin, *The self-dual point of the two-dimensional random-cluster model is critical for* $q \geqslant 1$, Probability Theory and Related Fields, (2011). To appear, preprint arXiv:1006:5073.

[6] ———, *Smirnov's fermionic observable away from criticality*, The Annals of Probability, (2011). To appear, preprint arXiv1010.0525.

[7] V. Beffara and P. Nolin, *Numerical estimates for monochromatic percolation exponents*. Unpublished, 2009.

[8] ———, *On monochromatic arm exponents for 2D critical percolation*, The Annals of Probability, 39 (2011), pp. 1286–1304.

[9] B. Bollobás and O. Riordan, *Percolation*, Cambridge University Press, 2006.

[10] S. R. Broadbent and J. M. Hammersley, *Percolation processes, I and II*, Proc. Cambridge Philos. Soc., 53 (1957), pp. 629–645.

[11] F. Camia and C. M. Newman, *Two-dimensional critical percolation: the full scaling limit*, Communications in Mathematical Physics, 268 (2005), pp. 1–38.

[12] J. Cardy, *Critical percolation in finite geometries*, Journal of Physics A: Mathematics and General, 25 (1992), pp. L201–L206.

[13] G. R. Grimmett, *Percolation*, vol. 321 of Grundlehren der Mathematischen Wissenschaften, Springer-Verlag, Berlin, second ed., 1999.

[14] ———, *The Random-Cluster Model*, vol. 333 of Grundlehren der Mathematischen Wissenschaften, Springer-Verlag, Berlin, 2006.

[15] A. Kemppainen, *On Random Planar Curves and Their Scaling Limits*, PhD thesis, University of Helsinki, 2009.

[16] A. Kemppainen and S. Smirnov, *Random curves, scaling limits and Loewner evolutions*, In preparation, (2009).

[17] H. Kesten, *The critical probability of bond percolation on the square lattice equals* 1/2, Communications in Mathematical Physics, 74 (1980), pp. 41–59.

[18] ———, *Percolation theory for mathematicians*, vol. 2 of Progress in Probability and Statistics, Birkhäuser, Boston, Mass., 1982.

[19] H. A. Kramers and G. H. Wannier, *Statistics of the two-dimensional ferromagnet I*, Physical Review, 60 (1941), pp. 252–262.

[20] G. F. Lawler, *Conformally Invariant Processes in the Plane*, vol. 114 of Mathematical Surveys and Monographs, American Mathematical Society, 2005.

[21] G. F. Lawler, O. Schramm, and W. Werner, *Values of Brownian intersection exponents I: Half-plane exponents*, Acta Mathematica, 187 (2001), pp. 237–273.

[22] ———, *Values of Brownian intersection exponents II: Plane exponents*, Acta Mathematica, 187 (2001), pp. 275–308.

[23] ———, *Values of Brownian intersection exponents III: Two-sided exponents*, Annales de l'Institut Henri Poincaré B: Probabilités et Statistiques, 38 (2002), pp. 109–123.

[24] ———, *Conformal restriction: The chordal case*, J. Amer. Math. Soc., 16 (2003), pp. 917–955.
[25] ———, *Conformal invariance of planar loop-erased random walks and uniform spanning trees*, The Annals of Probability, 32 (2004), pp. 939–995.
[26] D. E. Marshall and S. Rohde, *The Loewner differential equation and slit mappings*, J. Amer. Math. Soc., 18 (2005), pp. 763–778.
[27] M. V. Menshikov, *Coincidence of critical points in percolation problems*, Soviet Mathematics Doklady, 33 (1986), pp. 856–859.
[28] L. Onsager, *Crystal statistics. I. A two-dimensional model with an order-disorder transition*, Physical Review, 65 (1944), pp. 117–149.
[29] C. Pommerenke, *Boundary Behaviour of Conformal Maps*, Springer, 1992.
[30] S. Rohde and O. Schramm, *Basic properties of SLE*, Annals of Mathematics, 161 (2005), pp. 883–924.
[31] L. Russo, *A note on percolation*, Zeitschrift Wahrscheinlichkeitstheorie und verwandte Gebiete, 43 (1978), pp. 39–48.
[32] O. Schramm, *Scaling limits of loop-erased random walks and uniform spanning trees*, Israel Journal of Mathematics, 118 (2000), pp. 221–288.
[33] P. D. Seymour and D. J. A. Welsh, *Percolation probabilities on the square lattice*, Annals of Discrete Mathematics, 3 (1978), pp. 227–245. Advances in graph theory (Cambridge Combinatorial Conf., Trinity College, Cambridge, 1977).
[34] S. Sheffield and N. Sun, *Strong path convergence from Loewner driving convergence*, Preprint, (2010). arXiv:1003.4675.
[35] S. Smirnov, *Critical percolation in the plane: Conformal invariance, Cardy's formula, scaling limits*, Comptes Rendus de l'Académie des Sciences, Paris, Série I : Mathématiques, 333 (2001), pp. 239–244.
[36] ———, *Critical percolation in the plane. I. Conformal invariance and Cardy's formula. II. Continuum scaling limit.* `http://www.math.kth.se/~stas/papers/percol.ps`, 2001.
[37] ———, *Conformal invariance in random cluster models. I. Holomorphic fermions in the Ising model*, Annals of Mathematics, 172 (2010), pp. 1435–1467.
[38] S. Smirnov and H. Duminil-Copin, *Conformal invariance in lattice models*, in Lecture notes of the 2010 Clay summer school, Buzios, 2010.
[39] S. Smirnov and W. Werner, *Critical exponents for two-dimensional percolation*, Mathematical Research Letters, 8 (2001), pp. 729–744.
[40] W. Werner, *Random planar curves and Schramm-Loewner evolutions*, in Lectures on probability theory and statistics, vol. 1840 of Lecture Notes in Mathematics, Springer, Berlin, 2004, pp. 107–195. Lectures from the 32nd Probability Summer School held in Saint-Flour, July 7–24, 2002.
[41] ———, *Lectures on two-dimensional critical percolation*, Institute for Advanced Studies, Lectures given at the Park City summer school, 2007. arXiv:0710.0856.
[42] D. B. Wilson, *Generating random spanning trees more quickly than the cover time*, in Proceedings of the Twenty-eighth Annual ACM Symposium on the Theory of Computing (Philadelphia, PA, 1996), New York, 1996, ACM, pp. 296–303.
[43] D. V. Wood, *Average and probability, problem 5*, The American Mathematical Monthly, 1 (1894), pp. 211–212.

Unité de Mathématiques Pures et Appliquées, École Normale Supérieure de Lyon, 46, allée d'Italie, F-69364 Lyon cedex 07, France
E-mail address: `Vincent.Beffara@ens-lyon.fr`

Clay Mathematics Proceedings
Volume 15, 2012

Noise Sensitivity and Percolation

Christophe Garban and Jeffrey E. Steif

Contents

2010 *Mathematics Subject Classification.* 60K35, 06E30, 60J67.

Overview

The goal of this set of lectures is to combine two seemingly unrelated topics:

- The study of **Boolean functions**, a field particularly active in computer science
- Some models in statistical physics, mostly **percolation**

The link between these two fields can be loosely explained as follows: a percolation configuration is built out of a collection of i.i.d. "bits" which determines whether the corresponding edges, sites, or blocks are present or absent. In that respect, any event concerning percolation can be seen as a Boolean function whose inputs are precisely these "bits".

Over the last 20 years, mainly thanks to the computer science community, a very rich structure has emerged concerning the properties of Boolean functions. The first part of this course will be devoted to a description of some of the main achievements in this field.

In some sense one can say, although this is an exaggeration, that computer scientists are mostly interested in the *stability* or *robustness* of Boolean functions. As we will see later in this course, the Boolean functions which "encode" large scale properties of **critical** percolation will turn out to be very *sensitive* to small perturbations. This phenomenon corresponds to what we will call **noise sensitivity**. Hence, the Boolean functions one wishes to describe here are in some sense *orthogonal* to the Boolean functions one encounters, ideally, in computer science. Remarkably, it turns out that the tools developed by the computer science community to capture the properties and stability of Boolean functions are also suitable for the study of noise sensitive functions. This is why it is worth us first spending some time on the general properties of Boolean functions.

One of the main tools needed to understand properties of Boolean functions is Fourier analysis on the hypercube. Noise sensitivity will correspond to our Boolean function being of "high frequency" while stability will correspond to our Boolean function being of "low frequency". We will apply these ideas to some other models from statistical mechanics as well; namely, first passage percolation and dynamical percolation.

Some of the different topics here can be found (in a more condensed form) in [**Gar10**].

Acknowledgements

We wish to warmly thank the organizers David Ellwood, Charles Newman, Vladas Sidoravicius and Wendelin Werner for inviting us to give this course at the Clay summer school 2010 in Buzios. It was a wonderful experience for us to give this set of lectures. We also wish to thank Ragnar Freij who served as a very good teaching assistant for this course and for various comments on the manuscript.

Some standard notations

In the following table, $f(n)$ and $g(n)$ are any sequences of positive real numbers.

$f(n) \asymp g(n)$	there exists some constant $C > 0$ such that $$C^{-1} \le \frac{f(n)}{g(n)} \le C\,, \ \forall n \ge 1$$
$f(n) \le O(g(n))$	there exists some constant $C > 0$ such that $$f(n) \le Cg(n)\,, \ \forall n \ge 1$$
$f(n) \ge \Omega(g(n))$	there exists some constant $C > 0$ such that $$f(n) \ge Cg(n)\,, \ \forall n \ge 1$$
$f(n) = o(g(n))$	$$\lim_{n\to\infty} \frac{f(n)}{g(n)} = 0$$

Part I. Boolean functions and key concepts

1. Boolean functions

DEFINITION I.1. *A* **Boolean function** *is a function from the hypercube* $\Omega_n := \{-1,1\}^n$ *into either* $\{-1,1\}$ *or* $\{0,1\}$.

Ω_n will be endowed with the uniform measure $\mathbb{P} = \mathbb{P}^n = (\frac{1}{2}\delta_{-1} + \frac{1}{2}\delta_1)^{\otimes n}$ and $\mathbb{E}$ will denote the corresponding expectation. At various times, Ω_n will be endowed with the general product measure $\mathbb{P}_p = \mathbb{P}_p^n = ((1-p)\delta_{-1} + p\delta_1)^{\otimes n}$ but in such cases the p will be explicit. $\mathbb{E}_p$ will then denote the corresponding expectations.

An element of Ω_n will be denoted by either ω or ω_n and its n bits by $x_1, \ldots, x_n$ so that $\omega = (x_1, \ldots, x_n)$.

Depending on the context, concerning the range, it might be more pleasant to work with one of $\{-1,1\}$ or $\{0,1\}$ rather than the other and at some specific places in these lectures, we will even relax the Boolean constraint (i.e. taking only two possible values). In these cases (which will be clearly mentioned), we will consider instead real-valued functions $f : \Omega_n \to \mathbb{R}$.

A Boolean function f is canonically identified with a subset A_f of Ω_n via $A_f := \{\omega : f(\omega) = 1\}$.

REMARK I.1. Often, Boolean functions are defined on $\{0,1\}^n$ rather than $\Omega_n = \{-1,1\}^n$. This does not make any fundamental difference at all but, as we will see later, the choice of $\{-1,1\}^n$ turns out to be more convenient when one wishes to apply Fourier analysis on the hypercube.

2. Some Examples

We begin with a few examples of Boolean functions. Others will appear throughout this part.

EXAMPLE 1 (Dictatorship).

$$\mathbf{DICT}_n(x_1, \ldots, x_n) := x_1$$

The first bit determines what the outcome is.

EXAMPLE 2 (Parity).

$$\mathbf{PAR}_n(x_1, \ldots, x_n) := \prod_{i=1}^{n} x_i$$

This Boolean function tells whether the number of -1's is even or odd.

These two examples are in some sense trivial, but they are good to keep in mind since in many cases they turn out to be the "extreme cases" for properties concerning Boolean functions.

The next rather simple Boolean function is of interest in social choice theory.

EXAMPLE 3 (Majority function). Let n be odd and define

$$\mathbf{MAJ}_n(x_1, \ldots, x_n) := \mathrm{sign}(\sum_{i=1}^{n} x_i).$$

Following are two further examples which will also arise in our discussions.

EXAMPLE 4 (Iterated 3-Majority function). Let $n = 3^k$ for some integer k. The bits are indexed by the leaves of a rooted 3-ary tree (so the root has degree 3, the leaves have degree 1 and all others have degree 4) with depth k. One iteratively applies the previous example (with $n = 3$) to obtain values at the vertices at level $k-1$, then level $k-2$, etc. until the root is assigned a value. The root's value is then the output of f. For example when $k = 2$, $f(-1,1,1;1,-1,-1;-1,1,-1) = -1$. The recursive structure of this Boolean function will enable explicit computations for various properties of interest.

EXAMPLE 5 (Clique containment). If $r = \binom{n}{2}$ for some integer n, then Ω_r can be identified with the set of labelled graphs on n vertices. (x_i is 1 iff the ith edge is present.) Recall that a **clique** of size k of a graph $G = (V, E)$ is a complete graph on k vertices embedded in G.

Now for any $1 \le k \le \binom{n}{2} = r$, let $\mathbf{CLIQ}_n^k$ be the indicator function of the event that the random graph G_ω defined by $\omega \in \Omega_r$ contains a clique of size k. Choosing $k = k_n$ so that this Boolean function is non-degenerate turns out to be a rather delicate issue. The interesting regime is near $k_n \approx 2\log_2(n)$. See the exercises for this "tuning" of $k = k_n$. It turns out that for most values of n, the Boolean function $\mathbf{CLIQ}_n^k$ is degenerate (i.e. has small variance) for all values of k. However, there is a sequence of n for which there is some $k = k_n$ for which $\mathbf{CLIQ}_n^k$ is nondegerate.

3. Pivotality and Influence

This section contains our first fundamental concepts. We will abbreviate $\{1, \ldots, n\}$ by $[n]$.

DEFINITION I.2. *Given a Boolean function f from Ω_n into either $\{-1,1\}$ or $\{0,1\}$ and a variable $i \in [n]$, we say that i* **is pivotal for** f **for** ω *if $\{f(\omega) \neq f(\omega^i)\}$ where ω^i is ω but flipped in the ith coordinate. Note that this event is measurable with respect to $\{x_j\}_{j \neq i}$.*

DEFINITION I.3. *The* **pivotal set**, *$\mathcal{P}$, for f is the random set of $[n]$ given by*

$$\mathcal{P}(\omega) = \mathcal{P}_f(\omega) := \{i \in [n] : \; i \text{ is pivotal for } f \text{ for } \omega\}.$$

In words, it is the (random) set of bits with the property that if you flip the bit, then the function output changes.

DEFINITION I.4. *The* **influence** *of the ith bit, $\mathbf{I}_i(f)$, is defined by*

$$\mathbf{I}_i(f) := \mathbb{P}(\; i \text{ is pivotal for } f \;) = \mathbb{P}(i \in \mathcal{P}).$$

Let also the **influence vector**, $\mathbf{Inf}(f)$, *be the collection of all the influences: i.e. $\{\mathbf{I}_i(f)\}_{i \in [n]}$.*

In words, the influence of the ith bit, $\mathbf{I}_i(f)$, is the probability that, on flipping this bit, the function output changes.

DEFINITION I.5. *The* **total influence**, $\mathbf{I}(f)$, *is defined by*

$$\mathbf{I}(f) := \sum_i \mathbf{I}_i(f) \; = \|\mathbf{Inf}(f)\|_1 \; (= \mathbb{E}(|\mathcal{P}|)).$$

It would now be instructive to go and compute these quantities for examples 1–3. See the exercises.

Later, we will need the last two concepts in the context when our probability measure is $\mathbb{P}_p$ instead. We give the corresponding definitions.

DEFINITION I.6. *The* **influence vector at level** p, $\{\mathbf{I}_i^p(f)\}_{i\in[n]}$, *is defined by*

$$\mathbf{I}_i^p(f) := \mathbb{P}_p(\text{ } i \text{ is pivotal for } f \text{ }) = \mathbb{P}_p(i \in \mathcal{P}).$$

DEFINITION I.7. *The* **total influence at level** p, $\mathbf{I}^p(f)$, *is defined by*

$$\mathbf{I}^p(f) := \sum_i \mathbf{I}_i^p(f) \ (= \mathbb{E}_p(|\mathcal{P}|)).$$

It turns out that the total influence has a geometric-combinatorial interpretation as the size of the so-called edge-boundary of the corresponding subset of the hypercube. See the exercises.

REMARK I.2. Aside from its natural definition as well as its geometric interpretation as measuring the edge-boundary of the corresponding subset of the hypercube (see the exercises), the notion of *total influence* arises very naturally when one studies **sharp thresholds** for *monotone functions* (to be defined in Part III). Roughly speaking, as we will see in detail in Part III, for a monotone event A, one has that $d\mathbb{P}_p\big[A\big]/dp$ is the total influence at level p (this is the Margulis-Russo formula). This tells us that the speed at which one changes from the event A "almost surely" *not* occurring to the case where it "almost surely" *does* occur is very sudden if the Boolean function happens to have a large total influence.

4. The Kahn, Kalai, Linial Theorem

This section addresses the following question. Does there always exist some variable i with (reasonably) large influence? In other words, for large n, what is the smallest value (as we vary over Boolean functions) that the largest influence (as we vary over the different variables) can take on?

Since for the constant function all influences are 0, and the function which is 1 only if all the bits are 1 has all influences $1/2^{n-1}$, clearly one wants to deal with functions which are reasonably balanced (meaning having variances not so close to 0) or alternatively, obtain lower bounds on the maximal influence in terms of the variance of the Boolean function.

The first result in this direction is the following result. A sketch of the proof is given in the exercises.

THEOREM I.1 (Discrete Poincaré). *If f is a Boolean function mapping Ω_n into $\{-1,1\}$, then*

$$\mathrm{Var}(f) \le \sum_i \mathbf{I}_i(f).$$

It follows that there exists some i such that

$$\mathbf{I}_i(f) \ge \mathrm{Var}(f)/n.$$

This gives a first answer to our question. For reasonably balanced functions, there is some variable whose influence is at least of order $1/n$. *Can we find a better "universal" lower bound on the maximal influence?* Note that for Example 3 all the influences are of order $1/\sqrt{n}$ (and the variance is 1). In terms of our question, this universal lower bound one is looking for should lie somewhere between $1/n$ and $1/\sqrt{n}$. The following celebrated result improves by a logarithmic factor on the above $\Omega(1/n)$ bound.

THEOREM I.2 ([**KKL88**]). *There exists a universal* $c > 0$ *such that if* f *is a Boolean function mapping* Ω_n *into* $\{0, 1\}$, *then there exists some* i *such that*

$$\mathbf{I}_i(f) \geq c\text{Var}(f)(\log n)/n.$$

What is remarkable about this theorem is that this "logarithmic" lower bound on the maximal influence turns out to be *sharp*! This is shown by the following example by Ben-Or and Linial.

EXAMPLE 6 (Tribes). Partition $[n]$ into subsequent blocks of length $\log_2(n) - \log_2(\log_2(n))$ with perhaps some leftover debris. Define $f = f_n$ to be 1 if there exists at least one block which contains all 1's, and 0 otherwise.

It turns out that one can check that the sequence of variances stays bounded away from 0 and that all the influences (including of course those belonging to the debris which are equal to 0) are smaller than $c(\log n)/n$ for some $c < \infty$. See the exercises for this. Hence the above theorem is indeed sharp.

Our next result tells us that if all the influences are "small", then the total influence is large.

THEOREM I.3 ([**KKL88**]). *There exists a* $c > 0$ *such that if* f *is a Boolean function mapping* Ω_n *into* $\{0, 1\}$ *and* $\delta := \max_i \mathbf{I}_i(f)$ *then*

$$\mathbf{I}(f) \geq c\,\text{Var}(f)\log(1/\delta).$$

Or equivalently,

$$\|\mathbf{Inf}(f)\|_1 \geq c\,\text{Var}(f)\log\frac{1}{\|\mathbf{Inf}(f)\|_\infty}.$$

One can in fact talk about the influence of a set of variables rather than the influence of a single variable.

DEFINITION I.8. *Given* $S \subseteq [n]$, *the* **influence of** S, $\mathbf{I}_S(f)$, *is defined by*

$$\mathbf{I}_S(f) := \mathbb{P}(\ f \textit{ is not determined by the bits in } S^c).$$

It is easy to see that when S is a single bit, this corresponds to our previous definition. The following is also proved in [**KKL88**]. We will not indicate the proof of this result in these lecture notes.

THEOREM I.4 ([**KKL88**]). *Given a sequence* f_n *of Boolean functions mapping* Ω_n *into* $\{0, 1\}$ *such that* $0 < \inf_n \mathbb{E}_n(f) \leq \sup_n \mathbb{E}_n(f) < 1$ *and any sequence* a_n *going to* ∞ *arbitrarily slowly, then there exists a sequence of sets* $S_n \subseteq [n]$ *such that* $|S_n| \leq a_n n/\log n$ *and* $\mathbf{I}_{S_n}(f_n) \to 1$ *as* $n \to \infty$.

Theorems I.2 and I.3 will be proved in Part V.

5. Noise sensitivity and noise stability

This subsection introduces our second set of fundamental concepts.

Let ω be uniformly chosen from Ω_n and let ω_ϵ be ω but with each bit independently "rerandomized" with probability ϵ. This means that each bit, independently of everything else, rechooses whether it is 1 or -1, each with probability $1/2$. Note that ω_ϵ then has the same distribution as ω.

The following definition is *central* for these lecture notes. Let m_n be an increasing sequence of integers and let $f_n : \Omega_{m_n} \to \{\pm 1\}$ or $\{0, 1\}$.

DEFINITION I.9. *The sequence* $\{f_n\}$ *is* **noise sensitive** *if for every* $\epsilon > 0$,

$$\lim_{n\to\infty} \mathbb{E}[f_n(\omega)f_n(\omega_\epsilon)] - \mathbb{E}[f_n(\omega)]^2 = 0. \tag{I.1}$$

Since f_n just takes 2 values, this says that the random variables $f_n(\omega)$ and $f_n(\omega_\epsilon)$ are asymptotically independent for $\epsilon > 0$ fixed and n large. We will see later that (I.1) holds for one value of $\epsilon \in (0,1)$ if and only if it holds for all such ϵ. The following notion captures the opposite situation where the two events above are close to being the same event if ϵ is small, uniformly in n.

DEFINITION I.10. *The sequence* $\{f_n\}$ *is* **noise stable** *if*

$$\lim_{\epsilon\to 0} \sup_n \mathbb{P}(f_n(\omega) \neq f_n(\omega_\epsilon)) = 0.$$

It is an easy exercise to check that a sequence $\{f_n\}$ is both noise sensitive and noise stable if and only it is degenerate in the sense that the sequence of variances $\{\mathrm{Var}(f_n)\}$ goes to 0. Note also that a sequence of Boolean functions could be neither noise sensitive nor noise stable (see the exercises).

It is also an easy exercise to check that Example 1 (dictator) is noise stable and Example 2 (parity) is noise sensitive. We will see later, when Fourier analysis is brought into the picture, that these examples are the two opposite extreme cases. For the other examples, it turns out that Example 3 (Majority) is noise stable, while Examples 4–6 are all noise sensitive. See the exercises. In fact, there is a deep theorem (see [**MOO10**]) which says in some sense that, among all low influence Boolean functions, Example 3 (Majority) is the stablest.

In Figure I.1, we give a slightly *impressionistic* view of what "noise sensitivity" is.

6. Benjamini, Kalai and Schramm noise sensitivity Theorem

The following is the main theorem concerning noise sensitivity.

THEOREM I.5 ([**BKS99**]). *If*

$$\lim_n \sum_k \mathbf{I}_k(f_n)^2 = 0,$$

then $\{f_n\}$ *is noise sensitive.*

REMARK I.3. The converse is clearly false as shown by Example 2. However, it turns out that the converse is true for so-called **monotone functions** (see the next part for the definition of this) as we will see in Part IV.

This theorem will allow us to conclude noise sensitivity of many of the examples we have introduced in this first part. See the exercises. This theorem will also be proved in Part V.

7. Percolation crossings: our final and most important example

We have saved our most important example to the end. This set of notes would not be being written if it were not for this example and for the results that have been proved for it.

Let us consider percolation on $\mathbb{Z}^2$ at the critical point $p_c(\mathbb{Z}^2) = 1/2$. (See Part II for a fast review on the model.) At this critical point, there is no infinite cluster, but somehow clusters are 'large' (there are clusters at all scales). This can

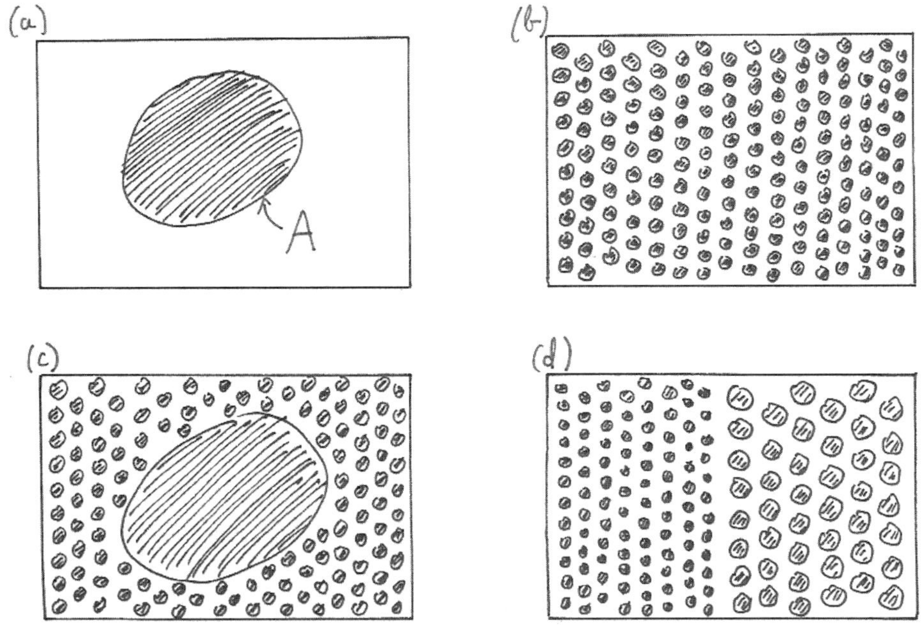

FIGURE I.1. Let us consider the following *"experiment"*: take a bounded domain in the plane, say a rectangle, and consider a measurable subset A of this domain. What would be an analogue of the above definitions of being *noise sensitive* or *noise stable* in this case? Start by sampling a point x uniformly in the domain according to Lebesgue measure. Then let us apply some noise to this position x so that we end up with a new position x_ϵ. One can think of many natural "noising" procedures here. For example, let x_ϵ be a uniform point in the ball of radius ϵ around x, conditioned to remain in the domain. (This is not quite perfect yet since this procedure does not exactly preserve Lebesgue measure, but let's not worry about this.) The natural analogue of the above definitions is to ask whether $1_A(x)$ and $1_A(x_\epsilon)$ are decorrelated or not. *Question:* According to this analogy, discuss the stability versus sensitivity of the sets A sketched in pictures (a) to (d) ? Note that in order to match with definitions I.9 and I.10, one should consider sequences of subsets $\{A_n\}$ instead, since noise sensitivity is an asymptotic notion.

be seen using duality or with the RSW Theorem II.1. In order to understand the geometry of the critical picture, the following large-scale *observables* turn out to be very useful: Let Ω be a piecewise smooth domain with two disjoint open arcs ∂_1 and ∂_2 on its boundary $\partial\Omega$. For each $n \geq 1$, we consider the scaled domain $n\Omega$. Let A_n be the event that there is an open path in ω from $n\partial_1$ to $n\partial_2$ which stays inside

$n\Omega$. Such events are called **crossing events**. They are naturally associated with Boolean functions whose entries are indexed by the set of edges inside $n\Omega$ (there are $O(n^2)$ such variables).

For simplicity, let us consider the particular case of rectangle crossings:

EXAMPLE 7 (Percolation crossings).

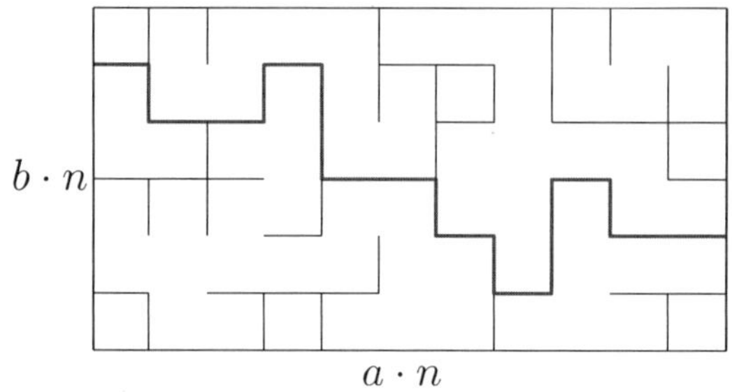

Let $a, b > 0$ and let us consider the rectangle $[0, a \cdot n] \times [0, b \cdot n]$. The left to right crossing event corresponds to the Boolean function $f_n : \{-1,1\}^{O(1)n^2} \to \{0,1\}$ defined as follows:

$$f_n(\omega) := \begin{cases} 1 & \text{if there is a left-right crossing} \\ 0 & \text{otherwise} \end{cases}$$

We will later prove that this sequence of Boolean functions $\{f_n\}$ is noise sensitive. This means that if a percolation configuration $\omega \sim \mathbb{P}_{p_c=1/2}$ is given to us, one cannot predict anything about the large scale clusters of the slightly perturbed percolation configuration ω_ϵ (where only an ϵ-fraction of the edges have been resampled).

REMARK I.4. The same statement holds for the above more general crossing events (i.e. in $(n\Omega, n\partial_1, n\partial_2)$).

Exercise sheet of Part I

EXERCISE I.1. Determine the pivotal set, the influence vector and the total influence for Examples 1–3.

EXERCISE I.2. Determine the influence vector for Example 4 and Example 6.

EXERCISE I.3. Show that in Example 6 the variances stay bounded away from 0. If the blocks are taken to be of size $\log_2 n$ instead, show that the influences would all be of order $1/n$. Why does this not contradict the KKL Theorem?

EXERCISE I.4. Ω_n has a graph structure where two elements are neighbors if they differ in exactly one location. The **edge boundary** of a subset $A \subseteq \Omega_n$, denoted by $\partial_E(A)$, is the set of edges where exactly one of the endpoints is in A.

Show that for any Boolean function, $\mathbf{I}(f) = |\partial_E(A_f)|/2^{n-1}$.

EXERCISE I.5. Prove Theorem I.1. This is a type of Poincaré inequality. Hint: use the fact that $\mathrm{Var}(f)$ can be written $2\mathbb{P}\big[f(\omega) \neq f(\widetilde{\omega})\big]$, where $\omega, \widetilde{\omega}$ are independent and try to "interpolate" from ω to $\widetilde{\omega}$.

EXERCISE I.6. Show that Example 3 (Majority) is noise stable.

EXERCISE I.7. Prove that Example 4 (iterated 3-majority) is noise sensitive directly without relying on Theorem I.5. Hint: use the recursive structure of this example in order to show that the criterion of noise sensitivity is satisfied.

EXERCISE I.8. Prove that Example 6 (tribes) is noise sensitive directly without using Theorem I.5. Here there is no recursive structure, so a more "probabilistic" argument is needed.

PROBLEM I.9. Recall Example 5 (clique containment).

(a) Prove that when $k_n = o(n^{1/2})$, $\mathbf{CLIQ}_n^{k_n}$ is asymptotically noise sensitive. Hint: start by obtaining an upper bound on the influences (which are identical for each edge) using Exercise I.4. Conclude by using Theorem I.5.

(b) Open exercise: Find a more direct proof of this fact (in the spirit of exercise I.8) which would avoid using Theorem I.5.

As pointed out after Example 5, for most values of $k = k_n$, the Boolean function $\mathbf{CLIQ}_n^{k_n}$ becomes degenerate. The purpose of the rest of this problem is to determine what the interesting regime is where $\mathbf{CLIQ}_n^{k_n}$ has a chance of being non-degenerate (i.e. variance bounded away from 0). The rest of this exercise is somewhat tangential to the course.

(c) If $1 \leq k \leq \binom{n}{2} = r$, what is the expected number of cliques in G_ω, $\omega \in \Omega_r$?

(d) Explain why there should be at most one choice of $k = k_n$ such that the variance of $\mathbf{CLIQ}_n^{k_n}$ remains bounded away from 0 ? (No rigorous proof required.) Describe this choice of k_n. Check that it is indeed in the regime $2\log_2(n)$.

(e) Note retrospectively that in fact, for any choice of $k = k_n$, $\mathbf{CLIQ}_n^{k_n}$ is noise sensitive.

EXERCISE I.10. Deduce from Theorem I.5 that both Example 4 (iterated 3-majority) and Example 6 (tribes) are noise sensitive.

EXERCISE I.11. Give a sequence of Boolean functions which is neither noise sensitive nor noise stable.

EXERCISE I.12. In the sense of Definition I.8, show that for the majority function and for fixed ϵ, any set of size $n^{1/2+\epsilon}$ has influence approaching 1 while any set of size $n^{1/2-\epsilon}$ has influence approaching 0.

PROBLEM I.13. Do you think a *"generic"* Boolean function would be stable or sensitive? Justify your intuition. Show that if f_n was a "randomly" chosen function, then a.s. $\{f_n\}$ is noise sensitive.

Part II. Percolation in a nutshell

In order to make these lecture notes as self-contained as possible, we review various aspects of the percolation model and give a short summary of the main useful results.

For a complete account of percolation, see [**Gri99**] and for a study of the 2-dimensional case, which we are concentrating on here, see the lecture notes [**Wer07**].

1. The model

Let us briefly start by introducing the model itself.

We will be concerned mainly with two-dimensional percolation and we will focus on two lattices: $\mathbb{Z}^2$ and the triangular lattice $\mathbb{T}$. (All the results stated for $\mathbb{Z}^2$ in these lecture notes are also valid for percolations on "reasonable" 2-d translation invariant graphs for which the RSW Theorem (see the next section) is known to hold at the corresponding critical point.)

Let us describe the model on the graph $\mathbb{Z}^2$ which has $\mathbb{Z}^2$ as its vertex set and edges between vertices having Euclidean distance 1. Let $\mathbb{E}^2$ denote the set of edges of the graph $\mathbb{Z}^2$. For any $p \in [0,1]$ we define a random subgraph of $\mathbb{Z}^2$ as follows: independently for each edge $e \in \mathbb{E}^2$, we keep this edge with probability p and remove it with probability $1-p$. Equivalently, this corresponds to defining a random configuration $\omega \in \{-1,1\}^{\mathbb{E}^2}$ where, independently for each edge $e \in \mathbb{E}^2$, we declare the edge to be open ($\omega(e) = 1$) with probability p or closed ($\omega(e) = -1$) with probability $1-p$. The law of the so-defined random subgraph (or configuration) is denoted by $\mathbb{P}_p$.

Percolation is defined similarly on the triangular grid $\mathbb{T}$, except that on this lattice we will instead consider *site* percolation (i.e. here we keep each site with probability p). The sites are the points $\mathbb{Z} + e^{i\pi/3}\mathbb{Z}$ so that neighboring sites have distance one from each other in the complex plane.

2. Russo-Seymour-Welsh

We will often rely on the following celebrated result known as the **RSW Theorem**.

THEOREM II.1 (RSW). *(see [**Gri99**]) For percolation on $\mathbb{Z}^2$ at $p = 1/2$, one has the following property concerning the crossing events. Let $a, b > 0$. There exists a constant $c = c(a,b) > 0$, such that for any $n \geq 1$, if A_n denotes the event that there is a left to right crossing in the rectangle $([0, a\cdot n] \times [0, b\cdot n]) \cap \mathbb{Z}^2$, then*

$$c < \mathbb{P}_{1/2}\big[A_n\big] < 1 - c\,.$$

In other words, this says that the Boolean functions f_n defined in Example 7 of Part I are non-degenerate.

The same result holds also in the case of site-percolation on $\mathbb{T}$ (also at $p = 1/2$).

FIGURE II.1. Pictures (by Oded Schramm) representing two percolation configurations respectively on $\mathbb{T}$ and on $\mathbb{Z}^2$ (both at $p = 1/2$). The sites of the triangular grid are represented by hexagons.

The parameter $p = 1/2$ plays a very special role for the two models under consideration. Indeed, there is a natural way to associate to each percolation configuration $\omega_p \sim \mathbb{P}_p$ a dual configuration ω_{p^*} on the so-called dual graph. In the case of $\mathbb{Z}^2$, its dual graph can be realized as $\mathbb{Z}^2 + (\frac{1}{2}, \frac{1}{2})$. In the case of the triangular lattice, $\mathbb{T}^* = \mathbb{T}$. The figure on the right illustrates this *duality* for percolation on $\mathbb{Z}^2$. It is easy to see that in both cases $p^* = 1 - p$. Hence, at $p = 1/2$, our two models happen to be *self-dual.*

This duality has the following very important consequence. For a domain in $\mathbb{T}$ with two specified boundary arcs, there is a 'left-right' crossing of white hexagons if and only if there is no 'top-bottom' crossing of black hexagons.

3. Phase transition

In percolation theory, one is interested in large scale connectivity properties of the random configuration $\omega = \omega_p$. In particular, as one raises the level p above a certain critical parameter $p_c(\mathbb{Z}^2)$, an infinite cluster (almost surely) emerges. This corresponds to the well-known *phase transition* of percolation. By a famous theorem of Kesten this transition takes place at $p_c(\mathbb{Z}^2) = \frac{1}{2}$. On the triangular grid, one also has $p_c(\mathbb{T}) = 1/2$. The event $\{0 \overset{\omega}{\longleftrightarrow} \infty\}$ denotes the event that there exists a self-avoiding path from 0 to ∞ consisting of open edges.

This phase transition can be measured with the *density function* $\theta_{\mathbb{Z}^2}(p) := \mathbb{P}_p(0 \overset{\omega}{\longleftrightarrow} \infty)$ which encodes important properties of the large scale connectivities of the random configuration ω: it corresponds to the density averaged over the space $\mathbb{Z}^2$ of the (almost surely unique) infinite cluster. The shape of the function $\theta_{\mathbb{Z}^2}$ is pictured on the right (notice the infinite derivative at p_c).

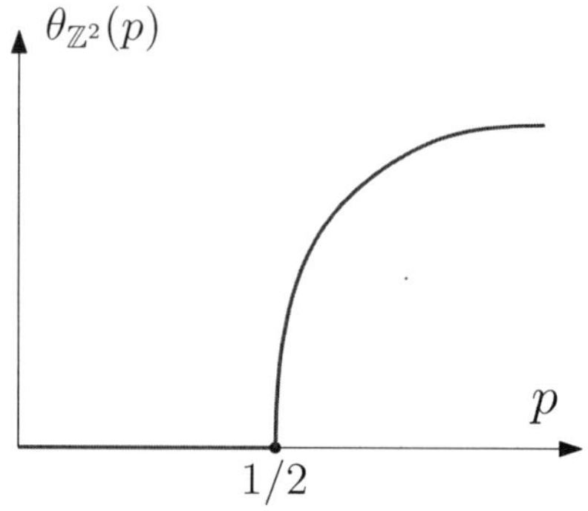

4. Conformal invariance at criticality and SLE processes

It has been conjectured for a long time that percolation should be *asymptotically* conformally invariant at the critical point. This should be understood in the same way as the fact that a Brownian motion (ignoring its time-parametrization) is a conformally invariant probabilistic object. One way to picture this conformal invariance is as follows: consider the 'largest' cluster C_δ surrounding 0 in $\delta\mathbb{Z}^2 \cap \mathbb{D}$ and such that $C_\delta \cap \partial\mathbb{D} = \emptyset$. Now consider some other simply connected domain Ω containing 0. Let $\hat{C}_\delta$ be the largest cluster surrounding 0 in a critical configuration in $\delta\mathbb{Z}^2 \cap \Omega$ and such that $\hat{C}_\delta \cap \partial\Omega = \emptyset$. Now let ϕ be the conformal map from $\mathbb{D}$ to Ω such that $\phi(0) = 0$ and $\phi'(0) > 0$. Even though the random sets $\phi(C_\delta)$ and $\hat{C}_\delta$ do not lie on the same lattice, the conformal invariance principle claims that when $\delta = o(1)$, these two random clusters are very close in law.

Over the last decade, two major breakthroughs have enabled a much better understanding of the critical regime of percolation:

- The invention of the SLE processes by Oded Schramm([**Sch00**]).
- The proof of conformal invariance on $\mathbb{T}$ by Stanislav Smirnov ([**Smi01**]).

The simplest precise statement concerning conformal invariance is the following. Let Ω be a bounded simply connected domain of the plane and let A, B, C and D be 4 points on the boundary of Ω in clockwise order. Scale the hexagonal lattice T by $1/n$ and perform critical percolation on this scaled lattice. Let $\mathbb{P}(\Omega, A, B, C, D, n)$ denote the probability that in the $1/n$ scaled hexagonal lattice there is an open path of hexagons in Ω going from the boundary of Ω between A and B to the boundary of Ω between C and D.

THEOREM II.2. *(Smirnov,* [**Smi01**]*)*
(i) For all Ω and A, B, C and D as above,

$$\mathbb{P}(\Omega, A, B, C, D, \infty) := \lim_{n\to\infty} \mathbb{P}(\Omega, A, B, C, D, n)$$

exists and is conformally invariant in the sense that if f is a conformal mapping, then $\mathbb{P}(\Omega, A, B, C, D, \infty) = \mathbb{P}(f(\Omega), f(A), f(B), f(C), f(D), \infty)$.
(ii) If Ω is an equilateral triangle (with side lengths 1), A, B and C the three corner points and D on the line between C and A having distance x from C, then the above limiting probability is x. (Observe, by conformal invariance, that this gives the limiting probability for all domains and 4 points.)

The first half was conjectured by M. Aizenman while J. Cardy conjectured the limit for the case of rectangles using the four corners. In this case, the formula is quite complicated involving hypergeometric functions but Lennart Carleson realized that this is then equivalent to the simpler formula given above in the case of triangles.

Note that, on $\mathbb{Z}^2$ at $p_c = 1/2$, proving the conformal invariance is still a challenging open problem.

We will not define the SLE processes in these notes. See the lecture notes by Vincent Beffara and references therein. The illustration below explains how SLE curves arise naturally in the percolation picture.

This celebrated picture (by Oded Schramm) represents an **exploration path** on the triangular lattice. This exploration path, which turns right when encountering black hexagons and left when encountering white ones, asymptotically converges towards SLE_6 (as the mesh size goes to 0).

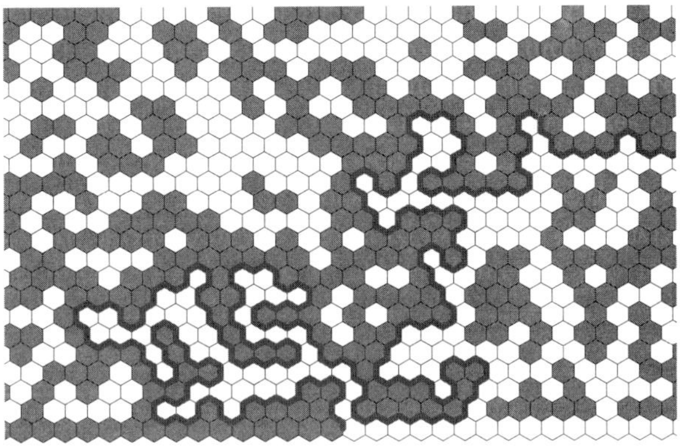

5. Critical exponents

The proof of conformal invariance combined with the detailed information given by the SLE_6 process enables one to obtain very precise information on the critical and *near-critical* behavior of site percolation on $\mathbb{T}$. For instance, it is known that on the triangular lattice the density function $\theta_{\mathbb{T}}(p)$ has the following behavior near $p_c = 1/2$:

$$\theta(p) = (p - 1/2)^{5/36+o(1)}\,,$$

when $p \to 1/2+$ (see [**Wer07**]).

In the rest of these lectures, we will often rely on three types of percolation events: namely the *one-arm*, *two-arm* and *four-arm* events. They are defined as follows: for any radius $R > 1$, let A_R^1 be the event that the site 0 is connected to distance R by some open path (one-arm). Next, let A_R^2 be the event that there are two "arms" of different colors from the site 0 (which itself can be of either color) to distance R away. Finally, let A_R^4 be the event that there are four "arms" of alternating color from the site 0 (which itself can be of either color) to distance R away (i.e. there are four connected paths, two open, two closed from 0 to radius R and the closed paths lie between the open paths). See Figure II.2 for a realization of two of these events.

It was proved in [**LSW02**] that the probability of the one-arm event decays as follows:

$$\mathbb{P}\big[A_R^1\big] := \alpha_1(R) = R^{-\frac{5}{48}+o(1)}\,.$$

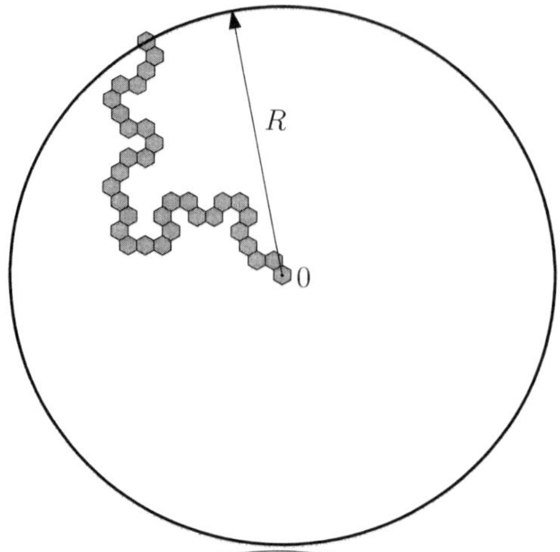

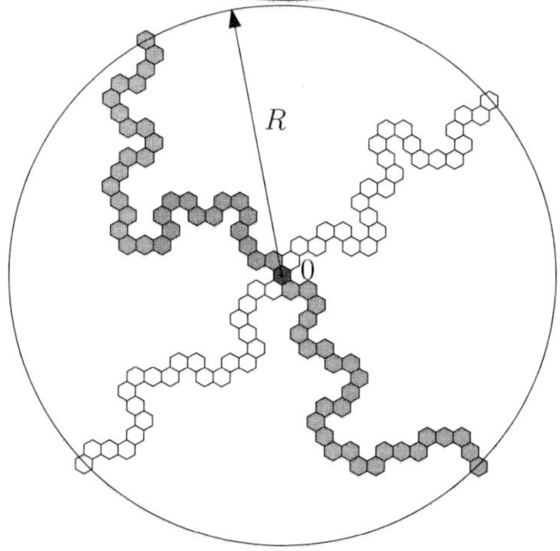

FIGURE II.2. A realization of the *one-arm* event is pictured on the top; the *four-arm* event is pictured on the bottom.

For the two-arms and four-arms events, it was proved by Smirnov and Werner in [**SW01**] that these probabilities decay as follows:

$$\mathbb{P}\big[A_R^2\big] := \alpha_2(R) = R^{-\frac{1}{4}+o(1)}$$

and

$$\mathbb{P}\big[A_R^4\big] := \alpha_4(R) = R^{-\frac{5}{4}+o(1)}\,.$$

REMARK II.1. Note the $o(1)$ terms in the above statements (which means of course goes to zero as $R \to \infty$). Its presence reveals that the above critical exponents are known so far only up to 'logarithmic' corrections. It is conjectured that there are no such 'logarithmic' corrections, but at the moment one has to deal with their possible existence. More specifically, it is believed that for the one-arm event,

$$\alpha_1(R) \asymp R^{-\frac{5}{48}}$$

where $\asymp$ means that the ratio of the two sides is bounded away from 0 and ∞ uniformly in R; similarly for the other arm events.

The four exponents we encountered concerning $\theta_{\mathbb{T}}$, α_1, α_2 and α_4 (i.e. $\frac{5}{36}$, $\frac{5}{48}$, $\frac{1}{4}$ and $\frac{5}{4}$) are known as *critical exponents*.

The *four-arm* event is clearly of particular relevance to us in these lectures. Indeed, if a point x is in the 'bulk' of a domain $(n\Omega, n\partial_1, n\partial_2)$, the event that this point is pivotal for the Left-Right crossing event A_n is intimately related to the four-arm event. See Part VI for more details.

6. Quasi-multiplicativity

Finally, let us end this overview by a type of scale invariance property of these arm events. More precisely, it is often convenient to "divide" these arm events into different scales. For this purpose, we introduce $\alpha_4(r, R)$ (with $r \le R$) to be the probability that the four-arm event holds from radius r to radius R ($\alpha_1(r, R)$, $\alpha_2(r, R)$ and $\alpha_3(r, R)$ are defined analogously). By independence on disjoint sets, it is clear that if $r_1 \le r_2 \le r_3$ then one has $\alpha_4(r_1, r_3) \le \alpha_4(r_1, r_2)\,\alpha_4(r_2, r_3)$. A very useful property known as **quasi-multiplicativity** claims that up to constants, these two expressions are the same (this makes the division into several scales practical). This property can be stated as follows.

PROPOSITION II.3 (**quasi-multiplicativity**, [**Kes87**]). *For any $r_1 \le r_2 \le r_3$, one has (both for $\mathbb{Z}^2$ and $\mathbb{T}$ percolations)*

$$\alpha_4(r_1, r_3) \asymp \alpha_4(r_1, r_2)\,\alpha_4(r_2, r_3)\,.$$

See [**Wer07, Nol09, SS10b**] for more details. Note also that the same property holds for the one-arm event. However, this is much easier to prove: it is an easy consequence of the RSW Theorem II.1 and the so-called FKG inequality which says that increasing events are positively correlated. The reader might consider doing this as an exercise.

Part III. Sharp thresholds and the critical point for 2-d percolation

1. Monotone functions and the Margulis-Russo formula

The class of so-called monotone functions plays a very central role in this subject.

DEFINITION III.1. *A function f is* **monotone** *if $x \leq y$ (meaning $x_i \leq y_i$ for each i) implies that $f(x) \leq f(y)$. An event is monotone if its indicator function is monotone.*

Recall that when the underlying variables are independent with 1 having probability p, we let $\mathbb{P}_p$ and $\mathbb{E}_p$ denote probabilities and expectations.

It is fairly obvious that for f monotone, $\mathbb{E}_p(f)$ should be increasing in p. The Margulis-Russo formula gives us an explicit formula for this (nonnegative) derivative.

THEOREM III.1. *Let A be an increasing event in Ω_n. Then*

$$d(\mathbb{P}_p(A))/dp = \sum_i \mathbf{I}_i^p(A).$$

Proof. Let us allow each variable x_i to have its own parameter p_i and let $\mathbb{P}_{p_1,\ldots,p_n}$ and $\mathbb{E}_{p_1,\ldots,p_n}$ be the corresponding probability measure and expectation. It suffices to show that

$$\partial(\mathbb{P}_{(p_1,\ldots,p_n)}(A))/\partial p_i = \mathbf{I}_i^{(p_1,\ldots,p_n)}(A)$$

where the definition of this latter term is clear. WLOG, take $i = 1$. Now

$$\mathbb{P}_{p_1,\ldots,p_n}(A) = \mathbb{P}_{p_1,\ldots,p_n}(A \backslash \{1 \in \mathcal{P}_A\}) + \mathbb{P}_{p_1,\ldots,p_n}(A \cap \{1 \in \mathcal{P}_A\}).$$

The event in the first term is measurable with respect to the other variables and hence the first term does not depend on p_1 while the second term is

$$p_1 \mathbb{P}_{p_2,\ldots,p_n}(\{1 \in \mathcal{P}_A\})$$

since $A \cap \{1 \in \mathcal{P}_A\}$ is the event $\{x_1 = 1\} \cap \{1 \in \mathcal{P}_A\}$. □

2. KKL away from the uniform measure case

Recall now Theorem I.2. For sharp threshold results, one needs lower bounds on the total influence not just at the special parameter 1/2 but at all p.

The following are the two main results concerning the KKL result for general p that we will want to have at our disposal. The proofs of these theorems will be outlined in the exercises in Part V.

THEOREM III.2 ([**BKK**$^+$**92**]). *There exists a universal $c > 0$ such that for any Boolean function f mapping Ω_n into $\{0, 1\}$ and, for any p, there exists some i such that*

$$\mathbf{I}_i^p(f) \geq c\mathrm{Var}_p(f)(\log n)/n$$

THEOREM III.3 ([**BKK**$^+$**92**]). *There exists a universal $c > 0$ such that for any Boolean function f mapping Ω_n into $\{0, 1\}$ and for any p,*

$$\mathbf{I}^p(f) \geq c\mathrm{Var}_p(f)\log(1/\delta_p)$$

where $\delta_p := \max_i \mathbf{I}_i^p(f)$.

3. Sharp thresholds in general : the Friedgut-Kalai Theorem

THEOREM III.4 ([**FK96**]). *There exists a $c_1 < \infty$ such that for any monotone event A on n variables where all the influences are the same, if $\mathbb{P}_{p_1}(A) > \epsilon$, then*

$$\mathbb{P}_{p_1 + \frac{c_1 \log(1/(2\epsilon))}{\log n}}(A) > 1 - \epsilon.$$

REMARK III.1. This says that for fixed ϵ, the probability of A moves from below ϵ to above $1-\epsilon$ in an interval of p of length of order at most $1/\log n$. The assumption of equal influences holds for example if the event is invariant under some transitive action, which is often the case. For example, it holds for Example 4 (iterated 3-majority) as well as for any graph property in the context of the random graphs $G(n,p)$.

Proof. Theorem III.2 and all the influences being the same tell us that

$$\mathbf{I}^p(A) \geq c \min\{\mathbb{P}_p(A), 1 - \mathbb{P}_p(A)\} \log n$$

for some $c > 0$. Hence Theorem III.1 yields

$$d(\log(\mathbb{P}_p(A)))/dp \geq c \log n$$

if $\mathbb{P}_p(A) \leq 1/2$. Letting $p* := p_1 + \frac{\log(1/2\epsilon)}{c \log n}$, an easy computation (using the fundamental theorem of calculus) yields

$$\log(P_{p^*}(A)) \geq \log(1/2).$$

Next, if $\mathbb{P}_p(A) \geq 1/2$, then

$$d(\log(1 - \mathbb{P}_p(A)))/dp \leq -c \log n$$

from which another application of the fundamental theorem yields

$$\log(1 - P_{p^{**}}(A)) \leq -\log(1/\epsilon)$$

where $p^{**} := p^* + \frac{\log(1/2\epsilon)}{c \log(n)}$. Letting $c_1 = 2/c$ gives the result. □

4. The critical point for percolation for $\mathbb{Z}^2$ and $\mathbb{T}$ is $\frac{1}{2}$

THEOREM III.5 ([**Kes80**]).

$$p_c(\mathbb{Z}^2) = p_c(\mathbb{T}) = \frac{1}{2}.$$

Proof. We first show that $\theta(1/2) = 0$. Let $\text{Ann}(\ell) := [-3\ell, 3\ell] \backslash [-\ell, \ell]$ and C_k be the event that there is a circuit in $\text{Ann}(4^k) + 1/2$ in the dual lattice around the origin consisting of closed edges. The C_k's are independent and RSW and FKG show that for some $c > 0$, $\mathbb{P}_{1/2}(C_k) \geq c$ for all k. This gives that $\mathbb{P}_{1/2}(C_k \text{ infinitely often}) = 1$ and hence $\theta(1/2) = 0$.

The next key step is a *finite size criterion* which implies percolation and which is interesting in itself. We outline its proof afterwards.

PROPOSITION III.6. *(Finite size criterion) Let J_n be the event that there is a crossing of a $2n \times (n-2)$ box. For any p, if there exists an n such that*

$$\mathbb{P}_p(J_n) \geq .98,$$

then a.s. there exists an infinite cluster.

Assume now that $p_c = 1/2 + \delta$ with $\delta > 0$. Let $I = [1/2, 1/2 + \delta/2]$. Since $\theta(1/2 + \delta/2) = 0$, it is easy to see that the maximum influence over all variables and over all $p \in I$ goes to 0 with n since being pivotal implies the existence of an open path from a neighbor of the given edge to distance $n/2$ away. Next, by RSW, $\inf_n \mathbb{P}_{1/2}(J_n) > 0$. If for all n, $\mathbb{P}_{1/2+\delta/2}(J_n) < .98$, then Theorems III.1 and III.3 would allow us to conclude that the derivative of $\mathbb{P}_p(J_n)$ goes to ∞ uniformly on I as $n \to \infty$, giving a contradiction. Hence $\mathbb{P}_{1/2+\delta/2}(J_n) \geq .98$ for some n implying, by Proposition III.6, that $\theta(1/2 + \delta/2) > 0$, a contradiction. □

Outline of proof of Proposition III.6.

The first step is to show that for any p and for any $\epsilon \leq .02$, if $\mathbb{P}_p(J_n) \geq 1 - \epsilon$, then $\mathbb{P}_p(J_{2n}) \geq 1 - \epsilon/2$. The idea is that by FKG and "glueing" one can show that one can cross a $4n \times (n-2)$ box with probability at least $1 - 5\epsilon$ and hence one obtains that $\mathbb{P}_p(J_{2n}) \geq 1 - \epsilon/2$ since, for this event to fail, it must fail in both the top and bottom halves of the box. It follows that if we place down a sequence of (possibly rotated and translated) boxes of sizes $2^{n+1} \times 2^n$ anywhere, then with probability 1, all but finitely many are crossed. Finally, one can place these boxes down in an intelligent way such that crossing all but finitely many of them necessarily entails the existence of an infinite cluster (see Figure III.1). □

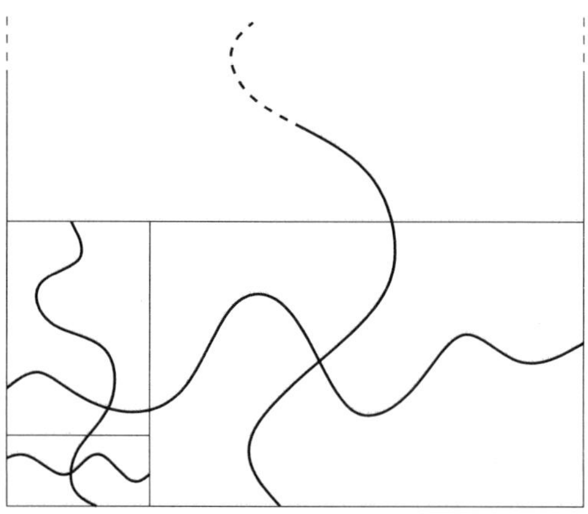

FIGURE III.1

5. Further discussion

The Margulis-Russo formula is due independently to Margulis [**Mar74**] and Russo [**Rus81**].

The idea to use the results from KKL to show that $p_c = 1/2$ is due to Bollobás and Riordan (see [**BR06**]). It was understood much earlier that obtaining a sharp threshold was the key step. Kesten (see [**Kes80**]) showed the necessary sharp threshold by obtaining a lower bound on the expected number of pivotals in a hands on fashion. Russo (see [**Rus82**]) had developed an earlier weaker, more qualitative, version of KKL and showed how it also sufficed to show that $p_c = 1/2$.

Exercise sheet of Part III

EXERCISE III.1. Develop an alternative proof of the Margulis-Russo formula using classical couplings.

EXERCISE III.2. Study, as best as you can, what the "threshold windows" are (i.e. where and how long does it take to go from a probability of order ϵ to a probability of order $1-\epsilon$) in the following examples:

(a) for $\mathbf{DICT}_n$
(b) for $\mathbf{MAJ}_n$
(c) for the tribes example
(d) for the iterated majority example.

Do not rely on [**KKL88**] type of results, but instead do hands-on computations specific to each case.

EXERCISE III.3. Write out the details of the proof of Proposition III.6.

PROBLEM III.4 (*What is the "sharpest" monotone event ?*). Show that among all monotone Boolean functions on Ω_n, $\mathbf{MAJ}_n$ is the one with largest total influence (at $p = 1/2$).
Hint: Use the Margulis-Russo formula.

EXERCISE III.5. A consequence of Problem III.4 is that the total influence at $p = 1/2$ of any monotone function is at most $O(\sqrt{n})$. A similar argument shows that for any p, there is a constant C_p so that the total influence at level p of any monotone function is at most $C_p\sqrt{n}$. Prove nonetheless that there exists $c > 0$ such for for any n, there exists a monotone function $f = f_n$ and a $p = p_n$ so that the total influence of f at level p is at least cn.

EXERCISE III.6. Find a monotone function $f : \Omega_n \to \{0, 1\}$ such that $d(\mathbb{E}_p(f))/dp$ is very large at $p = 1/2$, but nevertheless there is no sharp threshold for f (this means that a large total influence at some value of p is not in general a sufficient condition for sharp threshold).

Part IV. Fourier analysis of Boolean functions (first facts)

1. Discrete Fourier analysis and the energy spectrum

It turns out that in order to understand and analyze the concepts previously introduced, which are in some sense purely probabilistic, a critical tool is Fourier analysis on the hypercube.

Recall that we consider our Boolean functions as functions from the hypercube $\Omega_n := \{-1,1\}^n$ into $\{-1,1\}$ or $\{0,1\}$ where Ω_n is endowed with the uniform measure $\mathbb{P} = \mathbb{P}^n = (\frac{1}{2}\delta_{-1} + \frac{1}{2}\delta_1)^{\otimes n}$.

In order to apply Fourier analysis, the natural setup is to enlarge our discrete space of Boolean functions and to consider instead the larger space $L^2(\{-1,1\}^n)$ of real-valued functions on Ω_n endowed with the inner product:

$$\begin{aligned} \langle f, g \rangle &:= \sum_{x_1,\ldots,x_n} 2^{-n} f(x_1,\ldots,x_n) g(x_1,\ldots,x_n) \\ &= \mathbb{E}\big[fg\big] \quad \text{for all } f,g \in L^2(\Omega_n)\,, \end{aligned}$$

where $\mathbb{E}$ denotes expectation with respect to the uniform measure $\mathbb{P}$ on Ω_n.

For any subset $S \subseteq \{1,2\ldots,n\}$, let χ_S be the function on $\{-1,1\}^n$ defined for any $x = (x_1,\ldots,x_n)$ by

$$\chi_S(x) := \prod_{i \in S} x_i\,. \tag{IV.1}$$

(So $\chi_\emptyset \equiv 1$.) It is straightforward (check this!) to see that this family of 2^n functions forms an orthonormal basis of $L^2(\{-1,1\}^n)$. Thus, any function f on Ω_n (and a fortiori any Boolean function f) can be decomposed as

$$f = \sum_{S \subseteq \{1,\ldots,n\}} \hat{f}(S)\,\chi_S,$$

where $\{\hat{f}(S)\}_{S\subseteq[n]}$ are the so-called Fourier coefficients of f. They are also sometimes called the **Fourier-Walsh** coefficients of f and they satisfy

$$\hat{f}(S) := \langle f, \chi_S \rangle = \mathbb{E}\big[f\chi_S\big].$$

Note that $\hat{f}(\emptyset)$ is the average $\mathbb{E}\big[f\big]$. As in classical Fourier analysis, if f is some Boolean function, its Fourier(-Walsh) coefficients provide information on the "regularity" of f. We will sometimes use the term *spectrum* when referring to the set of Fourier coefficients.

Of course one may find many other orthonormal bases for $L^2(\{-1,1\}^n)$, but there are many situations for which this particular set of functions $\{\chi_S\}_{S\subseteq\{1,\ldots,n\}}$ arises naturally. First of all there is a well-known theory of Fourier analysis on groups, a theory which is particularly simple and elegant on Abelian groups (thus including our special case of $\{-1,1\}^n$, but also $\mathbb{R}/\mathbb{Z}$, $\mathbb{R}$ and so on). For Abelian groups, what turns out to be relevant for doing harmonic analysis is the set $\hat{G}$ of **characters** of G (i.e. the group homomorphisms from G to $\mathbb{C}^*$). In our case of $G = \{-1,1\}^n$, the characters are precisely our functions χ_S indexed by $S \subseteq \{1,\ldots,n\}$ since they satisfy $\chi_S(x \cdot y) = \chi_S(x)\chi_S(y)$. This background is not however needed and we won't talk in these terms.

These functions also arise naturally if one performs simple random walk on the hypercube (equipped with the Hamming graph structure), since they are the eigenfunctions of the corresponding Markov chain (heat kernel) on $\{-1,1\}^n$. Last but not least, we will see later in this part that the basis $\{\chi_S\}$ turns out to be particularly well adapted to our study of noise sensitivity.

We introduce one more concept here without motivation; it will be very well motivated later on in the chapter.

DEFINITION IV.1. *For any real-valued function* $f : \Omega_n \to \mathbb{R}$, *the* **energy spectrum** E_f *is defined by*

$$E_f(m) := \sum_{|S|=m} \hat{f}(S)^2, \ \forall m \in \{1, \dots, n\}.$$

2. Examples

First note that, from the Fourier point of view, Dictator and Parity have simple representations since they are χ_1 and $\chi_{[n]}$ respectively. Each of the two corresponding energy spectra are trivially concentrated on 1 point, namely 1 and n.

For Example 3, the Majority function, Bernasconi explicitly computed the Fourier coefficients and when n goes to infinity, one ends up with the following asymptotic formula for the energy spectrum:

$$E_{\mathbf{MAJ}_n}(m) = \sum_{|S|=m} \widehat{\mathbf{MAJ}}_n(S)^2 = \begin{cases} \frac{4}{\pi\, m 2^m}\binom{m-1}{\frac{m-1}{2}} + O(m/n) & \text{if } m \text{ is odd}, \\ 0 & \text{if } m \text{ is even}. \end{cases}$$

(The reader may think about why the "even" coefficients are 0.) See [**O'D03**] for a nice overview and references therein concerning the spectral behavior of the majority function.

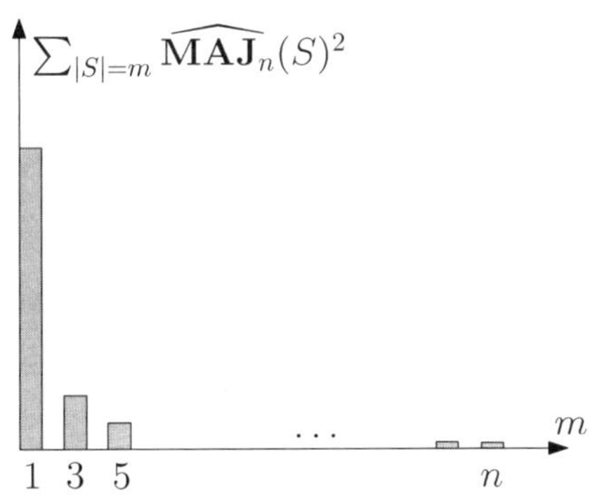

FIGURE IV.1. Shape of the energy spectrum for the Majority function

Picture IV.1 represents the shape of the energy spectrum of $\mathbf{MAJ}_n$: its spectrum is concentrated on low frequencies, which is typical of stable functions.

3. Noise sensitivity and stability in terms of the energy spectrum

In this section, we describe the concepts of noise sensitivity and noise stability in terms of the energy spectrum.

The first step is to note that, given any real-valued function $f : \Omega_n \to \mathbb{R}$, the correlation between $f(\omega)$ and $f(\omega_\epsilon)$ is nicely expressed in terms of the Fourier coefficients of f as follows:

$$\begin{aligned} \mathbb{E}\big[f(\omega)f(\omega_\epsilon)\big] &= \mathbb{E}\Big[\Big(\sum_{S_1} \hat{f}(S_1)\chi_{S_1}(\omega)\Big)\Big(\sum_{S_2} \hat{f}(S_2)\chi_{S_2}(\omega_\epsilon)\Big)\Big] \\ &= \sum_S \hat{f}(S)^2 \mathbb{E}\big[\chi_S(\omega)\,\chi_S(\omega_\epsilon)\big] \\ &= \sum_S \hat{f}(S)^2 (1-\epsilon)^{|S|}. \end{aligned} \tag{IV.2}$$

Moreover, we immediately obtain

$$\mathrm{Cov}(f(\omega), f(\omega_\epsilon)) = \sum_{m=1}^{n} E_f(m)(1-\epsilon)^m. \tag{IV.3}$$

Note that either of the last two expressions tell us that $\mathrm{Cov}(f(\omega), f(\omega_\epsilon))$ is nonnegative and decreasing in ϵ. Also, we see that the "level of noise sensitivity" of a Boolean function is naturally encoded in its energy spectrum. It is now an an easy exercise to prove the following proposition.

PROPOSITION IV.1 ([**BKS99**]). *A sequence of Boolean functions* $f_n : \{-1,1\}^{m_n} \to \{0,1\}$ *is noise sensitive if and only if, for any* $k \geq 1$,

$$\sum_{m=1}^{k} \sum_{|S|=m} \hat{f}_n(S)^2 = \sum_{m=1}^{k} E_{f_n}(m) \underset{n\to\infty}{\longrightarrow} 0\,.$$

Moreover, (I.1) holding does not depend on the value of $\epsilon \in (0,1)$ *chosen.*

There is a similar spectral description of noise stability which, given (IV.2), is an easy exercise.

PROPOSITION IV.2 ([**BKS99**]). *A sequence of Boolean functions* $f_n : \{-1,1\}^{m_n} \to \{0,1\}$ *is noise stable if and only if, for any* $\epsilon > 0$, *there exists* k *such that for all* n,

$$\sum_{m=k}^{\infty} \sum_{|S|=m} \hat{f}_n(S)^2 = \sum_{m=k}^{\infty} E_{f_n}(m) < \epsilon.$$

So, as argued in the introduction, a function of "high frequency" will be sensitive to noise while a function of "low frequency" will be stable.

4. Link between the spectrum and influence

In this section, we relate the notion of influence with that of the spectrum.

PROPOSITION IV.3. *If* $f : \Omega_n \to \{0,1\}$, *then for all* k,

$$\mathbf{I}_k(f) = 4 \sum_{S: k\in S} \hat{f}(S)^2$$

and

$$\mathbf{I}(f) = 4\sum_S |S|\hat{f}(S)^2.$$

Proof. If $f : \Omega_n \to \mathbb{R}$, we introduce the functions

$$\nabla_k f : \left\{ \begin{array}{ccc} \Omega_n & \to & \mathbb{R} \\ \omega & \mapsto & f(\omega) - f(\sigma_k(\omega)) \end{array} \right. \quad \text{for all } k \in [n]\,,$$

where σ_k acts on Ω_n by flipping the kth bit (thus $\nabla_k f$ corresponds to a discrete derivative along the kth bit).

Observe that

$$\nabla_k f(\omega) = \sum_{S\subseteq\{1,\ldots,n\}} \hat{f}(S)\,[\chi_S(\omega) - \chi_S(\sigma_k(\omega))] = \sum_{S\subseteq\{1,\ldots,n\},k\in S} 2\hat{f}(S)\,\chi_S(\omega),$$

from which it follows that for any $S \subseteq [n]$,

$$\widehat{\nabla_k f}(S) = \left\{ \begin{array}{ll} 2\hat{f}(S) & \text{if } k \in S \\ 0 & \text{otherwise} \end{array} \right. \tag{IV.4}$$

Clearly, if f maps into $\{0,1\}$, then $\mathbf{I}_k(f) := \|\nabla_k f\|_1$ and since $\nabla_k f$ takes values in $\{-1,0,1\}$ in this case, we have $\|\nabla_k f\|_1 = \|\nabla_k f\|_2^2$. Applying Parseval to $\nabla_k f$ and using (IV.4), one obtains the first statement of the proposition. The second is obtained by summing over k and exchanging the order of summation. □

REMARK IV.1. If f maps into $\{-1,1\}$ instead, then one can easily check that $\mathbf{I}_k(f) = \sum_{S:k\in S} \hat{f}(S)^2$ and $\mathbf{I}(f) = \sum_S |S|\hat{f}(S)^2$.

5. Monotone functions and their spectrum

It turns out that for monotone functions, there is an alternative useful spectral description of the influences.

PROPOSITION IV.4. *If $f : \Omega_n \to \{0,1\}$ is monotone, then for all k*

$$\mathbf{I}_k(f) = 2\hat{f}(\{k\})$$

If f maps into $\{-1,1\}$ instead, then one has that $\mathbf{I}_k(f) = \hat{f}(\{k\})$. (Observe that Parity shows that the assumption of monotonicity is needed here; note also that the proof shows that the weaker result with $=$ replaced by $\geq$ holds in general.)

Proof. We prove only the first statement; the second is proved in the same way.

$$\hat{f}(\{k\}) := \mathbb{E}\big[f\chi_{\{k\}}\big] = \mathbb{E}\big[f\chi_{\{k\}}I_{\{k\notin\mathcal{P}\}}\big] + \mathbb{E}\big[f\chi_{\{k\}}I_{\{k\in\mathcal{P}\}}\big]$$

It is easily seen that the first term is 0 (independent of whether f is monotone or not) and the second term is $\frac{\mathbf{I}_k(f)}{2}$ due to monotonicity. □

REMARK IV.2. This tells us that, for monotone functions mapping into $\{-1,1\}$, the sum in Theorem I.5 is exactly the total weight of the level 1 Fourier coefficients, that is, the energy spectrum at 1, $E_f(1)$. (If we map into $\{0,1\}$ instead, there is simply an extra irrelevant factor of 4.) So Theorem I.5 and Propositions IV.1 and IV.4 imply that for monotone functions, if the energy spectrum at 1 goes to 0, then this is true for any fixed level. In addition, Propositions IV.1 (with $k = 1$) and IV.4 easily imply that for monotone functions the converse of Theorem I.5 holds.

Another application of Proposition IV.4 gives a general upper bound for the total influence for monotone functions.

PROPOSITION IV.5. *If $f : \Omega_n \to \{-1,1\}$ or $\{0,1\}$ is monotone, then*

$$\mathbf{I}(f) \leq \sqrt{n}.$$

Proof. If the image is $\{-1,1\}$, then by Proposition IV.4, we have

$$\mathbf{I}(f) = \sum_{k=1}^{n} \mathbf{I}_k(f) = \sum_{k=1}^{n} \hat{f}(\{k\}).$$

By the Cauchy-Schwarz inequality, this is at most $(\sum_{k=1}^{n} \hat{f}^2(\{k\}))^{1/2}\sqrt{n}$. By Parseval's formula, the first term is at most 1 and we are done. If the image is $\{0,1\}$, the above proof can easily be modified or one can deduce it from the first case since the total influence of the corresponding ± 1-valued function is the same. □

REMARK IV.3. The above result with some universal c on the right hand side follows (for odd n) from an earlier exercise showing that Majority has the largest influence together with the known influences for Majority. However, the above argument yields a more direct proof of the $\sqrt{n}$ bound.

Exercise sheet of Part IV

EXERCISE IV.1. Prove the discrete Poincaré inequality, Theorem I.1, using the spectrum.

EXERCISE IV.2. Compute the Fourier coefficients for the indicator function that there are all 1's.

EXERCISE IV.3. Show that all even size Fourier coefficients for the Majority function are 0. Can you extend this result to a broader class of Boolean functions?

EXERCISE IV.4. For the Majority function $\mathbf{MAJ}_n$, find the limit (as the number of voters n goes to infinity) of the following quantity (total weight of the level-3 Fourier coefficients)

$$E_{\mathbf{MAJ}_n}(3) := \sum_{|S|=3} \widehat{\mathbf{MAJ}}_n(S)^2 .$$

EXERCISE IV.5. Let f_n be a sequence of Boolean functions which is noise sensitive and g_n be a sequence of Boolean functions which is noise stable. Show that f_n and g_n are asymptotically uncorrelated.

EXERCISE IV.6 (Another equivalent definition of noise sensitivity). Assume that $\{A_n\}$ is a noise sensitive sequence. (This of course means that the indicator functions of these events is a noise sensitive sequence.)

(a) Show for each $\epsilon > 0$, we have that $\mathbb{P}\big[\omega_\epsilon \in A_n \bigm| \omega\big] - \mathbb{P}\big[A_n\big]$ approaches 0 in probability.
Hint: use the Fourier representation.
(b) Can you show the above implication without using the Fourier representation?
(c) Discuss if this implication is surprising.
(d) Show that the condition in part (a) implies that the sequence is noise sensitive directly without the Fourier representation.

EXERCISE IV.7. How does the spectrum of a generic Boolean function look? Use this to give an alternative answer to the question asked in problem I.13 of Part I.

EXERCISE IV.8. *(Open exercise).* For Boolean functions, can one have ANY (reasonable) shape of the energy spectrum or are there restrictions?

For the next exercises, we introduce the following functional which measures the stability of Boolean functions. For any Boolean function $f : \Omega_n \to \{-1, 1\}$, let

$$\mathbb{S}_f \, : \epsilon \mapsto \mathbb{P}\big[f(\omega) \neq f(\omega_\epsilon)\big] \, .$$

Obviously, the smaller $\mathbb{S}_f$ is, the more stable f is.

EXERCISE IV.9. Express the functional $\mathbb{S}_f$ in terms of the Fourier expansion of f.

By a **balanced** Boolean function we mean one which takes its two possible values each with probability $1/2$.

EXERCISE IV.10. Among balanced Boolean functions, does there exist some function f^* which is "stablest" in the sense that for any balanced Boolean function f and any $\epsilon > 0$,

$$\mathbb{S}_{f^*}(\epsilon) \le \mathbb{S}_f(\epsilon)\,?$$

If yes, describe the set of these extremal functions and prove that these are the only ones.

PROBLEM IV.11. In this problem, we wish to understand the asymptotic shape of the energy spectrum for $\mathbf{MAJ}_n$.

(a) Show that for all $\epsilon \ge 0$,

$$\lim_{n\to\infty} \mathbb{S}_{\mathbf{MAJ}_n}(\epsilon) = \frac{1}{2} - \frac{\arcsin(1-\epsilon)}{\pi} = \frac{\arccos(1-\epsilon)}{\pi}.$$

Hint: The relevant limit is easily expressed as the probability that a certain 2-dimensional Gaussian variable (with a particular correlation structure) falls in a certain area of the plane. One can write down the corresponding density function and this probability as an explicit integral but this integral does not seem so easy to evaluate. However, this Gaussian probability can be computed directly by representing the joint distribution in terms of two independent Gaussians.

Note that the above limit immediately implies that for $f_n = \mathbf{MAJ}_n$,

$$\lim_{n\to\infty} \mathbb{E}(f_n(\omega) f_n(\omega_\epsilon)) = \frac{2\arcsin(1-\epsilon)}{\pi}.$$

(b) Deduce from (a) and the Taylor expansion for $\arcsin(x)$ the limiting value, as $n \to \infty$ of $E_{\mathbf{MAJ}_n}(k) = \sum_{|S|=k} \widehat{\mathbf{MAJ}}_n(S)^2$ for all $k \ge 1$. Check that the answer is consistent with the values obtained earlier for $k = 1$ and $k = 3$ (Exercise IV.4).

Part V. Hypercontractivity and its applications

In this lecture, we will prove the main theorems about influences stated in Part I. As we will see, these proofs rely on techniques imported from harmonic analysis, in particular *hypercontractivity*. As we will see later in this part and in Part VII, these types of proofs extend to other contexts which will be of interest to us: noise sensitivity and sub-Gaussian fluctuations.

1. Heuristics of proofs

All the subsequent proofs which will be based on *hypercontractivity* will have more or less the same flavor. Let us now explain in the particular case of Theorem I.2 what the overall scheme of the proof is.

Recall that we want to prove that there exists a universal constant $c > 0$ such that for any function $f : \Omega_n \to \{0, 1\}$, one of its variables has influence at least $c \frac{\log n \operatorname{Var}(f)}{n}$.

Let f be a Boolean function. Suppose all its influences $\mathbf{I}_k(f)$ are "small" (this would need to be made quantitative). This means that $\nabla_k f$ must have small support. Using the intuition coming from the Weyl-Heisenberg uncertainty, $\widehat{\nabla_k f}$ should then be quite spread out in the sense that most of its spectral mass should be concentrated on high frequencies.

This intuition, which is still vague at this point, says that having small influences pushes the spectrum of $\nabla_k f$ towards high frequencies. Now, summing up as we did in Section 4 of Part IV, but restricting ourselves only to frequencies S of size smaller than some large (well-chosen) $1 \ll M \ll n$, one easily obtains

$$
\begin{aligned}
\sum_{0<|S|<M} \hat{f}(S)^2 &\leq 4 \sum_{0<|S|<M} |S| \hat{f}(S)^2 \\
&= \sum_k \sum_{0<|S|<M} \widehat{\nabla_k f}(S)^2 \\
&\text{"}\ll\text{"} \sum_k \|\widehat{\nabla_k f}\|_2^2 \\
&= \mathbf{I}(f)\,, \qquad \text{(V.1)}
\end{aligned}
$$

where, in the third line, we used the informal statement that $\widehat{\nabla_k f}$ should be supported on high frequencies if f has small influences. Now recall (or observe) that

$$\sum_{|S|>0} \hat{f}(S)^2 = \operatorname{Var}(f)\,.$$

Therefore, in the above equation (V.1), if we are in the case where a positive fraction of the Fourier mass of f is concentrated below M, then (V.1) says that $\mathbf{I}(f)$ is much larger than $\operatorname{Var}(f)$. In particular, at least one of the influences has to be "large". If, on the other hand, we are in the case where most of the spectral mass of f is supported on frequencies of size higher than M, then we also obtain that $\mathbf{I}(f)$ is large by using the formula:

$$\mathbf{I}(f) = 4 \sum_S |S| \hat{f}(S)^2\,.$$

REMARK V.1. Note that these heuristics suggest that there is a subtle balance between $\sum_k \mathbf{I}_k(f) = \mathbf{I}(f)$ and $\sup_k \mathbf{I}_k(f)$. Namely, if influences are all small (i.e. $\| \cdot \|_\infty$ is small), then their sum on the other hand has to be "large". The right balance is exactly quantified by Theorem I.3.

Of course it now remains to convert the above sketch into a proof. The main difficulty in the above program is to obtain quantitative spectral information on functions with values in $\{-1, 0, 1\}$ knowing that they have small support. This is done ([**KKL88**]) using techniques imported from harmonic analysis, namely hypercontractivity.

2. About hypercontractivity

First, let us state what hypercontractivity corresponds to. Let $(K_t)_{t\geq 0}$ be the heat kernel on $\mathbb{R}^n$. Hypercontractivity is a statement which quantifies how functions are regularized under the heat flow. The statement, which goes back to a number of authors, can be simply stated as follows:

THEOREM V.1 (Hypercontractivity). *Consider $\mathbb{R}^n$ with standard Gaussian measure. If $1 < q < 2$, there is some $t = t(q) > 0$ (which does not depend on the dimension n) such that for any $f \in L^q(\mathbb{R}^n)$,*

$$\|K_t * f\|_2 \leq \|f\|_q .$$

The dependence $t = t(q)$ is explicit but will not concern us in the Gaussian case. Hypercontractivity is thus a regularization statement: if one starts with some initial "rough" L^q function f outside of L^2 and waits long enough ($t(q)$) under the heat flow, then we end up being in L^2 with a good control on its L^2 norm.

This concept has an interesting history as is nicely explained in O'Donnell's lecture notes (see [**O'D**]). It was originally invented by Nelson in [**Nel66**] where he needed regularization estimates on Free fields (which are the building blocks of quantum field theory) in order to apply these in "constructive field theory". It was then generalized by Gross in his elaboration of logarithmic Sobolev inequalities ([**Gro75**]), which is an important tool in analysis. Hypercontractivity is intimately related to these Log-Sobolev inequalities and thus has many applications in the theory of Semigroups, mixing of Markov chains and other topics.

We now state the result in the case which concerns us, namely the hypercube. For any $\rho \in [0, 1]$, let T_ρ be the following **noise operator** on the set of functions on the hypercube: recall from Part I that if $\omega \in \Omega_n$, we denote by ω_ϵ an ϵ-noised configuration of ω. For any $f : \Omega_n \to \mathbb{R}$, we define $T_\rho f : \omega \mapsto \mathbb{E}\big[f(\omega_{1-\rho}) \bigm| \omega\big]$. This noise operator acts in a very simple way on the Fourier coefficients, as the reader can check:

$$T_\rho : f = \sum_S \hat{f}(S)\, \chi_S \mapsto \sum_S \rho^{|S|} \hat{f}(S)\, \chi_S .$$

We have the following analogue of Theorem V.1.

THEOREM V.2 (Bonami-Gross-Beckner). *For any $f : \Omega_n \to \mathbb{R}$ and any $\rho \in [0, 1]$,*

$$\|T_\rho f\|_2 \leq \|f\|_{1+\rho^2} .$$

The analogy with the classical result V.1 is clear: the heat flow is replaced here by the random walk on the hypercube. You can find the proof of Theorem V.2 in the appendix attached to the present part.

REMARK V.2. The term *hypercontractive* refers here to the fact that one has an operator which maps L^q into L^2 ($q < 2$), which is a contraction.

Before going into the detailed proof of Theorem I.2, let us see why Theorem V.2 provides us with the type of spectral information we need. In the above sketch, we assumed that all influences were small. This can be written as

$$\mathbf{I}_k(f) = \|\nabla_k f\|_1 = \|\nabla_k f\|_2^2 \ll 1\,,$$

for any $k \in [n]$. Now if one applies the hypercontractive estimate to these functions $\nabla_k f$ for some fixed $0 < \rho < 1$, we obtain that

$$\|T_\rho(\nabla_k f)\|_2 \le \|\nabla_k f\|_{1+\rho^2} = \|\nabla_k f\|_2^{2/(1+\rho^2)} \ll \|\nabla_k f\|_2 \tag{V.2}$$

where, for the equality, we used once again that $\nabla_k f \in \{-1, 0, 1\}$. After squaring, this gives on the Fourier side,

$$\sum_S \rho^{2|S|} \widehat{\nabla_k f}(S)^2 \ll \sum_S \widehat{\nabla_k f}(S)^2\,.$$

This shows (under the assumption that $\mathbf{I}_k(f)$ is small) that the spectrum of $\nabla_k f$ is indeed mostly concentrated on high frequencies.

REMARK V.3. We point out that Theorem V.2 in fact tells us that any function with small support has its frequencies concentrated on large sets as follows. It is easy to see that given any $p < 2$, if a function h on a probability space has very small support, then its L_p norm is much smaller than its L_2 norm. Using Theorem V.2, we would then have for such a function that

$$\|T_\rho(h)\|_2 \le \|h\|_{1+\rho^2} \ll \|h\|_2\,,$$

yielding that

$$\sum_S \rho^{2|S|} \widehat{h}(S)^2 \ll \sum_S \widehat{h}(S)^2$$

which can only occur if h has its frequencies concentrated on large sets. From this point of view, one also sees that under the small influence assumption, one did not actually need the third term in (V.2) in the above outline.

3. Proof of the KKL Theorems on the influences of Boolean functions

We will start by proving Theorem I.2, and then Theorem I.3. In fact, it turns out that one can recover Theorem I.2 directly from Theorem I.3; see the exercises. Nevertheless, since the proof of Theorem I.2 is slightly simpler, we start with this one.

3.1. Proof of Theorem I.2. Let $f : \Omega_n \to \{0, 1\}$. Recall that we want to show that there is some $k \in [n]$ such that

$$\mathbf{I}_k(f) \ge c\mathrm{Var}(f)\frac{\log n}{n}\,, \tag{V.3}$$

for some universal constant $c > 0$.

We divide the analysis into the following two cases.

Case 1:

Suppose that there is some $k \in [n]$ such that $\mathbf{I}_k(f) \geq n^{-3/4}\,\mathrm{Var}(f)$. Then the bound V.3 is clearly satisfied for a small enough $c > 0$.

Case 2:

Now, if f does not belong to the first case, this means that for all $k \in [n]$,

$$\mathbf{I}_k(f) = \|\nabla_k f\|_2^2 \leq \mathrm{Var}(f) n^{-3/4}\,. \tag{V.4}$$

Following the above heuristics, we will show that under this assumption, most of the Fourier spectrum of f is supported on high frequencies. Let $M \geq 1$, whose value will be chosen later. We wish to bound from above the bottom part (up to M) of the Fourier spectrum of f.

$$\begin{aligned}\sum_{1\leq|S|\leq M} \hat{f}(S)^2 &\leq \sum_{1\leq|S|\leq M} |S|\hat{f}(S)^2 \\ &\leq 2^{2M}\sum_{|S|\geq 1}(1/2)^{2|S|}|S|\hat{f}(S)^2 \\ &= \frac{1}{4}\,2^{2M}\sum_k \|T_{1/2}(\nabla_k f)\|_2^2\,,\end{aligned}$$

(see Section 4 of Part IV). Now by applying hypercontractivity (Theorem V.2) with $\rho = 1/2$ to the above sum, we obtain

$$\begin{aligned}\sum_{1\leq|S|\leq M} \hat{f}(S)^2 &\leq \frac{1}{4}\,2^{2M}\sum_k \|\nabla_k f\|_{5/4}^2 \\ &\leq 2^{2M}\sum_k \mathbf{I}_k(f)^{8/5} \\ &\leq 2^{2M}\, n\,\mathrm{Var}(f)^{8/5} n^{\frac{-3}{4}\cdot\frac{8}{5}} \\ &\leq 2^{2M}\, n^{-1/5}\,\mathrm{Var}(f)\,,\end{aligned}$$

where we used the assumption V.4 and the obvious fact that $\mathrm{Var}(f)^{8/5} \leq \mathrm{Var}(f)$ (recall $\mathrm{Var}(f) \leq 1$ since f is Boolean). Now with $M := \lfloor \frac{1}{20}\log_2 n \rfloor$, this gives

$$\sum_{1\leq|S|\leq\frac{1}{20}\log_2 n} \hat{f}(S)^2 \leq n^{1/10-1/5}\,\mathrm{Var}(f) = n^{-1/10}\,\mathrm{Var}(f)\,.$$

This shows that under our above assumption, most of the Fourier spectrum is concentrated above $\Omega(\log n)$. We are now ready to conclude:

$$\begin{aligned}
\sup_k \mathbf{I}_k(f) \geq \frac{\sum_k \mathbf{I}_k(f)}{n} &= \frac{4\sum_{|S|\geq 1}|S|\hat{f}(S)^2}{n} \\
&\geq \frac{1}{n}\Big[\sum_{|S|>M}|S|\hat{f}(S)^2\Big] \\
&\geq \frac{M}{n}\Big[\sum_{|S|>M}\hat{f}(S)^2\Big] \\
&= \frac{M}{n}\Big[\mathrm{Var}(f) - \sum_{1\leq|S|\leq M}\hat{f}(S)^2\Big] \\
&\geq \frac{M}{n}\mathrm{Var}(f)\big[1-n^{-1/10}\big] \\
&\geq c_1\,\mathrm{Var}(f)\frac{\log n}{n}\,,
\end{aligned}$$

with $c_1 = \frac{1}{20\log 2}(1-2^{-1/10})$. By combining with the constant given in case 1, this completes the proof. □

REMARK V.4. We did not try here to optimize the proof in order to find the best possible universal constant $c > 0$. Note though, that even without optimizing at all, the constant we obtain is not that bad.

3.2. Proof of Theorem I.3. We now proceed to the proof of the stronger result, Theorem I.3, which states that there is a universal constant $c > 0$ such that for any $f : \Omega_n \to \{0,1\}$,

$$\|\mathbf{I}(f)\| = \|\mathbf{Inf}(f)\|_1 \geq c\,\mathrm{Var}(f)\log\frac{1}{\|\mathbf{Inf}(f)\|_\infty}\,.$$

The strategy is very similar. Let $f : \Omega_n \to \{0,1\}$ and let $\delta := \|\mathbf{Inf}(f)\|_\infty = \sup_k \mathbf{I}_k(f)$. Assume for the moment that $\delta \leq 1/1000$. As in the above proof, we start by bounding the bottom part of the spectrum up to some integer M (whose value will be fixed later). Exactly in the same way as above, one has

$$\begin{aligned}
\sum_{1\leq|S|\leq M}\hat{f}(S)^2 &\leq 2^{2M}\sum_k \mathbf{I}_k(f)^{8/5} \\
&\leq 2^{2M}\delta^{3/5}\sum_k \mathbf{I}_k(f) = 2^{2M}\delta^{3/5}\,\mathbf{I}(f)\,.
\end{aligned}$$

Now,

$$\begin{aligned}
\mathrm{Var}(f) = \sum_{|S|\geq 1}\hat{f}(S)^2 &\leq \sum_{1\leq|S|\leq M}\hat{f}(S)^2 + \frac{1}{M}\sum_{|S|>M}|S|\hat{f}(S)^2 \\
&\leq \big[2^{2M}\delta^{3/5} + \frac{1}{M}\big]\,\mathbf{I}(f)\,.
\end{aligned}$$

Choose $M := \frac{3}{10}\log_2(\frac{1}{\delta}) - \frac{1}{2}\log_2\log_2(\frac{1}{\delta})$. Since $\delta < 1/1000$, it is easy to check that $M \geq \frac{1}{10}\log_2(1/\delta)$ which leads us to

$$\mathrm{Var}(f) \quad \leq \quad \left[\frac{1}{\log_2(1/\delta)} + \frac{10}{\log_2(1/\delta)}\right]\mathbf{I}(f)$$

(V.5)

which gives

$$\mathbf{I}(f) = \|\mathbf{Inf}(f)\|_1 \geq \frac{1}{11\log 2}\mathrm{Var}(f)\log\frac{1}{\|\mathbf{Inf}(f)\|_\infty}.$$

This gives us the result for $\delta \leq 1/1000$.

Next the discrete Poincaré inequality, which says that $\mathbf{I}(f) \geq \mathrm{Var}(f)$, tells us that the claim is true for $\delta \geq 1/1000$ if we take c to be $1/\log 1000$. Since this is larger than $\frac{1}{11\log 2}$, we obtain the result with the constant $c = \frac{1}{11\log 2}$. □

4. KKL away from the uniform measure

In Part III (on sharp thresholds), we needed an extension of the above KKL Theorems to the p-biased measures $\mathbb{P}_p = (p\delta_1 + (1-p)\delta_{-1})^{\otimes n}$. These extensions are respectively Theorems III.2 and III.3.

A first natural idea in order to extend the above proofs would be to extend the hypercontractive estimate (Theorem V.2) to these p-biased measures $\mathbb{P}_p$. This extension of Bonami-Gross-Beckner is possible, but it turns out that the control it gives gets worse near the edges (p close to 0 or 1). This is problematic since both in Theorems III.2 and III.3, we need bounds which are uniform in $p \in [0,1]$.

Hence, one needs a different approach to extend the KKL Theorems. A nice approach was provided in [**BKK**$^+$**92**], where they prove the following general theorem.

THEOREM V.3 ([**BKK**$^+$**92**]). *There exists a universal $c > 0$ such that for any measurable function $f : [0,1]^n \to \{0,1\}$, there exists a variable k such that*

$$\mathbf{I}_k(f) \geq c\,\mathrm{Var}(f)\frac{\log n}{n}.$$

Here the 'continuous' hypercube is endowed with the uniform (Lebesgue) measure and for any $k \in [n]$, $\mathbf{I}_k(f)$ denotes the probability that f is not almost-surely constant on the fiber given by $(x_i)_{i\neq k}$.

In other words,

$$\mathbf{I}_k(f) = \mathbb{P}\big[\mathrm{Var}\big(f(x_1,\ldots,x_n)\bigm| x_i, i\neq k\big) > 0\big].$$

It is clear how to obtain Theorem III.2 from the above theorem. If $p \in [0,1]$ and $f : \Omega_n \to \{0,1\}$, consider $\bar{f}_p : [0,1]^n \to \{0,1\}$ defined by

$$\bar{f}_p(x_1,\ldots,x_n) = f((1_{x_i<p} - 1_{x_i\geq p})_{i\in[n]}).$$

Friedgut noticed in [**Fri04**] that one can recover Theorem V.3 from Theorem III.2. The first idea is to use a symmetrization argument in such a way that the problem reduces to the case of monotone functions. Then, the main idea is the approximate the uniform measure on $[0,1]$ by the dyadic random variable

$$X_M : (x_1,\ldots,x_M) \in \{-1,1\}^M \mapsto \sum_{m=1}^{M}\frac{x_m+1}{2}2^{-m}.$$

One can then approximate $f : [0,1]^n \to \{0,1\}$ by the Boolean function $\hat{f}_M$ defined on $\{-1,1\}^{M\times n}$ by

$$\hat{f}_M(x_1^1,\ldots,x_M^1,\ldots,x_1^n,\ldots,x_M^n) := f(X_M^1,\ldots,X_M^n)\,.$$

Still (as mentioned in the above heuristics) this proof requires two technical steps: a monotonization procedure and an "approximation" step (going from f to $\hat{f}_M$). Since in our applications to sharp thresholds we used Theorems III.2 and III.3 only in the case of monotone functions, for the sake of simplicity we will not present the monotonization procedure in these notes.

Furthermore, it turns out that for our specific needs (the applications in Part III), we do not need to deal with the approximation part either. The reason is that for any Boolean function f, the function $p \mapsto \mathbf{I}_k^p(f)$ is continuous. Hence it is enough to obtain uniform bounds on $\mathbf{I}_k^p(f)$ for dyadic values of p (i.e. $p \in \{m2^{-M}\}\cap[0,1]$).

See the exercises for the proof of Theorems III.2 and III.3 when f is assumed to be monotone (problem V.4).

REMARK V.5. We mentioned above that generalizing hypercontractivity would not allow us to obtain uniform bounds (with p taking any value in $[0,1]$) on the influences. It should be noted though that Talagrand obtained ([**Tal94**]) results similar to Theorems III.2 and III.3 by somehow generalizing hypercontractivity, but along a different line. Finally, let us point out that both Talagrand ([**Tal94**]) and Friedgut and Kalai ([**FK96**]) obtain sharper versions of Theorems III.2 and III.3 where the constant $c = c_p$ in fact improves (i.e. blows up) near the edges.

5. The noise sensitivity theorem

In this section, we prove the milestone Theorem I.5 from [**BKS99**]. Before recalling what the statement is, let us define the following functional on Boolean functions. For any $f : \Omega_n \to \{0,1\}$, let

$$\mathbf{H}(f) := \sum_k \mathbf{I}_k(f)^2 = \|\mathbf{Inf}(f)\|_2^2\,.$$

Recall the Benjamini-Kalai-Schramm Theorem.

THEOREM V.4 ([**BKS99**]). *Consider a sequence of Boolean functions $f_n : \Omega_{m_n} \to \{0,1\}$. If*

$$\mathbf{H}(f_n) = \sum_{k=1}^{m_n} \mathbf{I}_k(f)^2 \to 0$$

as $n \to \infty$, then $\{f_n\}_n$ is noise sensitive.

We will in fact prove this theorem under a stronger condition, namely that $\mathbf{H}(f_n) \le (m_n)^{-\delta}$ for some exponent $\delta > 0$. Without this assumption of "polynomial decay" on $\mathbf{H}(f_n)$, the proof is more technical and relies on estimates obtained by Talagrand. See the remark at the end of this proof. For our application to the *noise sensitivity of percolation* (see Part VI), this stronger assumption will be satisfied and hence we stick to this simpler case in these notes.

The assumption of polynomial decay in fact enables us to prove the following more quantitative result.

PROPOSITION V.5 ([**BKS99**]). *For any $\delta > 0$, there exists a constant $M = M(\delta) > 0$ such that if $f_n : \Omega_{m_n} \to \{0,1\}$ is any sequence of Boolean functions satisfying*

$$\mathbf{H}(f_n) \le (m_n)^{-\delta},$$

then

$$\sum_{1\le |S| \le M \log(m_n)} \widehat{f_n}(S)^2 \to 0 .$$

Using Proposition IV.1, this proposition obviously implies Theorem I.5 when $\mathbf{H}(f_n)$ decays as assumed. Furthermore, this gives a quantitative "logarithmic" control on the noise sensitivity of such functions.

Proof. The strategy will be very similar to the one used in the KKL Theorems (even though the goal is very different). The main difference here is that the regularization term ρ used in the hypercontractive estimate must be chosen in a more delicate way than in the proofs of KKL results (where we simply took $\rho = 1/2$).

Let $M > 0$ be a constant whose value will be chosen later.

$$\begin{aligned}
\sum_{1\le|S|\le M\log(m_n)} \widehat{f_n}(S)^2 &\le 4 \sum_{1\le|S|\le M\log(m_n)} |S|\widehat{f_n}(S)^2 \\
&= \sum_k \sum_{1\le|S|\le M\log(m_n)} \widehat{\nabla_k f_n}(S)^2 \\
&\le \sum_k \Big(\frac{1}{\rho^2}\Big)^{M\log(m_n)} \|T_\rho(\nabla_k f_n)\|_2^2 \\
&\le \sum_k \Big(\frac{1}{\rho^2}\Big)^{M\log(m_n)} \|\nabla_k f_n\|_{1+\rho^2}^2 .
\end{aligned}$$

by Theorem V.2.

Now, since f_n is Boolean, one has $\|\nabla_k f_n\|_{1+\rho^2} = \|\nabla_k f_n\|_2^{2/(1+\rho^2)}$, hence

$$\begin{aligned}
\sum_{0<|S|<M\log(m_n)} \hat{f}_n(S)^2 &\le \rho^{-2M\log(m_n)} \sum_k \|\nabla_k f_n\|_2^{4/(1+\rho^2)} \\
&= \rho^{-2M\log(m_n)} \sum_k \mathbf{I}_k(f_n)^{2/(1+\rho^2)} \\
&\le \rho^{-2M\log(m_n)} (m_n)^{\rho^2/(1+\rho^2)} \Big(\sum_k \mathbf{I}_k(f_n)^2\Big)^{\frac{1}{1+\rho^2}} \quad \text{(by Hölder)} \\
&= \rho^{-2M\log(m_n)} (m_n)^{\rho^2/(1+\rho^2)}\, \mathbf{H}(f_n)^{\frac{1}{1+\rho^2}} \\
&\le \rho^{-2M\log(m_n)} (m_n)^{\frac{\rho^2-\delta}{1+\rho^2}} .
\end{aligned}$$

Now by choosing $\rho \in (0,1)$ close enough to 0, and then by choosing $M = M(\delta)$ small enough, we obtain the desired logarithmic noise sensitivity. $\square$

We now give some indications of the proof of Theorem I.5 in the general case.

Recall that Theorem I.5 is true independently of the speed of convergence of $\mathbf{H}(f_n) = \sum_k \mathbf{I}_k(f_n)^2$. The proof of this general result is a bit more involved than the one we gave here. The main lemma is as follows:

LEMMA V.6 ([**BKS99**]). *There exist absolute constants C_k such that for any monotone Boolean function f and for any $k \geq 2$, one has*

$$\sum_{|S|=k} \hat{f}(S)^2 \leq C_k \mathbf{H}(f)\, (-\log \mathbf{H}(f))^{k-1}.$$

This lemma "mimics" a result from Talagrand's [**Tal96**]. Indeed, Proposition 2.3 in [**Tal96**] can be translated as follows: for any monotone Boolean function f, its level-2 Fourier weight (i.e. $\sum_{|S|=2} \hat{f}(S)^2$) is bounded by $O(1)\mathbf{H}(f)\log(1/\mathbf{H}(f))$. Lemma V.6 obviously implies Theorem I.5 in the monotone case, while the general case can be deduced by a monotonization procedure. It is worth pointing out that hypercontractivity is used in the proof of this lemma.

6. Appendix: proof of hypercontractivity

The purpose of this appendix is to show that we are not using a giant "hammer" but rather that this needed inequality arising from Fourier analysis is understandable from first principles. In fact, historically, the proof by Gross of the Gaussian case first looked at the case of the hypercube and so we have the tools to obtain the Gaussian case should we want to. Before starting the proof, observe that for $\rho = 0$ (where 0^0 is defined to be 1), this simply reduces to $|\int f| \leq \int |f|$.

Proof of Theorem V.2.

6.1. Tensorization.

In this first section, we show that it is sufficient, via a tensorization procedure, that the result holds for $n = 1$ in order for us to be able to conclude by induction the result for all n.

The key step of the argument is the following lemma.

LEMMA V.7. *Let $q \geq p \geq 1$, $(\Omega_1, \mu_1), (\Omega_2, \mu_2)$ be two finite probability spaces, $K_i : \Omega_i \times \Omega_i \to \mathbb{R}$ and assume that for $i = 1, 2$*

$$\|T_i(f)\|_{L_q(\Omega_i,\mu_i)} \leq \|f\|_{L_p(\Omega_i,\mu_i)}$$

where $T_i(f)(x) := \int_{\Omega_i} f(y) K_i(x,y) d\mu_i(y)$. Then

$$\|T_1 \otimes T_2(f)\|_{L_q((\Omega_1,\mu_1)\times(\Omega_2,\mu_2))} \leq \|f\|_{L_p((\Omega_1,\mu_1)\times(\Omega_2,\mu_2))}$$

where $T_1 \otimes T_2(f)(x_1, x_2) := \int_{\Omega_1\times\Omega_2} f(y_1,y_2) K_1(x_1,y_1) K_2(x_2,y_2) d\mu_1(y_1) \times d\mu_2(y_2)$.

Proof. One first needs to recall Minkowski's inequality for integrals, which states that, for $g \geq 0$ and $r \in [1, \infty)$, we have

$$\left(\int \left(\int g(x,y) d\nu(y)\right)^r d\mu(x)\right)^{1/r} \leq \int \left(\int g(x,y)^r d\mu(x)\right)^{1/r} d\nu(y).$$

(Note that when ν consists of 2 point masses each of size 1, then this reduces to the usual Minkowski inequality.)

One can think of T_1 acting on functions of both variables by leaving the second variable untouched and analogously for T_2. It is then easy to check that $T_1 \otimes T_2 = T_1 \circ T_2$. By thinking of x_2 as fixed, our assumption on T_1 yields

$$\|T_1 \otimes T_2(f)\|^q_{L_q((\Omega_1,\mu_1)\times(\Omega_2,\mu_2))} \leq \int_{\Omega_2} \left(\int_{\Omega_1} |T_2(f)|^p d\mu_1(x_1)\right)^{q/p} d\mu_2(x_2).$$

(It might be helpful here to think of $T_2(f)(x_1, x_2)$ as a function $g^{x_2}(x_1)$ where x_2 is fixed).

Applying Minkowski's integral inequality to $|T_2(f)|^p$ with $r = q/p$, this in turn is at most

$$\left[\int_{\Omega_1}\left(\int_{\Omega_2}|T_2(f)|^q d\mu_2(x_2)\right)^{p/q} d\mu_1(x_1)\right]^{q/p}.$$

Fixing now the x_1 variable and applying our assumption on T_2 gives that this is at most $\|f\|^q_{L_p((\Omega_1,\mu_1)\times(\Omega_2,\mu_2))}$, as desired. □

The next key observation, easily obtained by expanding and interchanging of summation, is that our operator T_ρ acting on functions on Ω_n corresponds to an operator of the type dealt with in the previous lemma with $K(x, y)$ being

$$\sum_{S\subseteq\{1,\ldots,n\}} \rho^{|S|}\chi_S(x)\chi_S(y).$$

In addition, it is easily checked that the function K for the Ω_n is simply an n-fold product of the function for the $n = 1$ case.

Assuming the result for the case $n = 1$, Lemma V.7 and the above observations allows us to conclude by induction the result for all n.

6.2. The $n = 1$ case. We now establish the case $n = 1$. We abbreviate T_ρ by T.

Since $f(x) = (f(-1)+f(1))/2+(f(1)-f(-1))/2\ x$, we have $Tf(x) = (f(-1)+f(1))/2+\rho(f(1)-f(-1))/2\ x$. Denoting $(f(-1)+f(1))/2$ by a and $(f(1)-f(-1))/2$ by b, it suffices to show that for all a and b, we have

$$(a^2 + \rho^2b^2)^{(1+\rho^2)/2} \le \frac{|a + b|^{1+\rho^2} + |a - b|^{1+\rho^2}}{2}.$$

Using $\rho \in [0, 1]$, the case $a = 0$ is immediate. For the case, $a \neq 0$, it is clear we can assume $a > 0$. Dividing both sides by $a^{1+\rho^2}$, we need to show that

$$(1 + \rho^2y^2)^{(1+\rho^2)/2} \le \frac{|1 + y|^{1+\rho^2} + |1 - y|^{1+\rho^2}}{2} \tag{V.6}$$

for all y and clearly it suffices to assume $y \ge 0$.

We first do the case that $y \in [0, 1)$. By the generalized Binomial formula, the right hand side of (V.6) is

$$\frac{1}{2}\left[\sum_{k=0}^{\infty}\binom{1+\rho^2}{k}y^k + \sum_{k=0}^{\infty}\binom{1+\rho^2}{k}(-y)^k\right] = \sum_{k=0}^{\infty}\binom{1+\rho^2}{2k}y^{2k}.$$

For the left hand side of (V.6), we first note the following. For $0 < \lambda < 1$, a simple calculation shows that the function $g(x) = (1 + x)^\lambda - 1 - \lambda x$ has a negative derivative on $[0, \infty)$ and hence $g(x) \le 0$ on $[0, \infty)$.

This yields that the left hand side of (V.6) is at most

$$1 + \left(\frac{1+\rho^2}{2}\right)\rho^2y^2$$

which is precisely the first two terms of the right hand side of (V.6). On the other hand, the binomial coefficients appearing in the other terms are nonnegative, since in the numerator there are an even number of terms with the first two terms being positive and all the other terms being negative. This verifies the desired inequality for $y \in [0,1)$.

The case $y = 1$ for (V.6) follows by continuity.

For $y > 1$, we let $z = 1/y$ and note, by multiplying both sides of (V.6) by $z^{1+\rho^2}$, we need to show

$$(z^2+\rho^2)^{(1+\rho^2)/2} \leq \frac{|1+z|^{1+\rho^2}+|1-z|^{1+\rho^2}}{2}. \tag{V.7}$$

Now, expanding $(1-z^2)(1-\rho^2)$, one sees that $z^2+\rho^2 \leq 1+z^2\rho^2$ and hence the desired inequality follows precisely from (V.6) for the case $y \in (0,1)$ already proved. This completes the $n = 1$ case and thereby the proof. □

Exercise sheet of Part V

EXERCISE V.1. Find a direct proof that Theorem I.3 implies Theorem I.2.

EXERCISE V.2. Is it true that the smaller the influences are, the more noise sensitive the function is?

EXERCISE V.3. Prove that Theorem V.3 indeed implies Theorem III.2.
Hint: use the natural projection.

PROBLEM V.4. In this problem, we prove Theorems III.2 and III.3 for the monotone case.

(1) Show that Theorem III.3 implies III.2 and hence one needs to prove only Theorem III.3 (This is the basically the same as Exercise V.1).
(2) Show that it suffices to prove the result when $p = k/2^\ell$ for integers k and ℓ.
(3) Let $\Pi : \{0,1\}^\ell \to \{0, 1/2^\ell, \ldots, (2^\ell-1)/2^\ell\}$ by $\Pi(x_1,\ldots,x_\ell) = \sum_{i=1}^{\ell} x_i/2^i$. Observe that if x is uniform, then $\Pi(x)$ is uniform on its range and that $\mathbb{P}(\Pi(x) \geq i/2^\ell) = (2^\ell - i)/2^\ell$.
(4) Define $g : \{0,1\}^\ell \to \{0,1\}$ by $g(x_1,\ldots,x_\ell) := I_{\{\Pi(x)\geq 1-p\}}$. Note that $\mathbb{P}(g(x)=1) = p$.
(5) Define $\tilde{f} : \{0,1\}^{n\ell} \to \{0,1\}$ by
$$\tilde{f}(x_1^1,\ldots,x_\ell^1,x_1^2,\ldots,x_\ell^2,\ldots,x_1^n,\ldots,x_\ell^n) =$$
$$f(g(x_1^1,\ldots,x_\ell^1),g(x_1^2,\ldots,x_\ell^2),\ldots,g(x_1^n,\ldots,x_\ell^n)).$$
Observe that $\tilde{f}$ (defined on $(\{0,1\}^{n\ell},\pi_{1/2})$) and f (defined on $(\{0,1\}^n,\pi_p)$) have the same distribution and hence the same variance.
(6) Show (or observe) that $\mathbf{I}_{(r,j)}(\tilde{f}) \leq \mathbf{I}_r^p(f)$ for each $r = 1,\ldots,n$ and $j = 1,\ldots,\ell$. Deduce from Theorem I.3 that
$$\sum_{r,j} \mathbf{I}_{(r,j)}(\tilde{f}) \geq c\mathrm{Var}(f)\log(1/\delta_p)$$
where $\delta_p := \max_i \mathbf{I}_i^p(f)$ where c comes from Theorem I.3.
(7) (Key step). Show that for each $r = 1,\ldots,n$ and $j = 1,\ldots,\ell$,
$$\mathbf{I}_{(r,j)}(\tilde{f}) \leq \mathbf{I}_r^p(f)/2^{j-1}.$$
(8) Combine parts 6 and 7 to complete the proof.

Part VI. First evidence of noise sensitivity of percolation

In this lecture, our goal is to collect some of the facts and theorems we have seen so far in order to conclude that percolation crossings are indeed noise sensitive. Recall from the "BKS" Theorem (Theorem I.5) that it is enough for this purpose to prove that influences are "small" in the sense that $\sum_k \mathbf{I}_k(f_n)^2$ goes to zero.

In the first section, we will deal with a careful study of influences in the case of percolation crossings on the triangular lattice. Then, we will treat the case of $\mathbb{Z}^2$, where conformal invariance is not known. Finally, we will speculate to what "extent" percolation is noise sensitive.

This whole part should be considered somewhat of a "pause" in our program, where we take the time to summarize what we have achieved so far in our understanding of the noise sensitivity of percolation, and what remains to be done if one wishes to prove things such as the existence of exceptional times in dynamical percolation.

1. Bounds on influences for crossing events in critical percolation on the triangular lattice

1.1. Setup. Fix $a, b > 0$, let us consider some rectangle $[0, a \cdot n] \times [0, b \cdot n]$, and let R_n be the set of of hexagons in $\mathbb{T}$ which intersect $[0, a \cdot n] \times [0, b \cdot n]$. Let f_n be the event that there is a left to right crossing event in R_n. (This is the same event as in Example 7 in Part I, but with $\mathbb{Z}^2$ replaced by $\mathbb{T}$). By the RSW Theorem II.1, we know that $\{f_n\}$ is non-degenerate. Conformal invariance tells us that $\mathbb{E}[f_n] = \mathbb{P}[f_n = 1]$ converges as $n \to \infty$. The limit is given by the so-called **Cardy's formula**.

In order to prove that this sequence of Boolean functions $\{f_n\}$ is noise sensitive, we wish to study its influence vector $\mathbf{Inf}(f_n)$ and we would like to prove that $\mathbf{H}(f_n) = \|\mathbf{Inf}(f_n)\|_2^2 = \sum \mathbf{I}_k(f_n)^2$ decays polynomially fast towards 0. (Recall that in these notes, we gave a complete proof of Theorem I.5 only in the case where $\mathbf{H}(f_n)$ decreases as an inverse polynomial of the number of variables.)

1.2. Study of the set of influences. Let x be a site (i.e. a hexagon) in the rectangle R_n. One needs to understand

$$\mathbf{I}_x(f_n) := \mathbb{P}[x \text{ is pivotal for } f_n]$$

It is easy but crucial to note that if x is at distance d from the boundary of R_n, in order for x to be pivotal, the *four-arm* event described in Part II (see Figure II.2) has to be satisfied in the ball $B(x, d)$ of radius d around the hexagon x. See the figure on the right.

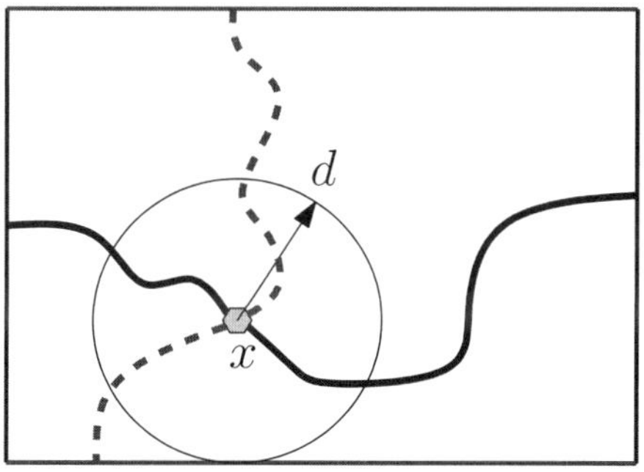

In particular, this implies (still under the assumption that $\mathrm{dist}(x, \partial R_n) = d$) that

$$\mathbf{I}_x(f_n) \le \alpha_4(d) = d^{-\frac{5}{4}+o(1)},$$

where $\alpha_4(d)$ denotes the probability of the four-arm event up to distance d. See Part II. The statement

$$\alpha_4(R) = R^{-5/4+o(1)}$$

implies that for any $\epsilon > 0$, there exists a constant $C = C_\epsilon$, such that for all $R \ge 1$,

$$\alpha_4(R) \le C\, R^{-5/4+\epsilon}.$$

The above bound gives us a very good control on the influences of the points in the *bulk* of the domain (i.e. the points far from the boundary). Indeed, for any fixed $\delta > 0$, let Δ_n^δ be the set of hexagons in R_n which are at distance at least δn from ∂R_n. Most of the points in R_n (except a proportion $O(\delta)$ of these) lie in Δ_n^δ, and for any such point $x \in \Delta_n^\delta$, one has by the above argument

$$\mathbf{I}_x(f_n) \le \alpha_4(\delta n) \le C\,(\delta n)^{-5/4+\epsilon} \le C\delta^{-5/4} n^{-5/4+\epsilon}. \tag{VI.1}$$

Therefore, the contribution of these points to $\mathbf{H}(f_n) = \sum_k \mathbf{I}_k(f_n)^2$ is bounded by $O(n^2)(C\delta^{-5/4}n^{-5/4+\epsilon})^2 = O(\delta^{-5/2}n^{-1/2+2\epsilon})$. As $n \to \infty$, this goes to zero polynomially fast. Since this estimate concerns "almost" all points in R_n, it seems we are close to proving the BKS criterion.

1.3. Influence of the boundary. Still, in order to complete the above analysis, one has to estimate what the influence of the points near the boundary is. The main difficulty here is that if x is close to the boundary, the probability for x to be pivotal is not related any longer to the above *four-arm* event. Think of the above figure when d gets very small compared to n. One has to distinguish two cases:

- x is close to a *corner*. This will correspond to a *two-arm* event in a quarter-plane.
- x is close to an *edge*. This involves the *three-arm* event in the half-plane $\mathbb{H}$.

Before detailing how to estimate the influence of points near the boundary, let us start by giving the necessary background on the involved critical exponents.

The two-arm and three-arm events in $\mathbb{H}$. For these particular events, it turns out that the critical exponents are known to be *universal*: they are two of the very few critical exponents which are known also on the square lattice $\mathbb{Z}^2$. The derivations of these types of exponents do not rely on SLE technology but are "elementary". Therefore, in this discussion, we will consider both lattices $\mathbb{T}$ and $\mathbb{Z}^2$.

The *three-arm* event in $\mathbb{H}$ corresponds to the event that there are three arms (two open arms and one 'closed' arm in the dual) going from 0 to distance R and such that they remain in the upper half-plane. See the figure for a self-explanatory definition. The *two-arm* event corresponds to just having one open and one closed arm.

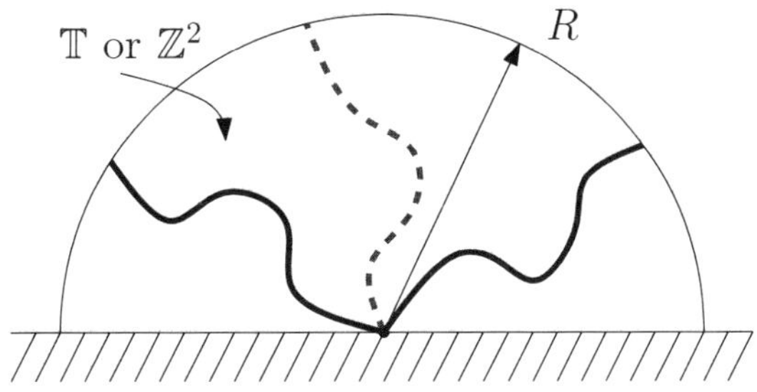

Let $\alpha_2^+(R)$ and $\alpha_3^+(R)$ denote the probabilities of these events. As in Part II, let $\alpha_2^+(r,R)$ and $\alpha_3^+(r,R)$ be the natural extensions to the annulus case (i.e. the probability that these events are satisfied in the annulus between radii r and R in the upper half-plane).

We will rely on the following result, which goes back as far as we know to M. Aizenman. See [**Wer07**] for a proof of this result.

PROPOSITION VI.1. *Both on the triangular lattice* $\mathbb{T}$ *and on* $\mathbb{Z}^2$, *one has that*

$$\alpha_2^+(r,R) \asymp (r/R)$$

and

$$\alpha_3^+(r,R) \asymp (r/R)^2 .$$

Note that, in these special cases, there are no $o(1)$ *correction terms in the exponent. The probabilities are in this case known up to constants.*

The two-arm event in the quarter-plane. In this case, the corresponding exponent is unfortunately not known on $\mathbb{Z}^2$, so we will need to do some work here in the next section, where we will prove noise sensitivity of percolation crossings on $\mathbb{Z}^2$.

The *two-arm* event in a corner corresponds to the event illustrated on the following picture. We will use the following proposition:

PROPOSITION VI.2 ([**SW01**]). *If* $\alpha_2^{++}(R)$ *denotes the probability of this event, then*

$$\alpha_2^{++}(R) = R^{-2+o(1)} ,$$

and with the obvious notations

$$\alpha_2^{++}(r,R) = (r/R)^{2+o(1)} .$$

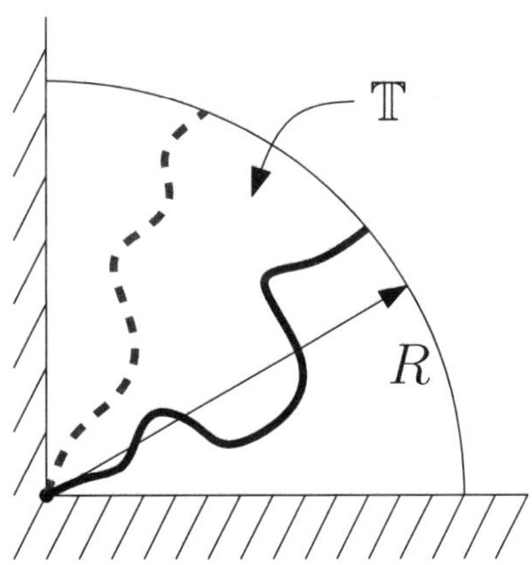

Now, back to our study of influences, we are in good shape (at least for the triangular lattice) since the two critical exponents arising from the boundary effects are larger than the *bulk* exponent $5/4$. This means that it is less likely for a point near the boundary to be pivotal than for a point in the bulk. Therefore in some sense the boundary helps us here.

More formally, summarizing the above facts, for any $\epsilon > 0$, there is a constant $C = C(\epsilon)$ such that for any $1 \le r \le R$,

$$\max\{\alpha_4(r,R), \alpha_3^+(r,R), \alpha_2^{++}(r,R)\} \le C(r/R)^{\frac{5}{4}-\epsilon}. \tag{VI.2}$$

Now, if x is some hexagon in R_n, let n_0 be the distance to the closest edge of ∂R_n and let x_0 be the point on ∂R_n such that $\mathrm{dist}(x, x_0) = n_0$. Next, let $n_1 \ge n_0$ be the distance from x_0 to the closest corner and let x_1 be this closest corner. It is easy to see that for x to be pivotal for f_n, the following events all have to be satisfied:

- The four-arm event in the ball of radius n_0 around x.
- The $\mathbb{H}$-three-arm event in the annulus centered at x_0 of radii $2n_0$ and n_1.
- The corner-two-arm event in the annulus centered at x_1 of radii $2n_1$ and n.

By independence on disjoint sets, one thus concludes that

$$\begin{aligned} \mathbf{I}_x(f_n) &\le \alpha_4(n_0)\,\alpha_3^+(2n_0, n_1)\,\alpha_2^{++}(2n_1, n) \\ &\le O(1) n^{-5/4+\epsilon}. \end{aligned}$$

1.4. Noise sensitivity of crossing events. This uniform bound on the influences over the whole domain R_n enables us to conclude that the BKS criterion is indeed verified. Indeed,

$$\mathbf{H}(f_n) = \sum_{x \in R_n} \mathbf{I}_x(f_n)^2 \le C n^2 (n^{-5/4+\epsilon})^2 = C n^{-1/2+2\epsilon}, \tag{VI.3}$$

where $C = C(a, b, \epsilon)$ is a universal constant. By taking $\epsilon < 1/4$, this gives us the desired polynomial decay on $\mathbf{H}(f_n)$, which by Proposition V.5) implies

THEOREM VI.3 ([**BKS99**]). *The sequence of percolation crossing events $\{f_n\}$ on $\mathbb{T}$ is noise sensitive.*

We will give some other consequences (for example, to sharp thresholds) of the above analysis on the influences of the crossing events in a later section.

2. The case of $\mathbb{Z}^2$ percolation

Let R_n denote similarly the $\mathbb{Z}^2$ rectangle closest to $[0, a \cdot n] \times [0, b \cdot n]$ and let f_n be the corresponding left-right crossing event (so here this corresponds exactly to example 7). Here one has to face two main difficulties:

- The main one is that due to the missing ingredient of *conformal invariance*, one does not have at our disposal the value of the *four-arm* critical exponent (which is of course believed to be $5/4$). In fact, even the *existence* of a critical exponent is an open problem.
- The second difficulty (also due to the lack of conformal invariance) is that it is now slightly harder to deal with boundary issues. Indeed, one can still use the above bounds on α_3^+ which are *universal*, but the exponent 2 for α_2^{++} is not known for $\mathbb{Z}^2$. So this requires some more analysis.

Let us start by taking care of the boundary effects.

2.1. Handling the boundary effect. What we need to do in order to carry through the above analysis for $\mathbb{Z}^2$ is to obtain a reasonable estimate on α_2^{++}. Fortunately, the following bound, which follows immediately from Proposition VI.1, is sufficient.

$$\alpha_2^{++}(r,R) \le O(1)\frac{r}{R}\,. \tag{VI.4}$$

Now let e be an edge in R_n. We wish to bound from above $\mathbf{I}_e(f_n)$. We will use the same notation as in the case of the triangular lattice: recall the definitions of n_0, x_0, n_1, x_1 there.

We obtain in the same way

$$\mathbf{I}_e(f_n) \le \alpha_4(n_0)\,\alpha_3^+(2n_0,n_1)\,\alpha_2^{++}(2n_1,n)\,. \tag{VI.5}$$

At this point, we need another *universal* exponent, which goes back also to M. Aizenman:

THEOREM VI.4 (M. Aizenman, see [**Wer07**]). *Let $\alpha_5(r,R)$ denote the probability that there are 5 arms (with four of them being of 'alternate colors'). Then there are some universal constants $c, C>0$ such that both for $\mathbb{T}$ and $\mathbb{Z}^2$, one has for all $1\le r\le R$,*

$$c\Big(\frac{r}{R}\Big)^2 \le \alpha_5(r,R) \le C\Big(\frac{r}{R}\Big)^2\,.$$

This result allows us to get a lower bound on $\alpha_4(r,R)$. Indeed, it is clear that

$$\alpha_4(r,R) \ge \alpha_5(r,R) \ge \Omega(1)\alpha_3^+(r,R)\,. \tag{VI.6}$$

In fact, one can obtain the following better lower bound on $\alpha_4(r,R)$ which we will need later.

LEMMA VI.5. *There exists some $\epsilon>0$ and some constant $c>0$ such that for any $1\le r\le R$,*

$$\alpha_4(r,R) \ge c(r/R)^{2-\epsilon}\,.$$

Proof. There are several ways to see why this holds, none of them being either very hard or very easy. One of them is to use **Reimer's inequality** (see [**Gri99**]) which in this case would imply that

$$\alpha_5(r,R) \le \alpha_1(r,R)\alpha_4(r,R)\,. \tag{VI.7}$$

The RSW Theorem II.1 can be used to show that

$$\alpha_1(r,R) \le (r/R)^\alpha$$

for some positive α. By Theorem VI.4, we are done. [See [[**GPS10**], Section 2.2 as well as the appendix] for more on these bounds.] □

Combining (VI.5) with (VI.6), one obtains

$$\begin{aligned}\mathbf{I}_e(f_n) &\le O(1)\alpha_4(n_0)\alpha_4(2n_0,n_1)\alpha_2^{++}(2n_1,n)\\ &\le O(1)\alpha_4(n_1)\,\frac{n_1}{n}\,,\end{aligned}$$

where in the last inequality we used quasi-multiplicativity (Proposition II.3) as well as the bound given by (VI.4).

Recall that we want an upper bound on $\mathbf{H}(f_n) = \sum \mathbf{I}_e(f_n)^2$. In this sum over edges $e \in R_n$, let us divide the set of edges into dyadic annuli centered around the 4 corners as in the next picture.

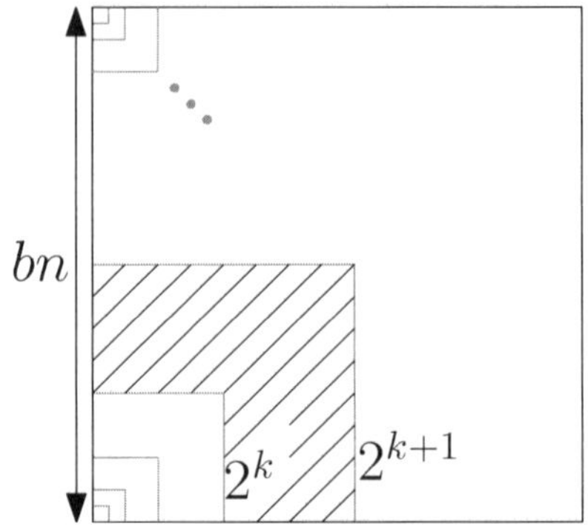

Notice that there are $O(1)2^{2k}$ edges in an annulus of radius 2^k. This enables us to bound $\mathbf{H}(f_n)$ as follows:

$$\begin{aligned} \sum_{e\in R_n} \mathbf{I}_e(f_n)^2 &\le O(1) \sum_{k=1}^{\log_2 n+O(1)} 2^{2k}\Big(\alpha_4(2^k)\frac{2^k}{n}\Big)^2 \\ &\le O(1)\,\frac{1}{n^2} \sum_{k\le \log_2 n+O(1)} 2^{4k}\,\alpha_4(2^k)^2\,. \end{aligned} \tag{VI.8}$$

It now remains to obtain a good upper bound on $\alpha_4(R)$, for all $R \ge 1$.

2.2. An upper bound on the four-arm event in $\mathbb{Z}^2$. This turns out to be a rather non-trivial problem. Recall that we obtained an easy lower bound on α_4 using α_5 (and Lemma VI.5 strengthens this lower bound). For an upper bound, completely different ideas are required. On $\mathbb{Z}^2$, the following estimate is available for the four-arm event.

PROPOSITION VI.6. *For critical percolation on $\mathbb{Z}^2$, there exists constants $\epsilon, C > 0$ such that for any $R \ge 1$, one has*

$$\alpha_4(1,R) \le C\Big(\frac{1}{R}\Big)^{1+\epsilon}.$$

Before discussing where such an estimate comes from, let us see that it indeed implies a polynomial decay for $\mathbf{H}(f_n)$.

Recall equation (VI.8). Plugging in the above estimate, this gives us

$$\begin{aligned} \sum_{e\in R_n} \mathbf{I}_e(f_n)^2 &\le O(1)\,\frac{1}{n^2} \sum_{k\le \log_2 n+O(1)} 2^{4k}\,(2^k)^{-2-2\epsilon} \\ &\le O(1)\frac{1}{n^2} n^{2-2\epsilon} = O(1) n^{-2\epsilon}\,, \end{aligned}$$

which implies the desired polynomial decay and thus the fact that $\{f_n\}$ is noise sensitive by Proposition V.5).

Let us now discuss different approaches which enable one to prove Proposition VI.6.

(a) Kesten proved implicitly this estimate in his celebrated paper [**Kes87**]. His main motivation for such an estimate was to obtain bounds on the corresponding critical exponent which governs the so-called *critical length*.

(b) In [**BKS99**], in order to prove noise sensitivity of percolation using their criterion on $\mathbf{H}(f_n)$, the authors referred to [**Kes87**], but they also gave a completely different approach which also yields this estimate.
Their alternative approach is very nice: finding an upper bound for $\alpha_4(R)$ is related to finding an upper bound for the influences for crossings of an $R \times R$ box. For this, they noticed the following nice phenomenon: if a monotone function f happens to be very little correlated with Majority, then its influences have to be small. The proof of this phenomenon uses for the first time in this context the concept of "randomized algorithms". For more on this approach, see Part VIII, which is devoted to these types of ideas.

(c) In [**SS10b**], the concept of randomized algorithms is used in a more powerful way. See again Part VIII. In this part, we provide a proof of this estimate in Proposition VIII.8.

REMARK VI.1. It turns out that that a multi-scale version of Proposition VI.6 stating that $\alpha_4(r,R) \le C\left(\frac{r}{R}\right)^{1+\epsilon}$ is also true. However, none of the three arguments given above seem to prove this stronger version. A proof of this stronger version is given in the appendix of [**SS10a**]. Since this multi-scale version is not needed until Part X, we stated here only the weaker version.

3. Some other consequences of our study of influences

In the previous sections, we handled the boundary effects in order to check that $\mathbf{H}(f_n)$ indeed decays polynomially fast. Let us list some related results implied by this analysis.

3.1. Energy spectrum of f_n. We start by a straightforward observation: since the f_n are monotone, we have by Proposition IV.4 that

$$\widehat{f_n}(\{x\}) = \frac{1}{2}\mathbf{I}_x(f_n)\,,$$

for any site x (or edge e) in R_n. Therefore, the bounds we obtained on $\mathbf{H}(f_n)$ imply the following control on the first layer of the energy spectrum of the crossing events $\{f_n\}$.

COROLLARY VI.7. *Let $\{f_n\}$ be the crossing events of the rectangles R_n.*

- *If we are on the triangular lattice $\mathbb{T}$, then we have the bound*

$$E_{f_n}(1) = \sum_{|S|=1} \widehat{f_n}(S)^2 \le n^{-1/2+o(1)}\,.$$

- *On the square lattice $\mathbb{Z}^2$, we end up with the weaker estimate*

$$E_{f_n}(1) \le C\,n^{-\epsilon}\,,$$

for some $\epsilon, C > 0$.

3.2. Sharp threshold of percolation. The above analysis gave an upper bound on $\sum_k \mathbf{I}_k(f_n)^2$. As we have seen in the first parts, the total influence $\mathbf{I}(f_n) = \sum_k \mathbf{I}_k(f_n)$ is also a very interesting quantity. Recall that, by Russo's formula, this is the quantity which shows "how sharp" the threshold is for $p \mapsto \mathbb{P}_p[f_n = 1]$.

The above analysis allows us to prove the following.

PROPOSITION VI.8. *Both on $\mathbb{T}$ and $\mathbb{Z}^2$, one has*

$$\mathbf{I}(f_n) \asymp n^2 \alpha_4(n)\,.$$

In particular, this shows that on $\mathbb{T}$ that

$$\mathbf{I}(f_n) \asymp n^{3/4+o(1)}\,.$$

REMARK VI.2. Since f_n is defined on $\{-1,1\}^{O(n^2)}$, note that the Majority function defined on the same hypercube has a much sharper threshold than the percolation crossings f_n.

Proof. We first derive an upper bound on the total influence. In the same vein (i.e., using dyadic annuli and quasi-multiplicativity) as we derived (VI.8) and with the same notation one has

$$\begin{aligned}\mathbf{I}(f_n) = \sum_e \mathbf{I}_e(f_n) &\le \sum_e O(1)\alpha_4(n_1)\frac{n_1}{n} \\ &\le O(1)\frac{1}{n} \sum_{k \le \log_2 n + O(1)} 2^{3k}\alpha_4(2^k)\,.\end{aligned}$$

Now, and this is the main step here, using quasi-multiplicativity one has $\alpha_4(2^k) \le O(1)\frac{\alpha_4(n)}{\alpha_4(2^k,n)}$, which gives us

$$\begin{aligned}\mathbf{I}(f_n) &\le O(1)\frac{\alpha_4(n)}{n} \sum_{k \le \log_2 n + O(1)} 2^{3k}\frac{1}{\alpha_4(2^k,n)} \\ &\le O(1)\frac{\alpha_4(n)}{n} \sum_{k \le \log_2 n + O(1)} 2^{3k}\frac{n^2}{2^{2k}} \quad \text{since } \alpha_4(r,R) \ge \alpha_5(r,R) \asymp (r/R)^{-2} \\ &\le O(1) n\, \alpha_4(n) \sum_{k \le \log_2 n + O(1)} 2^k \\ &\le O(1) n^2 \alpha_4(n)\end{aligned}$$

as desired.

For the lower bound on the total influence, we proceed as follows. One obtains a lower bound by just summing over the influences of points whose distance to the boundary is at least $n/4$. It would suffice if we knew that for such edges or hexagons, the influence is at least a constant times $\alpha_4(n)$. This is in fact known to be true. It is not very involved and is part of the *folklore* results in percolation. However, it still would lead us too far from our topic. The needed technique is known under the name of **separation of arms** and is closely related to the statement of quasi-multiplicativity. See [**Wer07**] for more details. □

4. Quantitative noise sensitivity

In this part, we have proved that the sequence of crossing events $\{f_n\}$ is noise sensitive. This can be roughly translated as follows: for any fixed level of noise $\epsilon > 0$, as $n \to \infty$, the large scale clusters of ω in the window $[0,n]^2$ are asymptotically independent of the large clusters of ω_ϵ.

REMARK VI.3. Note that this picture is correct, but in order to make it rigorous, this would require some work, since so far we only worked with left-right crossing events. The non-trivial step here is to prove that in some sense, in the scaling limit $n \to \infty$, any macroscopic property concerning percolation (e.g., diameter of clusters) is measurable with respect to the σ-algebra generated by the crossing events. This is a rather subtle problem since we need to make precise what kind of information we keep in what we call the "scaling limit" of percolation (or subsequential scaling limits in the case of $\mathbb{Z}^2$). An example of something which is not present in the scaling limit is whether one has more open sites than closed ones since by noise sensitivity we know that this is asymptotically uncorrelated with crossing events. We will not need to discuss these notions of scaling limits more in these lecture notes, since the focus is mainly on the discrete model itself including the model of dynamical percolation which is presented at the end of these lecture notes.

At this stage, a natural question to ask is to what extent the percolation picture is sensitive to noise. In other words, can we let the noise $\epsilon = \epsilon_n$ go to zero with the "size of the system" n, and yet keep this independence of large scale structures between ω and ω_{ϵ_n} ? If yes, can we give *quantitative estimates* on how fast the noise $\epsilon = \epsilon_n$ may go to zero? One can state this question more precisely as follows.

QUESTION VI.1. *If $\{f_n\}$ denote our left-right crossing events, for which sequences of noise-levels $\{\epsilon_n\}$ do we have*

$$\lim_{n\to\infty} \mathrm{Cov}[f_n(\omega), f_n(\omega_{\epsilon_n})] = 0\ ?$$

The purpose of this section is to briefly discuss this question based on the results we have obtained so far.

4.1. Link with the energy spectrum of $\{f_n\}$. It is an exercise to show that Question VI.1 is essentially equivalent to the following one.

QUESTION VI.2. *For which sequences $\{k_n\}$ going to infinity do we have*

$$\sum_{m=1}^{k_n} E_{f_n}(m) = \sum_{1\le |S|\le k_n} \widehat{f_n}(S)^2 \underset{n\to\infty}{\longrightarrow} 0\ ?$$

Recall that we have already obtained some relevant information on this question. Indeed, we have proved in this part that $\mathbf{H}(f_n) = \sum_x \mathbf{I}_x(f_n)^2$ decays polynomially fast towards 0 (both on $\mathbb{Z}^2$ and $\mathbb{T}$). Therefore Proposition V.5 tells us that for some constant $c > 0$, one has for both $\mathbb{T}$ and $\mathbb{Z}^2$ that

$$\sum_{1\le |S|\le c\log n} \widehat{f_n}(S)^2 \to 0\,. \tag{VI.9}$$

Therefore, back to our original question VI.1, this gives us the following quantitative statement: if the noise ϵ_n satisfies $\epsilon_n \gg \frac{1}{\log n}$, then $f_n(\omega)$ and $f_n(\omega_{\epsilon_n})$ are asymptotically independent.

4.2. Noise stability regime. Of course, one cannot be too demanding on the rate of decay of $\{\epsilon_n\}$. For example if $\epsilon_n \ll \frac{1}{n^2}$, then in the window $[0, n]^2$, with high probability, the configurations ω and ω_{ϵ_n} are identical. This brings us to the next natural question concerning the *noise stability regime* of crossing events.

QUESTION VI.3. *Let $\{f_n\}$ be our sequence of crossing events. For which sequences $\{\epsilon_n\}$ do we have*

$$\mathbb{P}\big[f_n(\omega) \neq f_n(\omega_{\epsilon_n})\big] \underset{n\to\infty}{\longrightarrow} 0\ ?$$

It is an exercise to show that this question is essentially equivalent to the following one.

For which sequences $\{k_n\}$ do we have

$$\sum_{|S|>k_n} \widehat{f_n}(S)^2 \to 0\ ?$$

Using the estimates of the present part, one can give the following non-trivial bound on the noise stability regime of $\{f_n\}$.

PROPOSITION VI.9. *Both on $\mathbb{Z}^2$ and $\mathbb{T}$, if*

$$\epsilon_n = o\Big(\frac{1}{n^2\alpha_4(n)}\Big)\,,$$

then

$$\mathbb{P}\big[f_n(\omega) \neq f_n(\omega_{\epsilon_n})\big] \underset{n\to\infty}{\longrightarrow} 0$$

On the triangular grid, using the critical exponent, this gives us a bound of $n^{-3/4}$ on the noise stability regime of percolation.

Proof. Let $\{\epsilon_n\}$ be a sequence satisfying the above assumption. There are $O(n^2)$ bits concerned. For simplicity, assume that there are exactly n^2 bits. Let us order these in some arbitrary way: $\{x_1, \ldots, x_{n^2}\}$ (or on $\mathbb{Z}^2$, $\{e_1, \ldots, e_{n^2}\}$).

Let $\omega = \omega_0 = (x_1, \ldots, x_{n^2})$ be sampled according to the uniform measure. Recall that the noised configuration ω_{ϵ_n} is produced as follows: for each $i \in [n^2]$, resample the bit x_i with probability ϵ_n, independently of everything else, obtaining the bit y_i. (In particular $y_i \neq x_i$ with probability $\epsilon_n/2$).

Now for each $i \in [n^2]$ define the intermediate configuration

$$\omega_i := (y_1, \ldots, y_i, x_{i+1}, \ldots, x_{n^2})$$

Notice that for each $i \in [n^2]$, ω_i is also sampled according to the uniform measure and one has for each $i \in \{1, \ldots, n^2\}$ that

$$\mathbb{P}\big[f_n(\omega_{i-1}) \neq f_n(\omega_i)\big] = (\epsilon_n/2)\,\mathbf{I}_{x_i}(f_n)\,.$$

Summing over all i, one obtains

$$\begin{aligned}
\mathbb{P}\big[f_n(\omega) \neq f_n(\omega_{\epsilon_n})\big] &= \mathbb{P}\big[f_n(\omega_0) \neq f_n(\omega_{n^2})\big] \\
&\leq \sum_{i=0}^{n^2-1} \mathbb{P}\big[f_n(\omega_i) \neq f_n(\omega_{i+1})\big] \\
&= (\epsilon_n/2) \sum_{i=1}^{n^2} \mathbf{I}_{x_i}(f_n) \\
&= (\epsilon_n/2)\,\mathbf{I}(f_n) \\
&\leq \epsilon_n O(1) n^2 \alpha_4(n) \quad \text{by Proposition VI.8,}
\end{aligned}$$

which concludes the proof. □

4.3. Where does the spectral mass lies? Proposition VI.9 (together with Exercise IX.2 in Part IX) implies that the Fourier coefficients of $\{f_n\}$ satisfy

$$\sum_{|S| \gg n^2 \alpha_4(n)} \widehat{f_n}(S)^2 \underset{n\to\infty}{\longrightarrow} 0\,. \tag{VI.10}$$

From Lemma VI.5, we know that even on $\mathbb{Z}^2$, $n^2\alpha_4(n)$ is larger than n^ϵ for some exponent $\epsilon > 0$. Combining the estimates on the spectrum that we achieved so far (equations (VI.9) and (VI.10)), we see that in order to localize the spectral mass of $\{f_n\}$, there is still a missing gap. See Figure VI.1.

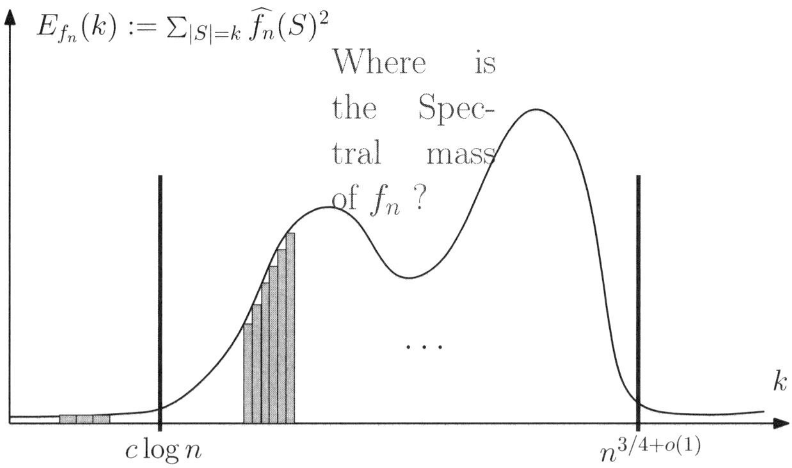

FIGURE VI.1. This picture summarizes our present knowledge of the energy spectrum of $\{f_n\}$ on the triangular lattice $\mathbb{T}$. Much remains to be understood to know where, in the range $[\Omega(\log n), n^{3/4+o(1)}]$, the spectral mass lies. This question will be analyzed in the following parts.

For our later applications to the model of dynamical percolation (in the last part of these lecture notes), a better understanding of the noise sensitivity of percolation than the "logarithmic" control we achieved so far will be needed.

Exercise sheet of Part VI

Instead of being the usual exercise sheet, this page will be devoted to a single Problem whose goal will be to do "hands-on" computations of the first layers of the energy spectrum of the percolation crossing events f_n. Recall from Proposition IV.1 that a sequence of Boolean functions $\{f_n\}$ is noise sensitive if and only if for any fixed $k \geq 1$,

$$\sum_{m=1}^{k} \sum_{|S|=m} \hat{f}_n(S)^2 = \sum_{m=1}^{k} E_{f_n}(m) \underset{n\to\infty}{\longrightarrow} 0\,.$$

In the present part, we obtained (using Proposition IV.4) that this is indeed the case for $k = 1$. The purpose here is to check by simple combinatorial arguments (without relying on hypercontractivity) that it is still the case for $k = 2$ and to convince ourselves that it works for all layers $k \geq 3$.

To start with, we will simplify our task by working on the torus $\mathbb{Z}^2/n\mathbb{Z}^2$. This has the very nice advantage that there are no boundary issues here.

Energy spectrum of crossing events on the torus (study of the first layers.)

Let T_n be either the square grid torus $\mathbb{Z}^2/n\mathbb{Z}^2$ or the triangular grid torus $\mathbb{T}/n\mathbb{T}$. Let f_n be the indicator of the event that there is an open circuit along the first coordinate of T_n.

(1) Using RSW, prove that there is a constant $c > 0$ such that for all $n \geq 1$,

$$c \leq \mathbb{P}\big[f_n = 1\big] \leq 1 - c\,.$$

(In other words, $\{f_n\}$ is non-degenerate.)

(2) Show that for all edges e (or sites x) in T_n

$$\mathbf{I}_e(f_n) \leq \alpha_4(\frac{n}{2})\,.$$

(3) Check that the BKS criterion (about $\mathbf{H}(f_n)$) is satisfied. Therefore $\{f_n\}$ is noise-sensitive

From now on, one would like to forget about the BKS Theorem and try to do some hands-on computations in order to get a feeling why most frequencies should be large.

(4) Show that if x, y are two sites of T_n (or similarly if e, e' are two edges of T_n), then

$$|\hat{f}(\{x, y\})| \leq 2\mathbb{P}\big[\ x \text{ and } y \text{ are pivotal points}\big]\,.$$

Does this result hold for general Boolean functions?

(5) Show that if $d := |x - y|$, then

$$\mathbb{P}\big[\ x \text{ and } y \text{ are pivotal points }\big] \leq O(1)\frac{\alpha_4(n/2)^2}{\alpha_4(\frac{d}{2}, \frac{n}{2})}\,.$$

(Hint: use Proposition II.3.)

(6) On the square lattice $\mathbb{Z}^2$, by carefully summing over all edges $e, e' \in T_n \times T_n$, show that

$$E_{f_n}(2) = \sum_{|S|=2} \widehat{f_n}(S)^2 \leq O(1) n^{-\epsilon}\,,$$

for some exponent $\epsilon > 0$.
Hint: you might decompose the sum in a dyadic way (as we did many times in the present section) depending on the mutual distance $d(e, e')$.

(7) On the triangular grid, what exponent does it give for the decay of $\mathbb{E}_{f_n}(2)$? Compare with the decay we found in Corollary VI.7 about the decay of the first layer $E_{f_n}(1)$ (i.e. $k = 1$). See also Lemma V.6 in this regard. Discuss this.

(8) For $\mathbb{T}$, what do you expect for higher (fixed) values of k? (I.e. for $E_{f_n}(k)$, $k \geq 3$)?

(9) *(Quite hard)* Try to obtain a nonrigorous combinatorial argument similar to the one above in the particular case $k = 2$, that for any fixed layer $k \geq 1$,

$$E_{f_n}(k) \underset{n\to\infty}{\longrightarrow} 0\,.$$

This would give us an alternative proof of noise sensitivity of percolation (at least in the case of the torus T_n) not relying on Theorem I.5.

Observe that one can do similar things for rectangles but then one has to deal with boundary issues.

Part VII. Anomalous fluctuations

In this lecture, our goal is to extend the technology we used to prove the KKL Theorems on influences and the BKS Theorem on noise sensitivity to a slightly different context: the study of fluctuations in **first passage percolation**.

1. The model of first passage percolation

Let us first explain what the model is. Let $0 < a < b$ be two positive numbers. We define a **random metric** on the graph $\mathbb{Z}^d$, $d \geq 2$ as follows. Independently for each edge $e \in \mathbb{E}^d$, fix its length τ_e to be a with probability $1/2$ and b with probability $1/2$. This is represented by a uniform configuration $\omega \in \{-1,1\}^{\mathbb{E}^d}$.

This procedure induces a well-defined (random) metric dist_ω on $\mathbb{Z}^d$ in the usual fashion. For any vertices $x, y \in \mathbb{Z}^d$, let

$$\mathrm{dist}_\omega(x,y) := \inf_{\substack{\text{paths } \gamma = \{e_1,\ldots,e_k\} \\ \text{connecting } x \to y}} \left\{ \sum \tau_{e_i}(\omega) \right\}.$$

REMARK VII.1. In greater generality, the lengths of the edges are i.i.d. non-negative random variables, but here, following [**BKS03**], we will restrict ourselves to the above uniform distribution on $\{a, b\}$ to simplify the exposition; see [**BR08**] for an extension to more general laws.

One of the main goals in first passage percolation is to understand the large-scale properties of this random metric space. For example, for any $T \geq 1$, one may consider the (random) ball

$$B_\omega(x,T) := \{y \in \mathbb{Z}^d : \mathrm{dist}_\omega(x,y) \leq T\}.$$

To understand the name *first passage percolation*, one can think of this model as follows. Imagine that water is pumped in at vertex x, and that for each edge e, it takes $\tau_e(\omega)$ units of time for the water to travel across the edge e. Then, $B_\omega(x,T)$ represents the region of space that has been wetted by time T.

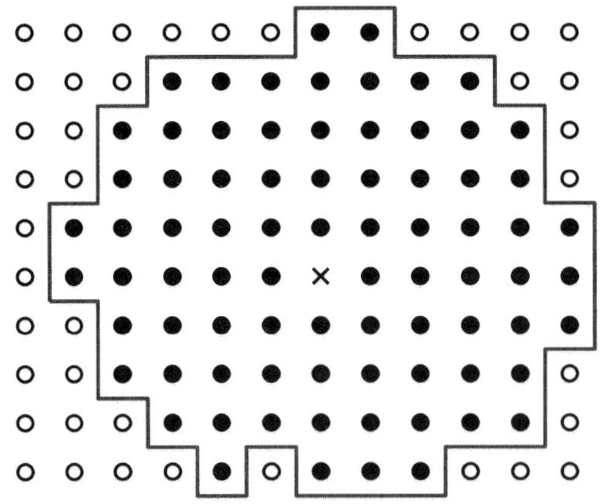

FIGURE VII.1. A sample of a wetted region at time T, i.e. $B_\omega(x,T)$, in first passage percolation.

An application of subadditivity shows that the renormalized ball $\frac{1}{T}B_\omega(0,T)$ converges as $T \to \infty$ towards a deterministic shape which can in certain cases be computed explicitly. This is a kind of "geometric law of large numbers". Whence the natural question:

QUESTION VII.1. *Describe the* fluctuations *of $B_\omega(0,T)$ around its asymptotic deterministic shape.*

This question has received tremendous interest in the last 15 years or so. It is widely believed that these fluctuations should be in some sense "universal". More precisely, the behavior of $B_\omega(0,T)$ around its limiting shape should not depend on the "microscopic" particularities of the model such as the law on the edges lengths but only on the dimension d of the underlying graph. The shape itself depends on the other hand of course on the microscopic parameters, in the same way as the critical point depends on the graph in percolation.

In the two-dimensional case, using very beautiful combinatorial bijections with random matrices, certain cases of *directed* last passage percolation (where the law on the edges is taken to be geometric or exponential) have been understood very deeply. For example, it is known (see [**Joh00**]) that the fluctuations of the ball of radius n (i.e. the points whose last passage times are below n) around n times its asymptotic deterministic shape are of order $n^{1/3}$ and the law of these fluctuations properly renormalized follows the Tracy-Widom distribution. Very interestingly, the fluctuations of the largest eigenvalue of GUE ensembles also follow this distribution.

2. State of the art

Returning to our initial model of (non-directed) first passage percolation, it is thus conjectured that, for dimension $d = 2$, fluctuations are of order $n^{1/3}$ following a Tracy-Widom Law. Still, the current state of understanding of this model is far from this conjecture.

Kesten first proved that the fluctuations of the ball of radius n are at most $\sqrt{n}$ (this did not yet exclude a possible Gaussian behavior with Gaussian scaling). Benjamini, Kalai and Schramm then strengthened this result by showing that the fluctuations are sub-Gaussian. This is still far from the conjectured $n^{1/3}$-fluctuations, but their approach has the great advantage of being very general; in particular their result holds in any dimension $d \geq 2$.

Let us now state their main theorem concerning the fluctuations of the metric dist.

THEOREM VII.1 ([**BKS03**]). *For all a, b, d, there exists an absolute constant $C = C(a,b,d)$ such that in $\mathbb{Z}^d$,*

$$\mathrm{Var}(\mathrm{dist}_\omega(0,v)) \leq C \frac{|v|}{\log |v|}$$

for any $v \in \mathbb{Z}^d$, $|v| \geq 2$.

To keep things simple in these notes, we will only prove the analogous statement on the torus where one has more symmetries and invariance to play with.

3. The case of the torus

Let $\mathbb{T}_m^d$ be the d-dimensional torus $(\mathbb{Z}/m\mathbb{Z})^d$. As in the above lattice model, independently for each edge of $\mathbb{T}_m^d$, we choose its length to be either a or b equally likely. We are interested here in the smallest length among all closed paths γ "winding" around the torus along the first coordinate $\mathbb{Z}/m\mathbb{Z}$ (i.e. those paths γ which when projected onto the first coordinate have winding number one). In [**BKS03**], this is called the shortest *circumference*. For any configuration $\omega \in \{a,b\}^{E(\mathbb{T}_m^d)}$, this shortest circumference is denoted by $\mathrm{Circ}_m(\omega)$.

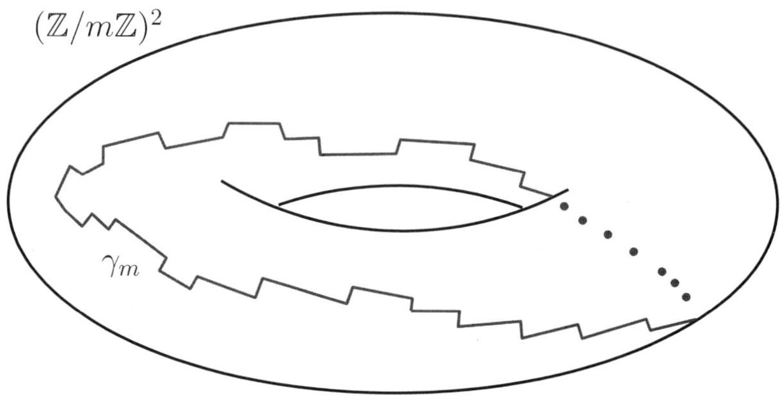

FIGURE VII.2. The shortest geodesic along the first coordinate for the random metric dist_ω on $(\mathbb{Z}/m\mathbb{Z})^2$.

THEOREM VII.2 ([**BKS03**]). *There is a constant $C = C(a,b)$ (which does not depend on the dimension d), such that*

$$\mathrm{var}(\mathrm{Circ}_m(\omega)) \le C\frac{m}{\log m}.$$

REMARK VII.2. A similar analysis as the one carried out below works in greater generality: if $G = (V,E)$ is some finite connected graph endowed with a random metric d_ω with $\omega \in \{a,b\}^{\otimes E}$, then one can obtain bounds on the fluctuations of the random diameter $D = D_\omega$ of (G, d_ω). See [**BKS03**, Theorem 2] for a precise statement in this more general context.

Proof.

For any edge e, let us consider the gradient along the edge e: $\nabla_e \mathrm{Circ}_m$. These gradient functions have values in $[-(b-a), b-a]$. By dividing our distances by the constant factor $b-a$, we can even assume without loss of generality that our gradient functions have values in $[-1,1]$. Doing so, we end up being in a setup similar to the one we had in Part V. The **influence** of an edge e corresponds here to $\mathbf{I}_e(\mathrm{Circ}_m) := \mathbb{P}\big[\nabla_e \mathrm{Circ}_m(\omega) \neq 0\big]$. We will prove later on that Circ_m has very small influences. In other words, we will show that the above gradient functions have small support, and hypercontractivity will imply the desired bound.

We have thus reduced the problem to the following general framework. Consider a real-valued function $f : \{-1,1\}^n \to \mathbb{R}$, such that for any variable k, $\nabla_k f \in [-1,1]$.

We are interested in $\mathrm{Var}(f)$ and we want to show that if "influences are small" then $\mathrm{Var}(f)$ is small. It is easy to check that the variance can be written

$$\mathrm{Var}(f) = \frac{1}{4}\sum_k \sum_{\emptyset \neq S \subseteq [n]} \frac{1}{|S|} \widehat{\nabla_k f}(S)^2 .$$

If all the variables have very small influence, then, as previously, $\nabla_k f$ should be of high frequency. Heuristically, this should then imply that

$$\begin{aligned}\mathrm{Var}(f) &\ll \sum_k \sum_{S \neq \emptyset} \widehat{\nabla_k f}(S)^2 \\ &= \sum_k \mathbf{I}_k(f) .\end{aligned}$$

This intuition is quantified by the following lemma on the link between the fluctuations of a real-valued function f on Ω_n and its influence vector.

LEMMA VII.3. *Let $f : \Omega_n \to \mathbb{R}$ be a (real-valued) function such that each of its discrete derivatives $\nabla_k f$, $k \in [n]$ have their values in $[-1, 1]$. Assume that the influences of f are small in the sense that there exists some $\alpha > 0$ such that for any $k \in \{1, \ldots, n\}$, $\mathbf{I}_k(f) \leq n^{-\alpha}$. Then there is some constant $C = C(\alpha)$ such that*

$$\mathrm{Var}(f) \leq \frac{C}{\log n} \sum_k \mathbf{I}_k(f) .$$

REMARK VII.3. If f is Boolean, then this follows from Theorem I.3 with $C(\alpha) = c/\alpha$ with c universal.

The proof of this lemma is postponed to the next section. In the meantime, let us show that in our special case of first passage percolation on the torus, the assumption on small influences is indeed verified. Since the edge lengths are in $\{a, b\}$, the smallest contour $\mathrm{Circ}_m(\omega)$ in $\mathbb{T}^d_m$ around the first coordinate lies somewhere in $[am, bm]$. Hence, if γ is a geodesic (a path in the torus with the required winding number) satisfying $\mathrm{length}(\gamma) = \mathrm{Circ}_m(\omega)$, then γ uses at most $\frac{b}{a}m$ edges. There might be several different geodesics minimizing the circumference. Let us choose randomly one of these in an "invariant" way and call it $\tilde{\gamma}$. For any edge $e \in E(\mathbb{T}^d_m)$, if, by changing the length of e, the circumference increases, then e has to be contained in any geodesic γ, and in particular in $\tilde{\gamma}$. This implies that $\mathbb{P}\big[\nabla_e \mathrm{Circ}_m(\omega) > 0\big] \leq \mathbb{P}\big[e \in \tilde{\gamma}\big]$. By symmetry we obtain that

$$\mathbf{I}_e(\mathrm{Circ}_m) = \mathbb{P}\big[\nabla_e \mathrm{Circ}_m(\omega) \neq 0\big] \leq 2\mathbb{P}\big[e \in \tilde{\gamma}\big] .$$

Now using the symmetries both of the torus $\mathbb{T}^d_m$ and of our observable Circ_m, if $\tilde{\gamma}$ is chosen in an appropriate invariant way (uniformly among all geodesics for instance), then it is clear that all the "vertical" edges (meaning those edges which, when projected onto the first coordinate, project onto a single vertex) have the same probability to lie in $\tilde{\gamma}$. The same is true for the "horizontal" edges. In particular we have that

$$\sum_{\text{"vertical" edges } e} \mathbb{P}\big[e \in \tilde{\gamma}\big] \leq \mathbb{E}\big[|\tilde{\gamma}|\big] \leq \frac{b}{a} m .$$

Since there are at least order m^d vertical edges, the influence of each of these is bounded by $O(1)m^{1-d}$. The same is true for the horizontal edges. All together this gives the desired assumption needed in Lemma VII.3. Applying this lemma, we indeed obtain that

$$\mathrm{Var}(\mathrm{Circ}_m(\omega)) \le O(1)\frac{m}{\log m},$$

where the constant does not depend on the dimension d; the dimension in fact helps us here, since it makes the influences smaller. □

REMARK VII.4. At this point, we know that for any edge e, $\mathbf{I}_e(\mathrm{Circ}_m) = O(\frac{m}{m^d})$. Hence, at least in the case of the torus, one easily deduces from Poincaré's inequality the theorem by Kesten which says that $\mathrm{Var}(\mathrm{Circ}_m) = O(m)$.

4. Upper bounds on fluctuations in the spirit of KKL

In this section, we prove Lemma VII.3.

Proof. Similarly as in the proofs of Part V, the proof relies on implementing hypercontractivity in the right way. We have that for any c,

$$\begin{aligned}
\mathrm{var}(f) &= \frac{1}{4}\sum_k \sum_{S\neq\emptyset} \frac{1}{|S|}\widehat{\nabla_k f}(S)^2 \\
&\le \frac{1}{4}\sum_k \sum_{0<|S|<c\log n} \widehat{\nabla_k f}(S)^2 + \frac{O(1)}{\log n}\sum_k \mathbf{I}_k(f)
\end{aligned}$$

where the $O(1)$ term depends on the choice of c.

Hence it is enough to bound the contribution of small frequencies, $0 < |S| < c\log n$, for some constant c which will be chosen later. As previously we have for any $\rho \in (0,1)$ and using hypercontractivity,

$$\begin{aligned}
\sum_k \sum_{0<|S|<c\log n} \widehat{\nabla_k f}(S)^2 &\le \rho^{-2c\log n}\sum_k \|T_\rho \nabla_k f\|_2^2 \\
&\le \rho^{-2c\log n}\sum_k \|\nabla_k f\|_{1+\rho^2}^2 \\
&\le \rho^{-2c\log n}\sum_k \mathbf{I}_k(f)^{2/(1+\rho^2)} \\
&\le \rho^{-2c\log n}\big(\sup_k \mathbf{I}_k(f)\big)^{\frac{1-\rho^2}{1+\rho^2}}\sum_k \mathbf{I}_k(f) \\
&\le \rho^{-2c\log n} n^{-\alpha\frac{1-\rho^2}{1+\rho^2}}\sum_k \mathbf{I}_k(f) \text{ by our assumption}.
\end{aligned} \tag{VII.1}$$

Now fixing any $\rho \in (0,1)$, and then choosing the constant c depending on ρ and α, the lemma follows. By optimizing on the choice of ρ, one could get better constants if one wants. □

5. Further discussion

Some words on the proof of Theorem VII.1

The main difficulty here is that the quantity of interest, $f(\omega) := \mathrm{dist}_\omega(0, v)$, is no longer invariant under a large class of graph automorphisms. This lack of symmetry makes the study of influences more difficult. For example, edges near the endpoints 0 or v have very high influence (of order one). To gain some more symmetry, the authors in [**BKS03**] rely on a very nice "averaging" procedure. We refer to this paper for more details.

Known lower bounds on the fluctuations

We discussed mainly here ways to obtain upper bounds on the fluctuations of the shapes in first passage percolation. It is worth pointing out that some non-trivial *lower* bounds on the fluctuations are known for $\mathbb{Z}^2$. See [**PP94, NP95**].

REMARK VII.5. We end by mentioning that the proof given in [**BKS03**] was based on an inequality by Talagrand. The proof given here avoids this inequality.

Exercise sheet of Part VII

PROBLEM VII.1. Let $n \geq 1$ and $d \geq 2$. Consider the random metric on the torus $\mathbb{Z}^d/n\mathbb{Z}^d$ as described in this part. For any $k \geq 1$, let $\mathcal{A}_n^k$ be the event that the shortest "horizontal" circuit is $\leq k$. If $d \geq 3$, show that for any choice of $k_n = k(n)$, the family of events $\mathcal{A}_n^{k_n}$ is noise sensitive. (Note that the situation here is similar to the Problem I.9 in Part I.) Finally, discuss the two-dimensional case, $d = 2$ (non-rigorously).

EXERCISE VII.2. Show that Lemma VII.3 is false if $\mathbf{I}_k(f)$ is taken to be the square of the L^2 norm of $\nabla_k f$ rather than the probability of its support (i.e. find a counterexample).

Part VIII. Randomized algorithms and noise sensitivity

In this part, we explain how the notion of **revealment** for so-called randomized algorithms can in some cases yield direct information concerning the energy spectrum which may allow not only noise sensitivity results but even quantitative noise sensitivity results.

1. BKS and randomized algorithms

In the previous part, we explained how Theorem I.5 together with bounds on the pivotal exponent for percolation yields noise sensitivity for percolation crossings. However, in [**BKS99**], a different approach was in fact used for showing noise sensitivity which, while still using Theorem I.5, did not use these bounds on the critical exponent. In that approach, one sees the first appearance of randomized algorithms. In a nutshell, the authors showed that (1) if a monotone function is very uncorrelated with all majority functions, then it is noise sensitive (in a precise quantitative sense) and (2) percolation crossings are very uncorrelated with all majority functions. The latter is shown by constructing a certain algorithm which, due to the RSW Theorem II.1, looks at very few bits but still looks at enough bits in order to be able to determine the output of the function.

2. The revealment theorem

An **algorithm** for a Boolean function f is an algorithm A which queries (asks the values of) the bits one by one, where the decision of which bit to ask can be based on the values of the bits previously queried, and stops once f is determined (being determined means that f takes the same value no matter how the remaining bits are set).

A **randomized algorithm** for a Boolean function f is the same as above but auxiliary randomness may also be used to decide the next value queried (including for the first bit). [In computer science, the term randomized decision tree would be used for our notion of randomized algorithm, but we will not use this terminology.]

The following definition of *revealment* will be crucial. Given a randomized algorithm A for a Boolean function f, we let J_A denote the random set of bits queried by A. (Note that this set depends both on the randomness corresponding to the choice of ω and the randomness inherent in running the algorithm, which are of course taken to be independent.)

DEFINITION VIII.1. *The* **revealment of a randomized algorithm** A *for a Boolean function* f*, denoted by* δ_A*, is defined by*

$$\delta_A := \max_{i\in[n]} \mathbb{P}(i \in J_A).$$

The **revealment of a Boolean function** f*, denoted by* δ_f*, is defined by*

$$\delta_f := \inf_A \delta_A$$

where the infimum is taken over all randomized algorithms A *for* f*.*

This section presents a connection between noise sensitivity and randomized algorithms. It will be used later to yield an alternative proof of noise sensitivity for percolation crossings which is not based upon Theorem I.5 (or Proposition V.5).

Two other advantages of the algorithmic approach of the present section over that mentioned in the previous section (besides the fact that it does not rest on Theorem I.5) is that it applies to nonmonotone functions and yields a more "quantitative" version of noise sensitivity.

We have only defined algorithms, randomized algorithms and revealment for Boolean functions but the definitions immediately extend to functions $f : \Omega_n \to \mathbb{R}$.

The main theorem of this section is the following.

THEOREM VIII.1 ([**SS10b**]). *For any function $f : \Omega_n \to \mathbb{R}$ and for each $k = 1, 2, \dots$, we have that*

$$E_f(k) = \sum_{S \subseteq [n],\, |S|=k} \hat{f}(S)^2 \le \delta_f\, k\, \|f\|^2, \tag{VIII.1}$$

where $\|f\|$ denotes the L^2 norm of f with respect to the uniform probability measure on Ω and δ_f is the revealment of f.

Before giving the proof, we make some comments to help the reader see what is happening and suggest why a result like this might be true. Our original function is a sum of monomials with coefficients given by the Fourier coefficients. Each time a bit is revealed by the algorithm, we obtain a new Boolean function obtained by just substituting in the value of the bit we obtained into the corresponding variable. On the algebraic side, those monomials which contain this bit go down by 1 in degree while the other monomials are unchanged. There might however be cancellation in the process which is what we hope for since when the algorithm stops, all the monomials (except the constant) must have been killed. The way cancellation occurs is illustrated as follows. The Boolean function at some stage might contain $(1/3)x_2x_4x_5 + (1/3)x_2x_4$ and then the bit x_5 might be revealed and take the value -1. When we substitute this value into the variable, the two terms cancel and disappear, thereby bringing us 1 step closer to a constant (and hence determined) function.

As far as why the result might be true, the intuition, very roughly speaking, is as follows. The theorem says that for a Boolean function we cannot, for example, have $\delta = 1/1000$ and $\sum_i \hat{f}(\{i\})^2 = 1/2$. If the level 1 monomials of the function were

$$a_1\omega_1 + a_2\omega_2 + \cdots + a_n\omega_n,$$

then it is clear that after the algorithm is over, then with high probability, the sum of the squares of the coefficients of the terms which have not been reduced to a constant is still reasonably large. Therefore, since the function at the end of the algorithm is constant, these remaining terms must necessarily have been cancelled by higher degree monomials which, after running the algorithm, have been "reduced to" degree 1 monomials. If, for the sake of this heuristic argument, we assume that each bit is revealed independently, then the probability that a degree $k \ge 2$ monomial is brought down to a degree 1 monomial (which is necessary for it to help to cancel the degree 1 terms described above) is at most δ^{k-1} and hence the expected sum of the squares of the coefficients from the degree $k \ge 2$ monomials which are brought down to degree 1 is at most δ^{k-1}. The total such sum for levels 2 to n is then at most

$$\sum_{k=2}^{n} \delta^{k-1} \le 2\delta$$

which won't be enough to cancel the (originally) degree 1 monomials which remained degree 1 after running the algorithm if δ is much less than $\sum_i \hat{f}(\{i\})^2$. A similar heuristic works for the other levels.

Proof. In the following, we let $\tilde{\Omega}$ denote the probability space that includes the randomness in the input bits of f and the randomness used to run the algorithm (which we assume to be independent) and we let $\mathbb{E}$ denote the corresponding expectation. Without loss of generality, elements of $\tilde{\Omega}$ can be represented as $\tilde{\omega} = (\omega, \tau)$ where ω are the random bits and τ represents the randomness necessary to run the algorithm.

Now, fix $k \geq 1$. Let

$$g(\omega) := \sum_{|S|=k} \hat{f}(S)\, \chi_S(\omega)\,, \qquad \omega \in \Omega.$$

The left hand side of (VIII.1) is equal to $\|g\|^2$.

Let $J \subseteq [n]$ be the random set of all bits examined by the algorithm. Let $\mathcal{A}$ denote the minimal σ-field for which J is measurable and every ω_i, $i \in J$, is measurable; this can be viewed as the relevant information gathered by the algorithm. For any function $h : \Omega \to \mathbb{R}$, let $h_J : \Omega \to \mathbb{R}$ denote the random function obtained by substituting the values of the bits in J. More precisely, if $\tilde{\omega} = (\omega, \tau)$ and $\omega' \in \Omega$, then $h_J(\tilde{\omega})(\omega')$ is $h(\omega'')$ where ω'' is ω on $J(\tilde{\omega})$ and is ω' on $[n]\backslash J(\tilde{\omega})$. In this way, h_J is a random variable on $\tilde{\Omega}$ taking values in the set of mappings from Ω to $\mathbb{R}$ and it is immediate that this random variable is $\mathcal{A}$-measurable. When the algorithm terminates, the unexamined bits in Ω are unbiased and hence $\mathbb{E}\big[h \,\big|\, \mathcal{A}\big] = \int h_J (= \hat{h}_J(\emptyset))$ where $\int$ is defined, as usual, to be integration with respect to uniform measure on Ω. It follows that $\mathbb{E}[h] = \mathbb{E}[\int h_J]$.

Similarly, for all h,

$$\text{(VIII.2)} \qquad \|h\|^2 = \mathbb{E}\big[h^2\big] = \mathbb{E}\Big[\int h_J^2\Big] = \mathbb{E}\big[\|h_J\|^2\big].$$

Since the algorithm determines f, it is $\mathcal{A}$ measurable, and we have

$$\|g\|^2 = \mathbb{E}[g\, f] = \mathbb{E}\Big[\mathbb{E}\big[g\, f \,\big|\, \mathcal{A}\big]\Big] = \mathbb{E}\Big[f\, \mathbb{E}\big[g \,\big|\, \mathcal{A}\big]\Big].$$

Since $\mathbb{E}\big[g \,\big|\, \mathcal{A}\big] = \hat{g}_J(\emptyset)$, Cauchy-Schwarz therefore gives

$$\text{(VIII.3)} \qquad \|g\|^2 \leq \sqrt{\mathbb{E}[\hat{g}_J(\emptyset)^2]}\; \|f\|\,.$$

We now apply Parseval to the (random) function g_J: this gives (for any $\tilde{\omega} = (\omega, \tau) \in \tilde{\Omega}$),

$$\hat{g}_J(\emptyset)^2 = \|g_J\|_2^2 - \sum_{|S|>0} \hat{g}_J(S)^2.$$

Taking the expectation over $\tilde{\omega} \in \tilde{\Omega}$, this leads to

$$\begin{aligned}
\mathbb{E}\big[\hat{g}_J(\emptyset)^2\big] &= \mathbb{E}\big[\|g_J\|_2^2\big] - \sum_{|S|>0} \mathbb{E}\big[\hat{g}_J(S)^2\big] \\
&= \|g\|_2^2 - \sum_{|S|>0} \mathbb{E}\big[\hat{g}_J(S)^2\big] \quad \text{by (VIII.2)} \\
&= \sum_{|S|=k} \hat{g}(S)^2 - \sum_{|S|>0} \mathbb{E}\big[\hat{g}_J(S)^2\big] \left\{ \begin{array}{l} \text{since } g \text{ is supported} \\ \text{on level-}k \text{ coefficients} \end{array} \right. \\
&\leq \sum_{|S|=k} \mathbb{E}\big[\hat{g}(S)^2 - \hat{g}_J(S)^2\big] \left\{ \begin{array}{l} \text{by restricting to} \\ \text{level-}k \text{ coefficients} \end{array} \right.
\end{aligned}$$

Now, since g_J is built randomly from g by fixing the variables in $J = J(\tilde{\omega})$, and since g by definition does not have frequencies larger than k, it is clear that for any S with $|S| = k$ we have

$$\hat{g}_J(S) = \left\{ \begin{array}{ll} \hat{g}(S) = \hat{f}(S), & \text{if } S \cap J(\tilde{\omega}) = \emptyset \\ 0, & \text{otherwise.} \end{array} \right.$$

Therefore, we obtain

$$\|\mathbb{E}[g \,|\, J]\|_2^2 = \mathbb{E}\big[\hat{g}_J(\emptyset)^2\big] \leq \sum_{|S|=k} \hat{g}(S)^2\, \mathbb{P}\big[S \cap J \neq \emptyset\big] \leq \|g\|_2^2\, k\, \delta\,.$$

Combining with (VIII.3) completes the proof. □

Proposition IV.1 and Theorem VIII.1 immediately imply the following corollary.

COROLLARY VIII.2. *If the revealments satisfy*

$$\lim_{n\to\infty} \delta_{f_n} = 0,$$

then $\{f_n\}$ *is noise sensitive.*

In the exercises, one is asked to show that certain sequences of Boolean functions are noise sensitive by applying the above corollary.

3. An application to noise sensitivity of percolation

In this section, we apply Corollary VIII.2 to prove noise sensitivity of percolation crossings. The following result gives the necessary assumption that the revealments approach 0.

THEOREM VIII.3 ([**SS10b**]). *Let* $f = f_n$ *be the indicator function for the event that critical site percolation on the triangular grid contains a left to right crossing of our* $n \times n$ *box. Then* $\delta_{f_n} \leq n^{-1/4+o(1)}$ *as* $n \to \infty$.

For critical bond percolation on the square grid, this holds with $1/4$ *replaced by some positive constant* $a > 0$.

Outline of Proof. We outline the argument only for the triangular lattice; the argument for the square lattice is similar. We first give a first attempt at a good algorithm. We consider from Part II the exploration path or interface from the bottom right of the square to the top left used to detect a left right crossing. This (deterministic) algorithm simply asks the bits that it needs to know in order to continue the interface. Observe that if a bit is queried, it is necessarily the case that there is both a black and white path from next to the hexagon to the boundary. It follows, from the exponent of 1/4 for the two-arm event in Part II, that, for hexagons far from the boundary, the probability that they are revealed is at most $R^{-1/4+o(1)}$ as desired. However, one cannot conclude that points near the boundary have small revealment and of course the right bottom point is always revealed.

The way that we modify the above algorithm so that all points have small revealment is as follows. We first choose a point x at random from the middle third of the right side. We then run two algorithms, the first one which checks whether there is a left right path from the right side *above* x to the left side and the second one which checks whether there is a left right path from the right side *below* x to the left side. The first part is done by looking at an interface from x to the top left corner as above. The second part is done by looking at an interface from x to the bottom left corner as above (but where the colors on the two sides of the interface need to be swapped.)

It can then be shown with a little work (but no new conceptual ideas) that this modified algorithm has the desired revealment of at most $R^{-1/4+o(1)}$ as desired. One of the things that one needs to use in this analysis is the so-called one-arm half-plane exponent, which has a known value of 1/3. See [**SS10b**] for details. □

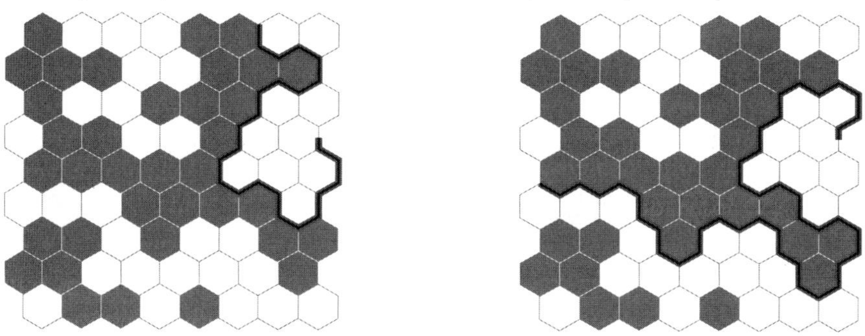

3.1. First quantitative noise sensitivity result. In this subsection, we give our first "polynomial bound" on the noise sensitivity of percolation. This is an important step in our understanding of quantitative noise sensitivity of percolation initiated in Part VI.

Recall that in the definition of noise sensitivity, ϵ is held fixed. However, as we have seen in Part VI, it is of interest to ask if the correlations can still go to 0 when $\epsilon = \epsilon_n$ goes to 0 with n but not so fast. The techniques of the present part imply the following result.

THEOREM VIII.4 ([**SS10b**]). *Let $\{f_n\}$ be as in Theorem VIII.3. Then, for the triangular lattice, for all $\gamma < 1/8$,*

$$\lim_{n\to\infty} \mathbb{E}[f_n(\omega)f_n(\omega_{1/n^\gamma})] - \mathbb{E}[f_n(\omega)]^2 = 0. \tag{VIII.4}$$

On the square lattice, there exists some $\gamma > 0$ with the above property.

Proof. We prove only the first statement; the square lattice case is handled similarly. First, (IV.3) gives us that every n and γ,

$$\mathbb{E}[f_n(\omega)f_n(\omega_{1/n^\gamma})] - \mathbb{E}[f_n(\omega)]^2 = \sum_{k=1} E_{f_n}(k)(1-1/n^\gamma)^k. \tag{VIII.5}$$

Note that there are order n^2 terms in the sum. Fix $\gamma < 1/8$. Choose $\epsilon > 0$ so that $\gamma + \epsilon < 1/8$. For large n, we have that $\delta_{f_n} \le 1/n^{1/4-\epsilon}$. The right hand side of (VIII.5) is at most

$$\sum_{k=1}^{n^{\gamma+\epsilon/2}} k/n^{1/4-\epsilon} + (1 - 1/n^{\gamma})^{n^{\gamma+\epsilon/2}}$$

by breaking up the sum at $n^{\gamma+\epsilon/2}$ and applying Theorems VIII.1 and VIII.3 to bound the $E_{f_n}(k)$ terms in the first part. The second term clearly goes to 0 while the first part also goes to 0 by the way ϵ was chosen. □

Observe that the *proof* of Theorem VIII.4 immediately yields the following general result.

COROLLARY VIII.5. *Let $\{f_n\}$ be a sequence of Boolean functions on m_n bits with $\delta(f_n) \le O(1)/n^{\beta}$ for all n. Then for all $\gamma < \beta/2$, we have that*

$$\lim_{n\to\infty} \mathbb{E}[f_n(\omega)f_n(\omega_{1/n^{\gamma}})] - \mathbb{E}[f_n(\omega)]^2 = 0. \tag{VIII.6}$$

4. Lower bounds on revealments

One of the goals of the present section is to show that one cannot hope to reach the conjectured 3/4-sensitivity exponent with Theorem VIII.1. Theorem VIII.4 told us that we obtain asymptotic decorrelation if the noise is $1/n^{\gamma}$ for $\gamma < 1/8$. Note that this differs from the conjectured "critical exponent" of 3/4 by a factor of 6. In this section, we investigate the degree to which the 1/8 could potentially be improved and in the discussion, we will bring up an interesting open problem. We will also derive an interesting general theorem giving a nontrivial lower bound on the revealment for monotone functions. We start with the following definition.

DEFINITION VIII.2. *Given a randomized algorithm A for a Boolean function f, let $C(A)$ (the cost of A) be the expected number of queries that the algorithm A makes. Let $C(f)$ (the cost of f) be the infimum of $C(A)$ over all randomized algorithms A for f.*

REMARK VIII.1. (i). It is easy to see that $C(f)$ is unchanged if we take the infimum over deterministic algorithms.
(ii). Clearly $n\delta_A \ge C(A)$ and hence $n\delta_f \ge C(f)$.
(iii). $C(f)$ is at least the total influence $\mathbf{I}(f)$ since for any algorithm A and any i, the event that i is pivotal necessarily implies that the bit i is queried by A.

The following result due to O'Donnell and Servedio ([**OS07**])is an essential improvement on the third part of the last remark.

THEOREM VIII.6. *Let f be a monotone Boolean function mapping Ω_n into $\{-1,1\}$. Then $C(f) \ge \mathbf{I}(f)^2$ and hence $\delta_f \ge \mathbf{I}(f)^2/n$.*

Proof. Fix any randomized algorithm A for f. Let $J = J_A$ be the random set of bits queried by A. We then have

$$\mathbf{I}(f) = \mathbb{E}[\sum_i f(\omega)\omega_i] = \mathbb{E}[f(\omega)\sum_i \omega_i I_{\{i\in J\}}] \le \sqrt{\mathbb{E}[f(\omega)^2]}\sqrt{\mathbb{E}[(\sum_i \omega_i I_{\{i\in J\}})^2]}$$

where the first equality uses monotonicity (recall Proposition IV.4) and then the Cauchy-Schwarz inequality is used. We now bound the first term by 1. For the second moment inside the second square root, the sum of the diagonal terms yields $\mathbb{E}[|J|]$ while the cross terms are all 0 since for $i \neq j$, $\mathbb{E}[\omega_i I_{\{i\in J\}}\omega_j I_{\{j\in J\}}] = 0$ as can be seen by breaking up the sum depending on whether i or j is queried first. This yields the result. □

Returning to our event f_n of percolation crossings, since the sum of the influences is $n^{3/4+o(1)}$, Theorem VIII.6 tells us that $\delta_{f_n} \geq n^{-1/2+o(1)}$. It follows from the method of proof in Theorem VIII.4 that Theorem VIII.1 cannot improve the result of Theorem VIII.4 past $\gamma = 1/4$ which is still a factor of 3 from the critical value 3/4. Of course, one could investigate the degree to which Theorem VIII.1 itself could be improved.

Theorem VIII.3 tells us that there are algorithms A_n for f_n such that $C(A_n) \leq n^{7/4+o(1)}$. On the other hand, Theorem VIII.6 tell us that it is necessarily the case that $C(A) \geq n^{6/4+o(1)}$.

Open Question: Find the smallest σ such that there are algorithms A_n for f_n with $C(A_n) \leq n^\sigma$. (We know $\sigma \in [6/4, 7/4]$.)

We mention another inequality relating revealment with influences which is a consequence of the results in [**OSSS05**].

THEOREM VIII.7. *Let f be a Boolean function mapping Ω_n into $\{-1,1\}$. Then* $\delta_f \geq \mathrm{Var}(f)/(n \max_i \mathbf{I}_i(f))$

It is interesting to compare Theorems VIII.6 and VIII.7. Assuming $\mathrm{Var}(f)$ is of order 1, and all the influences are of order $1/n^\alpha$, then it is easy to check that Theorem VIII.6 gives a better bound when $\alpha < 2/3$ and Theorem VIII.7 gives a better bound when $\alpha > 2/3$. For crossings of percolation, where α should be 5/8, it is better to use Theorem VIII.6 rather than VIII.7.

Finally, there are a number of interesting results concerning revealment obtained in the paper [**BSW05**]. Four results are as follows.

1. If f is reasonably balanced on n bits, then the revealment is at least of order $1/n^{1/2}$.
2. There is a reasonably balanced function on n bits whose revealment is at most $O(1)(\log n)/n^{1/2}$.
3. If f is reasonably balanced on n bits and is monotone, then the revealment is at least of order $1/n^{1/3}$.
4. There is a reasonably balanced monotone function on n bits whose revealment is at most $O(1)(\log n)/n^{1/3}$.

We finally end this section by giving one more reference which gives an interesting connection between percolation, algorithms and game theory; see [**PSSW07**].

5. An application to a critical exponent

In this section, we show how Theorem VIII.1 or in fact Theorem VIII.6 can be used to show that the four-arm exponent is strictly larger than 1; recall that with SLE technology, this can be shown for the triangular lattice.

PROPOSITION VIII.8. *Both on the triangular lattice $\mathbb{T}$ and on $\mathbb{Z}^2$, there exists $\epsilon_0 > 0$ such that*

$$\alpha_4(R) \leq 1/R^{1+\epsilon_0}$$

We will assume the separation of arms result mentioned earlier in Part VI which says that for the event f_R, the influence of any variable further than distance $R/10$ from the boundary, a set of variables that we will denote by B for bulk, is $\asymp \alpha_4(R)$.

Proof. Theorems VIII.3 and VIII.1 imply that for some $a > 0$,

$$\sum_i \hat{f}_R(\{i\})^2 \leq 1/R^a.$$

Next, using the separation of arms as explained above, we have

$$R^2\alpha_4^2(R) \leq O(1)\sum_{i\in B} \mathbf{I}_i^2. \tag{VIII.7}$$

Proposition IV.4 then yields

$$R^2\alpha_4^2(R) \leq O(1/R^a)$$

and the result follows. □

Observe that Theorem VIII.6 could also be used as follows. Theorem VIII.3 implies that $C(f_R) \leq R^{2-a}$ for some $a > 0$ and then Theorem VIII.6 yields $\mathbf{I}(f_R)^2 \leq R^{2-a}$.

Exactly as in (VIII.7), one has, again using separation of arms, that

$$R^2\alpha_4(R) \leq O(1)\sum_{i\in B} \mathbf{I}_i \leq O(1)\mathbf{I}(f_R). \tag{VIII.8}$$

Altogether this gives us

$$R^4\alpha_4^2(R) \leq O(1)R^{2-a},$$

again yielding the result.

We finally mention that it is not so strange that either of Theorems VIII.1 or VIII.6 can be used here since, as the reader can easily verify, for the case of monotone functions all of whose variables have the same influence, the case $k = 1$ in Theorem VIII.1 is equivalent to Theorem VIII.6.

REMARK VIII.2. We now mention that the proof for the multi-scale version of Proposition VI.6 is an extension of the approach of O'Donnell and Servedio above.

6. Does noise sensitivity imply low revealment?

As far as these lectures are concerned, this subsection will not connect to anything that follows and hence can be viewed as tangential.

It is natural to ask if the converse of Corollary VIII.2 might be true. A moment's thought reveals that example 2, Parity, provides a counterexample. However, it is more interesting perhaps that there is a monotone counterexample to the converse which is provided by example 5, Clique containment.

PROPOSITION VIII.9. *Clique containment provides an example showing that the converse of Corollary VIII.2 is false for monotone functions.*

Outline of Proof. We first explain more precisely the size of the clique that we are looking for. Given n and k, let $f(n,k) := \binom{n}{k}2^{-\binom{k}{2}}$, which is just the expected number of cliques of size k in a random graph. When k is around $2\log_2(n)$, it is easy to check that $f(n, k+1)/f(n,k)$ is $o(1)$ as $n \to \infty$. For such k, clearly if $f(n,k)$ is small, then with high probability there is no k-clique while it can be shown, via a second moment type argument, that if $f(n,k)$ is large, then with high probability there is a k-clique. One now takes k_n to be around $2\log_2(n)$ such that $f(n,k_n) \geq 1$ and $f(n,k_n+1) < 1$. Since $f(n,k+1)/f(n,k)$ is $o(1)$, it follows with some thought from the above that the clique number is concentrated on at most 2 points. Furthermore, if $f(n,k_n)$ is very large and $f(n,k_n+1)$ very small, then it is concentrated on one point. Again, see [**AS00**] for details.

Finally, we denote the event that the random graph on n vertices contains a clique of size k_n by A_n. We have already seen in one of the exercises that this example is noise sensitive. We will only consider a sequence of n's so that A_n is nondegenerate in the sense that the probabilities of this sequence stay bounded away from 0 and 1. An interesting point is that there is such a sequence. Again, see [**AS00**] for this. To show that the revealments do not go to 0, it suffices to show that the sequence of costs (see Definition VIII.2 and the remarks afterwards) is $\Omega(n^2)$. We prove something stronger but, to do this, we must first give a few more definitions.

DEFINITION VIII.3. *For a given Boolean function f, a* **witness** *for ω is any subset W of the variables such that the elements of ω in W determine f in the sense that for every ω' which agrees with ω on W, we have that $f(\omega) = f(\omega')$. The* **witness size** *of ω, denoted $w(\omega)$, is the size of the smallest witness for ω. The* **expected witness size**, *denoted by $w(f)$, is $\mathbb{E}(w(\omega))$.*

Observe that, for any Boolean function f, the bits revealed by any algorithm A for f and for any ω is always a witness for ω. It easily follows that the cost $C(f)$ satisfies $C(f) \geq w(f)$. Therefore, in order to prove the proposition, it suffices to show that

$$w(f_n) = \Omega(n^2). \tag{VIII.9}$$

REMARK VIII.3. (i). The above also implies that with a fixed uniform probability, $w(\omega)$ is $\Omega(n^2)$.
(ii). Of course when f_n is 1, there is always a (small) witness of size $\binom{k_n}{2} \ll n$ and so the large average witness size comes from when f_n is -1.
(iii). However, it is not deterministically true that when f_n is -1, $w(\omega)$ is necessarily of size $\Omega(n^2)$. For example, for $\omega \equiv -1$ (corresponding to the empty graph), the witness size is $o(n^2)$ as is easily checked. Clearly the empty graph has the smallest witness size among ω with $f_n = -1$.

LEMMA VIII.10. *Let E_n be the event that all sets of vertices of size at least $.97n$ contains C_{k_n-3}. Then $\lim_{n\to\infty} \mathbb{P}(E_n) = 1$.*

Proof. This follows, after some work, from the Janson inequalities. See [**AS00**] for details concerning these inequalities. □

LEMMA VIII.11. *Let U be any collection of at most $n^2/1000$ edges in C_n. Then there exist distinct v_1, v_2, v_3 such that no edge in U goes between any v_i and v_j and*

$$|\{e \in U : e \text{ is an edge between } \{v_1,v_2,v_3\} \text{ and } \{v_1,v_2,v_3\}^c\}| \leq n/50. \tag{VIII.10}$$

Proof. We use the probabilistic method where we choose $\{v_1,v_2,v_3\}$ to be a uniformly chosen 3-set. It is immediate that the probability that the first condition fails is at most $3|U|/\binom{n}{2} \leq 1/100$. Letting Y be the number of edges in the set appearing in (VIII.10) and Y' be the number of U edges touching v_1, it is easy to see that

$$\mathbb{E}(Y) \leq 3\mathbb{E}(Y') = 6|U|/n \leq n/100$$

where the equality follows from the fact that, for any graph, the number of edges is half the total degree. By Markov's inequality, the probability of the event in (VIII.10) holds with probably at least $1/2$. This shows that the random 3-set $\{v_1, v_2, v_3\}$ satisfies the two stated conditions with positive probability and hence such a 3-set exists. □

By Lemma VIII.10, we have $\mathbb{P}(A_n^c \cap E_n) \geq c > 0$ for all large n. To prove the theorem, it therefore suffices to show that if $A_n^c \cap E_n$ occurs, there is no witness of size smaller than $n^2/1000$. Assume U to be any set of edges of size smaller than $n^2/1000$. Choose $\{v_1, v_2, v_3\}$ from Lemma VIII.11. By the second condition in this lemma, there exists a set S of size at least $.97n$ which is disjoint from $\{v_1, v_2, v_3\}$ which has no U-edge to $\{v_1, v_2, v_3\}$. Since E_n occurs, S contains a C_{k_n-3}, whose vertices we denote by T. Since there are no U-edges between T and $\{v_1, v_2, v_3\}$ or within $\{v_1, v_2, v_3\}$ (by the first condition in Lemma VIII.11) and T is the complete graph, U cannot be a witness since A_n^c occured. □

The key step in the proof of Proposition VIII.9 is (VIII.9). This is stated without proof in [**FKW02**]; however, E. Friedgut provided us with the above proof.

Exercise sheet of Part VIII

EXERCISE VIII.1. Compute the revealment for Majority function on 3 bits.

EXERCISE VIII.2. Use Corollary VIII.2 to show that Examples 4 and 6, Iterated 3-Majority function and tribes, are noise sensitive.

EXERCISE VIII.3. For transitive monotone functions, is there a relationship between revealment and the minimal cost over all algorithms?

EXERCISE VIII.4. Show that for transitive monotone functions, Theorem VIII.6 yields the same result as Theorem VIII.1 does for the case $k = 1$.

EXERCISE VIII.5. What can you say about the sequence of revealments for the Iterated 3-Majority function? [It can be shown that the sequence of revealments decays like $1/n^\sigma$ for some σ but it is an open question what σ is.]

EXERCISE VIII.6. You are given a sequence of Boolean functions and told that it is not noise sensitive using noise $\epsilon_n = 1/n^{1/5}$. What, if anything, can you conclude about the sequence of revealments δ_n?

EXERCISE VIII.7. Note that a consequence of Corollary VIII.2 and the last line in Remark IV.2 is that if $\{f_n\}$ is a sequence of monotone functions, then, if the revealments of $\{f_n\}$ go to 0, the sums of the squared influences approach 0. Show that this implication is false without the monotonicity assumption.

Part IX. The spectral sample

It turns out that it is very useful to view the Fourier coefficients of a Boolean function as a random subset of the input bits where the "weight" or "probability" of a subset is its squared Fourier coefficient. It is our understanding that it was Gil Kalai who suggested that thinking of the spectrum as a random set could shed some light on the types of questions we are looking at here. The following is the crucial definition in this part.

1. Definition of the spectral sample

DEFINITION IX.1. *Given a Boolean function $f : \Omega_n \to \{\pm 1\}$ or $\{0,1\}$, we let the* **spectral measure** $\hat{\mathbb{Q}} = \hat{\mathbb{Q}}_f$ *of f be the measure on subsets $\{1,\dots,n\}$ given by*

$$\hat{\mathbb{Q}}_f(S) := \hat{f}(S)^2,\ S \subset \{1,\dots,n\}.$$

We let $\mathscr{S}_f = \mathscr{S}$ denote a subset of $\{1,\dots,n\}$ chosen according to this measure and call this the **spectral sample**. *We let $\hat{\mathbb{Q}}$ also denote the corresponding expectation (even when $\hat{\mathbb{Q}}$ is not a probability measure).*

By Parseval, the total mass of the so-defined spectral measure is

$$\sum_{S\subset\{1,\dots,n\}} \hat{f}(S)^2 = \mathbb{E}\big[f^2\big].$$

This makes the following definition natural.

DEFINITION IX.2. *Given a Boolean function $f : \Omega_n \to \{\pm 1\}$ or $\{0,1\}$, we let the* **spectral probability measure** $\hat{\mathbb{P}} = \hat{\mathbb{P}}_f$ *of f be the probability measure on subsets of $\{1,\dots,n\}$ given by*

$$\hat{\mathbb{P}}_f(S) := \frac{\hat{f}(S)^2}{\mathbb{E}[f^2]},\ S \subset \{1,\dots,n\}.$$

Since $\hat{\mathbb{P}}_f$ is just $\hat{\mathbb{Q}}_f$ up to a renormalization factor, the **spectral sample** $\mathscr{S}_f = \mathscr{S}$ *will denote as well a random subset of $[n]$ sampled according to $\hat{\mathbb{P}}_f$. We let $\hat{\mathbb{E}}_f = \hat{\mathbb{E}}$ denote its corresponding expectation.*

REMARK IX.1.

(i) Note that if f maps into $\{\pm 1\}$, then, by Parseval's formula, $\hat{\mathbb{Q}}_f = \hat{\mathbb{P}}_f$ while if it maps into $\{0,1\}$, $\hat{\mathbb{Q}}_f$ will be a subprobability measure.

(ii) Observe that if $(f_n)_n$ is a sequence of non-degenerate Boolean functions into $\{0,1\}$, then $\hat{\mathbb{P}}_{f_n} \asymp \hat{\mathbb{Q}}_{f_n}$.

(iii) There is no statistical relationship between ω and $\mathscr{S}_f$ as they are defined on different probability spaces. The spectral sample will just be a convenient point of view in order to understand the questions we are studying.

Some of the formulas and results we have previously derived in these notes have very simple formulations in terms of the spectral sample. For example, it is immediate to check that (IV.2) simply becomes

$$\mathbb{E}[f(\omega)f(\omega_\epsilon)] = \hat{\mathbb{Q}}_f[(1-\epsilon)^{|\mathscr{S}|}] \tag{IX.1}$$

or

$$\mathbb{E}[f(\omega)f(\omega_\epsilon)] - \mathbb{E}[f(\omega)]^2 = \hat{\mathbb{Q}}_f[(1-\epsilon)^{|\mathscr{S}|} I_{\mathscr{S}\neq\emptyset}]. \tag{IX.2}$$

Next, in terms of the spectral sample, Propositions IV.1 and IV.2 simply become the following proposition.

PROPOSITION IX.1. *If $\{f_n\}$ is a sequence of Boolean functions mapping into $\{\pm 1\}$, then we have the following.*
1. $\{f_n\}$ is noise sensitive if and only if $|\mathscr{S}_{f_n}| \to \infty$ in probability on the set $\{|\mathscr{S}_{f_n}| \neq 0\}$.
2. $\{f_n\}$ is noise stable if and only if the random variables $\{|\mathscr{S}_{f_n}|\}$ are tight.

There is also a nice relationship between the pivotal set $\mathcal{P}$ and the spectral sample. The following result, which is simply Proposition IV.3 (see also the remark after this proposition), tells us that the two random sets $\mathcal{P}$ and $\mathscr{S}$ have the same 1-dimensional marginals.

PROPOSITION IX.2. *If f is a Boolean function mapping into $\{\pm 1\}$, then for all $i \in [n]$ we have that*

$$\mathbb{P}(i \in \mathcal{P}) = \hat{\mathbb{Q}}(i \in \mathscr{S})$$

and hence $\mathbb{E}(|\mathcal{P}|) = \hat{\mathbb{Q}}(|\mathscr{S}|)$.

(This proposition is stated with $\hat{\mathbb{Q}}$ instead of $\hat{\mathbb{P}}$ since if f maps into $\{0, 1\}$ instead, then the reader can check that the above holds with an extra factor of 4 on the right hand side while if $\hat{\mathbb{P}}$ were used instead, then this would not be true for any constant.) Even though $\mathscr{S}$ and $\mathcal{P}$ have the same "1-dimensional" marginals, it is not however true that these two random sets have the same distribution. For example, it is easily checked that for $\mathbf{MAJ}_3$, these two distributions are different. Interestingly, as we will see in the next section, $\mathscr{S}$ and $\mathcal{P}$ also always have the same "2-dimensional" marginals. This will prove useful when applying second moment method arguments.

Before ending this section, let us give an alternative proof of Proposition VI.9 using this point of view of thinking of $\mathscr{S}$ as a random set.

Alternative proof of Proposition VI.9 The statement of the proposition when converted to the spectrum states (see the exercises in this part if this is not clear) that for any $a_n \to \infty$,

$$\lim_{n\to\infty} \hat{\mathbb{P}}(|\mathscr{S}_n| \geq a_n n^2 \alpha_4(n)) = 0.$$

However this immediately follows from Markov's inequality using Propositions VI.8 and IX.2. □

2. A way to sample the spectral sample in a sub-domain

In this section, we describe a method of "sampling" the spectral measure restricted to a subset of the bits. As an application of this, we show that $\mathscr{S}$ and $\mathcal{P}$ in fact have the same 2-dimensional marginals, namely that for all i and j, $\mathbb{P}(i, j \in \mathcal{P}) = \hat{\mathbb{Q}}(i, j \in \mathscr{S})$.

In order to first get a little intuition about the spectral measure, we start with an easy proposition.

PROPOSITION IX.3 ([**GPS10**]). *For a Boolean function f and $A \subseteq \{1, 2, \ldots, n\}$, we have*

$$\hat{\mathbb{Q}}(\mathscr{S}_f \subseteq A) = \mathbb{E}[|\mathbb{E}(f|A)|^2]$$

where conditioning on A means conditioning on the bits in A.

Proof. Noting that $\mathbb{E}(\chi_S|A)$ is χ_S if $S \subseteq A$ and 0 otherwise, we obtain by expanding that

$$\mathbb{E}(f|A) = \sum_{S \subseteq A} \hat{f}(S)\, \chi_S.$$

Now apply Parseval's formula. □

If we have a subset $A \subseteq \{1, 2, \ldots, n\}$, how do we "sample" from $A \cap \mathscr{S}$? A nice way to proceed is as follows: choose a random configuration outside of A, then look at the induced function on A and sample from the induced function's spectral measure. The following proposition justifies in precise terms this way of sampling. Its proof is just an extension of the proof of Proposition IX.3.

PROPOSITION IX.4 ([**GPS10**]). *Fix a Boolean function f on Ω_n. For $A \subseteq \{1, 2, \ldots, n\}$ and $y \in \{\pm 1\}^{A^c}$, that is a configuration on A^c, let g_y be the function defined on $\{\pm 1\}^A$ obtained by using f but fixing the configuration to be y outside of A. Then for any $S \subseteq A$, we have*

$$\hat{\mathbb{Q}}(\mathscr{S}_f \cap A = S) = \mathbb{E}[\hat{\mathbb{Q}}(\mathscr{S}_{g_y} = S)] = \mathbb{E}[\hat{g}_y^2(S)].$$

Proof. Using the first line of the proof of Proposition IX.3, it is easy to check that for any $S \subseteq A$, we have that

$$\mathbb{E}\big[f\, \chi_S \,|\, \mathcal{F}_{A^c}\big] = \sum_{S' \subseteq A^c} \hat{f}(S \cup S')\, \chi_{S'}\,.$$

This gives

$$\mathbb{E}\Big[\mathbb{E}\big[f\, \chi_S \,|\, \mathcal{F}_{A^c}\big]^2\Big] = \sum_{S' \subseteq A^c} \hat{f}(S \cup S')^2 = \hat{\mathbb{Q}}[\mathscr{S} \cap A = S]$$

which is precisely the claim. □

REMARK IX.2. Observe that Proposition IX.3 is a special case of Proposition IX.4 when S is taken to be $\emptyset$ and A is replaced by A^c.

The following corollary was first observed by Gil Kalai.

COROLLARY IX.5 ([**GPS10**]). *If f is a Boolean function mapping into $\{\pm 1\}$, then for all i and j,*

$$\mathbb{P}(i, j \in \mathcal{P}) = \hat{\mathbb{Q}}(i, j \in \mathscr{S}).$$

(The comment immediately following Proposition IX.2 holds here as well.)

Proof. Although it has already been established that $\mathcal{P}$ and $\mathscr{S}$ have the same 1-dimensional marginals, we first show how Proposition IX.4 can be used to establish this. This latter proposition yields, with $A = S = \{i\}$, that

$$\hat{\mathbb{Q}}(i \in \mathscr{S}) = \hat{\mathbb{Q}}(\mathscr{S} \cap \{i\} = \{i\}) = \mathbb{E}[\hat{g}_y^2(\{i\})].$$

Note that g_y is $\pm\omega_i$ if i is pivotal and constant if i is not pivotal. Hence the last term is $\mathbb{P}(i \in \mathcal{P})$.

For the 2-dimensional marginals, one first checks this by hand when $n = 2$. For general n, taking $A = S = \{i, j\}$ in Proposition IX.4, we have

$$\hat{\mathbb{Q}}(i, j \in \mathscr{S}) = \mathbb{P}(\mathscr{S} \cap \{i, j\} = \{i, j\}) = \mathbb{E}[\hat{g}_y^2(\{i, j\})].$$

For fixed y, the $n = 2$ case tells us that $\hat{g}_y^2(\{i, j\}) = \mathbb{P}(i, j \in \mathcal{P}_{g_y})$. Finally, a little thought shows that $\mathbb{E}[\mathbb{P}(i, j \in \mathcal{P}_{g_y})] = \mathbb{P}(i, j \in \mathcal{P})$, completing the proof. □

3. Nontrivial spectrum near the upper bound for percolation

We now return to our central event of percolation crossings of the rectangle R_n where f_n denotes this event. At this point, we know that for $\mathbb{Z}^2$, (most of) the spectrum lies between n^{ϵ_0} (for some $\epsilon_0 > 0$) and $n^2\alpha_4(n)$ while for $\mathbb{T}$ it sits between $n^{1/8+o(1)}$ and $n^{3/4+o(1)}$. In this section, we show that there is a nontrivial amount of spectrum near the upper bound $n^2\alpha_4(n)$. For $\mathbb{T}$, in terms of quantitative noise sensitivity, this tells us that if our noise sequence ϵ_n is equal to $1/n^{3/4-\delta}$ for fixed $\delta > 0$, then in the limit, the two variables $f(\omega)$ and $f(\omega_{\epsilon_n})$ are not perfectly correlated; i.e., there is some degree of independence. (See the exercises for understanding such arguments.) However, we cannot conclude that there is full independence since we don't know that "all" of the spectrum is near $n^{3/4+o(1)}$ (yet!).

THEOREM IX.6 ([**GPS10**]). *Consider our percolation crossing functions $\{f_n\}$ (with values into $\{\pm 1\}$) of the rectangles R_n for $\mathbb{Z}^2$ or $\mathbb{T}$. There exists $c > 0$ such that for all n,*

$$\hat{\mathbb{P}}\big[|\mathscr{S}_n| \geq cn^2\alpha_4(n)\big] \geq c.$$

The key lemma for proving this is the following second moment bound on the number of pivotals which we prove afterwards. It has a similar flavor to Exercise 6 in Part VI.

LEMMA IX.7 ([**GPS10**]). *Consider our percolation crossing functions $\{f_n\}$ above and let R'_n be the box concentric with R_n with half the radius. If $X_n = |\mathcal{P}_n \cap R'_n|$ is the cardinality of the set of pivotal points in R'_n, then there exists a constant C such that for all n we have that*

$$\mathbb{E}[|X_n|^2] \leq C\mathbb{E}[|X_n|]^2.$$

Proof of Theorem IX.6. Since $\mathcal{P}_n$ and $\mathscr{S}_n$ have the same 1 and 2-dimensional marginals, it follows fairly straightforward from Lemma IX.7 that we also have that for all n

$$\hat{\mathbb{P}}\big[|\mathscr{S}_n \cap R'_n|^2\big] \leq C\hat{\mathbb{P}}\big[|\mathscr{S}_n \cap R'_n|\big]^2.$$

Recall now the Paley-Zygmund inequality which states that if $Z \geq 0$, then for all $\theta \in (0,1)$,

$$\mathbb{P}(Z \geq \theta\,\mathbb{E}[Z]) \geq (1-\theta)^2\frac{\mathbb{E}[Z]^2}{\mathbb{E}[Z^2]}.$$

The two above inequalities (with $Z = |\mathscr{S}_n \cap R'_n|$ and $\theta = 1/2$) imply that for all n,

$$\hat{\mathbb{P}}\Big[|\mathscr{S}_n \cap R'_n| \geq \frac{\hat{\mathbb{E}}\big[|\mathscr{S}_n \cap R'_n|\big]}{2}\Big] \geq \frac{1}{4C}.$$

Now, by Proposition IX.2, one has that $\hat{\mathbb{E}}\big[|\mathscr{S}_n \cap R'_n|\big] = \mathbb{E}[X_n]$. Furthermore (a trivial modification of) Proposition VI.8 yields $\mathbb{E}[X_n] \asymp n^2\alpha_4(n)$ which thus completes the proof. □

We are now left with

Proof of Lemma IX.7. As indicated at the end of the proof of Theorem IX.6, we have that $\mathbb{E}(X_n) \asymp n^2\alpha_4(n)$. Next, for $x, y \in R'_n$, a picture shows that

$$\mathbb{P}(x, y \in \mathcal{P}_n) \leq \alpha_4^2(|x-y|/2)\alpha_4(2|x-y|, n/2)$$

since we need to have the four-arm event around x to distance $|x-y|/2$, the same for y, and the four-arm event in the annulus centered at $(x+y)/2$ from distance $2|x-y|$ to distance $n/2$ and finally these three events are independent. This is by quasi-multiplicity at most

$$O(1)\alpha_4^2(n)/\alpha_4(|x-y|,n)$$

and hence

$$\mathbb{E}[|X_n|^2] \le O(1)\alpha_4^2(n)\sum_{x,y}\frac{1}{\alpha_4(|x-y|,n)}.$$

Since, for a given x, there are at most $O(1)2^{2k}$ y's with $|x-y|\in[2^k,2^{k+1}]$, using quasi-multiplicity, the above sum is at most

$$O(1)n^2\alpha_4^2(n)\sum_{k=0}^{\log_2(n)}\frac{2^{2k}}{\alpha_4(2^k,n)}.$$

Using

$$\frac{1}{\alpha_4(r,R)} \le (R/r)^{2-\epsilon}$$

(this is the fact that the four-arm exponent is strictly less than 2), the sum becomes at most

$$O(1)n^{4-\epsilon}\alpha_4^2(n)\sum_{k=0}^{\log_2(n)}2^{k\epsilon}.$$

Since the last sum is at most $O(1)n^{\epsilon}$, we are done. □

In terms of the consequences for quantitative noise sensitivity, Theorem IX.6 implies the following corollary; see the exercises for similar implications. We state this only for the triangular lattice. An analogous result holds for $\mathbb{Z}^2$.

COROLLARY IX.8. *For $\mathbb{T}$, there exists $c>0$ so that if $\epsilon_n = 1/(n^2\alpha_4(n))$, then for all n,*

$$\mathbb{P}(f_n(\omega)\neq f_n(\omega_{\epsilon_n}))\ge c.$$

Note, importantly, this does *not* say that $f_n(\omega)$ and $f_n(\omega_{\epsilon_n})$ become asymptotically uncorrelated, only that they are not asymptotically completely correlated. To ensure that they are asymptotically uncorrelated is significantly more difficult and requires showing that "all" of the spectrum is near $n^{3/4}$. This much more difficult task is the subject of the next part.

Exercise sheet of Part IX

EXERCISE IX.1. Let $\{f_n\}$ be an arbitrary sequence of Boolean functions mapping into $\{\pm 1\}$ with corresponding spectral samples $\{\mathscr{S}_n\}$.
(i). Show that $\hat{\mathbb{P}}\big[0<|\mathscr{S}_n|\le A_n\big]\to 0$ implies that $\hat{\mathbb{E}}\big[(1-\epsilon_n)^{|\mathscr{S}_n|}I_{\mathscr{S}_n\neq\emptyset}\big]\to 0$ if $\epsilon_n A_n\to\infty$.
(ii). Show that $\hat{\mathbb{E}}\big[(1-\epsilon_n)^{|\mathscr{S}_n|}I_{\mathscr{S}_n\neq\emptyset}\big]\to 0$ implies that $\hat{\mathbb{P}}\big[0<|\mathscr{S}_n|\le A_n\big]\to 0$ if $\epsilon_n A_n = O(1)$.

EXERCISE IX.2. Let $\{f_n\}$ be an arbitrary sequence of Boolean functions mapping into $\{\pm 1\}$ with corresponding spectral samples $\{\mathscr{S}_n\}$.
(i). Show that $\mathbb{P}\big[f(\omega) \neq f(\omega_{\epsilon_n})\big] \to 0$ and $A_n \epsilon_n = \Omega(1)$ imply that $\hat{\mathbb{P}}\big[|\mathscr{S}_n| \geq A_n\big] \to 0$.
(ii). Show that $\hat{\mathbb{P}}\big[|\mathscr{S}_n| \geq A_n\big] \to 0$ and $A_n \epsilon_n = o(1)$ imply that $\mathbb{P}\big[f(\omega) \neq f(\omega_{\epsilon_n})\big] \to 0$.

EXERCISE IX.3. Prove Corollary IX.8.

EXERCISE IX.4. For the iterated 3-Majority sequence, recall that the total influence is n^α where $\alpha = 1 - \log 2/\log 3$. Show that for $\epsilon_n = 1/n^\alpha$, $\mathbb{P}(f_n(\omega) \neq f_n(\omega_{\epsilon_n}))$ does not tend to 0.

EXERCISE IX.5. Assume that $\{f_n\}$ is a sequence of monotone Boolean functions on n bits with total influence equal to $n^{1/2}$ up to constants. Show that the sequence cannot be noise sensitive. Is it necessarily noise stable as the Majority function is?

EXERCISE IX.6. Assume that $\{f_n\}$ is a sequence of monotone Boolean functions with mean 0 on n bits. Show that one cannot have noise sensitivity when using noise level $\epsilon_n = 1/n^{1/2}$.

EXERCISE IX.7. Show that $\mathcal{P}$ and $\mathscr{S}$ have the same 2-dimensional marginals using only Proposition IX.3 rather than Proposition IX.4.
Hint: It suffices to show that $\mathbb{P}(\{i,j\} \cap \mathcal{P} = \emptyset) = \hat{\mathbb{Q}}(\{i,j\} \cap \mathscr{S} = \emptyset)$.

EXERCISE IX.8. (Challenging problem) Do you expect that exercise IX.5 is sharp, meaning that, if $1/2$ is replaced by $\alpha < 1/2$, then one can find noise sensitive examples?

Part X. Sharp noise sensitivity of percolation

We will explain in this part the main ideas of the proof in [**GPS10**] that most of the "spectral mass" lies near $n^2\alpha_4(n) \approx n^{3/4+o(1)}$. This proof being rather long and involved, the content of this part will be far from a formal proof. Rather it should be considered as a (hopefully convincing) heuristic explanation of the main results, and possibly for the interested readers as a "reading guide" for the paper [**GPS10**].

Very briefly speaking, the idea behind the proof is to identify properties of the geometry of $\mathscr{S}_{f_n}$ which are reminiscent of a self-similar fractal structure. Ideally, $\mathscr{S}_{f_n}$ would behave like a spatial branching tree (or in other words a fractal percolation process), where distinct branches evolve independently of each other. This is conjecturally the case, but it turns out that it is very hard to control the dependency structure within $\mathscr{S}_{f_n}$. In [**GPS10**], only a tiny hint of spatial *independence* within $\mathscr{S}_{f_n}$ is proved. One of the main difficulties of the proof is to overcome the fact that one has very little independence to play with.

A substantial part of this part focuses on the much simpler case of fractal percolation. Indeed, this process can be seen as the simplest toy model for the spectral sample $\mathscr{S}_{f_n}$. Explaining the simplified proof adapted to this setting already enables us to convey some of the main ideas for handling $\mathscr{S}_{f_n}$.

1. State of the art and main statement

See Figure X.1 where we summarize what we have learned so far about the spectral sample $\mathscr{S}_{f_n}$ of a left to right crossing event f_n.

From this table, we see that the main question now is to prove that all the spectral mass indeed diverges at speed $n^2\alpha_4(n)$ which is $n^{3/4+o(1)}$ for the triangular lattice. This is the content of the following theorem.

THEOREM X.1 ([**GPS10**]).

$$\limsup_{n\to\infty} \hat{\mathbb{P}}\big[0 < |\mathscr{S}_{f_n}| < \lambda\, n^2\alpha_4(n)\big] \underset{\lambda\to 0}{\longrightarrow} 0\,.$$

On the triangular lattice $\mathbb{T}$, the rate of decay in λ is known explicitly. Namely:

THEOREM X.2 ([**GPS10**]). *On the triangular grid $\mathbb{T}$, the lower tail of $|\mathscr{S}_{f_n}|$ satisfies*

$$\limsup_{n\to\infty} \hat{\mathbb{P}}\big[0 < |\mathscr{S}_{f_n}| < \lambda\, \hat{\mathbb{E}}\big[|\mathscr{S}_{f_n}|\big]\big)\big] \underset{\lambda\to 0}{\asymp} \lambda^{2/3}\,.$$

This result deals with what one might call the "macroscopic" lower tail, i.e. with quantities which asymptotically are still of order $\hat{\mathbb{E}}\big[|\mathscr{S}_{f_n}|\big]$ (since λ remains fixed in the above lim sup). It turns out that in our later study of dynamical percolation in Part XI, we will need a sharp control on the full lower tail. This is the content of the following stronger theorem:

THEOREM X.3 ([**GPS10**]). *On $\mathbb{Z}^2$ and on the triangular grid $\mathbb{T}$, for all $1 \le r \le n$, one has*

$$\hat{\mathbb{P}}\big[0 < |\mathscr{S}_{f_n}| < r^2\alpha_4(r)\big] \asymp \frac{n^2}{r^2}\alpha_4(r,n)^2\,.$$

On the triangular grid, this translates into

$$\hat{\mathbb{P}}\big[0 < |\mathscr{S}_{f_n}| < u\big] \approx n^{-\frac{1}{2}} u^{\frac{2}{3}}\,,$$

	on the square lattice $\mathbb{Z}^2$	on the triangular lattice $\mathbb{T}$
The spectral mass diverges at polynomial speed	There is a positive exponent $\epsilon > 0$, s.t. $\hat{\mathbb{P}}\big[0 < \lvert\mathscr{S}_{f_n}\rvert < n^{\epsilon}\big] \to 0$	The same holds for all $\epsilon < 1/8$
Lower tail estimates	On both lattices, Theorem VIII.1 enables to obtain (non-sharp) lower tail estimates	
A positive fraction of the spectral mass lies "where it should"	There is some universal $c > 0$ s.t. $\hat{\mathbb{P}}\big[\lvert\mathscr{S}_{f_n}\rvert > c\,n^2\alpha_4(n)\big] > c$	$\hat{\mathbb{P}}\big[\lvert\mathscr{S}_{f_n}\rvert > c\,n^{3/4+o(1)}\big] > c$
May be summarized by the following picture	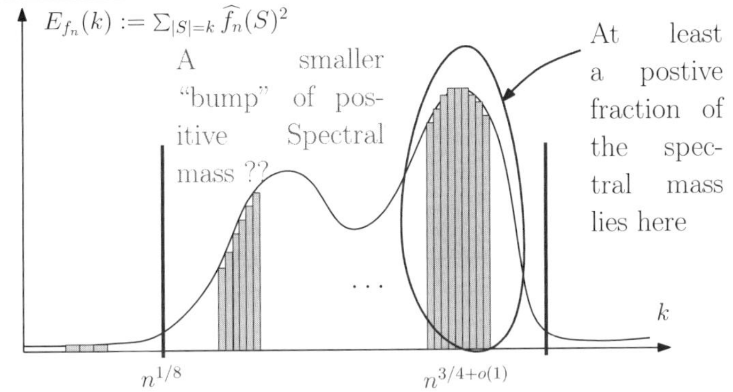	

FIGURE X.1. A summary of some of the results obtained so far for $\mathscr{S}_{f_n}$.

where we write $\approx$ to avoid relying on $o(1)$ terms in the exponents.

2. Overall strategy

In the above theorems, it is clear that we are mostly interested in the cardinality of $\mathscr{S}_{f_n}$. However, our strategy will consist in understanding as much as we can about the typical *geometry* of the random set $\mathscr{S}_{f_n}$ sampled according to the spectral probability measure $\hat{\mathbb{P}}_{f_n}$.

As we have seen so far, the random set $\mathscr{S}_{f_n}$ shares many properties with the set of pivotal points $\mathcal{P}_{f_n}$. A first possibility would be that they are asymptotically similar. After all, noise sensitivity is intimately related with pivotal points, so it is not unreasonable to hope for such a behavior. This scenario would be very convenient for us since the geometry of $\mathcal{P}_{f_n}$ is now well understood (at least on $\mathbb{T}$) thanks to the SLE processes. In particular, in the case of $\mathcal{P}_{f_n}$, one can "explore" $\mathcal{P}_{f_n}$ in a Markovian way by relying on exploration processes. Unfortunately, based on very convincing heuristics, it is conjectured that the scaling limits of $\frac{1}{n}\mathscr{S}_{f_n}$ and $\frac{1}{n}\mathcal{P}_{f_n}$ are singular random compact sets of the square. See Figure X.2 for a quick overview of the similarities and differences between these two random sets.

The conclusion of this table is that they indeed share many properties, but one cannot deduce lower tail estimates on $|\mathscr{S}_{f_n}|$ out of lower tail estimates on $|\mathcal{P}_{f_n}|$. Also, even worse, we will not be allowed to rely on spatial Markov properties for $\mathscr{S}_{f_n}$.

	Pivotal set $\mathcal{P}_{f_n}$	Spectral set $\mathscr{S}_{f_n}$
First moment	$\mathbb{E}\big[\lvert\mathcal{P}_{f_n}\rvert\big]$ =	$\hat{\mathbb{E}}\big[\lvert\mathscr{S}_{f_n}\rvert\big]$
Second moment	$\mathbb{E}\big[\lvert\mathcal{P}_{f_n}\rvert^2\big]$ =	$\hat{\mathbb{E}}\big[\lvert\mathscr{S}_{f_n}\rvert^2\big]$
Higher moments ($k \geq 3$)	In general, they differ !	
Methods for sampling these random sets	Easy (and fast) using two exploration paths: 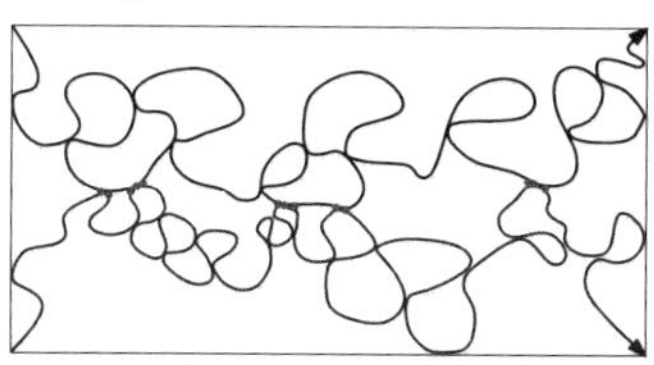	The spectral sample $\mathscr{S}_{f_n}$ is much harder to sample. In fact, the only known way to proceed is to compute the weights $\hat{f}_n(S)^2$, one at a time ...
Spatial correlation structure	Distant regions in $\mathcal{P}_{f_n}$ behave more or less independently of each other. Furthermore, one can use the very convenient spatial Markov property due to the i.i.d structure of the percolation picture.	? Much of the picture here remains unclear
Lower Tail behavior	$\mathbb{P}\big[\lvert\mathcal{P}_{f_n}\rvert = 1\big] \asymp n^{-\frac{11}{12}}$	$\hat{\mathbb{P}}\big[\lvert\mathscr{S}_{f_n}\rvert = 1\big] \asymp n^{-\frac{1}{2}}$

FIGURE X.2. Similarities and differences between $\mathscr{S}_{f_n}$ and $\mathcal{P}_{f_n}$.

However, even though $\mathcal{P}_{f_n}$ and $\mathscr{S}_{f_n}$ differ in many ways, they share at least one essential property: a seemingly *self-similar fractal behavior*. The main strategy in [**GPS10**] to control the lower-tail behavior of $|\mathscr{S}_{f_n}|$ is to prove that in some very weak sense, $\mathscr{S}_{f_n}$ behaves like the simplest model among self-similar fractal processes in $[0,n]^2$: i.e. a super-critical spatial Galton-Watson tree embedded in $[0,n]^2$, also called a *fractal percolation process*. The lower tail of this very simple toy model will be investigated in detail in the next section with a technique which will be suitable for $\mathscr{S}_{f_n}$. The main difficulty which arises in this program is the lack of knowledge of the independency structure within $\mathscr{S}_{f_n}$. In other words, when we try to compare

$\mathscr{S}_{f_n}$ with a fractal percolation process, the self-similarity already requires some work, but the hardest part is to deal with the fact that distinct "branches" (or rather their analogues) are not known to behave even slightly independently of each other. We will discuss these issues in Section 4 but will not give a complete proof.

3. Toy model: the case of fractal percolation

As we explained above, our main strategy is to exploit the fact that $\mathscr{S}_{f_n}$ has a certain self-similar fractal structure. Along this section, we will consider the simplest case of such a self-similar fractal object: namely *fractal percolation*, and we will detail in this simple setting what our later strategy will be. Deliberately, this strategy will not be optimal in this simplified case. In particular, we will not rely on the martingale techniques that one can use with fractal percolation or Galton-Watson trees, since such methods would not be available for our spectral sample $\mathscr{S}_{f_n}$.

3.1. Definition of the model and first properties. To make the analogy with $\mathscr{S}_{f_n}$ easier let

$$n := 2^h\,, h \geq 1\,,$$

and let's fix a parameter $p \in (0,1)$.

Now, *fractal percolation* on $[0,n]^2$ is defined inductively as follows: divide $[0,2^h]^2$ into 4 squares and retain each of them independently with probability p. Let $\mathcal{T}^1$ be the union of the retained 2^{h-1}-squares. The second-level tree $\mathcal{T}^2$ is obtained by reiterating the same procedure independently for each 2^{h-1}-square in $\mathcal{T}^1$. Continuing in the same fashion all the way to the squares of unit size, one obtains $\mathcal{T}_n = \mathcal{T} := \mathcal{T}^h$ which is a random subset of $[0,n]^2$. See [**LyP11**] for more on the definition of *fractal percolation*. See also Figure X.3 for an example of $\mathcal{T}^5$.

REMARK X.1. We thus introduced two different notations for the same random set ($\mathcal{T}_{n=2^h} \equiv \mathcal{T}^h$). The reason for this is that on the one hand the notation $\mathcal{T}_n$ defined on $[0,n]^2 = [0,2^h]^2$ makes the analogy with $\mathscr{S}_{f_n}$ (also defined on $[0,n]^2$) easier, while on the other hand inductive proofs will be more convenient with the notation $\mathcal{T}^h$.

In order to have a supercritical Galton-Watson tree, one has to choose $p \in (1/4,1)$. Furthermore, one can easily check the following easy proposition.

PROPOSITION X.4. *Let $p \in (1/4,1)$. Then*

$$\mathbb{E}\big[|\mathcal{T}_n|\big] = n^2 p^h = n^{2+\log_2 p}\,,$$

and

$$\mathbb{E}\big[|\mathcal{T}_n|^2\big] \leq O(1)\mathbb{E}\big[|\mathcal{T}_n|\big]^2\,.$$

In particular, by the second moment method (e.g. the Paley-Zygmund inequality), with positive probability, $\mathcal{T}_n$ is of order $n^{2+\log_2 p}$.

Let

$$\alpha := 2 + \log_2 p\,.$$

This parameter α corresponds to the "fractal dimension" of $\mathcal{T}_n$. To make the analogy with $\mathscr{S}_{f_n}$ even clearer, one could choose p in such a way that $\alpha = 2 + \log_2 p = 3/4$, but we will not need to.

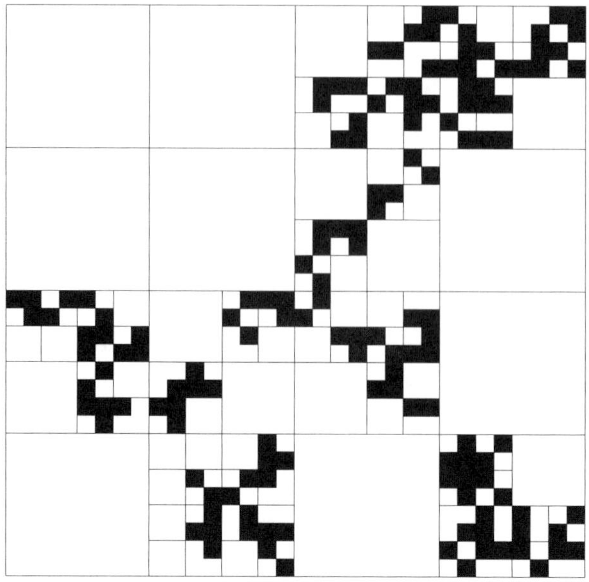

FIGURE X.3. A realization of a fractal percolation $\mathcal{T}_{2^5} = \mathcal{T}^5$

The above proposition implies that on the event $\mathcal{T}_n \neq \emptyset$, with positive conditional probability $|\mathcal{T}_n|$ is large (of order n^α). This is the exact analogue of Theorem IX.6 for the spectral sample $\mathscr{S}_{f_n}$.

Let us first analyze what would be the analogue of Theorem X.1 in the case of our toy model $\mathcal{T}_n$. We have the following.

PROPOSITION X.5.

$$\limsup_{n\to\infty} \mathbb{P}\big[0 < |\mathcal{T}_n| < \lambda\, n^\alpha)\big] \underset{\lambda\to 0}{\longrightarrow} 0\,.$$

REMARK X.2. If one could rely on martingale techniques, then this proposition is a corollary of standard results. Indeed, as is well-known

$$M_i := \frac{|\mathcal{T}^i|}{(4p)^i}\,,$$

is a positive martingale. Therefore it converges, as $n \to \infty$, to a non-negative random variable $W \geq 0$. Furthermore, the conditions of the Kesten-Stigum Theorem are fulfilled (see for example Section 5.1 in [**LyP11**]) and therefore W is positive on the event that there is no extinction. This implies the above proposition.

As we claimed above, we will intentionally follow a more hands-on approach in this section which will be more suitable to the random set $\mathscr{S}_{f_n}$ which we have in mind. Furthermore this approach will have the great advantage to provide the following much more precise result, which is the analogue of Theorem X.3 for $\mathcal{T}_n$.

PROPOSITION X.6. *For any* $1 \leq r \leq n$,

$$\mathbb{P}\big[0 < |\mathcal{T}_n| < r^\alpha\big] \asymp \big(\frac{r}{n}\big)^{\log_2 1/\mu}\,,$$

where μ *is an explicit constant in* $(0,1)$ *computed in Exercise X.2.*

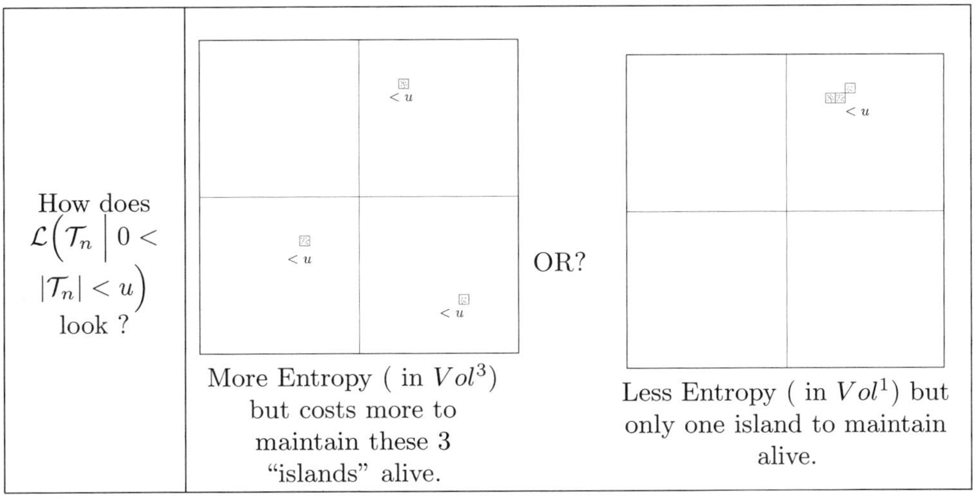

FIGURE X.4. Entropy v.s. Clustering effect

3.2. Strategy and heuristics. Letting $u \ll n^{\alpha}$, we wish to estimate $\mathbb{P}\big[0 < |\mathcal{T}_n| < u\big]$. Even though we are only interested in the size of $\mathcal{T}_n$, we will try to estimate this quantity by understanding the *geometry* of the conditional set:

$$\mathcal{T}_n^{|u} := \mathcal{L}\left(\mathcal{T}_n \,\middle|\, 0 < |\mathcal{T}_n| < u\right).$$

The first natural question to ask is whether this conditional random set is typically *localized* or not. See Figure X.4.

Intuitively, it is quite clear that the set $\mathcal{T}_n$ conditioned to be very small will tend to be localized. So it is the picture on the right in Figure X.4 which is more likely. This would deserve a proof of course, but we will come back to this later. The fact that it should look more and more localized tells us that as one shrinks u, this should make our conditional $\mathcal{T}_n^{|u}$ more and more singular with respect to the unconditional one. But how much localization should we see? This is again fairly easy to answer, at least on the intuitive level. Indeed, $\mathcal{T}_n^{|u}$ should tend to localize until it reaches a certain mesoscopic scale r such that $1 \ll r \ll n$. One can compute how much it costs to maintain a single branch (or $O(1)$ branches) alive until scale r, but once this is achieved, one should let the system evolve in a "natural" way. In particular, once the tree survives all the way to a mesoscopic square of size r, it will (by the second moment method) produce $\Omega(r^{\alpha})$ leaves there with positive probability.

To summarize, typically $\mathcal{T}_n^{|u}$ will maintain $O(1)$ many branches alive at scale $1 \ll r \ll n$, and then it will let the branching structure evolve in a basically unconditional way. The intermediate scale r is chosen so that $r^{\alpha} \asymp u$.

DEFINITION X.1. *If $1 \le r \le n = 2^h$ is such that $r = 2^l, 0 \le l \le h$, let $\mathcal{T}_{(r)}$ denote the set of branches that were still alive at scale $r = 2^l$ in the iterative construction of $\mathcal{T}_n$. In other words, $\mathcal{T}_{(r)} \equiv \mathcal{T}^{h-l}$ and $\mathcal{T}_n \subset \bigcup \mathcal{T}_{(r)}$. This random set $\mathcal{T}_{(r)}$ will be the analogue of the "r-smoothing" $\mathscr{S}_{(r)}$ of the spectral sample $\mathscr{S}_{f_n}$ defined later in Definition X.2.*

Returning to our problem, the above heuristics say that one expects to have for any $1 \ll u \ll n^\alpha$.

$$\begin{aligned}\mathbb{P}\big[0 < |\mathcal{T}_n| < u\big] &\asymp \mathbb{P}\big[0 < |\mathcal{T}_{(r)}| \le O(1)\big] \\ &\asymp \mathbb{P}\big[|\mathcal{T}_{(r)}| = 1\big]\,,\end{aligned}$$

where r is a dyadic integer chosen such that $r^\alpha \asymp u$. Or in other words, we expect that

$$\mathbb{P}\big[0 < |\mathcal{T}_n| < r^\alpha\big] \asymp \mathbb{P}\big[|\mathcal{T}_{(r)}| = 1\big]\,. \tag{X.1}$$

In the next subsection, we briefly explain how this heuristic can be implemented into a proof in the case of the tree $\mathcal{T}_n$ in a way which will be suitable to the study of $\mathscr{S}_{f_n}$. We will only skim through the main ideas for this tree case.

3.3. Setup of a proof for $\mathcal{T}_n$. Motivated by the above heuristics, we divide our system into two scales: above and below the mesoscopic scale r. One can write the lower tail event as follows (let $1 \ll r \ll n$):

$$\mathbb{P}\big[0 < |\mathcal{T}_n| < r^\alpha\big] = \sum_{k\ge 1} \mathbb{P}\big[|\mathcal{T}_{(r)}| = k\big]\, \mathbb{P}\big[0 < |\mathcal{T}_n| < r^\alpha \bigm| |\mathcal{T}_{(r)}| = k\big]\,. \tag{X.2}$$

It is not hard to estimate the second term $\mathbb{P}\big[0 < |\mathcal{T}_n| < r^\alpha \bigm| |\mathcal{T}_{(r)}| = k\big]$. Indeed, in this term we are conditioning on having exactly k branches alive at scale r. Independently of where they are, "below" r, these k branches evolve independently of each other. Furthermore, by the second moment method, there is a universal constant $c > 0$ such that each of them exceeds the fatal amount of r^α leaves with probability at least c (note that in the opposite direction, each branch could also go extinct with positive probability). This implies that

$$\mathbb{P}\big[0 < |\mathcal{T}_n| < r^\alpha \bigm| |\mathcal{T}_{(r)}| = k\big] \le (1-c)^k\,.$$

REMARK X.3. Note that one makes heavy use of the independence structure within $\mathcal{T}_n$ here. This aspect is much more nontrivial for the spectral sample $\mathscr{S}_{f_n}$. Fortunately it turns out, and this is a key fact, that in [**GPS10**] one can prove a weak independence statement which in some sense makes it possible to follow this route.

We are left with the following upper bound:

$$\mathbb{P}\big[0 < |\mathcal{T}_n| < r^\alpha\big] \le \sum_{k\ge 1} \mathbb{P}\big[|\mathcal{T}_{(r)}| = k\big](1-c)^k\,. \tag{X.3}$$

In order to prove our goal of (X.1), by exploiting the exponential decay given by $(1-c)^k$ (which followed from independence), it is enough to prove the following bound on the mesoscopic behavior of $\mathcal{T}$:

LEMMA X.7. *There is a* **sub-exponential** *function $k \mapsto g(k)$ such that for all $1 \le r \le n$,*

$$\mathbb{P}\big[|\mathcal{T}_{(r)}| = k\big] \le g(k)\, \mathbb{P}\big[|\mathcal{T}_{(r)}| = 1\big]\,.$$

Notice as we did in Definition X.1 that since $\mathcal{T}_{(r)}$ has the same law as $\mathcal{T}^{h-l}$, this is a purely Galton-Watson tree type of question.

The big advantage of our strategy so far is that initially we were looking for a sharp control on $\mathbb{P}\big[0 < |\mathcal{T}_n| < u\big]$ and now, using this "two-scales" argument, it only remains to prove a crude upper bound on the lower tail of $|\mathcal{T}_{(r)}|$. By scale invariance this is nothing else than obtaining a crude upper bound on the lower tail of $|\mathcal{T}_n|$. Hence this division into two scales greatly simplified our task.

3.4. Sub-exponential estimate on the lower-tail (Lemma X.7). The first step towards proving and understanding Lemma X.7 is to understand the term $\mathbb{P}\big[|\mathcal{T}_{(r)}| = 1\big]$. From now on, it will be easier to work with the "dyadic" notations instead, i.e. with $\mathcal{T}^i \equiv \mathcal{T}_{2^i}$ (see remark X.1). With these notations, the first step is equivalent to understanding the probabilities $p_i := \mathbb{P}\big[|\mathcal{T}^i| = 1\big]$. This aspect of the problem is very specific to the case of Galton-Watson trees and gives very little insight into the later study of the spectrum $\mathscr{S}_{f_n}$. Therefore we postpone the details to Exercise X.2. The conclusion of this (straightforward) exercise is that p_i behaves as $i \to \infty$ like

$$p_i \sim c\,\mu^i\,,$$

for an explicit exponent $\mu \in (0,1)$ (see Exercise X.2). In particular, in order to prove Proposition X.6, it is now enough to find a sub-exponential function $k \mapsto g(k)$ such that for any $i, k \geq 1$,

$$\mathbb{P}\big[|\mathcal{T}^i| = k\big] \leq g(k)\mu^i\,. \tag{X.4}$$

More precisely, we will prove the following lemma.

Lemma X.8. *Let $g(k) := 2^{\theta \log_2^2(k+2)}$, where θ is a fixed constant to be chosen later. Then for all $i, k \geq 1$, one has*

$$\mathbb{P}\big[|\mathcal{T}^i| = k\big] \leq g(k)\,\mu^i\,. \tag{X.5}$$

We provide the proof of this lemma here, since it can be seen as a "toy proof" of the corresponding sub-exponential estimate needed for the r-smoothed spectral samples $\mathscr{S}_{(r)}$, stated in the coming Theorem X.13. The proof of this latter theorem shares some similarities with the proof below but is much more technical since in the case of $\mathscr{S}_{(r)}$ one has to deal with a more complex structure than the branching structure of a Galton-Watson tree.

Proof. We proceed by double induction. Let $k \geq 2$ be fixed and assume that equation (X.5) is already satisfied for all pair (i', k') such that $k' < k$. Based on this assumption, let us prove by induction on i that all pairs (i, k) satisfy equation (X.5) as well.

First of all, if i is small enough, this is obvious by the definition of $g(k)$. Let

$$J = J_k := \sup\{i \geq 1 : g(k)\mu^i > 10\}\,.$$

Then, it is clear that equation (X.5) is satisfied for all (i, k) with $i \leq J_k$. Now let $i > J_k$.

If $\mathcal{T}^i$ is such that $|\mathcal{T}^i| = k \geq 1$, let $L = L(\mathcal{T}^i) \geq 0$ be the largest integer such that $\mathcal{T}^i$ intersects only one square of size 2^{i-L}. This means that below scale 2^{i-L}, the tree $\mathcal{T}^i$ splits into at least 2 live branches in distinct dyadic squares of size 2^{i-L-1}. Let $d \in \{2,3,4\}$ be the number of such live branches. By decomposing on the value of L, and using the above assumption, we get

$$\mathbb{P}\big[|\mathcal{T}^i| = k\big] \le \mathbb{P}\big[L(\mathcal{T}^i) > i - J_k\big] +$$

$$\frac{1}{1-q}\sum_{l=0}^{i-J_k} \mathbb{P}\big[L(\mathcal{T}^i) = l\big] \sum_{d=2}^{4} \binom{4}{d} (\mu^{i-l-1})^d \sum_{\substack{(k_j)_{1\le j\le d} \\ k_j \ge 1, \sum k_j = k}} \prod_j g(k_j)$$

where q is the probability that our Galton-Watson tree goes extinct.

Let us first estimate what $\mathbb{P}\big[L(\mathcal{T}^i) \ge m\big]$ is for $m \ge 0$. If $m \ge 1$, this means that among the 2^{2m} dyadic squares of size 2^{i-m}, only one will remain alive all the way to scale 1. Yet, it might be that some other such squares are still alive at scale 2^{i-m} but will go extinct by the time they reach scale 1. Let $p_{m,b}$ be the probability that the process $\mathcal{T}^{m+b}$, which lives in $[0, 2^{m+b}]^2$, is entirely contained in a dyadic square of size 2^b. With such notations, one has

$$\mathbb{P}\big[L(\mathcal{T}^i) \ge m\big] = p_{m,i-m}\,.$$

Furthermore, if $i = m$, one has $p_{i,0} = p_i \sim c\mu^i$. It is not hard to prove (see Exercise X.2) the following lemma.

LEMMA X.9. *For any value of $m, b \ge 0$, one has*

$$p_{m,b} \le \mu^m\,.$$

In particular, one has a universal upper bound in $b \ge 0$.

It follows from the lemma that $\mathbb{P}\big[L(\mathcal{T}^i) = l\big] \le \mathbb{P}\big[L(\mathcal{T}^i) \ge l\big] \le \mu^l$ and

$$\mathbb{P}\big[L(\mathcal{T}^i) > i - J_k\big] \le \mu^{i-J_k} \tag{X.6}$$

$$\le \frac{1}{10} g(k)\,\mu^i \text{ by the definition of } J_k\,. \tag{X.7}$$

This gives us that for some constant C

$$\mathbb{P}\big[|\mathcal{T}^i| = k\big] \le \frac{\mu^i}{10} g(k) + C\sum_{l=0}^{i-J_k} \mu^l \sum_{d=2}^{4} (\mu^{i-l})^d \sum_{\substack{(k_j)_{1\le j\le d} \\ k_j \ge 1, \sum k_j = k}} \prod_j g(k_j)$$

$$= \frac{\mu^i}{10} g(k) + C\mu^i \sum_{d=2}^{4} \sum_{l=0}^{i-J_k} (\mu^{i-l})^{d-1} \sum_{\substack{(k_j)_{1\le j\le d} \\ k_j \ge 1, \sum k_j = k}} \prod_j g(k_j)\,.$$

Let us deal with the $d = 2$ sum (the contributions coming from $d > 2$ being even smaller). By concavity of $k \mapsto \theta \log_2^2(k+2)$, one obtains that for any (k_1, k_2) such that $k_1 + k_2 = k$: $g(k_1)g(k_2) \le g(k/2)^2$. Since there are at most k^2 such pairs, this gives us the following bound on the $d = 2$ sum.

$$\sum_{l=0}^{i-J_k} (\mu^{i-l})^{2-1} \sum_{\substack{(k_j)_{1\le j\le 2} \\ k_j \ge 1, \sum k_j = k}} \prod_j g(k_j) \le \sum_{l=0}^{i-J_k} \mu^{i-l} k^2 g(k/2)^2$$

$$\le \frac{1}{1-\mu}\, \mu^{J_k}\, k^2\, g(k/2)^2$$

$$\le 10 \frac{1}{1-\mu}\, k^2\, g(k/2)^2\, (\mu g(k))^{-1}\,,$$

by definition of J_k.

Now, some easy analysis implies that if one chooses the constant $\theta > 0$ large enough, then for any $k \ge 2$, one has $C10\frac{1}{1-\mu}k^2\, g(k/2)^2\, (\mu g(k))^{-1} \le \frac{1}{10} g(k)$. Altogether (and taking into consideration the $d > 2$ contributions), this implies that

$$\mathbb{P}\big[|\mathcal{T}^i| = k\big] \le \frac{2}{5} g(k)\mu^i \le g(k)\mu^i\,,$$

as desired. □

REMARK X.4. Recall the initial question from Figure X.4 which asked whether the clustering effect wins over the entropy effect or not. This question enabled us to motivate the setup of the proof but in the end, we did not specifically address it. Notice that the above proof in fact solves the problem (see Exercise X.3).

4. Back to the spectrum: an exposition of the proof

4.1. Heuristic explanation. Let us now apply the strategy we developed for $\mathcal{T}_n$ to the case of the spectral sample $\mathscr{S}_{f_n}$. Our goal is to prove Theorem X.3 (of which Theorems X.1 and X.2 are straightforward corollaries). Let $\mathscr{S}_{f_n} \subset [0, n]^2$ be our spectral sample. We have seen (Theorem IX.6) that with positive probability $|\mathscr{S}_{f_n}| \asymp n^2 \alpha_4(n)$. For all $1 < u < n^2\alpha_4(n)$, we wish to understand the probability $\hat{\mathbb{P}}\big[0 < |\mathscr{S}_{f_n}| < u\big]$. Following the notations we used for $\mathcal{T}_n$, let $\mathscr{S}_{f_n}^{|u}$ be the spectral sample conditioned on the event $\{0 < |\mathscr{S}_{f_n}| < u\}$.

Question: How does $\mathscr{S}_{f_n}^{|u}$ typically look?

To answer this question, one has to understand whether $\mathscr{S}_{f_n}^{|u}$ tends to be *localized* or not. Recall from Figure X.4 the illustration of the competition between entropy and clustering effects in the case of $\mathcal{T}_n$. The same figure applies to the spectral sample $\mathscr{S}_{f_n}$. We will later state a **clustering lemma** (Lemma X.14) which will strongly support the localized behavior described in the next proposition.

Therefore we are guessing that our conditional set $\mathscr{S}_{f_n}^{|u}$ will tend to localize into $O(1)$ many squares of a certain scale r and will have a "normal" size within these r-squares. It remains to understand what this mesoscopic scale r as a function of u is.

By "scale invariance", one expects that if $\mathscr{S}_{f_n}$ is conditioned to live in a square of size r, then $|\mathscr{S}_{f_n}|$ will be of order $r^2\alpha_4(r)$ with positive conditional probability. More precisely, the following lemma will be proved in Problem X.6.

LEMMA X.10. *There is a universal* $c \in (0,1)$ *such that for any* n *and for any* r*-square* $B \subset [n/4, 3n/4]^2$ *in the "bulk" of* $[0,n]^2$*, one has*

$$\hat{\mathbb{P}}\Big[\frac{|\mathscr{S}_{f_n}|}{r^2\alpha_4(r)} \in (c, 1/c) \mid \mathscr{S}_{f_n} \neq \emptyset \text{ and } \mathscr{S}_{f_n} \subset B\Big] > c\,. \tag{X.8}$$

In fact this lemma holds uniformly in the position of the r-square B inside $[0,n]^2$, but we will not discuss this here.

What this lemma tells us is that for any $1 < u < n^2\alpha_4(n)$, if one chooses $r = r_u$ in such a way that $r^2\alpha_4(r) \asymp u$, then we expect to have the following estimate:

$$\begin{aligned}\hat{\mathbb{P}}\big[0 < |\mathscr{S}_{f_n}| < u\big] &\asymp \hat{\mathbb{P}}\big[\mathscr{S}_{f_n} \text{ intersects } O(1) \text{ } r\text{-square in } [0,n]^2\big]\\ &\asymp \hat{\mathbb{P}}\big[\mathscr{S}_{f_n} \text{ intersects a single } r\text{-square in } [0,n]^2\big]\end{aligned}$$

At this point, let us introduce a concept which will be very helpful in what follows.

DEFINITION X.2 ("r-smoothing"). *Let* $1 \le r \le n$*. Consider the domain* $[0,n]^2$ *and divide it into a grid of squares of edge-length* r*. (If* $1 \ll r \ll n$*, one can view this grid as a mesoscopic grid).*

If n *is not divisible by* r*, write* $n = mr + q$ *and consider the grid of* r*-squares covering* $[0,(m+1)r]^2$*.*

Now, for each subset $S \subset [0,n]^2$*, define* $S_{(r)}$ *to be the set of* $r \times r$ *squares in the above grid which intersect* S*. In particular* $|S_{(r)}|$ *will correspond to the number of such* r*-squares which intersect* S*. With a slight abuse of notation,* $S_{(r)}$ *will sometimes also denote the actual subset of* $[0,n]^2$ *consisting of the union of these* r*-squares.*

One can view the application $S \mapsto S_{(r)}$ *as an* r-**smoothing** *since all the details below the scale* r *are lost.*

REMARK X.5. Note that in Definition X.1, we relied on a slightly different notion of "r-smoothing" since in that case, $\mathcal{T}_{(r)}$ could also include r-branches which might go extinct by the time they reached scale one. The advantage of this choice was that there was an exact scale-invariance from $\mathcal{T}$ to $\mathcal{T}_{(r)}$ while in the case of $\mathscr{S}_{f_n}$, there is no such exact scale-invariance from $\mathscr{S}$ to $\mathscr{S}_{(r)}$.

With these notations, the above discussion leads us to believe that the following proposition should hold.

PROPOSITION X.11. *For all* $1 \le r \le n$*, one has*

$$\hat{\mathbb{P}}\big[0 < |\mathscr{S}_{f_n}| < r^2\alpha_4(r)\big] \asymp \hat{\mathbb{P}}_{f_n}\big[|\mathscr{S}_{(r)}| = 1\big]\,.$$

Before explaining the setup used in [**GPS10**] to prove such a result, let us check that it indeed implies Theorem X.3. By neglecting the boundary issues, one has

$$\hat{\mathbb{P}}_{f_n}\big[|\mathscr{S}_{(r)}| = 1\big] \asymp \sum_{\substack{r\text{-squares}\\ B \subset [n/4,3n/4]^2}} \hat{\mathbb{P}}\big[\mathscr{S}_{f_n} \neq \emptyset \text{ and } \mathscr{S}_{f_n} \subset B\big]\,. \tag{X.9}$$

There are $O(\frac{n^2}{r^2})$ such B squares, and for each of these, one can check (see Exercise X.5) that

$$\hat{\mathbb{P}}\big[\mathscr{S}_{f_n} \neq \emptyset \text{ and } \mathscr{S}_{f_n} \subset B\big] \asymp \alpha_4(r,n)^2\,.$$

Therefore, Proposition X.11 indeed implies Theorem X.3.

4.2. Setup and organization of the proof of Proposition X.11. To start with, assume we knew that disjoint regions in the spectral sample $\mathscr{S}_{f_n}$ behave more or less independently of each other in the following (vague) sense. For any $k \geq 1$ and any mesoscopic scale $1 \leq r \leq n$, if one conditions on $\mathscr{S}_{(r)}$ to be equal to $B_1 \cup \cdots \cup B_k$ for k disjoint r-squares, then the conditional law of $\mathscr{S}_{|\bigcup B_i}$ should be "similar" to an independent product of $\mathcal{L}\big[\mathscr{S}_{|B_i} \mid \mathscr{S} \cap B_i \neq \emptyset\big]$, $i \in \{1, \ldots, k\}$. Similarly as in the tree case (where the analogous property for $\mathcal{T}_n$ was an exact independence factorization), and assuming that the above comparison with an independent product could be made quantitative, this would potentially imply the following upper bound for a certain absolute constant $c > 0$:

$$\hat{\mathbb{P}}\big[0 < |\mathscr{S}_{f_n}| < r^2 \alpha_4(r)\big] \leq \sum_{k \geq 1} \hat{\mathbb{P}}\big[|\mathscr{S}_{(r)}| = k\big]\, (1-c)^k\,. \tag{X.10}$$

This means that even if one managed to obtain a good control on the dependency structure within $\mathscr{S}_{f_n}$ (in the above sense), one would still need to have a good estimate on $\hat{\mathbb{P}}\big[|\mathscr{S}_{(r)}| = k\big]$ in order to deduce Proposition X.11. This part of the program is achieved in [**GPS10**] without requiring any information on the dependency structure of $\mathscr{S}_{f_n}$. More precisely, the following result is proved:

THEOREM X.12 ([**GPS10**]). *There is a sub-exponential function $g \mapsto g(k)$, such that for any $1 \leq r \leq n$ and any $k \geq 1$,*

$$\hat{\mathbb{P}}\big[|\mathscr{S}_{(r)}| = k\big] \;\leq\; g(k)\, \hat{\mathbb{P}}\big[|\mathscr{S}_{(r)}| = 1\big]\,.$$

The proof of this result will be described briefly in the next subsection.

One can now describe how the proof of Theorem X.3 is organized in [**GPS10**]. It is divided into three main parts:

(1) The first part deals with proving the multi-scale sub-exponential bound on the lower-tail of $|\mathscr{S}_{(r)}|$ given by Theorem X.12.
(2) The second part consists in proving as much as we can on the dependency structure of $\mathscr{S}_{f_n}$. Unfortunately here, it seems to be very challenging to achieve a good understanding of all the "independence" that should be present within $\mathscr{S}_{f_n}$. The only hint of independence which was finally proved in [**GPS10**] is a very *weak* one (see subsection 4.4). In particular, it is too weak to readily imply a bound like (X.10).
(3) Since disjoint regions of the spectral sample $\mathscr{S}_{f_n}$ are not known to behave independently of each other, the third part of the proof consists in adapting the setup we used for the tree (where distinct branches evolve exactly independently of each other) into a setup where the weak hint of independence obtained in the second part of the program turns out to be enough to imply the bound given by (X.10) for an appropriate absolute constant $c > 0$. This final part of the proof will be discussed in subsection 4.5.

The next three subsections will be devoted to each of these 3 parts of the program.

4.3. Some words about the sub-exponential bound on the lower tail of $\mathscr{S}_{(r)}$. In this subsection, we turn our attention to the proof of the first part of the program, i.e. on Theorem X.12. In fact, as in the case of $\mathcal{T}_n$, the following more explicit statement is proved in [**GPS10**].

THEOREM X.13 ([**GPS10**]). *There exists an absolute constant $\theta > 0$ such that for any $1 \le r \le n$ and any $k \ge 1$,*

$$\hat{\mathbb{P}}\big[|\mathscr{S}_{(r)}| = k\big] \le 2^{\theta \log_2^2(k+2)}\, \hat{\mathbb{P}}\big[|\mathscr{S}_{(r)}| = 1\big]\,.$$

REMARK X.6. Note that the theorems from [**BKS99**] on the noise sensitivity of percolation are all particular cases ($r = 1$) of this intermediate result in [**GPS10**].

The main idea in the proof of this theorem is in some sense to assign a *tree structure* to each possible set $\mathscr{S}_{(r)}$. The advantage of working with a tree structure is that it is easier to work with inductive arguments. In fact, once a mapping $\mathscr{S}_{(r)} \mapsto$ "tree structure" has been designed, the proof proceeds similarly as in the case of $\mathcal{T}_{(r)}$ by double induction on the depth of the tree as well as on $k \ge 1$. Of course, this mapping is a delicate affair: it has to be designed in an "efficient" way so that it can compete against entropy effects caused by the exponential growth of the number of tree structures.

We will not give the details of how to define such a mapping, but let us describe informally how it works. More specifically than a tree structure, we will in fact assign an *annulus structure* to each set $\mathscr{S}_{(r)}$.

DEFINITION X.3. *Let $\mathcal{A}$ be a finite collection of disjoint (topological) annuli in the plane. We call this an* **annulus structure**. *Furthermore, we will say that a set $S \subset \mathbb{R}^2$ is* **compatible** *with $\mathcal{A}$ (or vice versa) if it is contained in $\mathbb{R}^2 \setminus \bigcup \mathcal{A}$ and intersects the inner disk of each annulus in $\mathcal{A}$. Note that it is allowed that one annulus is "inside" of another annulus.*

The mapping procedure in [**GPS10**] assigns to each $\mathscr{S}_{(r)}$ an annulus structure $\mathcal{A} \subset [0, n]^2$ in such a way that it is compatible with $\mathscr{S}_{(r)}$. See Figure X.5 for an example. Again, we will not describe this procedure nor discuss the obvious boundary issues which arise here, but let us state a crucial property satisfied by annulus structures.

LEMMA X.14 (**clustering Lemma**). *If $\mathcal{A}$ is an annulus structure contained in $[0, n]^2$, then*

$$\hat{\mathbb{P}}\big[\mathscr{S}_{(r)} \text{ is compatible with } \mathcal{A}\big] \le \prod_{A \in \mathcal{A}} \alpha_4(A)^2\,,$$

where $\alpha_4(A)$ denotes the probability of having a four-arm event in the annulus A.

REMARK X.7. To deal with boundary issues, one would also need to incorporate within our annulus structures half-annuli centered on the boundaries as well as quarter disks centered at the corners of $[0, n]^2$.

Let us briefly comment on this lemma.

- First of all, its proof is an elegant combination of linear algebra and percolation. It is a short and relatively elementary argument. See Lemma 4.3 in [**GPS10**].
- It is very powerful in dealing with the possible non-injectivity of the mapping $\mathcal{S}_{(r)} \mapsto \mathcal{A}$. Indeed, while describing the setup above, one might have

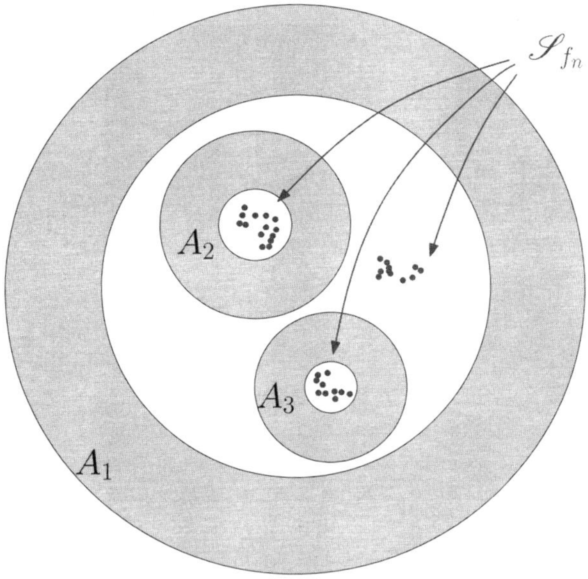

FIGURE X.5. An example of an **annulus structure** $\mathcal{A} := \{A_1, A_2, A_3\}$ compatible with a spectral sample $\mathscr{S}_{f_n}$.

objected that if the mapping were not injective enough, then the cardinality of the "fibers" above each annulus structure would have to be taken into account as well. Fortunately, the above lemma reads as follows: for any fixed annulus structure $\mathcal{A}$,

$$\sum_{\mathscr{S}_{(r)}:\,\mathscr{S}_{(r)}\mapsto\mathcal{A}} \hat{\mathbb{P}}\big[\mathscr{S}_{(r)}\big] \le \hat{\mathbb{P}}\big[\mathscr{S}_{(r)} \text{ is compatible with } \mathcal{A}\big] \le \prod_{A\in\mathcal{A}} \alpha_4(A)^2 .$$

- Another essential feature of this lemma is that it quantifies very efficiently the fact that the clustering effect wins over the entropy effect in the sense of Figure X.4. The mechanism responsible for this is that the probability of the *four-arm* event squared has an exponent (equal to $5/2$ on $\mathbb{T}$) larger than the *volume* exponent equal to 2. To illustrate this, let us analyze the situation when $k = 2$ (still neglecting boundary issues). The probability that the spectrum $\mathscr{S}_{f_n}$ intersects two and only two r-squares at macroscopic distance $\Omega(n)$ from each other can be easily estimated using the lemma. Indeed, in such a case, $\mathscr{S}_{(r)}$ would be compatible with an annulus structure consisting of two annuli, each being approximately of the type $A(r,n)$. There are $O(\frac{n^2}{r^2}) \times O(\frac{n^2}{r^2})$ such possible annulus structures. Using the lemma each of them costs (on $\mathbb{T}$) $(\frac{r}{n})^{5+o(1)}$. An easy exercise shows that this is much smaller than $\hat{\mathbb{P}}\big[|\mathscr{S}_{(r)}| = 2\big]$. In other words, if $|\mathscr{S}_{(r)}|$ is conditioned to be small, it tends to be localized. Also, the way that the lemma is stated makes it very convenient to work with higher values of k.

The details of the proof of Theorem X.13 can be found in [**GPS10**]. The double induction there is in some sense very close to the one we carried out in detail in subsection 3.4 in the case of the tree; this is the reason why we included this latter proof. For those who might read the proof in [**GPS10**], there is a notion of *overcrowded cluster* defined there; it exactly corresponds in the case of the tree to stopping the analysis above scale J_k instead of going all the way to scale 1 (note that without stopping at this scale J_k, the double induction in subsection 3.4 would have failed).

4.4. Some words on the weak independence property proved in [GPS10]. This part of the program is in some sense the main one. To introduce it, let us start by a naive but tempting strategy. What the first part of the program (Theorem X.13) tells us is that for any mesoscopic scale $1 \le r \le n$, if $\mathscr{S}_{f_n}$ is non-empty, it is very unlikely that it will intersect few squares of size r. In other words, it is very unlikely that $|\mathscr{S}_{(r)}|$ will be small. Let $B_1, \ldots, B_m$ denote the set of $O(n^2/r^2)$ r-squares which tile $[0,n]^2$. One might try the following *scanning procedure*: explore the spectral sample $\mathscr{S}_{f_n}$ inside the squares B_i one at a time. More precisely, before starting the scanning procedure, we consider our spectral sample $\mathscr{S}_{f_n}$ as a random subset of $[0,n]^2$ about which we do not know anything yet. Then, at step one, we reveal $\mathscr{S}_{|B_1}$. This gives us some partial information about $\mathscr{S}_{f_n}$. What we still have to explore is a random set of $[0,n]^2 \setminus B_1$ which follows the law of a spectral sample conditioned on what was seen in B_1 and we keep going in this way. By Theorem X.13, many of these squares will be non-empty. Now, it is not hard to prove the following lemma (using similar methods as in Problem X.6).

LEMMA X.15. *There is a universal constant $c > 0$ such that for any r-square B in the bulk $[n/4, 3n/4]^2$, one has*

$$\hat{\mathbb{P}}\big[|\mathscr{S}_{f_n} \cap B| > c\, r^2 \alpha_4(r) \;\big|\; \mathscr{S}_{f_n} \cap B \neq \emptyset\big] > c\,.$$

This lemma in fact holds uniformly in the position of B inside $[0,n]^2$.

If one could prove the following (much) stronger result: there exists a universal constant $c > 0$ such that uniformly on the sets $S \subset [0,n]^2 \setminus B$ one has

$$\hat{\mathbb{P}}\big[|\mathscr{S}_{f_n} \cap B| > c\, r^2 \alpha_4(r) \;\big|\; \mathscr{S}_{f_n} \cap B \neq \emptyset \text{ and } \mathscr{S}_{|B^c} = S\big] > c\,, \tag{X.11}$$

then it would not be hard to make the above scanning strategy work together with Theorem X.13 in order to obtain Theorem X.3. (Note that such a result would indeed give a strong hint of independence within $\mathscr{S}_{f_n}$.) However, as we discussed before, the current understanding of the independence within $\mathscr{S}_{f_n}$ is far from giving such a statement. Instead, the following result is proved in [**GPS10**]. We provide here a slightly simplified version.

THEOREM X.16 ([**GPS10**]). *There exists a uniform constant $c > 0$ such that for any set $W \subset [0,n]^2$ and any r-square B such that $B \cap W = \emptyset$, one has*

$$\hat{\mathbb{P}}\big[|\mathscr{S}_{f_n} \cap B| > c\, r^2 \alpha_4(r) \;\big|\; \mathscr{S}_{f_n} \cap B \neq \emptyset \textit{ and } \mathscr{S}_{f_n} \cap W = \emptyset\big] > c\,.$$

Note that this theorem in some sense interpolates between part of Lemma X.10 and Lemma X.15 which correspond respectively to the special cases $W = B^c$ and $W = \emptyset$. Yet it looks very weak compared to the expected (X.11) which is stated uniformly on the behavior of $\mathscr{S}_{f_n}$ outside of B.

Assuming this weak hint of independence (Theorem X.16), it seems we are in bad shape if we try to apply the above scanning procedure. Indeed, we face the following two obstacles:

(1) The first obstacle is that one would keep a good control only as far as one would not see any "spectrum". Namely, while revealing $\mathscr{S}_{|B_i}$ one at a time, the first time one finds a square B_i such that $\mathscr{S}_{|B_i} \neq \emptyset$, one would be forced to stop the scanning procedure there. In particular, if the size of the spectrum in this first non-trivial square does not exceed $r^2\alpha_4(r)$, then we cannot conclude anything.
(2) The second obstacle is that, besides the conditioning $\mathscr{S} \cap W = \emptyset$, our estimate is also conditioned on the event that $\mathscr{S} \cap B \neq \emptyset$. In particular, in the above "naive" scanning strategy where squares are revealed in a sequential way, at each step one would have to update the probability that $\mathscr{S} \cap B_{i+1} \neq \emptyset$ based on what was discovered so far.

It is the purpose of the third part of the program to adapt the above scanning strategy to these constraints. Before describing this third part in the next subsection, let us say a few words on how to prove Theorem X.16.

A crucial step in the proof of this theorem is to understand the following "one-point function" for any $x \in B$ at distance at least $r/3$ from the boundary:

$$\hat{\mathbb{P}}\big[x \in \mathscr{S}_{f_n} \text{ and } \mathscr{S}_{f_n} \cap W = \emptyset\big] .$$

A very useful observation is to rewrite this one-point function in terms of an explicit coupling of two iid percolation configurations. It works as follows: let (ω_1, ω_2) be a coupling of two i.i.d. percolations on $[0, n]^2$ which are such that

$$\begin{cases} \omega_1 = \omega_2 & \text{on } W^c \\ \omega_1, \omega_2 & \text{are independent on } W \end{cases}$$

One can check that the one-point function we are interested in is related to this coupling in the following simple way:

$$\hat{\mathbb{P}}\big[x \in \mathscr{S}_{f_n} \text{ and } \mathscr{S}_{f_n} \cap W = \emptyset\big] = \mathbb{P}\big[x \text{ is pivotal for } \omega_1 \text{ and } \omega_2\big] .$$

REMARK X.8. You may check this identity in the special cases where $W = \emptyset$ or $W = \{x\}^c$.

Thanks to this observation, the proof of Theorem X.16 proceeds by analyzing this W-coupling. See [**GPS10**] for the complete details.

4.5. Adapting the setup to the weak hint of independence. As we discussed in the previous subsection, one faces two main obstacles if, on the basis of the weak independence given by Theorem X.16, one tries to apply the naive sequential scanning procedure described earlier.

Let us start with the first obstacle. Assume that we scan the domain $[0, n]^2$ in a sequential way, i.e., we choose an increasing family of subsets $(W_l)_{l\geq 1} = (\{w_1, \dots, w_l\})_{l\geq 1}$. At each step, we reveal what $\mathscr{S}_{|\{w_{l+1}\}}$ is, conditioned on what was discovered so far (i.e., conditioned on $\mathscr{S}_{|W_l}$). From the weak independence Theorem X.16, it is clear that if we want this strategy to have any chance to be successful, we have to choose $(W_l)_{l\geq 1}$ in such a way that $(\mathscr{S}_{f_n} \cap W_l)_{l\geq 1}$ will remain empty for some time (so that we can continue to rely on our weak independence result); of course this cannot remain empty forever, so the game is to choose the increasing family $(W_l)_{l\geq 1}$ in such a way that the first time $\mathscr{S}_{f_n} \cap \{w_l\}$ will happen to be non-empty, it should give a strong indication that $\mathscr{S}_{f_n}$ is large in the r-neighborhood of w_l.

As we have seen, revealing the entire mesoscopic boxes B_i one at a time is not a successful idea. Here is a much better idea (which is not yet the right one due to the second obstacle, but we are getting close): in each r-square B_i, instead of revealing all the bits, let us reveal only a very small proportion δ_r of them. Lemma X.15 tells us that if $\mathscr{S} \cap B_i \neq \emptyset$, then each point $x \in B_i$ has probability of order $\alpha_4(r)$ to be in $\mathscr{S}_{f_n}$. Therefore if we choose $\delta_r \ll (r^2\alpha_4(r))^{-1}$, then with high probability, by revealing only a proportion δ_r of the points in B_i, we will "miss" the spectral sample $\mathscr{S}_{f_n}$. Hence, we have to choose $\delta_r \geq (r^2\alpha_4(r))^{-1}$. In fact choosing $\delta \asymp (r^2\alpha_4(r))^{-1}$ is exactly the right balance. Indeed, we know from Theorem X.13 that many r-squares B_i will be touched by the spectral sample; now, in this more sophisticated scanning procedure, if the first such square encountered happens to contain few points (i.e. $\ll r^2\alpha_4(r)$), then with the previous scanning strategy, we would "lose", but with the present one, due to our choice of δ_r, most likely we will keep $\mathscr{S}_{f_n} \cap W_l = \emptyset$ so that we can continue further on until we reach a "good" square (i.e. a square containing of order $r^2\alpha_4(r)$ points).

Now, Theorems X.13 and X.16 together tell us that with high probability, one will eventually reach such a good square. Indeed, suppose the m first r-squares touched by the spectral sample happened to contain few points; then, most likely, if W_{l_m} is the set of bits revealed so far, by our choice of δ_r we will still have $\mathscr{S} \cap W_{l_m} = \emptyset$. This allows us to still rely on Theorem X.16, which basically tells us that there is a positive conditional probability for the next one to be a "good" square (we are neglecting the second obstacle here). This says that the probability to visit m consecutive bad squares seems to decrease exponentially fast. Since m is typically very large (by Theorem X.13), we conclude that, with high probability, we will finally reach good squares. In the first good square encountered, by our choice of δ_r, there is now a positive probability to reveal a bit present in $\mathscr{S}_{f_n}$. In this case, the sequential scanning will have to stop, since we will not be able to use our weak independence result anymore, but this is not a big issue: indeed, assume you have some random set $S \subset B$. If by revealing each bit only with probability δ_r, you end up finding a point in S, most likely your set S is at least $\Omega(r^2\alpha_4(r))$ large. This is exactly the size we are looking for in Theorem X.3.

Now, only the second obstacle remains. It can be rephrased as follows: assume you applied the above strategy in $B_1, \ldots, B_h$ (i.e. you revealed each point in B_i, $i \in \{1, \ldots, h\}$ only with probability δ_r) and that you did not find any spectrum yet. In other words, if W_l denotes the set of points visited so far, then $\mathscr{S}_{f_n} \cap W_l = \emptyset$. Now if B_{h+1} is the next r-square to be scanned (still in a "dilute" way with intensity δ_r), we seem to be in good shape since we know how to control the conditioning $\mathscr{S}_{f_n} \cap W_l = \emptyset$. However, if we want to rely on the uniform control given by Theorem X.16, we also need to further condition on $\mathscr{S}_{f_n} \cap B_{h+1} \neq \emptyset$. In other words, we need to control the following conditional expectation:

$$\hat{\mathbb{P}}\big[\mathscr{S}_{f_n} \cap B_{h+1} \neq \emptyset \bigm| \mathscr{S}_{f_n} \cap W_l = \emptyset\big] .$$

It is quite involved to estimate such quantities. Fortunately, by changing our sequential scanning procedure into a slightly more "abstract" procedure, one can avoid dealing with such terms. More precisely, within each r-square B, we will still reveal only a δ_r proportion of the bits (so that the first obstacle is still taken care of), but instead of operating in a sequential way (i.e. scanning B_1, then B_2 and so on), we will gain a lot by considering the combination of Theorem X.13 and

Theorem X.16 in a more abstract fashion. Namely, the following large deviation lemma from [**GPS10**] captures exactly what we need in our present situation.

LEMMA X.17 ([**GPS10**]). *Let $X_i, Y_i \in \{0,1\}$, $i \in \{1,\dots,m\}$ be random variables such that for each i $Y_i \le X_i$ a.s. If $\forall J \subset [m]$ and $\forall i \in [m] \setminus J$, we have*

$$\mathbb{P}\big[Y_i = 1 \mid Y_j = 0,\ \forall j \in J\big] \ge c\,\mathbb{P}\big[X_i = 1 \mid Y_j = 0,\ \forall j \in J\big], \tag{X.12}$$

then if $X := \sum X_i$ and $Y := \sum Y_i$, one has that

$$\mathbb{P}\big[Y = 0 \mid X > 0\big] \le c^{-1}\mathbb{E}\big[e^{-(c/e)X} \mid X > 0\big].$$

Recall that $B_1, \dots, B_m$ denotes the set of r-squares which tile $[0,n]^2$. For each $i \in [m]$, let $X_i := 1_{\mathscr{S} \cap B_i \neq \emptyset}$ and $Y_i := 1_{\mathscr{S} \cap B_i \cap \mathcal{W} \neq \emptyset}$, where $\mathcal{W}$ is an independent uniform random subset of $[0,n]^2$ of intensity δ_r. Note that our set of bits (X_i, Y_i) are functions of the random set $\mathscr{S}_{f_n}$ plus some additional randomness (provided by the random dilute set $\mathcal{W}$).

This lemma enables us to combine our two main results, Theorems X.16 and X.13, in a very nice way: By our choice of the intensity δ_r, Theorem X.16, exactly states that the assumption (X.12) is satisfied for a certain constant $c > 0$. Lemma X.17 then implies that

$$\hat{\mathbb{P}}\big[Y = 0 \mid X > 0\big] \le c^{-1}\mathbb{E}\big[e^{-(c/e)X} \mid X > 0\big].$$

Now, notice that $X = \sum X_i$ exactly corresponds to $|\mathscr{S}_{(r)}|$ while the event $\{X > 0\}$ corresponds to $\{\mathscr{S}_{f_n} \neq \emptyset\}$ and the event $\{Y = 0\}$ corresponds to $\{\mathscr{S}_{f_n} \cap \mathcal{W} = \emptyset\}$. Therefore Theorem X.13 leads us to

$$\begin{aligned}
\hat{\mathbb{P}}\big[\mathscr{S}_{f_n} \cap \mathcal{W} = \emptyset,\ \mathscr{S}_{f_n} \neq \emptyset\big] &\le c^{-1}\mathbb{E}\big[e^{-(c/e)|\mathscr{S}_{(r)}|},\ \mathscr{S}_{f_n} \neq \emptyset\big] \\
&\le c^{-1}\sum_{k \ge 1}\hat{\mathbb{P}}\big[|\mathscr{S}_{(r)}| = k\big]e^{-(c/e)k} \\
&\le c^{-1}\Big(\sum_{k \ge 1} 2^{\theta \log_2^2(k+2)}e^{-(c/e)k)}\Big)\hat{\mathbb{P}}\big[|\mathscr{S}_{(r)}| = 1\big] \\
&\le C(\theta)\,\hat{\mathbb{P}}\big[|\mathscr{S}_{(r)}| = 1\big] \asymp \frac{n^2}{r^2}\alpha_4(r,n)^2\,. \qquad \text{(X.13)}
\end{aligned}$$

This shows that on the event that $\mathscr{S}_{f_n} \neq \emptyset$, it is very unlikely that we do not detect the spectral sample on the δ_r-dilute set $\mathcal{W}$. This is enough for us to conclude using the following identity:

$$\hat{\mathbb{P}}\big[\mathscr{S}_{f_n} \cap \mathcal{W} = \emptyset \mid \mathscr{S}_{f_n}\big] = (1 - \delta_r)^{|\mathscr{S}_{f_n}|} = (1 - \frac{1}{r^2\alpha_4(r)})^{|\mathscr{S}_{f_n}|}.$$

Indeed, by averaging this identity we obtain

$$\begin{aligned}
\hat{\mathbb{P}}\big[\mathscr{S}_{f_n} \cap \mathcal{W} = \emptyset, \mathscr{S}_{f_n} \neq \emptyset\big] &= \hat{\mathbb{E}}\big[\hat{\mathbb{P}}\big[\mathscr{S}_{f_n} \cap \mathcal{W} = \emptyset \mid \mathscr{S}_{f_n}\big]\,1_{\mathscr{S}_{f_n} \neq \emptyset}\big] \\
&= \hat{\mathbb{E}}\big[(1 - \frac{1}{r^2\alpha_4(r)})^{|\mathscr{S}_{f_n}|}\,1_{\mathscr{S}_{f_n} \neq \emptyset}\big] \\
&\ge \Omega(1)\hat{\mathbb{P}}\big[0 < |\mathscr{S}_{f_n}| < r^2\alpha_4(r)\big],
\end{aligned}$$

which, combined with (X.13) yields the desired upper bound in Theorem X.3. See Problem X.7 for the lower bound.

5. The radial case

The next part will focus on the existence of *exceptional times* in the model of dynamical percolation. A main tool in the study of these exceptional times is the spectral measure $\hat{\mathbb{Q}}_{g_R}$ where g_R is the Boolean function $g_R := \{-1,1\}^{O(R^2)} \to \{0,1\}$ defined to be the indicator function of the one-arm event $\{0 \longleftrightarrow \partial B(0,R)\}$. Note that by definition, g_R is such that $\|g_R\|_2^2 = \alpha_1(R)$.

In [**GPS10**], the following "sharp" theorem on the lower tail of $\mathscr{S}_{g_R}$ is proved.

THEOREM X.18 ([**GPS10**]). *Let g_R be the one-arm event in $B(0,R)$. Then for any $1 \le r \le R$, one has*

$$\hat{\mathbb{Q}}_{g_R}\big[0 < |\mathscr{S}_{g_R}| < r^2\alpha_4(r)\big] \asymp \frac{\alpha_1(R)^2}{\alpha_1(r)}\,. \tag{X.14}$$

The proof of this theorem is in many ways similar to the chordal case (Theorem X.3). An essential difference is that the "clustering v.s. entropy" mechanism is very different in this case. Indeed in the chordal left to right case, when $\mathscr{S}_{f_n}$ is conditioned to be very small, the proof of Theorem X.3 shows that typically $\mathscr{S}_{f_n}$ localizes in some r-square whose location is "uniform" in the domain $[0,n]^2$. In the radial case, the situation is very different: $\mathscr{S}_{g_R}$ conditioned to be very small will in fact tend to localize in the r-square centered at the origin. This means that the analysis of the mesoscopic behavior (i.e. the analogue of Theorem X.13) has to be adapted to the radial case. In particular, in the definition of an annulus structure, the annuli containing the origin play a distinguished role. See [**GPS10**] for complete details.

Exercise sheet of Part X

EXERCISE X.1. Prove Proposition X.4.

EXERCISE X.2. Consider the fractal percolation process $\mathcal{T}^i$, $i \ge 1$ introduced in this part. (Recall that $\mathcal{T}_{2^i} \equiv \mathcal{T}^i$). Recall that in Section 3, it was important to estimate the quantity $\mathbb{P}\big[|\mathcal{T}^i| = 1\big]$. This is one of the purposes of the present exercise.

(a) Let $p_i := \mathbb{P}\big[|\mathcal{T}^i| = 1\big]$. By recursion, show that there is a constant $c \in (0,1)$ so that, as $i \to \infty$
$$p_i \sim c\mu^i\,,$$
where $\mu := 4p(1-p+pq)^3$ and q is the probability of extinction for the Galton-Watson tree correponding to $(\mathcal{T}^i)_{i\ge1}$.

(b) Using the generating function $s \mapsto f(s)(= E(s^{\text{ number of offspring}})$ of this Galton-Watson tree, and by studying the behavior of its i-th iterates $f^{(i)}$, prove the same result with $\mu := f'(q)$. Check that it gives the same formula.

(c) Recall the definition of $p_{m,b}$ from Section 3. Let $p_{m,\infty}$ be the probability that exactly 1 person at generation m survives forever. Prove that
$$p_{m,\infty} = (1-q)\mu^m$$
for the same exponent μ. Prove Lemma X.9. Finally, prove that $\lim_{b\to\infty} p_{m,b} = p_{m,\infty}$.

EXERCISE X.3. Extract from the proof of Lemma X.8 the answer to the question asked in Figure X.4.

EXERCISE X.4. Prove that

$$\text{Theorem X.3} \Rightarrow \text{Theorem X.2} \Rightarrow \text{Theorem X.1}$$

EXERCISE X.5. Consider an r-square $B \subset [n/4, 3n/4]^2$ in the "bulk" of $[0, n]^2$.

(a) Prove using Proposition IX.3 that

$$\hat{\mathbb{P}}\big[\mathscr{S}_{f_n} \neq \emptyset \text{ and } \mathscr{S}_{f_n} \subset B\big] \asymp \alpha_4(r,n)^2$$

(b) Check that the clustering Lemma X.14 is consistent with this estimate.

PROBLEM X.6. The purpose of this exercise is to prove Lemma X.10.

(a) Using Proposition IX.3, prove that for any $x \in B$ at distance $r/3$ from the boundary,

$$\mathbb{P}\big[x \in \mathscr{S}_{f_n} \text{ and } \mathscr{S}_{f_n} \cap B^c = \emptyset\big] \asymp \alpha_4(r)\alpha_4(r,n)^2\,.$$

(b) Recover the same result using Proposition IX.4 instead.

(c) Conclude using Exercise X.5 that $\hat{\mathbb{E}}\big[|\mathscr{S}_{f_n} \cap \bar{B}| \mid \mathscr{S}_{f_n} \neq \emptyset \text{ and } \mathscr{S}_{f_n} \subset B\big] \asymp r^2\alpha_4(r)$, where $\bar{B} \subset B$ is the set of points $x \in B$ at distance at least $r/3$ from the boundary.

(d) Study the second-moment $\hat{\mathbb{E}}\big[|\mathscr{S}_{f_n} \cap \bar{B}|^2 \mid \mathscr{S}_{f_n} \neq \emptyset \text{ and } \mathscr{S}_{f_n} \subset B\big]$.

(e) Deduce Lemma X.10.

PROBLEM X.7. Most of this part was devoted to the explanation of the proof of Theorem X.3. Note that we in fact only discussed how to prove the upper bound. This is because the lower bound is much easier to prove and this is the purpose of this problem.

(a) Deduce from Lemma X.10 and Exercise X.5(a) that the lower bound on $\hat{\mathbb{P}}\big[0 < |\mathscr{S}_{f_n}| < r^2\alpha_4(r)\big]$ given in Theorem X.3 is correct. I.e., show that there exists a constant $c > 0$ such that

$$\hat{\mathbb{P}}\big[0 < |\mathscr{S}_{f_n}| < r^2\alpha_4(r)\big] > c\frac{n^2}{r^2}\alpha_4(r,n)^2\,.$$

(b) (Hard) In the same fashion, prove the lower bound part of Theorem X.18.

Part XI. Applications to dynamical percolation

In this section, we present a very natural model where percolation undergoes a time-evolution: this is the model of **dynamical percolation** described below. The study of the "dynamical" behavior of percolation as opposed to its "static" behavior turns out to be very rich: interesting phenomena arise especially at the phase transition point. We will see that in some sense, dynamical planar percolation at criticality is a very unstable or chaotic process. In order to understand this instability, sensitivity of percolation (and therefore its Fourier analysis) will play a key role. In fact, the original motivation for the paper [**BKS99**] on noise sensitivity was to solve a particular problem in the subject of dynamical percolation. [**Ste09**] provides a recent survey on the subject of dynamical percolation.

We mention that one can read all but the last section of the present part without having read Part X.

1. The model of dynamical percolation

This model was introduced by Häggström, Peres and Steif [**HPS97**] inspired by a question that Paul Malliavin asked at a lecture at the Mittag-Leffler Institute in 1995. This model was invented independently by Itai Benjamini.

In the general version of this model as it was introduced, given an arbitrary graph G and a parameter p, the edges of G switch back and forth according to independent 2-state continuous time Markov chains where closed switches to open at rate p and open switches to closed at rate $1-p$. Clearly, the product measure with density p, denoted by π_p in this part, is the unique stationary distribution for this Markov process. The general question studied in dynamical percolation is whether, when we start with the stationary distribution π_p, there exist atypical times at which the percolation structure looks markedly different than that at a fixed time. In almost all cases, the term "markedly different" refers to the existence or nonexistence of an infinite connected component. Dynamical percolation on site percolation models, which includes our most important case of the hexagonal lattice, is defined analogously.

We very briefly summarize a few early results in the area. It was shown in [**HPS97**] that below criticality, there are no times at which there is an infinite cluster and above criticality, there is an infinite cluster at all times. See the exercises. In [**HPS97**], examples of graphs which do not percolate at criticality but for which there exist exceptional times where percolation occurs were given. (Also given were examples of graphs which do percolate at criticality but for which there exist exceptional times where percolation does not occur.) A fairly refined analysis of the case of so-called *spherically symmetric* trees was given. See the exercises for some of these.

Given the above results, it is natural to ask what happens on the standard graphs that we work with. Recall that for $\mathbb{Z}^2$, we have seen that there is no percolation at criticality. It turns out that it is also known (see below) that for $d \geq 19$, there is no percolation at criticality for $\mathbb{Z}^d$. It is a major open question to prove that this is also the case for intermediate dimensions; the consensus is that this should be the case.

2. What's going on in high dimensions: $\mathbb{Z}^d, d \geq 19$?

For the high dimensional case, $\mathbb{Z}^d, d \geq 19$, it was shown in [**HPS97**] that there are no exceptional times of percolation at criticality.

THEOREM XI.1 ([**HPS97**]). *For the integer lattice $\mathbb{Z}^d$ with $d \geq 19$, dynamical critical percolation has no exceptional times of percolation.*

The key reason for this is a highly nontrivial result due to work of Hara and Slade ([**HS94**]), using earlier work of Barsky and Aizenman ([**BA91**]), that says that if $\theta(p)$ is the probability that the origin percolates when the parameter is p, then for $p \geq p_c$

$$\theta(p) = O(p - p_c)\,. \tag{XI.1}$$

(This implies in particular that there is no percolation at criticality.) In fact, this is the only thing which is used in the proof and hence the result holds whenever the percolation function satisfies this "finite derivative condition" at the critical point.

Outline of Proof. By countable additivity, it suffices to show that there are no times at which the origin percolates during $[0,1]$. We use a first moment argument. We break the time interval $[0,1]$ into m intervals each of length $1/m$. If we fix one of these intervals, the set of edges which are open at *some time* during this interval is i.i.d. with density about $p_c + 1/m$. Hence the probability that the origin percolates with respect to these set of edges is by (XI.1) at most $O(1/m)$. It follows that if N_m is the number of intervals where this occurs, then $\mathbb{E}[N_m]$ is at most $O(1)$. It is not hard to check that $N \leq \liminf_m N_m$, where N is the cardinality of the set of times during $[0,1]$ at which the origin percolates. Fatou's Lemma now yields that $\mathbb{E}(N) < \infty$ and hence there are at most finitely many exceptional times during $[0,1]$ at which the origin percolates. To go from here to having no exceptional times can either be done by using some rather abstract Markov process theory or by a more hands on approach as was done in [**HPS97**] and which we refer to for details. □

REMARK XI.1. It is known that (XI.1) holds for any homogeneous tree (see [**Gri99**] for the binary tree case) and hence there are no exceptional times of percolation in this case also.

REMARK XI.2. It is was proved by Kesten and Zhang [**KZ87**], that (XI.1) fails for $\mathbb{Z}^2$ and hence the proof method above to show that there are no exceptional times fails. This infinite derivative in this case might suggest that there are in fact exceptional times for critical dynamical percolation on $\mathbb{Z}^2$, an important question left open in [**HPS97**].

3. $d = 2$ and BKS

One of the questions posed in [**HPS97**] was whether there are exceptional times of percolation for $\mathbb{Z}^2$. It was this question which was one of the main motivations for the paper [**BKS99**]. While they did not prove the existence of exceptional times of percolation, they did obtain the following very interesting result which has a very similar flavor.

THEOREM XI.2 ([**BKS99**]). *Consider an $R \times R$ box on which we run critical dynamical percolation. Let S_R be the number of times during $[0,1]$ at which the configuration changes from having a percolation crossing to not having one. Then*

$$S_R \to \infty \text{ in probability as } R \to \infty.$$

Noise sensitivity of percolation as well as the above theorem tells us that certain large scale connectivity properties decorrelate very quickly. This suggests that in some vague sense $\omega_t^{p_c}$ "changes" very quickly as time goes on and hence there might be some chance that an infinite cluster appears since we are given many "chances".

In the next section, we begin our study of exceptional times for $\mathbb{Z}^2$ and the hexagonal lattice.

4. The second moment method and the spectrum

In this section, we reduce the question of exceptional times to a "second moment method" computation which in turn reduces to questions concerning the spectral behavior for specific Boolean functions involving percolation. Since $p = 1/2$, our dynamics can be equivalently defined by having each edge or hexagon be rerandomized at rate 1.

The key random variable which one needs to look at is

$$X = X_R := \int_0^1 1_{0 \overset{\omega_t}{\longleftrightarrow} R} \, dt$$

where $0 \overset{\omega_t}{\longleftrightarrow} R$ is of course the event that at time t there is an open path from the origin to distance R away. Note that the above integral is simply the Lebesgue measure of the set of times in $[0,1]$ at which this occurs.

We want to apply the second moment method here. We isolate the easy part of the argument so that the reader who is not familiar with this method understands it in a more general context. However, the reader should keep in mind that the difficult part is always to prove the needed bound on the second moments which in this case is (XI.2).

PROPOSITION XI.3. *If there exists a constant C such that for all R*

$$\mathbb{E}(X_R^2) \le C\mathbb{E}(X_R)^2, \tag{XI.2}$$

then a.s. there are exceptional times of percolation.

Proof. For any nonnegative random variable Y, the Cauchy-Schwarz inequality applied to $YI_{\{Y>0\}}$ yields

$$\mathbb{P}(Y > 0) \ge \mathbb{E}(Y)^2/\mathbb{E}(Y^2).$$

Hence by (XI.2), we have that for all R,

$$\mathbb{P}(X_R > 0) \ge 1/C$$

and hence by countable additivity (as we have a decreasing sequence of events)

$$\mathbb{P}(\cap_R\{X_R > 0\}) \ge 1/C.$$

Had the set of times that a fixed edge is on been a closed set, then the above would have yielded by compactness that there is an exceptional time of percolation with probability at least $1/C$. However, this is not a closed set. On the other hand, this point is very easily fixed by modifying the process so that the times each edge is on is a closed set and observing that a.s. no new times of percolation are introduced by this modification. The details are left to the reader. Once we have an exceptional time with positive probability, ergodicity immediately implies that this occurs a.s. □

The first moment of X_R is, due to Fubini's Theorem, simply the probability of our one-arm event, namely $\alpha_1(R)$. The second moment of X_R is easily seen to be (XI.3)

$$\mathbb{E}(X^2) = \mathbb{E}\Big(\int_0^1\int_0^1 1_{0\overset{\omega_s}{\longleftrightarrow}R}\, 1_{0\overset{\omega_t}{\longleftrightarrow}R}\, ds\, dt\Big) = \int_0^1\int_0^1 \mathbb{P}(0 \overset{\omega_s}{\longleftrightarrow} R, 0 \overset{\omega_t}{\longleftrightarrow} R)\, ds\, dt$$

which is, by time invariance, at most

$$2\int_0^1 \mathbb{P}(0 \overset{\omega_s}{\longleftrightarrow} R, 0 \overset{\omega_0}{\longleftrightarrow} R)\, ds. \tag{XI.4}$$

The key observation now, which brings us back to noise sensitivity, is that the integrand $\mathbb{P}(0 \overset{\omega_s}{\longleftrightarrow} R, 0 \overset{\omega_0}{\longleftrightarrow} R)$ is precisely $\mathbb{E}[f_R(\omega)f_R(\omega_\epsilon)]$ where f_R is the indicator of the event that there is an open path from the origin to distance R away and $\epsilon = 1 - e^{-s}$ since looking at our process at two different times is exactly looking at a configuration and a noisy version.

What we have seen in this subsection is that proving the existence of exceptional times comes down to proving a second moment estimate and furthermore that the integrand in this second moment estimate concerns noise sensitivity, something for which we have already developed a fair number of tools to handle.

5. Proof of existence of exceptional times for the hexagonal lattice via randomized algorithms

In [**SS10b**], exceptional times were shown to exist for the hexagonal lattice; this was the first transitive graph for which such a result was obtained. However, the methods in this paper did not allow the authors to prove that $\mathbb{Z}^2$ had exceptional times.

THEOREM XI.4 ([**SS10b**]). *For dynamical percolation on the hexagonal lattice $\mathbb{T}$ at the critical point $p_c = 1/2$, there exist almost surely exceptional times $t \in [0,\infty)$ such that ω_t has an infinite cluster.*

Proof. As we noted in the previous section, two different times of our model can be viewed as "noising" where the probability that a hexagon is rerandomized within t units of time is $1 - e^{-t}$. Hence, by (IV.2), we have that

$$\mathbb{P}\big[0 \overset{\omega_0}{\longleftrightarrow} R,\ 0 \overset{\omega_t}{\longleftrightarrow} R\big] = \mathbb{E}\big[f_R\big]^2 + \sum_{\emptyset \neq S \subseteq B(0,R)} \hat{f}_R(S)^2 \exp(-t|S|) \tag{XI.5}$$

where $B(0,R)$ are the set of hexagons involved in the event f_R. We see in this expression that, for small times t, the frequencies contributing in the correlation between $\{0 \overset{\omega_0}{\longleftrightarrow} R\}$ and $\{0 \overset{\omega_t}{\longleftrightarrow} R\}$ are of "small" size $|S| \lesssim 1/t$. Therefore, in order to detect the existence of exceptional times, one needs to achieve good control on the *lower tail* of the Fourier spectrum of f_R.

The approach of this section is to find an algorithm minimizing the revealment as much as possible and to apply Theorem VIII.1. However there is a difficulty here, since our algorithm might have to look near the origin, in which case it is difficult to keep the revealment small. There are other reasons for a potential problem. If R is very large and t very small, then if one conditions on the event $\{0 \overset{\omega_0}{\longleftrightarrow} R\}$, since few sites are updated, the open path in ω_0 from 0 to distance R will still be preserved in ω_t at least up to some distance $L(t)$ (further away, large scale connections start to decorrelate). In some sense the geometry associated to the event $\{0 \overset{\omega}{\longleftrightarrow} R\}$ is "frozen" on a certain scale between time 0 and time t. Therefore, it is natural to divide our correlation analysis into two scales: the ball of radius $r = r(t)$ and the annulus from $r(t)$ to R. Obviously the "frozen radius" $r = r(t)$ increases as $t \to 0$. We therefore proceed as follows instead. For any r, we have

$$\begin{aligned} \mathbb{P}\big[0 \overset{\omega_0}{\longleftrightarrow} R,\, 0 \overset{\omega_t}{\longleftrightarrow} R\big] &\le \mathbb{P}\big[0 \overset{\omega_0}{\longleftrightarrow} r\big]\,\mathbb{P}\big[r \overset{\omega_0}{\longleftrightarrow} R,\, r \overset{\omega_t}{\longleftrightarrow} R\big] \\ &\le \alpha_1(r)\,\mathbb{E}\big[f_{r,R}(\omega_0) f_{r,R}(\omega_t)\big], \end{aligned} \tag{XI.6}$$

where $f_{r,R}$ is the indicator function of the event, denoted by $r \overset{\omega}{\longleftrightarrow} R$, that there is an open path from distance r away to distance R away. Now, as above, we have

$$\mathbb{E}\big[f_{r,R}(\omega_0) f_{r,R}(\omega_t)\big] \le \mathbb{E}\big[f_{r,R}\big]^2 + \sum_{k=1}^{\infty} \exp(-tk) \sum_{|S|=k} \hat{f}_{r,R}(S)^2. \tag{XI.7}$$

The Boolean function $f_{r,R}$ somehow avoids the singularity at the origin, and it is possible to find algorithms for this function with small revealments. In any case, letting $\delta = \delta_{r,R}$ be the revealment of $f_{r,R}$, it follows from Theorem VIII.1 and the fact that $\sum_k k\exp(-tk) \le O(1)/t^2$ that

$$\mathbb{E}\big[f_{r,R}(\omega_0) f_{r,R}(\omega_t)\big] \le \alpha_1(r,R)^2 + O(1)\delta\alpha_1(r,R)/t^2. \tag{XI.8}$$

The following proposition gives a bound on δ. We will sketch why it is true afterwards.

Proposition XI.5 ([**SS10b**]). *Let $2 \le r < R$. Then*

$$\delta_{r,R} \le O(1)\alpha_1(r,R)\,\alpha_2(r)\,. \tag{XI.9}$$

Putting together (XI.6), (XI.8), Proposition XI.5 and using quasi-multiplicativity of α_1 yields

$$\mathbb{P}\big[0 \overset{\omega_0}{\longleftrightarrow} R,\, 0 \overset{\omega_t}{\longleftrightarrow} R\big] \le O(1)\frac{\alpha_1(R)^2}{\alpha_1(r)}\left(1 + \frac{\alpha_2(r)}{t^2}\right).$$

This is true for all r and t. If we choose $r = r(t) = (1/t)^8$ and ignore $o(1)$ terms in the critical exponents (which can easily be handled rigorously), we obtain, using the explicit values for the one and two-arm critical exponents, that

$$\mathbb{P}\big[0 \overset{\omega_0}{\longleftrightarrow} R,\, 0 \overset{\omega_t}{\longleftrightarrow} R\big] \le O(1)t^{-5/6}\alpha_1(R)^2\,. \tag{XI.10}$$

Now, since $\int_0^1 t^{-5/6}dt < \infty$, by integrating the above correlation bound over the unit interval, one obtains that $\mathbb{E}\big[X_R^2\big] \le C\mathbb{E}\big[X_R\big]^2$ for some constant C as desired. □

Outline of proof of Proposition XI.5.
We use an algorithm that mimics the one we used for percolation crossings except the present setup is "radial". As in the chordal case, we randomize the starting point of our exploration process by choosing a site uniformly on the 'circle' of radius R. Then, we explore the picture with an exploration path γ directed towards the origin; this means that as in the case of crossings, when the interface encounters

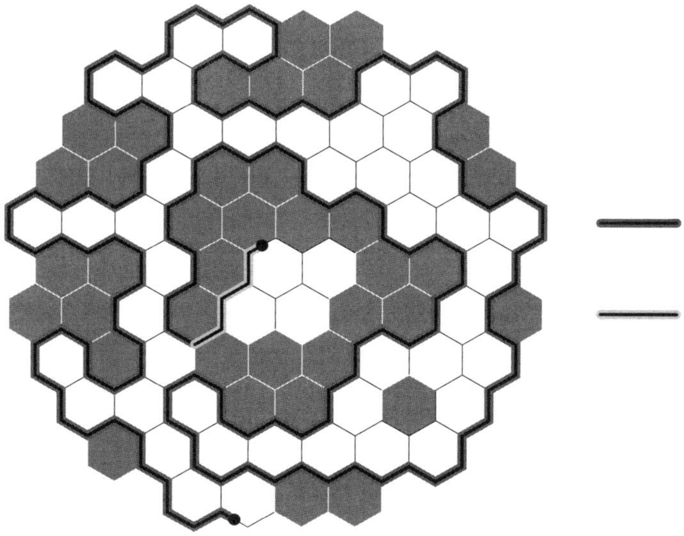

an open (resp. closed) site, it turns say to the left (resp. right), the only difference being that when the exploration path closes a loop around the origin, it continues its exploration inside the connected component of the origin. (It is known that this discrete curve converges towards *radial* SLE_6 on $\mathbb{T}$, when the mesh goes to zero.) It turns out that the so-defined exploration path gives all the information we need. Indeed, if the exploration path closes a clockwise loop around the origin, this means that there is a closed circuit around the origin making $f_{r,R}$ equal to zero. On the other hand, if the exploration path does not close any clockwise loop until it reaches radius r, it means that $f_{r,R} = 1$. Hence, we run the exploration path until either it closes a clockwise loop or it reaches radius r. This is our algorithm. Neglecting boundary issues (points near radius r or R), if x is a point at distance u from 0, with $2r < u < R/2$, in order for x to be examined by the algorithm, it is needed that there is an open path from $2u$ to R and the two-arm event holds in the ball centered at u with radius $u/2$. Hence for $|x| = u$, $\mathbb{P}\big[x \in J\big]$ is at most $O(1)\alpha_2(u)\alpha_1(u, R)$. Due to the explicit values of the one and two-arm exponents, this expression is decreasing in u. Hence, ignoring the boundary, the revealment is at most $O(1)\alpha_2(r)\alpha_1(r, R)$. See [**SS10b**] for more details. □

We now assume that the reader is familiar with the notion of Hausdorff dimension. We let $\mathcal{E} \subseteq [0, \infty]$ denote the (random) set of these exceptional times at which percolation occurs. It is an immediate consequence of Fubini's Theorem that $\mathcal{E}$ has Lebesgue measure zero and hence we should look at its Hausdorff dimension if we want to measure its "size". The first result is the following.

THEOREM XI.6 ([**SS10b**]). *The Hausdorff dimension of $\mathcal{E}$ is an almost sure constant in* $[1/6, 31/36]$.

It was conjectured there that the dimension of the set of exceptional times is a.s. $31/36$.

Outline of Proof. The fact that the dimension is an almost sure constant follows from easy 0-1 Laws. The lower bounds are obtained by placing a random measure on $\mathcal{E}$ with finite so-called α–energies for any $\alpha < 1/6$ and using a result called

Frostman's Theorem. (This is a standard technique once one has good control of the correlation structure.) Basically, the $1/6$ comes from the fact that for any $\alpha < 1/6$, one can multiply the integrand in $\int_0^1 t^{-5/6}dt$ by $(1/t)^\alpha$ and still be integrable. It is the amount of "room to spare" you have. If one could obtain better estimates on the correlations, one could thereby improve the lower bounds on the dimension. The upper bound is obtained via a first moment argument similar to the proof of Theorem XI.1 but now using (II.1). □

Before moving on to our final method of dealing with the spectrum, let us consider what we might have lost in the above argument. Using the above argument, we optimized things by taking $r(t) = (1/t)^8$. However, at time t compared to time 0, we have noise which is about t. Since we now know the exact noise sensitivity exponent, in order to obtain decorrelation, the noise level should be at least about the negative 3/4th power of the radius of the region we are looking at. So, events in our annulus should decorrelate if $r(t) >> (1/t)^{4/3}$. This suggests there might be potential for improvement. Note we used an inner radius which is 6 times larger than potentially necessary ($8 = 6 \times 4/3$). This 6 is the same 6 by which the result in Theorem VIII.4 differed by the true exponent ($3/4 = 6 \times 1/8$) and the same 6 explaining the gap in Theorem XI.6 $(1-1/6) = 6 \times (1-31/36)$. This last difference is also seen by comparing the exponents in (XI.10) and the last term in (XI.11) below.

6. Proof of existence of exceptional times via the geometric approach of the spectrum

Recall that our third approach for proving the noise sensitivity of percolation crossings was based on a geometrical analysis of the spectrum, viewing the spectrum as a random set. This approach yielded the exact noise sensitivity exponent for percolation crossings for the hexagonal lattice. This approach can also be used here as we will now explain. Two big advantages of this approach are that it succeeded in proving the existence of exceptional times for percolation crossings on $\mathbb{Z}^2$, something which [**SS10b**] was not able to do, as well as obtaining the exact Hausdorff dimension for the set of exceptional times, namely the upper bound of $31/36$ in the previous result.

THEOREM XI.7 ([**GPS10**]). *For the triangular lattice, the Hausdorff dimension of $\mathcal{E}$ is almost surely* $31/36$.

Proof. As explained in the previous section, it suffices to lower the $5/6$ in (XI.10) to $5/36$. (Note that (XI.10) was really only obtained for numbers strictly larger than $5/6$, with the $O(1)$ depending on this number; the same will be true for the $5/36$.)

Let $s(r)$ be the inverse of the map $r \to r^2\alpha_4(r) \sim r^{3/4}$. So more or less, $s(r) := r^{4/3}$. Using Theorem X.18, we obtain the following:

$$\begin{aligned}
\mathbb{E}\big[f_R(\omega_0)f_R(\omega_t)\big] &= \sum_S \exp(-t|S|)\hat{f}_R(S)^2 \\
&= \sum_{k=1}^{\infty} \sum_{S:|S|\in[(k-1)/t,k/t)} \exp(-t|S|)\hat{f}_R(S)^2 \\
&\le \sum_{k=1}^{\infty} \exp(-k)\hat{\mathbb{Q}}\big[|\mathscr{S}_{f_R}| < k/t\big] \\
&\le O(1)\sum_{k=1}^{\infty} \exp(-k)\frac{\alpha_1(R)^2}{\alpha_1(s(k/t))} \\
&\le O(1)\alpha_1(R)^2 \sum_{k=1}^{\infty} \exp(-k)(\frac{k}{t})^{4/3\times 5/48} \\
&\le O(1)\alpha_1(R)^2(\frac{1}{t})^{5/36}. \qquad \text{(XI.11)}
\end{aligned}$$

This completes the proof. (Of course, there are $o(1)$ terms in these exponents which we are ignoring.) □

We have done a lot of the work for proving that there are exceptional times also on $\mathbb{Z}^2$.

THEOREM XI.8 ([**GPS10**]). *For dynamical percolation on $\mathbb{Z}^2$ at the critical point $p_c = 1/2$, there exist almost surely exceptional times $t \in [0,\infty)$ such that ω_t has an infinite cluster.*

Proof. $s(r)$ is defined as it was before but now we cannot say that $s(r)$ is about $r^{4/3}$. However, we can say that for some fixed $\delta > 0$, we have that for all r,

$$s(r) \ge r^{\delta} \qquad \text{(XI.12)}$$

From the previous proof, we still have

$$\frac{\mathbb{E}\big[f_R(\omega_0)f_R(\omega_t)\big]}{\alpha_1(R)^2} \le O(1)\sum_{k=1}^{\infty}\exp(-k)\frac{1}{\alpha_1(s(k/t))}. \qquad \text{(XI.13)}$$

Exactly as in the proof of Theorem XI.4, we need to show that the right hand side is integrable near 0 in order to carry out the second moment argument.

Quasi-multiplicativity can be used to show that

$$\alpha_1(s(1/t)) \le k^{O(1)}\alpha_1(s(k/t)). \qquad \text{(XI.14)}$$

(Note that if things behaved exactly as power laws, this would be clear.)

Therefore the above sum is at most

$$O(1)\sum_{k=1}^{\infty}\exp(-k)\frac{k^{O(1)}}{\alpha_1(s(1/t))} \le O(1)\frac{1}{\alpha_1(s(1/t))} \qquad \text{(XI.15)}$$

V. Beffara has shown that there exists $\epsilon_0 > 0$ such that for all r,

$$\alpha_1(r)\alpha_4(r) \ge r^{\epsilon_0 - 2}. \qquad \text{(XI.16)}$$

Note that Theorem VI.4 and (VI.7) tell us that the left hand side is larger than $\Omega(1)r^{-2}$. The above tells us that we get an (important) extra power of r in (VI.7).

It follows that

$$\text{(XI.17)} \qquad \frac{1}{\alpha_1(s(1/t))} \leq \alpha_4(s(1/t))s(1/t)^{2-\epsilon_0} = (1/t)s(1/t)^{-\epsilon_0}.$$

(XI.12) tells us that the last factor is at most t^η for some $\eta > 0$ and hence the relevant integral converges as desired. The rest of the argument is the same. □

One can also consider exceptional times for other events, such as for example times at which there is an infinite cluster in the upper half-plane or times at which there are two infinite clusters in the whole plane, and consider the corresponding Hausdorff dimension. A number of results of this type, which are not sharp, are given in [**SS10b**] while various sharp results are given in [**GPS10**].

Exercise sheet of Part XI

EXERCISE XI.1. Prove that on any graph below criticality, there are no times at which there is an infinite cluster while above criticality, there is an infinite cluster at all times.

EXERCISE XI.2. Consider critical dynamical percolation on a general graph satisfying $\theta(p_c) = 0$. Show that a.s. $\{t : \omega_t \text{ percolates }\}$ has Lebesgue measure 0.

EXERCISE XI.3. (Somewhat hard). A *spherically symmetric* tree is one where all vertices at a given level have the same number of children, although this number may depend on the given level. Let T_n be the number of vertices at the nth level. Show that there is percolation at p if

$$\sum_n \frac{1}{p^{-n}T_n} < \infty$$

Hint: Let X_n be the number of vertices in the nth level which are connected to the root. Apply the second moment method to the sequence of X_n's.

The convergence of the sum is also necessary for percolation but this is harder and you are not asked to show this. This theorem is due to Russell Lyons.

EXERCISE XI.4. Show that if T_n is $n^2 2^n$ up to multiplicative constants, then the critical value of the graph is $1/2$ and we percolate at the critical value. (This yields a graph which percolates at the critical value.)

EXERCISE XI.5. (Quite a bit harder). Consider dynamical percolation on a spherically symmetric tree. Show that there for the parameter p, there are exceptional times at which percolation occurs if

$$\sum_n \frac{1}{np^{-n}T_n} < \infty.$$

Hint: Find an appropriate random variable X_n to which the second moment method can be applied.

EXERCISE XI.6. Find a spherically symmetric tree which does not percolate at criticality but for which there are exceptional times at which percolation occurs.

References

[AS00] Noga Alon and Joel H. Spencer. *The probabilistic method.* Wiley-Interscience Series in Discrete Mathematics and Optimization. Wiley-Interscience [John Wiley & Sons], New York, second edition, 2000. With an appendix on the life and work of Paul Erdos.

[BA91] D. J. Barsky and M. Aizenman. Percolation critical exponents under the triangle condition. *Ann. Probab.*, 19(4):1520–1536, 1991.

[BR08] Michel Benaïm and Raphaël Rossignol. Exponential concentration for first passage percolation through modified Poincaré inequalities. *Ann. Inst. Henri Poincaré Probab. Stat.*, 44(3):544–573, 2008.

[BKS99] Itai Benjamini, Gil Kalai, and Oded Schramm. Noise sensitivity of Boolean functions and applications to percolation. *Inst. Hautes Études Sci. Publ. Math.*, (90):5–43 (2001), 1999.

[BKS03] Itai Benjamini, Gil Kalai, and Oded Schramm. First passage percolation has sublinear distance variance. *Ann. Probab.*, 31(4):1970–1978, 2003.

[BSW05] Itai Benjamini, Oded Schramm, and David B. Wilson. Balanced Boolean functions that can be evaluated so that every input bit is unlikely to be read. In *STOC'05: Proceedings of the 37th Annual ACM Symposium on Theory of Computing*, pages 244–250. ACM, New York, 2005.

[BR06] Béla Bollobás and Oliver Riordan. A short proof of the Harris-Kesten theorem. *Bull. London Math. Soc.*, 38(3):470–484, 2006.

[BKK+92] Jean Bourgain, Jeff Kahn, Gil Kalai, Yitzhak Katznelson, and Nathan Linial. The influence of variables in product spaces. *Israel J. Math.*, 77(1-2):55–64, 1992.

[Fri04] Ehud Friedgut. Influences in product spaces: KKL and BKKKL revisited. *Combin. Probab. Comput.*, 13(1):17–29, 2004.

[FK96] Ehud Friedgut and Gil Kalai. Every monotone graph property has a sharp threshold. *Proc. Amer. Math. Soc.*, 124(10):2993–3002, 1996.

[FKW02] Ehud Friedgut, Jeff Kahn, and Avi Wigderson. Computing graph properties of randomized subcube partitions. In *Randomization and approximation techniques in computer science*, volume 2483 of *Lecture Notes in Comput. Sci.*, pages 105–113. Springer, Berlin, 2002.

[Gar10] Christophe Garban. Oded Schramm's contributions to noise sensitivity, 2010. *Ann. Probab.*, volume in honor of Oded Schramm, 39(5):1702–1767, 2011.

[GPS10] Christophe Garban, Gábor Pete, and Oded Schramm. The Fourier spectrum of critical percolation. *Acta Mathematica*, 205(1):19–104, 2010.

[Gri99] Geoffrey Grimmett. *Percolation.* Grundlehren der mathematischen Wissenschaften 321. Springer-Verlag, Berlin, second edition, 1999.

[Gro75] Leonard Gross. Logarithmic Sobolev inequalities. *Amer. J. Math.*, 97(4):1061–1083, 1975.

[HPS97] Olle Häggström, Yuval Peres, and Jeffrey E. Steif. Dynamical percolation. *Ann. Inst. H. Poincaré Probab. Statist.*, 33(4):497–528, 1997.

[HS94] Takashi Hara and Gordon Slade. Mean-field behaviour and the lace expansion. In *Probability and phase transition (Cambridge, 1993)*, volume 420 of *NATO Adv. Sci. Inst. Ser. C Math. Phys. Sci.*, pages 87–122. Kluwer Acad. Publ., Dordrecht, 1994.

[Joh00] Kurt Johansson. Shape fluctuations and random matrices. *Comm. Math. Phys.*, 209(2):437–476, 2000.

[KKL88] Jeff Kahn, Gil Kalai, and Nathan Linial. The influence of variables on Boolean functions. *29th Annual Symposium on Foundations of Computer Science*, (68–80), 1988.

[Kes80] Harry Kesten. The critical probability of bond percolation on the square lattice equals $\frac{1}{2}$. *Comm. Math. Phys.*, 74(1):41–59, 1980.

[Kes87] Harry Kesten. Scaling relations for 2D-percolation. *Comm. Math. Phys.*, 109(1):109–156, 1987.

[KZ87] Harry Kesten and Yu Zhang. Strict inequalities for some critical exponents in two-dimensional percolation. *J. Statist. Phys.*, 46(5-6):1031–1055, 1987.

[LSW02] Gregory F. Lawler, Oded Schramm, and Wendelin Werner. One-arm exponent for critical 2D percolation. *Electron. J. Probab.*, 7:no. 2, 13 pp. (electronic), 2002.

[LyP11] Russell Lyons with Yuval Peres. *Probability on Trees and Networks.* Cambridge University Press, 2011. In preparation. Current version available at `http://php.indiana.edu/~rdlyons/`.

[Mar74] G. A. Margulis. Probabilistic characteristics of graphs with large connectivity. *Problemy Peredači Informacii*, 10(2):101–108, 1974.

[MOO10] Elchanan Mossel, Ryan O'Donnell, and Krzysztof Oleszkiewicz. Noise stability of functions with low influences: invariance and optimality. *Ann. Math.*, 171(1):295–341, 2010.

[Nel66] Edward Nelson. A quartic interaction in two dimensions. In *Mathematical Theory of Elementary Particles (Proc. Conf., Dedham, Mass., 1965)*, pages 69–73. M.I.T. Press, Cambridge, Mass., 1966.

[NP95] Charles M. Newman and Marcelo S. T. Piza. Divergence of shape fluctuations in two dimensions. *Ann. Probab.*, 23(3):977–1005, 1995.

[Nol09] Pierre Nolin. Near-critical percolation in two dimensions. *Electron. J. Probab.*, 13(55):1562—1623, 2009.

[O'D] Ryan O'Donnell. History of the hypercontractivity theorem. `http://boolean-analysis.blogspot.com/`.

[O'D03] Ryan O'Donnell. *Computational Applications Of Noise Sensitivity.* PhD thesis, M.I.T., 2003.

[OSSS05] Ryan O'Donnell, Michael Saks, Oded Schramm, and Rocco Servedio. Every decision tree has an influential variable. *FOCS*, 2005.

[OS07] Ryan O'Donnell and Rocco A. Servedio. Learning monotone decision trees in polynomial time. *SIAM J. Comput.*, 37(3):827–844 (electronic), 2007.

[PP94] R. Pemantle and Y. Peres. Planar first-passage percolation times are not tight. In *Probability and phase transition (Cambridge, 1993)*, volume 420 of *NATO Adv. Sci. Inst. Ser. C Math. Phys. Sci.*, pages 261–264. Kluwer Acad. Publ., Dordrecht, 1994.

[PSSW07] Yuval Peres, Oded Schramm, Scott Sheffield, and David B. Wilson. Random-turn hex and other selection games. *Amer. Math. Monthly*, 114(5):373–387, 2007.

[Rus81] Lucio Russo. On the critical percolation probabilities. *Z. Wahrsch. Verw. Gebiete*, 56(2):229–237, 1981.

[Rus82] Lucio Russo. An approximate zero-one law. *Z. Wahrsch. Verw. Gebiete*, 61(1):129–139, 1982.

[Sch00] Oded Schramm. Scaling limits of loop-erased random walks and uniform spanning trees. *Israel J. Math.*, 118:221–288, 2000.

[SS10a] Oded Schramm and Stanislav Smirnov. On the scaling limits of planar percolation. *Ann. Probab.*, volume in honor of Oded Schramm, 39(5):1768–1814, 2011.

[SS10b] Oded Schramm and Jeffrey Steif. Quantitative noise sensitivity and exceptional times for percolation. *Ann. Math.*, 171(2):619–672, 2010.

[Smi01] Stanislav Smirnov. Critical percolation in the plane: conformal invariance, Cardy's formula, scaling limits. *C. R. Acad. Sci. Paris Sér. I Math.*, 333(3):239–244, 2001.

[SW01] Stanislav Smirnov and Wendelin Werner. Critical exponents for two-dimensional percolation. *Math. Res. Lett.*, 8(5-6):729–744, 2001.

[Ste09] Jeffrey Steif. A survey of dynamical percolation. *Fractal geometry and stochastics, IV, Birkhauser*, pages 145–174, 2009.

[Tal94] Michel Talagrand. On Russo's approximate zero-one law. *Ann. Probab.*, 22(3):1576–1587, 1994.

[Tal96] Michel Talagrand. How much are increasing sets positively correlated? *Combinatorica*, 16(2):243–258, 1996.

[Wer07] Wendelin Werner. *Lectures on two-dimensional critical percolation.* IAS Park City Graduate Summer School, 2007.

CNRS and UMPA, Ecole Normale Supérieure de Lyon, 46 allée d'Italie, 69364 Lyon Cedex 07 France

E-mail address: `christophe.garban@ens-lyon.fr`

Mathematical Sciences, Chalmers University of Technology and Mathematical Sciences, Göteborg University, SE-41296 Gothenburg, Sweden

E-mail address: `steif@math.chalmers.se`

Clay Mathematics Proceedings
Volume 15, 2012

Scaling Limits of Random Trees and Planar Maps

Jean-François Le Gall and Grégory Miermont

Contents

1. Introduction

The main goal of these lectures is to present some of the recent progress in the asymptotics for large random planar maps. Recall that a planar map is simply a graph drawn on the two-dimensional sphere and viewed up to direct homeomorphisms of the sphere. The faces of the map are the connected components of the complement of edges, or in other words the regions of the sphere delimited by the graph. Special cases of planar maps are triangulations, respectively quadrangulations, respectively p-angulations, where each face is adjacent to exactly 3, respectively 4, respectively p, edges (see Section 4 for more precise definitions).

Planar maps play an important role in several areas of mathematics and physics. They have been studied extensively in combinatorics since the pioneering work of Tutte (see in particular [**51**]), which was motivated by the famous four-color theorem. Graphs drawn on surfaces also have important algebraic and geometric applications; see the book [**27**]. In theoretical physics, the enumeration of planar maps (and of maps on surfaces of higher genus) has strong connections with matrix models, as shown by the work of 't Hooft [**24**] and Brézin et al [**10**]. More recently, graphs on surfaces have been used in physics as discrete models of random geometry in the so-called two-dimensional quantum gravity; see in particular the book [**3**] (a different mathematical approach to quantum gravity using the Gaussian free field appears in the work of Duplantier and Sheffield [**16**]). A nice account of the connections between planar maps and the statistical physics of random surfaces can be found in Bouttier's thesis [**7**]. From the probabilistic perspective, a planar map can be viewed as a discretization of a surface, and finding a continuous limit for large planar maps chosen at random in a suitable class should lead to an interesting model of a "Brownian surface". This is of course analogous to the well-known fact that Brownian motion appears as the scaling limit of long discrete random paths. In a way similar to the convergence of rescaled random walks to Brownian motion, one expects that the scaling limit of large random planar maps is universal in the sense that it should not depend on the details of the discrete model one is considering. These ideas appeared in the pioneering paper of Chassaing and Schaeffer [**12**] and in the subsequent work of Markert and Mokkadem [**37**] in the case of quadrangulations, and a little later in Schramm [**48**], who gave a precise form to the question of the existence of a scaling limit for large random triangulations of the sphere.

To formulate the latter question, consider a random planar map M_n which is uniformly distributed over a certain class of planar maps (for instance, triangulations, or quadrangulations) with n faces. Equip the vertex set $V(M_n)$ with the graph distance d_{gr}. It has been known for some time that the diameter of the resulting metric space is of order $n^{1/4}$ when n is large (see [**12**] for the case of quadrangulations). One then expects that the rescaled random metric spaces $(V(M_n), n^{-1/4}d_{gr})$ will converge in distribution as n tends to infinity towards a certain random metric space, which should be the same, up to trivial scaling factors, independently of the class of planar maps we started from. For the previous convergence to make sense, we need to say what it means for a sequence of metric spaces to converge. To this end we use the notion of the Gromov-Hausdorff distance, as it was suggested in [**48**]. Roughly speaking (see Section 2 for a more precise definition) a sequence (E_n) of compact metric spaces converges to a limiting space E_∞ if it is possible to embed isometrically all spaces E_n and E_∞ in the same "big" metric space E, in such a way that the Hausdorff distance between E_n and E_∞ tends to 0 as $n \to \infty$.

The preceding question of the existence of the scaling limit of large random planar maps is still open, but there has been significant progress in this direction, and our aim is to present some of the results that have been obtained in recent years.

Much of the recent progress in the understanding of asymptotic properties of large random planar maps was made possible by the use of bijections between different classes of planar maps and certain labeled trees. In the particular case of quadrangulations, such bijections were discovered by Cori and Vauquelin [**14**] and later popularized by Schaeffer [**47**] (see also Chassaing and Schaeffer [**12**]). The Cori-Vauquelin-Schaeffer bijection was extended to much more general planar maps by Bouttier, Di Francesco and Guitter [**9**]. In the case of bipartite planar maps, this extension takes a particularly simple form, which explains why some of the recent work [**36, 31, 32**] concentrates on the bipartite case. The reason why the bijections between maps and trees are interesting is the fact that properties of large (labeled) trees are often much easier to understand than those of large graphs. Indeed, it has been known for a long time and in particular since the work of Aldous [**1, 2**] that one can often describe the asymptotic properties of large random trees in terms of "continuous trees" whose prototype is the so-called CRT or Brownian continuum random tree. In the case of trees with labels, the relevant scaling limit for most of

FIGURE 1. Two planar quadrangulations, with respectively 2500 and 20000 vertices. These pictures represent the quadrangulations as graphs, and do not take account of the embedding in the sphere. Simulations by J.-F. Marckert.

the discrete models of interest is the CRT equipped with Brownian labels, which can conveniently be constructed and studied via the path-valued process called the Brownian snake (see e.g. [**28**]).

A key feature of the bijections between planar maps and labeled trees is the fact that, up to an appropriate translation, labels on the tree correspond to distances in the map from a distinguished vertex that plays a special role. Therefore, the known results about scaling limits of labeled trees immediately give much information about asymptotics of distances from this distinguished vertex. This idea was exploited by Chassaing and Schaeffer [**12**] in the case of quadrangulations and then by Marckert and Miermont [**36**] (for bipartite planar maps) and Miermont [**38**] (for general planar maps). In view of deriving the Gromov-Hausdorff convergence of rescaled planar maps, it is however not sufficient to control distances from a distinguished vertex. Still, a simple argument gives an effective bound on the distance between two arbitrary vertices in terms of quantities depending only on the labels on the tree, or equivalently on the distances from the distinguished vertex (see Proposition 5.9(i) below). This bound was used in [**31**] to show via a compactness argument that the scaling limit of rescaled uniformly distributed $2p$-angulations with n faces exists along suitable subsequences. Furthermore, this scaling limit is a quotient space of the CRT for an equivalence relation defined in terms of Brownian labels on the CRT: Roughly speaking, two vertices of the CRT need to be identified if they have the same label and if, when travelling from one vertex to the other one along the contour of the CRT, one only encounters vertices with larger label. The results of [**31**] are not completely satisfactory, because they require the extraction of suitable subsequences. The reason why this is necessary is the fact that the distance on the limiting space (that is, on the quotient of the CRT we have just described) has not been fully identified, even though lower and upper bounds are available. Still we call Brownian map any random metric space that arises as the scaling limit of uniformly distributed $2p$-angulations with n faces. This terminology is borrowed from Marckert and Mokkadem [**37**], who studied a weaker form of the convergence of rescaled random quadrangulations. Although the distribution of the Brownian map has not been fully characterized, it is possible to derive many properties of this random object (these properties will be common to any of the limiting random metric spaces that can arise in the scaling limit). In particular, it has been shown that the Brownian map has dimension 4 [**31**] and that it is homeomorphic to the 2-sphere [**34, 39**]. The latter fact is maybe not surprising since we started from larger and larger graphs drawn on the sphere: Still it implies that large random planar maps will have no "bottlenecks", meaning cycles whose length is small in comparison with the diameter of the graph but such that both connected components of the complement of the cycle have a macroscopic size.

In the subsequent sections, we discuss most of the preceding results in detail. We restrict our attention to the case of quadrangulations, because the bijections with trees are simpler in that case: The labeled trees corresponding to quadrangulations are just plane trees (rooted ordered trees) equipped with integer labels, such that the label of the root is 0 and the label can change by at most 1 in absolute value along each edge of the tree.

The first three sections below are devoted to asymptotics for random (labeled) trees, in view of our applications to random planar maps. In Section 1, we discuss asymptotics for uniformly distributed plane trees with n edges. We give a detailed proof of the fact that the suitably rescaled contour function of these discrete trees converges in distribution to the normalized Brownian excursion (this is a special case of the results of [**2**]). To this end, we first recall the basic facts of excursion theory that we need. In Section 2, we show that the convergence of rescaled contour functions can be restated as a convergence in the Gromov-Hausdorff sense of the trees viewed as random metric spaces for the graph distance. The limiting space is then the CRT, which we define precisely as the random real tree coded by a normalized Brownian excursion. Section 2 also contains basic facts about the Gromov-Hausdorff distance, and in particular its definition in terms of correspondences. In Section 3, we consider labeled trees and we give a detailed proof of the fact that rescaled labeled trees converge (in a suitable sense) towards the CRT equipped with Brownian labels.

The last four sections are devoted to planar maps and their scaling limits. Section 4 presents the combinatorial facts about planar maps that we need. In particular, we describe the Cori-Vauquelin-Schaeffer bijection between (rooted and pointed) quadrangulations and labeled trees. We also explain how labels on the tree give access to distances from the distinguished vertex in the map, and provide useful upper and lower bounds for other distances. In Section 5, we give the compactness argument that makes it possible to get sequential limits for rescaled uniformly distributed quadrangulations with n faces, in the Gromov-Hausdorff sense. The identification of the limit (or Brownian map) as a quotient space of the CRT for the equivalence relation described above is explained in Section 6. In that section, we are not able to give the full details of the proofs, but we try to present the main ideas. As a simple consequence of some of the estimates needed in the identification of the Brownian map, we also compute its Hausdorff dimension. Finally, Section 7 is devoted to the homeomorphism theorem. We follow the approach of [**39**], which consists in establishing the absence of "bottlenecks" in the Brownian map before proving via a theorem of Whyburn that this space is homeomorphic to the sphere.

To conclude this introduction, let us mention that, even though the key problem of the uniqueness of the Brownian map remains unsolved, many properties of this space have been investigated successfully. Often these results give insight into the properties of large planar maps. This is in particular the case for the results of [**32**], which give a complete description of all geodesics connecting an arbitrary point of the Brownian map to the distinguished point. Related results have been obtained in the paper [**40**], which deals with maps on surfaces of arbitrary genus. Very recently, the homeomorphism theorem of [**34**] has been extended by Bettinelli [**5**] to higher genus. As a final remark, one expects that the Brownian map should be the scaling limit for all random planar maps subject to some bound on the maximal degree of faces. One may ask what happens for random planar maps such that the distribution of the degree of a typical face has a heavy tail: This problem is discussed in [**33**], where it is shown that this case leads to different scaling limits.

2. Discrete trees and convergence towards the Brownian excursion

2.1. Plane trees. We will be interested in (finite) rooted ordered trees, which are called plane trees in combinatorics (see e.g. [**50**]). We set $\mathbb{N} = \{1, 2, \ldots\}$ and by convention $\mathbb{N}^0 = \{\varnothing\}$. We introduce the set

$$\mathcal{U} = \bigcup_{n=0}^{\infty} \mathbb{N}^n.$$

An element of $\mathcal{U}$ is thus a sequence $u = (u^1, \ldots, u^n)$ of elements of $\mathbb{N}$, and we set $|u| = n$, so that $|u|$ represents the "generation" of u. If $u = (u^1, \ldots, u^k)$ and $v = (v^1, \ldots, v^\ell)$ belong to $\mathcal{U}$, we write $uv = (u^1, \ldots, u^k, v^1, \ldots, v^\ell)$ for the concatenation of u and v. In particular $u\varnothing = \varnothing u = u$.

The mapping $\pi : \mathcal{U}\backslash\{\varnothing\} \longrightarrow \mathcal{U}$ is defined by $\pi((u^1, \ldots, u^n)) = (u^1, \ldots, u^{n-1})$ ($\pi(u)$ is the "parent" of u).

A plane tree τ is a finite subset of $\mathcal{U}$ such that:

(i) $\varnothing \in \tau$.
(ii) $u \in \tau\backslash\{\varnothing\} \Rightarrow \pi(u) \in \tau$.
(iii) For every $u \in \tau$, there exists an integer $k_u(\tau) \geq 0$ such that, for every $j \in \mathbb{N}$, $uj \in \tau$ if and only if $1 \leq j \leq k_u(\tau)$

The number $k_u(\tau)$ is interpreted as the "number of children" of u in τ.

We denote by $\mathbf{A}$ the set of all plane trees. In what follows, we see each vertex of the tree τ as an individual of a population whose τ is the family tree. By definition, the size $|\tau|$ of τ is the number of edges of τ, $|\tau| = \#\tau - 1$. For every integer $k \geq 0$, we put

$$\mathbf{A}_k = \{\tau \in \mathbf{A} : |\tau| = k\}.$$

EXERCISE 2.1. *Verify that the cardinality of $\mathbf{A}_k$ is the k-th Catalan number*

$$\#\mathbf{A}_k = \mathrm{Cat}_k := \frac{1}{k+1}\binom{2k}{k}.$$

A plane tree can be coded by its Dyck path or **contour function**. Suppose that the tree is embedded in the half-plane in such a way that edges have length one. Informally, we imagine the motion of a particle that starts at time $t = 0$ from the root of the tree and then explores the tree from the left to the right, moving continuously along the edges at unit speed (in the way explained by the arrows of Fig.2), until all edges have been explored and the particle has come back to the root. Since it is clear that each edge will be crossed twice in this evolution, the total time needed to explore the tree is $2|\tau|$. The value $C(s)$ of the contour function at time $s \in [0, 2|\tau|]$ is the distance (on the tree) between the position of the particle at time s and the root. By convention $C(s) = 0$ if $s \geq 2|\tau|$. Fig.2 explains the construction of the contour function better than a formal definition.

Let $k \geq 0$ be an integer. A Dyck path of length $2k$ is a sequence $(x_0, x_1, x_2, \ldots, x_{2k})$ of nonnegative integers such that $x_0 = x_{2k} = 0$, and $|x_i - x_{i-1}| = 1$ for every $i = 1, \ldots, 2k$. Clearly, if τ is a plane tree of size k, and $(C(s))_{s\geq 0}$ is its contour function, the sequence $(C(0), C(1), \ldots, C(2k))$ is a Dyck path of length $2k$. More precisely, we have the following easy result.

PROPOSITION 2.2. *The mapping $\tau \mapsto (C(0), C(1), \ldots, C(2k))$ is a bijection from $\mathbf{A}_k$ onto the set of all Dyck paths of length $2k$.*

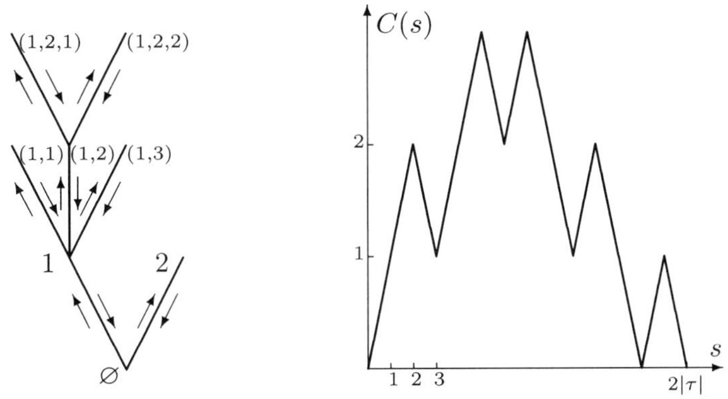

FIGURE 2. A tree and its contour function

2.2. Galton-Watson trees. Let μ be a critical or subcritical offspring distribution. This means that μ is a probability measure on $\mathbb{Z}_+$ such that

$$\sum_{k=0}^{\infty} k\mu(k) \leq 1.$$

We exclude the trivial case where $\mu(1) = 1$.

To define Galton-Watson trees, we let $(K_u, u \in \mathcal{U})$ be a collection of independent random variables with law μ, indexed by the set $\mathcal{U}$. Denote by θ the random subset of $\mathcal{U}$ defined by

$$\theta = \{u = (u^1, \ldots, u^n) \in \mathcal{U} : u^j \leq K_{(u^1,\ldots,u^{j-1})} \text{ for every } 1 \leq j \leq n\}.$$

PROPOSITION 2.3. *θ is a.s. a tree. Moreover, if*

$$Z_n = \#\{u \in \theta : |u| = n\},$$

$(Z_n, n \geq 0)$ is a Galton-Watson process with offspring distribution μ and initial value $Z_0 = 1$.

REMARK 2.4. *Clearly $k_u(\theta) = K_u$ for every $u \in \theta$.*

The tree θ, or any random tree with the same distribution, will be called a Galton-Watson tree with offspring distribution μ, or in short a μ-Galton-Watson tree. We also write Π_μ for the distribution of θ on the space $\mathbf{A}$.

We leave the easy proof of the proposition to the reader. The finiteness of the tree θ comes from the fact that the Galton-Watson process with offspring distribution μ becomes extinct a.s., so that $Z_n = 0$ for n large.

If τ is a tree and $1 \leq j \leq k_\varnothing(\tau)$, we write $T_j\tau$ for the tree τ shifted at j:

$$T_j\tau = \{u \in \mathcal{U} : ju \in \tau\}.$$

Note that $T_j\tau$ is a tree.

Then Π_μ may be characterized by the following two properties (see e.g. [**44**] for more general statements):

(i) $\Pi_\mu(k_\varnothing = j) = \mu(j), \quad j \in \mathbb{Z}_+$.
(ii) For every $j \geq 1$ with $\mu(j) > 0$, the shifted trees $T_1\tau, \ldots, T_j\tau$ are independent under the conditional probability $\Pi_\mu(d\tau \mid k_\varnothing = j)$ and their conditional distribution is Π_μ.

Property (ii) is often called the branching property of the Galton-Watson tree. We now give an explicit formula for Π_μ.

PROPOSITION 2.5. *For every $\tau \in \mathbf{A}$,*

$$\Pi_\mu(\tau) = \prod_{u \in \tau} \mu(k_u(\tau)).$$

PROOF. We can easily check that

$$\{\theta = \tau\} = \bigcap_{u \in \tau} \{K_u = k_u(\tau)\},$$

so that

$$\Pi_\mu(\tau) = P(\theta = \tau) = \prod_{u \in \tau} P(K_u = k_u(\tau)) = \prod_{u \in \tau} \mu(k_u(\tau)).$$

□

We will be interested in the particular case when $\mu = \mu_0$ is the (critical) geometric offspring distribution, $\mu_0(k) = 2^{-k-1}$ for every $k \in \mathbb{Z}_+$. In that case, the proposition gives

$$\Pi_{\mu_0}(\tau) = 2^{-2|\tau|-1}$$

(note that $\sum_{u \in \tau} k_u(\tau) = |\tau|$ for every $\tau \in \mathbf{A}$).

In particular $\Pi_{\mu_0}(\tau)$ only depends on $|\tau|$. As a consequence, for every integer $k \geq 0$, the conditional probability distribution $\Pi_{\mu_0}(\cdot \mid |\tau| = k)$ is just the uniform probability measure on $\mathbf{A}_k$. This fact will be important later.

2.3. The contour function in the geometric case. In general, the Dyck path of a Galton-Watson tree does not have a "nice" probabilistic structure (see however Section 1 of [**29**]). In this section we restrict our attention to the case when $\mu = \mu_0$ is the critical geometric offspring distribution.

First recall that $(S_n)_{n \geq 0}$ is a simple random walk on $\mathbb{Z}$ (started from 0) if it can be written as

$$S_n = X_1 + X_2 + \cdots + X_n$$

where $X_1, X_2, \ldots$ are i.i.d. random variables with distribution $P(X_n = 1) = P(X_n = -1) = \frac{1}{2}$.

Set $T = \inf\{n \geq 0 : S_n = -1\} < \infty$ a.s. The random finite path

$$(S_0, S_1, \ldots, S_{T-1})$$

(or any random path with the same distribution) is called an excursion of simple random walk. Obviously this random path is a random Dyck path of length $T - 1$.

PROPOSITION 2.6. *Let θ be a μ_0-Galton-Watson tree. Then the Dyck path of θ is an excursion of simple random walk.*

PROOF. Since plane trees are in one-to-one correspondence with Dyck paths (Proposition 2.2), the statement of the proposition is equivalent to saying that the random plane tree θ coded by an excursion of simple random walk is a μ_0-Galton-Watson tree. To see this, introduce the upcrossing times of the random walk S from 0 to 1:

$$U_1 = \inf\{n \geq 0 : S_n = 1\} \ , \ V_1 = \inf\{n \geq U_1 : S_n = 0\}$$

and by induction, for every $j \geq 1$,

$$U_{j+1} = \inf\{n \geq V_j : S_n = 1\} \ , \ V_{j+1} = \inf\{n \geq U_{j+1} : S_n = 0\}.$$

Let $K = \sup\{j : U_j \le T\}$ ($\sup\varnothing = 0$). From the relation between a plane tree and its associated Dyck path, one easily sees that $k_\varnothing(\theta) = K$, and that for every $i = 1, \ldots, K$, the Dyck path associated with the subtree $T_i\theta$ is the path ω_i, with

$$\omega_i(n) := S_{(U_i+n)\wedge(V_i-1)} - 1 \quad , \; 0 \le n \le V_i - U_i - 1.$$

A simple application of the Markov property now shows that K is distributed according to μ_0 and that conditionally on $K = k$, the paths $\omega_1, \ldots, \omega_k$ are k independent excursions of simple random walk. The characterization of Π_{μ_0} by properties (i) and (ii) listed before Proposition 2.5 now shows that θ is a μ_0-Galton-Watson-tree. □

2.4. Brownian excursions. Our goal is to prove that the (suitably rescaled) contour function of a tree uniformly distributed over $\mathbf{A}_k$ converges in distribution as $k \to \infty$ towards a normalized Brownian excursion. We first need to recall some basic facts about Brownian excursions.

We consider a standard linear Brownian motion $B = (B_t)_{t\ge0}$ starting from the origin. The process $\beta_t = |B_t|$ is called reflected Brownian motion. We denote by $(L_t^0)_{t\ge0}$ the local time process of B (or of β) at level 0, which can be defined by the approximation

$$L_t^0 = \lim_{\varepsilon\to0} \frac{1}{2\varepsilon} \int_0^t ds\, \mathbf{1}_{[-\varepsilon,\varepsilon]}(B_s) = \lim_{\varepsilon\to0} \frac{1}{2\varepsilon} \int_0^t ds\, \mathbf{1}_{[0,\varepsilon]}(\beta_s),$$

for every $t \ge 0$, a.s.

Then $(L_t^0)_{t\ge0}$ is a continuous increasing process, and the set of increase points of the function $t \to L_t^0$ coincides with the set

$$\mathcal{Z} = \{t \ge 0 : \beta_t = 0\}$$

of all zeros of β. Consequently, if we introduce the right-continuous inverse of the local time process,

$$\sigma_\ell := \inf\{t \ge 0 : L_t^0 > \ell\} \;, \quad \text{for every } \ell \ge 0,$$

we have

$$\mathcal{Z} = \{\sigma_\ell : \ell \ge 0\} \cup \{\sigma_{\ell-} : \ell \in D\}$$

where D denotes the countable set of all discontinuity times of the mapping $\ell \to \sigma_\ell$.

The connected components of the open set $\mathbb{R}_+\backslash\mathcal{Z}$ are called the *excursion intervals* of β away from 0. The preceding discussion shows that, with probability one, the excursion intervals of β away from 0 are exactly the intervals $(\sigma_{\ell-}, \sigma_\ell)$ for $\ell \in D$. Then, for every $\ell \in D$, we define the excursion $e_\ell = (e_\ell(t))_{t\ge0}$ associated with the interval $(\sigma_{\ell-}, \sigma_\ell)$ by setting

$$e_\ell(t) = \begin{cases} \beta_{\sigma_{\ell-}+t} & \text{if } 0 \le t \le \sigma_\ell - \sigma_{\ell-}\,, \\ 0 & \text{if } t > \sigma_\ell - \sigma_{\ell-}\,. \end{cases}$$

We view e_ℓ as an element of the excursion space E, which is defined by

$$E = \{e \in C(\mathbb{R}_+, \mathbb{R}_+) : e(0) = 0 \text{ and } \zeta(e) := \sup\{s > 0 : e(s) > 0\} \in (0, \infty)\},$$

where $\sup\varnothing = 0$ by convention. Note that we require $\zeta(e) > 0$, so that the zero function does not belong to E. The space E is equipped with the metric d defined by

$$d(e, e') = \sup_{t\ge0} |e(t) - e'(t)| + |\zeta(e) - \zeta(e')|$$

and with the associated Borel σ-field. Notice that $\zeta(e_\ell) = \sigma_\ell - \sigma_{\ell-}$ for every $\ell \in D$. The following theorem is the basic result of excursion theory in our particular setting.

THEOREM 2.7. *The point measure*

$$\sum_{\ell \in D} \delta_{(\ell, e_\ell)}(ds\, de)$$

is a Poisson measure on $\mathbb{R}_+ \times E$, *with intensity*

$$2ds \otimes \mathbf{n}(de)$$

where $\mathbf{n}(de)$ *is a* σ*-finite measure on* E.

The measure $\mathbf{n}(de)$ is called the Itô measure of positive excursions of linear Brownian motion, or simply the Itô excursion measure (our measure $\mathbf{n}$ corresponds to the measure n_+ in Chapter XII of [**46**]). The next corollary follows from standard properties of Poisson measures.

COROLLARY 2.8. *Let* A *be a measurable subset of* E *such that* $0 < \mathbf{n}(A) < \infty$, *and let* $T_A = \inf\{\ell \in D : e_\ell \in A\}$. *Then,* T_A *is exponentially distributed with parameter* $\mathbf{n}(A)$, *and the distribution of* e_{T_A} *is the conditional measure*

$$\mathbf{n}(\cdot \mid A) = \frac{\mathbf{n}(\cdot \cap A)}{\mathbf{n}(A)}.$$

Moreover, T_A *and* e_{T_A} *are independent.*

This corollary can be used to calculate various distributions under the Itô excursion measure. The distribution of the height and the length of the excursion are given as follows: For every $\varepsilon > 0$,

$$\mathbf{n}\Big(\max_{t \geq 0} e(t) > \varepsilon\Big) = \frac{1}{2\varepsilon}$$

and

$$\mathbf{n}(\zeta(e) > \varepsilon) = \frac{1}{\sqrt{2\pi\varepsilon}}.$$

The Itô excursion measure enjoys the following scaling property. For every $\lambda > 0$, define a mapping $\Phi_\lambda : E \longrightarrow E$ by setting $\Phi_\lambda(e)(t) = \sqrt{\lambda}\, e(t/\lambda)$, for every $e \in E$ and $t \geq 0$. Then we have $\Phi_\lambda(\mathbf{n}) = \sqrt{\lambda}\,\mathbf{n}$.

This scaling property is useful when defining conditional versions of the Itô excursion measure. We discuss the conditioning of $\mathbf{n}(de)$ with respect to the length $\zeta(e)$. There exists a unique collection $(\mathbf{n}_{(s)}, s > 0)$ of probability measures on E such that the following properties hold:

(i) For every $s > 0$, $\mathbf{n}_{(s)}(\zeta = s) = 1$.
(ii) For every $\lambda > 0$ and $s > 0$, we have $\Phi_\lambda(\mathbf{n}_{(s)}) = \mathbf{n}_{(\lambda s)}$.
(iii) For every measurable subset A of E,

$$\mathbf{n}(A) = \int_0^\infty \mathbf{n}_{(s)}(A)\, \frac{ds}{2\sqrt{2\pi s^3}}.$$

We may and will write $\mathbf{n}_{(s)} = \mathbf{n}(\cdot \mid \zeta = s)$. The measure $\mathbf{n}_{(1)} = \mathbf{n}(\cdot \mid \zeta = 1)$ is called the law of the normalized Brownian excursion.

There are many different descriptions of the Itô excursion measure: See in particular [**46**, Chapter XII]. We state the following proposition, which emphasizes the Markovian properties of $\mathbf{n}$. For every $t > 0$ and $x > 0$, we set

$$q_t(x) = \frac{x}{\sqrt{2\pi t^3}} \exp(-\frac{x^2}{2t}).$$

Note that the function $t \mapsto q_t(x)$ is the density of the first hitting time of x by B. For $t > 0$ and $x, y \in \mathbb{R}$, we also let

$$p_t(x,y) = \frac{1}{\sqrt{2\pi t}} \exp(-\frac{(y-x)^2}{2t})$$

be the usual Brownian transition density.

PROPOSITION 2.9. *The Itô excursion measure* $\mathbf{n}$ *is the only* σ*-finite measure on* E *that satisfies the following two properties:*

(i) *For every* $t > 0$*, and every* $f \in C(\mathbb{R}_+, \mathbb{R}_+)$*,*

$$\mathbf{n}(f(e(t))\,\mathbf{1}_{\{\zeta > t\}}) = \int_0^\infty f(x)\, q_t(x)\, dx.$$

(ii) *Let* $t > 0$*. Under the conditional probability measure* $\mathbf{n}(\cdot \mid \zeta > t)$*, the process* $(e(t+r))_{r \geq 0}$ *is Markov with the transition kernels of Brownian motion stopped upon hitting* 0*.*

This proposition can be used to establish absolute continuity properties of the conditional measures $\mathbf{n}_{(s)}$ with respect to $\mathbf{n}$. For every $t \geq 0$, let $\mathcal{F}_t$ denote the σ-field on E generated by the mappings $r \mapsto e(r)$, for $0 \leq r \leq t$. Then, if $0 < t < 1$, the measure $\mathbf{n}_{(1)}$ is absolutely continuous with respect to $\mathbf{n}$ on the σ-field $\mathcal{F}_t$, with Radon-Nikodým density

$$\frac{d\mathbf{n}_{(1)}}{d\mathbf{n}}\Big|_{\mathcal{F}_t}(e) = 2\sqrt{2\pi}\, q_{1-t}(e(t)).$$

This formula provides a simple derivation of the finite-dimensional marginals under $\mathbf{n}_{(1)}$, noting that the finite-dimensional marginals under $\mathbf{n}$ are easily obtained from Proposition 2.9. More precisely, for every integer $p \geq 1$, and every choice of $0 < t_1 < t_2 < \cdots < t_p < 1$, we get that the distribution of $(e(t_1), \ldots, e(t_p))$ under $\mathbf{n}_{(1)}(de)$ has density

$$(1) \qquad 2\sqrt{2\pi}\, q_{t_1}(x_1)\, p^*_{t_2-t_1}(x_1,x_2)\, p^*_{t_3-t_2}(x_2,x_3) \cdots p^*_{t_p-t_1}(x_{p-1},x_p)\, q_{1-t_p}(x_p)$$

where

$$p^*_t(x,y) = p_t(x,y) - p_t(x,-y)\ , t > 0\ ,\ \ x,y > 0$$

is the transition density of Brownian motion killed when it hits 0. As a side remark, formula (1) shows that the law of $(e(t))_{0 \leq t \leq 1}$ under $\mathbf{n}_{(1)}$ is invariant under time-reversal.

2.5. Convergence of contour functions to the Brownian excursion. The following theorem can be viewed as a special case of the results in Aldous [**2**]. The space of all continuous functions from $[0,1]$ into $\mathbb{R}_+$ is denoted by $C([0,1], \mathbb{R}_+)$, and is equipped with the topology of uniform convergence.

THEOREM 2.10. *For every integer $k \geq 1$, let θ_k be a random tree that is uniformly distributed over $\mathbf{A}_k$, and let $(C_k(t))_{t\geq 0}$ be its contour function. Then*

$$\Big(\frac{1}{\sqrt{2k}} C_k(2k\,t)\Big)_{0\leq t\leq 1} \xrightarrow[k\to\infty]{\rm (d)} (\mathbb{e}_t)_{0\leq t\leq 1}$$

where $\mathbb{e}$ is distributed according to $\mathbf{n}_{(1)}$ (i.e. $\mathbb{e}$ is a normalized Brownian excursion) and the convergence holds in the sense of weak convergence of the laws on the space $C([0,1],\mathbb{R}_+)$.

PROOF. We already noticed that $\Pi_{\mu_0}(\cdot \mid |\tau| = k)$ coincides with the uniform distribution over $\mathbf{A}_k$. By combining this with Proposition 2.6, we get that $(C_k(0), C_k(1), \ldots, C_k(2k))$ is distributed as an excursion of simple random walk conditioned to have length $2k$. Recall our notation $(S_n)_{n\geq 0}$ for simple random walk on $\mathbb{Z}$ starting from 0, and $T = \inf\{n \geq 0 : S_n = -1\}$. To get the desired result, we need to verify that the law of

$$\Big(\frac{1}{\sqrt{2k}} S_{\lfloor 2kt \rfloor}\Big)_{0\leq t\leq 1}$$

under $P(\cdot \mid T = 2k+1)$ converges to $\mathbf{n}_{(1)}$ as $k \to \infty$. This result can be seen as a conditional version of Donsker's theorem (see Kaigh [**26**] for similar statements). We will provide a detailed proof, because this result plays a major role in what follows, and because some of the ingredients of the proof will be needed again in Section 3 below. As usual, the proof is divided into two parts: We first check the convergence of finite-dimensional marginals, and then establish the tightness of the sequence of laws.

Finite-dimensional marginals. We first consider one-dimensional marginals. So we fix $t \in (0,1)$, and we will verify that

$$\lim_{k\to\infty} \sqrt{2k}\, P\Big(S_{\lfloor 2kt \rfloor} = \lfloor x\sqrt{2k} \rfloor \text{ or } \lfloor x\sqrt{2k} \rfloor + 1 \,\Big|\, T = 2k+1\Big) = 4\sqrt{2\pi}\, q_t(x)\, q_{1-t}(x), \tag{2}$$

uniformly when x varies over a compact subset of $(0,\infty)$. Comparing with the case $p = 1$ of formula (1), we see that the law of $(2k)^{-1/2} S_{\lfloor 2kt \rfloor}$ under $P(\cdot \mid T = 2k+1)$ converges to the law of $e(t)$ under $n_{(1)}(de)$ (we even get a local version of this convergence).

In order to prove (2), we will use two lemmas. The first one is a very special case of classical local limit theorems (see e.g. Chapter 2 of Spitzer [**49**]).

LEMMA 2.11. *For every $\varepsilon > 0$,*

$$\lim_{n\to\infty} \sup_{x\in\mathbb{R}} \sup_{s\geq\varepsilon} \Big|\sqrt{n} P\Big(S_{\lfloor ns \rfloor} = \lfloor x\sqrt{n} \rfloor \text{ or } \lfloor x\sqrt{n} \rfloor + 1\Big) - 2\,p_s(0,x)\Big| = 0.$$

In our special situation, the result of the lemma is easily obtained by direct calculations using the explicit form of the law of S_n and Stirling's formula.

The next lemma is (a special case of) a famous formula of Kemperman (see e.g. [**45**] Chapter 6). For every integer $\ell \in \mathbb{Z}$, we use P_ℓ for a probability measure under which the simple random walk S starts from ℓ.

LEMMA 2.12. *For every $\ell \in \mathbb{Z}_+$ and every integer $n \geq 1$,*

$$P_\ell(T = n) = \frac{\ell+1}{n}\, P_\ell(S_n = -1).$$

PROOF. It is easy to see that

$$P_\ell(T=n)=\frac{1}{2}P_\ell(S_{n-1}=0,\,T>n-1).$$

On the other hand,

$$\begin{aligned}P_\ell(S_{n-1}=0,\,T>n-1)&=P_\ell(S_{n-1}=0)-P_\ell(S_{n-1}=0,\,T\le n-1)\\&=P_\ell(S_{n-1}=0)-P_\ell(S_{n-1}=-2,\,T\le n-1)\\&=P_\ell(S_{n-1}=0)-P_\ell(S_{n-1}=-2),\end{aligned}$$

where the second equality is a simple application of the reflection principle. So we have

$$P_\ell(T=n)=\frac{1}{2}\Big(P_\ell(S_{n-1}=0)-P_\ell(S_{n-1}=-2)\Big)$$

and an elementary calculation shows that this is equivalent to the statement of the lemma. □

Let us turn to the proof of (2). We first write for $i\in\{1,\dots,2k\}$ and $\ell\in\mathbb{Z}_+$,

$$P(S_i=\ell\mid T=2k+1)=\frac{P(\{S_i=\ell\}\cap\{T=2k+1\})}{P(T=2k+1)}.$$

By an application of the Markov property of S,

$$P(\{S_i=\ell\}\cap\{T=2k+1\})=P(S_i=\ell,T>i)\,P_\ell(T=2k+1-i).$$

Furthermore, a simple time-reversal argument (we leave the details to the reader) shows that

$$P(S_i=\ell,T>i)=2\,P_\ell(T=i+1).$$

Summarizing, we have obtained

$$\begin{aligned}(3)\qquad P(S_i=\ell\mid T=2k+1)&=\frac{2P_\ell(T=i+1)P_\ell(T=2k+1-i)}{P(T=2k+1)}\\&=\frac{2(2k+1)(\ell+1)^2}{(i+1)(2k+1-i)}\;\frac{P_\ell(S_{i+1}=-1)P_\ell(S_{2k+1-i}=-1)}{P(S_{2k+1}=-1)}\end{aligned}$$

using Lemma 2.12 in the second equality.

We apply this identity with $i=\lfloor 2kt\rfloor$ and $\ell=\lfloor x\sqrt{2k}\rfloor$ or $\ell=\lfloor x\sqrt{2k}\rfloor+1$. Using Lemma 2.11, we have first

$$\frac{2(2k+1)(\lfloor x\sqrt{2k}\rfloor+1)^2}{(\lfloor 2kt\rfloor+1)(2k+1-\lfloor 2kt\rfloor)}\times\frac{1}{P(S_{2k+1}=-1)}\approx 2\sqrt{2\pi}\,(k/2)^{1/2}\,\frac{x^2}{t(1-t)}$$

and, using Lemma 2.11 once again,

$$\begin{aligned}&P_{\lfloor x\sqrt{2k}\rfloor}(S_{\lfloor 2kt\rfloor+1}=-1)P_{\lfloor x\sqrt{2k}\rfloor}(S_{2k+1-\lfloor 2kt\rfloor}=-1)\\&\quad+\;P_{\lfloor x\sqrt{2k}\rfloor+1}(S_{\lfloor 2kt\rfloor+1}=-1)P_{\lfloor x\sqrt{2k}\rfloor+1}(S_{2k+1-\lfloor 2kt\rfloor}=-1)\\&\quad\approx 2\,k^{-1}\,p_t(0,x)p_{1-t}(0,x).\end{aligned}$$

Putting these estimates together, and noting that $q_t(x)=(x/t)p_t(0,x)$, we arrive at (2).

Higher order marginals can be treated in a similar way. Let us sketch the argument in the case of two-dimensional marginals. We observe that, if $0<i<j<2k$ and if $\ell,m\in\mathbb{Z}_+$, we have, by the same arguments as above,

$$\begin{aligned}&P(S_i=\ell,S_j=m,T=2k+1)\\&=\;2\,P_\ell(T=i+1)\,P_\ell(S_{j-i}=m,T>j-i)\,P_m(T=k+1-j).\end{aligned}$$

Only the middle term $P_\ell(S_{j-i} = m, T > j-i)$ requires a different treatment than in the case of one-dimensional marginals. However, by an application of the reflection principle, one has

$$P_\ell(S_{j-i} = m, T > j-i) = P_\ell(S_{j-i} = m) - P_\ell(S_{j-i} = -m-2).$$

Hence, using Lemma 2.11, we easily obtain that for $x, y > 0$ and $0 < s < t < 1$,

$$\begin{aligned} P_{\lfloor x\sqrt{2k}\rfloor}(S_{\lfloor 2kt\rfloor - \lfloor 2ks\rfloor} = \lfloor y\sqrt{2k}\rfloor) + P_{\lfloor x\sqrt{2k}\rfloor+1}(S_{\lfloor 2kt\rfloor - \lfloor 2ks\rfloor} = \lfloor y\sqrt{2k}\rfloor) \\ \approx (2k)^{-1/2}\, p^*_{t-s}(x,y), \end{aligned}$$

and the result for two-dimensional marginals follows in a straightforward way.

Tightness. We start with some combinatorial considerations. We fix $k \geq 1$. Let $(x_0, x_1, \ldots, x_{2k})$be a Dyck path with length $2k$, and let $i \in \{0, 1, \ldots, 2k-1\}$. We set, for every $j \in \{0, 1, \ldots, 2k\}$,

$$x_j^{(i)} = x_i + x_{i\oplus j} - 2 \min_{i\wedge(i\oplus j) \leq n \leq i\vee(i\oplus j)} x_n$$

with the notation $i \oplus j = i + j$ if $i + j \leq 2k$, and $i \oplus j = i + j - 2k$ if $i + j > 2k$. It is elementary to see that $(x_0^{(i)}, x_1^{(i)}, \ldots, x_{2k}^{(i)})$ is again a Dyck path with length $2k$. Moreover, the mapping $\Phi_i : (x_0, x_1, \ldots, x_{2k}) \longrightarrow (x_0^{(i)}, x_1^{(i)}, \ldots, x_{2k}^{(i)})$ is a bijection from the set of all Dyck paths with length $2k$ onto itself. To see this, one may check that the composition $\Phi_{2k-i} \circ \Phi_i$ is the identity mapping. This property is easily verified by viewing Φ_i as a mapping defined on plane trees with $2k$ edges (using Proposition 2.2): The plane tree corresponding to the image under Φ_i of the Dyck path associated with a tree τ is the "same" tree τ re-rooted at the corner corresponding to the i-th step of the contour exploration of τ. From this observation it is obvious that the composition $\Phi_{2k-i} \circ \Phi_i$ leads us back to the original plane tree.

To simplify notation, we set for every $i, j \in \{0, 1, \ldots, 2k\}$,

$$\check{C}_k^{i,j} = \min_{i\wedge j \leq n \leq i\vee j} C_k(n).$$

The preceding discussion then gives the identity in distribution

$$\Big(C_k(i) + C_k(i\oplus j) - 2\check{C}_k^{i,i\oplus j}\Big)_{0\leq j\leq 2k} \overset{(d)}{=} (C_k(j))_{0\leq j\leq 2k}. \tag{4}$$

LEMMA 2.13. *For every integer $p \geq 1$, there exists a constant K_p such that, for every $k \geq 1$ and every $i \in \{0, 1, \ldots, 2k\}$,*

$$E[C_k(i)^{2p}] \leq K_p\, i^p.$$

Assuming that the lemma holds, the proof of tightness is easily completed. Using the identity (4), we get for $0 \leq i < j \leq 2k$,

$$\begin{aligned} E[(C_k(j) - C_k(i))^{2p}] &\leq E[(C_k(i) + C_k(j) - 2\check{C}_k^{i,j})^{2p}] \\ &= E[C_k(j-i)^{2p}] \\ &\leq K_p(j-i)^p. \end{aligned}$$

It readily follows that the bound

$$E\Big[\Big(\frac{C_k(2kt) - C_k(2ks)}{\sqrt{2k}}\Big)^{2p}\Big] \leq K_p\,(t-s)^p.$$

holds at least if s and t are of the form $s = i/2k$, $t = j/2k$, with $0 \leq i < j \leq 2k$. Since the function C_k is 1-Lipschitz, a simple argument shows that the same bound holds (possibly with a different constant K_p) whenever $0 \leq s < t \leq 1$. This gives the desired tightness, but we still have to prove the lemma.

Proof of Lemma 2.13. Clearly, we may restrict our attention to the case $1 \leq i \leq k$ (note that $(C_k(2k-i))_{0\leq i\leq 2k}$ has the same distribution as $(C_k(i))_{0\leq i\leq 2k}$). Recall that $C_k(i)$ has the same distribution as S_i under $P(\cdot \mid T = 2k+1)$. By formula (3), we have thus, for every integer $\ell \geq 0$,

$$P(C_k(i) = \ell) = \frac{2(2k+1)(\ell+1)^2}{(i+1)(2k+1-i)} \, \frac{P_\ell(S_{i+1} = -1)P_\ell(S_{2k+1-i} = -1)}{P(S_{2k+1} = -1)}.$$

From Lemma 2.11 (and our assumption $i \leq k$), we can find two positive constants c_0 and c_1 such that

$$P(S_{2k+1} = -1) \geq c_0(2k)^{-1/2}\,, \quad P_\ell(S_{2k+1-i} = -1) \leq c_1(2k)^{-1/2}.$$

It then follows that

$$\begin{aligned} P(C_k(i) = \ell) &\leq 4c_1(c_0)^{-1}\, \frac{(\ell+1)^2}{i+1}\, P_\ell(S_{i+1} = -1) \\ &= 4c_1(c_0)^{-1}\, \frac{(\ell+1)^2}{i+1}\, P(S_{i+1} = \ell+1). \end{aligned}$$

Consequently,

$$\begin{aligned} E[C_k(i)^{2p}] &= \sum_{\ell=0}^{\infty} \ell^{2p} P(C_k(i) = \ell) \\ &\leq \frac{4c_1(c_0)^{-1}}{i+1} \sum_{\ell=0}^{\infty} \ell^{2p}(\ell+1)^2\, P(S_{i+1} = \ell+1) \\ &\leq \frac{4c_1(c_0)^{-1}}{i+1}\, E[(S_{i+1})^{2p+2}]. \end{aligned}$$

However, it is well known and easy to prove that $E[(S_{i+1})^{2p+2}] \leq K'_p(i+1)^{p+1}$, with some constant K'_p independent of i. This completes the proof of the lemma and of Theorem 2.10. □

Extensions and variants of Theorem 2.10 can be found in [**2**], [**17**] and [**18**]. To illustrate the power of this theorem, let us give a typical application. The height $H(\tau)$ of a plane tree τ is the maximal generation of a vertex of τ.

COROLLARY 2.14. *Let θ_k be uniformly distributed over $\mathbf{A}_k$. Then*

$$\frac{1}{\sqrt{2k}} H(\theta_k) \xrightarrow[k\to\infty]{(d)} \max_{0\leq t\leq 1} \mathbb{e}_t.$$

Since

$$\frac{1}{\sqrt{2k}} H(\theta_k) = \max_{0\leq t\leq 1} \Big(\frac{1}{\sqrt{2k}} C_k(2k\,t)\Big)$$

the result of the corollary is immediate from Theorem 2.10.

The limiting distribution in Corollary 2.14 is known in the form of a series: For every $x > 0$,

$$P\Big(\max_{0\leq t\leq 1} \mathbb{e}_t > x\Big) = 2\sum_{k=1}^{\infty} (4k^2x^2 - 1)\exp(-2k^2x^2).$$

See Chung [**13**].

3. Real trees and the Gromov-Hausdorff convergence

Our main goal in this section is to interpret the convergence of contour functions in Theorem 2.10 as a convergence of discrete random trees towards a "continuous random tree" which is coded by the Brownian excursion in the same sense as a plane tree is coded by its contour function. We need to introduce a suitable notion of a continuous tree, and then to explain in which sense the convergence takes place.

3.1. Real trees. We start with a formal definition. In these notes, we consider only *compact* real trees, and so we include this compactness property in the definition.

DEFINITION 3.1. *A compact metric space* $(\mathcal{T}, d)$ *is a real tree if the following two properties hold for every* $a, b \in \mathcal{T}$.

(i) *There is a unique isometric map* $f_{a,b}$ *from* $[0, d(a,b)]$ *into* $\mathcal{T}$ *such that* $f_{a,b}(0) = a$ *and* $f_{a,b}(d(a,b)) = b$.
(ii) *If* q *is a continuous injective map from* $[0,1]$ *into* $\mathcal{T}$, *such that* $q(0) = a$ *and* $q(1) = b$, *we have*

$$q([0,1]) = f_{a,b}([0, d(a,b)]).$$

A rooted real tree is a real tree $(\mathcal{T}, d)$ *with a distinguished vertex* $\rho = \rho(\mathcal{T})$ *called the root. In what follows, real trees will always be rooted, even if this is not mentioned explicitly.*

Informally, one should think of a (compact) real tree as a connected union of line segments in the plane with no loops. Asssume for simplicity that there are finitely many segments in the union. Then, for any two points a and b in the tree, there is a unique path going from a to b in the tree, which is the concatentation of finitely many line segments. The distance between a and b is then the length of this path.

Let us consider a rooted real tree $(\mathcal{T}, d)$. The range of the mapping $f_{a,b}$ in (i) is denoted by $[\![a, b]\!]$ (this is the "line segment" between a and b in the tree). In particular, $[\![\rho, a]\!]$ is the path going from the root to a, which we will interpret as the ancestral line of vertex a. More precisely, we can define a partial order on the tree by setting $a \preccurlyeq b$ (a is an ancestor of b) if and only if $a \in [\![\rho, b]\!]$.

If $a, b \in \mathcal{T}$, there is a unique $c \in \mathcal{T}$ such that $[\![\rho, a]\!] \cap [\![\rho, b]\!] = [\![\rho, c]\!]$. We write $c = a \wedge b$ and call c the most recent common ancestor to a and b.

By definition, the multiplicity of a vertex $a \in \mathcal{T}$ is the number of connected components of $\mathcal{T} \backslash \{a\}$. Vertices of $\mathcal{T}$ which have multiplicity 1 are called leaves.

3.2. Coding real trees. In this subsection, we describe a method for constructing real trees, which is well-suited to our forthcoming applications to random trees. This method is nothing but a continuous analog of the coding of discrete trees by contour functions.

We consider a (deterministic) continuous function $g : [0,1] \longrightarrow [0, \infty)$ such that $g(0) = g(1) = 0$. To avoid trivialities, we will also assume that g is not identically zero. For every $s, t \in [0,1]$, we set

$$m_g(s,t) = \inf_{r \in [s \wedge t, s \vee t]} g(r),$$

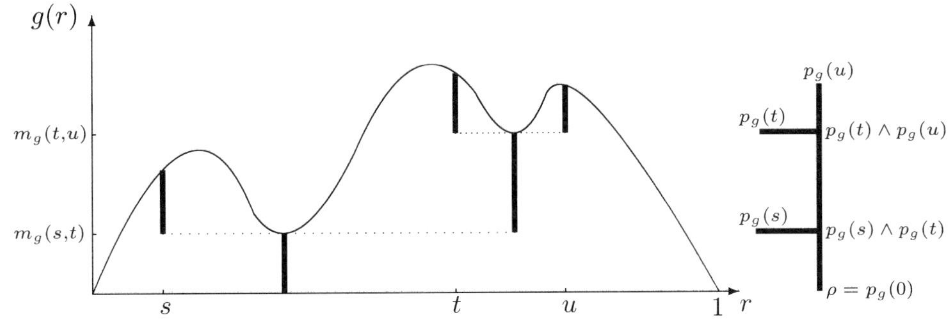

FIGURE 3. Coding a tree by a continuous function

and

$$d_g(s,t) = g(s) + g(t) - 2m_g(s,t).$$

Clearly $d_g(s,t) = d_g(t,s)$ and it is also easy to verify the triangle inequality

$$d_g(s,u) \leq d_g(s,t) + d_g(t,u)$$

for every $s, t, u \in [0,1]$. We then introduce the equivalence relation $s \sim t$ iff $d_g(s,t) = 0$ (or equivalently iff $g(s) = g(t) = m_g(s,t)$). Let $\mathcal{T}_g$ be the quotient space

$$\mathcal{T}_g = [0,1]/\sim .$$

Obviously the function d_g induces a distance on $\mathcal{T}_g$, and we keep the notation d_g for this distance. We denote by $p_g : [0,1] \longrightarrow \mathcal{T}_g$ the canonical projection. Clearly p_g is continuous (when $[0,1]$ is equipped with the Euclidean metric and $\mathcal{T}_g$ with the metric d_g), and the metric space $(\mathcal{T}_g, d_g)$ is thus compact.

THEOREM 3.1. *The metric space $(\mathcal{T}_g, d_g)$ is a real tree. We will view $(\mathcal{T}_g, d_g)$ as a rooted tree with root $\rho = p_g(0) = p_g(1)$.*

REMARK 3.2. *It is also possible to prove that any (rooted) real tree can be represented in the form $\mathcal{T}_g$. We will leave this as an exercise for the reader.*

To get an intuitive understanding of Theorem 3.1, the reader should have a look at Fig.3. This figure shows how to construct a simple subtree of $\mathcal{T}_g$, namely the "reduced tree" consisting of the union of the ancestral lines in $\mathcal{T}_g$ of three vertices $p_g(s), p_g(t), p_g(u)$ corresponding to three (given) times $s, t, u \in [0,1]$. This reduced tree is the union of the five bold line segments that are constructed from the graph of g in the way explained on the left part of the figure. Notice that the lengths of the horizontal dotted lines play no role in the construction, and that the reduced tree should be viewed as pictured on the right part of Fig.3. The ancestral line of $p_g(s)$ (resp. $p_g(t), p_g(u)$) is a line segment of length $g(s)$ (resp. $g(t), g(u)$). The ancestral lines of $p_g(s)$ and $p_g(t)$ share a common part, which has length $m_g(s,t)$ (the line segment at the bottom in the left or the right part of Fig.3), and of course a similar property holds for the ancestral lines of $p_g(s)$ and $p_g(u)$, or of $p_g(t)$ and $p_g(u)$.

The following re-rooting lemma, which is of independent interest, is a useful ingredient of the proof of Theorem 3.1 (a discrete version of this lemma already appeared at the beginning of the proof of tightness in Theorem 2.10).

LEMMA 3.3. *Let $s_0 \in [0,1)$. For any real $r \geq 0$, denote the fractional part of r by $\overline{r} = r - \lfloor r \rfloor$. Set*

$$g'(s) = g(s_0) + g(\overline{s_0 + s}) - 2m_g(s_0, \overline{s_0 + s}),$$

for every $s \in [0,1]$. Then, the function g' is continuous and satisfies $g'(0) = g'(1) = 0$, so that we can define $\mathcal{T}_{g'}$. Furthermore, for every $s, t \in [0,1]$, we have

$$d_{g'}(s,t) = d_g(\overline{s_0 + s}, \overline{s_0 + t}) \tag{5}$$

and there exists a unique isometry R from $\mathcal{T}_{g'}$ onto $\mathcal{T}_g$ such that, for every $s \in [0,1]$,

$$R(p_{g'}(s)) = p_g(\overline{s_0 + s}). \tag{6}$$

Assuming that Theorem 3.1 is proved, we see that $\mathcal{T}_{g'}$ coincides with the real tree $\mathcal{T}_g$ re-rooted at $p_g(s_0)$. Thus the lemma tells us which function codes the tree $\mathcal{T}_g$ re-rooted at an arbitrary vertex.

Proof. It is immediately checked that g' satisfies the same assumptions as g, so that we can make sense of $\mathcal{T}_{g'}$. Then the key step is to verify the relation (5). Consider first the case where $s, t \in [0, 1 - s_0)$. Then two possibilities may occur.

If $m_g(s_0 + s, s_0 + t) \geq m_g(s_0, s_0 + s)$, then $m_g(s_0, s_0 + r) = m_g(s_0, s_0 + s) = m_g(s_0, s_0 + t)$ for every $r \in [s,t]$, and so

$$m_{g'}(s,t) = g(s_0) + m_g(s_0 + s, s_0 + t) - 2m_g(s_0, s_0 + s).$$

It follows that

$$\begin{aligned}
d_{g'}(s,t) &= g'(s) + g'(t) - 2m_{g'}(s,t) \\
&= g(s_0 + s) - 2m_g(s_0, s_0 + s) + g(s_0 + t) \\
&\qquad - 2m_g(s_0, s_0 + t) - 2(m_g(s_0 + s, s_0 + t) - 2m_g(s_0, s_0 + s)) \\
&= g(s_0 + s) + g(s_0 + t) - 2m_g(s_0 + s, s_0 + t) \\
&= d_g(s_0 + s, s_0 + t).
\end{aligned}$$

If $m_g(s_0 + s, s_0 + t) < m_g(s_0, s_0 + s)$, then the minimum in the definition of $m_{g'}(s,t)$ is attained at r_1 defined as the first $r \in [s,t]$ such that $g(s_0 + r) = m_g(s_0, s_0 + s)$ (because for $r \in [r_1, t]$ we will have $g(s_0 + r) - 2m_g(s_0, s_0 + r) \geq -m_g(s_0, s_0 + r) \geq -m_g(s_0, s_0 + r_1)$). Therefore,

$$m_{g'}(s,t) = g(s_0) - m_g(s_0, s_0 + s),$$

and

$$\begin{aligned}
d_{g'}(s,t) &= g(s_0 + s) - 2m_g(s_0, s_0 + s) + g(s_0 + t) \\
&\quad -2m_g(s_0, s_0 + t) + 2m_g(s_0, s_0 + s) \\
&= d_g(s_0 + s, s_0 + t).
\end{aligned}$$

The other cases are treated in a similar way and are left to the reader.

By (5), if $s, t \in [0,1]$ are such that $d_{g'}(s,t) = 0$, then $d_g(\overline{s_0 + s}, \overline{s_0 + t}) = 0$ so that $p_g(\overline{s_0 + s}) = p_g(\overline{s_0 + t})$. Noting that $\mathcal{T}_{g'} = p_{g'}([0,1])$, we can define R in a unique way by the relation (6). From (5), R is an isometry, and it is also immediate that R takes $\mathcal{T}_{g'}$ onto $\mathcal{T}_g$. □

Thanks to the lemma, the fact that $\mathcal{T}_g$ verifies property (i) in the definition of a real tree is obtained from the particular case when $a = \rho$ and $b = p_g(s)$ for some

$s \in [0,1]$. In that case however, the isometric mapping $f_{\rho,b}$ is easily constructed by setting

$$f_{\rho,b}(t) = p_g(\sup\{r \le s : g(r) = t\}) , \quad \text{for every } 0 \le t \le g(s) = d_g(\rho, b).$$

The remaining part of the argument is straightforward: See Section 2 in [**19**].

REMARK 3.4. *A short proof of Theorem 3.1 using the characterization of real trees via the so-called four-point condition can be found in* [**20**].

The following simple observation will be useful in Section 7: If $s,t \in [0,1]$, the line segment $[\![p_g(s), p_g(t)]\!]$ in the tree $\mathcal{T}_g$ coincides with the collection of the vertices $p_g(r)$, for all $r \in [0,1]$ such that either $g(r) = m_g(r,s) \ge m_g(s,t)$ or $g(r) = m_g(r,t) \ge m_g(s,t)$. This easily follows from the construction of the distance d_g.

3.3. The Gromov-Hausdorff convergence. In order to make sense of the convergence of discrete trees towards real trees, we will use the Gromov-Hausdorff distance between compact metric spaces, which has been introduced by Gromov (see e.g. [**22**]) in view of geometric applications.

If (E,δ) is a metric space, the notation $\delta_{Haus}(K,K')$ stands for the usual Hausdorff metric between compact subsets of E :

$$\delta_{Haus}(K,K') = \inf\{\varepsilon > 0 : K \subset U_\varepsilon(K') \text{ and } K' \subset U_\varepsilon(K)\},$$

where $U_\varepsilon(K) := \{x \in E : \delta(x,K) \le \varepsilon\}$.

A pointed metric space is just a pair consisting of a metric space E and a distinguished point ρ of E. We often write E instead of (E,ρ) to simplify notation.

Then, if (E_1,ρ_1) and (E_2,ρ_2) are two pointed compact metric spaces, we define the distance $d_{GH}(E_1,E_2)$ by

$$d_{GH}(E_1,E_2) = \inf\{\delta_{Haus}(\varphi_1(E_1),\varphi_2(E_2)) \vee \delta(\varphi_1(\rho_1),\varphi_2(\rho_2))\}$$

where the infimum is over all possible choices of the metric space (E,δ) and the isometric embeddings $\varphi_1 : E_1 \longrightarrow E$ and $\varphi_2 : E_2 \longrightarrow E$ of E_1 and E_2 into E.

Two pointed compact metric spaces E_1 and E_2 are called equivalent if there is an isometry that maps E_1 onto E_2 and preserves the distinguished points. Obviously $d_{GH}(E_1,E_2)$ only depends on the equivalence classes of E_1 and E_2. We denote by $\mathbb{K}$ the space of all equivalence classes of pointed compact metric spaces.

THEOREM 3.5. *d_{GH} defines a metric on the set $\mathbb{K}$. Furthermore the metric space $(\mathbb{K}, d_{GH})$ is separable and complete.*

A proof of the fact that d_{GH} is a metric on the set $\mathbb{K}$ can be found in [**11**, Theorem 7.3.30]. This proof is in fact concerned with the non-pointed case, but the argument is easily adapted to our setting. The separability of the space $(\mathbb{K}, d_{GH})$ follows from the fact that finite metric spaces are dense in $\mathbb{K}$. Finally the completeness of $(\mathbb{K}, d_{GH})$ can be obtained as a consequence of the compactness theorem in [**11**, Theorem 7.4.15].

In our applications, it will be important to have the following alternative definition of d_{GH}. First recall that if (E_1,d_1) and (E_2,d_2) are two compact metric spaces, a correspondence between E_1 and E_2 is a subset $\mathcal{R}$ of $E_1 \times E_2$ such that for every $x_1 \in E_1$ there exists at least one $x_2 \in E_2$ such that $(x_1,x_2) \in \mathcal{R}$ and conversely for every $y_2 \in E_2$ there exists at least one $y_1 \in E_1$ such that $(y_1,y_2) \in \mathcal{R}$.

The distortion of the correspondence $\mathcal{R}$ is defined by

$$\mathrm{dis}(\mathcal{R}) = \sup\{|d_1(x_1, y_1) - d_2(x_2, y_2)| : (x_1, x_2), (y_1, y_2) \in \mathcal{R}\}.$$

PROPOSITION 3.6. *Let (E_1, ρ_1) and (E_2, ρ_2) be two pointed compact metric spaces. Then,*

$$d_{GH}(E_1, E_2) = \frac{1}{2} \inf_{\mathcal{R} \in \mathcal{C}(E_1, E_2),\, (\rho_1, \rho_2) \in \mathcal{R}} \mathrm{dis}(\mathcal{R}), \tag{7}$$

where $\mathcal{C}(E_1, E_2)$ denotes the set of all correspondences between E_1 and E_2.

See [**11**, Theorem 7.3.25] for a proof of this proposition in the non-pointed case, which is easily adapted.

The following consequence of Proposition 3.6 will be very useful. Notice that a rooted real tree can be viewed as a pointed compact metric space, whose distinguished point is the root.

COROLLARY 3.7. *Let g and g' be two continuous functions from $[0, 1]$ into $\mathbb{R}_+$, such that $g(0) = g(1) = g'(0) = g'(1) = 0$. Then,*

$$d_{GH}(\mathcal{T}_g, \mathcal{T}_{g'}) \leq 2\|g - g'\|,$$

where $\|g - g'\| = \sup_{t \in [0,1]} |g(t) - g'(t)|$ is the supremum norm of $g - g'$.

Proof. We rely on formula (7). We can construct a correspondence between $\mathcal{T}_g$ and $\mathcal{T}_{g'}$ by setting

$$\mathcal{R} = \{(a, a') : \exists t \in [0, 1] \text{ such that } a = p_g(t) \text{ and } a' = p_{g'}(t)\}.$$

Note that $(\rho, \rho') \in \mathcal{R}$, if $\rho = p_g(0)$, resp. $\rho' = p_{g'}(0)$, is the root of $\mathcal{T}_g$, resp. the root of $\mathcal{T}_{g'}$. In order to bound the distortion of $\mathcal{R}$, let $(a, a') \in \mathcal{R}$ and $(b, b') \in \mathcal{R}$. By the definition of $\mathcal{R}$ we can find $s, t \geq 0$ such that $p_g(s) = a$, $p_{g'}(s) = a'$ and $p_g(t) = b$, $p_{g'}(t) = b'$. Now recall that

$$d_g(a, b) = g(s) + g(t) - 2m_g(s, t),$$
$$d_{g'}(a', b') = g'(s) + g'(t) - 2m_{g'}(s, t),$$

so that

$$|d_g(a, b) - d_{g'}(a', b')| \leq 4\|g - g'\|.$$

Thus we have $\mathrm{dis}(\mathcal{R}) \leq 4\|g - g'\|$ and the desired result follows from (7). □

3.4. Convergence towards the CRT. As in subsection 2.5, we use the notation $\mathbb{e}$ for a normalized Brownian excursion. We view $\mathbb{e} = (\mathbb{e}_t)_{0 \leq t \leq 1}$ as a (random) continuous function over the interval $[0, 1]$, which satisfies the same assumptions as the function g in subsection 3.2.

DEFINITION 3.2. *The Brownian continuum random tree, also called the CRT, is the random real tree $\mathcal{T}_{\mathbb{e}}$ coded by the normalized Brownian excursion.*

The CRT $\mathcal{T}_{\mathbb{e}}$ is thus a random variable taking values in the set $\mathbb{K}$. Note that the measurability of this random variable follows from Corollary 3.7.

REMARK 3.8. *Aldous* [**1**],[**2**] *uses a different method to define the CRT. The preceding definition then corresponds to Corollary 22 in* [**2**]. *Note that our normalization differs by an unimportant scaling factor 2 from the one in Aldous' papers: The CRT there is the tree $\mathcal{T}_{2\mathbb{e}}$ instead of $\mathcal{T}_{\mathbb{e}}$.*

We will now restate Theorem 2.10 as a convergence in distribution of discrete random trees towards the CRT in the space $(\mathbb{K}, d_{GH})$.

THEOREM 3.9. *For every $k \geq 1$, let θ_k be uniformly distributed over $\mathbf{A}_k$, and equip θ_k with the usual graph distance d_{gr}. Then*

$$(\theta_k, (2k)^{-1/2} d_{gr}) \xrightarrow[k\to\infty]{\text{(d)}} (\mathcal{T}_{\mathbb{e}}, d_{\mathbb{e}})$$

in the sense of convergence in distribution for random variables with values in $(\mathbb{K}, d_{GH})$.

PROOF. As in Theorem 2.10, let C_k be the contour function of θ_k, and define a rescaled version of C_k by setting

$$\widetilde{C}_k(t) = (2k)^{-1/2} C_k(2k\,t)$$

for every $t \in [0,1]$. Note that the function $\widetilde{C}_k$ is continuous and nonnegative over $[0,1]$ and vanishes at 0 and at 1. Therefore we can define the real tree $\mathcal{T}_{\widetilde{C}_k}$.

Now observe that this real tree is very closely related to the (rescaled) discrete tree θ_k. Indeed $\mathcal{T}_{\widetilde{C}_k}$ is (isometric to) a finite union of line segments of length $(2k)^{-1/2}$ in the plane, with genealogical structure prescribed by θ_k, in the way suggested in the left part of Fig.2. From this observation, and the definition of the Gromov-Hausdorff distance, we easily get

$$d_{GH}\Big((\theta_k, (2k)^{-1/2} d_{gr}), (\mathcal{T}_{\widetilde{C}_k}, d_{\widetilde{C}_k})\Big) \leq (2k)^{-1/2}. \tag{8}$$

On the other hand, by combining Theorem 2.10 and Corollary 3.7, we have

$$(\mathcal{T}_{\widetilde{C}_k}, d_{\widetilde{C}_k}) \xrightarrow[k\to\infty]{\text{(d)}} (\mathcal{T}_{\mathbb{e}}, d_{\mathbb{e}}).$$

The statement of Theorem 3.9 now follows from the latter convergence and (8). □

REMARK 3.10. *Theorem 3.9 contains in fact less information than Theorem 2.10, because the lexicographical ordering that is inherent to the notion of a plane tree (and also to the coding of real trees by functions) disappears when we look at a plane tree as a metric space. Still, Theorem 3.9 is important from the conceptual viewpoint: It is crucial to think of the CRT as a continuous limit of rescaled discrete random trees.*

There are analogs of Theorem 3.9 for other classes of combinatorial trees. For instance, if τ_n is distributed uniformly among all rooted Cayley trees with n vertices, then $(\tau_n, (4n)^{-1/2} d_{gr})$ converges in distribution to the CRT $\mathcal{T}_{\mathbb{e}}$, in the space $\mathbb{K}$. Similarly, discrete random trees that are uniformly distributed over binary trees with $2k$ edges converge in distribution (modulo a suitable rescaling) towards the CRT. All these results can be derived from a general statement of convergence of conditioned Galton-Watson trees due to Aldous [**2**] (see also [**29**]). A recent work of Haas and Miermont [**23**] provides further extensions of Theorem 3.9 to Pólya trees (unordered rooted trees).

4. Labeled trees and the Brownian snake

4.1. Labeled trees. In view of forthcoming applications to random planar maps, we now introduce labeled trees. A labeled tree is a pair $(\tau, (\ell(v))_{v\in\tau})$ that consists of a plane tree τ (see subsection 2.1) and a collection $(\ell(v))_{v\in\tau}$ of integer labels assigned to the vertices of τ – in our formalism for plane trees, the tree τ coincides with the set of all its vertices. We assume that labels satisfy the following three properties:

(i) for every $v \in \tau$, $\ell(v) \in \mathbb{Z}$;
(ii) $\ell(\varnothing) = 0$;
(iii) for every $v \in \tau \backslash \{\varnothing\}$, $\ell(v) - \ell(\pi(v)) = 1, 0,$ or -1,

where we recall that $\pi(v)$ denotes the parent of v. Condition (iii) just means that when crossing an edge of τ the label can change by at most 1 in absolute value.

The motivation for introducing labeled trees comes from the fact that (rooted and pointed) planar quadrangulations can be coded by such trees (see Section 4 below). Our goal in the present section is to derive asymptotics for large labeled trees chosen uniformly at random, in the same way as Theorem 2.10, or Theorem 3.9, provides asymptotics for large plane trees. For every integer $k \geq 0$, we denote by $\mathbf{T}_k$ the set of all labeled trees with k edges. It is immediate that

$$\#\mathbf{T}_k = 3^k \#\mathbf{A}_k = \frac{3^k}{k+1}\binom{2k}{k}$$

simply because for each edge of the tree there are three possible choices for the label increment along this edge.

Let $(\tau, (\ell(v))_{v \in \tau})$ be a labeled tree with k edges. As we saw in subsection 2.1, the plane tree τ is coded by its contour function $(C_t)_{t \geq 0}$. We can similarly encode the labels by another function $(V_t)_{t \geq 0}$, which is defined as follows. If we explore the tree τ by following its contour, in the way suggested by the arrows of Fig.2, we visit successively all vertices of τ (vertices that are not leaves are visited more than once). Write $v_0 = \varnothing, v_1, v_2, \ldots, v_{2k} = \varnothing$ for the successive vertices visited in this exploration. For instance, in the particular example of Fig.1 we have

$$v_0 = \varnothing, v_1 = 1, v_2 = (1,1), v_3 = 1, v_4 = (1,2), v_5 = (1,2,1), v_6 = (1,2), \ldots$$

The finite sequence $v_0, v_1, v_2, \ldots, v_{2k}$ will be called the contour exploration of the vertices of τ.

Notice that $C_i = |v_i|$, for every $i = 0, 1, \ldots, 2k$, by the definition of the contour function. We similarly set

$$V_i = \ell(v_i) \quad \text{for every } i = 0, 1, \ldots, 2k.$$

To complete this definition, we set $V_t = 0$ for $t \geq 2k$ and, for every $i = 1, \ldots, 2k$, we define V_t for $t \in (i-1, i)$ by using linear interpolation. We will call $(V_t)_{t \geq 0}$ the "label contour function" of the labeled tree $(\tau, (\ell(v))_{v \in \tau})$ Clearly $(\tau, (\ell(v))_{v \in \tau})$ is determined by the pair $(C_t, V_t)_{t \geq 0}$.

Our goal is now to describe the scaling limit of this pair when the labeled tree $(\tau, (\ell(v))_{v \in \tau})$ is chosen uniformly at random in $\mathbf{T}_k$ and $k \to \infty$. As an immediate consequence of Theorem 2.10 (and the fact that the number of possible labelings is the same for every plane tree with k edges), the scaling limit of $(C_t)_{t \geq 0}$ is the normalized Brownian excursion. To describe the scaling limit of $(V_t)_{t \geq 0}$ we need to introduce the Brownian snake.

4.2. The snake driven by a deterministic function. Consider a continuous function $g : [0,1] \longrightarrow \mathbb{R}_+$ such that $g(0) = g(1) = 0$ (as in subsection 3.2). We also assume that g is Hölder continuous: There exist two positive constants K and

γ such that, for every $s, t \in [0,1]$,

$$|g(s) - g(t)| \le K\,|s-t|^{\gamma}.$$

As in subsection 3.2, we also set, for every $s, t \in [0,1]$,

$$m_g(s,t) = \min_{r \in [s\wedge t, s\vee t]} g(r).$$

LEMMA 4.1. *The function $(m_g(s,t))_{s,t\in[0,1]}$ is nonnegative definite in the sense that, for every integer $n \ge 1$, for every $s_1, \ldots, s_n \in [0,1]$ and every $\lambda_1, \ldots, \lambda_n \in \mathbb{R}$, we have*

$$\sum_{i=1}^{n}\sum_{j=1}^{n} \lambda_i \lambda_j \, m_g(s_i, s_j) \ge 0.$$

PROOF. Fix $s_1, \ldots, s_n \in [0,1]$, and let $t \ge 0$. For $i, j \in \{1, \ldots, n\}$, put $i \approx j$ if $m_g(s_i, s_j) \ge t$. Then $\approx$ is an equivalence relation on $\{i : g(s_i) \ge t\} \subset \{1, \ldots, n\}$. By summing over the different classes of this equivalence relation, we get that

$$\sum_{i=1}^{n}\sum_{j=1}^{n} \lambda_i\lambda_j \mathbf{1}_{\{t \le m_g(s_i,s_j)\}} = \sum_{\mathcal{C} \text{ class of } \approx} \Big(\sum_{i\in\mathcal{C}} \lambda_i\Big)^2 \ge 0.$$

Now integrate with respect to dt to get the desired result. □

By Lemma 4.1 and a standard application of the Kolmogorov extension theorem, there exists a centered Gaussian process $(Z^g_s)_{s\in[0,1]}$ whose covariance is

$$E[Z^g_s Z^g_t] = m_g(s,t)$$

for every $s, t \in [0,1]$. Consequently we have

$$\begin{aligned} E[(Z^g_s - Z^g_t)^2] &= E[(Z^g_s)^2] + E[(Z^g_t)^2] - 2E[Z^g_s Z^g_t] \\ &= g(s) + g(t) - 2m_g(s,t) \\ &\le 2K\,|s-t|^{\gamma}, \end{aligned}$$

where the last bound follows from our Hölder continuity assumption on g (this calculation also shows that $E[(Z^g_s - Z^g_t)^2] = d_g(s,t)$, in the notation of subsection 3.2). From the previous bound and an application of the Kolmogorov continuity criterion, the process $(Z^g_s)_{s\in[0,1]}$ has a modification with continuous sample paths. This leads us to the following definition.

DEFINITION 4.1. *The snake driven by the function g is the centered Gaussian process $(Z^g_s)_{s\in[0,1]}$ with continuous sample paths and covariance*

$$E[Z^g_s Z^g_t] = m_g(s,t)\ , \quad s,t \in [0,1].$$

Notice that we have in particular $Z^g_0 = Z^g_1 = 0$. More generally, for every $t \in [0,1]$, Z^g_t is normal with mean 0 and variance $g(t)$.

REMARK 4.2. *Recall from subsection 3.2 the definition of the equivalence relation $\sim$ associated with g: $s \sim t$ iff $d_g(s,t) = 0$. Since we have $E[(Z^g_s - Z^g_t)^2] = d_g(s,t)$, a simple argument shows that almost surely for every $s, t \in [0,1]$, the condition $s \sim t$ implies that $Z^g_s = Z^g_t$. In other words we may view Z^g as a process indexed by the quotient $[0,1]/\sim$, that is by the tree $\mathcal{T}_g$. Indeed, it is then very natural to interpret Z^g as Brownian motion indexed by the tree $\mathcal{T}_g$: In the particular case when $\mathcal{T}_g$ is a finite union of segments (which holds if g is piecewise monotone), Z^g can be constructed by running independent Brownian motions along the branches of*

$\mathcal{T}_g$. *It is however more convenient to view* Z^g *as a process indexed by* $[0,1]$ *because later the function* g *(and thus the tree* $\mathcal{T}_g$*) will be random and we avoid considering a random process indexed by a random set.*

4.3. Convergence towards the Brownian snake. Let $\mathbb{e}$ be as previously a normalized Brownian excursion. By standard properties of Brownian paths, the function $t \mapsto \mathbb{e}_t$ is a.s. Hölder continuous (with exponent $\frac{1}{2} - \varepsilon$ for any $\varepsilon > 0$), and so we can apply the construction of the previous subsection to (almost) every realization of $\mathbb{e}$.

In other words, we can construct a pair $(\mathbb{e}_t, Z_t)_{t\in[0,1]}$ of continuous random processes, whose distribution is characterized by the following two properties:

(i) $\mathbb{e}$ is a normalized Brownian excursion;

(ii) conditionally given $\mathbb{e}$, Z is distributed as the snake driven by $\mathbb{e}$.

The process Z will be called the Brownian snake (driven by $\mathbb{e}$). This terminology is a little different from the usual one: Usually, the Brownian snake is viewed as a path-valued process (see e.g. [**28**]) and Z_t would correspond only to the terminal point of the value at time t of this path-valued process.

We can now answer the question raised at the end of subsection 4.1. The following theorem is due to Chassaing and Schaeffer [**12**]. More general results can be found in [**25**].

THEOREM 4.3. *For every integer* $k \geq 1$, *let* $(\theta_k, (\ell^k(v))_{v\in\theta_k})$ *be distributed uniformly over the set* $\mathbf{T}_k$ *of all labeled trees with* k *edges. Let* $(C_k(t))_{t\geq 0}$ *and* $(V_k(t))_{t\geq 0}$ *be respectively the contour function and the label contour function of the labeled tree* $(\theta_k, (\ell^k(v))_{v\in\theta_k})$. *Then,*

$$\Big(\frac{1}{\sqrt{2k}}C_k(2k\,t), \Big(\frac{9}{8k}\Big)^{1/4} V_k(2k\,t)\Big)_{t\in[0,1]} \xrightarrow[k\to\infty]{\text{(d)}} (\mathbb{e}_t, Z_t)_{t\in[0,1]}$$

where the convergence holds in the sense of weak convergence of the laws on the space $C([0,1], \mathbb{R}_+^2)$.

PROOF. From Theorem 2.10 and the Skorokhod representation theorem, we may assume without loss of generality that

$$\sup_{0\leq t\leq 1} |(2k)^{-1/2} C_k(2kt) - \mathbb{e}_t| \xrightarrow[k\to\infty]{\text{a.s.}} 0. \tag{9}$$

We first discuss the convergence of finite-dimensional marginals: We prove that for every choice of $0 \leq t_1 < t_2 < \cdots < t_p \leq 1$, we have

$$\Big(\frac{1}{\sqrt{2k}}C_k(2k\,t_i), \Big(\frac{9}{8k}\Big)^{1/4} V_k(2k\,t_i)\Big)_{1\leq i\leq p} \xrightarrow[k\to\infty]{\text{(d)}} (\mathbb{e}_{t_i}, Z_{t_i})_{1\leq i\leq p}. \tag{10}$$

Since for every $i \in \{1, \ldots, n\}$,

$$|C_k(2kt_i) - C_k(\lfloor 2kt_i \rfloor)| \leq 1\ , \quad |V_k(2kt_i) - V_k(\lfloor 2kt_i \rfloor)| \leq 1$$

we may replace $2kt_i$ by its integer part $\lfloor 2kt_i \rfloor$ in (10).

Consider the case $p = 1$. We may assume that $0 < t_1 < 1$, because otherwise the result is trivial. It is immediate that conditionally on θ_k, the label increments $\ell^k(v) - \ell^k(\pi(v))$, $v \in \theta_k \backslash \{\emptyset\}$, are independent and uniformly distributed over $\{-1, 0, 1\}$. Consequently, we may write

$$(C_k(\lfloor 2kt_1 \rfloor), V_k(\lfloor 2kt_1 \rfloor)) \overset{\text{(d)}}{=} \Big(C_k(\lfloor 2kt_1 \rfloor), \sum_{i=1}^{C_k(\lfloor 2kt_1 \rfloor)} \eta_i\Big)$$

where the variables $\eta_1, \eta_2, \dots$ are independent and uniformly distributed over $\{-1, 0, 1\}$, and are also independent of the trees θ_k. By the central limit theorem,

$$\frac{1}{\sqrt{n}} \sum_{i=1}^{n} \eta_i \xrightarrow[n\to\infty]{\text{(d)}} \Big(\frac{2}{3}\Big)^{1/2} N$$

where N is a standard normal variable. Thus if we set for $\lambda \in \mathbb{R}$,

$$\Phi(n,\lambda) = E\Big[\exp\Big(i\frac{\lambda}{\sqrt{n}}\sum_{i=1}^{n}\eta_i\Big)\Big]$$

we have $\Phi(n,\lambda) \longrightarrow \exp(-\lambda^2/3)$ as $n \to \infty$.

Then, for every $\lambda, \lambda' \in \mathbb{R}$, we get by conditioning on θ_k

$$\begin{aligned}
&E\Big[\exp\Big(i\frac{\lambda}{\sqrt{2k}}C_k(\lfloor 2kt_1\rfloor) + i\frac{\lambda'}{\sqrt{C_k(\lfloor 2kt_1\rfloor)}}\sum_{i=1}^{C_k(\lfloor 2kt_1\rfloor)}\eta_i\Big)\Big]\\
&\quad = E\Big[\exp\Big(i\frac{\lambda}{\sqrt{2k}}C_k(\lfloor 2kt_1\rfloor)\Big) \times \Phi(C_k(\lfloor 2kt_1\rfloor),\lambda')\Big]\\
&\quad \xrightarrow[k\to\infty]{} E[\exp(i\lambda \mathbb{e}_{t_1})] \times \exp(-\lambda'^2/3)
\end{aligned}$$

using the (almost sure) convergence of $(2k)^{-1/2}C_k(\lfloor 2kt_1\rfloor)$ towards $\mathbb{e}_{t_1} > 0$. In other words we have obtained the joint convergence in distribution

$$\Big(\frac{C_k(\lfloor 2kt_1\rfloor)}{\sqrt{2k}}, \frac{1}{\sqrt{C_k(\lfloor 2kt_1\rfloor)}}\sum_{i=1}^{C_k(\lfloor 2kt_1\rfloor)}\eta_i\Big) \xrightarrow[k\to\infty]{\text{(d)}} (\mathbb{e}_{t_1}, (2/3)^{1/2}N), \tag{11}$$

where the normal variable N is independent of $\mathbb{e}$.

From the preceding observations, we have

$$\begin{aligned}
&\Big(\frac{C_k(\lfloor 2kt_1\rfloor)}{\sqrt{2k}}, \Big(\frac{9}{8k}\Big)^{1/4} V_k(\lfloor 2kt_1\rfloor)\Big)\\
&\overset{\text{(d)}}{=} \Big(\frac{C_k(\lfloor 2kt_1\rfloor)}{\sqrt{2k}}, \Big(\frac{3}{2}\Big)^{1/2}\Big(\frac{C_k(\lfloor 2kt_1\rfloor)}{\sqrt{2k}}\Big)^{1/2}\frac{1}{\sqrt{C_k(\lfloor 2kt_1\rfloor)}}\sum_{i=1}^{C_k(\lfloor 2kt_1\rfloor)}\eta_i\Big)
\end{aligned}$$

and from (11) we get

$$\Big(\frac{C_k(\lfloor 2kt_1\rfloor)}{\sqrt{2k}}, \Big(\frac{9}{8k}\Big)^{1/4} V_k(\lfloor 2kt_1\rfloor)\Big) \xrightarrow[k\to\infty]{\text{(d)}} (\mathbb{e}_{t_1}, \sqrt{\mathbb{e}_{t_1}}\,N).$$

This gives (10) in the case $p = 1$, since by construction it holds that $(\mathbb{e}_{t_1}, Z_{t_1}) \overset{\text{(d)}}{=} (\mathbb{e}_{t_1}, \sqrt{\mathbb{e}_{t_1}}\,N)$.

Let us discuss the case $p = 2$ of (10). We fix t_1 and t_2 with $0 < t_1 < t_2 < 1$. Recall the notation

$$\check{C}_k^{i,j} = \min_{i\wedge j \le n \le i\vee j} C_k(n)\,, \qquad i,j \in \{0,1,\dots,2k\}$$

introduced in Section 1. Write $v_0^k = \varnothing, v_1^k, \dots, v_{2k}^k = \varnothing$ for the contour exploration of vertices of θ_k (see the end of subsection 4.1). Then we know that

$$\begin{aligned}
&C_k(\lfloor 2kt_1\rfloor) = |v^k_{\lfloor 2kt_1\rfloor}|,\ C_k(\lfloor 2kt_2\rfloor) = |v^k_{\lfloor 2kt_2\rfloor}|,\\
&V_k(\lfloor 2kt_1\rfloor) = \ell^k(v^k_{\lfloor 2kt_1\rfloor}),\ V_k(\lfloor 2kt_2\rfloor) = \ell^k(v^k_{\lfloor 2kt_2\rfloor}),
\end{aligned}$$

and furthermore $\check{C}_k^{\lfloor 2kt_1\rfloor,\lfloor 2kt_2\rfloor}$ is the generation in θ_k of the last common ancestor to $v^k_{\lfloor 2kt_1\rfloor}$ and $v^k_{\lfloor 2kt_2\rfloor}$. From the properties of labels on the tree θ_k, we now see that conditionally on θ_k,

$$(V_k(\lfloor 2kt_1\rfloor), V_k(\lfloor 2kt_2\rfloor)) \overset{\text{(d)}}{=}$$

$$\Big(\sum_{i=1}^{\check{C}_k^{\lfloor 2kt_1\rfloor,\lfloor 2kt_2\rfloor}} \eta_i + \sum_{i=\check{C}_k^{\lfloor 2kt_1\rfloor,\lfloor 2kt_2\rfloor}+1}^{C_k(\lfloor 2kt_1\rfloor)} \eta'_i\,,\ \sum_{i=1}^{\check{C}_k^{\lfloor 2kt_1\rfloor,\lfloor 2kt_2\rfloor}} \eta_i + \sum_{i=\check{C}_k^{\lfloor 2kt_1\rfloor,\lfloor 2kt_2\rfloor}+1}^{C_k(\lfloor 2kt_2\rfloor)} \eta''_i\Big) \tag{12}$$

where the variables $\eta_i, \eta'_i, \eta''_i$ are independent and uniformly distributed over $\{-1, 0, 1\}$.

From (9), we have

$$\Big((2k)^{-1/2}C_k(\lfloor 2kt_1\rfloor), (2k)^{-1/2}C_k(\lfloor 2kt_2\rfloor), (2k)^{-1/2}\check{C}_k^{\lfloor 2kt_1\rfloor,\lfloor 2kt_2\rfloor}\Big) \xrightarrow[k\to\infty]{\text{a.s.}} (\mathbb{e}_{t_1}, \mathbb{e}_{t_2}, m_{\mathbb{e}}(t_1,t_2)).$$

By arguing as in the case $p = 1$, we now deduce from (12) that

$$\Big(\frac{C_k(\lfloor 2kt_1\rfloor)}{\sqrt{2k}}, \frac{C_k(\lfloor 2kt_2\rfloor)}{\sqrt{2k}}, \Big(\frac{9}{8k}\Big)^{1/4} V_k(\lfloor 2kt_1\rfloor), \Big(\frac{9}{8k}\Big)^{1/4} V_k(\lfloor 2kt_2\rfloor)\Big)$$
$$\xrightarrow[k\to\infty]{\text{(d)}} (\mathbb{e}_{t_1}, \mathbb{e}_{t_2}, \sqrt{m_{\mathbb{e}}(t_1,t_2)}\, N + \sqrt{\mathbb{e}_{t_1} - m_{\mathbb{e}}(t_1,t_2)}\, N'\,,$$
$$\sqrt{m_{\mathbb{e}}(t_1,t_2)}\, N + \sqrt{\mathbb{e}_{t_2} - m_{\mathbb{e}}(t_1,t_2)}\, N'')$$

where N, N', N'' are three independent standard normal variables, which are also independent of $\mathbb{e}$. The limiting distribution in the last display is easily identified with that of $(\mathbb{e}_{t_1}, \mathbb{e}_{t_2}, Z_{t_1}, Z_{t_2})$, and this gives the case $p = 2$ in (10). The general case is proved by similar arguments and we leave details to the reader.

To complete the proof of Theorem 4.3, we need a tightness argument. The laws of the processes

$$\Big(\frac{1}{\sqrt{2k}} C_k(2k\,t)\Big)_{t\in[0,1]}$$

are tight by Theorem 2.10, and so we need only verify the tightness of the processes

$$\Big(\Big(\frac{9}{8k}\Big)^{1/4} V_k(2k\,t)\Big)_{t\in[0,1]}.$$

This is a consequence of the following lemma, which therefore completes the proof of Theorem 4.3. □

LEMMA 4.4. *For every integer $p \geq 1$, there exists a constant $K_p < \infty$ such that, for every $k \geq 1$ and every $s, t \in [0,1]$,*

$$E\Big[\Big(\frac{V_k(2kt) - V_k(2ks)}{k^{1/4}}\Big)^{4p}\Big] \leq K_p\, |t-s|^p.$$

PROOF. Simple arguments show that we may restrict our attention to the case when $s = i/(2k)$, $t = j/(2k)$, with $i, j \in \{0, 1, \ldots, 2k\}$. By using the same decomposition as in (12), we have

$$V_k(j) - V_k(i) \overset{\text{(d)}}{=} \sum_{n=1}^{d_{gr}(v_i^k, v_j^k)} \eta_n \tag{13}$$

where the random variables η_n are independent and uniform over $\{-1,0,1\}$ (and independent of θ_k) and

$$d_{gr}(v_i^k, v_j^k) = C_k(i) + C_k(j) - 2\check{C}_k^{i,j}$$

is the graph distance in the tree θ_k between vertices v_i^k and v_j^k. From (13) and by conditioning with respect to θ_k, we get the existence of a constant K'_p such that

$$E[(V_k(i) - V_k(j))^{4p}] \leq K'_p\, E[(d_{gr}(v_i^k, v_j^k))^{2p}].$$

So the lemma will be proved if we can verify the bound

$$E[(C_k(i) + C_k(j) - 2\check{C}_k^{i,j})^{2p}] \leq K''_p\, |j - i|^p \tag{14}$$

with a constant K''_p independent of k. By the identity (4), it is enough to prove that this bound holds for $i = 0$. However, the case $i = 0$ is exactly Lemma 2.13. This completes the proof.

5. Planar maps

5.1. Definitions. A map is a combinatorial object, which can be best visualized as a class of graphs embedded in a surface. In these lectures, we will exclusively focus on the case of *plane* (or *planar*) maps, where the surface is the 2-dimensional sphere $\mathbb{S}^2$.

Let us first formalize the notion of map. We will not enter into details, referring the reader to the book by Mohar and Thomassen [**42**] for a very complete exposition. Another useful reference, discussing in depth the different equivalent ways to define maps (in particular through purely algebraic notions) is the book by Lando and Zvonkin [**27**, Chapter 1].

An *oriented edge* in $\mathbb{S}^2$ is a mapping $e : [0,1] \to \mathbb{S}^2$ that is continuous, and such that either e is injective, or the restriction of e to $[0,1)$ is injective and $e(0) = e(1)$. In the latter case, e is also called a loop. An oriented edge will always be considered up to reparametrization by a continuous increasing function from $[0,1]$ to $[0,1]$, and we will always be interested in properties of edges that do not depend on a particular parameterization. The origin and target of e are the points $e^- = e(0)$ and $e^+ = e(1)$. The reversal of e is the oriented edge $\overline{e} = e(1-\cdot)$. An *edge* is a pair $\mathbf{e} = \{e, \overline{e}\}$, where e is an oriented edge. The *interior* of $\mathbf{e}$ is defined as $e((0,1))$.

An *embedded graph* in $\mathbb{S}^2$ is a graph[1] $G = (V, E)$ such that

- V is a (finite) subset of $\mathbb{S}^2$
- E is a (finite) set of edges in $\mathbb{S}^2$
- the vertices incident to $\mathbf{e} = \{e, \overline{e}\} \in E$ are $e^-, e^+ \in V$
- the interior of an edge $\mathbf{e} \in E$ does not intersect V nor the edges of E distinct from $\mathbf{e}$

The support of an embedded graph $G = (V, E)$ is

$$\mathrm{supp}\,(G) = V \cup \bigcup_{\mathbf{e}=\{e,\overline{e}\}\in E} e([0,1])\,.$$

A *face* of the embedding is a connected component of the set $\mathbb{S}^2 \setminus \mathrm{supp}\,(G)$.

[1]all the graphs considered here are finite, and are multigraphs in which multiple edges and loops are allowed

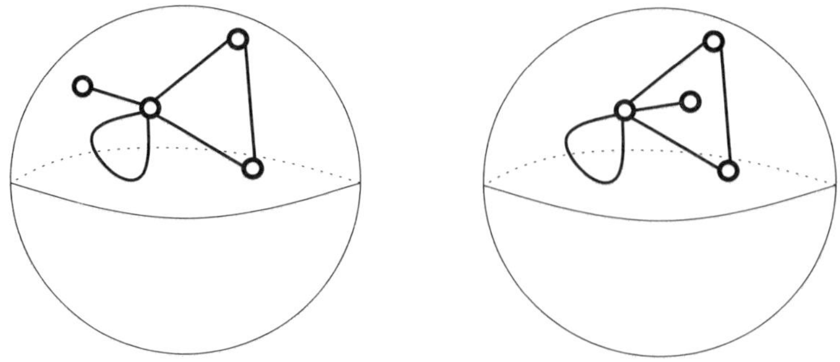

FIGURE 4. Two planar maps, with 4 vertices and 3 faces of degrees 1,3,6 and 1,4,5 respectively

DEFINITION 5.1. *A (planar) map is a connected embedded graph. Equivalently, a map is an embedded graph whose faces are all homeomorphic to the Euclidean unit disk in* $\mathbb{R}^2$.

Topologically, one would say that a map is the 1-skeleton of a CW-complex decomposition of $\mathbb{S}^2$. We will denote maps using bold characters $\mathbf{m}, \mathbf{q}, \ldots$

Let $\mathbf{m} = (V, E)$ be a map, and let $\vec{E} = \{e \in \mathbf{e} : \mathbf{e} \in E\}$ be the set of all oriented edges of $\mathbf{m}$. Since $\mathbb{S}^2$ is oriented, it is possible to define, for every oriented edge $e \in \vec{E}$, a unique face f_e of $\mathbf{m}$, located to the left of the edge e. We call f_e the face *incident* to e. Note that the edges incident to a given face form a closed curve in $\mathbb{S}^2$, but not necessarily a Jordan curve (it can happen that $f_e = f_{\bar{e}}$ for some e). The degree of a face f is defined as

$$\deg(f) = \#\{e \in \vec{E} : f_e = f\}.$$

The oriented edges incident to a given face f, are arranged cyclically in counterclockwise order around the face in what we call the *facial ordering.* With every oriented edge e, we can associate a *corner* incident to e, which is a small simply connected neighborhood of e^- intersected with f_e. Then the corners of two different oriented edges do not intersect.

Of course, the degree of a vertex $u \in V$ is the usual graph-theoretical notion

$$\deg(u) = \#\{e \in \vec{E} : e^- = u\}.$$

Similarly as for faces, the outgoing edges from u are organized cyclically in counterclockwise order around u.

A *rooted* map is a pair $(\mathbf{m}, e)$ where $\mathbf{m} = (V, E)$ is a map and $e \in \vec{E}$ is a distinguished oriented edge, called the root. We often omit the mention of e in the notation.

5.2. Euler's formula. An important property of maps is the so-called *Euler formula.* If $\mathbf{m}$ is a map, $V(\mathbf{m}), E(\mathbf{m}), F(\mathbf{m})$ denote respectively the sets of all vertices, edges and faces of $\mathbf{m}$. Then,

$$\#V(\mathbf{m}) - \#E(\mathbf{m}) + \#F(\mathbf{m}) = 2\,. \tag{15}$$

This is a relatively easy result in the case of interest (the planar case): One can remove the edges of the graph one by one until a spanning tree $\mathbf{t}$ of the graph is obtained, for which the result is trivial (it has one face, and $\#V(\mathbf{t}) = \#E(\mathbf{t}) + 1$).

5.3. Isomorphism, automorphism and rooting. In the sequel, we will always consider maps "up to deformation" in the following sense.

DEFINITION 5.2. *The maps* $\mathbf{m}, \mathbf{m}'$ *on* $\mathbb{S}^2$ *are isomorphic if there exists an orientation-preserving homeomorphism* h *of* $\mathbb{S}^2$ *onto itself, such that* h *induces a graph isomorphism of* $\mathbf{m}$ *with* $\mathbf{m}'$.

The rooted maps $(\mathbf{m}, e)$ *and* $(\mathbf{m}', e')$ *are isomorphic if* $\mathbf{m}$ *and* $\mathbf{m}'$ *are isomorphic through a homeomorphism* h *that maps* e *to* e'.

In the sequel, we will almost always identify two isomorphic maps $\mathbf{m}, \mathbf{m}'$. This of course implies that the (non-embedded, combinatorial) graphs associated with $\mathbf{m}, \mathbf{m}'$ are isomorphic, but this is stronger: For instance the two maps of Fig.4 are not isomorphic, since a map isomorphism preserves the degrees of faces.

An *automorphism* of a map $\mathbf{m}$ is an isomorphism of $\mathbf{m}$ with itself. It should be interpreted as a *symmetry* of the map. An important fact is the following.

PROPOSITION 5.1. *An automorphism of* $\mathbf{m}$ *that fixes an oriented edge fixes all the oriented edges.*

Loosely speaking, the only automorphism of a rooted map is the identity. This explains why rooting is an important tool in the combinatorial study of maps, as it "kills the symmetries". The idea of the proof of the previous statement is to see that if e is fixed by the automorphism, then all the edges incident to e^- should also be fixed (since an automorphism preserves the orientation). One can thus progress in the graph (by connectedness) and show that all the edges are fixed.

In a rooted map, the face f_e incident to the root edge e is often called the *external face*, or root face. The other faces are called *internal*. The vertex e^- is called the root vertex.

From now on, unless otherwise specified, all maps will be rooted.

We end this presentation by introducing the notion of *graph distance* in a map $\mathbf{m}$. A *chain* of length $k \geq 1$ is a sequence $e_{(1)}, \ldots, e_{(k)}$ of oriented edges in $\vec{E}(\mathbf{m})$, such that $e^+_{(i)} = e^-_{(i+1)}$ for $1 \leq i \leq k-1$, and we say that the chain links the vertices $e^-_{(1)}$ and $e^+_{(k)}$. We also allow, for every vertex $u \in V(\mathbf{m})$, a chain with length 0, starting and ending at u. The *graph distance* $d_{\mathbf{m}}(u, v)$ between two vertices $u, v \in V(\mathbf{m})$ is the minimal k such that there exists a chain with length k linking u and v. A chain with minimal length between two vertices is called a *geodesic chain*.

5.4. The Cori-Vauquelin-Schaeffer bijection. Via the identification of maps up to isomorphisms the set of all maps becomes a countable set. For instance, the set $\mathbf{M}_n$ of all rooted maps with n edges is a finite set: The $2n$ oriented edges should be organized around a finite family of polygons (the faces of the map), and the number of ways to associate the boundary edges of these polygons is finite. A natural question to ask is "what is the cardinality of $\mathbf{M}_n$?".

Tutte answered this question (and many other counting problems for maps), motivated in part by the 4-color problem. He developed a powerful method, the "quadratic method", to solve the apparently ill-defined equations for the generating functions of maps. For recent developments in this direction, see the article by Bousquet-Mélou and Jehanne [**6**]. The method, however, is a kind of "black box" which solves such counting problems without giving much extra information about the structure of maps. One obtains

$$\#\mathbf{M}_n = \frac{2}{n+2} 3^n \mathrm{Cat}_n\,,$$

where $\mathrm{Cat}_n = \frac{1}{n+1}\binom{2n}{n}$ is the n-th Catalan number. We also mention the huge literature on the enumeration of maps using matrix integrals, initiating in [**24, 10**], which is particularly popular in the physics literature. See [**27**, Chapter 4] for an introduction to this approach.

Motivated by the very simple form of the formula enumerating $\mathbf{M}_n$, Cori and Vauquelin [**14**] gave in 1981 a bijective approach to this formula. These approaches reached their full power with the work of Schaeffer starting in his 1998 thesis [**47**]. We now describe the bijective approach in the case of quadrangulations.

5.4.1. *Quadrangulations.* A map $\mathbf{q}$ is a quadrangulation if all its faces are of degree 4. We let $\mathbf{Q}_n$ be the set of all (rooted) quadrangulations with n faces. Quadrangulations are a very natural family of maps to consider, in virtue of the fact that there exists a "trivial" bijection between $\mathbf{M}_n$ and $\mathbf{Q}_n$, which can be described as follows.

Let $\mathbf{m}$ be a map with n edges, and imagine that the vertices of $\mathbf{m}$ are colored in black. We then create a new map by adding inside each face of $\mathbf{m}$ a white vertex, and joining this white vertex to every corner of the face f it belongs to, by non-intersecting edges inside the face f. In doing so, notice that some black vertices may be joined to the same white vertex with several edges. Lastly, we erase the interiors of the edges of the map $\mathbf{m}$. We end up with a map $\mathbf{q}$, which is a plane quadrangulation with n faces, each face containing exactly one edge of the initial map. We adopt a rooting convention, for instance, we root $\mathbf{q}$ at the first edge coming after e in counterclockwise order around e^-, where e is the root of $\mathbf{m}$.

Notice that $\mathbf{q}$ also comes with a bicoloration of its vertices in black and white, in which two adjacent vertices have different colors. This says that $\mathbf{q}$ is *bipartite*, and as a matter of fact, every (planar!) quadrangulation is bipartite. So this coloring is superfluous: One can recover it by declaring that the black vertices are those at even distance from the root vertex of $\mathbf{q}$, and the white vertices are those at odd distance from the root vertex.

Conversely, starting from a rooted quadrangulation $\mathbf{q}$, we can recover a bipartite coloration as above, by declaring that the vertices at even distance from the root edge are black. Then, we draw the diagonal linking the two black corners incident to every face of $\mathbf{q}$. Finally, we remove the interior of the edges of $\mathbf{q}$ and root the resulting map $\mathbf{m}$ at the first outgoing diagonal from e^- in clockwise order from the root edge e of $\mathbf{q}$. One checks that this is indeed a left- and right-inverse of the previous mapping from $\mathbf{M}_n$ to $\mathbf{Q}_n$. See Fig.5 below for an illustration of these bijections.

For the record, we state the following useful fact.

PROPOSITION 5.2. *A (planar) map is bipartite if and only if its faces all have even degree.*

5.4.2. *The CVS bijection.* Recall that $\mathbf{Q}_n$ is the set of all rooted quadrangulations with n faces. A simple application of Euler's formula shows that any element of $\mathbf{Q}_n$ has $2n$ edges ($4n$ oriented edges, 4 for each face) and $n+2$ vertices.

Let $\mathbf{T}_n$ be the set of all labeled trees with n edges, as defined in Section 3. If $(\tau, (\ell(u))_{u\in\tau}) \in \mathbf{T}_n$, then τ is a plane tree with n edges, and $\ell : \tau \to \mathbb{Z}$ is a label

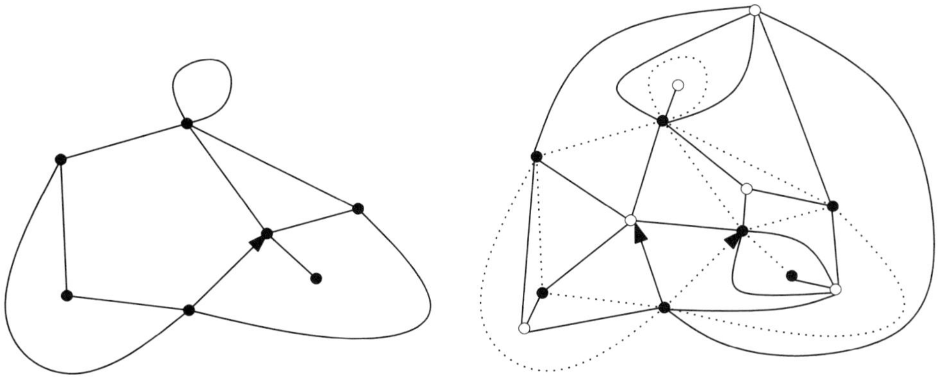

FIGURE 5. The so-called "trivial" bijection

function on τ, such that $\ell(\varnothing) = 0$ and

$$|\ell(u) - \ell(\pi(u))| \leq 1 , \qquad \text{for every } u \in \tau \setminus \{\varnothing\} .$$

In order to avoid trivialities, we now assume that $n \geq 1$. It will be convenient here to view a plane tree τ as a planar map, by embedding it in $\mathbb{S}^2$, and rooting it at the edge going from $\varnothing$ to the vertex 1. Let $\varnothing = v_0, v_1, \ldots, v_{2n} = \varnothing$ be the contour exploration of the vertices of the tree τ (see the end of subsection 4.1). For $i \in \{0, 1, \ldots, 2n-1\}$, we let e_i be the oriented edge from v_i to v_{i+1}, and extend the sequences (v_i) and (e_i) to infinite sequences by $2n$-periodicity. With each oriented edge e_i, we can associate a corner around e_i^-, as explained in subsection 4.1. In the remaining part of Section 4, we will often identify the oriented edge e_i with the associated corner, and we adopt the notation $\ell(e_i) = \ell(e_i^-)$. In particular, note that $\ell(e_i) = V_i, 0 \leq i \leq 2n$ is the label contour sequence as defined in Section 3.

For every $i \geq 0$, we define the *successor* of i by

$$s(i) = \inf\{j > i : \ell(e_j) = \ell(e_i) - 1\} ,$$

with the convention that $\inf \emptyset = \infty$. Note that $s(i) = \infty$ if and only if $\ell(e_i)$ equals $\min\{\ell(v) : v \in \tau\}$. This is a simple consequence of the fact that the integer-valued sequence $(\ell(e_i), i \geq 0)$ can decrease only by taking unit steps.

Consider a point v_* in $\mathbb{S}^2$ that does not belong to the support of τ, and denote by e_∞ a corner around v_*, i.e. a small neighborhood of v_* with v_* excluded, not intersecting the corners $e_i, i \geq 0$. By convention, we set

$$\ell(v_*) = \ell(e_\infty) = \min\{\ell(u) : u \in \tau\} - 1.$$

For every $i \geq 0$, the successor of the corner e_i is then defined by

$$s(e_i) = e_{s(i)} .$$

The CVS construction consists in drawing, for every $i \in \{0, 1, \ldots, 2n-1\}$, an *arc*, which is an edge from the corner e_i to the corner $s(e_i)$ inside $\mathbb{S}^2 \setminus (\{v_*\} \cup \operatorname{supp}(\tau))$. See Fig.6 for an illustration of the CVS construction.

LEMMA 5.3. *It is possible to draw the arcs in such a way that the graph with vertex-set $\tau \cup \{v_*\}$ and edge-set consisting of the edges of τ and the arcs is an embedded graph.*

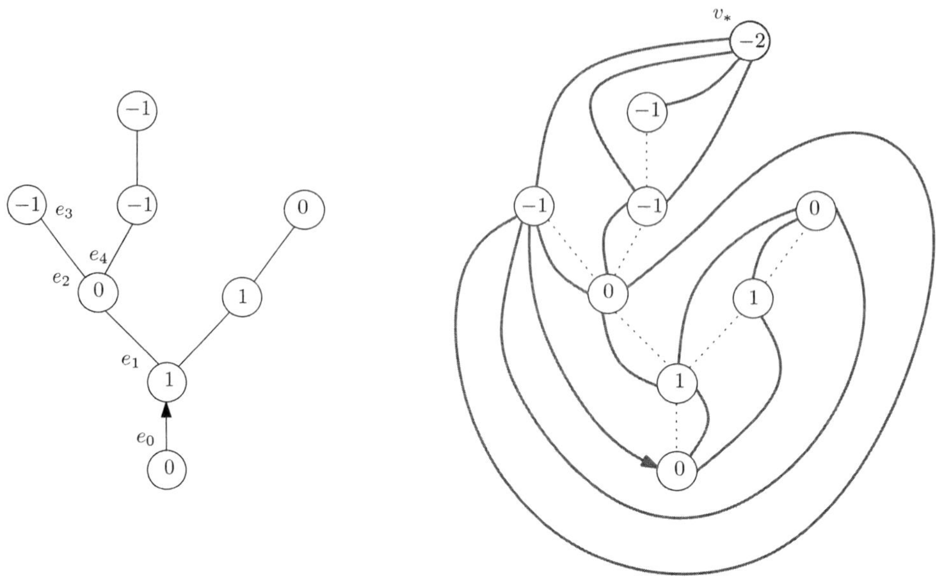

FIGURE 6. Illustration of the Cori-Vauquelin-Schaeffer bijection, in the case $\epsilon = 1$. For instance, e_3 is the successor of e_0, e_2 the successor of e_1, and so on.

PROOF. Since τ is a tree, we can see it as a map with a unique face $\mathbb{S}^2 \setminus \mathrm{supp}\,(\tau)$. The latter can in turn be seen as an open polygon, bounded by the edges $e_0, e_1, \ldots, e_{2n-1}$ in counterclockwise order. Hence, the result will follow if we can show that the arcs do not cross, i.e. that it is not possible to find pairwise distinct corners $e^{(1)}, e^{(2)}, e^{(3)}, e^{(4)}$ that arise in this order in the cyclic order induced by the contour exploration, and such that $e^{(3)} = s(e^{(1)})$ and $e^{(4)} = s(e^{(2)})$.

If this were the case, then we would have $\ell(e^{(2)}) \geq \ell(e^{(1)})$, as otherwise the successor of $e^{(1)}$ would be between $e^{(1)}$ and $e^{(2)}$. Similarly, $\ell(e^{(3)}) \geq \ell(e^{(2)})$. But by definition, $\ell(e^{(3)}) = \ell(e^{(1)}) - 1$, giving $\ell(e^{(2)}) \geq \ell(e^{(3)}) + 1 \geq \ell(e^{(2)}) + 1$, which is a contradiction. □

We call $\mathbf{q}$ the graph with vertex-set $V(\tau) \cup \{v_*\}$ and edge-set formed by the arcs, now excluding the (interiors of the) edges of τ.

LEMMA 5.4. *The embedded graph* $\mathbf{q}$ *is a quadrangulation with* n *faces.*

PROOF. First we check that $\mathbf{q}$ is connected, and hence is a map. But this is obvious since the consecutive successors of any given corner e, given by $e, s(e), s(s(e)), \ldots$, form a finite sequence ending at e_∞. Hence, every vertex in $\mathbf{q}$ can be joined by a chain to v_*, and the graph is connected.

To check that $\mathbf{q}$ is a quadrangulation, let us consider an edge of τ, corresponding to two oriented edges $e, \overline{e}$. Let us first assume that $\ell(e^+) = \ell(e^-) - 1$. Then, the successor of e is incident to e^+ and the preceding construction gives an arc starting from e^- (more precisely from the corner associated with e) and ending at e^+. Next, let e' be the corner following $\overline{e}$ in the contour exploration around τ. Then $\ell(e') = \ell(e^-) = \ell(\overline{e}) + 1$, giving that $s(\overline{e}) = s(s(e'))$. Indeed, $s(e')$ is the first corner coming after e' in contour order and with label $\ell(e') - 1 = \ell(e) - 1$, while $s(s(e'))$

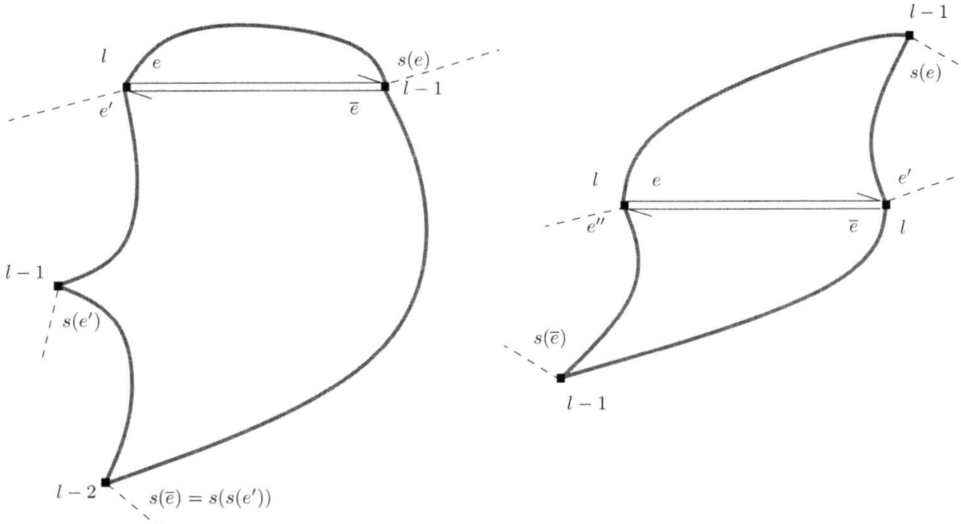

FIGURE 7. Illustration of the proof of Lemma 5.4. In this figure, $l = \ell(e)$

is the first corner coming after e' with label $\ell(e) - 2$. Therefore, it has to be the first corner coming after $\overline{e}$, with label $\ell(e) - 2 = \ell(\overline{e}) - 1$.

We deduce that the arcs joining the corners e to $s(e)$, resp. $\overline{e}$ to $s(\overline{e})$, resp. e' to $s(e')$, resp. $s(e')$ to $s(s(e')) = s(\overline{e})$, form a quadrangle, that contains the edge $\{e, \overline{e}\}$, and no other edge of τ.

If $\ell(e^+) = \ell(e^-) + 1$, the situation is the same by interchanging the roles of e and $\overline{e}$.

The only case that remains is when $\ell(e^+) = \ell(e^-)$. In this case, if e' and e'' are the corners following e and $\overline{e}$ respectively in the contour exploration of τ, then $\ell(e) = \ell(e') = \ell(\overline{e}) = \ell(e'')$, so that $s(e) = s(e')$ on the one hand and $s(\overline{e}) = s(e'')$ on the other hand. We deduce that the edge $\{e, \overline{e}\}$ is the diagonal of a quadrangle formed by the arcs linking e to $s(e)$, e' to $s(e') = s(e)$, $\overline{e}$ to $s(\overline{e})$ and e'' to $s(e'') = s(\overline{e})$. The different cases are summed up in Fig.7.

Now, notice that $\mathbf{q}$ has $2n$ edges (one per corner of τ) and $n + 2$ vertices, so it must have n faces by Euler's formula. So all the faces must be of the form described above. This completes the proof. □

Note that the quadrangulation $\mathbf{q}$ has a distinguished vertex v_*, but for now it is not a rooted quadrangulation. To fix this root, we will need an extra parameter $\epsilon \in \{-1, 1\}$. If $\epsilon = 1$ we let the root edge of $\mathbf{q}$ be the arc linking e_0 with $s(e_0)$, and oriented from $s(e_0)$ from e_0. If $\epsilon = -1$, the root edge is this same arc, but oriented from e_0 to $s(e_0)$.

In this way, we have defined a mapping Φ, from $\mathbf{T}_n \times \{-1, 1\}$ to the set $\mathbf{Q}_n^\bullet$ of pairs $(\mathbf{q}, v_*)$, where $\mathbf{q} \in \mathbf{Q}_n$ and $v_* \in V(\mathbf{q})$. We call such pairs *pointed quadrangulations*.

THEOREM 5.5. *For every $n \geq 1$, the mapping Φ is a bijection from $\mathbf{T}_n \times \{-1, 1\}$ onto $\mathbf{Q}_n^\bullet$.*

We omit the proof of this result. See Chassaing and Schaeffer [**12**, Theorem 4].

COROLLARY 5.6. *We have the following formula for every $n \geq 1$:*

$$\#\mathbf{M}_n = \#\mathbf{Q}_n = \frac{2}{n+2} 3^n \mathrm{Cat}_n$$

PROOF. We first notice that $\#\mathbf{Q}_n^\bullet = (n+2)\#\mathbf{Q}_n$, since every quadrangulation $\mathbf{q} \in \mathbf{Q}_n$ has $n+2$ vertices, each of which induces a distinct element of $\mathbf{Q}_n^\bullet$. On the other hand, it is obvious that

$$\#\mathbf{T}_n \times \{-1, 1\} = 2 \cdot 3^n \#\mathbf{A}_n = 2 \cdot 3^n \mathrm{Cat}_n \,.$$

The result follows from Theorem 5.5. □

The probabilistic counterpart of this can be stated as follows.

COROLLARY 5.7. *Let Q_n be a uniform random element in $\mathbf{Q}_n$, and conditionally given Q_n, let v_* be chosen uniformly at random in $V(Q_n)$. On the other hand, let θ_n be chosen uniformly at random in $\mathbf{T}_n$, and let ϵ be independent of θ_n and uniformly distributed in $\{-1, 1\}$. Then $\Phi(\theta_n, \epsilon)$ has the same distribution as (Q_n, v_*).*

The proof is obvious, since the probability that (Q_n, v_*) equals some particular $(\mathbf{q}, v) \in \mathbf{Q}_n^\bullet$ equals $((n+2)\#\mathbf{Q}_n)^{-1} = (\#\mathbf{Q}_n^\bullet)^{-1}$.

5.4.3. *Interpretation of the labels.* The CVS bijection will be of crucial importance to us when we will deal with metric properties of random elements of $\mathbf{Q}_n$, because the labels on $\mathbf{q}$ that are inherited from a labeled tree through the CVS construction turn out to measure certain distances in $\mathbf{q}$. Recall that the set τ is identified with $V(\mathbf{q}) \setminus \{v_*\}$ if (τ, ℓ) and $\mathbf{q}$ are associated through the CVS bijection (the choice of ϵ is irrelevant here). Hence, the function ℓ is also a function on $V(\mathbf{q}) \setminus \{v_*\}$, and we extend it by letting, as previously, $\ell(v_*) = \min\{\ell(u) : u \in \tau\} - 1$. For simplicity, we write

$$\min \ell = \min\{\ell(u) : u \in \tau\} \,.$$

PROPOSITION 5.8. *For every $v \in V(\mathbf{q})$, we have*

$$d_{\mathbf{q}}(v, v_*) = \ell(v) - \min \ell + 1 \,, \tag{16}$$

where $d_{\mathbf{q}}$ is the graph distance on $\mathbf{q}$.

PROOF. Let $v \in V(\mathbf{q}) \setminus \{v_*\} = \tau$, and let e be a corner (in τ) incident to v. Then the chain of arcs

$$e \to s(e) \to s^2(e) \to \ldots \to e_\infty$$

is a chain of length $\ell(e) - \ell(e_\infty) = \ell(v) - \ell(v_*)$ between v and v_*. Therefore, $d_{\mathbf{q}}(v, v_*) \leq \ell(v) - \ell(v_*)$. On the other hand, if $v = v_0, v_1, \ldots, v_d = v_*$ are the consecutive vertices of any chain linking v to v_*, then since $|\ell(e) - \ell(s(e))| = 1$ by definition for any corner e and since the edges of $\mathbf{q}$ all connect a corner to its successor, we get

$$d = \sum_{i=1}^{d} |\ell(v_i) - \ell(v_{i-1})| \geq |\ell(v_0) - \ell(v_d)| = \ell(v) - \ell(v_*) \,,$$

as desired. □

Remark. The preceding proof also shows that the chain of arcs $e \to s(e) \to s^2(e) \to \ldots \to e_\infty$ is a geodesic chain linking e^- to v_*. Such a geodesic chain, or more generally a chain of the form $e \to s(e) \to s^2(e) \to \ldots \to s^k(e)$, will be called a successor geodesic chain.

The triangle inequality for $d_{\mathbf{q}}$ (or the second part of the proof) gives the useful bound

$$d_{\mathbf{q}}(u,v) \geq |\ell(u) - \ell(v)|\,, \tag{17}$$

This bound will be improved in the next subsection.

As a consequence of the proposition, we obtain for instance that the "volume of spheres" around v_* can be interpreted in terms of ℓ: for every $k \geq 0$,

$$|\{v \in V(\mathbf{q}) : d_{\mathbf{q}}(v, v_*) = k\}| = |\{u \in \tau : \ell(u) - \min \ell + 1 = k\}|\,.$$

5.4.4. *Two useful bounds.* The general philosophy in the forthcoming study of random planar maps is then the following: Information about labels in a random labeled tree, which follows from the results of subsection 3.3 if this tree is uniformly distributed over $\mathbf{T}_n$, allows one to obtain information about distances in the associated quadrangulation. One major problem with this approach is that exact information will only be available for distances to a distinguished vertex v_*. There is no simple expression for the distances between two vertices distinct from v_* in terms of the labels in the tree. However, more advanced properties of the CVS bijection allow to get useful bounds on these distances. Recall that $e_0, e_1, e_2, \ldots$ is the contour sequence of corners (or oriented edges) around a tree $\tau \in \mathbf{A}_n$, starting from the root (see the beginning of subsection 5.4.2). We view $(e_i, i \geq 0)$ as cyclically ordered, and for any two corners e, e' of τ, we let $[e, e']$ be the set of all corners encountered when starting from e, following the cyclic contour order, and stopping when visiting e'.

PROPOSITION 5.9. *Let $((\tau,\ell),\epsilon)$ be an element in $\mathbf{T}_n \times \{-1,1\}$, and $(\mathbf{q}, v_*) = \Phi(((\tau,\ell),\epsilon))$. Let u, v be two vertices in $V(\mathbf{q}) \setminus \{v_*\}$, and let e, e' be two corners of τ such that $e^- = u, (e')^- = v$.*

(i) *There holds that*

$$d_{\mathbf{q}}(u,v) \leq \ell(u) + \ell(v) - 2 \min_{e'' \in [e,e']} \ell(e'') + 2\,,$$

(ii) *There holds that*

$$d_{\mathbf{q}}(u,v) \geq \ell(u) + \ell(v) - 2 \min_{w \in [[u,v]]} \ell(w)\,,$$

where $[[u,v]]$ is the set of all vertices lying on the geodesic path from u to v in the tree τ.

PROOF. For simplicity, let $m = \min_{e'' \in [e,e']} \ell(e'')$. Let e'' be the first corner in $[e,e']$ such that $\ell(e'') = m$. The corner $s^k(e)$, whenever it is well defined (i.e. whenever $d_{\mathbf{q}}(e^-, v_*) \geq k$), is called the k-th successor of e. Then e'' is the $(\ell(e) - m)$-th successor of e. Moreover, by definition, $s(e'')$ does not belong to $[e,e']$ since it has lesser label than e'', and necessarily, $s(e'')$ is also the $(\ell(e') - m + 1)$-st successor of e'. Hence, the successor geodesic chain $e \to s(e) \to s^2(e) \to \cdots \to s(e'')$ from $u = e^-$ to $s(e'')^+$, concatenated with the similar geodesic chain from v to $s(e'')^+$ is a path of length

$$\ell(u) + \ell(v) - 2m + 2\,,$$

and the distance $d_{\mathbf{q}}(u,v)$ is less than or equal to this quantity. This proves (i).

Let us prove (ii). Let $w \in [[u,v]]$ be such that $\ell(w) = \min\{\ell(w') : w' \in [[u,v]]\}$. If $w = u$ or $w = v$ then the statement follows trivially from (17). So we exclude this case. We can then write τ as the union $\tau = \tau_1 \cup \tau_2$ of two connected subgraphs of τ such that $\tau_1 \cap \tau_2 = \{w\}$, τ_1 contains u but not v and τ_2 contains v but not u. There may be several such decompositions, so we just choose one. We consider a geodesic path γ from u to v in $\mathbf{q}$. If v_* belongs to this path, then this means that $d_{\mathbf{q}}(u,v) = d_{\mathbf{q}}(v_*,u) + d_{\mathbf{q}}(v_*,v)$ and the desired lower bound immediately follows from (16). So we may assume that v_* does not belong to γ. From our choice of τ_1 and τ_2, we can then find two corners $e_{(1)}$ and $e_{(2)}$ of τ such that $e_{(1)}^-$ belongs to τ_1 and $e_{(2)}^-$ belongs to τ_2, $e_{(1)}^-$ and $e_{(2)}^-$ are consecutive points on γ, and the corners $e_{(1)}$ and $e_{(2)}$ are connected by an edge of $\mathbf{q}$. From the latter property, we must have $e_{(2)} = s(e_{(1)})$ or $e_{(1)} = s(e_{(2)})$. Consider only the first case for definiteness (the other one is treated in a similar fashion). Since the contour exploration of vertices of τ must visit w between any visit of $u = e_{(1)}^-$ and any visit of $v = e_{(2)}^-$, the definition of the successor ensures that $\ell(w) \geq \ell(e_{(2)})$ (with equality only possible if $w = e_{(2)}^-$). Then, using (17) once again, we have

$$\begin{aligned} d_{\mathbf{q}}(u,v) &= d_{\mathbf{q}}(u,e_{(2)}^-) + d_{\mathbf{q}}(e_{(2)}^-,v) \\ &\geq \ell(u) - \ell(e_{(2)}^-) + \ell(v) - \ell(e_{(2)}^-) \\ &\geq \ell(u) + \ell(v) - 2\ell(w), \end{aligned}$$

giving the desired result. □

6. Basic convergence results for uniform quadrangulations

For the remaining part of this course, our main goal will be to study the scaling limits of random planar quadrangulations chosen according to the uniform probability measure on $\mathbf{Q}_n$. Thanks to Corollary 5.7, the CVS bijection and the study of scaling limits of random labeled trees will turn out to be useful tools to study this problem. Ultimately, the question we would like to address is to study the convergence in distribution of an appropriately rescaled version of the random metric space $(V(Q_n), d_{Q_n})$, in the sense of the Gromov-Hausdorff topology.

One of the motivations for this problem comes from physics, and we refer the interested reader to [**3**] for an extensive discussion. In the past 15 years or so, physicists have been starting to view random maps as possible discrete models for a continuum model of random surfaces (called the Euclidean 2-dimensional quantum gravity model), which is still ill-defined from a mathematical point of view. We thus want to investigate whether the scaling limit of Q_n exists in the above sense, and does define a certain random surface. One can also ask the natural question of whether this limiting random surface is *universal*, in the sense that it also arises as the scaling limit of many different models of random maps, for instance, maps chosen uniformly at random in the set of all p-angulations with n faces:

$$\mathbf{M}_n^p = \{\mathbf{m} : \deg(f) = p \text{ for every } f \in F(\mathbf{m}), \#F(\mathbf{m}) = n\}, \qquad p \geq 3\,.$$

Indeed, most of the results that we will describe in the sequel do have analogs in this more general setting [**36, 38, 41, 31**], thanks to nice generalizations of the CVS bijection that are due to Bouttier, Di Francesco and Guitter [**9**].

This is of course analogous to the celebrated Donsker Theorem, according to which Brownian motion is the universal scaling limit of discrete random walks, as well as to the fact that the Brownian CRT is the scaling limit of many different models of random trees (see the remarks at the end of subsection 3.4).

6.1. Radius and profile. We will first address a simpler question than the one raised above, which is to determine by what factor we should rescale the distance d_{Q_n} in order to get an interesting scaling limit as $n \to \infty$.

Let $\mathbf{q} \in \mathbf{Q}_n$ be a rooted planar quadrangulation, and v be a vertex of $\mathbf{q}$. As before, let $d_{\mathbf{q}}$ denote the graph distance on the vertex set of $\mathbf{q}$. We define the *radius* of $\mathbf{q}$ seen from v as

$$\mathcal{R}(\mathbf{q}, v) = \max_{u \in V(\mathbf{q})} d_{\mathbf{q}}(u, v)\,,$$

and the *profile* of $\mathbf{q}$ seen from v as the sequence

$$I_{\mathbf{q},v}(k) = \mathrm{Card}\,\{u \in V(\mathbf{q}) : d_{\mathbf{q}}(u,v) = k\}\,, \qquad k \geq 0$$

which measures the 'volumes' of the spheres centered at v in the graph metric. The profile can be seen as a measure on $\mathbb{Z}_+$ with total volume $n+2$. Our first limit theorem is the following.

THEOREM 6.1. *Let Q_n be uniformly distributed over $\mathbf{Q}_n$, and conditionally on Q_n, let v_* be chosen uniformly among the $n+2$ vertices of Q_n. Let also $(\mathbb{e}, Z)$ be as in subsection 4.3.*

(i) We have

$$\left(\frac{9}{8n}\right)^{1/4} \mathcal{R}(Q_n, v_*) \xrightarrow[n\to\infty]{(d)} \sup Z - \inf Z\,.$$

(ii) If v_{**} is another vertex chosen uniformly in $V(Q_n)$ and independent of v_*,

$$\left(\frac{9}{8n}\right)^{1/4} d_{Q_n}(v_*, v_{**}) \xrightarrow[n\to\infty]{(d)} \sup Z\,.$$

(iii) Finally, the following convergence in distribution holds for the weak topology on probability measures on $\mathbb{R}_+$:

$$\frac{I_{Q_n,v_*}((8n/9)^{1/4}\cdot)}{n+2} \xrightarrow[n\to\infty]{(d)} \mathcal{I}\,,$$

where $\mathcal{I}$ is the occupation measure of Z above its infimum, defined as follows: For every non-negative, measurable $g : \mathbb{R}_+ \to \mathbb{R}_+$,

$$\langle \mathcal{I}, g\rangle = \int_0^1 \mathrm{d}t\, g(Z_t - \inf Z)\,.$$

The points (i) and (iii) are due to Chassaing and Schaeffer [**12**], and (ii) is due to Le Gall [**30**], although these references state these properties in a slightly different context, namely, in the case where v_* is the root vertex rather than a uniformly chosen vertex. This indicates that as $n \to \infty$, the root vertex plays no particular role. Some information about the limiting distributions in (i) and (ii) can be found in Delmas [**15**].

Property (ii) identifies the so-called 2-point function of the Brownian map. An important generalization of this result has been obtained by Bouttier and Guitter [**8**], who were able to compute the 3-point function, namely the joint asymptotic distribution of the mutual distances between three vertices chosen uniformly at random in $V(Q_n)$.

PROOF. Let $((T_n, L_n), \epsilon)$ be a uniform random element in $\mathbf{T}_n \times \{-1, 1\}$. Then by Corollary 5.7 we may assume that (Q_n, v_*) equals $\Phi(((T_n, L_n), \epsilon))$, where Φ is the CVS bijection.

Let C_n and V_n be respectively the contour function and the label contour function of (T_n, L_n) (cf. subsections 2.1 and 4.1), and let $u_i^n, 0 \le i \le 2n$ be the contour exploration of vertices of T_n as defined in subsection 4.1 (so that $C_n(i) = |u_i^n|$ and $V_n(i) = L_n(u_i^n)$).

By Proposition 5.8, the radius of Q_n viewed from v_* then equals

$$\mathcal{R}(Q_n, v_*) = \max L_n - \min L_n + 1 = \max V_n - \min V_n + 1 .$$

Property (i) immediately follows from this equality and Theorem 4.3.

As for (ii), we first observe that we may slightly change the hypothesis on the distribution of v_{**}. It clearly suffices to prove the desired convergence when v_{**} is replaced by a vertex that is uniformly chosen among the n vertices of Q_n that are distinct from both v_* and the vertex $\varnothing$ of T_n (recall that $V(Q_n) \setminus \{v_*\} = V(T_n)$).

Now, for $s \in [0, 2n)$, we let $\langle s \rangle = \lceil s \rceil$ if C_n has slope $+1$ immediately after s, and $\langle s \rangle = \lfloor s \rfloor$ otherwise. Then, if $u \in T_n$, we have $u^n_{\langle s \rangle} = u$ if and only if $u \neq \varnothing$ and s is a time when the contour exploration around T_n explores either of the two oriented edges between u and its parent $\pi(u)$. Therefore, for every $u \in T_n \setminus \{\varnothing\}$, the Lebesgue measure of $\{s \in [0, 2n) : u^n_{\langle s \rangle} = u\}$ equals 2. Consequently, if U is a uniform random variable in $[0, 1)$, independent of (T_n, L_n), then $u^n_{\langle 2nU \rangle}$ is uniform in $T_n \setminus \{\varnothing\}$. Hence, it suffices to prove the desired result with $u^n_{\langle 2nU \rangle}$ instead of v_{**}.

Since $|s - \langle s \rangle| \le 1$, Theorem 4.3 entails that

$$\begin{aligned}\left(\frac{8n}{9}\right)^{-1/4} d_{Q_n}(v_*, u^n_{\langle 2nU \rangle}) &= \left(\frac{8n}{9}\right)^{-1/4} (L_n(u^n_{\langle 2nU \rangle}) - \min L_n + 1) \\ &= \left(\frac{8n}{9}\right)^{-1/4} (V_n(\langle 2nU \rangle) - \min V_n + 1) ,\end{aligned}$$

converges in distribution to $Z_U - \inf Z$ (here U is also assumed to be independent of $(\mathbb{e}, Z)$). The fact that $Z_U - \inf Z$ has the same distribution as $\sup Z$, or equivalently as $-\inf Z$, can be derived from the invariance of the CRT under uniform re-rooting, see e.g. [**35**]. This completes the proof of (ii).

Finally, for (iii) we just note that, for every bounded continuous $g : \mathbb{R}_+ \to \mathbb{R}$,

$$\begin{aligned}\frac{1}{n+2} \sum_{k \in \mathbb{Z}_+} & I_{Q_n, v_*}(k)\, g(((8n/9)^{-1/4} k) \\ &= \frac{1}{n+2} \sum_{v \in Q_n} g((8n/9)^{-1/4} d_{Q_n}(v_*, v)) \\ &= E_{**}[g((8n/9)^{-1/4} d_{Q_n}(v_*, v_{**}))] \\ &\underset{n\to\infty}{\longrightarrow} E_U[g(Z_U - \inf Z)] \\ &= \int_0^1 \mathrm{d}t\, g(Z_t - \inf Z) ,\end{aligned}$$

where E_{**} and E_U means that we take the expectation only with respect to v_{**} and U in the corresponding expressions (these are conditional expectations given (Q_n, v_*) and $(\mathbb{e}, Z)$ respectively). In the penultimate step, we used the convergence established in the course of the proof of (ii). □

6.2. Convergence as metric spaces. We would like to be able to understand the full scaling limit picture for random maps, in a similar way as it was done for trees, where we showed, using Theorem 2.10, that the distances in discrete trees, once rescaled by $\sqrt{2n}$, converge to the distances in the CRT $(\mathcal{T}_{\mathbb{e}}, d_{\mathbb{e}})$. We thus ask if there is an analog of the CRT that arises as the limit of the properly rescaled metric spaces (Q_n, d_{Q_n}). In view of Theorem 6.1, the correct normalization for the distance should be $n^{1/4}$.

Assume that (T_n, L_n) is uniformly distributed over $\mathbb{T}_n$, let ϵ be uniform in $\{-1, 1\}$ and independent of (T_n, L_n), and let Q_n be the random uniform quadrangulation with n faces and with a uniformly chosen vertex v_*, which is obtained from $((T_n, L_n), \epsilon)$ via the CVS bijection. We now follow Le Gall [**31**][2]. Recall our notation $u_0^n, u_1^n, \ldots, u_{2n}^n$ for the contour exploration of the vertices of T_n, and recall that in the CVS bijection these vertices are also viewed as elements of $V(Q_n) \setminus \{v_*\}$. Define a pseudo-metric on $\{0, \ldots, 2n\}$ by letting $d_n(i,j) = d_{Q_n}(u_i^n, u_j^n)$. A major problem comes from the fact that $d_n(i,j)$ cannot be expressed as a simple functional of (C_n, V_n). The only distances that we are able to handle in an easy way are distances to v_*, through the following rewriting of (16):

$$d_{Q_n}(v_*, u_i^n) = V_n(i) - \min V_n + 1 . \tag{18}$$

We also define, for $i, j \in \{0, 1, \ldots, 2n\}$,

$$d_n^0(i,j) = V_n(i) + V_n(j) - 2 \max\Big(\min_{i \le k \le j} V_n(k), \min_{j \le k \le i} V_n(k) \Big) + 2 .$$

Here, if $j < i$, the condition $i \le k \le j$ means that $k \in \{i, i+1, \ldots, 2n\} \cup \{0, 1, \ldots, j\}$ and similarly for the condition $j \le k \le i$ if $i < j$.

As a consequence of Proposition 5.9(i), we have the bound $d_n \le d_n^0$.

We now extend the function d_n to $[0, 2n]^2$ by letting

$$\begin{aligned} d_n(s,t) \;=\; & (\lceil s \rceil - s)(\lceil t \rceil - t) d_n(\lfloor s \rfloor, \lfloor t \rfloor) + (\lceil s \rceil - s)(t - \lfloor t \rfloor) d_n(\lfloor s \rfloor, \lceil t \rceil) \\ & + (s - \lfloor s \rfloor)(\lceil t \rceil - t) d_n(\lceil s \rceil, \lfloor t \rfloor) + (s - \lfloor s \rfloor)(t - \lfloor t \rfloor) d_n(\lceil s \rceil, \lceil t \rceil) , \end{aligned} \tag{19}$$

recalling that $\lfloor s \rfloor = \sup\{k \in \mathbb{Z}_+ : k \le s\}$ and $\lceil s \rceil = \lfloor s \rfloor + 1$. The function d_n^0 is extended to $[0, 2n]^2$ by the obvious similar formula.

It is easy to check that d_n thus extended is continuous on $[0, 2n]^2$ and satisfies the triangle inequality (although this is not the case for d_n^0), and that the bound $d_n \le d_n^0$ still holds. We define a rescaled version of these functions by letting

$$D_n(s,t) = \left(\frac{9}{8n} \right)^{1/4} d_n(2ns, 2nt) , \qquad 0 \le s, t \le 1 .$$

We define similarly the functions D_n^0 on $[0,1]^2$. Then, as a consequence of Theorem 4.3, we have

$$(D_n^0(s,t), 0 \le s, t \le 1) \xrightarrow[n \to \infty]{(d)} (D^0(s,t), 0 \le s, t \le 1) , \tag{20}$$

for the uniform topology on $C([0,1]^2, \mathbb{R})$, where by definition

$$D^0(s,t) = Z_s + Z_t - 2 \max\Big(\min_{s \le r \le t} Z_r, \min_{t \le r \le s} Z_r \Big) , \tag{21}$$

where if $t < s$ the condition $s \le r \le t$ means that $r \in [s, 1] \cup [0, t]$.

We can now state

[2] At this point, it should be noted that [**31, 34, 32**] consider another version of Schaeffer's bijection, where no distinguished vertex v_* has to be considered. This results in considering pairs (T_n, L_n) in which L_n is conditioned to be positive. The scaling limits of such pairs are still tractable, and in fact, are simple functionals of $(\mathbb{e}, Z)$, as shown in [**35, 30**]. So there will be some differences from our exposition, but these turn out to be unimportant.

PROPOSITION 6.2. *The family of laws of $(D_n(s,t), 0 \le s,t \le 1)$, as n varies, is relatively compact for the weak topology on probability measures on $C([0,1]^2,\mathbb{R})$.*

PROOF. Let $s,t,s',t' \in [0,1]$. Then by a simple use of the triangle inequality, and the fact that $D_n \le D_n^0$,

$$|D_n(s,t) - D_n(s',t')| \le D_n(s,s') + D_n(t,t') \le D_n^0(s,s') + D_n^0(t,t'),$$

which allows one to estimate the modulus of continuity at a fixed $\delta > 0$:

$$\sup_{\substack{|s-s'|\le\delta \\ |t-t'|\le\delta}} |D_n(s,t) - D_n(s',t')| \le 2 \sup_{|s-s'|\le\delta} D_n^0(s,s'). \tag{22}$$

However, the convergence in distribution (20) entails that for every $\varepsilon > 0$,

$$\limsup_{n\to\infty} P\left(\sup_{|s-s'|\le\delta} D_n^0(s,s') \ge \varepsilon\right) \le P\left(\sup_{|s-s'|\le\delta} D^0(s,s') \ge \varepsilon\right),$$

and the latter quantity goes to 0 when $\delta \to 0$ (for any fixed value of $\epsilon > 0$) by the continuity of D^0 and the fact that $D^0(s,s) = 0$. Hence, taking $\eta > 0$ and letting $\varepsilon = \varepsilon_k = 2^{-k}$, we can choose $\delta = \delta_k$ (tacitly depending also on η) such that

$$\sup_{n\ge1} P\left(\sup_{|s-s'|\le\delta_k} D_n^0(s,s') \ge 2^{-k}\right) \le \eta 2^{-k}, \qquad k \ge 1,$$

entailing

$$P\left(\bigcap_{k\ge1}\left\{\sup_{|s-s'|\le\delta_k} D_n^0(s,s') \le 2^{-k}\right\}\right) \ge 1 - \eta,$$

for all $n \ge 1$. Together with (22), this shows that with probability at least $1-\eta$, the function D_n belongs to the set of all functions f from $[0,1]^2$ into $\mathbb{R}$ such that $f(0,0) = 0$ and, for every $k \ge 1$,

$$\sup_{\substack{|s-s'|\le\delta_k \\ |t-t'|\le\delta_k}} |f(s,t) - f(s',t')| \le 2^{-k}.$$

The latter set is compact by the Arzelà-Ascoli theorem. The conclusion then follows from Prokhorov's theorem. □

At this point, we are allowed to say that the random distance functions D_n admit a limit in distribution, up to taking $n \to \infty$ along a subsequence:

$$(D_n(s,t), 0 \le s,t \le 1) \xrightarrow{(d)} (D(s,t), 0 \le s,t \le 1) \tag{23}$$

for the uniform topology on $C([0,1]^2,\mathbb{R})$. In fact, we are going to need a little more than the convergence of D_n. From the relative compactness of the components, we see that the closure of the collection of laws of the triplets

$$((2n)^{-1}C_n(2n\cdot), (9/8n)^{1/4}V_n(2n\cdot), D_n), \quad n \ge 1$$

is compact in the space of all probability measures on $C([0,1],\mathbb{R})^2 \times C([0,1]^2,\mathbb{R})$. Therefore, it is possible to choose a subsequence $(n_k, k \ge 1)$ so that this triplet converges in distribution to a limit, which is denoted by $(\mathbb{e}, Z, D)$ (from Theorem 4.3, this is of course consistent with the preceding notation). The joint convergence to the triplet $(\mathbb{e}, Z, D)$ gives a coupling of D, D^0 such that $D \le D^0$, since $D_n \le D_n^0$ for every n.

Define a random equivalence relation on $[0,1]$ by letting $s \approx t$ if $D(s,t) = 0$. We let $M = [0,1]/\approx$ be the associated quotient space, endowed with the quotient distance, which we still denote by D. The canonical projection $[0,1] \to M$ is denoted by $\mathbf{p}$.

Finally, let $s_* \in [0,1]$ be such that $Z_{s_*} = \inf Z$. It can be proved that s_* is unique a.s., see [**37**] or [**35**], and we will admit this fact (although it is not really needed for the next statement). We set $\rho_* = \mathbf{p}(s_*)$. We can now state the main result of this section.

THEOREM 6.3. *The random pointed metric space (M, D, ρ_*) is the limit in distribution of the spaces $(V(Q_n), (9/8n)^{1/4} d_{Q_n}, v_*)$, for the Gromov-Hausdorff topology, along the subsequence $(n_k, k \geq 1)$. Moreover, we have a.s. for every $x \in M$ and $s \in [0,1]$ such that $\mathbf{p}(s) = x$,*

$$D(\rho_*, x) = D(s_*, s) = Z_s - \inf Z\,.$$

Note that, in the discrete model, a point at which the minimal label in T_n is attained lies at distance 1 from v_*. Therefore, the point ρ_* should be seen as the continuous analog of the distinguished vertex v_*. The last identity in the statement of the theorem is then of course the continuous analog of (16) and (18).

PROOF. For the purposes of this proof, it is useful to assume, using the Skorokhod representation theorem, that the convergence

$$((2n)^{-1/2} C_n(2n\cdot), (9/8n)^{1/4} V_n(2n\cdot), D_n) \longrightarrow (\mathbb{e}, Z, D)$$

holds a.s. along the subsequence (n_k). In what follows we restrict our attention to values of n in this sequence.

For every n, let $i_*^{(n)}$ be any index in $\{0, 1, \ldots, 2n\}$ such that $V_n(i_*^{(n)}) = \min V_n$. Then for every $v \in V(Q_n)$, it holds that

$$|d_{Q_n}(v_*, v) - d_{Q_n}(u^n_{i_*^{(n)}}, v)| \leq 1$$

because $d_{Q_n}(v_*, u^n_{i_*^{(n)}}) = 1$ (v_* and $u^n_{i_*^{(n)}}$ are linked by an arc in the CVS bijection). Moreover, since $(8n/9)^{-1/4} V_n(2n\cdot)$ converges to Z uniformly on $[0,1]$, and since we know[3] that Z attains its overall infimum at a unique point s_*, it is easy to obtain that $i_*^{(n)}/2n$ converges as $n \to \infty$ towards s_*.

For every integer n, we construct a correspondence $\mathcal{R}_n$ between $V(Q_n)$ and M_n, by putting:

- $(v_*, \rho_*) \in \mathcal{R}_n$;
- $(u^n_{\lfloor 2ns \rfloor}, \mathbf{p}(s)) \in \mathcal{R}_n$, for every $s \in [0,1]$.

We then verify that the distortion of $\mathcal{R}_n$ (with respect to the metrics $(9/8n)^{1/4} d_{Q_n}$ on $V(Q_n)$ and D on M) converges to 0 a.s. as $n \to \infty$. We first observe that

$$\begin{aligned}
&\sup_{s \in [0,1]} |(9/8n)^{1/4} d_{Q_n}(v_*, u^n_{\lfloor 2ns \rfloor}) - D(\rho_*, \mathbf{p}(s))| \\
&\quad \leq (9/8n)^{1/4} + \sup_{s \in [0,1]} |(9/8n)^{1/4} d_{Q_n}(u^n_{i_*^{(n)}}, u^n_{\lfloor 2ns \rfloor}) - D(\rho_*, \mathbf{p}(s))| \\
&\quad = (9/8n)^{1/4} + \sup_{s \in [0,1]} |D_n(i_*^{(n)}/2n, \lfloor 2ns \rfloor/2n) - D(s_*, s)|,
\end{aligned}$$

[3] We could also perform the proof without using this fact, but it makes things a little easier

which tends to 0 as $n \to \infty$, by the a.s. uniform convergence of D_n to D, and the fact that $i_*^{(n)}/2n$ converges to s_*. Similarly, we have

$$\sup_{s,t\in[0,1]} |(9/8n)^{1/4} d_{Q_n}(u^n_{\lfloor 2ns\rfloor}, u^n_{\lfloor 2nt\rfloor}) - D(\mathbf{p}(s), \mathbf{p}(t))|$$
$$= \sup_{s,t\in[0,1]} |D_n(\lfloor 2ns\rfloor/2n, \lfloor 2nt\rfloor/2n) - D(s,t)|$$

which tends to 0 as $n \to \infty$. We conclude that the distortion of $\mathcal{R}_n$ converges to 0 a.s. and that the pointed metric spaces $(V(Q_n), (9/8n)^{-1/4} d_{Q_n}, v_*)$ also converge a.s. to (M, D, ρ_*) in the Gromov-Hausdorff topology.

Let us prove the last statement of the theorem. Using once again the uniform convergence of D_n to D, we obtain that for every $s \in [0,1]$,

$$\begin{aligned} D(s_*, s) &= \lim_{n\to\infty} D_n(i_*^{(n)}/2n, \lfloor 2ns\rfloor/2n) \\ &= \lim_{n\to\infty} \left(\frac{8n}{9}\right)^{-1/4} d_{Q_n}(v_*, u^n_{\lfloor 2ns\rfloor}) \\ &= \lim_{n\to\infty} \left(\frac{8n}{9}\right)^{-1/4} (V_n(\lfloor 2ns\rfloor) - \min V_n + 1) \\ &= Z_s - \inf Z\,, \end{aligned}$$

as desired. □

It is tempting to call (M, D) the "Brownian map", or the "Brownian continuum map", by analogy with the fact that the "Brownian continuum random tree" is the scaling limit of uniformly distributed plane trees with n edges. However, the choice of the subsequence in Theorem 6.3 poses a problem of uniqueness of the limit. As we see in the previous statement, only the distances to ρ_* are *a priori* defined as simple functionals of the process Z. Distances between other points in M seem to be harder to handle. The following conjecture is however very appealing.

CONJECTURE 6.1. *The spaces $(V(Q_n), n^{-1/4} d_{Q_n})$ converge in distribution, for the Gromov-Hausdorff topology.*

Marckert and Mokkadem [**37**] and Le Gall [**31**] give a natural candidate for the limit (called the Brownian map in [**37**]) but until now the convergence result in the above conjecture has not been proved.

7. Identifying the Brownian map

7.1. The Brownian map as a quotient of the CRT. In the previous section, we wrote the scaling limit of rescaled random quadrangulations (along a suitable subsequence) as a quotient space $M = [0,1]/\approx$ where the equivalence relation $\approx$ is defined by $s \approx t$ iff $D(s,t) = 0$. In this section, we provide a more explicit description of this quotient.

Recall the notation of the previous section. In particular, $((T_n, L_n), \epsilon)$ is uniformly distributed over $\mathbf{T}_n \times \{-1, 1\}$, and (Q_n, v_*) is the pointed quadrangulation that is the image of $((T_n, L_n), \epsilon)$ under the CVS bijection. For every $n \geq 1$, $u^n_0, u^n_1, \ldots, u^n_{2n}$ is the contour exploration of the vertices of T_n. Thus, $C_n(k) = |u^n_k|$ and $V_n(k) = L_n(u^n_k)$ for $0 \leq k \leq 2n$.

As in the proof of Theorem 6.3, we may assume that, along the sequence (n_k) we have the almost sure convergence

$$((2n)^{-1/2}C_n(2ns),(9/8n)^{1/4}V_n(2ns),D_n(s,t))_{s,t\in[0,1]} \underset{n\to\infty}{\longrightarrow} (\mathbb{e}_s,Z_s,D(s,t))_{s,t\in[0,1]} \tag{24}$$

uniformly over $[0,1]^2$. Recall from the proof of Theorem 6.3 that this implies the almost sure convergence

$$\left(V(Q_n),\left(\frac{9}{8}\right)^{1/4}d_{Q_n}\right)\underset{n\to\infty}{\longrightarrow}(M,D)$$

in the Gromov-Hausdorff sense, along the sequence (n_k).

As in Section 2 above, introduce the random equivalence relation $\sim_{\mathbb{e}}$ on $[0,1]$ by

$$s\sim_{\mathbb{e}} t \text{ iff } \mathbb{e}_s=\mathbb{e}_t=\min_{s\wedge t\le r\le s\vee t}\mathbb{e}_r$$

and recall that the CRT $\mathcal{T}_{\mathbb{e}}$ is defined as the quotient space $[0,1]/\sim_{\mathbb{e}}$ equipped with the distance $d_{\mathbb{e}}$.

LEMMA 7.1. *We have almost surely for every $s,t\in[0,1]$,*

$$s\sim_{\mathbb{e}} t\Rightarrow D(s,t)=0 \quad (\Leftrightarrow s\approx t).$$

PROOF. We can use the convergence of the first components in (24) to see that if $s\sim_{\mathbb{e}} t$ and $s<t$ we can find integers $i_n<j_n$ such that $i_n/2n\longrightarrow s$, $j_n/2n\longrightarrow t$, and, for every sufficiently large n (belonging to the sequence (n_k)),

$$C_n(i_n)=C_n(j_n)=\min_{i_n\le k\le j_n}C_n(k).$$

Then, from the definition of the contour function, we must have $u^n_{i_n}=u^n_{j_n}$, and thus $d_n(i_n,j_n)=0$. Using the convergence (24) again, we conclude that $D(s,t)=0$.

Consequence. Recall that $p_{\mathbb{e}}:[0,1]\longrightarrow\mathcal{T}_{\mathbb{e}}$ denotes the canonical projection. Then $D(s,t)$ only depends on $p_{\mathbb{e}}(s)$ and $p_{\mathbb{e}}(t)$. We can therefore put for every $a,b\in\mathcal{T}_{\mathbb{e}}$,

$$D(a,b)=D(s,t)$$

where s, resp. t, is an arbitrary representative of a, resp. of b, in $[0,1]$. Then D is (again) a pseudo-distance on $\mathcal{T}_{\mathbb{e}}$. With a slight abuse of notation we keep writing $a\approx b$ iff $D(a,b)=0$, for $a,b\in\mathcal{T}_{\mathbb{e}}$. Then the Brownian map M can be written as

$$M=[0,1]/\approx\;=\mathcal{T}_{\mathbb{e}}/\approx$$

where the first equality was the definition of M and the second one corresponds to the fact that there is an obvious canonical isometry between the two quotient spaces.

One may wonder why it is more interesting to write the Brownian map M as a quotient space of the CRT $\mathcal{T}_{\mathbb{e}}$ rather than as a quotient space of $[0,1]$. The point is that it will be possible to give a simple intuitive description of $\approx$ viewed as an equivalence relation on $\mathcal{T}_{\mathbb{e}}$. This is indeed the main goal of the next section.

7.2. Identifying the equivalence relation $\approx$. We noticed in subsection 4.2 that the process Z (the Brownian snake driven by $\mathbb{e}$) can be viewed as indexed by $\mathcal{T}_{\mathbb{e}}$. This will be important in what follows: For $a \in \mathcal{T}_{\mathbb{e}}$, we will write $Z_a = Z_t$ for any choice of t such that $a = p_{\mathbb{e}}(t)$. We also set $a_* = p_{\mathbb{e}}(s_*)$: a_* is thus the unique vertex of $\mathcal{T}_{\mathbb{e}}$ such that

$$Z_{a_*} = \min_{a \in \mathcal{T}_{\mathbb{e}}} Z_a.$$

We first need to define intervals on the tree $\mathcal{T}_{\mathbb{e}}$. For simplicity we consider only leaves of $\mathcal{T}_{\mathbb{e}}$. Recall that a point a of $\mathcal{T}_{\mathbb{e}}$ is a leaf if $\mathcal{T}_{\mathbb{e}} \backslash \{a\}$ is connected. Equivalently a vertex a distinct from the root ρ is a leaf if and only $p_{\mathbb{e}}^{-1}(a)$ is a singleton. Note in particular that a_* is a leaf of $\mathcal{T}_{\mathbb{e}}$.

Let a and b be two (distinct) leaves of $\mathcal{T}_{\mathbb{e}}$, and let s and t be the unique elements of $[0, 1)$ such that $p_{\mathbb{e}}(s) = a$ and $p_{\mathbb{e}}(t) = b$. Assume that $s < t$ for definiteness. We then set

$$[a, b] = p_{\mathbb{e}}([s, t])$$
$$[b, a] = p_{\mathbb{e}}([t, 1] \cup [0, s]).$$

It is easy to verify that $[a, b] \cap [b, a] = [\![a, b]\!]$ is the line segment between a and b in $\mathcal{T}_{\mathbb{e}}$.

THEOREM 7.2. *Almost surely, for every distinct $a, b \in \mathcal{T}_{\mathbb{e}}$,*

$$a \approx b \quad \Leftrightarrow \begin{cases} a, b \text{ are leaves of } \mathcal{T}_{\mathbb{e}} \text{ and} \\ Z_a = Z_b = \max\Big(\min_{c \in [a,b]} Z_c, \min_{c \in [b,a]} Z_c\Big) \end{cases}$$

REMARK 7.3. *We know that the minimum of Z over $\mathcal{T}_{\mathbb{e}}$ is attained at the unique vertex a_*. If a and b are (distinct) leaves of $\mathcal{T}_{\mathbb{e}} \backslash \{a_*\}$, exactly one of the two intervals $[a, b]$ and $[b, a]$ contains the vertex a_*. Obviously the minimum of Z over this interval is equal to Z_{a_*} and thus cannot be equal to Z_a or Z_b.*

The proof of the implication $\Leftarrow$ in the theorem is easy. Suppose that $a = p_{\mathbb{e}}(s)$ and $b = p_{\mathbb{e}}(t)$ with $s < t$ (for definiteness). If

$$Z_a = Z_b = \max\Big(\min_{c \in [a,b]} Z_c, \min_{c \in [b,a]} Z_c\Big)$$

this means that

$$Z_s = Z_t = \max\Big(\min_{r \in [s,t]} Z_r, \min_{r \in [t,1] \cup [0,s]} Z_r\Big).$$

The last identity is equivalent to saying that $D^0(s, t) = 0$, and since $D \le D^0$ we have also $D(s, t) = 0$, or equivalently $a \approx b$.

Unfortunately, the proof of the converse implication is much harder, and we will only give some key ideas of the proof, referring to [**31**] for additional details.

We start with a preliminary lemma. We denote by $\mathrm{vol}(\cdot)$ the mass measure on $\mathcal{T}_{\mathbb{e}}$, which is simply the image of the Lebesgue measure on $[0, 1]$ under the projection $p_{\mathbb{e}} : [0, 1] \longrightarrow \mathcal{T}_{\mathbb{e}}$.

LEMMA 7.4. *Almost surely, for every $\delta \in (0, 1)$, there exists a (random) constant $C_\delta(\omega)$ such that, for every $r > 0$ and every $a \in \mathcal{T}_{\mathbb{e}}$,*

$$\mathrm{vol}(\{b \in \mathcal{T}_{\mathbb{e}} : D(a, b) \le r\}) \le C_\delta \, r^{4-\delta}.$$

We omit the proof of this lemma. The first ingredient of the proof is a "re-rooting invariance property" of random planar maps, which makes it possible to reduce the proof to the case $a = a_*$. In that case we can use the formula $D(a_*, b) = Z_b - \min Z$ and explicit moment calculations for the Brownian snake (see Corollary 6.2 in [**32**] for a detailed proof).

Let us come to the proof of the implication $\Rightarrow$ in Theorem 7.2. For simplicity we consider only the case when a and b are leaves of $\mathcal{T}_e$ (it would be necessary to show also that the equivalence class of any vertex of $\mathcal{T}_e$ that is not a leaf is a singleton – this essentially follows from Lemma 2.2 in [**31**]). We let $s, t \in [0, 1]$ be such that $a = p_e(s)$ and $b = p_e(t)$, and assume for definiteness that $0 \leq s_* < s < t \leq 1$.

We assume that $a \approx b$, and our goal is to prove that

$$Z_a = Z_b = \min_{c \in [a,b]} Z_c.$$

We already know that $Z_a = Z_b$, because

$$Z_a - \min Z = D(a_*, a) = D(a_*, b) = Z_b - \min Z.$$

First step. We first establish that

$$Z_a = Z_b = \min_{c \in [\![a,b]\!]} Z_c. \tag{25}$$

To see this, we go back to the discrete picture. We can find $a_n, b_n \in T_n$ such that $a_n \longrightarrow a$ and $b_n \longrightarrow b$ as $n \to \infty$ (strictly speaking these convergences make no sense: What we mean is that $a_n = u^n_{i_n}$, $b_n = u^n_{j_n}$ with $i_n/2n \longrightarrow s$ and $j_n/2n \longrightarrow t$). Then the condition $D(a, b) = 0$ implies that

$$n^{-1/4}\, d_{Q_n}(a_n, b_n) \longrightarrow 0. \tag{26}$$

Recall, from Proposition 5.9, the notation $[\![a_n, b_n]\!]$ for the set of vertices lying on the geodesic path from a_n to b_n in the tree T_n. By Proposition 5.9(ii), we have

$$d_{Q_n}(a_n, b_n) \geq L_n(a_n) + L_n(b_n) - 2 \min_{c \in [\![a_n,b_n]\!]} L_n(c).$$

We multiply both sides of this inequality by $n^{-1/4}$ and let n tend to ∞, using (26). Modulo some technical details that we omit (essentially one needs to check that any vertex of $\mathcal{T}_e$ belonging to $[\![a, b]\!]$ is of the form $p_e(r)$, where $r = \lim k_n/2n$ and the integers k_n are such that $u^n_{k_n}$ belongs to $[\![a_n, b_n]\!]$), we get that

$$Z_a + Z_b - 2 \min_{c \in [\![a,b]\!]} Z_c \leq 0$$

from which (25) immediately follows.

Second step. We argue by contradiction, assuming that

$$\min_{c \in [a,b]} Z_c < Z_a = Z_b.$$

Let γ_n be a discrete geodesic from a_n to b_n in the quadrangulation Q_n (here we view a_n and b_n as vertices of the quadrangulation Q_n, and this geodesic is of course different from the geodesic from a_n to b_n in the tree T_n). From (26) the maximal distance between a_n (or b_n) and a vertex visited by γ_n is $o(n^{1/4})$ as $n \to \infty$. As a consequence, using the triangle inequality and (16), we have

$$\sup_{u \in \gamma_n} |L_n(u) - L_n(a_n)| = o(n^{1/4})$$

as $n \to \infty$.

To simplify the presentation of the argument, we assume that, for infinitely many values of n, the geodesic path γ_n from a_n to b_n stays in the lexicographical interval $[a_n, b_n]$. This lexicographical interval is defined, analogously to the continuous setting, as the set of all vertices visited by the contour exploration sequence $(u_i^n)_{0\leq i\leq 2n}$ between its last visit of a_n and its first visit of b_n. Note that the preceding assumption may not hold, and so the real argument is slightly more complicated than what follows.

We use the previous assumption to prove the following claim. If $x \in [a, b]$, we denote by $\phi_{a,b}(x)$ the last ancestor of x that belongs to $[\![a, b]\!]$ (the condition $x \in [a, b]$ ensures that the ancestral line $[\![\rho, x]\!]$ intersects $[\![a, b]\!]$). Alternatively, $\phi_{a,b}(x)$ is the point of $[\![a, b]\!]$ at minimal $d_{\mathbb{e}}$-distance of x in the tree $\mathcal{T}_{\mathbb{e}}$.

Claim. Let $\varepsilon > 0$. For every $c \in [a, b]$ such that

$$\begin{cases} Z_c < Z_a + \varepsilon & \\ Z_x > Z_a + \varepsilon/2 & \forall x \in [\![\phi_{a,b}(c), c]\!] \end{cases}$$

we have $D(a, c) \leq \varepsilon$.

The claim eventually leads to the desired contradiction: Using the first step of the proof (which ensures that $Z_c \geq Z_a$ for $c \in [\![a, b]\!]$) and the properties of the Brownian snake, one can check that, under the condition

$$\min_{c\in[a,b]} Z_c < Z_a = Z_b,$$

the volume of the set of all vertices c that satisfy the assumptions of the claim is bounded below by a (random) positive constant times ε^2, at least for sufficiently small $\varepsilon > 0$ (see Lemma 2.4 in [**31**] for a closely related statement). The desired contradiction follows since Lemma 7.4 implies that, for every $\delta \in (0, 1)$,

$$\mathrm{vol}(\{c : D(a, c) \leq \varepsilon\}) \leq C_\delta\, \varepsilon^{4-\delta}.$$

To complete this sketch, we explain why the claim holds. Again, we need to go back to the discrete setting. We consider a vertex $u \in [a_n, b_n]$ such that

(i) $L_n(u) < L_n(a_n) + \varepsilon n^{1/4}$;
(ii) $L_n(v) > L_n(a_n) + \frac{\varepsilon}{2}\, n^{1/4}$, $\quad \forall v \in [\![\phi^n_{a_n,b_n}(u), u]\!]$

where $\phi^n_{a_n,b_n}(u)$ is the last ancestor of u in the tree T_n that belongs to $[\![a_n, b_n]\!]$.

Condition (ii) guarantees that the vertex u lies "between" $[\![a_n, b_n]\!]$ and the geodesic γ_n: If this were not the case, the geodesic γ_n would contain a point in $[\![\phi^n_{a_n,b_n}(u), u]\!]$, which is impossible by (ii) (we already noticed that the label of a vertex of the geodesic γ_n must be $L_n(a_n) + o(n^{1/4})$.

Consider the geodesic path from u to v_* in Q_n that is obtained from the successor geodesic chain $e \to s(e) \to s^2(e) \to \cdots$ starting from any corner e of u in T_n. Since arcs in the CVS bijection do not cross edges of the tree and since we know that the vertex u lies in the area between $[\![a_n, b_n]\!]$ and the geodesic γ_n, the geodesic we have just constructed cannot "cross" $[\![a_n, b_n]\!]$ and so it must intersect γ_n at a vertex w. This vertex w is such that

$$L_n(u) - L_n(w) = d_{Q_n}(u, w).$$

Since w belongs to γ_n, we have $d_{Q_n}(w, a_n) = o(n^{1/4})$, and therefore

$$L_n(u) - L_n(a_n) = d_{Q_n}(u, a_n) + o(n^{1/4}).$$

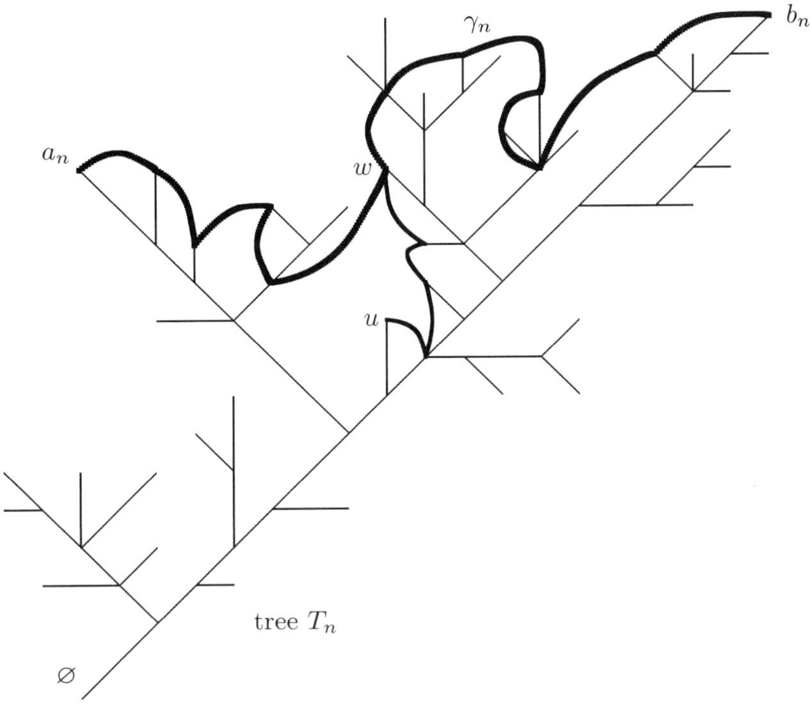

FIGURE 8. Illustration of the proof: The geodesic path γ_n from a_n to b_n is represented by the thick curves. The thin curves correspond to the beginning of the successor geodesic chain starting from u. This chain does not cross the line segment $[\![a_n, b_n]\!]$ and thus has to meet the path γ_n at some point w.

By (i), we now get

$$d_{Q_n}(u, a_n) \leq \varepsilon n^{1/4} + o(n^{1/4}).$$

We have thus obtained a discrete analog of the claim. To get the continuous version as stated above, we just need to do a careful passage to the limit $n \to \infty$. □

7.3. Hausdorff dimension. The limit in distribution (along a suitable subsequence) in Theorem 6.3 can be written as $(\mathcal{T}_{\mathbb{e}}/\approx, D)$, and the space $\mathcal{T}_{\mathbb{e}}/\approx$ is completely identified: Roughly speaking two vertices a and b of the CRT $\mathcal{T}_{\mathbb{e}}$ are identified if and only if they have the same label $Z_a = Z_b$ and if one can go from a to b following the "contour" of the tree $\mathcal{T}_{\mathbb{e}}$ and visiting only vertices with larger label. In order to prove Conjecture 6.1, it would be necessary to characterize the distance D. Much is known about D (in particular Theorem 6.1 characterizes the distribution of the profile of distances from the distinguished point ρ_*, and one can show that this profile has the same distribution if one replaces ρ_* by a "typical" point of M). Still the characterization of D remains an open problem.

Nevertheless, one can show that the "Brownian map" $(\mathcal{T}_{\mathbb{e}}/\approx, D)$, that is, any of the random metric spaces that can arise as the limit in Theorem 6.3, has Hausdorff dimension 4 and is homeomorphic to the 2-sphere. This was proved in [**31**] and [**34**]. The remainder of these notes will be devoted to the proof of these two results.

THEOREM 7.5. *Almost surely, the space* (M, D) *has Hausdorff dimension* 4.

The lower bound is an easy consequence of Lemma 7.4. Recall that vol is the image measure of Lebesgue measure on $[0,1]$ under $p_{\mathbb{e}}$. We let Vol be the induced measure on (M, D), that is, the image of Lebesgue measure on $[0,1]$ under the projection $\mathbf{p} : [0,1] \to M$. Then Lemma 7.4 implies that a.s., for every $\delta \in (0,1)$, and every $x \in M$, it holds that

$$\limsup_{r \downarrow 0} \frac{\mathrm{Vol}(B_D(x,r))}{r^{4-\delta}} = 0\,,$$

where $B_D(x,r) = \{y \in M : D(x,y) < r\}$ is the open ball centered at x with radius r. This last fact, combined with standard density theorems for Hausdorff measures, implies that a.s. the Hausdorff dimension of (M, D) is greater than or equal to $4-\delta$, for every $\delta \in (0,1)$.

For the upper bound, we rely on the following easy lemma.

LEMMA 7.6. *Almost surely, for every* $\alpha \in (0, 1/4)$, *the label process* Z *is Hölder continuous with exponent* α.

PROOF. This is obtained by the classical Kolmogorov continuity criterion, and moment estimates for Z. Let s,t be such that $0 \le s < t \le 1$, and recall that conditionally given $\mathbb{e}$, $Z_s - Z_t$ is a Gaussian random variable with variance $d_{\mathbb{e}}(s,t)$. Consequently, for every $p > 0$, there exists $C_p \in (0,\infty)$ such that

$$E[|Z_s - Z_t|^p \mid \mathbb{e}] = C_p d_{\mathbb{e}}(s,t)^{p/2}\,,$$

and since $\mathbb{e}$ is a.s. Hölder continuous with exponent 2α, we deduce the existence of a (random) $C'_p \in (0,\infty)$ such that

$$E[|Z_s - Z_t|^p \mid \mathbb{e}] \le C'_p |s-t|^{p\alpha}\,.$$

The desired Hölder continuity property then follows from an application of the classical Kolmogorov lemma. □

From this, we deduce that the projection $\mathbf{p} : [0,1] \to M$ is a.s. Hölder continuous with index $\alpha \in (0,1/4)$ as well. Indeed, using the fact that $D \le D^0$, where D^0 is defined in (21), we get

$$\begin{aligned} D(\mathbf{p}(s), \mathbf{p}(t)) &= D(s,t) \\ &\le Z_s + Z_t - 2 \inf_{s\wedge t \le u \le s \vee t} Z_u \\ &\le 2 \sup_{s\wedge t \le u,v \le s\vee t} |Z_u - Z_v| \\ &\le C''_p |s-t|^{\alpha}\,, \end{aligned}$$

for some $C''_p \in (0,\infty)$. The fact that the Hausdorff dimension of (M, D) is bounded above by $1/\alpha$ is then a classical consequence of this last property. This completes the proof of the theorem.

8. The homeomorphism theorem

THEOREM 8.1. *Almost-surely, the Brownian map* (M, D) *is homeomorphic to the 2-sphere* $\mathbb{S}^2$.

This result was first obtained by Le Gall and Paulin [**34**], by arguing directly on the quotient space $M = \mathcal{T}_{\mathbb{e}}/\approx$. More precisely, Le Gall and Paulin observe that the equivalence relations $\sim_{\mathbb{e}}$ and $\approx$ may be viewed as equivalence relations on the sphere $\mathbb{S}^2$. Upon showing that the associated classes are closed, arcwise connected, and have connected complements, one can then apply a theorem due to Moore [**43**], showing that under these hypotheses, the quotient $\mathbb{S}^2/\approx$ is itself homeomorphic to $\mathbb{S}^2$. Here, we will adopt a different approach, introduced in Miermont [**39**], which relies more on the discrete approximations described in these notes. The idea is roughly as follows: Even though the property of being homeomorphic to $\mathbb{S}^2$ is not preserved under Gromov-Hausdorff convergence, this preservation can be deduced under an additional property, called *regular convergence*, introduced by Whyburn. This property says heuristically that the spaces under consideration do not have small bottlenecks, i.e. cycles of vanishing diameters that separate the spaces into two macroscopic components.

In this section, when dealing with elements of the space $\mathbb{K}$ of isometry classes of pointed compact metric spaces, we will often omit to mention the distinguished point, as its role is less crucial than it was in Sections 6 and 7.

8.1. Geodesic spaces and regular convergence. A metric space (X, d) is said to be a *geodesic* metric space if for every $x, y \in X$, there exists an isometry $f : [0, d(x, y)] \to X$ such that $f(0) = x$ and $f(d(x, y)) = y$. Any such f is called a geodesic path between x and y. For instance, real trees are geodesic metric spaces by Definition 3.1. The set $\mathbb{K}_{\mathrm{geo}}$ of isometry classes of (rooted) compact geodesic metric spaces is closed in $(\mathbb{K}, d_{GH})$, as shown in [**11**].

DEFINITION 8.1. *Let $((X_n, d_n), n \geq 1)$ be a sequence of compact geodesic metric spaces, converging to (X, d) in $(\mathbb{K}, d_{GH})$. We say that the convergence is* regular *if for every $\varepsilon > 0$, one can find $\delta > 0$ and $N \in \mathbb{N}$ such that, for every $n > N$, every closed path γ in X_n with diameter at most δ is homotopic to 0 in its ε-neighborhood.*

For instance, let Y_n be the complement in the unit sphere $\mathbb{S}^2 \subset \mathbb{R}^3$ of the open $1/n$-neighborhood of the North pole, and endow Y_n with the intrinsic distance induced from the usual Euclidean metric on $\mathbb{R}^3$ (so that the distance between $x, y \in Y_n$ is the minimal length of a path from x to y in Y_n). Let X_n be obtained by gluing two (disjoint) copies of Y_n along their boundaries, and endow it with the natural intrinsic distance. Then X_n converges in the Gromov-Hausdorff sense to a bouquet of two spheres, i.e. two (disjoint) copies of $\mathbb{S}^2$ whose North poles have been identified. However, the convergence is not regular, because the path γ that consists in the boundary of (either copy of) Y_n viewed as a subset of X_n has vanishing diameter as $n \to \infty$, but is not homotopic to 0 in its ε-neighborhood for any $\varepsilon \in (0, 1)$ and for any n. Indeed, such an ε-neighborhood is a cylinder, around which γ makes one turn.

THEOREM 8.2. *Let $((X_n, d_n), n \geq 1)$ be a sequence of $\mathbb{K}_{\mathrm{geo}}$ that converges regularly to a limit (X, d) that is not reduced to a point. If (X_n, d_n) is homeomorphic to $\mathbb{S}^2$ for every $n \geq 1$, then so is (X, d).*

This theorem is an easy reformulation of a result of Whyburn in the context of Gromov-Hausdorff convergence; see the paper by Begle [**4**]. In the latter, it is assumed that every X_n should be a compact subset of a compact metric space (Z, δ), independent of n, and that X_n converges in the Hausdorff sense to X. This transfers to our setting, because, if (X_n, d_n) converges to (X, d) in the Gromov-Hausdorff sense, then one can find a compact metric space (Z, δ) containing isometric copies $X'_n, n \geq 1$ and X' of $X_n, n \geq 1$ and X, such that X'_n converges in the Hausdorff sense to X', see for instance [**21**, Lemma A.1]. In [**4**], it is also assumed in the definition of regular convergence that for every $\varepsilon > 0$, there exist $\delta > 0$ and $N \in \mathbb{N}$ such that, for every $n \geq N$, any two points of X_n that lie at distance $\leq \delta$ are in a connected subset of X_n of diameter $\leq \varepsilon$. This condition is tautologically satisfied for geodesic metric spaces, which is the reason why we work in this context.

8.2. Quadrangulations seen as geodesic spaces. Theorem 8.2 gives a natural method to prove Theorem 8.1, using the convergence of quadrangulations to the Brownian map, as stated in Theorem 6.3. However, the finite space $(V(Q_n), d_{Q_n})$ is certainly not a geodesic space, nor homeomorphic to the 2-sphere. Hence, we have to modify a little these spaces so that they satisfy the hypotheses of Theorem 8.2. We will achieve this by constructing a particular[4] graphical representation of $\mathbf{q}$.

Let $(X_f, d_f), f \in F(\mathbf{q})$ be disjoint copies of the emptied unit cube "with bottom removed"

$$\mathcal{C} = [0,1]^3 \setminus \left((0,1)^2 \times [0,1)\right) ,$$

endowed with the intrinsic metric d_f inherited from the Euclidean metric (the distance between two points of X_f is the minimal Euclidean length **of a path in** X_f). Obviously each (X_f, d_f) is a geodesic metric space homeomorphic to a closed disk of $\mathbb{R}^2$. We will write elements of X_f in the form $(s,t,r)_f$, where $(s,t,r) \in \mathcal{C}$ and the subscript f is used to differentiate points of the different spaces X_f. The boundary ∂X_f is then the collection of all points $(s,t,r)_f$ for $(s,t,r) \in ([0,1]^2 \setminus (0,1)^2) \times \{0\}$.

Let $f \in F(\mathbf{q})$ and let e_1, e_2, e_3, e_4 be the four oriented edges incident to f enumerated in a way consistent with the counterclockwise order on the boundary (here the labeling of these edges is chosen arbitrarily among the 4 possible labelings preserving the cyclic order). We then define

$$\begin{array}{lcl} c_{e_1}(t) = (t,0,0)_f & , & 0 \leq t \leq 1 \\ c_{e_2}(t) = (1,t,0)_f & , & 0 \leq t \leq 1 \\ c_{e_3}(t) = (1-t,1,0)_f & , & 0 \leq t \leq 1 \\ c_{e_4}(t) = (0,1-t,0)_f & , & 0 \leq t \leq 1 \,. \end{array}$$

In this way, for every oriented edge e of the map $\mathbf{q}$, we have defined a path c_e which goes along one of the four edges of the square ∂X_f, where f is the face located to the left of e.

We define an equivalence relation $\equiv$ on the disjoint union $\amalg_{f \in F(\mathbf{q})} X_f$, as the coarsest equivalence relation such that, for every oriented edge e of $\mathbf{q}$, and every $t \in [0,1]$, we have $c_e(t) \equiv c_{\overline{e}}(1-t)$. By identifying points of the same equivalence class, we glue the oriented sides of the squares ∂X_f pairwise, in a way that is consistent with the map structure. More precisely, the topological quotient $\mathcal{S}_{\mathbf{q}} := \amalg_{f \in F(\mathbf{q})} X_f / \equiv$ is a surface which has a 2-dimensional cell complex structure, whose

[4]The way we do this is by no means canonical. For instance, the emptied cubes X_f used to fill the faces of $\mathbf{q}$ below could be replaced by unit squares for the l^1 metric. However, our choice avoids the existence of too many geodesic paths between vertices of the map in the surface where it is embedded.

1-skeleton $\mathcal{E}_{\mathbf{q}} := \amalg_{f\in F(\mathbf{q})}\partial X_f/\equiv$ is a representative of the map $\mathbf{q}$, with faces (2-cells) $X_f \setminus \partial X_f$. In particular, $\mathcal{S}_{\mathbf{q}}$ is homeomorphic to $\mathbb{S}^2$ by [**42**, Lemma 3.1.4]. With an oriented edge e of $\mathbf{q}$ one associates an edge of the graph drawing $\mathcal{E}_{\mathbf{q}}$ in $\mathcal{S}_{\mathbf{q}}$, more simply called an edge of $\mathcal{S}_{\mathbf{q}}$, made of the equivalence classes of points in $c_e([0,1])$ (or $c_{\overline{e}}([0,1])$). We also let $\mathcal{V}_{\mathbf{q}}$ be the 0-skeleton of this complex, i.e. the vertices of the graph — these are the equivalent classes of the corners of the squares ∂X_f. We call them the vertices of $\mathcal{S}_{\mathbf{q}}$ for simplicity.

We next endow the disjoint union $\amalg_{f\in F(\mathbf{q})}X_f$ with the largest pseudo-metric $D_{\mathbf{q}}$ that is compatible with $d_f, f\in F(\mathbf{q})$ and with $\equiv$, in the sense that $D_{\mathbf{q}}(x,y)\leq d_f(x,y)$ for $x,y\in X_f$, and $D_{\mathbf{q}}(x,y) = 0$ for $x\equiv y$. Therefore, the function $D_{\mathbf{q}} : \amalg_{f\in F(\mathbf{q})}X_f \times \amalg_{f\in F(\mathbf{q})}X_f \to \mathbb{R}_+$ is compatible with the equivalence relation $\equiv$, and its quotient mapping defines a pseudo-metric on the quotient space $\mathcal{S}_{\mathbf{q}}$, which is still denoted by $D_{\mathbf{q}}$.

PROPOSITION 8.3. *The space $(\mathcal{S}_{\mathbf{q}}, D_{\mathbf{q}})$ is a geodesic metric space homeomorphic to $\mathbb{S}^2$. Moreover, the space $(\mathcal{V}_{\mathbf{q}}, D_{\mathbf{q}})$ is isometric to $(V(\mathbf{q}), d_{\mathbf{q}})$, and any geodesic path in $\mathcal{S}_{\mathbf{q}}$ between two elements of $\mathcal{V}_{\mathbf{q}}$ is a concatenation of edges of $\mathcal{S}_{\mathbf{q}}$. Last,*

$$d_{GH}((V(\mathbf{q}), d_{\mathbf{q}}), (\mathcal{S}_{\mathbf{q}}, D_{\mathbf{q}})) \leq 3\,.$$

PROOF. We first check that $D_{\mathbf{q}}$ is a true metric on $\mathcal{S}_{\mathbf{q}}$, i.e. that it separates points. To see this, we use the fact [**11**, Theorem 3.1.27] that $D_{\mathbf{q}}$ admits the constructive expression:

$$D_{\mathbf{q}}(a,b) = \inf\left\{\sum_{i=0}^{n} d(x_i,y_i) : n\geq 0, x_0 = a, y_n = b, y_i\equiv x_{i+1} \text{ for } 0\leq i\leq n-1\right\},$$

where we have set $d(x,y) = d_f(x,y)$ if $x,y\in X_f$ for some f, and $d(x,y)=\infty$ otherwise. It follows that, for $a\in X_f\backslash\partial X_f$ and $b\neq a$, $D_{\mathbf{q}}(a,b) > \min(d(a,b), d_f(a,\partial X_f)) > 0$, so a and b are separated.

To verify that $D_{\mathbf{q}}$ is a a true metric on $\mathcal{S}_{\mathbf{q}}$, it remains to treat the case where $a\in\partial X_f, b\in\partial X_{f'}$ for some $f,f'\in F(\mathbf{q})$. The crucial observation is that a shortest path in X_f between two points of ∂X_f is entirely contained in ∂X_f. It is then a simple exercise to check that if a,b are in distinct equivalence classes, the distance $D_{\mathbf{q}}(a,b)$ will be larger than the length of some fixed non-trivial path with values in $\mathcal{E}_{\mathbf{q}}$. More precisely, if (the equivalence classes of) a,b belong to the same edge of $\mathcal{S}_{\mathbf{q}}$, then we can find representatives a',b' in the same X_f and we will have $D_{\mathbf{q}}(a,b)\geq d_f(a',b')$. If the equivalence class of a is not a vertex of $\mathcal{S}_{\mathbf{q}}$ but that of b is, then $D_{\mathbf{q}}(a,b)$ is at least equal to the distance of $a\in X_f$ to the closest corner of the square ∂X_f. Finally, if the (distinct) equivalence classes of a,b are both vertices, then $D_{\mathbf{q}}(a,b)\geq 1$. One deduces that $D_{\mathbf{q}}$ is a true distance on $\mathcal{S}_{\mathbf{q}}$, which makes it a geodesic metric space by [**11**, Corollary 3.1.24]. Since $\mathcal{S}_{\mathbf{q}}$ is a compact topological space, the metric $D_{\mathbf{q}}$ induces the quotient topology on $\mathcal{S}_{\mathbf{q}}$ by [**11**, Exercise 3.1.14], hence $(\mathcal{S}_{\mathbf{q}}, D_{\mathbf{q}})$ is homeomorphic to $\mathbb{S}^2$.

From the observations in the last paragraph, a shortest path between vertices of $\mathcal{S}_{\mathbf{q}}$ takes values in $\mathcal{E}_{\mathbf{q}}$. Since an edge of $\mathcal{S}_{\mathbf{q}}$ is easily checked to have length 1 for the distance $D_{\mathbf{q}}$, such a shortest path will have the same length as a geodesic path for the (combinatorial) graph distance between the two vertices. Hence $(\mathcal{V}_{\mathbf{q}}, D_{\mathbf{q}})$ is indeed isometric to $(V(\mathbf{q}), d_{\mathbf{q}})$. The last statement follows immediately from this and the fact that diam $(X_f, d_f) \leq 3$, entailing that $\mathcal{V}_{\mathbf{q}}$ is 3-dense in $(\mathcal{S}_{\mathbf{q}}, D_{\mathbf{q}})$, i.e. its 3-neighborhood in $(\mathcal{S}_{\mathbf{q}}, D_{\mathbf{q}})$ equals $\mathcal{S}_{\mathbf{q}}$. □

In view of the proposition, we can view $D_{\mathbf{q}}$ as an extension to $\mathcal{S}_{\mathbf{q}}$ of the graph distance $d_{\mathbf{q}}$ on $V(\mathbf{q})$. For this reason, we will denote $D_{\mathbf{q}}$ by $d_{\mathbf{q}}$ from now on, which should not set any ambiguity.

8.3. Proof of the homeomorphism theorem. We now work in the setting of the beginning of subsection 7.1. Recall that the uniform pointed quadrangulation (Q_n, v_*) is encoded by a uniform random element (T_n, L_n) of $\mathbf{T}_n$ via the CVS bijection (the parameter $\epsilon \in \{-1, 1\}$ will play no role here), and that C_n and V_n are the contour and label processes of (T_n, L_n). We assume that the amost sure convergence (24) holds uniformly on $[0,1]^2$, along the sequence (n_k), which is fixed. In what follows, all convergences as $n \to \infty$ hold along this sequence, or along some further subsequence.

We can also assume that $(V(Q_n), d_{Q_n})$ is actually the (isometric) space $(\mathcal{V}_{Q_n}, d_{Q_n})$, i.e. the subspace of vertices of the space $(\mathcal{S}_{Q_n}, d_{Q_n})$ constructed in the previous subsection. Recalling from subsection 5.4.2 that, in the CVS bijection, each edge of the tree T_n lies in exactly one face of Q_n, we may and will assume that T_n is also embedded in the surface $\mathcal{S}_{Q_n}$, in such a way that the set of its vertices is $\mathcal{V}_{Q_n} \setminus \{v_*\}$, where $v_* \in V(Q_n)$ is identified with its counterpart in $\mathcal{V}_{Q_n}$, and that each edge of T_n lies entirely in the corresponding face of $\mathcal{S}_{Q_n}$ via the CVS bijection.

We will rely on the following lemma. Let $\mathrm{Sk}(\mathcal{T}_{\mathbb{e}})$ be the complement of the set of leaves in the CRT $\mathcal{T}_{\mathbb{e}}$. Equivalently, $\mathrm{Sk}(\mathcal{T}_{\mathbb{e}})$ is the set of all points $a \in \mathcal{T}_{\mathbb{e}}$ such that $\mathcal{T}_{\mathbb{e}} \setminus \{a\}$ is disconnected, and it also coincides with the set of all $a \in \mathcal{T}_{\mathbb{e}}$ that can be written $a = p_{\mathbb{e}}(s) = p_{\mathbb{e}}(s')$ for some $0 \leq s < s' < 1$. The set $\mathrm{Sk}(\mathcal{T}_{\mathbb{e}})$ is called the *skeleton* of $\mathcal{T}_{\mathbb{e}}$.

LEMMA 8.4. *The following property is true with probability* 1. *Let* $a \in \mathrm{Sk}(\mathcal{T}_{\mathbb{e}})$, *and let* $s \in (0,1)$ *be such that* $a = p_{\mathbb{e}}(s)$. *Then for every* $\varepsilon > 0$, *there exists* $t \in (s, (s+\varepsilon) \wedge 1)$ *such that* $Z_t < Z_s$.

This lemma is a consequence of [**34**, Lemma 3.2] (see also [**31**, Lemma 2.2] for a slightly weaker statement). The proof relies on a precise study of the label function Z, and we refer the interested reader to [**34**]. Note that this result (and the analogous statement derived by time-reversal) implies that a.s., if $a \in \mathrm{Sk}(\mathcal{T}_{\mathbb{e}})$, then in each component of $\mathcal{T}_{\mathbb{e}} \setminus \{a\}$, one can find points b that are arbitrarily close to a and such that $Z_b < Z_a$.

LEMMA 8.5. *Almost surely, for every* $\varepsilon > 0$, *there exists* $\delta \in (0, \varepsilon)$ *such that, for* n *large enough, any simple loop* γ_n *made of edges of* $\mathcal{S}_{Q_n}$, *with diameter* $\leq n^{1/4}\delta$, *splits* $\mathcal{S}_{Q_n}$ *in two Jordan domains, one of which has diameter* $\leq n^{1/4}\varepsilon$.

PROOF. We argue by contradiction. Assume that, with positive probability, along some (random) subsequence of (n_k) there exist simple loops γ_n made of edges of $\mathcal{S}_{Q_n}$, with diameters $o(n^{1/4})$ as $n \to \infty$, such that the two Jordan domains bounded by γ_n are of diameters $\geq n^{1/4}\varepsilon$, where $\varepsilon > 0$ is some fixed constant. From now on we argue on this event. By abuse of notation we will sometimes identify the chain γ_n with the set of vertices it visits, or with the union of its edges, in a way that should be clear from the context.

By the Jordan curve theorem, the path γ_n splits $\mathcal{S}_{Q_n}$ into two Jordan domains, which we denote by $\mathcal{D}_n$ and $\mathcal{D}'_n$. Since the diameters of both these domains are at least $n^{1/4}\varepsilon$, and since every point in $\mathcal{S}_{Q_n}$ is at distance at most 3 from some vertex, we can find vertices y_n and y'_n belonging to $\mathcal{D}_n$ and $\mathcal{D}'_n$ respectively, and which lie at distance at least $n^{1/4}\varepsilon/4$ from γ_n. Since $V(Q_n) = T_n \cup \{v_*\}$, we can always assume that y_n and y'_n are distinct from v_*. Now, consider the geodesic path from y_n to y'_n in T_n, and let x_n be the first vertex of this path that belongs to γ_n.

In the contour exploration around T_n, the vertex x_n is visited at least once in the interval between y_n and y'_n, and another time in the interval between y'_n and y_n. More precisely, let j_n and j'_n be such that $y_n = u^n_{j_n}, y'_n = u^n_{j'_n}$, and assume first that $j_n < j'_n$ for infinitely many n. For such n, we can find integers $i_n \in (j_n, j'_n)$ and $i'_n \in (0, j_n) \cup (j'_n, 2n)$ such that $x_n = u^n_{i_n} = u^n_{i'_n}$. Up to further extraction, we may and will assume that

$$\frac{i_n}{2n} \to s\,, \qquad \frac{i'_n}{2n} \to s'\,, \qquad \frac{j_n}{2n} \to t\,, \qquad \frac{j'_n}{2n} \to t'\,, \tag{27}$$

for some $s, s', t, t' \in [0,1]$ such that $t \leq s \leq t'$ and $s' \in [0,t] \cup [t,1]$. Since

$$d_{Q_n}(x_n, y_n) \wedge d_{Q_n}(x_n, y'_n) \geq n^{1/4}\varepsilon/4\,,$$

we deduce from (24) that $D(s,t), D(s',t), D(s,t'), D(s',t') > 0$, and in particular, s, s', t, t' are all distinct. Since $u^n_{i_n} = u^n_{i'_n}$, we conclude that $s \sim_{\mathbb{e}} s'$, so that $p_{\mathbb{e}}(s) \in \mathrm{Sk}(\mathcal{T}_{\mathbb{e}})$. One obtains the same conclusion by a similar argument if $j_n > j'_n$ for every n large. We let $x = p_{\mathbb{e}}(s)$ and $y = p_{\mathbb{e}}(t)$. Note that $y \neq x$ because $D(s,t) > 0$ (recall Lemma 7.1).

Since $x \in \mathrm{Sk}(\mathcal{T}_{\mathbb{e}})$, by Theorem 7.2 we deduce that $D(a_*, x) = D(s_*, s) > 0$, where $a_* = p_{\mathbb{e}}(s_*)$ is as before the a.s. unique leaf of $\mathcal{T}_{\mathbb{e}}$ where Z attains its minimum. In particular, we obtain by (18), (24) and the fact that $\mathrm{diam}\,(\gamma_n) = o(n^{1/4})$ that

$$\liminf_{n\to\infty} n^{-1/4} d_{Q_n}(v_*, \gamma_n) = \liminf_{n\to\infty} n^{-1/4} d_{Q_n}(v_*, x_n) > 0\,.$$

Therefore, for n large enough, v_* does not belong to γ_n, and for definiteness, we will assume that for such n, $\mathcal{D}_n$ is the component of $\mathcal{S}_{Q_n} \setminus \gamma_n$ that does not contain v_*.

Now, we let $L^+_n = L_n - \min L_n + 1$, and in the rest of this proof, we call $L^+_n(v) = d_{Q_n}(v_*, v)$ the *label of the vertex* v in Q_n. Let $l_n = d_{Q_n}(v_*, \gamma_n) = \min_{v\in\gamma_n} L^+_n(v)$ be the minimal distance from v_* to a point visited by γ_n. Note that, for every vertex $v \in \mathcal{D}_n$, the property $L^+_n(v) \geq l_n$ holds, since any geodesic chain from v_* to v in Q_n has to cross γ_n.

Recalling that the vertex x_n was chosen so that the simple path in T_n from x_n to y_n lies entirely in $\mathcal{D}_n$, we conclude that the labels of vertices on this path are all greater than or equal to l_n. By passing to the limit, one concludes that for every c in the path $[\![x, y]\!]$ in $\mathcal{T}_{\mathbb{e}}$, there holds that $Z_c \geq Z_x$. Since the process Z evolves like Brownian motion along line segments of the tree $\mathcal{T}_{\mathbb{e}}$, we deduce that for every $c \in [\![x, y]\!]$ close enough to x, we have in fact $Z_c > Z_x$. From the interpretation of line segments in $\mathcal{T}_{\mathbb{e}}$ in terms of the coding function $\mathbb{e}$ (see the end of subsection 3.2), we can find $\overline{s} \in (0,1)$ such that $p_{\mathbb{e}}(\overline{s}) = x$, and such that, for every $u > \overline{s}$ sufficiently close to $\overline{s}$, the intersection of $[\![x, p_{\mathbb{e}}(u)]\!]$ with $[\![x, y]\!]$ will be of the form $[\![x, p_{\mathbb{e}}(r)]\!]$ for some $r \in (\overline{s}, u]$. By Lemma 8.4, and the fact that $Z_c \geq Z_x$ for every $c \in [\![x, y]\!]$ close enough to x, we can find $u > \overline{s}$ encoding a point $a = p_{\mathbb{e}}(u)$ and some $\eta > 0$ such that $Z_a \leq Z_x - (9/8)^{1/4}\eta$, and such that $[\![x, a]\!] \cap [\![x, y]\!] = [\![x, b]\!]$ for some $b \neq x$ such that $Z_b \geq Z_x + (9/8)^{1/4}\eta$.

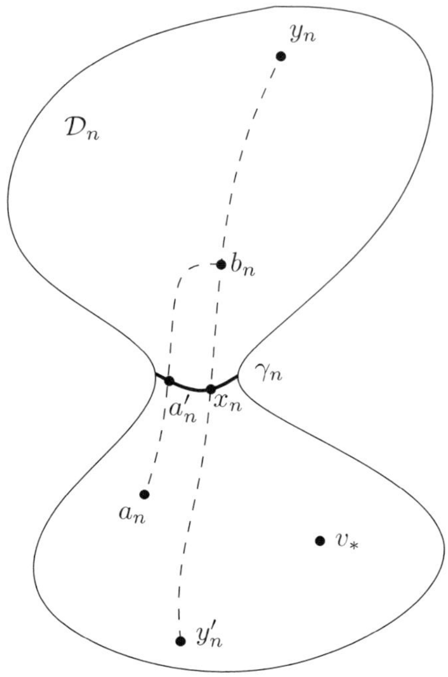

FIGURE 9. Illustration of the proof. The surface $\mathcal{S}_{Q_n}$ is depicted as a sphere with a bottleneck circled by γ_n (thick line). The dashed lines represent paths of T_n that are useful in the proof: One enters the component $\mathcal{D}_n$, and the other goes out after entering, identifying in the limit a point of the skeleton with another.

We then go back once again to the discrete approximations of the Brownian map, by considering k_n such that $k_n/2n$ converges to u. From the fact that $Z_a < Z_x$, we deduce that the vertex $a_n = u^n_{k_n}$ has label $L^+_n(a_n) < l_n$ for every n large enough. Indeed, the convergence (24) and the fact that $\mathrm{diam}\,(\gamma_n) = o(n^{1/4})$ imply that $(9/8n)^{1/4} l_n \to Z_x - \inf Z$. Consequently, the point a_n does not belong to $\mathcal{D}_n$. Moreover, the path in T_n from a_n to x_n meets the path from x_n to y_n at a point b_n such that $L^+_n(b_n) \geq l_n + \eta n^{1/4}$. The path from a_n to b_n has to cross the loop γ_n at some vertex, and we let a'_n be the first such vertex. By letting $n \to \infty$ one last time, we find a vertex $a' \in \mathcal{T}_{\mathbf{e}}$, which in the appropriate sense is the limit of a'_n as $n \to \infty$, such that $[\![a', x]\!]$ meets $[\![x, y]\!]$ at b. In particular, $a' \neq x$. But since a'_n and x_n are both on γ_n, we deduce that $D(a', x) = 0$. This contradicts Theorem 7.2 because x is not a leaf of $\mathcal{T}_{\mathbf{e}}$. This contradiction completes the proof of the lemma. □

We claim that Lemma 8.5 suffices to verify that the convergence of $(V(Q_n), (9/8n)^{1/4} d_{Q_n})$ to (M, D) is regular, and hence to conclude by Theorem 8.2 that the limit (M, D) is a topological sphere. To see this, we first choose $\varepsilon < \mathrm{diam}\,(M)/3$ to avoid trivialities. Let γ_n be a loop in $\mathcal{S}_{Q_n}$ with diameter $\leq n^{1/4}\delta$. Consider the union of the closures of faces of $\mathcal{S}_{Q_n}$ that are visited by γ_n. The boundary of this union is a collection $\mathcal{L}$ of pairwise disjoint simple loops made of edges of $\mathcal{S}_{Q_n}$. If x, y belong to the preceding union of faces, the fact that a face of $\mathcal{S}_{Q_n}$ has diameter less than 3 implies that there exist points x' and y' of γ_n at distance at most 3 from x and y respectively. Therefore, the diameters of the loops in $\mathcal{L}$ all are $\leq n^{1/4}\delta + 6$.

By the Jordan Curve Theorem, each of these loops splits $\mathcal{S}_{Q_n}$ into two simply connected components. By definition, one of these two components contains γ_n entirely. By Lemma 8.5, one of the two components has diameter $\leq n^{1/4}\varepsilon$. If we show that the last two properties hold simultaneously for one of the two components associated with (at least) one of the loops in $\mathcal{L}$, then obviously γ_n will be homotopic to 0 in its ε-neighborhood in $(\mathcal{S}_{Q_n}, n^{-1/4}d_{Q_n})$. So assume the contrary: The component not containing γ_n associated with every loop of $\mathcal{L}$ is of diameter $\leq n^{1/4}\varepsilon$. If this holds, then any point in $\mathcal{S}_{Q_n}$ must be at distance at most $n^{1/4}\varepsilon + 3$ from some point in γ_n. Take x, y such that $d_{Q_n}(x, y) = \mathrm{diam}\,(\mathcal{S}_{Q_n})$. Then there exist points x' and y' in γ_n at distance at most $n^{1/4}\varepsilon + 3$ respectively from x and y, and we conclude that $d_{Q_n}(x', y') \geq \mathrm{diam}\,(\mathcal{S}_{Q_n}) - 6 - 2n^{1/4}\varepsilon > n^{1/4}\delta \geq \mathrm{diam}\,(\gamma_n)$ for n large enough by our choice of ε. This contradiction completes the proof.

Note added in proof. The uniqueness problem for the Brownian map has been solved in two very recent papers of the authors: See the preprints arxiv:1104.1606 and arxiv:1105.4842. Consequently, Conjecture 6.1 is now a theorem, and analogs of this result hold for more general random planar maps such as triangulations.

References

[1] D. Aldous. The continuum random tree I. *Ann. Probab.* 19:1–28, 1991.

[2] D. Aldous. The continuum random tree III. *Ann. Probab.* 21:248–289, 1993.

[3] J. Ambjørn, B. Durhuus, and T. Jonsson. *Quantum geometry. A statistical field theory approach.* Cambridge Monographs on Mathematical Physics. Cambridge University Press, Cambridge, 1997.

[4] E.G. Begle. Regular convergence. *Duke Math. J.*, 11: 441–450, 1944.

[5] J. Bettinelli. The topology of scaling limits of positive genus random quadrangulations. Preprint, 2010, arXiv:1012.3726

[6] M. Bousquet-Mélou and A. Jehanne. Polynomial equations with one catalytic variable, algebraic series and map enumeration. *J. Combin. Theory Ser. B*, 96(5):623–672, 2006.

[7] J. Bouttier. *Physique Statistique des Surfaces Aléatoires et Combinatoire Bijective des Cartes Planaires.* PhD thesis, Université Paris 6, 2005.

[8] J. Bouttier and E. Guitter. The three-point function of planar quadrangulations. *J. Stat. Mech. Theory Exp.* 2008, no. 7, P07020, 39 pp.

[9] J. Bouttier, P. Di Francesco, and E. Guitter. Planar maps as labeled mobiles. *Electron. J. Combin.*, 11:no. 69, 27 pp. (electronic), 2004.

[10] E. Brézin, C. Itzykson, G. Parisi, and J. B. Zuber. Planar diagrams. *Comm. Math. Phys.*, 59(1):35–51, 1978.

[11] D. Burago, Y. Burago, and S. Ivanov. *A Course in Metric Geometry.* Graduate Studies in Mathematics, vol. 33. AMS, Boston, 2001.

[12] P. Chassaing and G. Schaeffer. Random planar lattices and integrated superBrownian excursion. *Probab. Theory Related Fields*, 128(2):161–212, 2004.

[13] K.L. Chung. Excursions in Brownian motion. *Ark. Mat.* 14:155–177, 1976.

[14] R. Cori and B. Vauquelin. Planar maps are well labeled trees. *Canad. J. Math.*, 33(5):1023–1042, 1981.

[15] J.-F. Delmas. Computation of moments for the length of the one dimensional ISE support. *Electron. J. Probab.*, 8:no. 17, 15 pp. (electronic), 2003.

[16] B. Duplantier and S. Sheffield. Liouville quantum gravity and KPZ. *Invent. Math.*, 185:333–393, 2011.

[17] T. Duquesne. A limit theorem for the contour process of conditioned Galton-Watson trees. *Ann. Probab.*, 31:996–1027, 2003.

[18] T. Duquesne and J.-F. Le Gall. Random Trees, Lévy Processes and Spatial Branching Processes. *Astérisque*, 281, 2002.

[19] T. Duquesne and J.-F. Le Gall. Probabilistic and fractal aspects of Lévy trees. *Probab. Th. Rel. Fields*, 131:553–603, 2005.
[20] S.N. Evans and A. Winter. Subtree prune and re-graft: A reversible tree-valued Markov process. *Ann. Probab.*, 34:918–961, 2006.
[21] A. Greven, P. Pfaffelhuber, and A. Winter. Convergence in distribution of random metric measure spaces (Λ-coalescent measure trees). *Probab. Theory Related Fields*, 145: 285–322, 2009.
[22] M. Gromov. *Metric Structures for Riemannian and Non-Riemannian Spaces.* Birkhäuser 1999.
[23] B. Haas and G. Miermont. Scaling limits of Markov branching trees, with applications to Galton-Watson and random unordered trees. Preprint, arXiv:1003.3632
[24] G. 't Hooft. A planar diagram theory for strong interactions. *Nucl. Phys. B*, 72:461–473, 1974.
[25] S. Janson and J.-F. Marckert. Convergence of discrete snakes. *J. Theoret. Probab.*, 18:615–647, 2005.
[26] W.D. Kaigh. An invariance principle for random walk conditioned by a late return to zero. *Ann. Probab.*, 4:115–121, 1976.
[27] S. K. Lando and A. K. Zvonkin. *Graphs on surfaces and their applications*, volume 141 of *Encyclopaedia of Mathematical Sciences.* Springer-Verlag, Berlin, 2004.
[28] J.-F. Le Gall. *Spatial Branching Processes, Random Snakes and Partial Differential Equations.* Birkhäuser 1999.
[29] J.-F. Le Gall. Random trees and applications. *Probab. Surveys*, 2:245-311, 2005.
[30] J.-F. Le Gall. A conditional limit theorem for tree-indexed random walk. *Stochastic Process. Appl.*, 116(4):539–567, 2006.
[31] J.-F. Le Gall. The topological structure of scaling limits of large planar maps. *Invent. Math.*, 169:621–670, 2007.
[32] J.-F. Le Gall. Geodesics in large planar maps and in the Brownian map. *Acta Mathematica*, 205:287–360, 2010.
[33] J.-F. Le Gall and G. Miermont. Scaling limits of random planar maps with large faces. *Ann. Probab.*, 39(1):1–69, 2011.
[34] J.-F. Le Gall and F. Paulin. Scaling limits of bipartite planar maps are homeomorphic to the 2-sphere. *Geom. Funct. Anal.*, 18(3):893–918, 2008.
[35] J.-F. Le Gall and M. Weill. Conditioned Brownian trees. *Ann. Inst. H. Poincaré Probab. Statist.*, 42(4):455–489, 2006.
[36] J.-F. Marckert and G. Miermont. Invariance principles for random bipartite planar maps. *Ann. Probab.*, 35(5):1642–1705, 2007.
[37] J.-F. Marckert and A. Mokkadem. Limit of normalized random quadrangulations: the Brownian map. *Ann. Probab.*, 34(6):2144–2202, 2006.
[38] G. Miermont. An invariance principle for random planar maps. In *Fourth Colloquium on Mathematics and Computer Sciences CMCS'06*, Discrete Math. Theor. Comput. Sci. Proc., AG, p. 39–58 (electronic). Nancy, 2006.
[39] G. Miermont. On the sphericity of scaling limits of random planar quadrangulations. *Electron. Commun. Probab.*, 13:248–257, 2008.
[40] G. Miermont. Tessellations of random maps of arbitrary genus. *Ann. Sci. Éc. Norm. Supér.*, 42(5):725–781, 2009.
[41] G. Miermont and M. Weill. Radius and profile of random planar maps with faces of arbitrary degrees. *Electron. J. Probab.*, 13: 79–106, 2008.
[42] B. Mohar and C. Thomassen. *Graphs on surfaces.* Johns Hopkins Studies in the Mathematical Sciences. Johns Hopkins University Press, Baltimore, 2001.
[43] R.L. Moore. Concerning upper-semicontinuous collections of continua. *Trans. Amer. Math. Soc.* 25:416-428, 1925.
[44] J. Neveu. Arbres et processus de Galton-Watson. *Ann. Inst. Henri Poincaré Probab. Stat.* 22:199–207, 1986.
[45] J.W. Pitman. Combinatorial stochastic processes. Ecole d'été de probabilités de Saint-Flour 2002. *Lecture Notes Math.* 1875. Springer, Berlin, 2006.
[46] D. Revuz and M. Yor. *Continuous Martingales and Brownian Motion.* Springer 1991.
[47] G. Schaeffer. *Conjugaison d'arbres et cartes combinatoires aléatoires.* PhD thesis, Université Bordeaux I, 1998.

[48] O. Schramm. Conformally invariant scaling limits: an overview and a collection of problems. In Proceedings of the International Congress of Mathematicians (Madrid 2006), Vol.I, pp. 513–543. European Math. Soc., Zürich, 2007.
[49] F. Spitzer. *Principles of Random Walk.* Van Nostrand 1963.
[50] R.P. Stanley. *Enumerative Combinatorics, Vol.2.* Cambridge University Press, Cambridge, 1999.
[51] W.T. Tutte. A census of planar maps. *Canad. J. Math.*, 15:249–271, 1963.

MATHÉMATIQUES, BAT.425, UNIVERSITÉ PARIS-SUD, 91405 ORSAY CEDEX FRANCE
E-mail address: jean-francois.legall@math.u-psud.fr

MATHÉMATIQUES, BAT.425, UNIVERSITÉ PARIS-SUD, 91405 ORSAY CEDEX FRANCE
E-mail address: gregory.miermont@math.u-psud.fr

Clay Mathematics Proceedings
Volume 15, 2012

Conformal Invariance of Lattice Models

Hugo Duminil-Copin and Stanislav Smirnov

ABSTRACT. These lecture notes provide an (almost) self-contained account on conformal invariance of the planar critical Ising and FK-Ising models. They present the theory of discrete holomorphic functions and its applications to planar statistical physics (more precisely to the convergence of fermionic observables). Convergence to SLE is discussed briefly. Many open questions are included.

CONTENTS

2010 *Mathematics Subject Classification.* Primary: 60K35; Secondary: 60J67, 82B23.

1. Introduction

The celebrated Lenz-Ising model is one of the simplest models of statistical physics exhibiting an order-disorder transition. It was introduced by Lenz in **[Len20]** as an attempt to explain Curie's temperature for ferromagnets. In the model, iron is modeled as a collection of atoms with fixed positions on a crystalline lattice. Each atom has a magnetic spin, pointing in one of two possible directions. We will set the spin to be equal to 1 or −1. Each configuration of spins has an intrinsic energy, which takes into account the fact that neighboring sites prefer to be aligned (meaning that they have the same spin), exactly like magnets tend to attract or repel each other. Fix a box $\Lambda \subset \mathbb{Z}^2$ of size n. Let $\sigma \in \{-1,1\}^\Lambda$ be a configuration of spins 1 or −1. The energy of the configuration σ is given by the Hamiltonian

$$E_\Lambda(\sigma) := -\sum_{x\sim y} \sigma_x \sigma_y$$

where $x \sim y$ means that x and y are neighbors in Λ. The energy is, up to an additive constant, twice the number of disagreeing neighbors. Following a fundamental principle of physics, the spin-configuration is sampled proportionally to its Boltzmann weight: at an inverse-temperature β, the probability $\mu_{\beta,\Lambda}$ of a configuration σ satisfies

$$\mu_{\beta,\Lambda}(\sigma) := \frac{\mathrm{e}^{-\beta E_\Lambda(\sigma)}}{Z_{\beta,\Lambda}}$$

where

$$Z_{\beta,\Lambda} := \sum_{\tilde\sigma\in\{-1,1\}^\Lambda} \mathrm{e}^{-\beta E_\Lambda(\tilde\sigma)}$$

is the so-called *partition function* defined in such a way that the sum of the weights over all possible configurations equals 1. Above a certain *critical inverse-temperature* β_c, the model has a *spontaneous magnetization* while below β_c does not (this phenomenon will be described in more detail in the next section). When β_c lies strictly between 0 and ∞, the Ising model is said to undergo a phase transition between an *ordered* and a *disordered* phase. The fundamental question is to study the phase transition between the two regimes.

Lenz's student Ising proved the absence of phase transition in dimension one (meaning $\beta_c = \infty$) in his PhD thesis **[Isi25]**, wrongly conjecturing the same picture in higher dimensions. This belief was widely shared, and motivated Heisenberg to introduce his famous model **[Hei28]**. However, some years later Peierls **[Pei36]** used estimates on the length of interfaces between spin clusters to disprove the conjecture, showing a phase transition in the two-dimensional case. Later, Kramers and Wannier **[KW41a, KW41b]** derived nonrigorously the value of the critical temperature.

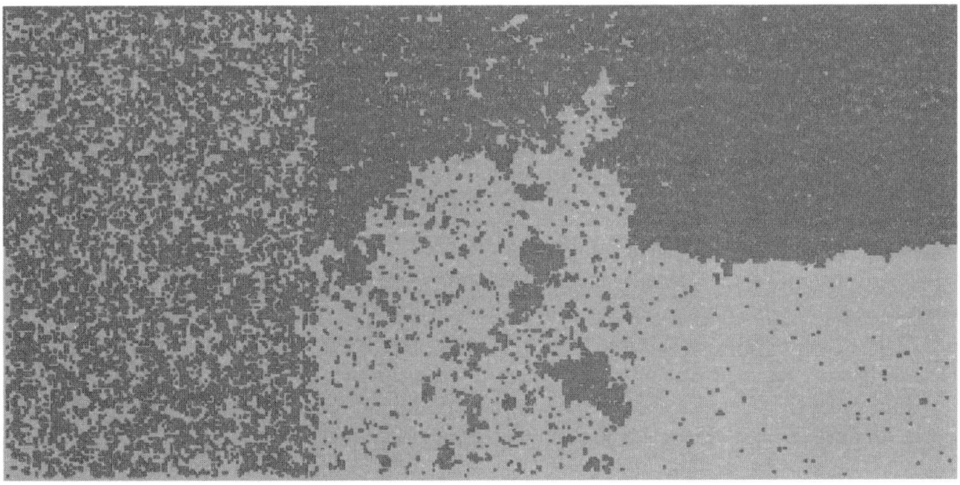

FIGURE 1. Ising configurations at $\beta < \beta_c$, at $\beta = \beta_c$, and $\beta > \beta_c$ respectively.

In 1944, Onsager [**Ons44**] computed the partition function of the model, followed by further computations with Kaufman, see [**KO50**] for instance[1]. In the physical approach to statistical models, the computation of the partition function is the first step towards a deep understanding of the model, enabling for instance the computation of the free energy. The formula provided by Onsager led to an explosion in the number of results on the 2D Ising model (papers published on the Ising model can now be counted in the thousands). Among the most noteworthy results, Yang derived rigorously the spontaneous magnetization [**Yan52**] (the result was derived nonrigorously by Onsager himself). McCoy and Wu [**MW73**] computed many important quantities of the Ising model, including several critical exponents, culminating with the derivation of two-point correlations between sites $(0,0)$ and (n,n) in the whole plane. See the more recent book of Palmer for an exposition of these and other results [**Pal07**].

The computation of the partition function was accomplished later by several other methods and the model became the most prominent example of an exactly solvable model. The most classical techniques include the transfer-matrices technique developed by Lieb and Baxter [**Lie67, Bax89**], the Pfaffian method, initiated by Fisher and Kasteleyn, using a connection with dimer models [**Fis66, Kas61**], and the combinatorial approach to the Ising model, initiated by Kac and Ward [**KW52**] and then developed by Sherman [**She60**] and Vdovichenko [**Vdo65**]; see also the more recent [**DZM**$^+$**99, Cim10**].

Despite the number of results that can be obtained using the partition function, the impossibility of computing it explicitly enough in finite volume made the geometric study of the model very hard to perform while using the classical methods. The lack of understanding of the geometric nature of the model remained mathematically unsatisfying for years.

The arrival of the renormalization group formalism (see [**Fis98**] for a historical exposition) led to a better physical and geometrical understanding, albeit mostly non-rigorous. It suggests that the block-spin renormalization transformation (coarse-graining, *e.g.* replacing a block of neighboring sites by one site having a spin equal to the dominant spin in the block) corresponds to appropriately changing the scale and the

[1]This result represented a shock for the community: it was the first mathematical evidence that the mean-field behavior was inaccurate in low dimensions.

temperature of the model. The Kramers-Wannier critical point then arises as the fixed point of the renormalization transformations. In particular, under simple rescaling the Ising model at the critical temperature should converge to a scaling limit, a continuous version of the originally discrete Ising model, corresponding to a quantum field theory. This leads to the idea of universality: the Ising models on different regular lattices or even more general planar graphs belong to the same renormalization space, with a unique critical point, and so at criticality the scaling limit and the scaling dimensions of the Ising model should always be the same (it should be independent of the lattice whereas the critical temperature depends on it).

Being unique, the scaling limit at the critical point must satisfy translation, rotation and scale invariance, which allows one to deduce some information about correlations [**PP66, Kad66**]. In seminal papers [**BPZ84b, BPZ84a**], Belavin, Polyakov and Zamolodchikov suggested a much stronger invariance of the model. Since the scaling-limit quantum field theory is a local field, it should be invariant by any map which is locally a composition of translation, rotation and homothety. Thus it becomes natural to postulate full conformal invariance (under all conformal transformations[2] of subregions). This prediction generated an explosion of activity in conformal field theory, allowing nonrigorous explanations of many phenomena; see [**ISZ88**] for a collection of the original papers of the subject.

To summarize, Conformal Field Theory asserts that the Ising model admits a scaling limit at criticality, and that this scaling limit is a conformally invariant object. From a mathematical perspective, this notion of conformal invariance of a model is ill-posed, since the meaning of scaling limit is not even clear. The following solution to this problem can be implemented: the scaling limit of the model could simply retain the information given by interfaces only. There is no reason why all the information of a model should be encoded into information on interfaces, yet one can hope that most of the relevant quantities can be recovered from it. The advantage of this approach is that there exists a mathematical setting for families of continuous curves.

In the Ising model, there is a canonical way to isolate macroscopic interfaces. Consider a simply-connected domain Ω with two points a and b on the boundary and approximate it by a discrete graph $\Omega_\delta \subset \delta\mathbb{Z}^2$. The boundary of Ω_δ determines two arcs ∂_{ab} and ∂_{ba} and we can fix the spins to be $+1$ on the arc ∂_{ab} and -1 on the arc ∂_{ba} (this is called Dobrushin boundary conditions). In this case, there exists an interface[3] separating $+1$ and -1 going from a to b and the prediction of Conformal Field Theory then translates into the following predictions for models: interfaces in Ω_δ converge when δ goes to 0 to a random continuous non-selfcrossing curve $\gamma_{(\Omega,a,b)}$ between a and b in Ω which is conformally invariant in the following way:

For any (Ω, a, b) and any conformal map $\psi : \Omega \to \mathbb{C}$, the random curve $\psi \circ \gamma_{(\Omega,a,b)}$ has the same law as $\gamma_{(\psi(\Omega),\psi(a),\psi(b))}$.

In 1999, Schramm proposed a natural candidate for the possible conformally invariant families of continuous non-selfcrossing curves. He noticed that interfaces of models further satisfy the *domain Markov property*, which, together with the assumption of conformal invariance, determine the possible families of curves. In [**Sch00**], he introduced the *Schramm-Loewner Evolution* (SLE for short): for $\kappa > 0$, the SLE(κ) is the random Loewner Evolution with driving process $\sqrt{\kappa}B_t$, where (B_t) is a standard Brownian motion (see Beffara's course in this volume). In our case, it implies that the random continuous curve $\gamma_{(\Omega,a,b)}$ described previously should be an SLE.

---

[2] *i.e.* one-to-one holomorphic maps.

[3] In fact the interface is not unique. In order to solve this issue, consider the closest interface to ∂_{ab}.

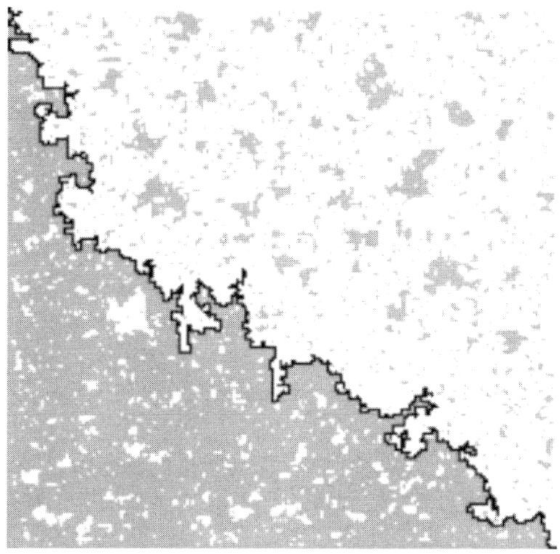

FIGURE 2. An interface between + and – in the Ising model.

Proving convergence of interfaces to an SLE is fundamental. Indeed, SLE processes are now well-understood and their path properties can be related to fractal properties of the critical phase. Critical exponents can then be deduced from these properties via the so-called *scaling relations*. These notes provide an (almost) self-contained proof of convergence to SLE for the two-dimensional Ising model and its random-cluster representation the FK-Ising model (see Section 3 for a formal definition).

Main result 1 *(Theorem 2.10) The law of interfaces of the critical Ising model converges in the scaling limit to a conformally invariant limit described by the Schramm-Loewner Evolution of parameter* $\kappa = 3$.
Main result 2 *(Theorem 3.13) The law of interfaces of the critical FK-Ising model converges in the scaling limit to a conformally invariant limit described by the Schramm-Loewner Evolution of parameter* $\kappa = 16/3$.

Even though we now have a mathematical framework for conformal invariance, it remains difficult to prove convergence of interfaces to SLEs. Observe that working with interfaces offers a further simplification: properties of these interfaces should also be conformally invariant. Therefore, one could simply look at a discrete *observable* of the model and try to prove that it converges in the scaling limit to a conformally covariant object. Of course, it is not clear that this observable would tell us anything about critical exponents, yet it already represents a significant step toward conformal invariance.

In 1994, Langlands, Pouliot and Saint-Aubin [**LPSA94**] published a number of numerical values in favor of conformal invariance (in the scaling limit) of crossing probabilities in the percolation model. More precisely, they checked that, taking different topological rectangles, the probability $C_\delta(\Omega, A, B, C, D)$ of having a path of adjacent open edges from AB to CD converges when δ goes to 0 towards a limit which is the same for (Ω, A, B, C, D) and $(\Omega', A', B', C', D')$ if they are images of each other by a conformal map. The paper [**LPSA94**], while only numerical, attracted many mathematicians to the domain. The same year, Cardy [**Car92**] proposed an explicit formula for the limit of percolation crossing probabilities. In 2001, Smirnov proved Cardy's formula rigorously for critical site percolation on the triangular lattice [**Smi01**], hence rigorously providing a concrete example of a conformally invariant property of the model. A somewhat incredible consequence of this theorem is that the mechanism can be reversed: even though Cardy's formula seems much weaker than convergence to SLE, they are actually equivalent. In other words, conformal covariance of one well-chosen observable of the model can be sufficient to prove conformal invariance of interfaces.

It is also possible to find an observable with this property in the Ising case (see Definition 2.9). This observable, called the *fermionic observable*, is defined in terms of the so-called high temperature expansion of the Ising model. Specific combinatorial properties of the Ising model translate into local relations for the fermionic observable. In particular, the observable can be proved to converge when taking the scaling limit. This convergence result (Theorem 2.11) is the main step in the proof of conformal invariance. Similarly, a fermionic observable can be defined in the FK-Ising case, and its convergence implies the convergence of interfaces.

Archetypical examples of conformally covariant objects are holomorphic solutions to boundary value problems such as Dirichlet or Riemann problems. It becomes natural to expect that discrete observables which are conformally covariant in the scaling limit are naturally preharmonic or preholomorphic functions, *i.e.* relevant discretizations of harmonic and holomorphic functions. Therefore, the proofs of conformal invariance harness *discrete complex analysis* in a substantial way. The use of discrete holomorphicity appeared first in the case of dimers [**Ken00**] and has been extended to several statistical physics models since then. Other than being interesting in themselves, preholomorphic functions have found several applications in geometry, analysis, combinatorics, and probability. We refer the interested reader to the expositions by Lovász [**Lov04**], Stephenson [**Ste05**], Mercat [**Mer01**], Bobenko and Suris [**BS08**]. Let us finish by mentioning that the previous discussion sheds a new light on both approaches described above: combinatorial properties of the discrete Ising model allow us to prove the convergence of discrete observables to conformally covariant objects. In other words, exact integrability and Conformal Field Theory are connected via the proof of the conformal invariance of the Ising model.

Acknowledgments These notes are based on a course on conformal invariance of lattice models given in Búzios, Brazil, in August 2010, as part of the Clay Mathematics Institute Summer School. The course consisted of six lectures by the second author. The authors wish to thank the organisers of both the Clay Mathematics Institute Summer School and the XIV Brazilian Probability School for this milestone event. We are particularly grateful to Vladas Sidoravicius for his incredible energy and the constant effort put into the organization of the school. We thank Stéphane Benoist, David Cimasoni and Alan Hammond for a careful reading of previous versions of this manuscript. The two authors were supported by the EU Marie-Curie RTN CODY, the ERC AG CONFRA, as well as by the Swiss FNS. The research of the second author is supported by the Chebyshev Laboratory (Department of Mathematics and Mechanics, St. Petersburg State University) under RF Government grant 11.G34.31.0026.

1.1. Organization of the notes. Section 2 presents the necessary background on the spin Ising model. In the first subsection, we recall general facts on the Ising model. In the second subsection, we introduce the low and high temperature expansions, as well as Kramers-Wannier duality. In the last subsection, we use the high-temperature expansion in spin Dobrushin domains to define the spin fermionic observable. Via the Kramers-Wannier duality, we explain how it relates to interfaces of the Ising model at criticality and we state the conformal invariance result for Ising.

Section 3 introduces the FK-Ising model. We start by defining general FK percolation models and we discuss planar duality. Then, we explain the Edwards-Sokal coupling, an important tool relating the spin Ising and FK-Ising models. Finally, we introduce the loop representation of the FK-Ising model in FK Dobrushin domains. It allows us to define the FK fermionic observable and to state the conformal invariance result for the FK-Ising model.

Section 4 is a brief survey of discrete complex analysis. We first deal with preharmonic functions and a few of their elementary properties. These properties will be used in Section 6. In the second subsection, we present a brief historic of preholomorphic functions. The third subsection is the most important, it contains the definition and several properties of s-holomorphic (or spin-holomorphic) functions. This notion is crucial in the proof of conformal invariance: the fermionic observables will be proved to be s-holomorphic, a fact which implies their convergence in the scaling limit. We also include a brief discussion on complex analysis on general graphs.

Section 5 is devoted to the convergence of the fermionic observables. First, we show that the FK fermionic observable is s-holomorphic and that it converges in the scaling limit. Second, we deal with the spin fermionic observable. We prove its s-holomorphicity and sketch the proof of its convergence.

Section 6 shows how to harness the convergence of fermionic observables in order to prove conformal invariance of interfaces in the spin and FK-Ising models. It mostly relies on tightness results and certain properties of Loewner chains.

Section 7 is intended to present several other applications of the fermionic observables. In particular, we provide an elementary derivation of the critical inverse-temperature.

Section 8 contains a discussion on generalizations of this approach to lattice models. It includes a subsection on the Ising model on general planar graphs. It also gathers conjectures regarding models more general than the Ising model.

1.2. Notations.

1.2.1. *Primal, dual and medial graphs.* We mostly consider the (**rotated**) **square lattice** $\mathbb{L}$ with vertex set $e^{i\pi/4}\mathbb{Z}^2$ and edges between nearest neighbors. An edge with end-points x and y will be denoted by $[xy]$. If there exists an edge e such that $e = [xy]$, we write $x \sim y$. Finite graphs G will always be subgraphs of $\mathbb{L}$ and will be called **primal graphs**. The **boundary** of G, denoted by ∂G, will be the set of sites of G with fewer than four neighbors in G.

The **dual graph** $G^\star$ of a planar graph G is defined as follows: sites of $G^\star$ correspond to faces of G (for convenience, the infinite face will not correspond to a dual site), edges of $G^\star$ connect sites corresponding to two adjacent faces of G. The **dual lattice** of $\mathbb{L}$ is denoted by $\mathbb{L}^\star$.

The **medial lattice** $\mathbb{L}^\diamond$ is the graph with vertex set being the centers of edges of $\mathbb{L}$, and edges connecting nearest vertices, see Fig. 6. The **medial graph** $G^\diamond$ is the subgraph of $\mathbb{L}^\diamond$ composed of all the vertices of $\mathbb{L}^\diamond$ corresponding to edges of G. Note that $\mathbb{L}^\diamond$ is a rotated and rescaled (by a factor $1/\sqrt{2}$) version of $\mathbb{L}$, and that it is the usual square lattice. We will often use the connection between the faces of $\mathbb{L}^\diamond$ and the sites of $\mathbb{L}$ and $\mathbb{L}^\star$. We say that a face of the medial lattice is *black* if it corresponds to a vertex of $\mathbb{L}$, and *white* otherwise. Edges of $\mathbb{L}^\diamond$ are oriented counterclockwise around black faces.

1.2.2. *Approximations of domains.* We will be interested in finer and finer graphs approximating continuous domains. For $\delta > 0$, the square lattice $\sqrt{2}\delta\mathbb{L}$ of mesh-size $\sqrt{2}\delta$ will be denoted by $\mathbb{L}_\delta$. The definitions of dual and medial lattices extend to this context. Note that the medial lattice $\mathbb{L}_\delta^\diamond$ has mesh-size δ.

For a simply connected domain Ω in the plane, we set $\Omega_\delta = \Omega \cap \mathbb{L}_\delta$. The edges connecting sites of Ω_δ are those included in Ω. The graph Ω_δ should be thought of as a discretization of Ω (we avoid technicalities concerning the regularity of the domain). More generally, when no continuous domain Ω is specified, Ω_δ stands for a finite simply connected (meaning that the complement is connected) subgraph of $\mathbb{L}_\delta$.

We will be considering sequences of functions on Ω_δ for δ going to 0. In order to make functions live in the same space, we implicitly perform the following operation: for a function f on Ω_δ, we choose for each square a diagonal and extend the function to

Ω in a piecewise linear way on every triangle (any reasonable way would do). Since no confusion will be possible, we denote the extension by f as well.

1.2.3. *Distances and convergence.* Points in the plane will be denoted by their complex coordinates, $\mathrm{Re}(z)$ and $\mathrm{Im}(z)$ will be the real and imaginary parts of z respectively. The **norm** will be the usual complex modulus $|\cdot|$. Unless otherwise stated, distances between points (even if they belong to a graph) are distances in the plane. The **distance** between a point z and a closed set F is defined by

$$d(z,F) := \inf_{y\in F} |z-y|. \tag{1.1}$$

Convergence of random parametrized curves (say with time-parameter in $[0,1]$) is in the sense of the **weak topology** inherited from the following distance on curves:

$$d(\gamma_1,\gamma_2) = \inf_{\phi} \sup_{u\in[0,1]} |\gamma_1(u)-\gamma_2(\phi(u))|, \tag{1.2}$$

where the infimum is taken over all reparametrizations (*i.e.* strictly increasing continuous functions $\phi:[0,1]\to[0,1]$ with $\phi(0)=0$ and $\phi(1)=1$).

2. Two-dimensional Ising model

2.1. Boundary conditions, infinite-volume measures and phase transition. The (spin) Ising model can be defined on any graph. However, we will restrict ourselves to the (rotated) square lattice. Let G be a finite subgraph of $\mathbb{L}$, and $b\in\{-1,+1\}^{\partial G}$. The Ising model with *boundary conditions* b is a random assignment of spins $\{-1,+1\}$ (or simply $-/+$) to vertices of G such that $\sigma_x = b_x$ on ∂G, where σ_x denotes the spin at site x. The partition function of the model is denoted by

$$Z^b_{\beta,G} = \sum_{\sigma\in\{-1,1\}^G:\ \sigma=b \text{ on } \partial G} \exp\Big[\beta \sum_{x\sim y} \sigma_x\sigma_y\Big], \tag{2.1}$$

where β is the inverse-temperature of the model and the second summation is over all pairs of neighboring sites x,y in G. The probability of a configuration σ is then equal to

$$\mu^b_{\beta,G}(\sigma) = \frac{1}{Z^b_{\beta,G}} \exp\Big[\beta \sum_{x\sim y} \sigma_x\sigma_y\Big]. \tag{2.2}$$

Equivalently, one can define the Ising model without boundary conditions, also called free boundary conditions (it is the one defined in the introduction). The measure with free boundary conditions is denoted by $\mu^f_{\beta,G}$.

We will not offer a complete exposition on the Ising model and we rather focus on crucial properties. The following result belongs to the folklore (see **[FKG71]** for the original paper). An event is called *increasing* if it is preserved by switching some spins from $-$ to $+$.

THEOREM 2.1 (Positive association at every temperature). *The Ising model on a finite graph G at temperature $\beta>0$ satisfies the following properties:*

- ***FKG inequality:*** *For any boundary conditions b and any increasing events A,B,*

$$\mu^b_{\beta,G}(A\cap B) \ge \mu^b_{\beta,G}(A)\mu^b_{\beta,G}(B). \tag{2.3}$$

- ***Comparison between boundary conditions:*** *For boundary conditions $b_1 \le b_2$ (meaning that spins $+$ in b_1 are also $+$ in b_2) and an increasing event A,*

$$\mu^{b_1}_{\beta,G}(A) \le \mu^{b_2}_{\beta,G}(A). \tag{2.4}$$

If (2.4) is satisfied for every increasing event, we say that $\mu_{\beta,G}^{b_2}$ *stochastically dominates* $\mu_{\beta,G}^{b_1}$ (denoted by $\mu_{\beta,G}^{b_1} \le \mu_{\beta,G}^{b_2}$). Two boundary conditions are extremal for the stochastic ordering: the measure with all + (resp. all −) boundary conditions, denoted by $\mu_{\beta,G}^{+}$ (resp. $\mu_{\beta,G}^{-}$) is the largest (resp. smallest).

Theorem 2.1 enables us to define infinite-volume measures as follows. Consider the nested sequence of boxes $\Lambda_n = [-n,n]^2$. For any $N > 0$ and any increasing event A depending only on spins in Λ_N, the sequence $(\mu_{\beta,\Lambda_n}^{+}(A))_{n \ge N}$ is decreasing[4]. The limit, denoted by $\mu_\beta^+(A)$, can be defined and verified to be independent on N.

In this way, μ_β^+ is defined for increasing events depending on a finite number of sites. It can be further extended to a probability measure on the σ-algebra spanned by cylindrical events (events measurable in terms of a finite number of spins). The resulting measure, denoted by μ_β^+, is called the infinite-volume Ising model with + boundary conditions.

Observe that one could construct (a priori) different infinite-volume measures, for instance with − boundary conditions (the corresponding measure is denoted by μ_β^-). If infinite-volume measures are defined from a property of compatibility with finite volume measures, then μ_β^+ and μ_β^- are extremal among infinite-volume measures of parameter β. In particular, if $\mu_\beta^+ = \mu_\beta^-$, there exists a unique infinite volume measure.

The Ising model in infinite-volume exhibits a phase transition at some critical inverse-temperature β_c:

THEOREM 2.2. *Let* $\beta_c = \frac{1}{2}\ln(1+\sqrt{2})$. *The* magnetization $\mu_\beta^+[\sigma_0]$ *at the origin is strictly positive for* $\beta > \beta_c$ *and equal to 0 when* $\beta < \beta_c$.

In other words, when $\beta > \beta_c$, there is long range memory, the phase is *ordered*. When $\beta < \beta_c$, the phase is called *disordered*. The existence of a critical temperature separating the ordered from the disordered phase is a relatively easy fact **[Pei36]** (although at the time it was quite unexpected). Its computation is more difficult. It was identified without proof by Kramers and Wannier **[KW41a, KW41b]** using the duality between low and high temperature expansions of the Ising model (see the argument in the next section). The first rigorous derivation is due to Yang **[Yan52]**. He uses Onsager's exact formula for the (infinite-volume) partition function to compute the spontaneous magnetization of the model. This quantity provides one criterion for localizing the critical point. The first probabilistic computation of the critical inverse-temperature is due to Aizenman, Barsky and Fernández **[ABF87]**. In Subsection 7.1, we present a short alternative proof of Theorem 2.2, using the fermionic observable.

The critical inverse-temperature has also an interpretation in terms of infinite-volume measures (these measures are called Gibbs measures). For $\beta < \beta_c$ there exists a unique Gibbs measure, while for $\beta > \beta_c$ there exist several. The classification of Gibbs measures in the ordered phase is interesting: in dimension two, any infinite-volume measure is a convex combination of μ_β^+ and μ_β^- (see **[Aiz80, Hig81]** or the recent proof **[CV10]**). This result is no longer true in higher dimension: non-translational-invariant Gibbs measures can be constructed using 3D Dobrushin domains **[Dob72]**.

When $\beta > \beta_c$, spin-correlations $\mu_\beta^+[\sigma_0\sigma_x]$ do not go to 0 when x goes to infinity. There is long range memory. At β_c, spin-correlations decay to 0 following a power law **[Ons44]**:

$$\mu_{\beta_c}^+[\sigma_0\sigma_x] \approx |x|^{-1/4}$$

when $x \to \infty$. When $\beta < \beta_c$, spin-correlations decay exponentially fast in $|x|$. More precisely, we will show the following result first due to **[MW73]**:

---

[4]Indeed, for any configuration of spins in $\partial\Lambda_n$ being smaller than all +, the restriction of $\mu_{\beta,\Lambda_{n+1}}^+$ to Λ_n is stochastically dominated by μ_{β,Λ_n}^+.

THEOREM 2.3. *For $\beta < \beta_c$, and $a = e^{i\pi/4}(x+iy) \in \mathbb{C}$,*

$$\tau_\beta(a) = \lim_{n\to\infty} -\frac{1}{n}\ln \mu_\beta^+[\sigma_0\sigma_{[na]}] = x \operatorname{arcsinh}(sx) + y \operatorname{arcsinh}(sy)$$

where $[na]$ is the site of $\mathbb{L}$ closest to na, and s solves the equation

$$\sqrt{1+s^2x^2} + \sqrt{1+s^2y^2} = \sinh(2\beta) + (\sinh(2\beta))^{-1}.$$

The quantity $\tau_\beta(z)$ is called the *correlation length* in direction z. When getting closer to the critical point, the correlation length goes to infinity and becomes isotropic (it does not depend on the direction, thus giving a glimpse of rotational invariance at criticality):

THEOREM 2.4 (see *e.g.* [**Mes06**]). *For $z \in \mathbb{C}$, the correlation length satisfies the following equality*

$$\lim_{\beta\nearrow\beta_c} \frac{\tau_\beta(z)}{(\beta_c-\beta)} = 4|z|. \tag{2.5}$$

2.2. Low and high temperature expansions of the Ising model. The *low temperature expansion* of the Ising model is a graphical representation on the dual lattice. Fix a spin configuration σ for the Ising model on G with + boundary conditions. The *collection of contours* of a spin configuration σ is the set of interfaces (edges of the dual graph) separating + and − clusters. In a collection of contours, an even number of dual edges automatically emanates from each dual vertex. Reciprocally, any family of dual edges with an even number of edges emanating from each dual vertex is the collection of contours of exactly one spin configuration (since we fix + boundary conditions).

The interesting feature of the low temperature expansion is that properties of the Ising model can be restated in terms of this graphical representation. We only give the example of the partition function on G but other quantities can be computed similarly. Let $\mathcal{E}_{G^\star}$ be the set of possible collections of contours, and let $|\omega|$ be the number of edges of a collection of contours ω, then

$$Z_{\beta,G}^+ = e^{\beta \#\text{ edges in } G^\star} \sum_{\omega\in\mathcal{E}_{G^\star}} \left(e^{-2\beta}\right)^{|\omega|}. \tag{2.6}$$

The *high temperature expansion* of the Ising model is a graphical representation on the primal lattice itself. It is not a geometric representation since one cannot map a spin configuration σ to a subset of configurations in the graphical representation, but rather a convenient way to represent correlations between spins using statistics of contours. It is based on the following identity:

$$e^{\beta\sigma_x\sigma_y} = \cosh(\beta) + \sigma_x\sigma_y\sinh(\beta) = \cosh(\beta)\left[1 + \tanh(\beta)\sigma_x\sigma_y\right] \tag{2.7}$$

PROPOSITION 2.5. *Let G be a finite graph and a, b be two sites of G. At inverse-temperature $\beta > 0$,*

$$Z_{\beta,G}^f = 2^{\#\text{ vertices } G}\cosh(\beta)^{\#\text{ edges in } G} \sum_{\omega\in\mathcal{E}_G} \tanh(\beta)^{|\omega|} \tag{2.8}$$

$$\mu_{\beta,G}^f[\sigma_a\sigma_b] = \frac{\sum_{\omega\in\mathcal{E}_G(a,b)} \tanh(\beta)^{|\omega|}}{\sum_{\omega\in\mathcal{E}_G} \tanh(\beta)^{|\omega|}}, \tag{2.9}$$

where $\mathcal{E}_G$ (resp. $\mathcal{E}_G(a,b)$) is the set of families of edges of G such that an even number of edges emanates from each vertex (resp. except at a and b, where an odd number of edges emanates).

The notation $\mathcal{E}_G$ coincides with the definition $\mathcal{E}_{G^\star}$ in the low temperature expansion for the dual lattice.

Proof. Let us start with the partition function (2.8). Let E be the set of edges of G. We know

$$\begin{aligned} Z^f_{\beta,G} &= \sum_{\sigma}\prod_{[xy]\in E} e^{\beta\sigma_x\sigma_y} \\ &= \cosh(\beta)^{\#\text{ edges in }G}\sum_{\sigma}\prod_{[xy]\in E}[1+\tanh(\beta)\sigma_x\sigma_y] \\ &= \cosh(\beta)^{\#\text{ edges in }G}\sum_{\sigma}\sum_{\omega\subset E}\tanh(\beta)^{|\omega|}\prod_{e=[xy]\in\omega}\sigma_x\sigma_y \\ &= \cosh(\beta)^{\#\text{ edges in }G}\sum_{\omega\subset E}\tanh(\beta)^{|\omega|}\sum_{\sigma}\prod_{e=[xy]\in\omega}\sigma_x\sigma_y \end{aligned}$$

where we used (2.7) in the second equality. Notice that $\sum_\sigma\prod_{e=[xy]\in\omega}\sigma_x\sigma_y$ equals $2^{\#\text{ vertices }G}$ if ω is in $\mathcal{E}_G$, and 0 otherwise, hence proving (2.8).

Fix $a,b\in G$. By definition,

$$\mu^f_{\beta,G}[\sigma_a\sigma_b] = \frac{\sum_\sigma\sigma_a\sigma_b e^{-\beta H(\sigma)}}{\sum_\sigma e^{-\beta H(\sigma)}} = \frac{\sum_\sigma\sigma_a\sigma_b e^{-\beta H(\sigma)}}{Z^f_{\beta,G}}, \tag{2.10}$$

where $H(\sigma) = -\sum_{i\sim j}\sigma_i\sigma_j$. The second identity boils down to proving that the right hand terms of (2.9) and (2.10) are equal, *i.e.*

$$\sum_\sigma\sigma_a\sigma_b e^{-\beta H(\sigma)} = 2^{\#\text{ vertices }G}\cosh(\beta)^{\#\text{ edges in }G}\sum_{\omega\in\mathcal{E}_G(a,b)}\tanh(\beta)^{|\omega|}. \tag{2.11}$$

The first lines of the computation for the partition function are the same, and we end up with

$$\begin{aligned} \sum_\sigma\sigma_a\sigma_b e^{-\beta H(\sigma)} &= \cosh(\beta)^{\#\text{ edges in }G}\sum_{\omega\subset E}\tanh(\beta)^{|\omega|}\sum_\sigma\sigma_a\sigma_b\prod_{e=[xy]\in\omega}\sigma_x\sigma_y \\ &= 2^{\#\text{ vertices }G}\cosh(\beta)^{\#\text{ edges in }G}\sum_{\omega\in\mathcal{E}_G(a,b)}\tanh(\beta)^{|\omega|} \end{aligned}$$

since $\sum_\sigma\sigma_a\sigma_b\prod_{e=[xy]\in\omega}\sigma_x\sigma_y$ equals $2^{\#\text{ vertices }G}$ if $\omega\in\mathcal{E}_G(a,b)$, and 0 otherwise. □

The set $\mathcal{E}_G$ is the set of collections of loops on G when forgetting the way we draw loops (since some elements of $\mathcal{E}_G$, like a figure eight, can be decomposed into loops in several ways), while $\mathcal{E}_G(a,b)$ is the set of collections of loops on G together with one curve from a to b.

PROPOSITION 2.6 (Kramers-Wannier duality). *Let $\beta>0$ and define $\beta^\star\in(0,\infty)$ such that* $\tanh(\beta^\star)=e^{-2\beta}$, *then for every graph G,*

$$2^{\#\text{ vertices }G^\star}\cosh(\beta^\star)^{\#\text{ edges in }G^\star}Z^+_{\beta,G} = \left(e^{\beta}\right)^{\#\text{ edges in }G^\star}Z^f_{\beta^\star,G^\star}. \tag{2.12}$$

Proof. When writing the contour of connected components for the Ising model with + boundary conditions, the only edges of $\mathbb{L}^\star$ used are those of $G^\star$. Indeed, edges between boundary sites cannot be present since boundary spins are +. Thus, the right and left-hand side terms of (2.12) both correspond to the sum on $\mathcal{E}_{G^\star}$ of $(e^{-2\beta})^{|\omega|}$ or equivalently of $\tanh(\beta^\star)^{|\omega|}$, implying the equality (see Fig. 3). □

We are now in a position to present the argument of Kramers and Wannier. Physicists expect the partition function to exhibit only one singularity, localized at the critical point. If $\beta_c^\star\neq\beta_c$, there would be at least two singularities, at β_c and $\beta_c^\star$, thanks to the previous relation between partition functions at these two temperatures. Thus, β_c must equal $\beta_c^\star$, which implies $\beta_c=\frac{1}{2}\ln(1+\sqrt{2})$. Of course, the assumption that there is a unique singularity is hard to justify.

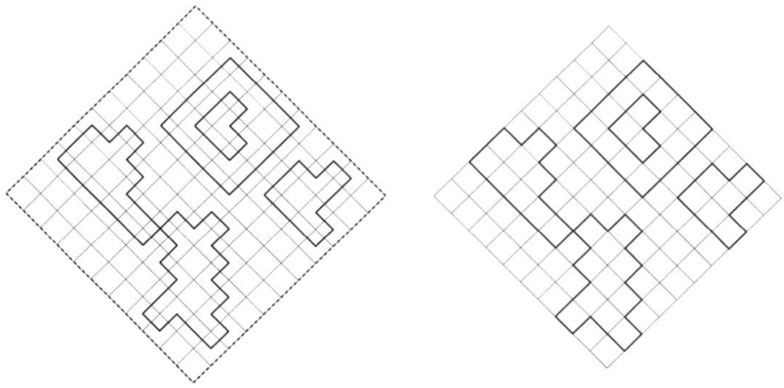

FIGURE 3. The possible collections of contours for + boundary conditions in the low-temperature expansion do not contain edges between boundary sites of G. Therefore, they correspond to collections of contours in $\mathcal{E}_{G^\star}$, which are exactly the collection of contours involved in the high-temperature expansion of the Ising model on $G^\star$ with free boundary conditions.

EXERCISE 2.7. *Extend the low and high temperature expansions to free and + boundary conditions respectively. Extend the high-temperature expansion to n-point spin correlations.*

EXERCISE 2.8 (Peierls argument). *Use the low and high temperature expansions to show that $\beta_c \in (0,\infty)$, and that correlations between spins decay exponentially fast when β is small enough.*

2.3. Spin-Dobrushin domain, fermionic observable and results on the Ising model. In this section we discuss the scaling limit of a single interface between + and − at criticality. We introduce the fundamental notions of Dobrushin domains and the so-called fermionic observable.

Let (Ω, a, b) be a simply connected domain with two marked points on the boundary. Let $\Omega_\delta^\diamond$ be the medial graph of Ω_δ composed of all the vertices of $\mathbb{L}_\delta^\diamond$ bordering a black face associated to Ω_δ, see Fig 4. This definition is non-standard since we include medial vertices not associated to edges of Ω_δ. Let a_δ and b_δ be two vertices of $\partial\Omega_\delta^\diamond$ close to a and b. We further require that b_δ is the southeast corner of a black face. We call the triplet $(\Omega_\delta^\diamond, a_\delta, b_\delta)$ a *spin-Dobrushin domain.*

Let $z_\delta \in \Omega_\delta^\diamond$. Mimicking the high-temperature expansion of the Ising model on Ω_δ, let $\mathcal{E}(a_\delta, z_\delta)$ be the set of collections of contours drawn on Ω_δ composed of loops and one interface from a_δ to z_δ, see Fig. 4. For a loop configuration ω, $\gamma(\omega)$ denotes the unique curve from a_δ to z_δ turning always left when there is an ambiguity. With these notations, we can define the spin-Ising fermionic observable.

DEFINITION 2.9. *On a spin Dobrushin domain $(\Omega_\delta^\diamond, a_\delta, b_\delta)$, the* spin-Ising fermionic observable *at $z_\delta \in \Omega_\delta^\diamond$ is defined by*

$$F_{\Omega_\delta, a_\delta, b_\delta}(z_\delta) = \frac{\sum_{\omega\in\mathcal{E}(a_\delta,z_\delta)} \mathrm{e}^{-\frac{1}{2}iW_{\gamma(\omega)}(a_\delta,z_\delta)}(\sqrt{2}-1)^{|\omega|}}{\sum_{\omega\in\mathcal{E}(a_\delta,b_\delta)} \mathrm{e}^{-\frac{1}{2}iW_{\gamma(\omega)}(a_\delta,b_\delta)}(\sqrt{2}-1)^{|\omega|}},$$

where the winding $W_\gamma(a_\delta, z_\delta)$ is the (signed) total rotation in radians of the curve γ between a_δ and z_δ.

The complex modulus of the denominator of the fermionic observable is connected to the partition function of a conditioned critical Ising model. Indeed, fix $b_\delta \in \partial\Omega_\delta^\diamond$. Even though $\mathcal{E}(a_\delta, b_\delta)$ is not exactly a high-temperature expansion (since there are two

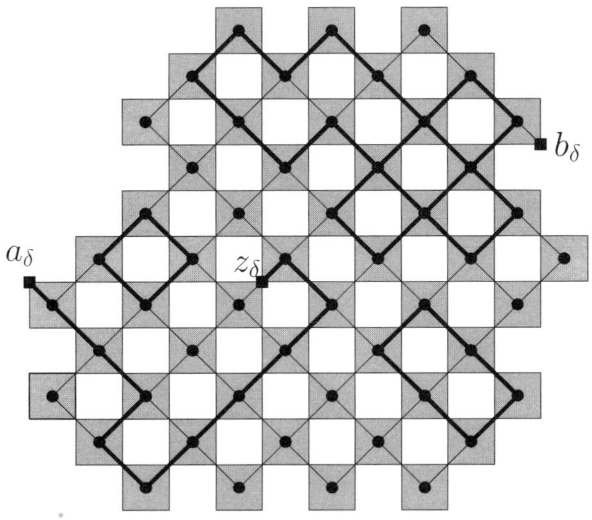

FIGURE 4. An example of collection of contours in $\mathcal{E}(a_\delta, z_\delta)$ on the lattice $\Omega_\diamond$.

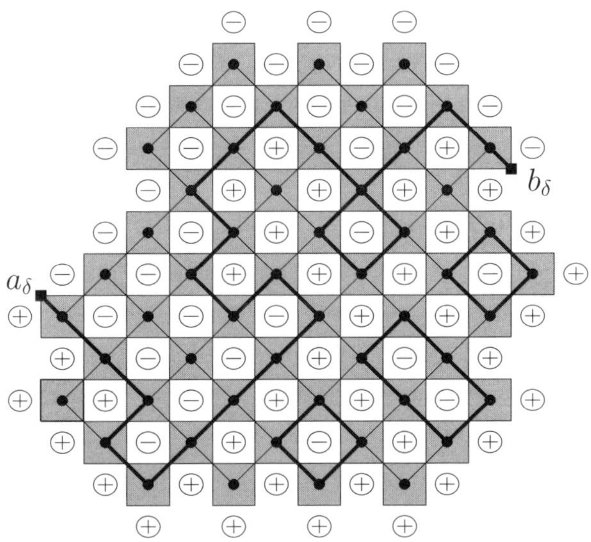

FIGURE 5. A high temperature expansion of an Ising model on the primal lattice together with the corresponding configuration on the dual lattice. The constraint that a_δ is connected to b_δ corresponds to the partition function of the Ising model with +/− boundary conditions on the domain.

half-edges starting from a_δ and b_δ respectively), it is in bijection with the set $\mathcal{E}(a, b)$. Therefore, (2.11) can be used to relate the denominator of the fermionic observable to the partition function of the Ising model on the primal graph with free boundary conditions conditioned on the fact that a and b have the same spin. Let us mention that the numerator of the observable also has an interpretation in terms of disorder operators of the critical Ising model.

The weights of edges are critical (since $\sqrt{2}-1 = e^{-2\beta_c}$). Therefore, the Kramers-Wannier duality has an enlightening interpretation here. The high-temperature expansion can be thought of as the low-temperature expansion of an Ising model on the dual graph, where the dual graph is constructed by adding one layer of dual vertices around ∂G, see Fig. 5. Now, the existence of a curve between a_δ and b_δ is equivalent to the existence of an interface between pluses and minuses in this new Ising model. Therefore, it corresponds to a model with Dobrushin boundary conditions on the dual graph. This fact is not surprising since the dual boundary conditions of the free boundary conditions conditioned on $\sigma_a = \sigma_b$ are the Dobrushin ones.

From now on, the Ising model on a spin Dobrushin domain is the critical Ising model on $\Omega_\delta^\star$ with Dobrushin boundary conditions. The previous paragraph suggests a connection between the fermionic observable and the interface in this model. In fact, Section 6 will show that the fermionic observable is crucial in the proof that the unique interface γ_δ going from a_δ to b_δ between the + component connected to the arc $\partial_{ab}^\star$ and the − component connected to $\partial_{ba}^\star$ (preserve the convention that the interface turns left every time there is a choice) is conformally invariant in the scaling limit. Figures 1 (center picture) and 2 show two interfaces in domains with Dobrushin boundary conditions.

THEOREM 2.10. *Let (Ω, a, b) be a simply connected domain with two marked points on the boundary. Let γ_δ be the interface of the* critical *Ising model with Dobrushin boundary conditions on the spin Dobrushin domain $(\Omega_\delta^\diamond, a_\delta, b_\delta)$. Then $(\gamma_\delta)_{\delta>0}$ converges weakly as $\delta \to 0$ to the (chordal) Schramm-Loewner Evolution with parameter $\kappa = 3$.*

The proof of Theorem 2.10 follows the program below, see Section 6:

- Prove that the family of interfaces $(\gamma_\delta)_{\delta>0}$ is tight.
- Prove that $M_t^{z_\delta} = F_{\Omega_\delta^\diamond \setminus \gamma_\delta[0,t], \gamma_\delta(t), b_\delta}(z_\delta)$ is a martingale for the discrete curve γ_δ.
- Prove that these martingales are converging when δ goes to 0. This provides us with a continuous martingale $(M_t^z)_t$ for any sub-sequential limit of the family $(\gamma_\delta)_{\delta>0}$.
- Use the martingales $(M_t^z)_t$ to identify the possible sub-sequential limits. Actually, we will prove that the (chordal) Schramm-Loewner Evolution with parameter $\kappa = 3$ is the only possible limit, thus proving the convergence.

The third step (convergence of the observable) will be crucial for the success of this program. We state it as a theorem on its own. The connection with the other steps will be explained in detail in Section 6.

THEOREM 2.11 (**[CS09]**). *Let Ω be a simply connected domain and a, b two marked points on its boundary, assuming that the boundary is smooth in a neighborhood of b. We have that*

$$F_{\Omega_\delta, a_\delta, b_\delta}(\cdot) \to \sqrt{\frac{\psi'(\cdot)}{\psi'(b)}} \quad \text{when } \delta \to 0 \tag{2.13}$$

uniformly on every compact subset of Ω, where ψ is any conformal map from Ω to the upper half-plane $\mathbb{H}$, mapping a to ∞ and b to 0.

The fermionic observable is a powerful tool to prove conformal invariance, yet it is also interesting in itself. Being defined in terms of the high-temperature expansion of the Ising model, it expresses directly quantities of the model. For instance, we will explain in Section 6 how a more general convergence result for the observable enables us to compute the energy density.

THEOREM 2.12 ([**HS10**]). *Let Ω be a simply connected domain and $a \in \Omega$. If $e_\delta = [xy]$ denotes the edge of $\Omega \cap \delta\mathbb{Z}^2$ closest to a, then the following equality holds:*

$$\mu^f_{\beta_c,\Omega\cap\delta\mathbb{Z}^2}\,[\sigma_x\sigma_y] \;=\; \frac{\sqrt{2}}{2} - \frac{\phi_a'(a)}{\pi}\,\delta + o(\delta),$$

where $\mu^f_{\beta_c,\Omega_\delta}$ is the Ising measure at criticality and ϕ_a is the unique conformal map from Ω to the disk $\mathbb{D}$ sending a to 0 and such that $\phi_a'(a) > 0$.

3. Two-dimensional FK-Ising model

In this section, another graphical representation of the Ising model, called the *FK-Ising model*, is presented in detail. Its properties will be used to describe properties of the Ising model in the following sections.

3.1. FK percolation. We refer to [**Gri06**] for a complete study on FK percolation (invented by Fortuin and Kasteleyn [**FK72**]). A *configuration* ω on G is a random subgraph of G, composed of the same sites and a subset of its edges. The edges belonging to ω are called *open*, the others *closed*. Two sites x and y are said to be *connected* (denoted by $x \leftrightarrow y$), if there is an *open path* — a path composed of open edges — connecting them. The maximal connected components are called *clusters*.

Boundary conditions ξ are given by a partition of ∂G. Let $o(\omega)$ (resp. $c(\omega)$) denote the number of open (resp. closed) edges of ω and $k(\omega,\xi)$ the number of connected components of the graph obtained from ω by identifying (or *wiring*) the vertices in ξ that belong to the same class of ξ.

The FK percolation $\phi^\xi_{p,q,G}$ on a finite graph G with parameters $p \in [0,1]$, and $q \in (0,\infty)$ and boundary conditions ξ is defined by

$$\phi^\xi_{p,q,G}(\omega) := \frac{p^{o(\omega)}(1-p)^{c(\omega)}q^{k(\omega,\xi)}}{Z^\xi_{p,q,G}}, \tag{3.1}$$

for any subgraph ω of G, where $Z^\xi_{p,q,G}$ is a normalizing constant called the *partition function* for the FK percolation. Here and in the following, we drop the dependence on ξ in $k(\omega,\xi)$.

The FK percolations with parameter $q < 1$ and $q \geq 1$ behave very differently. For now, we restrict ourselves to the second case. When $q \geq 1$, the FK percolation is *positively correlated*: an event is called *increasing* if it is preserved by addition of open edges.

THEOREM 3.1. *For $q \geq 1$ and $p \in [0,1]$, the FK percolation on G satisfies the following two properties:*

- ***FKG inequality:*** *For any boundary conditions ξ and any increasing events A, B,*

$$\phi^\xi_{p,q,G}(A\cap B) \;\geq\; \phi^\xi_{p,q,G}(A)\phi^\xi_{p,q,G}(B). \tag{3.2}$$

- ***Comparison between boundary conditions:*** *for any ξ refinement of ψ and any increasing event A,*

$$\phi^\psi_{p,q,G}(A) \;\geq\; \phi^\xi_{p,q,G}(A). \tag{3.3}$$

The previous result is very similar to Theorem 2.1. As in the Ising model case, one can define a notion of stochastic domination. Two boundary conditions play a special role in the study of FK percolation: the *wired* boundary conditions, denoted by $\xi = 1$, are specified by the fact that all the vertices on the boundary are pairwise connected. The *free* boundary conditions, denoted by $\xi = 0$, are specified by the absence of wirings between boundary sites. The free and wired boundary conditions are extremal among all boundary conditions for stochastic ordering.

Infinite-volume measures can be defined as limits of measures on nested boxes. In particular, we set $\phi^1_{p,q}$ for the infinite-volume measure with wired boundary conditions and $\phi^0_{p,q}$ for the infinite-volume measure with free boundary conditions. Like the Ising model, the model exhibits a phase transition in the infinite-volume limit.

THEOREM 3.2. *For any $q \geq 1$, there exists $p_c(q) \in (0,1)$ such that for any infinite volume measure $\phi_{p,q}$,*

- *if $p < p_c(q)$, there is almost surely no infinite cluster under $\phi_{p,q}$,*
- *if $p > p_c(q)$, there is almost surely a* unique *infinite cluster under $\phi_{p,q}$.*

Note that $q = 1$ is simply bond percolation. In this case, the existence of a phase transition is a well-known fact. The existence of a critical point in the general case $q \geq 1$ is not much harder to prove: a coupling between two measures $\phi_{p_1,q,G}$ and $\phi_{p_2,q,G}$ can be constructed in such a way that $\phi_{p_1,q,G}$ stochastically dominates $\phi_{p_2,q,G}$ if $p_1 \geq p_2$ (this coupling is not as straightforward as in the percolation case, see *e.g.* [**Gri06**]). The determination of the critical value is a much harder task.

A natural notion of duality also exists for the FK percolation on the square lattice (and more generally on any planar graph). We present duality in the simplest case of wired boundary conditions. Construct a model on $G^\star$ by declaring any edge of the dual graph to be open (resp. closed) if the corresponding edge of the primal graph is closed (resp. open) for the initial FK percolation model.

PROPOSITION 3.3. *The dual model of the FK percolation with parameters (p,q) with wired boundary conditions is the FK percolation with parameters $(p^\star,q)$ and free boundary conditions on $G^\star$, where*

$$p^\star = p^\star(p,q) := \frac{(1-p)q}{(1-p)q+p} \tag{3.4}$$

Proof. Note that the state of edges between two sites of ∂G is not relevant when boundary conditions are wired. Indeed, sites on the boundary are connected via boundary conditions anyway, so that the state of each boundary edge does not alter the connectivity properties of the subgraph, and is independent of other edges. For this reason, forget about edges between boundary sites and consider only inner edges (which correspond to edges of $G^\star$): $o(\omega)$ and $c(\omega)$ then denote the number of open and closed inner edges.

Set $e^\star$ for the dual edge of $G^\star$ associated to the (inner) edge e. From the definition of the dual configuration $\omega^\star$ of ω, we have $o(\omega^\star) = a - o(\omega)$ where a is the number of edges in $G^\star$ and $o(\omega^\star)$ is the number of open dual edges. Moreover, connected components of $\omega^\star$ correspond exactly to faces of ω, so that $f(\omega) = k(\omega^\star)$, where $f(\omega)$ is the number of faces (counting the infinite face). Using Euler's formula

$$\#\text{ edges} + \#\text{ connected components} + 1 = \#\text{sites} + \#\text{ faces},$$

which is valid for any planar graph, we obtain, with s being the number of sites in G,

$$k(\omega) = s - 1 + f(\omega) - o(\omega) = s - 1 + k(\omega^\star) - a + o(\omega^\star).$$

The probability of $\omega^\star$ is equal to the probability of ω under $\phi^1_{G,p,q}$, *i.e.*

$$\begin{aligned}\phi^1_{G,p,q}(\omega) &= \frac{1}{Z^1_{G,p,q}}p^{o(\omega)}(1-p)^{c(\omega)}q^{k(\omega)}\\ &= \frac{(1-p)^a}{Z^1_{G,p,q}}[p/(1-p)]^{o(\omega)}q^{k(\omega)}\\ &= \frac{(1-p)^a}{Z^1_{G,p,q}}[p/(1-p)]^{a-o(\omega^\star)}q^{s-1-a+k(\omega^\star)+o(\omega^\star)}\\ &= \frac{p^a q^{s-1-a}}{Z^1_{G,p,q}}[q(1-p)/p]^{o(\omega^\star)}q^{k(\omega^\star)} = \phi^0_{p^\star,q,G^\star}(\omega^\star)\end{aligned}$$

since $q(1-p)/p = p^\star/(1-p^\star)$, which is exactly the statement. □

It is then natural to define the self-dual point $p_{sd} = p_{sd}(q)$ solving the equation $p^\star_{sd} = p_{sd}$, which gives

$$p_{sd} = p_{sd}(q) := \frac{\sqrt{q}}{1+\sqrt{q}}.$$

Note that, mimicking the Kramers-Wannier argument, one can give a simple heuristic justification in favor of $p_c(q) = p_{sd}(q)$. Recently, the computation of $p_c(q)$ was performed for every $q \geq 1$:

THEOREM 3.4 ([**BDC10**]). *The critical parameter $p_c(q)$ of the FK percolation on the square lattice equals $p_{sd}(q) = \sqrt{q}/(1+\sqrt{q})$ for every $q \geq 1$.*

EXERCISE 3.5. *Describe the dual of a FK percolation with parameters (p,q) and free boundary conditions. What is the dual model of the FK percolation in infinite-volume with wired boundary conditions?*

EXERCISE 3.6 (Zhang's argument for FK percolation, [**Gri06**]). *Consider the FK percolation with parameters $q \geq 1$ and $p = p_{sd}(q)$. We suppose known the fact that infinite clusters are unique, and that the probability that there is an infinite cluster is 0 or 1.*

Assume that there is a.s. an infinite cluster for the measure $\phi^0_{p_{sd},q}$.

1) Let $\varepsilon < 1/100$. Show that there exists $n > 0$ such that the $\phi^0_{p_{sd},q}$-probability that the infinite cluster touches $[-n,n]^2$ is larger than $1-\varepsilon$. Using the FKG inequality for decreasing events (one can check that the FKG inequality holds for decreasing events as well), show that the $\phi^0_{p_{sd},q}$-probability that the infinite cluster touches $\{n\}\times[-n,n]$ from the outside of $[-n,n]^2$ is larger than $1-\varepsilon^{\frac{1}{4}}$.

2) Using the uniqueness of the infinite cluster and the fact that the probability that there exists an infinite cluster equals 0 or 1 (can you prove these facts?), show that a.s. there is no infinite cluster for the FK percolation with free boundary conditions at the self-dual point.

3) Is the previous result necessarily true for the FK percolation with wired boundary conditions at the self-dual point? What can be said about $p_c(q)$?

EXERCISE 3.7. *Prove Euler's formula.*

3.2. FK-Ising model and Edwards-Sokal coupling. The Ising model can be coupled to the FK percolation with cluster-weight $q = 2$ [**ES88**]. For this reason, the $q = 2$ FK percolation model will be called the FK-Ising model. We now present this coupling, called the Edwards-Sokal coupling, along with some consequences for the Ising model.

Let G be a finite graph and let ω be a configuration of open and closed edges on G. A spin configuration σ can be constructed on the graph G by assigning independently to each cluster of ω a + or − spin with probability 1/2 (note that all the sites of a cluster receive the same spin).

PROPOSITION 3.8. *Let $p \in (0,1)$ and G a finite graph. If the configuration ω is distributed according to a FK measure with parameters $(p,2)$ and free boundary conditions, then the spin configuration σ is distributed according to an Ising measure with inverse-temperature $\beta = -\frac{1}{2}\ln(1-p)$ and free boundary conditions.*

Proof. Consider a finite graph G, let $p \in (0,1)$. Consider a measure P on pairs (ω,σ), where ω is a FK configuration with free boundary conditions and σ is the corresponding random spin configuration, constructed as explained above. Then, for (ω,σ), we have:

$$P[(\omega,\sigma)] = \frac{1}{Z^0_{p,2,G}} p^{o(\omega)}(1-p)^{c(\omega)}2^{k(\omega)} \cdot 2^{-k(\omega)} = \frac{1}{Z^0_{p,2,G}} p^{o(\omega)}(1-p)^{c(\omega)}.$$

Now, we construct another measure $\tilde{P}$ on pairs of percolation configurations and spin configurations as follows. Let $\tilde{\sigma}$ be a spin configuration distributed according to an Ising model with inverse-temperature β satisfying $e^{-2\beta} = 1-p$ and free boundary conditions. We deduce $\tilde{\omega}$ from $\tilde{\sigma}$ by closing all edges between neighboring sites with different spins, and by independently opening with probability p edges between neighboring sites with same spins. Then, for any $(\tilde{\omega},\tilde{\sigma})$,

$$\tilde{P}[(\tilde{\omega},\tilde{\sigma})] = \frac{e^{-2\beta r(\tilde{\sigma})}p^{o(\tilde{\omega})}(1-p)^{a-o(\tilde{\omega})-r(\tilde{\sigma})}}{Z^f_{\beta,p}} = \frac{p^{o(\tilde{\omega})}(1-p)^{c(\tilde{\omega})}}{Z^f_{\beta,p}}$$

where a is the number of edges of G and $r(\tilde{\sigma})$ the number of edges between sites with different spins.

Note that the two previous measures are in fact defined on the same set of compatible pairs of configurations: if σ has been obtained from ω, then ω can be obtained from σ via the second procedure described above, and the same is true in the reverse direction for $\tilde{\omega}$ and $\tilde{\sigma}$. Therefore, $P = \tilde{P}$ and the marginals of P are the FK percolation with parameters $(p,2)$ and the Ising model at inverse-temperature β, which is the claim. □

The coupling gives a randomized procedure to obtain a spin-Ising configuration from a FK-Ising configuration (it suffices to assign random spins). The proof of Proposition 3.8 provides a randomized procedure to obtain a FK-Ising configuration from a spin-Ising configuration.

If one considers wired boundary conditions for the FK percolation, the Edwards-Sokal coupling provides us with an Ising configuration with + boundary conditions (or −, the two cases being symmetric). We do not enter into details, since the generalization is straightforward.

An important consequence of the Edwards-Sokal coupling is the relation between Ising correlations and FK connectivity properties. Indeed, two sites which are connected in the FK percolation configuration must have the same spin, while sites which are not have independent spins. This implies:

COROLLARY 3.9. *For $p \in (0,1)$, G a finite graph and $\beta = -\frac{1}{2}\ln(1-p)$, we obtain*

$$\mu^f_{\beta,G}[\sigma_x\sigma_y] = \phi^0_{p,2,G}(x \leftrightarrow y),$$
$$\mu^+_{\beta,G}[\sigma_x] = \phi^1_{p,2,G}(x \leftrightarrow \partial G).$$

In particular, $\beta_c = -\frac{1}{2}\ln[1-p_c(2)]$.

Proof. We leave the proof as an exercise. □

The uniqueness of Ising infinite-volume measures was discussed in the previous section. The same question can be asked in the case of the FK-Ising model. First, it can be proved that $\phi^1_{p,2}$ and $\phi^0_{p,2}$ are extremal among all infinite-volume measures. Therefore, it is sufficient to prove that $\phi^1_{p,2} = \phi^0_{p,2}$ to prove uniqueness. Second, the absence of an infinite cluster for $\phi^1_{p,2}$ can be shown to imply the uniqueness of the infinite-volume measure. Using the equality $p_c = p_{sd}$, the measure is necessarily unique whenever $p < p_{sd}$ since $\phi^1_{p,2}$ has no infinite cluster. Planar duality shows that the only value of p for which uniqueness could eventually fail is the (critical) self-dual point $\sqrt{2}/(1+\sqrt{2})$. It turns out that even for this value, there exists a unique infinite volume measure. Since this fact will play a role in the proof of conformal invariance, we now sketch an elementary proof due to W. Werner (the complete proof can be found in [**Wer09**]).

PROPOSITION 3.10. *There exists a unique infinite-volume FK-Ising measure with parameter* $p_c = \sqrt{2}/(1+\sqrt{2})$ *and there is almost surely no infinite cluster under this measure. Correspondingly, there exists a unique infinite-volume spin Ising measure at* β_c.

Proof. As described above, it is sufficient to prove that $\phi^0_{p_{sd},2} = \phi^1_{p_{sd},2}$. First note that there is no infinite cluster for $\phi^0_{p_{sd},2}$ thanks to Exercise 3.6. Via the Edwards-Sokal coupling, the infinite-volume Ising measure with free boundary conditions, denoted by $\mu^f_{\beta_c}$, can be constructed by coloring clusters of the measure $\phi^0_{p_{sd},2}$. Since there is no infinite cluster, this measure is obviously symmetric by global exchange of +/−. In particular, the argument of Exercise 3.6 can be applied to prove that there are neither + nor − infinite clusters. Therefore, fixing a box, there exists a + star-connected circuit surrounding the box with probability one (two vertices x and y are said to be *star-connected* if y is one of the eight closest neighbors to x).

One can then argue that the configuration inside the box stochastically dominates the Ising configuration for the infinite-volume measure with + boundary conditions (roughly speaking, the circuit of spin + behaves like + boundary conditions). We deduce that $\mu^f_{\beta_c}$ restricted to the box (in fact to any box) stochastically dominates $\mu^+_{\beta_c}$. This implies that $\mu^f_{\beta_c} \geq \mu^+_{\beta_c}$. Since the other inequality is obvious, $\mu^f_{\beta_c}$ and $\mu^+_{\beta_c}$ are equal.

Via Edwards-Sokal's coupling again, $\phi^0_{p_{sd},2} = \phi^1_{p_{sd},2}$ and there is no infinite cluster at criticality. Moreover, $\mu^-_{\beta_c} = \mu^f_{\beta_c} = \mu^+_{\beta_c}$ and there is a unique infinite-volume Ising measure at criticality. □

REMARK 3.11. *More generally, the FK percolation with integer parameter* $q \geq 2$ *can be coupled with Potts models. Many properties of Potts models are derived using FK percolation, since we have the FKG inequality at our disposal, while there is no equivalent of the spin-Ising FKG inequality for Potts models.*

3.3. Loop representation of the FK-Ising model and fermionic observable. Let (Ω, a, b) be a simply connected domain with two marked points on the boundary. Let Ω_δ be an approximation of Ω, and let ∂_{ab} and ∂_{ba} denote the counterclockwise arcs in the boundary $\partial\Omega_\delta$ joining a to b (resp. b to a). We consider a FK-Ising measure with wired boundary conditions on ∂_{ba} – all the edges are pairwise connected – and free boundary conditions on the arc ∂_{ab}. These boundary conditions are called the *Dobrushin boundary conditions.* We denote by $\phi^{a,b}_{\Omega_\delta,p}$ the associated FK-Ising measure with parameter p.

The *dual boundary arc* $\partial^\star_{ba}$ is the set of sites of $\Omega^\star_\delta$ adjacent to ∂_{ba} while the *dual boundary arc* $\partial^\star_{ab}$ is the set of sites of $\mathbb{L}^\star_\delta \setminus \Omega^\star_\delta$ adjacent to ∂_{ab}, see Fig. 6. A *FK-Dobrushin domain* $(\Omega^\diamond_\delta, a_\delta, b_\delta)$ is given by

- a medial graph $\Omega^\diamond_\delta$ defined as the set of medial vertices associated to edges of Ω_δ and to dual edges of $\partial^\star_{ab}$,
- medial sites $a_\delta, b_\delta \in \Omega^\diamond_\delta$ between arcs ∂_{ba} an $\partial^\star_{ab}$, see Fig. 6 again,

with the additional condition that b_δ is the southeast corner of a black face belonging to the domain.

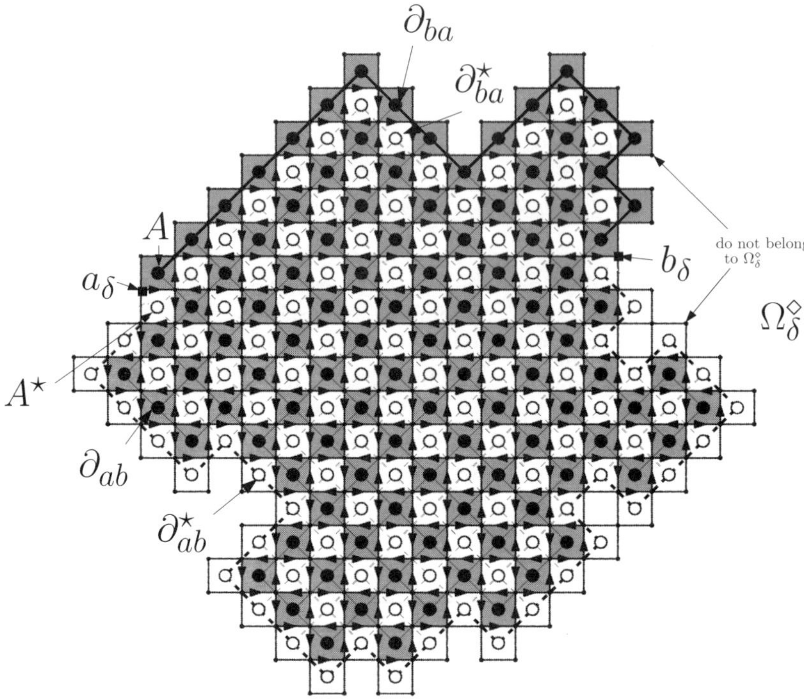

FIGURE 6. A domain Ω_δ with Dobrushin boundary conditions: the vertices of the primal graph are black, the vertices of the dual graph $\Omega^\star_\delta$ are white, and between them lies the medial graph $\Omega^\diamond_\delta$. The arcs ∂_{ba} and $\partial^\star_{ab}$ are the two outermost arcs. Moreover, arcs $\partial^\star_{ba}$ and ∂_{ab} are the arcs bordering ∂_{ba} and $\partial^\star_{ab}$ from the inside. The arcs ∂_{ab} and ∂_{ba} (resp. $\partial^\star_{ab}$ and $\partial^\star_{ba}$) are drawn in solid lines (resp. dashed lines)

REMARK 3.12. *Note that the definition of $\Omega^\diamond_\delta$ is not the same as in Section 1.2.1 since we added medial vertices associated to dual edges of $\partial^\star_{ab}$. We chose this definition to make sites of the dual and the primal lattices play symmetric roles. The condition that b_δ is the south corner of a black face belonging to the domain is a technical condition.*

Let $(\Omega^\diamond_\delta, a_\delta, b_\delta)$ be a FK-Dobrushin domain. For any FK-Ising configuration with Dobrushin boundary conditions on Ω_δ, we construct a loop configuration on $\Omega^\diamond_\delta$ as follows: The interfaces between the primal clusters and the dual clusters (*i.e* clusters in the dual model) form a family of loops together with a path from a_δ to b_δ. The loops are drawn as shown in Figure 7 following the edges of the medial lattice. The orientation of the medial lattice naturally gives an orientation to the loops, so that we are working with a model of oriented loops on the medial lattice.

The curve from a_δ to b_δ is called the *exploration path* and denoted by $\gamma = \gamma(\omega)$. It is the interface between the open cluster connected to ∂_{ba} and the dual-open cluster connected to $\partial^\star_{ab}$. As in the Ising model case, one can study its scaling limit when the mesh size goes to 0:

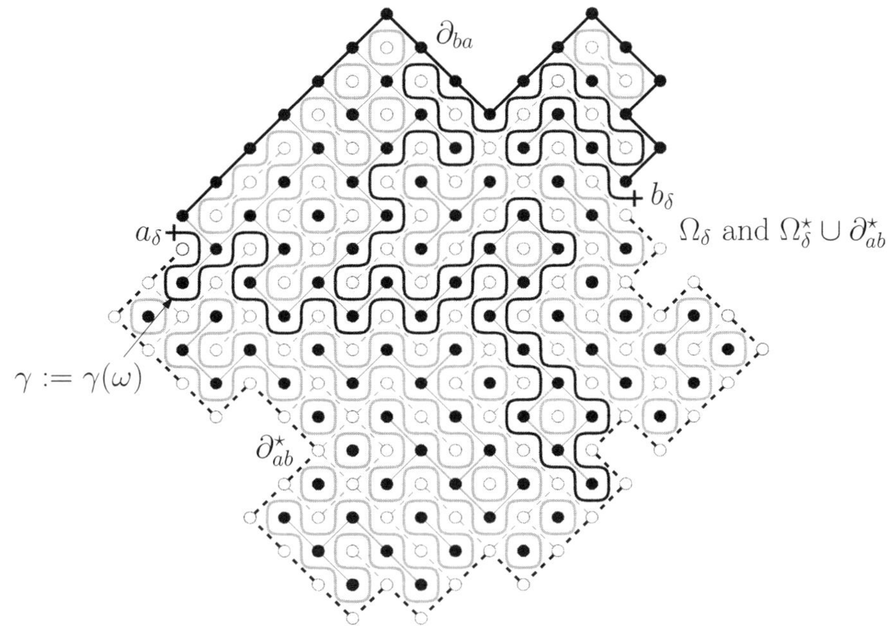

FIGURE 7. A FK percolation configuration in the Dobrushin domain $(\Omega_\delta, a_\delta, b_\delta)$, together with the corresponding interfaces on the medial lattice: the loops are grey, and the exploration path γ from a_δ to b_δ is black. Note that the exploration path is the interface between the open cluster connected to the wired arc and the dual-open cluster connected to the white faces of the free arc.

THEOREM 3.13 (Conformal invariance of the FK-Ising model, [**KS10, CDHKS12**]). *Let Ω be a simply connected domain with two marked points a, b on the boundary. Let γ_δ be the interface of the* critical *FK-Ising with Dobrushin boundary conditions on $(\Omega_\delta, a_\delta, b_\delta)$. Then the law of γ_δ converges weakly, when $\delta \to 0$, to the chordal Schramm-Loewner Evolution with $\kappa = 16/3$.*

As in the Ising model case, the proof of this theorem also involves a discrete observable, which converges to a conformally invariant object. We define it now.

DEFINITION 3.14. *The edge FK fermionic observable is defined on edges of $\Omega_\delta^\diamond$ by*

$$F_{\Omega_\delta^\diamond, a_\delta, b_\delta, p}(e) = \mathbb{E}_{\Omega_\delta, p}^{a_\delta, b_\delta}\left[e^{\frac{1}{2}\cdot \mathrm{i} W_\gamma(e, b_\delta)} 1_{e\in\gamma}\right], \tag{3.5}$$

where $W_\gamma(e, b_\delta)$ denotes the winding between the center of e and b_δ.

The vertex FK fermionic observable *is defined on vertices of $\Omega_\delta^\diamond \setminus \partial\Omega_\delta^\diamond$ by*

$$F_{\Omega_\delta^\diamond, a_\delta, b_\delta, p}(v) = \frac{1}{2}\sum_{e\sim v} F_{\Omega_\delta^\diamond, a_\delta, b_\delta, p}(e) \tag{3.6}$$

where the sum is over the four medial edges having v as an endpoint.

When we consider the observable at criticality (which will be almost always the case), we drop the dependence on p in the notation. More generally, if (Ω, a, b) is fixed, we simply denote the observable on $(\Omega_\delta^\diamond, a_\delta, b_\delta, p_{sd})$ by F_δ.

The quantity $F_\delta(e)$ is a complexified version of the probability that e belongs to the exploration path. The complex weight makes the link between F_δ and probabilistic properties less explicit. Nevertheless, the vertex fermionic observable F_δ converges when δ goes to 0:

THEOREM 3.15. **[Smi10a]** *Let (Ω, a, b) be a simply connected domain with two marked points on the boundary. Let F_δ be the* vertex *fermionic observable in $(\Omega_\delta^\diamond, a_\delta, b_\delta)$. Then, we have*

$$\frac{1}{\sqrt{2\delta}} F_\delta(\cdot) \;\to\; \sqrt{\phi'(\cdot)} \quad \text{when } \delta \to 0 \tag{3.7}$$

uniformly on any compact subset of Ω, where ϕ is any conformal map from Ω to the strip $\mathbb{R}\times(0,1)$ mapping a to $-\infty$ and b to ∞.

As in the case of the spin Ising model, this statement is the heart of the proof of conformal invariance. Yet, the observable itself can be helpful for the understanding of other properties of the FK-Ising model. For instance, it enables us to prove a statement equivalent to the celebrated Russo-Seymour-Welsh Theorem for percolation. This result will be central for the proof of compactness of exploration paths (an important step in the proof of Theorems 2.10 and 3.13).

THEOREM 3.16 (RSW-type crossing bounds, **[DCHN10]**). *There exists a constant $c > 0$ such that for any rectangle R of size $4n \times n$, one has*

$$\phi^0_{p_{sd},2,R}(\text{there exists an open path from left to right}) \geq c. \tag{3.8}$$

Before ending this section, we present a simple yet crucial result: we show that it is possible to compute rather explicitly the distribution of the loop representation. In particular, at criticality, the weight of a loop configuration depends only on the number of loops.

PROPOSITION 3.17. *Let $p \in (0,1)$ and let $(\Omega_\delta^\diamond, a_\delta, b_\delta)$ be a* FK Dobrushin domain, then for any configuration ω,

$$\phi^{a_\delta,b_\delta}_{\Omega_\delta,p}(\omega) \;=\; \frac{1}{Z}\, x^{o(\omega)} \sqrt{2}^{\ell(\omega)} \tag{3.9}$$

where $x = p/[\sqrt{2}(1-p)]$, $\ell(\omega)$ is the number of loops in the loop configuration associated to ω, $o(\omega)$ is the number of open edges, and Z is the normalization constant.

Proof. Recall that

$$\phi^{a_\delta,b_\delta}_{\Omega_\delta,p}(\omega) \;=\; \frac{1}{Z}[p/(1-p)]^{o(\omega)} 2^{k(\omega)}.$$

Using arguments similar to Proposition 3.3, the dual of $\phi^{a_\delta,b_\delta}_{\Omega_\delta,p}$ can be proved to be $\phi^{b_\delta,a_\delta}_{\Omega_\delta^\star,p^\star}$ (in this sense, Dobrushin boundary conditions are self-dual). With $\omega^\star$ being the dual configuration of ω, we find

$$\begin{aligned}
\phi_{\Omega_\delta,p}^{a_\delta,b_\delta}(\omega) &= \sqrt{\phi_{\Omega_\delta,p}^{a_\delta,b_\delta}(\omega)\, \phi_{\Omega_\delta^\star,p^\star}^{b_\delta,a_\delta}(\omega^\star)} \\
&= \frac{1}{\sqrt{ZZ^\star}} \sqrt{p/(1-p)}^{\,o(\omega)} \sqrt{2}^{\,k(\omega)} \sqrt{p^\star/(1-p^\star)}^{\,o(\omega^\star)} \sqrt{2}^{\,k(\omega^\star)} \\
&= \frac{1}{\sqrt{ZZ^\star}} \sqrt{\frac{p(1-p^\star)}{(1-p)p^\star}}^{\,o(\omega)} \sqrt{p^\star/(1-p^\star)}^{\,o(\omega^\star)+o(\omega)} \sqrt{2}^{\,k(\omega)+k(\omega^\star)} \\
&= \frac{\sqrt{2}\sqrt{p^\star/(1-p^\star)}^{\,o(\omega)+o(\omega^\star)}}{\sqrt{ZZ^\star}} x^{o(\omega)} \sqrt{2}^{\,k(\omega)+k(\omega^\star)-1}
\end{aligned}$$

where the definition of $p^\star$ was used to prove that $\frac{p(1-p^\star)}{(1-p)p^\star} = x^2$. Note that $\ell(\omega) = k(\omega) + k(\omega^\star) - 1$ and

$$\tilde{Z} = \frac{\sqrt{ZZ^\star}}{\sqrt{2}\sqrt{p^\star/(1-p^\star)}^{\,o(\omega)+o(\omega^\star)}}$$

does not depend on the configuration (the sum $o(\omega) + o(\omega^\star)$ being equal to the total number of edges). Altogether, this implies the claim. □

4. Discrete complex analysis on graphs

Complex analysis is the study of harmonic and holomorphic functions in complex domains. In this section, we shall discuss how to discretize harmonic and holomorphic functions, and what are the properties of these discretizations.

There are many ways to introduce discrete structures on graphs which can be developed in parallel to the usual complex analysis. We need to consider scaling limits (as the mesh of the lattice tends to zero), so we want to deal with discrete structures which converge to the continuous complex analysis as finer and finer graphs are taken.

4.1. Preharmonic functions.

4.1.1. *Definition and connection with random walks.* Introduce the (non-normalized) discretization of the Laplacian operator $\Delta := \frac{1}{4}(\partial_{xx}^2 + \partial_{yy}^2)$ in the case of the square lattice $\mathbb{L}_\delta$. For $u \in \mathbb{L}_\delta$ and $f : \mathbb{L}_\delta \to \mathbb{C}$, define

$$\Delta_\delta f(u) = \frac{1}{4} \sum_{v \sim u} \big(f(v) - f(u)\big).$$

The definition extends to rescaled square lattices in a straightforward way (for instance to $\mathbb{L}_\delta^\diamond$).

DEFINITION 4.1. *A function $h : \Omega_\delta \to \mathbb{C}$ is* preharmonic *(resp.* pre-superharmonic, pre-subharmonic*) if $\Delta_\delta h(x) = 0$ (resp. ≤ 0, ≥ 0) for every $x \in \Omega_\delta$.*

One fundamental tool in the study of preharmonic functions is the classical relation between preharmonic functions and simple random walks:

Let (X_n) be a simple random walk killed at the first time it exits Ω_δ; then h is preharmonic on Ω_δ if and only if $(h(X_n))$ is a martingale.

Using this fact, one can prove that harmonic functions are determined by their value on $\partial\Omega_\delta$, that they satisfy Harnack's principle, etc. We refer to [**Law91**] for a deeper study on preharmonic functions and their link to random walks. Also note that the set of preharmonic functions is a complex vector space. As in the continuum, it is easy to see that preharmonic functions satisfy the maximum and minimum principles.

4.1.2. *Derivative estimates and compactness criteria.* For general functions, a control on the gradient provides regularity estimates on the function itself. It is a well-known fact that harmonic functions satisfy the reverse property: controlling the function allows us to control the gradient. The following lemma shows that the same is true for preharmonic functions.

PROPOSITION 4.2. *There exists $C > 0$ such that, for any preharmonic function $h : \Omega_\delta \to \mathbb{C}$ and any two neighboring sites $x, y \in \Omega_\delta$,*

$$|h(x) - h(y)| \leq C\delta \frac{\sup_{z\in\Omega_\delta} |h(z)|}{d(x, \Omega^c)}. \tag{4.1}$$

Proof. Let $x, y \in \Omega_\delta$. The preharmonicity of h translates to the fact that $h(X_n)$ is a martingale (where X_n is a simple random walk killed at the first time it exits Ω_δ). Therefore, for x, y two neighboring sites of Ω_δ, we have

$$h(x) - h(y) = \mathbb{E}[h(X_\tau) - h(Y_{\tau'})] \tag{4.2}$$

where under $\mathbb{E}$, X and Y are two simple random walks starting respectively at x and y, and τ, τ' are any stopping times. Let $2r = d(x, \Omega^c) > 0$, so that $U = x + [-r, r]^2$ is included in Ω_δ. Fix τ and τ' to be the hitting times of ∂U_δ and consider the following coupling of X and Y (one has complete freedom in the choice of the joint law in (4.2)): (X_n) is a simple random walk and Y_n is constructed as follows,

- if $X_1 = y$, then $Y_n = X_{n+1}$ for $n \geq 0$,
- if $X_1 \neq y$, then $Y_n = \sigma(X_{n+1})$, where σ is the orthogonal symmetry with respect to the perpendicular bisector ℓ of $[X_1, y]$, whenever X_{n+1} does not reach ℓ. As soon as it does, set $Y_n = X_{n+1}$.

It is easy to check that Y is also a simple random walk. Moreover, we have

$$|h(x) - h(y)| \leq \mathbb{E}[|h(X_\tau) - h(Y_{\tau'})| 1_{X_\tau \neq Y_{\tau'}}] \leq 2\Big(\sup_{z\in\partial U_\delta} |h(z)|\Big)\ \mathbb{P}(X_\tau \neq Y_{\tau'})$$

Using the definition of the coupling, the probability on the right is known: it is equal to the probability that X does not touch ℓ before exiting the ball and is smaller than $\frac{C'}{r}\delta$ (with C' a universal constant), since U_δ is of radius r/δ for the graph distance. We deduce that

$$|h(x) - h(y)| \leq 2\Big(\sup_{z\in\partial U_\delta} |h(z)|\Big)\frac{C'}{r}\delta \leq 2\Big(\sup_{z\in\Omega_\delta} |h(z)|\Big)\frac{C'}{r}\delta$$

□

Recall that functions on Ω_δ are implicitly extended to Ω.

PROPOSITION 4.3. *A family $(h_\delta)_{\delta>0}$ of preharmonic functions on the graphs Ω_δ is precompact for the uniform topology on compact subsets of Ω if* one *of the following properties holds:*

(1) $(h_\delta)_{\delta>0}$ is uniformly bounded on any compact subset of Ω,

or

(2) for any compact subset K of Ω, there exists $M = M(K) > 0$ such that for any $\delta > 0$,

$$\delta^2 \sum_{x\in K_\delta} |h_\delta(x)|^2 \leq M.$$

Proof. Let us prove that the proposition holds under the first hypothesis and then that the second hypothesis implies the first one.

We are faced with a family of continuous maps $h_\delta : \Omega \to \mathbb{C}$ and we aim to apply the Arzelà-Ascoli theorem. It is sufficient to prove that the functions h_δ are uniformly Lipschitz on any compact subset since they are uniformly bounded on any compact subset of Ω. Let K be a compact subset of Ω. Proposition 4.2 shows that $|h_\delta(x) - h_\delta(y)| \leq C_K \delta$ for any two neighbors $x, y \in K_\delta$, where

$$C_K = C \frac{\sup_{\delta>0} \sup_{x\in\Omega:d(x,K)\leq r/2} |h_\delta(x)|}{d(K,\Omega^c)},$$

implying that $|h_\delta(x) - h_\delta(y)| \leq 2C_K|x-y|$ for any $x, y \in K_\delta$ (not necessarily neighbors). The Arzelá-Ascoli theorem concludes the proof.

Now assume that the second hypothesis holds, and let us prove that $(h_\delta)_{\delta>0}$ is bounded on any compact subset of Ω. Take $K \subset \Omega$ compact, let $2r = d(K, \Omega^c) > 0$ and consider $x \in K_\delta$. Using the second hypothesis, there exists $k := k(x)$ such that $\frac{r}{2\delta} \leq k \leq \frac{r}{\delta}$ and

$$\delta \sum_{y\in\partial U_{k\delta}} |h_\delta(y)|^2 \leq 2M/r, \tag{4.3}$$

where $U_{k\delta} = x + [-\delta k, \delta k]^2$ is the box of size k (for the graph distance) around x and $M = M(y + [-r,r]^2)$. Exercise 4.4 implies

$$h_\delta(x) = \sum_{y\in\partial U_{k\delta}} h_\delta(y) H_{U_{k\delta}}(x,y) \tag{4.4}$$

for every $x \in U_{\delta k}$. Using the Cauchy-Schwarz inequality, we find

$$\begin{aligned} h_\delta(x)^2 &= \left(\sum_{y\in\partial U_{k\delta}} h_\delta(y) H_{U_{k\delta}}(x,y) \right)^2 \\ &\leq \left(\delta \cdot \sum_{y\in\partial U_{k\delta}} |h_\delta(y)|^2 \right) \left(\frac{1}{\delta} \cdot \sum_{y\in\partial U_{k\delta}} H_{U_{k\delta}}(x,y)^2 \right) \leq 2M/r \cdot C \end{aligned}$$

where C is a uniform constant. The last inequality used Exercise 4.5 to affirm that $H_{U_{k\delta}}(x,y) \leq C\delta$ for some $C = C(r) > 0$. □

EXERCISE 4.4. *The discrete harmonic measure $H_{\Omega_\delta}(\cdot, y)$ of $y \in \partial\Omega_\delta$ is the unique harmonic function on $\Omega_\delta \setminus \partial\Omega_\delta$ vanishing on the boundary $\partial\Omega_\delta$, except at y, where it equals 1. Equivalently, $H_{\Omega_\delta}(x,y)$ is the probability that a simple random walk starting from x exits $\Omega_\delta \setminus \partial\Omega_\delta$ through y. Show that for any harmonic function $h : \Omega_\delta \to \mathbb{C}$,*

$$h = \sum_{y\in\partial\Omega_\delta} h(y) H_{\Omega_\delta}(\cdot, y).$$

EXERCISE 4.5. *Prove that there exists $C > 0$ such that $H_{Q_\delta}(0,y) \leq C\delta$ for every $\delta > 0$ and $y \in \partial Q_\delta$, where $Q = [-1,1]^2$.*

4.1.3. *Discrete Dirichlet problem and convergence in the scaling limit.* Preharmonic functions on square lattices of smaller and smaller mesh size were studied in a number of papers in the early twentieth century (see e.g. [**PW23, Bou26, Lus26**]), culminating in the seminal work of Courant, Friedrichs and Lewy. It was shown in [**CFL28**] that solutions to the Dirichlet problem for a discretization of an elliptic operator converge to the solution of the analogous continuous problem as the mesh of the lattice tends to zero. A first interesting fact is that the limit of preharmonic functions is indeed harmonic.

PROPOSITION 4.6. *Any limit of a sequence of preharmonic functions on Ω_δ converging uniformly on any compact subset of Ω is harmonic in Ω.*

Proof. Let (h_δ) be a sequence of preharmonic functions on Ω_δ converging to h. Via Propositions 4.2 and 4.3, $(\frac{1}{\delta}[h_\delta(\cdot+\delta)-h_\delta])_{\delta>0}$ is precompact. Since $\partial_x h$ is the only possible sub-sequential limit of the sequence, $(\frac{1}{\sqrt{2}\delta}[h_\delta(\cdot+\delta)-h_\delta])_{\delta>0}$ converges (indeed its discrete primitive converges to h). Similarly, one can prove convergence of discrete derivatives of any order. In particular, $0=\frac{1}{2\delta^2}\Delta_\delta h_\delta$ converges to $\frac{1}{4}[\partial_{xx}h+\partial_{yy}h]$. Therefore, h is harmonic. □

In particular, preharmonic functions with a given boundary value problem converge in the scaling limit to a harmonic function with the same boundary value problem in a rather strong sense, including convergence of all partial derivatives. The finest result of convergence of discrete Dirichlet problems to the continuous ones will not be necessary in our setting and we state the minimal required result:

THEOREM 4.7. *Let Ω be a simply connected domain with two marked points a and b on the boundary, and f a bounded continuous function on the boundary of Ω. Let $f_\delta:\partial\Omega_\delta\to\mathbb{C}$ be a sequence of uniformly bounded functions converging uniformly away from a and b to f. Let h_δ be the unique preharmonic map on Ω_δ such that $(h_\delta)_{|\partial\Omega_\delta}=f_\delta$. Then*

$$h_\delta \longrightarrow h \quad \textit{when } \delta\to 0$$

uniformly on compact subsets of Ω, where h is the unique harmonic function on Ω, continuous on $\overline{\Omega}$, satisfying $h_{|\partial\Omega}=f$.

Proof. Since $(f_\delta)_{\delta>0}$ is uniformly bounded by some constant M, the minimum and maximum principles imply that $(h_\delta)_{\delta>0}$ is bounded by M. Therefore, the family (h_δ) is precompact (Proposition 4.3). Let $\tilde{h}$ be a sub-sequential limit. Necessarily, $\tilde{h}$ is harmonic inside the domain (Proposition 4.6) and bounded. To prove that $\tilde{h}=h$, it suffices to show that $\tilde{h}$ can be continuously extended to the boundary by f.

Let $x\in\partial\Omega\smallsetminus\{a,b\}$ and $\varepsilon>0$. There exists $R>0$ such that for δ small enough,

$$|f_\delta(x')-f_\delta(x)|<\varepsilon \qquad \text{for every } x'\in\partial\Omega\cap Q(x,R),$$

where $Q(x,R)=x+[-R,R]^2$. For $r<R$ and $y\in Q(x,r)$, we have

$$|h_\delta(y)-f_\delta(x)| \;=\; \mathbb{E}_y[f_\delta(X_\tau)-f_\delta(x)]$$

for X a random walk starting at y, and τ its hitting time of the boundary. Decomposing between walks exiting the domain inside $Q(x,R)$ and others, we find

$$|h_\delta(y)-f_\delta(x)| \;\le\; \varepsilon \;+\; 2M\mathbb{P}_y[X_\tau\notin Q(x,R)]$$

Exercise 4.8 guarantees that $\mathbb{P}_y[X_\tau\notin Q(x,R)]\le(r/R)^\alpha$ for some independent constant $\alpha>0$. Taking $r=R(\varepsilon/2M)^{1/\alpha}$ and letting δ go to 0, we obtain $|\tilde{h}(y)-f(x)|\le 2\varepsilon$ for every $y\in Q(x,r)$. □

EXERCISE 4.8. *Show that there exists $\alpha>0$ such that for any $1\gg r>\delta>0$ and any curve γ inside $\mathbb{D}:=\{z:|z|<1\}$ from $C=\{z:|z|=1\}$ to $\{z:|z|=r\}$, the probability for a random walk on $\mathbb{D}_\delta$ starting at 0 to exit $(\mathbb{D}\smallsetminus\gamma)_\delta$ through C is smaller than r^α. To prove this, one can show that in any annulus $\{z:x\le|z|\le 2x\}$, the random walk trajectory has a uniformly positive probability to close a loop around the origin.*

4.1.4. *Discrete Green functions.* This paragraph concludes the section by mentioning the important example of discrete Green functions. For $y\in\Omega_\delta\smallsetminus\partial\Omega_\delta$, let $G_{\Omega_\delta}(\cdot,y)$ be the *discrete Green function* in the domain Ω_δ with singularity at y, *i.e.* the unique function on Ω_δ such that

- its Laplacian on $\Omega_\delta\smallsetminus\partial\Omega_\delta$ equals 0 except at y, where it equals 1,
- $G_{\Omega_\delta}(\cdot,y)$ vanishes on the boundary $\partial\Omega_\delta$.

The quantity $-G_{\Omega_\delta}(x,y)$ is the number of visits at x of a random walk started at y and stopped at the first time it reaches the boundary. Equivalently, it is also the number of visits at y of a random walk started at x stopped at the first time it reaches the boundary. Green functions are very convenient, in particular because of the Riesz representation formula for (not necessarily harmonic) functions:

PROPOSITION 4.9 (Riesz representation formula). *Let $f : \Omega_\delta \to \mathbb{C}$ be a function vanishing on $\partial\Omega_\delta$. We have*

$$f = \sum_{y\in\Omega_\delta} \Delta_\delta f(y) G_{\Omega_\delta}(\cdot, y).$$

Proof. Note that $f - \sum_{y\in\Omega_\delta} \Delta_\delta f(y) G_{\Omega_\delta}(\cdot,y)$ is harmonic and vanishes on the boundary. Hence, it equals 0 everywhere. □

Finally, a regularity estimate on discrete Green functions will be needed. This proposition is slightly technical. In the following, $aQ_\delta = [-a,a]^2 \cap \mathbb{L}_\delta$ and $\nabla_x f(x) = (f(x+\delta) - f(x), f(x+i\delta) - f(x))$.

PROPOSITION 4.10. *There exists $C > 0$ such that for any $\delta > 0$ and $y \in 9Q_\delta$,*

$$\sum_{x\in Q_\delta} |\nabla_x G_{9Q_\delta}(x,y)| \le C\delta \sum_{x\in Q_\delta} G_{9Q_\delta}(x,y).$$

Proof. In the proof, C_1,...,C_6 denote universal constants. First assume $y \in 9Q_\delta \setminus 3Q_\delta$. Using random walks, one can easily show that there exists $C_1 > 0$ such that

$$\frac{1}{C_1} G_{9Q_\delta}(x,y) \le G_{9Q_\delta}(x',y) \le C_1 G_{9Q_\delta}(x,y)$$

for every $x, x' \in 2Q_\delta$ (this is a special application of Harnack's principle). Using Proposition 4.2, we deduce

$$\sum_{x\in Q_\delta} |\nabla_x G_{9Q_\delta}(x,y)| \le \sum_{x\in Q_\delta} C_2\delta \max_{x\in 2Q_\delta} G_{9Q_\delta}(x,y) \le C_1C_2\delta \sum_{x\in Q_\delta} G_{9Q_\delta}(x,y)$$

which is the claim for $y \in 9Q_\delta \setminus 3Q_\delta$.

Assume now that $y \in 3Q_\delta$. Using the fact that $G_{9Q_\delta}(x,y)$ is the number of visits of x for a random walk starting at y (and stopped on the boundary), we find

$$\sum_{x\in Q_\delta} G_{9Q_\delta}(x,y) \ge C_3/\delta^2.$$

Therefore, it suffices to prove $\sum_{x\in Q_\delta} |\nabla G_{9Q_\delta}(x,y)| \le C_4/\delta$. Let $G_{\mathbb{L}_\delta}$ be the Green function in the whole plane, *i.e.* the function with Laplacian equal to $\delta_{x,y}$, normalized so that $G_{\mathbb{L}_\delta}(y,y) = 0$, and with sublinear growth. This function has been widely studied, it was proved in **[MW40]** that

$$G_{\mathbb{L}_\delta}(x,y) = \frac{1}{\pi}\ln\left(\frac{|x-y|}{\delta}\right) + C_5 + o\left(\frac{\delta}{|x-y|}\right).$$

Now, $G_{\mathbb{L}_\delta}(\cdot,y) - G_{9Q_\delta}(\cdot,y) - \frac{1}{\pi}\ln\left(\frac{1}{\delta}\right)$ is harmonic and has bounded boundary conditions on $\partial 9Q_\delta$. Therefore, Proposition 4.2 implies

$$\sum_{x\in Q_\delta} \left|\nabla_x (G_{\mathbb{L}_\delta}(x,y) - G_{9Q_\delta}(x,y))\right| \le C_6\delta\cdot 1/\delta^2 = C_6/\delta.$$

Moreover, the asymptotic of $G_{\mathbb{L}_\delta}(\cdot,y)$ leads to

$$\sum_{x\in Q_\delta} \left|\nabla_x G_{\mathbb{L}_\delta}(x,y)\right| \le C_7/\delta.$$

Summing the two inequalities, the result follows readily. □

4.2. Preholomorphic functions.

4.2.1. *Historical introduction.* Preholomorphic functions appeared implicitly in Kirchhoff's work **[Kir47]**, in which a graph is modeled as an electric network. Assume every edge of the graph is a unit resistor and for $u \sim v$, let $F(uv)$ be the current from u to v. The first and the second Kirchhoff's laws of electricity can be restated:

- the sum of currents flowing from a vertex is zero:

$$\sum_{v\sim u} F(uv) = 0, \tag{4.5}$$

- the sum of the currents around any oriented closed contour γ is zero:

$$\sum_{[uv]\in\gamma} F(uv) = 0. \tag{4.6}$$

Different resistances amount to putting weights into (4.5) and (4.6). The second law is equivalent to saying that F is given by the gradient of a potential function H, and the first equivalent to H being preharmonic.

Besides the original work of Kirchhoff, the first notable application of preholomorphic functions is perhaps the famous article **[BSST40]** of Brooks, Smith, Stone and Tutte, where preholomorphic functions were used to construct tilings of rectangles by squares.

Preholomorphic functions distinctively appeared for the first time in the papers **[Isa41, Isa52]** of Isaacs, where he proposed two definitions (and called such functions mono-diffric). Both definitions ask for a discrete version of the Cauchy-Riemann equations $\partial_{i\alpha}F = i\partial_\alpha F$ or equivalently that the $\bar z$-derivative is 0. In the first definition, the equation that the function must satisfy is

$$i[f(E)-f(S)] \;=\; f(W)-f(S)$$

while in the second, it is

$$i[f(E)-f(W)] \;=\; f(N)-f(S),$$

where N, E, S and W are the four corners of a face. A few papers of his and other mathematicians followed, studying the first definition, which is asymmetric on the square lattice. The second (symmetric) definition was reintroduced by Ferrand, who also discussed the passage to the scaling limit and gave new proofs of Riemann uniformization and the Courant-Friedrichs-Lewy theorems **[Fer44, LF55]**. This was followed by extensive studies of Duffin and others, starting with **[Duf56]**.

4.3. Isaacs's definition of preholomorphic functions. We will be working with Isaacs's second definition (although the theories based on both definitions are almost the same). The definition involves the following discretization of the $\bar\partial = \partial_x + i\partial_y$ operator. For a complex valued function f on $\mathbb{L}_\delta$ (or on a finite subgraph of it), and $x \in \mathbb{L}_\delta^\star$, define

$$\bar\partial_\delta f(x) \;=\; \frac{1}{2}[f(E)-f(W)] + \frac{i}{2}[f(N)-f(S)]$$

where N, E, S and W denote the four vertices adjacent to the dual vertex x indexed in the obvious way.

Remark 4.11. *When defining derivation, one uses duality between a graph and its dual. Quantities related to the derivative of a function on G are defined on the dual graph $G^\star$. Similarly, notions related to the second derivative are defined on the graph G again, whereas a primitive would be defined on $G^\star$.*

Definition 4.12. *A function $f : \Omega_\delta \to \mathbb{C}$ is called* preholomorphic *if $\bar\partial_\delta f(x) = 0$ for every $x \in \Omega_\delta^\star$. For $x \in \Omega_\delta^\star$, $\bar\partial_\delta f(x) = 0$ is called the* discrete Cauchy-Riemann equation *at x.*

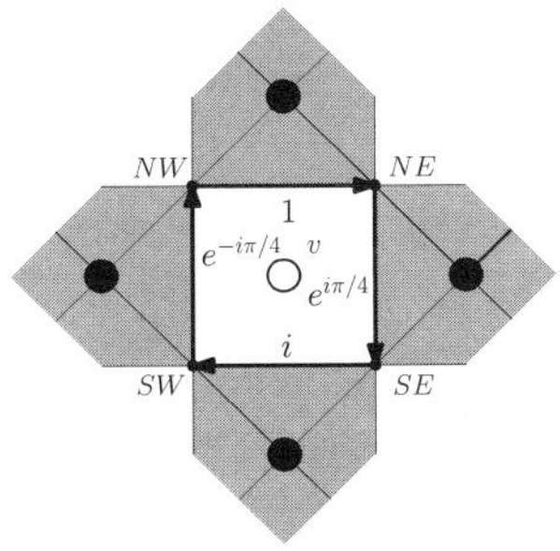

FIGURE 8. Lines $\ell(e)$ for medial edges around a white face.

The theory of preholomorphic functions starts much like the usual complex analysis. Sums of preholomorphic functions are also preholomorphic, discrete contour integrals vanish, primitive (in a simply-connected domain) and derivative are well-defined and are preholomorphic functions on the dual square lattice, etc. In particular, the (discrete) gradient of a preharmonic function is preholomorphic (this property has been proposed as a suitable generalization in higher dimensions).

EXERCISE 4.13. *Prove that the restriction of a continuous holomorphic function to $\mathbb{L}_\delta$ satisfies discrete Cauchy-Riemann equations up to $O(\delta^3)$.*

EXERCISE 4.14. *Prove that any preholomorphic function is preharmonic for a slightly modified Laplacian (the average over edges at distance $\sqrt{2}\delta$ minus the value at the point). Prove that the (discrete) gradient of a preharmonic function is preholomorphic (this property has been proposed as a suitable generalization in higher dimensions). Prove that the limit of preholomorphic functions is holomorphic.*

EXERCISE 4.15. *Prove that the integral of a preholomorphic function along a discrete contour vanishes. Prove that the primitive and the differential of preholomorphic functions are preholomorphic.*

EXERCISE 4.16. *Prove that $\frac{1}{\sqrt{2}\delta}\bar{\partial}_\delta$ and $\frac{1}{2\delta^2}\Delta_\delta$ converge (when $\delta \to 0$) to ∂, $\bar{\partial}$ and Δ in the sense of distributions.*

4.4. s-holomorphic functions. As explained in the previous sections, the theory of preholomorphic functions starts like the continuum theory. Unfortunately, problems arrive quickly. For instance, the square of a preholomorphic function is no longer preholomorphic in general. This makes the theory of preholomorphic functions significantly harder than the usual complex analysis, since one cannot transpose proofs from continuum to discrete in a straightforward way. In order to partially overcome this difficulty, we introduce s-holomorphic functions (for *spin*-holomorphic), a notion that will be central in the study of the spin and FK fermionic observables.

4.4.1. *Definition of s-holomorphic functions.* To any edge of the medial lattice e, we associate a line $\ell(e)$ passing through the origin and $\sqrt{\bar{e}}$ (the choice of the square root is not important, and recall that e being oriented, it can be thought of as a complex number). The different lines associated with medial edges on $\mathbb{L}_\delta^\diamond$ are $\mathbb{R}$, $e^{i\pi/4}\mathbb{R}$, $i\mathbb{R}$ and $e^{3i\pi/4}\mathbb{R}$, see Fig. 8.

DEFINITION 4.17. *A function $f : \Omega_\delta^\diamond \to \mathbb{C}$ is s-holomorphic if for any edge e of $\Omega_\delta^\diamond$, we have*

$$P_{\ell(e)}[f(x)] = P_{\ell(e)}[f(y)]$$

where x, y are the endpoints of e and P_ℓ is the orthogonal projection on ℓ.

The definition of s-holomorphicity is not rotationally invariant. Nevertheless, f is s-holomorphic if and only if $e^{i\pi/4}f(i\cdot)$ (resp. $if(-\cdot)$) is s-holomorphic.

PROPOSITION 4.18. *Any s-holomorphic function $f : \Omega_\delta^\diamond \to \mathbb{C}$ is preholomorphic on $\Omega_\delta^\diamond$.*

Proof. Let $f : \Omega_\delta^\diamond \to \mathbb{C}$ be a s-holomorphic function. Let v be a vertex of $\mathbb{L}_\delta \cup \mathbb{L}_\delta^\star$ (this is the vertex set of the dual of the medial lattice). Assume that $v \in \Omega_\delta^\star$, the other case is similar. We aim to show that $\bar\partial_\delta f(v) = 0$. Let NW, NE, SE and SW be the four vertices around v as illustrated in Fig. 8. Next, let us write relations provided by the s-holomorphicity, for instance

$$P_{\mathbb{R}}[f(NW)] = P_{\mathbb{R}}[f(NE)].$$

Expressed in terms of f and its complex conjugate $\bar f$ only, we obtain

$$f(NW) + \overline{f(NW)} = f(NE) + \overline{f(NE)}.$$

Doing the same with the other edges, we find

$$\begin{aligned} f(NE) + i\overline{f(NE)} &= f(SE) + i\overline{f(SE)} \\ f(SE) - \overline{f(SE)} &= f(SW) - \overline{f(SW)} \\ f(SW) - i\overline{f(SW)} &= f(NW) - i\overline{f(NW)} \end{aligned}$$

Multiplying the second identity by $-i$, the third by -1, the fourth by i, and then summing the four identities, we obtain

$$0 = (1-i)\left[f(NW) - f(SE) + if(SW) - if(NE)\right] = 2(1-i)\bar\partial_\delta f(v)$$

which is exactly the discrete Cauchy-Riemann equation in the medial lattice. □

4.4.2. *Discrete primitive of F^2.* One might wonder why s-holomorphicity is an interesting concept, since it is more restrictive than preholomorphicity. The answer comes from the fact that a relevant discretization of $\frac{1}{2}\mathrm{Im}\left(\int^z f^2\right)$ can be defined for s-holomorphic functions f.

THEOREM 4.19. *Let $f : \Omega_\delta^\diamond \to \mathbb{C}$ be an s-holomorphic function on the discrete simply connected domain $\Omega_\delta^\diamond$, and $b_0 \in \Omega_\delta$. Then, there exists a unique function $H : \Omega_\delta \cup \Omega_\delta^\star \to \mathbb{C}$ such that*

$$\begin{aligned} H(b_0) &= 1 \quad \text{and} \\ H(b) - H(w) &= \delta\left|P_{\ell(e)}[f(x)]\right|^2 \ \left(= \delta\left|P_{\ell(e)}[f(y)]\right|^2\right) \end{aligned}$$

for every edge $e = [xy]$ of $\Omega_\delta^\diamond$ bordered by a black face $b \in \Omega_\delta$ and a white face $w \in \Omega_\delta^\star$.

An elementary computation shows that for two neighboring sites $b_1, b_2 \in \Omega_\delta$, with v being the medial vertex at the center of $[b_1b_2]$,

$$H(b_1) - H(b_2) = \frac{1}{2}\mathrm{Im}\left[f(v)^2\cdot(b_1 - b_2)\right],$$

the same relation holding for sites of $\Omega_\delta^\star$. This legitimizes the fact that H is a discrete analogue of $\frac{1}{2}\mathrm{Im}\left(\int^z f^2\right)$.

Proof. The uniqueness of H is straightforward since $\Omega_\delta^\diamond$ is simply connected. To obtain the existence, construct the value at some point by summing increments along an arbitrary path from b_0 to this point. The only thing to check is that the value obtained does not depend on the path chosen to define it. Equivalently, we must check the second Kirchhoff's law. Since the domain is simply connected, it is sufficient to check it for elementary square contours around each medial vertex v (these are the simplest closed contours). Therefore, we need to prove that

$$\left|P_{\ell(n)}[f(v)]\right|^2 - \left|P_{\ell(e)}[f(v)]\right|^2 + \left|P_{\ell(s)}[f(v)]\right|^2 - \left|P_{\ell(w)}[f(v)]\right|^2 = 0, \tag{4.7}$$

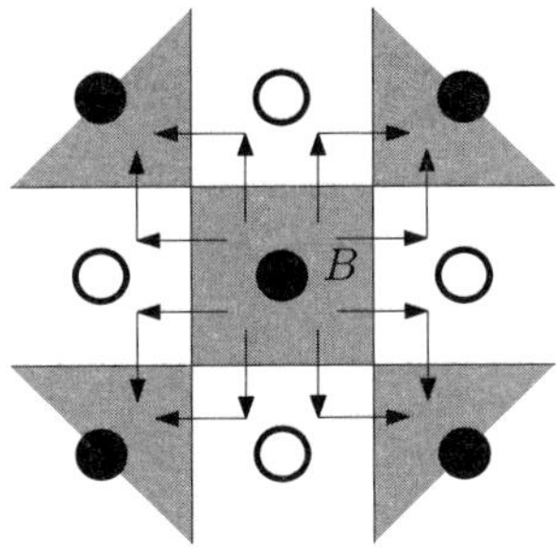

FIGURE 9. Arrows corresponding to contributions to $2\Delta H^\bullet$. Note that arrows from black to white contribute negatively, those from white to black positively.

where n, e, s and w are the four medial edges with endpoint v, indexed in the obvious way. Note that $\ell(n)$ and $\ell(s)$ (resp. $\ell(e)$ and $\ell(w)$) are orthogonal. Hence, (4.7) follows from

$$|P_{\ell(n)}[f(v)]|^2 + |P_{\ell(s)}[f(v)]|^2 = |f(v)|^2 = |P_{\ell(e)}[f(v)]|^2 + |P_{\ell(w)}[f(v)]|^2. \tag{4.8}$$

□

Even if the primitive of f is preholomorphic and thus preharmonic, this is not the case for H in general[5]. Nonetheless, H satisfies subharmonic and superharmonic properties. Denote by $H^\bullet$ and H° the restrictions of $H : \Omega_\delta \cup \Omega_\delta^\star \to \mathbb{C}$ to Ω_δ (black faces) and $\Omega_\delta^\star$ (white faces).

PROPOSITION 4.20. *If $f : \Omega_\delta^\diamond \to \mathbb{C}$ is s-holomorphic, then $H^\bullet$ and H° are respectively subharmonic and superharmonic.*

Proof. Let B be a vertex of $\Omega_\delta \smallsetminus \partial\Omega_\delta$. We aim to show that the sum of increments of $H^\bullet$ between B and its four neighbors is positive. In other words, we need to prove that the sum of increments along the sixteen arrows drawn in Fig. 9 is positive. Let a, b, c and d be the four values of $\sqrt{\delta}P_{\ell(e)}[f(y)]$ for every vertex $y \in \Omega_\delta^\diamond$ around B and any edge $e = [yz]$ bordering B (there are only four different values thanks to the definition of s-holomorphicity). An easy computation shows that the eight interior increments are thus $-a^2$, $-b^2$, $-c^2$, $-d^2$ (each appearing twice). Using the s-holomorphicity of f on vertices of $\Omega_\delta^\diamond$ around B, we can compute the eight exterior increments in terms of a, b, c and d: we obtain $(a\sqrt{2}-b)^2$, $(b\sqrt{2}-a)^2$, $(b\sqrt{2}-c)^2$, $(c\sqrt{2}-b)^2$, $(c\sqrt{2}-d)^2$, $(d\sqrt{2}-c)^2$, $(d\sqrt{2}+a)^2$, $(a\sqrt{2}+d)^2$. Hence, the sum S of increments equals

$$S = 4(a^2+b^2+c^2+d^2) - 4\sqrt{2}(ab+bc+cd-da) \tag{4.9}$$

$$= 4\left|e^{-i\pi/4}a - b + e^{i3\pi/4}c - id\right|^2 \geq 0. \tag{4.10}$$

The proof for H° follows along the same lines. □

REMARK 4.21. *A subharmonic function in a domain is smaller than the harmonic function with the same boundary conditions. Therefore, $H^\bullet$ is smaller than the harmonic function solving the same boundary value problem while H° is bigger than the harmonic function solving the same boundary value problem. Moreover, $H^\bullet(b)$ is larger than $H^\circ(w)$ for two neighboring faces. Hence, if $H^\bullet$ and H° are close to each other on the boundary, then they are sandwiched between two harmonic functions with roughly the*

[5] H is roughly (the imaginary part of) the primitive of *the square* of f.

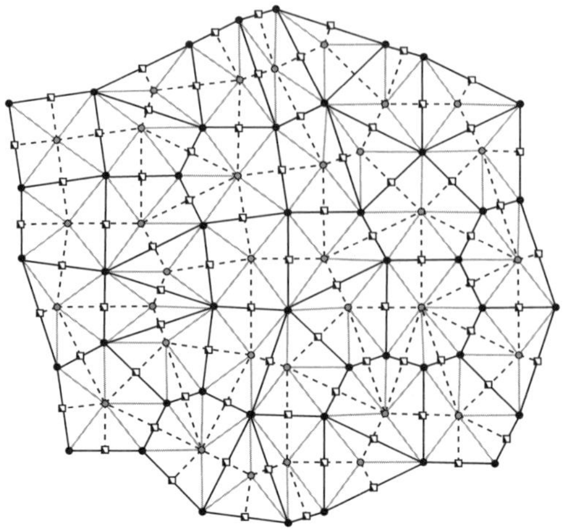

FIGURE 10. The black graph is the isoradial graph. Grey vertices are the vertices on the dual graph. There exists a radius $r > 0$ such that all faces can be put into an incircle of radius r. Dual vertices have been drawn in such a way that they are the centers of these circles.

same boundary conditions. In this case, they are almost harmonic. This fact will be central in the proof of conformal invariance.

4.5. Isoradial graphs and circle packings. Duffin [**Duf68**] extended the definition of preholomorphic functions to isoradial graphs. *Isoradial graphs* are planar graphs that can be embedded in such a way that there exists $r > 0$ so that each face has a circumcircle of same radius $r > 0$, see Fig. 10. When the embedding satisfies this property, it is said to be an isoradial embedding. We would like to point out that isoradial graphs form a rather large family of graphs. While not every topological quadrangulation (graph all of whose faces are quadrangles) admits a isoradial embedding, Kenyon and Schlenker [**KS05**] gave a simple necessary and sufficient topological condition for its existence. It seems that the first appearance of a related family of graphs in the probabilistic context was in the work of Baxter [**Bax89**], where the eight-vertex model and the Ising model were considered on Z-invariant graphs, arising from planar line arrangements. These graphs are topologically the same as the isoradial ones, and though they are embedded differently into the plane, by [**KS05**] they always admit isoradial embeddings. In [**Bax89**], Baxter was not considering scaling limits, and so the actual choice of embedding was immaterial for his results. However, weights in his models would suggest an isoradial embedding, and the Ising model was so considered by Mercat [**Mer01**], Boutilier and de Tilière [**BdT11, BdT10**], Chelkak and Smirnov [**CS08**] (see the last section for more details). Additionally, the dimer and the uniform spanning tree models on such graphs also have nice properties, see e.g. [**Ken02**]. Today, isoradial graphs seem to be the largest family of graphs for which certain lattice models, including the Ising model, have nice integrability properties (for instance, the star-triangle relation works nicely). A second reason to study isoradial graphs is that it is perhaps the largest family of graphs for which the Cauchy-Riemann operator admits a nice discretization. In particular, restrictions of holomorphic functions to such graphs are preholomorphic to higher orders. The fact that isoradial graphs are natural graphs both for discrete analysis and statistical physics sheds yet another light on the connection between the two domains.

In [**Thu86**], Thurston proposed circle packings as another discretization of complex analysis. Some beautiful applications were found, including yet another proof of the Riemann uniformization theorem by Rodin and Sullivan [**RS87**]. More interestingly, circle packings were used by He and Schramm [**HS93**] in the best result so far on the Koebe uniformization conjecture, stating that any domain can be conformally uniformized to a domain bounded by circles and points. In particular, they established the conjecture for domains with countably many boundary components. More about circle packings can be learned from Stephenson's book [**Ste05**]. Note that unlike the discretizations discussed above, the circle packings lead to non-linear versions of the Cauchy-Riemann equations, see *e.g.* the discussion in [**BMS05**].

5. Convergence of fermionic observables

In this section, we prove the convergence of fermionic observables *at criticality* (Theorems 2.11 and 3.15). We start with the easier case of the FK-Ising model. We present the complete proof of the convergence, the main tool being the discrete complex analysis that we developed in the previous section. We also sketch the proof of the convergence for the spin Ising model.

5.1. Convergence of the FK fermionic observable. In this section, fix a simply connected domain (Ω, a, b) with two points on the boundary. For $\delta > 0$, always consider a discrete FK Dobrushin domain $(\Omega_\delta^\diamond, a_\delta, b_\delta)$ and the critical FK-Ising model with Dobrushin boundary conditions on it. Since the domain is fixed, set $F_\delta = F_{\Omega_\delta^\diamond, a_\delta, b_\delta, p_{sd}}$ for the FK fermionic observable.

The proof of convergence is in three steps:

- First, prove the s-holomorphicity of the observable.
- Second, prove the convergence of the function H_δ naturally associated to the s-holomorphic functions $F_\delta/\sqrt{2\delta}$.
- Third, prove that $F_\delta/\sqrt{2\delta}$ converges to $\sqrt{\phi'}$.

5.1.1. *s-holomorphicity of the (vertex) fermionic observable for FK-Ising.* The next two lemmata deal with the edge fermionic observable. They are the key steps of the proof of the s-holomorphicity of the vertex fermionic observable.

LEMMA 5.1. *For an edge $e \in \Omega_\delta^\diamond$, $F_\delta(e)$ belongs to $\ell(e)$.*

Proof. The winding at an edge e can only take its value in the set $W + 2\pi\mathbb{Z}$ where W is the winding at e of an arbitrary interface passing through e. Therefore, the winding weight involved in the definition of $F_\delta(e)$ is always proportional to $e^{iW/2}$ with a real coefficient, thus $F_\delta(e)$ is proportional to $e^{iW/2}$. In any FK Dobrushin domain, b_δ is the southeast corner and the last edge is thus going to the right. Therefore $e^{iW/2}$ belongs to $\ell(e)$ for any e and so does $F_\delta(e)$. □

Even though the proof is finished, we make a short parenthetical remark: the definition of s-holomorphicity is not rotationally invariant, nor is the definition of FK Dobrushin domains, since the medial edge pointing to b_δ has to be oriented southeast. The latter condition has been introduced in such a way that this lemma holds true. Even though this condition seems arbitrary, it has no influence on the convergence result, meaning that one could perform a (slightly modified) proof with another orientation.

LEMMA 5.2. *Consider a medial vertex v in $\Omega_\delta^\diamond \setminus \partial\Omega_\delta^\diamond$. We have*

$$F_\delta(N) - F_\delta(S) = i[F_\delta(E) - F_\delta(W)] \tag{5.1}$$

where N, E, S and W are the adjacent edges indexed in clockwise order.

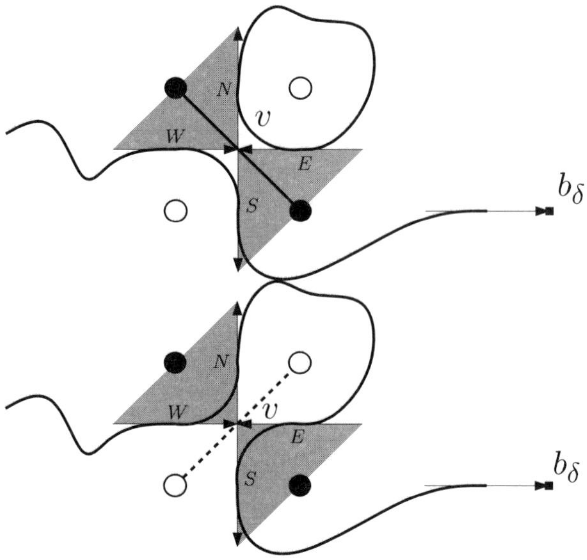

FIGURE 11. Two associated configurations, one with one exploration path and a loop, one without the loop. One can go from one to the other by switching the state of the edge.

Proof. Let us assume that v corresponds to a primal edge pointing SE to NW, see Fig. 11. The case NE to SW is similar.

We consider the involution s (on the space of configurations) which switches the state (open or closed) of the edge of the primal lattice corresponding to v. Let e be an edge of the medial graph and set

$$e_\omega \quad := \quad \phi_{\Omega^\diamond_\delta, a_\delta, b_\delta, p_{sd}}(\omega)\, \mathrm{e}^{\frac{\mathrm{i}}{2} W_\gamma(e, b_\delta)} 1_{e \in \gamma}$$

the contribution of the configuration ω to $F_\delta(e)$. Since s is an involution, the following relation holds:

$$F_\delta(e) = \sum_\omega e_\omega = \frac{1}{2} \sum_\omega \left[e_\omega + e_{s(\omega)}\right].$$

In order to prove (5.1), it suffices to prove the following for any configuration ω:

$$N_\omega + N_{s(\omega)} - S_\omega - S_{s(\omega)} = i\left[E_\omega + E_{s(\omega)} - W_\omega - W_{s(\omega)}\right]. \tag{5.2}$$

There are three possibilities:

Case 1: the exploration path $\gamma(\omega)$ does not go through any of the edges adjacent to v. It is easy to see that neither does $\gamma(s(\omega))$. All the terms then vanish and (5.2) trivially holds.

Case 2: $\gamma(\omega)$ goes through two edges around v. Note that it follows the orientation of the medial graph, and thus enters v through either W or E and leaves through N or S. We assume that $\gamma(\omega)$ enters through the edge W and leaves through the edge S (*i.e.* that the primal edge corresponding to v is open). The other cases are treated similarly. It is then possible to compute the contributions of all the edges adjacent to v of ω and $s(\omega)$ in terms of W_ω. Indeed,

- The probability of $s(\omega)$ is equal to $1/\sqrt{2}$ times the probability of ω (due to the fact that there is one less open edge of weight 1 – we are at the self-dual point – and one less loop of weight $\sqrt{2}$, see Proposition 3.17);

- Windings of the curve can be expressed using the winding of W. For instance, the winding of N in the configuration ω is equal to the winding of W minus a $\pi/2$ turn.

The contributions are given as:

configuration	W	E	N	S
ω	W_ω	0	0	$e^{i\pi/4}W_\omega$
$s(\omega)$	$W_\omega/\sqrt{2}$	$e^{i\pi/2}W_\omega/\sqrt{2}$	$e^{-i\pi/4}W_\omega/\sqrt{2}$	$e^{i\pi/4}W_\omega/\sqrt{2}$

Using the identity $e^{i\pi/4} - e^{-i\pi/4} = i\sqrt{2}$, we deduce (5.2) by summing (with the right weight) the contributions of all the edges around v.

Case 3: $\gamma(\omega)$ goes through the four medial edges around v. Then the exploration path of $s(\omega)$ goes through only two, and the computation is the same as in the second case.

In conclusion, (5.2) is always satisfied and the claim is proved. □

Recall that the FK fermionic observable is defined on medial edges as well as on medial vertices. Convergence of the observable means convergence of the *vertex* observable. The edge observable is just a very convenient tool in the proof. The two previous properties of the edge fermionic observable translate into the following result for the vertex fermionic observable.

PROPOSITION 5.3. *The vertex fermionic observable F_δ is s-holomorphic.*

Proof. Let v be a medial vertex and let N, E, S and W be the four medial edges around it. Using Lemmata 5.1 and 5.2, one can see that (5.1) can be rewritten (by taking the complex conjugate) as:

$$F_\delta(N) + F_\delta(S) = F_\delta(E) + F_\delta(W).$$

In particular, from (3.6),

$$F_\delta(v) := \frac{1}{2}\sum_{e \text{ adjacent}} F_\delta(e) = F_\delta(N) + F_\delta(S) = F_\delta(E) + F_\delta(W).$$

Using Lemma 5.1 again, $F_\delta(N)$ and $F_\delta(S)$ are orthogonal, so that $F_\delta(N)$ is the projection of $F_\delta(v)$ on $\ell(N)$ (and similarly for other edges). Therefore, for a medial edge $e = [xy]$, $F_\delta(e)$ is the projection of $F_\delta(x)$ and $F_\delta(y)$ with respect to $\ell(e)$, which proves that the vertex fermionic observable is s-holomorphic. □

The function $F_\delta/\sqrt{2\delta}$ is preholomorphic for every $\delta > 0$. Moreover, Lemma 5.1 identifies the boundary conditions of $F_\delta/\sqrt{2\delta}$ (its argument is determined) so that this function solves a discrete Riemann-Hilbert boundary value problem. These problems are significantly harder to handle than the Dirichlet problems. Therefore, it is more convenient to work with a discrete analogue of $\text{Im}\left(\int^z [F_\delta(z)/\sqrt{2\delta}]^2 dz\right)$, which should solve an approximate Dirichlet problem.

5.1.2. *Convergence of* $(H_\delta)_{\delta>0}$. Let A be the black face (vertex of Ω_δ) bordering a_δ, see Fig. 6. Since the FK fermionic observable $F_\delta/\sqrt{2\delta}$ is s-holomorphic, Theorem 4.19 defines a function $H_\delta : \Omega_\delta \cup \Omega_\delta^\star \to \mathbb{R}$ such that

$$\begin{aligned} H_\delta(A) &= 1 \quad \text{and} \\ H_\delta(B) - H_\delta(W) &= \left|P_{\ell(e)}[F_\delta(x)]\right|^2 = \left|P_{\ell(e)}[F_\delta(y)]\right|^2 \end{aligned}$$

for the edge $e = [xy]$ of $\Omega_\delta^\diamond$ bordered by a black face $B \in \Omega_\delta$ and a white face $W \in \Omega_\delta^\star$. Note that its restriction $H^\bullet$ to Ω_δ is subharmonic and its restriction H_δ° to $\Omega_\delta^\star$ is superharmonic.

Let us start with two lemmata addressing the question of boundary conditions for H_δ.

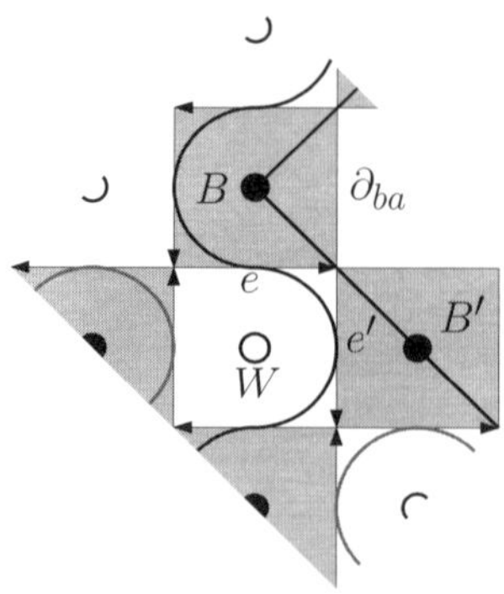

FIGURE 12. Two adjacent sites B and B' on ∂_{ba} together with the notation needed in the proof of Lemma 5.4.

LEMMA 5.4. *The function $H_\delta^\bullet$ is equal to 1 on the arc ∂_{ba}. The function H_δ° is equal to 0 on the arc $\partial_{ab}^\star$.*

Proof. We first prove that $H_\delta^\bullet$ is constant on ∂_{ba}. Let B and B' be two adjacent consecutive sites of ∂_{ba}. They are both adjacent to the same dual vertex $W \in \Omega_\delta^\star$, see Fig. 12. Let e (resp. e') be the edge of the medial lattice between W and B (resp. B'). We deduce

$$H_\delta^\bullet(B) - H_\delta^\bullet(B') = |F_\delta(e)|^2 - |F_\delta(e')|^2 = 0 \tag{5.3}$$

The second equality is due to $|F_\delta(e)| = \phi^{a_\delta,b_\delta}_{\Omega_\delta^\diamond,p_{sd}}(W \overset{\star}{\leftrightarrow} \partial^\star_{ab})$ (see Lemma 7.3). Hence, $H_\delta^\bullet$ is constant along the arc. Since $H_\delta^\bullet(A) = 1$, the result follows readily.

Similarly, H_δ° is constant on the arc $\partial^\star_{ab}$. Moreover, the dual white face $A^\star \in \partial^\star_{ab}$ bordering a_δ (see Fig. 6) satisfies

$$H_\delta^\circ(A^\star) \;=\; H_\delta^\bullet(A) - |F_\delta(e)|^2 \;=\; 1 - 1 \;=\; 0 \tag{5.4}$$

where e is the edge separating A and $A^\star$, which necessarily belongs to γ. Therefore $H_\delta^\circ = 0$ on $\partial^\star_{ab}$. □

LEMMA 5.5. *The function $H_\delta^\bullet$ converges to 0 on the arc ∂_{ab} uniformly away from a and b, H_δ° converges to 1 on the arc $\partial^\star_{ba}$ uniformly away from a and b.*

Proof. Once again, we prove the result for $H_\delta^\bullet$. The same reasoning then holds for H_δ°. Let B be a site of ∂_{ab} at distance r of ∂_{ba} (and therefore at graph distance r/δ of ∂_{ba} in Ω_δ). Let W be an adjacent site of B on $\partial^\star_{ab}$. Lemma 5.4 implies $H_\delta^\circ(W) = 0$. From the definition of H_δ, we find

$$H_\delta^\bullet(B) \;=\; H_\delta^\circ(W) + \left|P_{\ell(e)}[F_\delta(e)]\right|^2 \;=\; \left|P_{\ell(e)}[F_\delta(e)]\right|^2 \;=\; \phi^{a_\delta,b_\delta}_{\Omega_\delta,p_{sd}}(e \in \gamma)^2.$$

Note that $e \in \gamma$ if and only if B is connected to the wired arc ∂_{ba}. Therefore, $\phi^{a_\delta,b_\delta}_{\Omega_\delta,p_{sd}}(e \in \gamma)$ is equal to the probability that there exists an open path from B to ∂_{ba} (the winding is deterministic, see Lemma 7.3 for details). Since the boundary conditions on ∂_{ab} are free, the comparison between boundary conditions shows that the latter probability is smaller than the probability that there exists a path from B to ∂U_δ in the box $U_\delta = (B + [-r, r]^2) \cap \mathbb{L}_\delta$ with wired boundary conditions. Therefore,

$$H_\delta^\bullet(B) \;=\; \phi^{a_\delta,b_\delta}_{\Omega_\delta,p_{sd}}(e \in \gamma)^2 \;\le\; \phi^1_{U_\delta,p_{sd}}(B \leftrightarrow \partial U_\delta)^2.$$

Proposition 3.10 implies that the right hand side converges to 0 (there is no infinite cluster for $\phi^1_{p_{sd},2}$), which gives a uniform bound for B away from a and b. □

The two previous lemmata assert that the boundary conditions for $H_\delta^\bullet$ and H_δ° are roughly 0 on the arc ∂_{ab} and 1 on the arc ∂_{ba}. Moreover, $H_\delta^\bullet$ and H_δ° are almost harmonic. This should imply that $(H_\delta)_{\delta>0}$ converges to the solution of the Dirichlet problem, which is the subject of the next proposition.

PROPOSITION 5.6. *Let (Ω, a, b) be a simply connected domain with two points on the boundary. Then, $(H_\delta)_{\delta>0}$ converges to Im(ϕ) uniformly on any compact subsets of Ω when δ goes to 0, where ϕ is any conformal map from Ω to $\mathbb{T} = \mathbb{R} \times (0,1)$ sending a to $-\infty$ and b to ∞.*

Before starting, note that $\mathrm{Im}(\phi)$ is the solution of the Dirichlet problem on (Ω, a, b) with boundary conditions 1 on ∂_{ba} and 0 on ∂_{ab}.

Proof. From the definition of H, $H_\delta^\bullet$ is subharmonic, let $h_\delta^\bullet$ be the preharmonic function with same boundary conditions as $H_\delta^\bullet$ on $\partial\Omega_\delta$. Note that $H_\delta^\bullet \le h_\delta^\bullet$. Similarly, h_δ° is defined to be the preharmonic function with same boundary conditions as H_δ° on $\partial\Omega_\delta^\star$. If $K \subset \Omega$ is fixed, where K is compact, let $b_\delta \in K_\delta$ and $w_\delta \in K_\delta^\star$ any neighbor of b_δ, we have

$$h_\delta^\circ(w_\delta) \;\le\; H_\delta^\circ(w_\delta) \;\le\; H_\delta^\bullet(b_\delta) \;\le\; h_\delta^\bullet(b_\delta). \tag{5.5}$$

Using Lemmata 5.4 and 5.5, boundary conditions for $H_\delta^\bullet$ (and therefore $h_\delta^\bullet$) are uniformly converging to 0 on ∂_{ab} and 1 on ∂_{ba} away from a and b. Moreover, $|h_\delta^\bullet|$ is bounded by 1 everywhere. This is sufficient to apply Theorem 4.7: $h_\delta^\bullet$ converges to $\mathrm{Im}(\phi)$ on any compact subset of Ω when δ goes to 0. The same reasoning applies to h_δ°. The convergence for $H_\delta^\bullet$ and H_δ° follows easily since they are sandwiched between $h_\delta^\bullet$ and h_δ°. □

5.1.3. *Convergence of FK fermionic observables* $(F_\delta/\sqrt{2\delta})_{\delta>0}$. This section contains the proof of Theorem 3.15. The strategy is straightforward: $(F_\delta/\sqrt{2\delta})_{\delta>0}$ is proved to be a precompact family for the uniform convergence on compact subsets of Ω. Then, the possible sub-sequential limits are identified using H_δ.

Proof of Theorem 3.15. First assume that the precompactness of the family $(F_\delta/\sqrt{2\delta})_{\delta>0}$ has been proved. Let $(F_{\delta_n}/\sqrt{2\delta_n})_{n\in\mathbb{N}}$ be a convergent subsequence and denote its limit by f. Note that f is holomorphic as it is a limit of preholomorphic functions. For two points $x, y \in \Omega$, we have:

$$H_{\delta_n}(y) - H_{\delta_n}(x) = \frac{1}{2}\mathrm{Im}\Big(\int_x^y \frac{1}{\delta_n} F_{\delta_n}^2(z)dz\Big)$$

(for simplicity, also denote the closest points of x, y in Ω_{δ_n} by x, y). On the one hand, the convergence of $(F_{\delta_n}/\sqrt{2\delta_n})_{n\in\mathbb{N}}$ being uniform on any compact subset of Ω, the right hand side converges to $\mathrm{Im}\big(\int_x^y f(z)^2 dz\big)$. On the other hand, the left-hand side converges to $\mathrm{Im}(\phi(y) - \phi(x))$. Since both quantities are holomorphic functions of y, there exists $C \in \mathbb{R}$ such that $\phi(y) - \phi(x) = C + \int_x^y f(z)^2 dz$ for every $x, y \in \Omega$. Therefore f equals $\sqrt{\phi'}$. Since this is true for any converging subsequence, the result follows.

Therefore, the proof boils down to the precompactness of $(F_\delta/\sqrt{2\delta})_{\delta>0}$. We will use the second criterion in Proposition 4.3. Note that it is sufficient to prove this result for squares $Q \subset \Omega$ such that a bigger square $9Q$ (with same center) is contained in Ω.

Fix $\delta > 0$. When jumping diagonally over a medial vertex v, the function H_δ changes by $\mathrm{Re}(F_\delta^2(v))$ or $\mathrm{Im}(F_\delta^2(v))$ depending on the direction, so that

$$\delta^2 \sum_{v\in Q_\delta^\diamond} \left|F_\delta(v)/\sqrt{2\delta}\right|^2 \;=\; \delta \sum_{x\in Q_\delta} |\nabla H_\delta^\bullet(x)| \;+\; \delta \sum_{x\in Q_\delta^\star} |\nabla H_\delta^\circ(x)| \tag{5.6}$$

where $\nabla H_\delta^\bullet(x) = (H_\delta^\bullet(x+\delta) - H_\delta^\bullet(x), H_\delta^\bullet(x+i\delta) - H_\delta^\bullet(x))$, and ∇H_δ° is defined similarly for H_δ°. It follows that it is enough to prove uniform boundedness of the right hand side in (5.6). We only treat the sum involving $H_\delta^\bullet$. The other sum can be handled similarly.

Write $H_\delta^\bullet = S_\delta + R_\delta$ where S_δ is a harmonic function with the same boundary conditions on $\partial 9Q_\delta$ as $H_\delta^\bullet$. Note that $R_\delta \le 0$ is automatically subharmonic. In order to prove that the sum of $|\nabla H_\delta^\bullet|$ on Q_δ is bounded by C/δ, we deal separately with $|\nabla S_\delta|$ and $|\nabla R_\delta|$. First,

$$\sum_{x\in Q_\delta} |\nabla S_\delta(x)| \le \frac{C_1}{\delta^2}\cdot C_2\delta\left(\sup_{x\in\partial Q_\delta}|S_\delta(x)|\right) \le \frac{C_3}{\delta}\left(\sup_{x\in 9Q_\delta}|H_\delta^\bullet(x)|\right) \le \frac{C_4}{\delta},$$

where in the first inequality we used Proposition 4.2 and the maximum principle for S_δ, and in the second the fact that S_δ and $H_\delta^\bullet$ share the same boundary conditions on $9Q_\delta$. The last inequality comes from the fact that $H_\delta^\bullet$ converges, hence remains bounded uniformly in δ.

Second, recall that $G_{9Q_\delta}(\cdot,y)$ is the Green function in $9Q_\delta$ with singularity at y. Since R_δ equals 0 on the boundary, Proposition 4.9 implies

$$R_\delta(x) = \sum_{y\in 9Q_\delta} \Delta R_\delta(y) G_{9Q_\delta}(x,y), \tag{5.7}$$

thus giving

$$\nabla R_\delta(x) = \sum_{y\in 9Q_\delta} \Delta R_\delta(y) \nabla_x G_{9Q_\delta}(x,y)$$

Therefore,

$$\begin{aligned}
\sum_{x\in Q_\delta} |\nabla R_\delta(x)| &= \sum_{x\in Q_\delta}\Big|\sum_{y\in 9Q_\delta} \Delta R_\delta(y)\nabla_x G_{9Q_\delta}(x,y)\Big| \\
&\le \sum_{y\in 9Q_\delta} \Delta R_\delta(y) \sum_{x\in Q_\delta} |\nabla_x G_{9Q_\delta}(x,y)| \\
&\le \sum_{y\in 9Q_\delta} \Delta R_\delta(y)\; C_5\delta \sum_{x\in Q_\delta} G_{9Q_\delta}(x,y) \\
&= C_5\delta \sum_{x\in Q_\delta}\sum_{y\in 9Q_\delta} \Delta R_\delta(y) G_{9Q_\delta}(x,y) \\
&= C_5\delta \sum_{x\in Q_\delta} R_\delta(x) = C_6/\delta
\end{aligned}$$

The second line uses the fact that $\Delta R_\delta \ge 0$, the third Proposition 4.10, the fifth Proposition 4.9 again, and the last inequality the facts that Q_δ contains of order $1/\delta^2$ sites and that R_δ is bounded uniformly in δ (since H_δ and S_δ are).

Thus, $\delta\sum_{x\in Q_\delta}|\nabla H_\delta^\bullet|$ is uniformly bounded. Since the same result holds for H_δ°, $(F_\delta/\sqrt{2\delta})_{\delta>0}$ is precompact on Q (and more generally on any compact subset of Ω) and the proof is completed. □

5.2. Convergence of the spin fermionic observable. We now turn to the proof of convergence for the spin fermionic observable. Fix a simply connected domain (Ω, a, b) with two points on the boundary. For $\delta > 0$, always consider the spin fermionic observable on the discrete spin Dobrushin domain $(\Omega_\delta^\diamond, a_\delta, b_\delta)$. Since the domain is fixed, we set $F_\delta = F_{\Omega_\delta^\diamond,a_\delta,b_\delta}$. We follow the same three steps as before, beginning with the s-holomorphicity. The other two steps are only sketched, since they are more technical than in the FK-Ising case, see **[CS09]**.

PROPOSITION 5.7. *For $\delta > 0$, F_δ is s-holomorphic on $\Omega_\delta^\diamond$.*

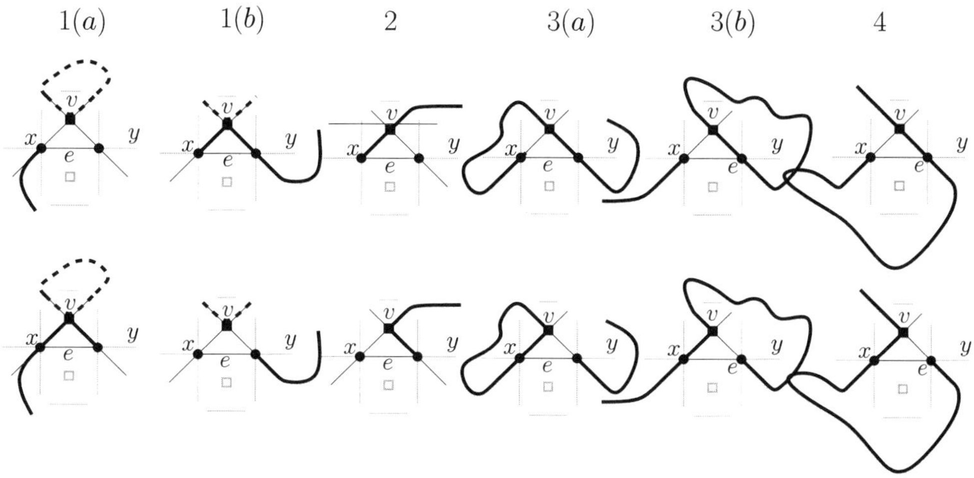

FIGURE 13. The different possible cases in the proof of Proposition 5.7: ω is depicted on the top, and ω' on the bottom.

Proof. Let x, y two adjacent medial vertices connected by the edge $e = [xy]$. Let v be the vertex of Ω_δ bordering the (medial) edge e. As before, set x_ω (resp. y_ω) for the contribution of ω to $F_\delta(x)$ (resp. $F_\delta(y)$). We wish to prove that

$$\sum_\omega P_{\ell(e)}(x_\omega) = \sum_\omega P_{\ell(e)}(y_\omega). \tag{5.8}$$

Note that the curve $\gamma(\omega)$ finishes at x_ω or at y_ω so that ω cannot contribute to $F_\delta(x)$ and $F_\delta(y)$ at the same time. Thus, it is sufficient to partition the set of configurations into pairs of configurations (ω, ω'), one contributing to y, the other one to x, such that $P_{\ell(e)}(x_\omega) = P_{\ell(e)}(y_{\omega'})$.

Without loss of generality, assume that e is pointing southeast, thus $\ell(e) = \mathbb{R}$ (other cases can be done similarly). First note that

$$x_\omega = \frac{1}{Z} \mathrm{e}^{-i\frac{1}{2}[W_{\gamma(\omega)}(a_\delta, x_\delta) - W_{\gamma'}(a_\delta, b_\delta)]} (\sqrt{2}-1)^{|\omega|},$$

where $\gamma(\omega)$ is the interface in the configuration ω, γ' is any curve from a_δ to b_δ (recall that $W_{\gamma'}(a_\delta, b_\delta)$ does not depend on γ'), and Z is a normalizing real number not depending on the configuration. There are six types of pairs that one can create, see Fig. 13 depicting the four main cases. Case 1 corresponds to the case where the interface reaches x or y and then extends by one step to reach the other vertex. In Case 2, γ reaches v before x and y, and makes an additional step to x or y. In Case 3, γ reaches x or y and sees a loop preventing it from being extended to the other vertex (in contrast to Case 1). In Case 4, γ reaches x or y, then goes away from v and comes back to the other vertex. Recall that the curve must always go to the left: in cases 1(a), 1(b), and 2 there can be a loop or even the past of γ passing through v. However, this does not change the computation.

We obtain the following table for x_ω and $y_{\omega'}$ (we always express $y_{\omega'}$ in terms of x_ω). Moreover, one can compute the argument modulo π of contributions x_ω since the orientation of e is known. When upon projecting on $\mathbb{R}$, the result follows.

configuration	Case 1(a)	Case 1(b)	Case 2	Case 3(a)	Case 3(b)	Case 4
x_ω	x_ω	x_ω	x_ω	x_ω	x_ω	x_ω
$y_{\omega'}$	$(\sqrt{2}-1)\mathrm{e}^{i\pi/4}x_\omega$	$\frac{\mathrm{e}^{i\pi/4}}{\sqrt{2}-1}x_\omega$	$\mathrm{e}^{-i\pi/4}x_\omega$	$\mathrm{e}^{3i\pi/4}x_\omega$	$\mathrm{e}^{3i\pi/4}x_\omega$	$\mathrm{e}^{-5i\pi/4}x_\omega$
arg. x_ω mod π	$5\pi/8$	$\pi/8$	$\pi/8$	$5\pi/8$	$5\pi/8$	$5\pi/8$

□

Proof of Theorem 2.11 (Sketch). The proof is roughly sketched. We refer to **[CS09]** for a complete proof.

Since F_δ is s-harmonic, one can define the observable H_δ as in Theorem 4.19, with the requirement that it is equal to 0 on the white face adjacent to b. Then, H_δ° is constant equal to 0 on the boundary as in the FK-Ising case. Note that H_δ should not converge to 0, even if boundary conditions are 0 away from a. Firstly, H_δ° is superharmonic and not harmonic, even though it is expected to be almost harmonic (away from a, $H_\delta^\bullet$ and H_δ° are close), this will not be true near a. Actually, H_δ should not remain bounded around a.

The main difference compared to the previous section is indeed the unboundedness of H_δ near a_δ which prevents us from the immediate use of Proposition 4.3. It is actually possible to prove that away from a, H_δ remains bounded, see **[CS09]**. This uses more sophisticated tools, among which are the boundary modification trick (see **[DCHN10]** for a quick description in the FK-Ising case, and **[CS09]** for the Ising original case). As before, boundedness implies precompactness (and thus boundedness) of $(F_\delta)_{\delta>0}$ away from a via Proposition 4.3. Since H_δ can be expressed in terms of F_δ, it is easy to deduce that H_δ is also precompact.

Now consider a convergent subsequence $(f_{\delta_n}, H_{\delta_n})$ converging to (f, H). One can check that H is equal to 0 on $\partial\Omega \setminus \{a\}$. Moreover, the fact that H_δ° equals 0 on the boundary and is superharmonic implies that H_δ° is greater than or equal to 0 everywhere, implying $H \geq 0$ in Ω. This property of harmonic functions in a domain almost determines them. There is only a one-parameter family of positive harmonic functions equal to 0 on the boundary. These functions are exactly the imaginary parts of conformal maps from Ω to the upper half-plane $\mathbb{H}$ mapping a to ∞. We can further assume that b is mapped to 0, since we are interested only in the imaginary part of these functions.

Fix one conformal map ψ from Ω to $\mathbb{H}$, mapping a to ∞ and b to 0. There exists $\lambda > 0$ such that $H = \lambda \mathrm{Im}\psi$. As in the case of the FK-Ising model, one can prove that $\mathrm{Im}\left(\int^z f^2\right) = H$, implying that $f^2 = \lambda\psi'$. Since $f(b) = 1$ (it is obvious from the definition that $F_\delta(b_\delta) = 1$), λ equals $\frac{1}{\psi'(b)}$. In conclusion, $f(z) = \sqrt{\psi'(z)/\psi'(b)}$ for every $z \in \Omega$. □

Note that some regularity hypotheses on the boundary near b are needed to ensure that the sequence $(f_{\delta_n}, H_{\delta_n})$ also converges near b. This is the reason for assuming that the boundary near b is smooth. We also mention that there is no normalization here. The normalization from the point of view of b was already present in the definition of the observable.

6. Convergence to chordal SLE(3) and chordal SLE(16/3)

The strategy to prove that a family of parametrized curves converges to SLE(κ) follows three steps:

- First, prove that the family of curves is tight.
- Then, show that any sub-sequential limit is a time-changed Loewner chain with a continuous driving process (see Beffara's course for details on Loewner chains and driving processes).
- Finally, show that the only possible driving processes for the sub-sequential limits is $\sqrt{\kappa}B_t$ where B_t is a standard Brownian motion.

The conceptual step is the third one. In order to identify the Brownian motion as being the only possible driving process for the curve, we find computable martingales expressed in terms of the limiting curve. These martingales will be the limits of fermionic observables. The fact that these (explicit) functions are martingales allows us to deduce martingale properties of the driving process. More precisely, we aim to use Lévy's

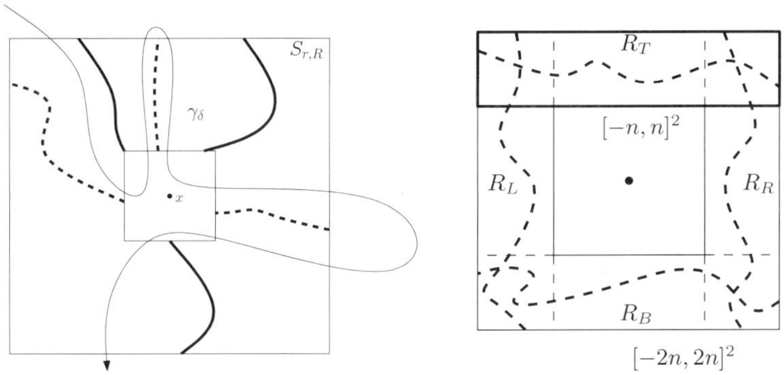

FIGURE 14. **Left:** The event $A_6(x,r,R)$. In the case of exploration paths, it implies the existence of alternating open and closed paths. **Right:** Rectangles R_T, R_R, R_B and R_L crossed by closed paths in the longer direction. The combination of these closed paths prevents the existence of a crossing from the inner to the outer boundary of the annulus.

theorem: a continuous real-valued process X such that X_t and $X_t^2 - at$ are martingales is necessarily $\sqrt{a}B_t$.

6.1. Tightness of interfaces for the FK-Ising model. In this section, we prove the following theorem:

THEOREM 6.1. *Fix a domain (Ω, a, b). The family $(\gamma_\delta)_{\delta>0}$ of random interfaces for the critical FK-Ising model in (Ω, a, b) is tight for the topology associated to the curve distance.*

The question of tightness for curves in the plane has been studied in the groundbreaking paper [**AB99**]. In that paper, it is proved that a sufficient condition for tightness is the absence, at every scale, of annuli crossed back and forth an unbounded number of times.

More precisely, for $x \in \Omega$ and $r < R$, let $S_{r,R}(x) = (x + [-R,R]^2) \setminus (x + [-r,r]^2)$ and define $\mathcal{A}_k(x;r,R)$ to be the event that there exist k crossings of the curve γ_δ between outer and inner boundaries of $S_{r,R}(x)$.

THEOREM 6.2 (Aizenman-Burchard [**AB99**]). *Let Ω be a simply connected domain and let a and b be two marked points on its boundary. Denote by $\mathbb{P}_\delta$ the law of a random curve $\tilde{\gamma}_\delta$ on Ω_δ from a_δ to b_δ. If there exist $k \in \mathbb{N}$, $C_k < \infty$ and $\Delta_k > 2$ such that for all $\delta < r < R$ and $x \in \Omega$,*

$$\mathbb{P}_\delta(\mathcal{A}_k(x;r,R)) \le C_k\Big(\frac{r}{R}\Big)^{\Delta_k},$$

then the family of curves $(\tilde{\gamma}_\delta)$ is tight.

We now show how to exploit this theorem in order to prove Theorem 6.1. The main tool is Theorem 3.16.

LEMMA 6.3 (Circuits in annuli). *Let $\mathcal{E}(x,n,N)$ be the probability that there exists an open path connecting the boundaries of $S_{n,N}(x)$. There exists a constant $c < 1$ such that for all $n > 0$,*

$$\phi^1_{p_{sd},S_{n,2n}(x)}(\mathcal{E}(x;n,2n)) \le c.$$

Note that the boundary conditions on the boundary of the annulus are wired. Via comparison between boundary conditions, this implies that the probability of an open path from the inner to the outer boundary is bounded uniformly on the configuration outside of the annulus. This uniform bound allows us to decouple what is happening inside the annulus with what is happening outside of it.

Proof. Assume $x = 0$. The result follows from Theorem 3.16 (proved in Section 7.2) applied in the four rectangles $R_B = [-2n, 2n] \times [-n, -2n]$, $R_L = [-2n, -n] \times [-2n, 2n]$, $R_T = [-2n, 2n] \times [n, 2n]$ and $R_R = [n, 2n] \times [-2n, 2n]$, see Fig. 14. Indeed, if there exists a closed path crossing each of these rectangles in the longer direction, one can construct from them a closed circuit in $S_{n,2n}$. Now, consider any of these rectangles, R_B for instance. Its aspect ratio is 4, so that Theorem 3.16 implies that there is a closed path crossing in the longer direction with probability at least $c_1 > 0$ (the wired boundary conditions are the dual of the free boundary conditions). The FKG inequality (3.2) implies that the probability of a circuit is larger than $c_1^4 > 0$. Therefore, the probability of a crossing is at most $c = 1 - c_1^4 < 1$. □

We are now in a position to prove Theorem 6.1.

Proof of Theorem 6.1. Fix $x \in \Omega$, $\delta < r < R$ and recall that we are on a lattice of mesh size δ. Let k to be fixed later. We first prove that

$$\phi_{\Omega_\delta, p_{sd}}^{a_\delta, b_\delta}(\mathcal{A}_{2k}(x; r, 2r)) \leq c^k \tag{6.1}$$

for some constant $c < 1$ uniform in x, k, r, δ and the configuration outside of $S_{r,2r}(x)$.

If $\mathcal{A}_{2k}(x; r, 2r)$ holds, then there are (at least) k open paths, alternating with k dual paths, connecting the inner boundary of the annulus to its outer boundary. Since the paths are alternating, one can deduce that there are k open crossings, each one being surrounded by closed crossings. Hence, using successive conditionings and the comparison between boundary conditions, the probability for each crossing is smaller than the probability that there is a crossing in the annulus with wired boundary conditions (since these boundary conditions maximize the probability of $\mathcal{E}(x; r, 2r)$). We obtain

$$\phi_{\Omega_\delta, p_{sd}}^{a_\delta, b_\delta}(\mathcal{A}_{2k}(x; r, 2r)) \leq \left[\phi_{p_{sd}, S_{r,2r}(x)}^1(\mathcal{E}(x; r, 2r))\right]^k.$$

Using Lemma 6.3, $\phi_{p_{sd}, S_{r,2r}(x)}^1(\mathcal{E}(x; r, 2r)) \leq c < 1$ and and (6.1) follows.

One can further fix k large enough so that $c^k < \frac{1}{8}$. Now, one can decompose the annulus $S_{r,R}(x)$ into roughly $\ln_2(R/r)$ annuli of the form $S_{r,2r}(x)$, so that for the previous k,

$$\phi_{\Omega_\delta, p_{sd}}^{a_\delta, b_\delta}(\mathcal{A}_{2k}(x; r, R)) \leq \left(\frac{r}{R}\right)^3. \tag{6.2}$$

Hence, Theorem 6.2 implies that the family (γ_δ) is tight. □

6.2. sub-sequential limits of FK-Ising interfaces are Loewner chains. This subsection requires basic knowledge of Loewner chains and we refer to Beffara's course in this volume for an overview on the subject. In the previous subsection, traces of interfaces in Dobrushin domains were shown to be tight. The natural discrete parametrization does not lead to a suitable continuous parametrization. We would prefer our sub-sequential limits to be parametrized as Loewner chains. In other words, we would like to parametrize the curve by its so-called h-capacity. In this case, we say that the curve is a *time-changed Loewner chain.*

THEOREM 6.4. *Any sub-sequential limit of the family $(\gamma_\delta)_{\delta>0}$ of FK-Ising interfaces is a time-changed Loewner chain.*

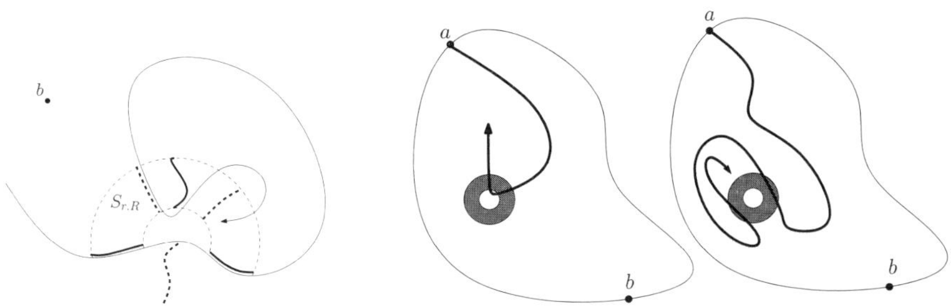

FIGURE 15. **Left:** An example of a fjord. Seen from b, the h-capacity (roughly speaking, the size) of the hull does not grow much while the curve is in the fjord. The event involves six alternating open and closed crossings of the annulus. **Right:** Conditionally on the beginning of the curve, the crossing of the annulus is unforced on the left, while it is forced on the right (it must go ultimately to b).

Not every continuous curve is a time-changed Loewner chain. In the case of FK interfaces, the limiting curve is fractal-like and has many double points, so that the following theorem is not a trivial statement. A general characterization for a parametrized non-selfcrossing curve in (Ω, a, b) to be a time-changed Loewner chain is the following:

- its h-capacity must be continuous,
- its h-capacity must be strictly increasing.
- the curve grows locally seen from infinity in the following sense: for any $t \geq 0$ and for any $\varepsilon > 0$, there exists $\delta > 0$ such that for any $s \leq t$, the diameter of $g_s(\Omega_s \smallsetminus \Omega_{s+\delta})$ is smaller than ε, where Ω_s is the connected component of $\Omega \smallsetminus \gamma[0,s]$ containing b and g_s is the conformal map from Ω_s to $\mathbb{H}$ with hydrodynamical renormalization (see Beffara's course).

The first condition is automatically satisfied by continuous curves. The third one usually follows from the two others when the curve is continuous, so that the crucial condition to check is the second one. This condition can be understood as being the fact that the tip of the curve is visible from b at every time. In other words, the family of hulls created by the curve (*i.e.* the complement of the connected component of $\Omega \smallsetminus \gamma_t$ containing b) is strictly increasing. This is the case if the curve does not enter long fjords created by its past at every scale, see Fig. 15.

In the case of FK interfaces, this corresponds to so-called six arm event, and it boils down to proving that $\Delta_6 > 2$. A general belief in statistical physics is that many exponents, called universal exponents, do not depend on the model. For instance, the so-called 5-arm exponent should equal 2. This would imply that $\Delta_6 > \Delta_5 = 2$. In general, proving that the 5-arm exponent equals 2 is very hard. Therefore, we need to invoke a stronger structural theorem to prove that sub-sequential limits are Loewner chains. Recently, Kemppainen and the second author proved the required theorem, and we describe it now.

For a family of parametrized curves $(\gamma_\delta)_{\delta>0}$, define Condition $(\star)$ by:

Condition $(\star)$*: There exist $C > 1$ and $\Delta > 0$ such that for any $0 < \delta < r < R/C$, for any stopping time τ and for any annulus $S_{r,R}(x)$ not containing γ_τ, the probability that γ_δ crosses the annulus $S_{r,R}(x)$ (from the outside to the inside) after time τ* ***while it is not forced to enter*** $S_{r,R}(x)$ **again** *is smaller than $C(r/R)^\Delta$, see Fig. 15.*

Roughly speaking, the previous condition is a uniform bound on unforced crossings. Note that it is necessary to assume the fact that the crossing is unforced.

THEOREM 6.5 ([**KS10**]). *If a family of curves* (γ_δ) *satisfies Condition* $(\star)$, *then it is tight for the topology associated to the curve distance. Moreover, any sub-sequential limit* (γ_{δ_n}) *is to a time-changed Loewner chain.*

Tightness is almost obvious, since Condition $(\star)$ implies the hypothesis in Aizenman-Burchard's theorem. The hard part is the proof that Condition $(\star)$ guarantees that the h-capacity of sub-sequential limits is strictly increasing and that they create Loewner chains. The reader is referred to [**KS10**] for a proof of this statement. We are now in a position to prove Theorem 6.4:

Proof of Theorem 6.4. Lemma 6.3 allows us to prove Condition $(\star)$ without difficulty. □

6.3. Convergence of FK-Ising interfaces to SLE(16/3). The FK fermionic observable is now proved to be a martingale for the discrete curves and to identify the driving process of any sub-sequential limit of FK-Ising interfaces.

LEMMA 6.6. *Let* $\delta > 0$. *The FK fermionic observable* $M_n^\delta(z) = F_{\Omega_\delta \smallsetminus \gamma[0,n],\gamma_n,b_\delta}(z)$ *is a martingale with respect to* $(\mathcal{F}_n)$, *where* $\mathcal{F}_n$ *is the* σ*-algebra generated by the* FK *interface* $\gamma[0,n]$.

Proof. For a Dobrushin domain $(\Omega_\delta^\diamond, a_\delta, b_\delta)$, the slit domain created by "removing" the first n steps of the exploration path is again a Dobrushin domain. Conditionally on $\gamma[0,n]$, the law of the FK-Ising model in this new domain is exactly $\phi^{\gamma_n,b_\delta}_{\Omega_\delta^\diamond \smallsetminus \gamma[0,n]}$. This observation implies that $M_n^\delta(z)$ is the random variable $1_{z\in\gamma_\delta} e^{\frac{1}{2}iW_{\gamma_\delta}(z,b)}$ conditionally on $\mathcal{F}_n$, therefore it is automatically a martingale. □

PROPOSITION 6.7. *Any sub-sequential limit of* $(\gamma_\delta)_{\delta>0}$ *which is a Loewner chain is the (chordal) Schramm-Loewner Evolution with parameter* $\kappa = 16/3$.

Proof. Consider a sub-sequential limit γ in the domain (Ω, a, b) which is a Loewner chain. Let ϕ be a map from (Ω, a, b) to $(\mathbb{H}, 0, \infty)$. Our goal is to prove that $\tilde{\gamma} = \phi(\gamma)$ is a chordal SLE(16/3) in the upper half-plane.

Since γ is assumed to be a Loewner chain, $\tilde{\gamma}$ is a growing hull from 0 to ∞ parametrized by its h-capacity. Let W_t be its continuous driving process. Also, define g_t to be the conformal map from $\mathbb{H} \smallsetminus \tilde{\gamma}[0,t]$ to $\mathbb{H}$ such that $g_t(z) = z + 2t/z + O(1/z^2)$ when z goes to ∞.

Fix $z' \in \Omega$. For $\delta > 0$, recall that $M_n^\delta(z')$ is a martingale for γ_δ. Since the martingale is bounded, $M_{\tau_t}^\delta(z')$ is a martingale with respect to $\mathcal{F}_{\tau_t}$, where τ_t is the first time at which $\phi(\gamma_\delta)$ has an h-capacity larger than t. Since the convergence is uniform, $M_t(z') := \lim_{\delta\to 0} M_{\tau_t}^\delta(z')$ is a martingale with respect to $\mathcal{G}_t$, where $\mathcal{G}_t$ is the σ-algebra generated by the curve $\tilde{\gamma}$ up to the first time its h-capacity exceeds t. By definition, this time is t, and $\mathcal{G}_t$ is the σ-algebra generated by $\tilde{\gamma}[0,t]$.

Recall that $M_t(z')$ is related to $\phi(z')$ via the conformal map from $\mathbb{H} \smallsetminus \tilde{\gamma}[0,t]$ to $\mathbb{R}\times(0,1)$, normalized to send $\tilde{\gamma}_t$ to $-\infty$ and ∞ to ∞. This last map is exactly $\frac{1}{\pi}\ln(g_t - W_t)$. Setting $z = \phi(z')$, we obtain that

$$\sqrt{\pi} M_t^z \ := \ \sqrt{\pi} M_t(z') \ = \ \sqrt{[\ln(g_t(z) - W_t)]'} \ = \ \sqrt{\frac{g_t'(z)}{g_t(z) - W_t}} \tag{6.3}$$

is a martingale. Recall that, when z goes to infinity,

$$g_t(z) \ = \ z + \frac{2t}{z} + O\Big(\frac{1}{z^2}\Big) \quad \text{and} \quad g_t'(z) = 1 - \frac{2t}{z^2} + O\Big(\frac{1}{z^3}\Big) \tag{6.4}$$

For $s \le t$,

$$\begin{aligned}\sqrt{\pi}\cdot\mathbb{E}[M_t^z|\mathcal{G}_s] &= \mathbb{E}\left[\sqrt{\frac{1-2t/z^2+O(1/z^3)}{z-W_t+2t/z+O(1/z^2)}}\ \Big|\ \mathcal{G}_s\right]\\ &= \frac{1}{\sqrt{z}}\,\mathbb{E}\left[1+\frac{1}{2}W_t/z+\frac{1}{8}\left(3W_t^2-16t\right)/z^2+O\left(1/z^3\right)\ \Big|\ \mathcal{G}_s\right]\\ &= \frac{1}{\sqrt{z}}\left(1+\frac{1}{2}\mathbb{E}[W_t|\mathcal{G}_s]/z+\frac{1}{8}\mathbb{E}[3W_t^2-16t|\mathcal{G}_s]/z^2+O\left(1/z^3\right)\right).\end{aligned}$$

Taking $s = t$ yields

$$\sqrt{\pi}\cdot M_s^z = \frac{1}{\sqrt{z}}\left(1+\frac{1}{2}W_s/z+\frac{1}{8}(3W_s^2-16s)/z^2+O(1/z^3)\right).$$

Since $\mathbb{E}[M_t^z|\mathcal{G}_s] = M_s^z$, terms in the previous asymptotic development can be matched together so that $\mathbb{E}[W_t|\mathcal{G}_s] = W_s$ and $\mathbb{E}[W_t^2 - \frac{16}{3}t|\mathcal{G}_s] = W_s^2 - \frac{16}{3}s$. Since W_t is continuous, Lévy's theorem implies that $W_t = \sqrt{\frac{16}{3}}B_t$ where B_t is a standard Brownian motion.

In conclusion, γ is the image by ϕ^{-1} of the chordal Schramm-Loewner Evolution with parameter $\kappa = 16/3$ in the upper half-plane. This is exactly the definition of the chordal Schramm-Loewner Evolution with parameter $\kappa = 16/3$ in the domain (Ω, a, b). □

Proof of Theorem 3.13. By Theorem 4.3, the family of curves is tight. Using Theorem 6.4, any sub-sequential limit is a time-changed Loewner chain. Consider such a sub-sequential limit and parametrize it by its h-capacity. Proposition 6.7 then implies that it is the Schramm-Loewner Evolution with parameter $\kappa = 16/3$. The possible limit being unique, the claim is proved. □

6.4. Convergence to SLE(3) for spin Ising interfaces. The proof of Theorem 2.10 is very similar to the proof of Theorem 3.13, except that we work with the spin Ising fermionic observable instead of the FK-Ising model one. The only point differing from the previous section is the proof that the spin fermionic observable is a martingale for the curve. We prove this fact now and leave the remainder of the proof as an exercise. Let γ be the interface in the critical Ising model with Dobrushin boundary conditions.

LEMMA 6.8. *Let $\delta > 0$, the spin fermionic observable $M_n^\delta(z) = F_{\Omega_\delta^\diamond\setminus\gamma[0,n],\gamma(n),b_\delta}(z)$ is a martingale with respect to $(\mathcal{F}_n)$, where $\mathcal{F}_n$ is the σ-algebra generated by the exploration process $\gamma[0,n]$.*

Proof. It is sufficient to check that $F_\delta(z)$ has the martingale property when $\gamma = \gamma(\omega)$ makes one step γ_1. In this case $\mathcal{F}_0$ is the trivial σ-algebra, so that we wish to prove

$$\mu_{\beta_c,\Omega}^{a,b}\left[F_{\Omega_\delta^\diamond\setminus[a_\delta\gamma_1],\gamma_1,b_\delta}(z)\right] = F_{\Omega_\delta^\diamond,a_\delta,b_\delta}(z), \tag{6.5}$$

where $\mu_{\beta_c,\Omega}^{a,b}$ is the critical Ising measure with Dobrushin boundary conditions in Ω. Write $Z_{\Omega_\delta^\diamond,a_\delta,b_\delta}$ (resp. $Z_{\Omega^\diamond\setminus[a_\delta x],x,b_\delta}$) for the partition function of the Ising model with Dobrushin boundary conditions on $(\Omega_\delta^\diamond, a_\delta, b_\delta)$ (resp. $(\Omega^\diamond\setminus[a_\delta x], x, b_\delta)$), *i.e.* $Z_{\Omega^\diamond\setminus[a_\delta x],x,b_\delta} = \sum_\omega(\sqrt{2}-1)^{|\omega|}$. Note that $Z_{\Omega^\diamond\setminus[a_\delta x],x,b_\delta}$ is almost the denominator of

$F_{\Omega_\delta^\diamond\setminus[a_\delta x],x,b_\delta}(z_\delta)$. By definition,

$$\begin{aligned}Z_{\Omega_\delta^\diamond,a_\delta,b_\delta}\,\mu^{a,b}_{\beta_c,\Omega}(\gamma_1=x) &= (\sqrt{2}-1)Z_{\Omega^\diamond\setminus[a_\delta x],x,b_\delta}\\ &= (\sqrt{2}-1)\mathrm{e}^{i\frac{1}{2}W_\gamma(x,b_\delta)}\frac{\sum_{\omega\in\mathcal{E}_{\Omega^\diamond\setminus[a_\delta x]}(x,z_\delta)}\mathrm{e}^{-i\frac{1}{2}W_\gamma(x,z_\delta)}(\sqrt{2}-1)^{|\omega|}}{F_{\Omega_\delta^\diamond\setminus[a_\delta x],x,b_\delta}(z_\delta)}\\ &= \mathrm{e}^{i\frac{1}{2}W_\gamma(a_\delta,b_\delta)}\frac{\sum_{\omega\in\mathcal{E}_{\Omega_\delta^\diamond}(a_\delta,z_\delta)}\mathrm{e}^{-i\frac{1}{2}W_\gamma(a_\delta,z_\delta)}(\sqrt{2}-1)^{|\omega|}1_{\{\gamma_1=x\}}}{F_{\Omega_\delta^\diamond\setminus[a_\delta x],x,b_\delta}(z_\delta)}\end{aligned}$$

In the second equality, we used the fact that $\mathcal{E}_{\Omega_\delta^\diamond\setminus[a_\delta x]}(x,z_\delta)$ is in bijection with configurations of $\mathcal{E}_{\Omega_\delta^\diamond}(a_\delta,z_\delta)$ such that $\gamma_1=x$ (there is still a difference of weight of $\sqrt{2}-1$ between two associated configurations). This gives

$$\mu^{a,b}_{\beta_c,\Omega}(\gamma_1=x)\,F_{\Omega_\delta^\diamond\setminus[a_\delta x],x,b_\delta}(z_\delta)=\frac{\sum_{\omega\in\mathcal{E}(a_\delta,z_\delta)}\mathrm{e}^{-i\frac{1}{2}W_\gamma(a_\delta,z_\delta)}(\sqrt{2}-1)^{|\omega|}1_{\{\gamma_1=x\}}}{\mathrm{e}^{-i\frac{1}{2}W_\gamma(a_\delta,b_\delta)}Z_{\Omega_\delta^\diamond,a_\delta,b_\delta}}.$$

The same holds for all possible first steps. Summing over all possibilities, we obtain the expectation on one side of the equality and $F_{\Omega_\delta^\diamond,a_\delta,b_\delta}(z_\delta)$ on the other side, thus proving (6.5). □

EXERCISE 6.9. *Prove that spin Ising interfaces converge to* SLE(3). *For tightness and the fact that sub-sequential limits are Loewner chains, it is sufficient to check Condition* ($\star$). *To do so, try to use Theorem 3.16 and the Edwards-Sokal coupling to prove an intermediate result similar to Lemma 6.3.*

7. Other results on the Ising and FK-Ising models

7.1. Massive harmonicity away from criticality. In this subsection, we consider the fermionic observable F for the FK-Ising model away from criticality. The Ising model is still solvable and the observable becomes massive harmonic (*i.e.* $\Delta f=\lambda^2 f$). We refer to [**BDC11**] for details on this paragraph. We start with a lemma which extends Lemma 5.2 to $p\neq p_{sd}=\sqrt{2}/(1+\sqrt{2})$.

LEMMA 7.1. *Let* $p\in(0,1)$. *Consider a vertex* $v\in\Omega^\diamond\setminus\partial\Omega^\diamond$,

$$F(A)-F(C) = i\mathrm{e}^{i\alpha}\left[F(B)-F(D)\right] \tag{7.1}$$

where A is an adjacent (to v) medial edge pointing towards v and B, C and D are indexed in such a way that A, B, C and D are found in counterclockwise order. The parameter α is defined by

$$\mathrm{e}^{i\alpha}=\frac{\mathrm{e}^{-i\pi/4}(1-p)\sqrt{2}+p}{\mathrm{e}^{-i\pi/4}p+(1-p)\sqrt{2}}.$$

The proof of this statement follows along the same lines as the proof of Lemma 5.2.

PROPOSITION 7.2. *For $p<\sqrt{2}/(1+\sqrt{2})$, there exists $\xi=\xi(p)>0$ such that for every n,*

$$\phi_p(0\leftrightarrow in) \le \mathrm{e}^{-\xi n}, \tag{7.2}$$

where the mesh size of the lattice $\mathbb{L}$ is 1.

In this proof, the lattices are rotated by an angle $\pi/4$. We will be able to estimate the connectivity probabilities using the FK fermionic observable. Indeed, the observable on the free boundary is related to the probability that sites are connected to the wired arc. More precisely:

LEMMA 7.3. *Fix (G,a,b) a Dobrushin domain and $p \in (0,1)$. Let $u \in G$ be a site on the free arc, and e be a side of the black diamond associated to u which borders a white diamond of the free arc. Then,*

$$|F(e)| = \phi_{p,G}^{a,b}(u \leftrightarrow \text{wired arc}). \tag{7.3}$$

Proof. Let u be a site of the free arc and recall that the exploration path is the interface between the open cluster connected to the wired arc and the dual open cluster connected to the free arc. Since u belongs to the free arc, u is connected to the wired arc if and only if e is on the exploration path, so that

$$\phi_{p,G}^{a,b}(u \leftrightarrow \text{wired arc}) = \phi_{p,G}^{a,b}(e \in \gamma).$$

The edge e being on the boundary, the exploration path cannot wind around it, so that the winding (denoted W_1) of the curve is deterministic (and easy to write in terms of that of the boundary itself). We deduce from this remark that

$$\begin{aligned}|F(e)| = |\phi_{p,G}^{a,b}(\mathrm{e}^{\frac{\mathrm{i}}{2}\mathrm{W}_1}1_{e\in\gamma})| &= |\mathrm{e}^{\frac{\mathrm{i}}{2}\mathrm{W}_1}\phi_{p,G}^{a,b}(e\in\gamma)| \\ &= \phi_{p,G}^{a,b}(e\in\gamma) = \phi_{p,G}^{a,b}(u \leftrightarrow \text{wired arc}).\end{aligned}$$

□

We are now in a position to prove Proposition 7.2. We first prove exponential decay in a strip with Dobrushin boundary conditions, using the observable. Then, we use classical arguments of FK percolation to deduce exponential decay in the bulk. We present the proof quickly (see **[BDC11]** for a complete proof).

Proof. Let $p < p_{sd}$, and consider the FK-Ising model of parameter p in the strip of height ℓ, with free boundary conditions on the top and wired boundary conditions on the bottom (the measure is denoted by $\phi_{p,\mathcal{S}_\ell}^{\infty,-\infty}$). It is easy to check that one can define the FK fermionic observable F in this case, by using the unique interface from $-\infty$ to ∞. This observable is the limit of finite volume observables, therefore it also satisfies Lemma 7.1.

Let e_k be the medial edge with center $ik + \frac{1+i}{\sqrt{2}}$, see Fig. 16. A simple computation using Lemmata 7.1 and 5.1 plus symmetries of the strip (via translation and horizontal reflection) implies

$$F(e_{k+1}) = \frac{[1+\cos(\pi/4-\alpha)]\cos(\pi/4-\alpha)}{[1+\cos(\pi/4+\alpha)]\cos(\pi/4+\alpha)}F(e_k). \tag{7.4}$$

Using the previous equality inductively, we find for every $\ell > 0$,

$$|F(e_\ell)| = \mathrm{e}^{-\xi\ell}|F(e_0)| \le \mathrm{e}^{-\xi\ell}$$

with

$$\xi := -\ln\frac{[1+\cos(\pi/4-\alpha)]\cos(\pi/4-\alpha)}{[1+\cos(\pi/4+\alpha)]\cos(\pi/4+\alpha)}. \tag{7.5}$$

Since e_ℓ is adjacent to the free arc, Lemma 7.3 implies

$$\phi_{p,\mathcal{S}_\ell}^{\infty,-\infty}[i\ell \leftrightarrow \mathbb{Z}] = |F(e_\ell)| \le \mathrm{e}^{-\xi\ell}$$

Now, let $N \in \mathbb{N}$ and recall that $\phi_{p,N}^0 := \phi_{p,2,[-N,N]^2}^0$ converges to the infinite-volume measure with free boundary conditions ϕ_p^0 when N goes to infinity.

Consider a configuration in the box $[-N,N]^2$, and let $A_{\max}$ be the site of the cluster of the origin which maximizes the ℓ^∞-norm $\max\{|x_1|,|x_2|\}$ (it could be equal to N). If there is more than one such site, we consider the greatest one in lexicographical order. Assume that $A_{\max}$ equals $a = a_1 + ia_2$ with $a_2 \ge |a_1|$ (the other cases can be treated the same way by symmetry, using the rotational invariance of the lattice).

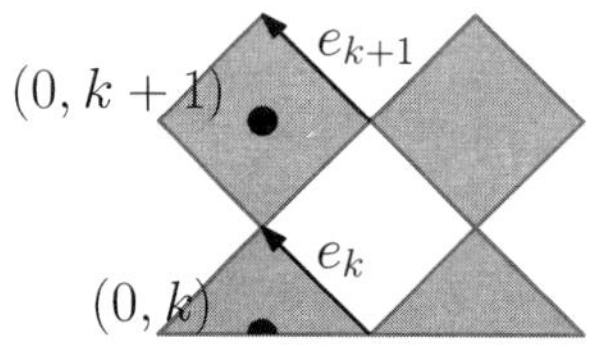

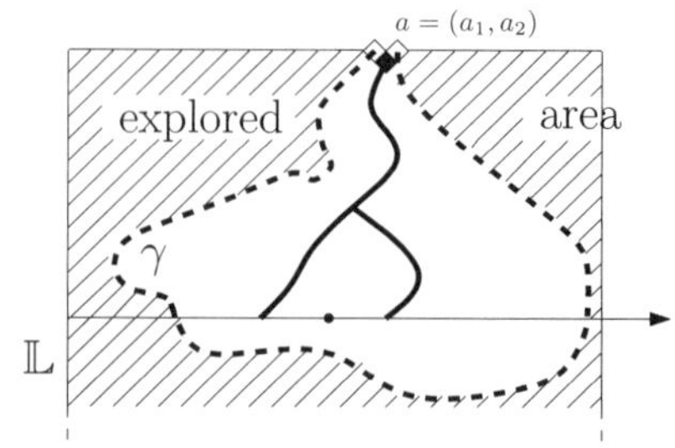

FIGURE 16. **Left:** Edges e_k and e_{k+1}. **Right:** A dual circuit surrounding an open path in the box $[-a_2, a_2]^2$. Conditioning on to the most exterior such circuit gives no information on the state of the edges inside it.

By definition, if $A_{\max}$ equals a, a is connected to 0 in $[-a_2, a_2]^2$. In addition to this, because of our choice of the free boundary conditions, there exists a dual circuit starting from $a + i/2$ in the dual of $[-a_2, a_2]^2$ (which is the same as $\mathbb{L}^* \cap [-a_2 - 1/2, a_2 + 1/2]^2$) and surrounding both a and 0. Let Γ be the outermost such dual circuit: we get

$$\phi^0_{p,N}(A_{\max} = a) = \sum_\gamma \phi^0_{p,N}(a \leftrightarrow 0 | \Gamma = \gamma)\phi^0_{p,N}(\Gamma = \gamma), \tag{7.6}$$

where the sum is over contours γ in the dual of $[-a_2, a_2]^2$ that surround both a and 0.

The event $\{\Gamma = \gamma\}$ is measurable in terms of edges outside or on γ. In addition, conditioning on this event implies that the edges of γ are dual-open. Therefore, from the domain Markov property, the conditional distribution of the configuration inside γ is a FK percolation model with free boundary conditions. Comparison between boundary conditions implies that the probability of $\{a \leftrightarrow 0\}$ conditionally on $\{\Gamma = \gamma\}$ is smaller than the probability of $\{a \leftrightarrow 0\}$ in the strip $\mathcal{S}_{a_2}$ with free boundary conditions on the top and wired boundary conditions on the bottom. Hence, for any such γ, we get

$$\phi^0_{p,N}(a \leftrightarrow 0 | \Gamma = \gamma) \le \phi^{\infty,-\infty}_{p,\mathcal{S}_{a_2}}(a \leftrightarrow 0) = \phi^{\infty,-\infty}_{p,\mathcal{S}_{a_2}}(a \leftrightarrow \mathbb{Z}) \le e^{-\xi a_2}$$

(observe that for the second measure, $\mathbb{Z}$ is wired, so that $\{a \leftrightarrow 0\}$ and $\{a \leftrightarrow \mathbb{Z}\}$ have the same probability). Plugging this into (7.6), we obtain

$$\phi^0_{p,N}(A_{\max} = a) \le \sum_\gamma e^{-\xi \max\{a_1,a_2\}} \phi^0_{p,N}(\Gamma = \gamma) \le e^{-\xi a_2} = e^{-\xi \max\{a_1,a_2\}}.$$

Fix $n \le N$. We deduce from the previous inequality that there exists a constant $0 < c < \infty$ such that

$$\phi^0_{p,N}(0 \leftrightarrow \mathbb{Z}^2 \smallsetminus [-n,n]^2) \le \sum_{a \in [-N,N]^2 \smallsetminus [-n,n]^2} \phi^0_{p,N}(A_{\max} = a) \le cne^{-\xi n}.$$

Since the estimate is uniform in N, we deduce that

$$\phi^0_p(0 \leftrightarrow in) \le \phi^0_p(0 \leftrightarrow \mathbb{Z}^2 \smallsetminus [-n,n]^2) \le cne^{-\xi n}. \tag{7.7}$$

□

THEOREM 7.4. *The critical parameter for the* FK-*Ising model is* $\sqrt{2}/(1+\sqrt{2})$. *The critical inverse-temperature for the Ising model is* $\frac{1}{2}\ln(1+\sqrt{2})$.

Proof. The inequality $p_c \geq \sqrt{2}/(1+\sqrt{2})$ follows from Proposition 7.2 since there is no infinite cluster for $\phi^0_{p,2}$ when $p < p_{sd}$ (the probability that 0 and in are connected converges to 0). In order to prove that $p_c \leq \sqrt{2}/(1+\sqrt{2})$, we harness the following standard reasoning.

Let A_n be the event that the point $n \in \mathbb{N}$ is in an open circuit which surrounds the origin. Notice that this event is included in the event that the point $n \in \mathbb{N}$ is in a cluster of radius larger than n. For $p < \sqrt{2}/(1+\sqrt{2})$, a modification of (7.7) implies that the probability of A_n decays exponentially fast. The Borel-Cantelli lemma shows that there is almost surely a finite number of n such that A_n occurs. In other words, there is a.s. only a finite number of open circuits surrounding the origin, which enforces the existence of an infinite dual cluster whenever $p < \sqrt{2}/(1+\sqrt{2})$. Using duality, the primal model is supercritical whenever $p > \sqrt{2}/(1+\sqrt{2})$, which implies $p_c \leq \sqrt{2}/(1+\sqrt{2})$. □

In fact, the FK fermionic observable F_δ in a Dobrushin domain $(\Omega_\delta^\diamond, a_\delta, b_\delta)$ is massive harmonic when $p \neq p_{sd}$. More precisely,

PROPOSITION 7.5. *Let $p \neq p_{sd}$,*

$$\Delta_\delta F_\delta(v) = (\cos 2\alpha - 1) F_\delta(v) \tag{7.8}$$

for every $v \in \Omega_\delta^\diamond \setminus \partial\Omega_\delta^\diamond$, where Δ_δ is the average on sites at distance $\sqrt{2}\delta$ minus the value at the point.

When δ goes to 0, one can perform two scaling limits. If $p = p_{sd}(1-\lambda\delta)$ goes to p_{sd} as δ goes to 0, $\frac{1}{\delta^2}(\Delta_\delta + [1-\cos 2\alpha]I)$ converges to $\Delta + \lambda^2 I$. Then F_δ (properly normalized) should converge to a function f satisfying $\Delta f + \lambda^2 f = 0$ inside the domain. Except for $\lambda = 0$, the limit will not be holomorphic, but massive harmonic. Discrete curves should converge to a limit which is not absolutely continuous with respect to SLE(16/3). The study of this regime, connected to massive SLEs, is a very interesting subject.

If we fix $p < p_{sd}$, one can interpret massive harmonicity in terms of killed random walks. Roughly speaking, $F_\delta(v)$ is the probability that a killed random walk starting at v visits the wired arc ∂_{ba}. Large deviation estimates on random walks allow to compute the asymptotic of F_δ inside the domain. In [**BDC11**], a surprising link (first noticed by Messikh [**Mes06**]) between correlation lengths of the Ising model and large deviations estimates of random walks is presented. We state the result in the following theorem:

THEOREM 7.6. *Fix $\beta < \beta_c$ (and α associated to it) and set*

$$m(\beta) := \cos(2\alpha).$$

For any $x \in \mathbb{L}$,

$$-\lim_{n\to\infty} \frac{1}{n} \ln \mu_\beta[\sigma_0\sigma(nx)] = -\lim_{n\to\infty} \frac{1}{n} \ln G_{m(\beta)}(0, nx). \tag{7.9}$$

Above, $G_m(0,x) := \mathbb{E}^x[m^\tau]$ for any $x \in \mathbb{L}$ and $m < 1$, where τ is the hitting time of the origin and $\mathbb{P}^x$ is the law of a simple random walk starting at x.

The massive Green function $G_m(0,x)$ on the right of (7.9) has been widely studied. In particular, we can compute the rate of decay in any direction and deduce Theorem 2.3 and Theorem 2.4 (see *e.g.* [**Mes06**]).

EXERCISE 7.7. *Prove Lemma 7.1 and the fact that F is massive harmonic inside the domain.*

7.2. Russo-Seymour-Welsh Theorem for FK-Ising. In this section, we sketch the proof of Theorem 3.16; see **[DCHN10]** for details. This theorem was improved in **[CDH12]**. We would like to emphasize that this result does not make use of scaling limits. Therefore, it is mostly independent of Sections 5 and 6.

We start by presenting a link between discrete harmonic measures and the probability for a point on the free arc ∂_{ab} of a FK Dobrushin domain to be connected to the wired arc ∂_{ba}.

Let us first define a notion of discrete harmonic measure in a FK Dobrushin domain Ω_δ which is slightly different from the usual one. First extend $\Omega_\delta \cup \Omega_\delta^\star$ by adding two extra layers of vertices: one layer of white faces adjacent to $\partial_{ab}^\star$, and one layer of black faces adjacent to ∂_{ba}. We denote the extended domains by $\tilde{\Omega}_\delta$ and $\tilde{\Omega}_\delta^\star$.

Define $(X_t^\bullet)_{t\geq 0}$ to be the continuous-time random walk on the black faces that jumps with rate 1 on neighbors, *except* for the faces on the extra layer adjacent to ∂_{ab} onto which it jumps with rate $\rho := 2/(\sqrt{2}+1)$. For $B \in \Omega_\delta$, we denote by $\tilde{H}^\bullet(B)$ the probability that the random walk $X_t^\bullet$ starting at B hits $\partial\tilde{\Omega}_\delta$ on the wired arc ∂_{ba} (in other words, if the random walk hits ∂_{ba} before hitting the extra layer adjacent to $\partial_{ab}^\star$). This quantity is called the (modified) harmonic measure of ∂_{ba} seen from B. Similarly, one can define a modified random walk X_t° and the associated harmonic measure of $\partial_{ab}^\star$ seen from w. We denote it by $H^\circ(w)$.

Proposition 7.8. *Consider a* FK *Dobrushin domain* $(\Omega_\delta, a_\delta, b_\delta)$, *for any site* B *on the free arc* ∂_{ab},

$$\sqrt{\tilde{H}^\circ(W)} \leq \phi_{\Omega_\delta,p_{sd}}^{a_\delta,b_\delta}[B \leftrightarrow \partial_{ba}] \leq \sqrt{\tilde{H}^\bullet(B)}, \tag{7.10}$$

where W *is any dual neighbor of* B not on $\partial_{ab}^\star$.

This proposition raises a connection between harmonic measure and connectivity properties of the FK-Ising model. To study connectivity probabilities for the FK-Ising model, it suffices to estimate events for simple random walks (slightly repelled on the boundary). The proof makes use of a variant of the "boundary modification trick". This trick was introduced in **[CS09]** to prove Theorem 2.11. It can be summarized as follows: one can extend the function H by 0 or 1 on the two extra layers, then $H^\bullet$ (resp. H°) is subharmonic (resp. superharmonic) for the Laplacian associated to the random walk $X^\bullet$ (resp. X°). Interestingly, $H^\bullet$ is not subharmonic for the usual Laplacian. This trick allows us to fix the boundary conditions (0 or 1), at the cost of a slightly modified notion of harmonicity.

We can now give the idea of the proof of Theorem 3.16 (we refer to **[DCHN10]** for details). The proof is a second moment estimate on the number of pairs of connected sites on opposite edges of the rectangle. We mention that another road to Theorem 3.16 has been proposed in **[KS10]**.

Proof of Theorem 3.16 (Sketch). Let $R_n = [0,4n]\times[0,n]$ be a rectangle and let N be the number of pairs (x,y), $x \in \{0\}\times[0,n]$ and $y \in \{4n\}\times[0,n]$ such that x is connected to y by an open path. The expectation of N is easy to obtain using the previous proposition. Indeed, it is the sum over all pairs x,y of the probability of $\{x \leftrightarrow y\}$ when the boundary conditions are free. Free boundary conditions can be thought of as a degenerate case of a Dobrushin domain, where $a = b = y$. In other words, we want to estimate the probability that x is connected to the wired arc $\partial_{ba} = \{y\}$. Except when x and y are close to the corners, the harmonic measure of y seen from x is of order $1/n^2$, so that the probability of $\{x \leftrightarrow y\}$ is of order $1/n$. Therefore, there exists a universal constant $c > 0$ such that $\phi_{p_{sd},R_n}^f[N] \geq cn$.

The second moment estimate is harder to obtain, as usual. Nevertheless, it can be proved, using successive conditioning and Proposition 7.8, that $\phi_{p_{sd},R_n}^f[N^2] \leq Cn^2$ for some universal $C > 0$; see **[DCHN10]** for a complete proof. Using the Cauchy-Schwarz inequality, we find

$$\phi^f_{p_{sd},R_n}[N > 0]\, \phi^f_{p_{sd},R_n}[N^2] \ \geq\ \phi^f_{p_{sd},R_n}[N]^2 \tag{7.11}$$

which implies

$$\phi^f_{R_n,p_{sd}}[\exists \text{ open crossing}] \ =\ \phi^f_{R_n,p_{sd}}[N > 0] \ \geq\ c^2/C \tag{7.12}$$

uniformly in n. □

We have already seen that Theorem 3.16 is central for proving tightness of interfaces. We would also like to mention an elementary consequence of Theorem 3.16.

PROPOSITION 7.9. *There exist constants $0 < c, C, \delta, \Delta < \infty$ such that for any sites $x, y \in \mathbb{L}$,*

$$\frac{c}{|x-y|^\delta} \ \leq\ \mu_{\beta_c}[\sigma_x\sigma_y] \ \leq\ \frac{C}{|x-y|^\Delta} \tag{7.13}$$

where μ_{β_c} is the unique infinite-volume measure at criticality.

Proof. Using the Edwards-Sokal coupling, (7.13) can be rephrased as

$$\frac{c}{|x-y|^\delta} \ \leq\ \phi_{p_{sd},2}[x \leftrightarrow y] \ \leq\ \frac{C}{|x-y|^\Delta},$$

where $\phi_{p_{sd},2}$ is the unique FK-Ising infinite-volume measure at criticality. In order to get the upper bound, it suffices to prove that $\phi_{p_{sd},2}(0 \leftrightarrow \partial\Lambda_k)$ decays polynomially fast, where Λ_k is the box of size $k = |x-y|$ centered at x. We consider the annuli $A_n = S_{2^{n-1},2^n}(x)$ for $n \leq \ln_2 k$, and $\mathcal{E}(A_n)$ the event that there is an open path crossing A_n from the inner to the outer boundary. We know from Corollary 6.3 (which is a direct application of Theorem 3.16) that there exists a constant $c < 1$ such that

$$\phi^1_{A_n,p_{sd},2}(\mathcal{E}(A_n)) \leq c$$

for all $n \geq 1$. By successive conditionings, we then obtain

$$\phi_{p_{sd},2}(0 \leftrightarrow \partial\Lambda_k) \ \leq\ \prod_{n=1}^{\ln_2 k} \phi^1_{A_n,p_{sd},2}(\mathcal{E}(A_n)) \leq c^N,$$

and the desired result follows. The lower bound can be done following the same kind of arguments (we leave it as an exercise). □

Therefore, the behavior at criticality (power law decay of correlations) is very different from the subcritical phase (exponential decay of correlations). Actually, the previous result is far from optimal. One can compute correlations between spins of a domain very explicitly. In particular, $\mu_{\beta_c}[\sigma_x\sigma_y]$ behaves like $|x-y|^{-\alpha}$, where $\alpha = 1/4$. We mention that α is one example of *critical exponent*. Even though we did not discuss how compute critical exponents, we mention that the technology developed in these notes has for its main purpose their computation.

To conclude this section, we mention that Theorem 3.16 leads to ratio mixing properties (see Exercise 7.10) of the Ising model. Recently, Lubetzky and Sly **[LS10]** used these spatial mixing properties in order to prove an important conjecture on the mixing time of the Glauber dynamics of the Ising model at criticality.

EXERCISE 7.10 (Spatial mixing). *Prove that there exist $c, \Delta > 0$ such that for any $r \leq R$,*

$$\left|\phi_{p_{sd},2}(A \cap B) - \phi_{p_{sd},2}(A)\phi_{p_{sd},2}(B)\right| \leq c\left(\frac{r}{R}\right)^\Delta \phi_{p_{sd},2}(A)\phi_{p_{sd},2}(B) \tag{7.14}$$

for any event A (resp. B) depending only on the edges in the box $[-r,r]^2$ (resp. outside $[-R,R]^2$).

7.3. Discrete singularities and energy density of the Ising model. In this subsection, we would like to emphasize the fact that slight modifications of the spin fermionic observable can be used directly to compute interesting quantities of the model. Now, we briefly present the example of the energy density between two neighboring sites x and y (Theorem 2.12).

So far, we considered observables depending on a point a on the boundary of a domain, but we could allow more flexibility and move a inside the domain: for $a_\delta \in \Omega_\delta^\diamond$, we define the fermionic observable $F_{\Omega_\delta}^{a_\delta}(z_\delta)$ for $z_\delta \neq a_\delta$ by

$$F_{\Omega_\delta}^{a_\delta}(z_\delta) = \lambda \frac{\sum_{\omega\in\mathcal{E}(a_\delta,z_\delta)} e^{-\frac{1}{2}iW_{\gamma(\omega)}(a_\delta,z_\delta)}(\sqrt{2}-1)^{|\omega|}}{\sum_{\omega\in\mathcal{E}}(\sqrt{2}-1)^{|\omega|}} \tag{7.15}$$

where λ is a well-chosen explicit complex number. Note that the denominator of the observable is simply the partition function for free boundary conditions $Z_{\beta_c,G}^f$. Actually, using the high-temperature expansion of the Ising model and local rearrangements, the observable can be related to spin correlations **[HS10]**:

LEMMA 7.11. *Let $[xy]$ be an horizontal edge of Ω_δ. Then*

$$\lambda\,\mu_{\beta_c,\Omega_\delta}^f[\sigma_x\sigma_y] = P_{\ell(ac)}[F_{\Omega_\delta}^a(c)] + P_{\ell(ad)}[F_{\Omega_\delta}^a(d)]$$

where a is the center of $[xy]$, $c = a + \delta\frac{1+i}{\sqrt{2}}$ and $d = a - \delta\frac{1+i}{\sqrt{2}}$.

If λ is chosen carefully, the function $F_\delta^{a_\delta}$ is s-holomorphic on $\Omega_\delta \setminus \{a_\delta\}$. Moreover, its complex argument is fixed on the boundary of the domain. Yet, the function is not s-holomorphic at a_δ (meaning that there is no way of defining $F_{\Omega_\delta}^a(a_\delta)$ so that the function is s-holomorphic at a_δ). In other words, there is a *discrete singularity* at a, whose behavior is related to the spin-correlation.

We briefly explain how one can address the problem of discrete singularities, and we refer to **[HS10]** for a complete study of this case. In the continuum, singularities are removed by subtracting Green functions. In the discrete context, we will do the same. We thus need to construct a discrete s-holomorphic Green function. Preholomorphic Green functions[6] were already constructed in **[Ken00]**. These functions are not s-holomorphic but relevant linear combinations of them are, see **[HS10]**. We mention that the s-holomorphic Green functions are very explicit and their convergence when the mesh size goes to 0 can be studied.

Proof of Theorem 2.12 (Sketch). The function $F_\delta^{a_\delta}/\delta$ converges uniformly on any compact subset of $\Omega \setminus \{a\}$. This fact is not helpful, since the interesting values of $F_\delta^{a_\delta}$ are located at neighbors of the singularity. It can be proved that, subtracting a well-chosen s-holomorphic Green function $g_{\Omega_\delta}^{a_\delta}$, one can erase the singularity at a_δ. More precisely, one can show that $[F_\delta^{a_\delta} - g_{\Omega_\delta}^{a_\delta}]/\delta$ converges uniformly on Ω towards an explicit conformal map. The value of this map at a is $\frac{\lambda}{\pi}\phi_a'(a)$. Now, $\mu_{\beta_c,\Omega_\delta}^f[\sigma_x\sigma_y]$ can be expressed in terms of $F_\delta^{a_\delta}$ for neighboring vertices of a_δ. Moreover, values of $g_{\Omega_\delta}^{a_\delta}$ for neighbors of a_δ can be computed explicitly. Using the fact that

$$F_\delta^{a_\delta} = g_{\Omega_\delta}^{a_\delta} + \delta\cdot\frac{1}{\delta}[F_\delta^{a_\delta} - g_{\Omega_\delta}^{a_\delta}],$$

and Lemma 7.11, the convergence result described above translates into the following asymptotics for the spin correlation of two neighbors

$$\mu_{\beta_c,\Omega_\delta}^f[\sigma_x\sigma_y] = \frac{\sqrt{2}}{2} - \delta\,\frac{1}{\pi}\phi_a'(a) + o(\delta).$$

□

---

[6] *i.e.* satisfying the Cauchy-Riemann equation except at a certain point.

8. Many questions and a few answers

8.1. Universality of the Ising model. Until now, we considered only the square lattice Ising model. Nevertheless, normalization group theory predicts that the scaling limit should be universal. In other words, the limit of critical Ising models on planar graphs should always be the same. In particular, the scaling limit of interfaces in spin Dobrushin domains should converge to SLE(3).

Of course, one should be careful about the way the graph is drawn in the plane. For instance, the isotropic spin Ising model of Section 2, when considered on a stretched square lattice (every square is replaced by a rectangle), is not conformally invariant (it is not invariant under rotations). Isoradial graphs form a large family of graphs possessing a natural embedding on which a critical Ising model is expected to be conformally invariant. More details are now provided about this fact.

DEFINITION 8.1. *A* rhombic embedding *of a graph G is a planar quadrangulation satisfying the following properties:*

- *the vertices of the quadrangulation are the vertices of G and $G^\star$,*
- *the edges connect vertices of G to vertices of $G^\star$ corresponding to adjacent faces of G,*
- *all the edges of the quadrangulation have equal length, see Fig. 10.*

A graph which admits a rhombic embedding is called isoradial.

Isoradial graphs are fundamental for two reasons. First, discrete complex analysis on isoradial graphs was extensively studied (see *e.g.* [**Mer01, Ken02, CS08**]) as explained in Section 4. Second, the Ising model on isoradial graphs satisfies very specific integrability properties and a natural critical point can be defined as follows. Let $J_{xy} = \operatorname{arctanh}[\tan(\theta/2)]$ where θ is the half-angle at the corner x (or equivalently y) made by the rhombus associated to the edge $[xy]$. One can define the *critical Ising model* with Hamiltonian

$$H(\sigma) = -\sum_{x\sim y} J_{xy}\sigma_x\sigma_y.$$

This Ising model on isoradial graphs (with rhombic embedding) is critical and conformally invariant in the following sense:

THEOREM 8.2 (Chelkak, Smirnov [**CS09**]). *The interfaces of the critical Ising model on isoradial graphs converge, as the mesh size goes to 0, to the chordal Schramm-Loewner Evolution with $\kappa = 3$.*

Note that the previous theorem is uniform on any rhombic graph discretizing a given domain (Ω, a, b), as soon as the edge-length of rhombi is small enough. This provides a first step towards universality for the Ising model.

QUESTION 8.3. *Since not every topological quadrangulation admits a rhombic embedding* [**KS05**]*, can another embedding with a sufficiently nice version of discrete complex analysis always be found?*

QUESTION 8.4. *Is there a more general discrete setup where one can get similar estimates, in particular convergence of preholomorphic functions to the holomorphic ones in the scaling limit?*

In another direction, consider a biperiodic lattice $\mathcal{L}$ (one can think of the universal cover of a finite graph on the torus), and define a Hamiltonian with periodic correlations (J_{xy}) by setting $H(\sigma) = -\sum_{x\sim y} J_{xy}\sigma_x\sigma_y$. The Ising model with this Hamiltonian makes perfect sense and there exists a critical inverse temperature separating the disordered phase from the ordered phase.

QUESTION 8.5. *Prove that there always exists an embedding of $\mathcal{L}$ such that the Ising model on $\mathcal{L}$ is conformally invariant.*

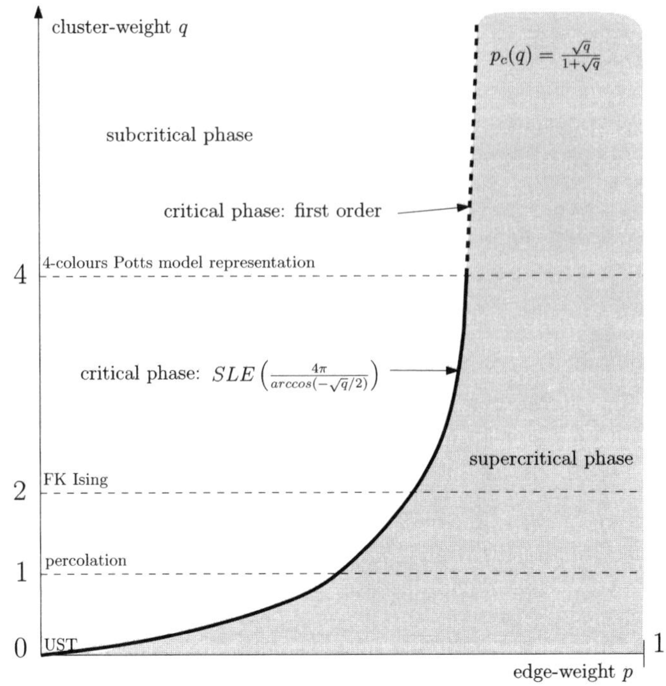

FIGURE 17. The phase diagram of the FK percolation model on the square lattice.

8.2. Full scaling limit of critical Ising model. It has been proved in **[KS10]** that the scaling limit of Ising interfaces in Dobrushin domains is SLE(3). The next question is to understand the full scaling limit of the interfaces. This question raises interesting technical problems. Consider the Ising model with free boundary conditions. Interfaces now form a family of loops. By consistency, each loop should look like a SLE(3). In **[HK11]**, Hongler and Kytolä made one step towards the complete picture by studying interfaces with $+/-/$free boundary conditions.

Sheffield and Werner **[SW10a, SW10b]** introduced a one-parameter family of processes of non-intersecting loops which are conformally invariant – called the Conformal Loop Ensembles CLE(κ) for $\kappa > 8/3$. Not surprisingly, loops of CLE(κ) are locally similar to SLE(κ), and these processes are natural candidates for the scaling limits of planar models of statistical physics. In the case of the Ising model, the limits of interfaces all together should be a CLE(3).

8.3. FK percolation for general cluster-weight $q \geq 0$. The FK percolation with cluster-weight $q \in (0,\infty)$ is conjectured to be critical for $p_c(q) = \sqrt{q}/(1+\sqrt{q})$ (see **[BDC10]** for the case $q \geq 1$). Critical FK percolation is expected to exhibit a very rich phase transition, whose properties depend strongly on the value of q (see Fig. 17). We use generalizations of the FK fermionic observable to predict the critical behavior for general q.

8.3.1. *Case* $0 \leq q \leq 4$. The critical FK percolation in Dobrushin domains can be associated to a loop model exactly like the FK-Ising model: each loop receives a weight $\sqrt{q}$. In this context, one can define a natural generalization of the fermionic observable on medial edges, called a *parafermionic observable*, by the formula

$$F(e) = \mathbb{E}_{\Omega_\delta^\diamond,a_\delta,b_\delta,p,q}\left[e^{\sigma\cdot iW_\gamma(e,b_\delta)}1_{e\in\gamma}\right], \tag{8.1}$$

where $\sigma = \sigma(q)$ is called the *spin* (σ takes a special value described below). Lemma 5.2 has a natural generalization to any $q \in [0, \infty)$:

PROPOSITION 8.6. *For $q \le 4$ and any FK Dobrushin domain, consider the observable F at criticality with spin $\sigma = 1 - \frac{2}{\pi}\arccos(\sqrt{q}/2)$. For any medial vertex inside the domain,*

$$F(N) - F(S) = i[F(E) - F(W)] \tag{8.2}$$

where N, E, S and W are the four medial edges adjacent to the vertex.

These relations can be understood as Cauchy-Riemann equations around some vertices. Importantly, F is not determined by these relations for general q (the number of variables exceeds the number of equations). For $q = 2$, which corresponds to $\sigma = 1/2$, the complex argument modulo π of the observable offers additional relations (Lemma 5.1) and it is then possible to obtain the preholomophicity (Proposition 5.3).

Parafermionic observables can be defined on medial vertices by the formula

$$F(v) = \frac{1}{2}\sum_{e\sim v} F(e)$$

where the summation is over medial edges with v as an endpoint. Even though they are only weakly-holomorphic, one still expects them to converge to a holomorphic function. The natural candidate for the limit is not hard to find:

CONJECTURE 8.7. *Let $q \le 4$ and (Ω, a, b) be a simply connected domain with two points on its boundary. For every $z \in \Omega$,*

$$\frac{1}{(2\delta)^{\sigma}} F_\delta(z) \to \phi'(z)^{\sigma} \quad \text{when } \delta \to 0 \tag{8.3}$$

where $\sigma = 1 - \frac{2}{\pi}\arccos(\sqrt{q}/2)$, F_δ is the observable (at $p_c(q)$) in discrete domains with spin σ, and ϕ is any conformal map from Ω to $\mathbb{R} \times (0,1)$ sending a to $-\infty$ and b to ∞.

Being mainly interested in the convergence of interfaces, one could try to follow the same program as in Section 6:

- Prove compactness of the interfaces.
- Show that sub-sequential limits are Loewner chains (with unknown random driving process W_t).
- Prove the convergence of discrete observables (more precisely martingales) of the model.
- Extract from the limit of these observables enough information to evaluate the conditional expectation and quadratic variation of increments of W_t (in order to harness the Lévy theorem). This would imply that W_t is the Brownian motion with a particular speed κ and so curves converge to SLE(κ).

The third step, corresponding to Conjecture 8.7, should be the most difficult. Note that the first two steps are also open for $q \ne 0, 1, 2$. Even though the convergence of observables is still unproved, one can perform a computation similar to the proof of Proposition 6.7 in order to identify the possible limiting curves (this is the fourth step). The following conjecture is thus obtained:

CONJECTURE 8.8. *For $q \le 4$, the law of critical FK interfaces converges to the Schramm-Loewner Evolution with parameter $\kappa = 4\pi/\arccos(-\sqrt{q}/2)$.*

The conjecture was proved by Lawler, Schramm and Werner [**LSW04a**] for $q = 0$, when they showed that the perimeter curve of the uniform spanning tree converges to SLE(8). Note that the loop representation with Dobrushin boundary conditions still makes sense for $q = 0$ (more precisely for the model obtained by letting $q \to 0$ and $p/q \to 0$). In fact, configurations have no loops, just a curve running from a to b (which then necessarily passes through all the edges), with all configurations being equally probable. The $q = 2$ case corresponds to Theorem 3.13. All other cases are wide open. The $q = 1$ case is particularly interesting, since it is actually bond percolation on the square lattice.

8.3.2. *Case* $q > 4$. The picture is very different and no conformal invariance is expected to hold. The phase transition is conjectured to be of first order : there are multiple infinite-volume measures at criticality. In particular, the critical FK percolation with wired boundary conditions should possess an infinite cluster almost surely while the critical FK percolation with free boundary conditions should not (in this case, the connectivity probabilities should even decay exponentially fast). This result is known only for $q \geq 25.72$ (see [**Gri06**] and references therein).

The observable still makes sense in the $q > 4$ case, providing σ is chosen so that $2\sin(\pi\sigma/2) = \sqrt{q}$. Interestingly, σ becomes purely imaginary in this case. A natural question is to relate this change of behavior for σ with the transition between conformally invariant critical behavior and first order critical behavior. Let us mention that the observable was used to compute the critical point on isoradial graphs for $q \geq 4$ [**BDS12**] and to show that the phase transition is second order for $1 \leq q \leq 4$ [**Dum12**].

8.4. $O(n)$ models on the hexagonal lattice. The Ising fermionic observable was introduced in [**Smi06**] in the setting of general $O(n)$ models on the hexagonal lattice. This model, introduced in [**DMNS81**] on the hexagonal lattice, is a lattice gas of non-intersecting loops. More precisely, consider configurations of non-intersecting simple loops on a finite subgraph of the hexagonal lattice and introduce two parameters: a loop-weight $n \geq 0$ (in fact $n \geq -2$) and an edge-weight $x > 0$, and ask the probability of a configuration to be proportional to $n^{\#\text{ loops}} x^{\#\text{ edges}}$.

Alternatively, an interface between two boundary points could be added: in this case configurations are composed of non-intersecting simple loops and one self-avoiding interface (avoiding all the loops) from a to b.

The $O(0)$ model is the self-avoiding walk, since no loop is allowed (there is still a self-avoiding path from a to b). The $O(1)$ model is the high-temperature expansion of the Ising model on the hexagonal lattice. For integers n, the $O(n)$-model is an approximation of the high-temperature expansion of spin $O(n)$-models (models for which spins are n-dimensional unit vectors).

The physicist Bernard Nienhuis [**Nie82, Nie84**] conjectured that $O(n)$-models in the range $n \in (0,2)$ (after certain modifications $n \in (-2,2)$ would work) exhibit a Berezinsky-Kosterlitz-Thouless phase transition [**Ber72, KT73**]:

CONJECTURE 8.9. *Let $x_c(n) = 1/\sqrt{2+\sqrt{2-n}}$. For $x < x_c(n)$ (resp. $x \geq x_c(n)$) the probability that two points are on the same loop decays exponentially fast (as a power law).*

The conjecture was rigorously established for two cases only. When $n = 1$, the critical value is related to the critical temperature of the Ising model. When $n = 0$, it was recently proved in [**DCS10**] that $\sqrt{2+\sqrt{2}}$ is the connective constant of the hexagonal lattice.

It turns out that the model exhibits one critical behavior at $x_c(n)$ and another on the interval $(x_c(n), +\infty)$, corresponding to dilute and dense phases (when in the limit the loops are simple and non-simple respectively), see Fig. 18. In addition to this, the two critical regimes are expected to be conformally invariant.

Exactly as in the case of FK percolation, the definition of the spin fermionic observable can be extended. For a discrete domain Ω with two points on the boundary a and

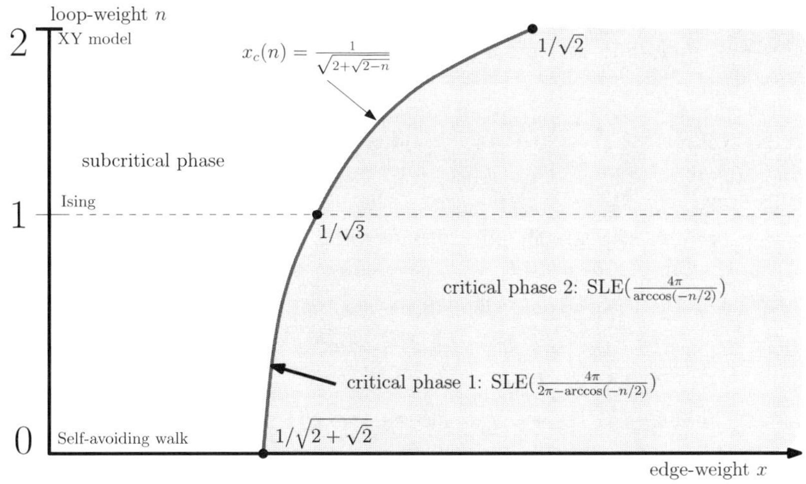

FIGURE 18. The phase diagram of the $O(n)$ model on the hexagonal lattice.

b, the *parafermionic observable* is defined on middle of edges by

$$F(z) = \frac{\sum_{\omega\in\mathcal{E}(a,z)} e^{-\sigma i W_\gamma(a,z)} x^{\#\text{ edges in }\omega} n^{\#\text{ loops in }\omega}}{\sum_{\omega\in\mathcal{E}(a,b)} e^{-\sigma i W_\gamma(a,b)} x^{\#\text{ edges in }\omega} n^{\#\text{ loops in }\omega}} \tag{8.4}$$

where $\mathcal{E}(a,z)$ is the set of configurations of loops with one interface from a to z. One can easily prove that the observable satisfies local relations at the (conjectured) critical value if σ is chosen carefully.

PROPOSITION 8.10. *If $x = x_c(n) = 1/\sqrt{2+\sqrt{2-n}}$, let F be the parafermionic observable with spin $\sigma = \sigma(n) = 1 - \frac{3}{4\pi}\arccos(-n/2)$; then*

$$(p-v)F(p) + (q-v)F(q) + (r-v)F(r) = 0 \tag{8.5}$$

where p, q and r are the three mid-edges adjacent to a vertex v.

This relation can be seen as a discrete version of the Cauchy-Riemann equation on the triangular lattice. Once again, the relations do not determine the observable for general n. Nonetheless, if the family of observables is precompact, then the limit should be holomorphic and it is natural to conjecture the following:

CONJECTURE 8.11. *Let $n \in [0,2]$ and (Ω, a, b) be a simply connected domain with two points on the boundary. For $x = x_c(n)$,*

$$F_\delta(z) \to \left(\frac{\psi'(z)}{\psi'(b)}\right)^\sigma \tag{8.6}$$

where $\sigma = 1 - \frac{3}{4\pi}\arccos(-n/2)$, F_δ is the observable in the discrete domain with spin σ and ψ is any conformal map from Ω to the upper half-plane sending a to ∞ and b to 0.

A conjecture on the scaling limit for the interface from a to b in the $O(n)$ model can also be deduced from these considerations:

CONJECTURE 8.12. *For $n \in [0,2)$ and $x_c(n) = 1/\sqrt{2+\sqrt{2-n}}$, as the mesh size goes to zero, the law of $O(n)$ interfaces converges to the chordal Schramm-Loewner Evolution with parameter $\kappa = 4\pi/(2\pi - \arccos(-n/2))$.*

This conjecture is only proved in the case $n = 1$ (Theorem 2.10). The other cases are open. The case $n = 0$ is especially interesting since it corresponds to self-avoiding walks. Proving the conjecture in this case would pave the way to the computation of many quantities, including the mean-square displacement exponent; see **[LSW04b]** for further details on this problem.

The phase $x < x_c(n)$ is subcritical and not conformally invariant (the interface converges to the shortest curve between a and b for the Euclidean distance). The critical phase $x \in (x_c(n), \infty)$ should be conformally invariant, and universality is predicted: the interfaces are expected to converge to the same SLE. The edge-weight $\tilde{x}_c(n) = 1/\sqrt{2-\sqrt{2-n}}$, which appears in Nienhuis's works **[Nie82, Nie84]**, seems to play a specific role in this phase. Interestingly, it is possible to define a parafermionic observable at $\tilde{x}_c(n)$ with a spin $\tilde{\sigma}(n)$ other than $\sigma(n)$:

PROPOSITION 8.13. *If $x = \tilde{x}_c(n)$, let F be the parafermionic observable with spin $\tilde{\sigma} = \tilde{\sigma}(n) = -\frac{1}{2} - \frac{3}{4\pi}\arccos(-n/2)$; then*

$$(p-v)F(p) + (q-v)F(q) + (r-v)F(r) = 0 \tag{8.7}$$

where p, q and r are the three mid-edges adjacent to a vertex v.

A convergence statement corresponding to Conjecture 8.11 for the observable with spin $\tilde{\sigma}$ enables to predict the value of κ for $\tilde{x}_c(n)$, and thus for every $x > x_c(n)$ thanks to universality.

CONJECTURE 8.14. *For $n \in [0,2)$ and $x \in (x_c(n), \infty)$, as the lattice step goes to zero, the law of $O(n)$ interfaces converges to the chordal Schramm-Loewner Evolution with parameter $\kappa = 4\pi/\arccos(-n/2)$.*

The case $n = 1$ corresponds to the subcritical high-temperature expansion of the Ising model on the hexagonal lattice, which also corresponds to the supercritical Ising model on the triangular lattice via Kramers-Wannier duality. The interfaces should converge to SLE(6). In the case $n = 0$, the scaling limit should be SLE(8), which is space-filling. For both cases, a (slightly different) model is known to converge to the corresponding SLE (site percolation on the triangular lattice for SLE(6), and the perimeter curve of the uniform spanning tree for SLE(8)). Yet, the known proofs do not extend to this context. Proving that the whole critical phase $(x_c(n), \infty)$ has the same scaling limit would be an important example of universality (not on the graph, but on the parameter this time).

The two previous sections presented a program to prove convergence of discrete curves towards the Schramm-Loewner Evolution. It was based on discrete martingales converging to continuous SLE martingales. One can study directly SLE martingales (*i.e.* with respect to $\sigma(\gamma[0,t])$). In particular, $g_t'(z)^\alpha[g_t(z)-W_t]^\beta$ is a martingale for SLE(κ) where $\kappa = 4(\alpha-\beta)/[\beta(\beta-1)]$. All the limits in these notes are of the previous forms, see *e.g.* Proposition 6.7. Therefore, the parafermionic observables are discretizations of very simple SLE martingales.

QUESTION 8.15. *Can new preholomorphic observables be found by looking at discretizations of more complicated SLE martingales?*

Conversely, in **[SS05]**, the harmonic explorer is constructed in such a way that a natural discretization of a SLE(4) martingale is a martingale of the discrete curve. This fact implied the convergence of the harmonic explorer to SLE(4).

QUESTION 8.16. *Can this reverse engineering be done for other values of κ in order to find discrete models converging to* SLE*?*

8.5. Discrete observables in other models. The study can be generalized to a variety of lattice models, see the work of Cardy, Ikhlef, Riva, Rajabpour **[IC09, RC07, RC06]**. Unfortunately, the observable is only partially preholomorphic (satisfying only

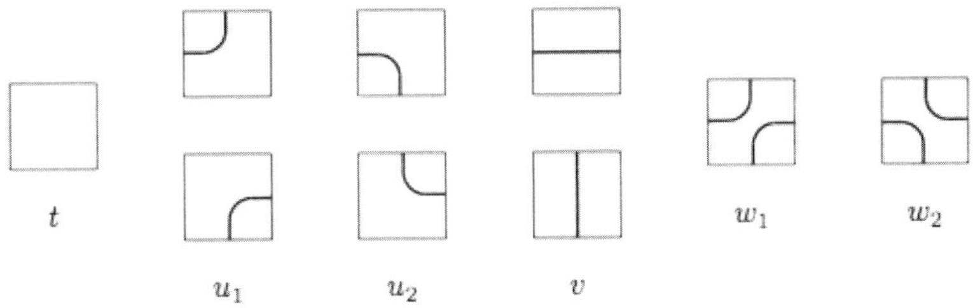

FIGURE 19. Different possible plaquettes with their associated weights.

some of the Cauchy-Riemann equations) except for the Ising case. Interestingly, weights for which there exists a half-holomorphic observable which is not degenerate in the scaling limit always correspond to weights for which the famous Yang-Baxter equality holds.

QUESTION 8.17. *The approach to two-dimensional integrable models described here is in several aspects similar to the older approaches based on the Yang-Baxter relations* [**Bax89**]. *Can one find a direct link between the two approaches?*

Let us give the example of the $O(n)$ model on the square lattice. We refer to [**IC09**] for a complete study of the following.

It is tempting to extend the definition of $O(n)$ models to the square lattice in order to obtain a family of models containing self-avoiding walks on $\mathbb{Z}^2$ and the high-temperature expansion of the Ising model. Nevertheless, difficulties arise when dealing with $O(n)$ models on non-trivalent graphs. Indeed, the indeterminacy when counting intersecting loops prevents us from defining the model as in the previous subsection.

One can still define a model of loops on $G \subset \mathbb{L}$ by distinguishing between local configurations: faces of $G^\star \subset \mathbb{L}^\star$ are filled with one of the nine plaquettes in Fig. 19. A weight p_v is associated to every face $v \in G^\star$ depending on the type of the face (meaning its plaquette). The probability of a configuration is then proportional to $n^{\#\text{ loops}} \prod_{v\in\mathbb{L}^\star} p_v$.

REMARK 8.18. *The case* $u_1 = u_2 = v = x$, $t = 1$ *and* $w_1 = w_2 = n = 0$ *corresponds to vertex self-avoiding walks on the square lattice. The case* $u_1 = u_2 = v = \sqrt{w_1} = \sqrt{w_2} = x$ *and* $n = t = 1$ *corresponds to the high-temperature expansion of the Ising model. The case* $t = u_1 = u_2 = v = 0$, $w_1 = w_2 = 1$ *and* $n > 0$ *corresponds to the FK percolation at criticality with* $q = n$.

A parafermionic observable can also be defined on the medial lattice:

$$F(z) = \frac{\sum_{\omega\in\mathcal{E}(a,z)} e^{-i\sigma W_\gamma(a,z)} n^{\#\text{ loops}} \prod_{v\in\mathbb{L}^\star} p_v}{\sum_{\omega\in\mathcal{E}} n^{\#\text{ loops}} \prod_{v\in\mathbb{L}^\star} p_v} \tag{8.8}$$

where $\mathcal{E}$ corresponds to all the configurations of loops on the graph, and $\mathcal{E}(a,z)$ corresponds to configurations with loops and one interface from a to z.

One can then look for a local relation for F around a vertex v, which would be a discrete analogue of the Cauchy-Riemann equation:

$$F(N) - F(S) = i[F(E) - F(W)], \tag{8.9}$$

An additional geometric degree of freedom can be added: the lattice can be stretched, meaning that each rhombus is not a square anymore, but a rhombus with inside angle α.

As in the case of FK percolations and spin Ising, one can associate configurations by pairs, and try to check (8.9) for each of these pairs, thus leading to a certain number of complex equations. We possess degrees of freedom in the choice of the weights of the model, of the spin σ and of the geometric parameter α. Very generally, one can thus try to solve the linear system and look for solutions. This leads to the following discussion:

Case $v = 0$ and $n = 1$: There exists a non-trivial solution for every spin σ, which is in bijection with a so-called six-vertex model in the disordered phase. The height function associated with this model should converge to the Gaussian free field. This is an example of a model for which interfaces cannot converge to SLE (in **[IC09]**; it is conjectured that the limit is described by SLE$(4,\rho)$).

Case $v = 0$ and $n \neq 1$: There exist unique weights associated to an observable with spin -1. This solution is in bijection with the FK percolation at criticality with $\sqrt{q} = n+1$. Nevertheless, physical arguments tend to show that the observable with this spin should have a trivial scaling limit. It would not provide any information on the scaling limit of the model itself; see **[IC09]** for additional details.

Case $v \neq 0$: Fix n. There exists a solution for $\sigma = \frac{3\eta}{2\pi} - \frac{1}{2}$ where $\eta \in [-\pi, \pi]$ satisfies $-\frac{n}{2} = \cos 2\eta$. Note that there are a priori four possible choices for σ. In general the following weights can be found:

$$\begin{cases} t &= -\sin(2\phi - 3\eta/2) + \sin(5\eta/2) - \sin(3\eta/2) + \sin(\eta/2) \\ u_1 &= -2\sin(\eta)\cos(3\eta/2 - \phi) \\ u_2 &= -2\sin(\eta)\sin(\phi) \\ v &= -2\sin(\phi)\cos(3\eta/2 - \phi) \\ w_1 &= -2\sin(\phi - \eta)\cos(3\eta/2 - \phi) \\ w_2 &= 2\cos(\eta/2 - \phi)\sin(\phi) \end{cases}$$

where $\phi = (1+\sigma)\alpha$. We now interpret these results:

When $\eta \in [0, \pi]$, the scaling limit has been argued to be described by a Coulomb gas with a coupling constant $2\eta/\pi$. In other words, the scaling limit should be the same as the corresponding $O(n)$ model on the hexagonal lattice. In particular, interfaces should converge to the corresponding Schramm-Loewner Evolution.

When $\eta \in [-\pi, 0]$, the scaling limit curve cannot be described by SLE, and it provides yet another example of a two-dimensional model for which the scaling limit is not described via SLE.

References

[AB99] M. Aizenman and A. Burchard, *Hölder regularity and dimension bounds for random curves*, Duke Math. J. **99** (1999), no. 3, 419–453.

[ABF87] M. Aizenman, D. J. Barsky, and R. Fernández, *The phase transition in a general class of Ising-type models is sharp*, J. Statist. Phys. **47** (1987), no. 3-4, 343–374.

[Aiz80] M. Aizenman, *Translation invariance and instability of phase coexistence in the two-dimensional Ising system*, Comm. Math. Phys. **73** (1980), no. 1, 83–94.

[Bax89] R.J. Baxter, *Exactly solved models in statistical mechanics*, Academic Press Inc. [Harcourt Brace Jovanovich Publishers], London, 1989, Reprint of the 1982 original.

[BDC10] V. Beffara and H. Duminil-Copin, *The self-dual point of the two-dimensional random-cluster model is critical for $q \geq 1$*, to appear in PTRF (2010), online first 10.1007/s00440-011-0353-8, 25 pages.

[BDC11] ———, *Smirnov's fermionic observable away from criticality*, to appear in Ann. Probab. (2011), 19 pages.

[BDS12] V. Beffara, H. Duminil-Copin and S. Smirnov, *On the critical parameters of the $q \geq 4$ random-cluster model on isoradial graphs*, preprint, 20 pages, 2012.

[BdT10] C. Boutillier and B. de Tilière, *The critical Z-invariant Ising model via dimers: the periodic case*, PTRF **147** (2010), no. 3-4, 379–413.

[BdT11] ———, *The critical Z-invariant Ising model via dimers: locality property*, Comm. Math. Phys. **301** (2011), no. 2, 473–516.

[Ber72] V.L. Berezinskii, *Destruction of long-range order in one-dimensional and two-dimensional systems possessing a continuous symmetry group. ii. quantum systems*, Soviet Journal of Experimental and Theoretical Physics **34** (1972), 610 pages.

[BMS05] A.I. Bobenko, C. Mercat, and Y. B. Suris, *Linear and nonlinear theories of discrete analytic functions. Integrable structure and isomonodromic Green's function*, J. Reine Angew. Math. **583** (2005), 117–161.

[Bou26] G. Bouligand, *Sur le problème de Dirichlet.*, Ann. Soc. Pol. Math. **4** (1926), 59–112.

[BPZ84a] A. A. Belavin, A. M. Polyakov, and A. B. Zamolodchikov, *Infinite conformal symmetry in two-dimensional quantum field theory*, Nuclear Phys. B **241** (1984), no. 2, 333–380.

[BPZ84b] ———, *Infinite conformal symmetry of critical fluctuations in two dimensions*, J. Statist. Phys. **34** (1984), no. 5-6, 763–774.

[BS08] A. I. Bobenko and Y. B. Suris, *Discrete differential geometry*, Graduate Studies in Mathematics, vol. 98, American Mathematical Society, Providence, RI, 2008, Integrable structure.

[BSST40] R. L. Brooks, C. A. B. Smith, A. H. Stone, and W. T. Tutte, *The dissection of rectangles into squares*, Duke Math. J. **7** (1940), 312–340.

[Car92] J. L. Cardy, *Critical percolation in finite geometries*, J. Phys. A **25** (1992), no. 4, L201–L206.

[CFL28] R. Courant, K. Friedrichs, and H. Lewy, *Über die partiellen Differenzengleichungen der mathematischen Physik*, Math. Ann. **100** (1928), no. 1, 32–74.

[Cim10] D. Cimasoni, *A generalized Kac-Ward formula*, to appear J. Stat. Mech. Theory Exp. (2010), 23 pages.

[CDH12] D. Chelkak, H. Duminil-Copin and C. Hongler *Crossing probabilities in topological rectangles for the critical planar FK-Ising model*, preprint (2012), 30 pages.

[CDHKS12] D. Chelkak, H. Duminil-Copin, C. Hongler, A. Kemppainen and S. Smirnov, *Convergence of Ising interfaces to Schramm's SLEs*, preprint (2012), 5 pages.

[CS08] D. Chelkak and S. Smirnov, *Discrete complex analysis on isoradial graphs*, to appear in Adv. in Math. (2008), 35 pages.

[CS09] ———, *Universality in the 2D Ising model and conformal invariance of fermionic observables*, to appear in Inv. Math. (2009), 52 pages.

[CV10] L. Coquille and Y. Velenik, *A finite-volume version of Aizenman-Higuchi theorem for the 2D Ising model*, to appear in PTRF (2010), 16 pages.

[Dum12] H. Duminil-Copin *Second-order phase transition for planar FK percolation with $1 \leq q \leq 4$ via parafermionic observables,* preprint, 2012.

[DCHN10] H. Duminil-Copin, C. Hongler, and P. Nolin, *Connection probabilities and RSW-type bounds for the two-dimensional FK Ising model*, Comm. in Pure and Applied Math., **64(9)** (2011), 1165–1198.

[DCS10] H. Duminil-Copin and S. Smirnov, *The connective constant of the honeycomb lattice equals* $\sqrt{2+\sqrt{2}}$, Annals of Math., **175(3)**, 1653–1665 (2012).

[DMNS81] D. Domany, D. Mukamel, B. Nienhuis, and A. Schwimmer, *Duality relations and equivalences for models with O(N) and cubic symmetry*, Nuclear Physics B **190** (1981), no. 2, 279–287.

[Dob72] R. L. Dobrushin, *Gibbs state, describing the coexistence of phases in the three-dimensional ising model.*, PTRF **17** (1972), 582–60.

[Duf56] R. J. Duffin, *Basic properties of discrete analytic functions*, Duke Math. J. **23** (1956), 335–363.

[Duf68] ———, *Potential theory on a rhombic lattice*, J. Combinatorial Theory **5** (1968), 258–272.

[DZM+99] N.P. Dolbilin, Y.M. Zinovév, A. S. Mishchenko, M.A. Shtańko, and M.I. Shtogrin, *The two-dimensional Ising model and the Kac-Ward determinant*, Izv. Ross. Akad. Nauk Ser. Mat. **63** (1999), no. 4, 79–100.

[ES88] R. G. Edwards and A. D. Sokal, *Generalization of the Fortuin-Kasteleyn-Swendsen-Wang representation and Monte Carlo algorithm*, Phys. Rev. D (3) **38** (1988), no. 6, 2009–2012.

[Fer44] J. Ferrand, *Fonctions préharmoniques et fonctions préholomorphes.*, Bull. Sci. Math. (2) **68** (1944), 152–180.

[Fis66] M. Fisher, *On the dimer solution of planar Ising models*, Journal of Mathematical Physics **7** (1966), no. 10, 1776–1781.

[Fis98] ———, *Renormalization group theory: its basis and formulation in statistical physics*, Rev. Modern Phys. **70** (1998), no. 2, 653–681.

[FK72] C. M. Fortuin and P. W. Kasteleyn, *On the random-cluster model. I. Introduction and relation to other models*, Physica **57** (1972), 536–564.

[FKG71] C. M. Fortuin, P. W. Kasteleyn, and J. Ginibre, *Correlation inequalities on some partially ordered sets*, Comm. Math. Phys. **22** (1971), 89–103.

[Gri06] G.R. Grimmett, *The random-cluster model*, Grundlehren der Mathematischen Wissenschaften [Fundamental Principles of Mathematical Sciences], vol. 333, Springer-Verlag, Berlin, 2006.

[Hei28] W. Heisenberg, *Zür theorie des ferromagnetismus*, Zeitschrift für Physik A Hadrons and Nuclei **49** (1928), no. 9, 619–636.

[Hig81] Y. Higuchi, *On the absence of non-translation invariant Gibbs states for the two-dimensional Ising model*, Random fields, Vol. I, II (Esztergom, 1979), Colloq. Math. Soc. János Bolyai, vol. 27, North-Holland, Amsterdam, 1981, pp. 517–534.

[HK11] C. Hongler and K. Kytolä, *Dipolar SLE in the Ising model with plus/minus/free boundary conditions*, 2011, 82 pages.

[HS93] Z.-X. He and O. Schramm, *Fixed points, Koebe uniformization and circle packings*, Ann. of Math. (2) **137** (1993), no. 2, 369–406.

[HS10] C. Hongler and S. Smirnov, *The energy density in the planar ising model*, to appear in PTRF (2010).

[IC09] Y. Ikhlef and J.L. Cardy, *Discretely holomorphic parafermions and integrable loop models*, J. Phys. A **42** (2009), no. 10, 102001, 11 pages.

[Isa41] R.P. Isaacs, *A finite difference function theory*, Univ. Nac. Tucumán. Revista A. **2** (1941), 177–201.

[Isa52] ———, *Monodiffric functions. Construction and applications of conformal maps*, Proceedings of a symposium (Washington, D. C.), National Bureau of Standards, Appl. Math. Ser., No. 18, U. S. Government Printing Office, 1952, pp. 257–266.

[Isi25] E. Ising, *Beitrag zur theorie des ferromagnetismus.*, Z. Phys. **31** (1925), 253–258.

[ISZ88] C. Itzykson, H. Saleur, and J.-B. Zuber (eds.), *Conformal invariance and applications to statistical mechanics*, World Scientific Publishing Co. Inc., Teaneck, NJ, 1988.

[Kad66] L. P. Kadanoff, *Scaling laws for Ising model near t_c*, Physics **2** (1966), no. 263.

[Kas61] P. W. Kasteleyn, *The statistics of dimers on a lattice.*, PhysicA **27** (1961), 1209–1225.

[Ken00] R. Kenyon, *Conformal invariance of domino tiling*, Ann. Probab. **28** (2000), no. 2, 759–795.

[Ken02] ———, *The Laplacian and Dirac operators on critical planar graphs*, Invent. Math. **150** (2002), no. 2, 409–439.

[Kir47] G. Kirchhoff, *über die auflösung der gleichungen, auf welche man bei der untersuchung der linearen vertheilung galvanischer ströme geführt wird.*, Annalen der Physik und Chemie **148** (1847), no. 12, 497–508.

[KO50] B. Kaufman and L. Onsager, *Crystal statistics. IV. long-range order in a binary crystal*, 1950.

[KS05] R. Kenyon and J.-M. Schlenker, *Rhombic embeddings of planar quad-graphs*, Trans. Amer. Math. Soc. **357** (2005), no. 9, 3443–3458 (electronic).

[KS10] A. Kemppainen and S. Smirnov, *Random curves, scaling limits and loewner evolutions*, in preparation (2010).

[KT73] J.M. Kosterlitz and D.J. Thouless, *Ordering, metastability and phase transitions in two-dimensional systems*, Journal of Physics C: Solid State Physics **6** (1973), 1181.

[KW41a] H.A. Kramers and G.H. Wannier, *Statistics of the two-dimensional ferromagnet, I*, Phys. Rev. **60** (1941), 252–262.

[KW41b] ———, *Statistics of the two-dimensional ferromagnet, II*, Phys. Rev. **60** (1941), 263–276.

[KW52] M. Kac and J. C. Ward, *A combinatorial solution of the two-dimensional ising model*, Phys. Rev **88** (1952), 1332–1337.

[Law91] G. F. Lawler, *Intersections of random walks*, Probability and its Applications, Birkhäuser Boston Inc., Boston, MA, 1991.

[Len20] W. Lenz, *Beitrag zum verständnis der magnetischen eigenschaften in festen körpern.*, Phys. Zeitschr. **21** (1920), 613–615.

[LF55] J. Lelong-Ferrand, *Représentation conforme et transformations à intégrale de Dirichlet bornée*, Gauthier-Villars, Paris, 1955.

[Lie67] E.H. Lieb, *Exact solution of the problem of the entropy of two-dimensional ice*, Physical Review Letters **18** (1967), no. 17, 692–694.

[Lov04] L. Lovász, *Discrete analytic functions: an exposition*, Surveys in differential geometry. Vol. IX, Surv. Differ. Geom., IX, Int. Press, Somerville, MA, 2004, pp. 241–273.

[LPSA94] R. Langlands, P. Pouliot, and Y. Saint-Aubin, *Conformal invariance in two-dimensional percolation*, Bull. Amer. Math. Soc. (N.S.) **30** (1994), no. 1, 1–61.

[LS10] E. Lubetzky and A. Sly, *Critical Ising on the square lattice mixes in polynomial time*, 2010, 26 pages.

[LSW04a] G. Lawler, O. Schramm, and W. Werner, *Conformal invariance of planar loop-erased random walks and uniform spanning trees*, Ann. Probab. **32** (2004), no. 1B, 939–995.

[LSW04b] G. F. Lawler, O. Schramm, and W. Werner, *On the scaling limit of planar self-avoiding walk*, Fractal geometry and applications: a jubilee of Benoît Mandelbrot, Part 2, Proc. Sympos. Pure Math., vol. 72, Amer. Math. Soc., Providence, RI, 2004, pp. 339–364.

[Lus26] L. Lusternik, *über einege Anwendungen der direkten Methoden in Variationsrechnung.*, Recueil de la Société Mathématique de Moscou, 1926, pp. 173–201.

[Mer01] C. Mercat, *Discrete Riemann surfaces and the Ising model*, Comm. Math. Phys. **218** (2001), no. 1, 177–216.

[Mes06] R.J. Messikh, *Surface tension near criticality of the 2d-Ising model.*, arXiv:math/0610636 (2006), 24 pages.

[MW40] W. H. McCrea and F. J. W. Whipple, *Random paths in two and three dimensions*, Proc. Roy. Soc. Edinburgh **60** (1940), 281–298.

[MW73] B.M. McCoy and T.T. Wu, *The two-dimensional Ising model*, Harvard University Press, Cambridge, MA, 1973.

[Nie82] B. Nienhuis, *Exact critical point and critical exponents of $o(n)$ models in two dimensions.*, Phys. Rev. Lett. **49** (1982), 1062–1065.

[Nie84] ———, *Coulomb gas description of 2D critical behaviour*, J. Statist. Phys. **34** (1984), 731–761.

[Ons44] L. Onsager, *Crystal statistics. i. a two-dimensional model with an order-disorder transition.*, Phys. Rev. (2) **65** (1944), 117–149.

[Pal07] J. Palmer, *Planar Ising correlations*, Progress in Mathematical Physics, vol. 49, Birkhäuser Boston Inc., Boston, MA, 2007.

[Pei36] R. Peierls, *On Ising's model of ferromagnetism.*, Math. Proc. Camb. Phil. Soc. **32** (1936), 477–481.

[PP66] A.Z. Patashinskii and V.L. Pokrovskii, *Behavior of ordered systems near the transition point*, Soviet Physics JETP **23** (1966), no. 292, 292.

[PW23] H.B. Phillips and N. Wiener, *Nets and the Dirichlet problem.*, Math. J. of Math. **2** (1923), 105–124.

[RC06] V. Riva and J.L. Cardy, *Holomorphic parafermions in the Potts model and stochastic Loewner evolution*, J. Stat. Mech. Theory Exp. (2006), no. 12, P12001, 19 pp. (electronic).

[RC07] M. A. Rajabpour and J.L. Cardy, *Discretely holomorphic parafermions in lattice Z_N models*, J. Phys. A **40** (2007), no. 49, 14703–14713.

[RS87] B. Rodin and D. Sullivan, *The convergence of circle packings to the Riemann mapping*, J. Differential Geom. **26** (1987), no. 2, 349–360.

[Sch00] O. Schramm, *Scaling limits of loop-erased random walks and uniform spanning trees*, Israel J. Math. **118** (2000), 221–288.

[She60] S. Sherman, *Combinatorial aspects of the Ising model for ferromagnetism. I. A conjecture of Feynman on paths and graphs*, J. Mathematical Phys. **1** (1960), 202–217.

[Smi01] S. Smirnov, *Critical percolation in the plane: conformal invariance, Cardy's formula, scaling limits*, C. R. Acad. Sci. Paris Sér. I Math. **333** (2001), no. 3, 239–244.

[Smi06] ———, *Towards conformal invariance of 2D lattice models*, International Congress of Mathematicians. Vol. II, Eur. Math. Soc., Zürich, 2006, pp. 1421–1451.

[Smi10a] ———, *Conformal invariance in random cluster models. I. Holomorphic fermions in the Ising model*, Ann. of Math. (2) **172** (2010), no. 2, 1435–1467.

[SS05] O. Schramm and S. Sheffield, *Harmonic explorer and its convergence to* SLE_4, Ann. Probab. **33** (2005), no. 6, 2127–2148.

[Ste05] K. Stephenson, *Introduction to circle packing*, Cambridge University Press, Cambridge, 2005, The theory of discrete analytic functions.

[SW10a] S. Sheffield and W. Werner, *Conformal loop ensembles: Construction via loop-soups*, Arxiv preprint arXiv:1006.2373 (2010), 21 pages.

[SW10b] ———, *Conformal loop ensembles: The Markovian characterization*, Arxiv preprint arXiv:1006.2374 (2010), 60 pages.

[Thu86] W. P. Thurston, *Zippers and univalent functions*, The Bieberbach conjecture (West Lafayette, Ind., 1985), Math. Surveys Monogr., vol. 21, Amer. Math. Soc., Providence, RI, 1986, pp. 185–197.

[Vdo65] N. V. Vdovichenko, *A calculation of the partition function for a plane dipole lattice*, Soviet Physics JETP **20** (1965), 477–488.

[Wer09] W. Werner, *Percolation et modèle d'Ising*, Cours Spécialisés [Specialized Courses], vol. 16, Société Mathématique de France, Paris, 2009.

[Yan52] C.N. Yang, *The spontaneous magnetization of a two-dimensional Ising model*, Phys. Rev. (2) **85** (1952), 808–816.

Département de Mathématiques, Université de Genève, Genève, Switzerland
E-mail address: hugo.duminil@unige.ch

Département de Mathématiques, Université de Genève, Genève, Switzerland. Chebyshev Laboratory, St. Petersburg State University, 14th Line, 29b, Saint Petersburg, 199178 Russia
E-mail address: stanislav.smirnov@unige.ch

Clay Mathematics Proceedings
Volume 15, 2012

Fractal and Multifractal Properties of Schramm-Loewner Evolution

Gregory F. Lawler

Introduction

This is a slightly expanded version of my lectures at the 2010 Clay Mathematics Institute summer (winter) school in Buzios, Brazil. The theme is the fine properties of Schramm-Loewner evolution (SLE) curves with an emphasis on recent work I have done with a number of co-authors on fractal and multifractal properties. I assume basic knowledge of SLE at the level of the foundational course presented by Vincent Beffara. I will try to discuss both results and ideas of proofs. Although discrete models motivate SLE, I will focus only on SLE itself and will not discuss the convergence of discrete models.

The basic theme tying the results together is the SLE curve. Fine analysis of the curve requires estimates of moments of the derivatives, and in turn leads to studying martingales and local martingales. In the process, I will discuss existence of the curve, Hausdorff dimension of the curve, and a number of more recent results that I have obtained with a number of co-authors.

The five sections correspond roughly to the five lectures that I gave. Here is a quick summary.

- Section 1 proves a basic result of Rohde and Schramm [**15**] on the existence of the SLE curve for $\kappa \neq 8$. Many small steps are left to the reader; one can treat this as an exercise in the deterministic Loewner equation and classical properties of univalent functions such as the distortion theorem. Two main ingredients go into the proof: the modulus of continuity of Brownian motion and an estimate of the moments of the derivative of the reverse map. By computing the moment, we can determine the optimal Hölder exponent and see why $\kappa = 8$ is the delicate case. The estimation of the moment is left to the next section.
- Section 2 discusses how to use the reverse Loewner flow to estimate the exponent. This was the tool in [**15**] to get their estimate. Here we expand signficantly on their work because finer analysis is needed to derive "two-point" or "second moment" estimates which are required to establish

2010 *Mathematics Subject Classification.* Primary 60J67; Secondary 28A80.
Research supported by National Science Foundation grant DMS-0907143.

fractal and multifractal behavior with probability one. Although there is a fair amount of calculation involved, there are a few general tools:
  - Use scaling and a good choice of reparametrization to reduce the problem to analysis of a one-variable SDE.
  - Find an appropriate martingale and use the Girsanov theorem to understand the measure obtained by weighting by the martingale.
  - For exceptional events on the path, find an event of high probability *in the weighted measure* that is contained in the exceptional event. Choose it appropriately so that two-point estimates can be obtained.

These are standard methods in stochastic analysis. One of the most fundamental techniques in large deviation theory is to study a new measure (sometimes called a "tilting") on a space on which an exceptional set has large probability. If this new measure arises from a martingale, then the Girsanov theorem is the tool for studying probabilities in the new measure.

The latter part of this section, starting with Section 2.6, contains some more advanced topics that were not covered in the lectures. I have included them in these notes because they are part of the reverse flow picture, but the material from this part is not used later. Readers should feel free to skip these and move to Section 3.

- Section 3 is essentially independent of Section 2 and considers the forward Loewner flow. The Hausdorff dimension of the SLE curve was analyzed in [**15**] and [**2**]. The basic questions are: how close does the SLE curve get to a $z \in \mathbb{H}$ and what does the path look like if it does get close to z? There is a fundamental local martingale in terms of the SLE Green's function, and if one uses a radial parametrization (depending on z), one gets a simple one-variable SLE. By weighting by this local martingale, one gets another process, two-sided radial SLE, which corresponds to SLE conditioned to hit a point. Here we use the Girsanov theorem to give very sharp estimates of the probability that the SLE gets near z. Finally, we discuss why trying to prove lower bounds for the Hausdorff dimension leads to studying a two-point estimate for the probability of getting close to two different points.
- Section 4 continues Section 3 by discussing the two-point estimate first proved by Beffara [**2**]. We only give a sketch of part of the argument as rederived in [**9**] and then we define an appropriate multi-point Green's function and corresponding two-point local martingale. The estimate and the two-point local martingale are used in the next section.
- Section 5 is devoted to the natural parametrization or length for SLE_κ, $\kappa < 8$. The usual parametrization for an SLE_κ is by capacity, which does not correspond to the "natural" scaled parametrization one would give to discrete models. I start by giving the intuition for a definition, which leads to an expression of the type analyzed in Section 2, and then give a precise definition as developed in [**8, 11**]. The proof of existence in [**11**] uses the ideas from Section 4.

There are many exercises interspersed throughout the notes. I warn you that I use facts from the exercises later on. Therefore, a reader should read them whether or not he or she chooses to actually do them. I have an additional section at the end with one more exercise on Brownian motion which can be considered as an easier example of some of the ideas from Sections 4 and 5. I assume the reader knows the basics of SLE and univalent function theory. Possible references are the notes from Beffara's course and my book [**7**].

These notes have been improved by questions and remarks by the participants of the school, and I thank all the participants. A particular note of thanks goes to Brent Werness for his work as a TA and his comments on these notes.

0.1. Basic definitions and notation. To set some basic definitions, I let g_t denote the conformal maps of chordal SLE_κ from 0 to ∞ in $\mathbb{H}$ parametrized so that the half-plane capacity grows at rate

$$a = \frac{2}{\kappa}.$$

I will use a throughout these notes because it makes formulas somewhat easier. It is always equal to $2/\kappa$ and the reader can make this replacement at any time! Under this parametrization, g_t satisfies the chordal Loewner equation

$$\partial_t g_t(z) = \frac{a}{g_t(z) - U_t}, \quad g_0(z) = z, \tag{1}$$

where $U_t = -B_t$ is a standard Brownian motion. (We choose this parametrization so that the driving function has variance parameter 1.) The equation is valid for all $z \in \mathbb{C} \setminus \{0\}$ up to time $T_z \in (0, \infty]$ and g_t is the unique conformal transformation of

$$H_t := \{z \in \mathbb{H} : T_z > t\}$$

onto $\mathbb{H}$ with $g_t(z) = z + o(1)$ as $z \to \infty$. Note that $T_{\overline{z}} = T_z$ and $g_t(\overline{z}) = \overline{g_t(z)}$, so we restrict to z in the upper half plane $\mathbb{H}$ or its closure $\overline{\mathbb{H}}$. If $z \in \overline{\mathbb{H}} \setminus \{0\}$, and we let

$$Z_t = Z_t(z) = g_t(z) - U_t,$$

then the Loewner equation (1) can be written as the SDE

$$dZ_t = \frac{a}{Z_t}\, dt + dB_t, \quad Z_0 = z.$$

If $z \in \mathbb{H}$, one should note that the process Z_t takes values in $\mathbb{H}$, but the Brownian motion B_t is a real Brownian motion. If $z \in \mathbb{R} \setminus \{0\}$, this equation becomes the usual real-valued Bessel SDE.

The word *curve* in these notes always means a continuous function of time. If $\gamma : [0, \infty) \to \mathbb{C}$ is a curve, we write γ_t for the image or trace up to time t,

$$\gamma_t = \{\gamma(s) : 0 \leq s \leq t\}.$$

1. The existence of the SLE curve

In this section, we will present a proof of the following theorem first proved by Rohde and Schramm [**15**]. We start with a definition.

Definition The conformal maps g_t are *generated by the curve* γ if $\gamma : [0, \infty) \to \mathbb{H}$ is a curve such that for each t, H_t is the unbounded component of $\mathbb{H} \setminus \gamma_t$.

THEOREM 1. [**15**] *If $\kappa \neq 8$, then with probability one the conformal maps g_t of chordal SLE_κ are generated by a curve.*

This theorem is also true for $\kappa = 8$, but the only current proof in this case comes from taking a limit of discrete processes [**12**]. We will consider only the $\kappa \neq 8$ case and along the way explain why $\kappa = 8$ is the hardest case. The curve is called the *SLE_κ curve.*

Let

$$\tilde{f}_t = g_t^{-1}, \qquad f_t = \tilde{f}_t(z + U_t) = g_t^{-1}(z + U_t). \tag{2}$$

(In some earlier work what we call $\tilde{f}_t$ is denoted f_t and what we call f_t is denoted $\hat{f}_t$. I have chosen the notation in (2) because f_t will be used more often than $\tilde{f}_t$ in this paper, and hence it will make the formulas nicer.) Heuristically, we would like to define

$$\gamma(t) = \tilde{f}_t(U_t) = f_t(0) = \lim_{y \to 0} f_t(iy), \tag{3}$$

so that $g_t(\gamma(t)) = U_t$. However, all that we know at the moment is that f_t is a conformal transformation of $\mathbb{H}$ onto H_t. One can give examples of conformal transformations such that the limit in (3) does not exist. In fact [**14**], one can give examples for solutions of the Loewner equation (1) with continuous U_t.

It is also possible to give examples for which the limit in (3) exists for all t, but for which the function γ is not continuous in t. However, if the limit (3) exists and γ is continuous, then it is not too difficult to see that H_t is the unbounded component of $\gamma(0,t]$. Indeed, since $\gamma(t) \in \partial H_t$, we know that $\gamma(0,t] \cap H_t = \emptyset$. Since H_t is simply connected (it is a conformal image of $\mathbb{H}$ under g_t^{-1}), it is connected and hence the bounded components of $\mathbb{H} \setminus \gamma_t$ cannot intersect H_t. Also, if we define $\hat{H}_t$ to be the unbounded component of $\mathbb{H} \setminus \gamma_t$ and $\hat{g}_t$ the conformal transformation of $\hat{H}_t$ onto $\mathbb{H}$ with $\hat{g}_t(z) - z = o(1)$ as $z \to \infty$, one can show that $\hat{g}_t$ satisfies (1) for $z \in \hat{H}_t$ and hence $\hat{g}_t(z) = g_t(z)$. In particular, $T_z > t$ and $z \in H_t$.

Notational convention. We will use ψ to denote a *(continuous, increasing) subpower function*, that is an continuous, increasing function $\psi : [0,\infty) \to (0,\infty)$ such that

$$\lim_{t \to \infty} \frac{\log \psi(t)}{\log t} = 0.$$

Different occurences of ψ indicate different subpower functions. Note that if ψ_1, ψ_2 are subpower functions, so are $\psi_1 \psi_2$, $\psi_1 + \psi_2$, and $\psi(t) = \psi_1(t^r)$ for $r > 0$.

Lévy's theorem on the modulus of continuity of Brownian motion shows that with probability one, the driving function is *weakly Hölder*-1/2, by which we mean that for some subpower function ψ,

$$|U_{t+s} - U_t| \leq \sqrt{s}\, \psi(1/s), \qquad 0 \leq t \leq 1, 0 \leq s \leq 1. \tag{4}$$

(In fact, we can choose $\psi(x) = c\sqrt{\log x}$ but we do not need this.) This condition is not sufficient to show existence of the curve. In fact, there are examples [**14**] with ψ constant for which the curve does not exist. To guarantee existence of the curve, we will bound $|f_t'(iy)|$ for y near zero. Let

$$\mathcal{D}_n = \left\{ \frac{j}{2^n} : j = 0, 1, \ldots, 2^n \right\}$$

denote the set of dyadic rationals in $[0,1]$ at level n.

LEMMA 2. *For $SLE_\kappa, \kappa \neq 8$, there exists $\theta = \theta_\kappa > 0$ such that with probability one there exists $C < \infty$ such that*

$$|f_t'(2^{-n}\,i)| \leq C\,2^{n(1-\theta)}, \quad t \in \mathcal{D}_{2n}. \tag{5}$$

PROOF. By the Borel-Cantelli lemma, it suffices to show that

$$\mathbb{P}\left\{|f_t'(2^{-n}\,i)| \geq C\,2^{n(1-\theta)}\right\} \leq c\,2^{-n(2+\epsilon)}$$

for some c, ϵ. See Theorem 11 and the comments following for the proof of this estimate. □

In this section, we will use a series of exercises to conclude the following deterministic result.

THEOREM 3. *Suppose $U_t, 0 \leq t \leq 1$ is a driving function satisfying* (4) *and* (5). *Then the corresponding maps are generated by a curve γ. Moreover,*

$$|\gamma(t+s) - \gamma(t)| \leq s^{\theta/2}\,\psi(1/s), \quad 0 \leq t < s+t \leq 1,$$

for some subpower function ψ.

♦In particular, it follows that γ is Hölder continuous of order α for all $\alpha < \theta/2$. In other words, γ is weakly Hölder-$(\theta/2)$

♦We recall our convention that the subpower function ψ takes different values in different places. The function ψ in Theorem 3 is not meant to be the same ψ as in (4). A careful reader can go through the proof and find how the ψ in the theorem depends on the ψ in (4).

♦Lemma 2 is not true for $\kappa = 8$. I would expect that one can give a direct proof of the existence of the curve for $\kappa = 8$, but it would require very careful analysis. In particular, we could not get away with being so cavalier about the subpower functions ψ.

We use the distortion theorem to construct the function γ. The first ingredient of the proof is a version of the distortion theorem that we leave as an exercise.

EXERCISE 4. *There exist C, r such if $f : \mathbb{H} \to \mathbb{C}$ is a conformal transformation, then for all $x \in \mathbb{R}, y > 0$,*

$$C^{-1}\,|f'(iy)| \leq |f'(iys)| \leq C\,|f'(iy)|, \quad \frac{1}{2} \leq s \leq 2,$$

$$C^{-1}\,(x^2+1)^{-r}\,|f'(iy)| \leq |f'(xy+iy)| \leq C\,(x^2+1)^{r}\,|f'(iy)|,$$

Hint: The distortion and growth theorems (see, e.g., [**7**]*) solve the equivalent problem in the unit disk $\mathbb{D}$. Although we do not need it here, you may wish to find the smallest possible r such that this holds.*

To extend the estimate to times that are not dyadic, we use the Loewner equation for the inverse. If g_t satisfies (1) and $\tilde{f}_t = g_t^{-1}$, then using $\tilde{f}_t(g_t(z)) = z$, we get the equation

$$\partial_t \tilde{f}_t(z) = \tilde{f}_t'(z)\,\frac{a}{U_t - z}. \tag{6}$$

Differentiating this, we get

$$\partial_t \tilde{f}_t'(z) = \tilde{f}_t''(z)\,\frac{a}{U_t - z} + \tilde{f}_t'(z)\,\frac{a}{(U_t - z)^2}.$$

Hence, if $z = x + iy$,

$$|\partial_t \tilde{f}_t'(z)| \leq a \left[|\tilde{f}_t''(z)| \, y^{-1} + |\tilde{f}_t'(z)| \, y^{-2} \right].$$

EXERCISE 5. *Show that there exists $c < \infty$ such that if $f : \mathbb{H} \to \mathbb{C}$ is a conformal transformation, then*

$$|f''(z)| \leq \frac{c}{\operatorname{Im}(z)} \, |f'(z)|.$$

Hint: Look up Bieberbach's theorem on the second coefficient of univalent functions on the disk. If you do this, you will find the optimal c.

Using the exercise, we now have

$$|\partial_t \tilde{f}_t'(x+iy)| \leq \frac{c}{y^2} \, |\tilde{f}_t'(x+iy)|. \tag{7}$$

It is now not difficult to show that the limit in (3) exists for all $t \in [0,1]$. We leave the steps as exercises.

EXERCISE 6. *Use (7) to prove the following. There exists $c < \infty$ such that if $\tilde{f}_t$ satisfies (6) and $s \leq y^2$, then*

$$c^{-1} \, |\tilde{f}_t'(x+iy)| \leq |\tilde{f}_{t+s}'(x+iy)| \leq c \, |\tilde{f}_t'(x+iy)|,$$

$$|\tilde{f}_{t+s}(x+iy) - \tilde{f}_t(x+iy)| \leq c \, y^2 \, |\tilde{f}_t'(x+iy)|. \tag{8}$$

EXERCISE 7. *Suppose $U_t, 0 \leq t \leq 1$, is a driving function satisfying (4) and (5). Then there exists a subpower function ψ such that for all $0 \leq t \leq 1$ and all $0 < y \leq 1$,*

$$|f_t'(iy)| \leq y^{\theta - 1} \, \psi(1/y).$$

EXERCISE 8. *Suppose $U_t, 0 \leq t \leq 1$, is a driving function satisfying (4) and (5). Then there exists a subpower function ψ such that for all $0 \leq t \leq 1$, the limit*

$$\gamma(t) = \lim_{y \to 0+} f_t(iy) \tag{9}$$

exists and for $0 < y \leq 1$,

$$|\gamma(t) - f_t(iy)| \leq y^{\theta} \, \psi(1/y). \tag{10}$$

We still have to show that γ is a continuous function of t and estimate its modulus of continuity. It suffices to estimate

$$|\gamma(t+s) - \gamma(t)|$$

where $t \in \mathcal{D}_{2n}$ and $0 \leq s \leq 2^{-2n}$. We use the triangle inequality. For every $y > 0$,

$$|\gamma(t+s) - \gamma(t)| \leq |\gamma(t+s) - f_{t+s}(iy)| + |f_{t+s}(iy) - f_t(iy)| + |\gamma(t) - f_t(iy)|.$$

Setting $y = 2^{-n}$ and using (10), we get

$$|\gamma(t+s) - \gamma(t)| \leq \, 2^{-n\theta} \, \psi(2^n) + |f_{t+s}(i2^{-n}) - f_t(i2^{-n})|.$$

We now write

$$|f_{t+s}(i2^{-n}) - f_t(i2^{-n})| = |\tilde{f}_{t+s}(U_{t+s} + i2^{-n}) - \tilde{f}_t(U_t + i2^{-n})|$$

$$\leq |\tilde{f}_{t+s}(U_{t+s} + i2^{-n}) - \tilde{f}_{t+s}(U_t + i2^{-n})| + |\tilde{f}_{t+s}(U_t + i2^{-n}) - \tilde{f}_t(U_t + i2^{-n})|.$$

The difference

$$|\tilde{f}_{t+s}(U_{t+s} + i2^{-n}) - \tilde{f}_{t+s}(U_t + i2^{-n})|$$

is bounded above by $|U_{t+s} - U_t|$ times the maximum of $|\tilde{f}'_{t+s}(z)|$ over all z on the interval connecting $U_{t+s} + i2^{-n}$ and $U_t + i2^{-n}$. Using Exercises 4 and 7, we see that this maximum is bounded above by $2^{n(1-\theta)}\,\psi(2^n)$ and (4) implies that

$$|U_{t+s} - U_t| \leq 2^{-n}\,\psi(2^n).$$

Therefore,

$$|\tilde{f}_{t+s}(U_{t+s} + i2^{-n}) - \tilde{f}_{t+s}(U_t + i2^{-n})| \leq 2^{-\theta n}\,\psi(2^n).$$

For the second term, we use (8) to get

$$\begin{aligned}|\tilde{f}_{t+s}(U_t + i2^{-n}) - \tilde{f}_t(U_t + i2^{-n})| &\leq c\,2^{-2n}\,|\tilde{f}'_t(U_t + i2^{-n})| \\ &\leq 2^{-n}\,2^{n(1-\theta)}\,\psi(2^n) \leq 2^{-n\theta}\,\psi(2^n).\end{aligned}$$

Combining all of the estimates, we have

$$|\gamma(s+t) - \gamma(t)| \leq s^{\theta/2}\,\psi(1/s), \qquad t \in \mathcal{D}_{2n}, \quad 0 \leq s \leq 2^{2n},$$

from which Theorem 3 follows.

1.1. Converse and Hölder continuity. We have seen that (4) and (5) imply that the curve γ is weakly $(\theta/2)$-Hölder.

PROPOSITION 9. *Suppose U_t satisfies* (4). *Then there exists a subpower function ψ such that for $t \in \mathcal{D}_{2n}$,*

$$\max_{0 \leq s \leq 2^{-2n}} |\gamma(s+t) - \gamma(t)| \geq 2^{-n}\,|f'_t(i2^{-n})|\,\psi(2^n)^{-1}.$$

SKETCH OF PROOF. We write ψ for $\psi(2^n)$ and allow ψ to change from line to line. Using (4), we can see that the image of $\gamma(t, t+2^{-2n}]$ under g_t has diameter at most $2^{-n}\,\psi$. Since it has half-plane capacity $a\,2^{-n}$, it must include at least one point $z = x + iy = \gamma(s)$ with $|x| \leq 2^{-n}\,\psi$ and $y \geq 2^{-n}/\psi$. (Why?) Distortion estimates imply that $|f'_t(z)| \geq |f'_t(i2^{-n})|\,\psi^{-1}$. The Koebe-1/4 theorem applied to the map f_t on the disk of radius y about z shows that $\gamma(t)$, which is $f_t(0)$, is not in the disk of radius $2^{-n}\,|f'_t(i2^{-n})|\,\psi^{-1}$ about $f_t(z) = \gamma(t+s)$. □

The methods of the next section allow us to determine the critical value of θ for which there exist n, y with $|f'_t(i2^{-n})| \geq 2^{(1-\theta)n}$. This is the basic idea of the following theorem, which we do not prove. One direction was proved in [**13**] and the other direction in [**4**]

THEOREM 10. *Let*

$$\alpha_* = \alpha_*(\kappa) = 1 - \frac{\kappa}{24 + 2\kappa - 8\sqrt{8+\kappa}}.$$

If γ is an SLE_κ curve and $\epsilon > 0$, then with probability one $\gamma(t), \epsilon \leq t \leq 1$, is weakly α_*-continuous, but not Hölder continuous of any order $\alpha > \alpha_*$.

♦The behavior of the curve at $t = 0$ is different than for positive t because we are starting with the initial condition for H_t of $\mathbb{H}$. This is why we restrict the curve to times $\epsilon \leq t \leq 1$ in the statement of the theorem.

♦Note that the theorem implies that for $\kappa = 8$, the curve is not Hölder-α for any $\alpha > 0$. This indicates why $\kappa = 8$ is the hardest value to show the existence of the curve.

♦This is a statement about the modulus of continuity of $\gamma(t)$ as a function of t in the capacity parametrization. Sometimes "Hölder continuity of SLE" refers to the properties of the function $z \mapsto f_t(z)$ for fixed t. This is discussed in [15]. For this problem $\kappa = 4$ is the value for which the function is not Hölder continuous of any order $\alpha > 0$.

---

2. The moments of $|f'|$

Using the existence of the curve as one motivation, we now proceed to discuss how one estimates

$$\mathbb{E}\left[|f_t'(z)|^\lambda\right],$$

where

$$f_t(z) = g_t^{-1}(z + U_t).$$

We summarize the main result here. Most of what we discuss here is proved in [6], but the final piece of the theorem as we state it here was done as Theorem 5.4 of [4]. A weaker form of this was in [15].

THEOREM 11. *If*

$$\lambda < \lambda_c = a + \frac{3}{16a} + 1 = \frac{2}{\kappa} + \frac{3\kappa}{32} + 1,$$

then as $t \to \infty$,

$$\mathbb{E}\left[|f_1'(i/t)|^\lambda\right] = \mathbb{E}\left[|f_{t^2}'(i)|^\lambda\right] \asymp t^{-\zeta},$$

where

$$\zeta = \zeta(\lambda) = \lambda + \frac{1}{2a}\sqrt{(2a+1)^2 - 4a\lambda} - 1 - \frac{1}{2a}.$$

Moreover, the expectation is carried on an event on which

$$|f_{t^2}'(i)| \approx t^\beta, \tag{11}$$

where

$$\beta = \beta(\lambda) = -\zeta'(\lambda) = \frac{1}{\sqrt{(2a+1)^2 - 4a\lambda}} - 1.$$

Roughly speaking,

$$\mathbb{P}\left\{|f_1'(i/t)| \approx t^\beta\right\} \approx t^{-(\zeta + \lambda\beta)}.$$

Note that $\zeta(\lambda_c) = a - \frac{1}{16a}, \beta(\lambda_c) = 1$ and hence

$$\zeta(\lambda_c) + \lambda_c\,\beta(\lambda_c) = 2a + \frac{1}{8a} + 1 = \frac{4}{\kappa} + \frac{\kappa}{16} + 1.$$

The right-hand side is minimized when $\kappa = 8$, at which it takes the value 2. For $\kappa \neq 8$, we can find $\beta < 1$ such that

$$\zeta(\lambda) + \lambda\beta(\lambda) > 2,$$

from which we can deduce Lemma 2 for $\theta = 1 - \beta$.

2.1. The reverse Loewner flow. We will use the *reverse Loewner flow* in studying the derivative. The reverse Loewner equation is the usual Loewner equation run backwards in time. It takes the form

$$\partial_t h_t(z) = -\frac{a}{h_t(z) - V_t} = \frac{a}{V_t - h_t(z)}, \qquad h_0(z) = z. \tag{12}$$

For each t, h_t is a conformal transformation of $\mathbb{H}$ onto a subdomain $h_t(\mathbb{H})$ satisfying $h_t(z) - z \to 0$ as $z \to \infty$. A relationship between the forward and reverse Loewner equations is given in the following exercise.

EXERCISE 12. *Suppose g_t, $0 \le t \le s$, is the solution to* (1) *and h_t is the solution to* (12) *with $V_t = U_{s-t} - U_s$. Then*

$$h_s(z) = f_s(z) - U_s.$$

If $U_t, 0 \le t \le s$, is a standard Brownian motion, then

$$V_t = U_{s-t} - U_s, \quad 0 \le t \le s,$$

is also a standard Brownian motion. Hence the following holds.

- If $g_t, 0 \le t \le s$, is the solution to (1) where U_t is a standard Brownian motion, and $h_t, 0 \le t \le s$, is the solution to (12) where V_t is a Brownian motion, then the random conformal transformations

$$z \longmapsto f_s(z) - U_s \quad \text{and} \quad z \longmapsto h_s(z)$$

  have the same distribution. In particular, f_s' and h_s' are identically distributed and

$$\mathbb{E}\left[|f_s'(z)|^\lambda\right] = \mathbb{E}\left[|h_s'(z)|^\lambda\right].$$

The joint distribution of $\{f_t' : 0 \le t \le s\}$ is *not* the same as that of $\{h_t' : 0 \le t \le s\}$. However, we can give the joint distributions. For second moment estimates, we need to consider two times simultaneously. We state the relationship here; the interested reader may wish to verify this.

- Suppose $f_t, 0 \le t \le s+u$ is the solution to (1) where U_t is a standard Brownian motion. Let $h_t, 0 \le t \le s+u$, be the solution to (12) with $V_t = U_{s+u-t} - U_{s+u}$. Let $\tilde{h}_t, 0 \le t \le s$, be the solution to (12) with $V_t = U_{s-t} - U_s$. Let

$$Z_t(z) = h_t(z) - V_t.$$

  Then,
  - $h_{s+u}(z) = \tilde{h}_s(Z_u(z)) - V_u$.
  - h_u and $\tilde{h}_s$ are independent.
  - $f_s'(w)\, f_{s+u}'(z) = h_u'(z)\, \tilde{h}_s'(Z_u(z))\, \tilde{h}_s'(w)$
  - $f_{s+u}(z) - f_s(w) = \tilde{h}_s(Z_u(z)) - \tilde{h}_s(w)$.

2.2. Some computations. For this section we assume that h_t satisfies (12) with $V_t = -B_t$ being a standard Brownian motion.

EXERCISE 13. *Use the scaling property of Brownian motion to show that if $r > 0$, $h_t(z)$ has the same distribution as $h_{r^2t}(rz)/r$, and hence $h_t'(z)$ has the same distribution at $h_{r^2t}'(rz)$.*

EXERCISE 14. *Prove the following "parabolic Harnack inequality". For every compact* $V \subset \mathbb{H}$ *and every* $s_0 \geq 1$, *there exist* c_1, c_2 *(depending on* V, s_0 *but not on* λ *or* κ*) such that for all* $t \geq 1$,

$$c_1^\lambda \, \mathbb{E}\left[|h_t'(i)|^\lambda\right] \leq \mathbb{E}\left[|h_{st}'(z)|^\lambda\right] \leq c_2^\lambda \, \mathbb{E}\left[|h_t'(i)|^\lambda\right], \qquad s_0^{-1} \leq s \leq s_0, \qquad z \in V.$$

Hint: Use scaling and the distortion theorem.

Let $z \in \mathbb{H}$ and define

$$Z_t = Z_t(z) = X_t + iY_t = h_t(z) - V_t = h_t(z) + B_t.$$

we will define a number of other quantities in this section. Even though we omit it in the notation, it is important to remember that there is a z dependence. The equation (12) can be written as

$$dX_t = -\frac{a\,X_t}{X_t^2 + Y_t^2}\,dt + dB_t, \qquad \partial_t Y_t = \frac{a\,Y_t}{X_t^2 + Y_t^2}. \tag{13}$$

Differentiating (12) with respect to z gives

$$\partial_t[\log h_t'(z)] = \frac{a}{Z_t^2},$$

and by taking real parts, we get

$$\partial_t \,|h_t'(z)| = |h_t'(z)|\,\frac{a(X_t^2 - Y_t^2)}{(X_t^2 + Y_t^2)^2}.$$

Let

$$S_t = \sin[\arg Z_t] = \frac{Y_t}{\sqrt{X_t^2 + Y_t^2}}, \qquad \Upsilon_t = \frac{|h_t'(z)|}{Y_t}.$$

The chain rule gives

$$\partial_t \Upsilon_t = -\Upsilon_t\,\frac{2aY_t^2}{(X_t^2 + Y_t^2)^2},$$

and an exercise in Itô's formula gives

$$dS_t^r = S_t^r\left[\frac{(2ar + \frac{r^2}{2} + \frac{r}{2})\,X_t^2 - \frac{r}{2}\,Y_t^2}{(X_t^2 + Y_t^2)^2}\,dt - \frac{r\,X_t}{X_t^2 + Y_t^2}\,dB_t\right]. \tag{14}$$

PROPOSITION 15. *Suppose* $r \in \mathbb{R}$ *and*

$$\lambda = \lambda(r) = r\left(1 + \frac{1}{2a}\right) - \frac{r^2}{4a}, \quad \zeta = \zeta(r) = r - \frac{r^2}{4a} = \lambda - \frac{r}{2a}.$$

If $z \in \mathbb{H}$, *let*

$$M_t = M_{t,r}(z) = |h_t'(z)|^\lambda\, Y_t^\zeta\, S_t^{-r}. \tag{15}$$

Then M_t *is a martingale satisfying*

$$dM_t = \frac{r\,X_t}{X_t^2 + Y_t^2}\,M_t\,dB_t. \tag{16}$$

EXERCISE 16. *Verify as many of the calculations above as you want. Also, establish the following deterministic estimates if* $z = x + iy$:

$$y^2 \leq Y_t^2 \leq y^2 + 2at,$$

$$|h_t'(z)| \leq \frac{Y_t}{Y_0} \leq \sqrt{1 + 2a(t/y^2)}.$$

♦The parameters λ, ζ are the same parameters as in Theorem 11. However, it is useful to include the extra parameter r. There is only a "one real variable" amount of randomness (nontrivial quadratic variation) in the martingale M_t. For convenience we have written it in terms of the sine, S_t; earlier versions of these computations chose to write it in terms of $(X^2 + Y^2)$. Either way, one must choose the appropriate "compensator" terms which turn out to be in terms of Y_t and $|h_t'(z)|$, both of which are differentiable in t.

We can consider a new measure $\mathbb{P}^*$ obtained by weighting by the martingale M. To be more precise, if E is an event in the σ-algebra $\mathcal{F}_t = \sigma\{B_s : 0 \le s \le t\}$, then

$$\mathbb{P}^*(E) = M_0^{-1}\, \mathbb{E}\left[M_t\, 1_E\right].$$

The Girsanov theorem tells us that

$$dB_t = \frac{r\, X_t}{X_t^2 + Y_t^2}\, dt + dW_t,$$

where W_t is a Brownian motion with respect to $\mathbb{P}^*$. In other words,

$$dX_t = \frac{(r-a)X_t}{X_t^2 + Y_t^2}\, dt + dW_t. \tag{17}$$

♦Suppose M_t is a continuous, positive process satisfying

$$dM_t = A_t\, M_t\, dB_t.$$

Then M_t is a local martingale, but not necessarily a martingale. If one chooses stopping times τ_n by

$$\tau_n = \inf\{t : M_t \ge n \text{ or } |A_t| \ge n\},$$

then $M_t^{(n)} := M_{t\wedge\tau_n}$ is a martingale satisfying

$$dM_t^{(n)} = A_t\, 1_{\{\tau_n > t\}}\, M_t^{(n)}\, dB_t.$$

The Girsanov theorem tells us that it we weight by the martingale, then B_t satisfies

$$dB_t = A_t\, dt + dW_t, \quad t < \tau_n,$$

where W_t is a Brownian motion in the new measure, which we denote by $\mathbb{P}^*$. At the moment, this is only valid for $t < \tau_n$. However, if

$$\lim_{n\to\infty} \mathbb{P}^*\{\tau_n \le t\} = 0, \tag{18}$$

then we can conclude that the process is actually a martingale.

In our particular case, one can see from (17) and the bounds on Y_t that X_t does not blow up in finite time. Since $|h_t'(z)|$ is expressed in a differential equation involving X_t, Y_t, it also does not blow up. This is how one verifies (18) and shows that M_t is a martingale.

Let us now choose $z = i$ so that

$$M_0 = 1.$$

Since M_t is a martingale, we have for all r,

$$\mathbb{E}[M_t] = \mathbb{E}\left[|h_t'(i)|^{\lambda}\, Y_t^{\zeta}\, S_t^{-r}\right] = 1.$$

Typically we expect for large t that

$$Y_t \asymp t^{1/2}, \quad S_t \asymp 1, \tag{19}$$

and hence we might want to conclude that

$$\mathbb{E}\left[|h_t'(z)|^{\lambda}\right] \asymp t^{-\zeta/2}.$$

This is a hand-waving argument, and, in fact, it is not valid for all values of λ. As we will see below, the values of λ for which it will be valid are those values for which (19) holds typically *when we weight by the martingale* M_t. These values which we call *good* r satisfy

$$r < r_c := 2a + \frac{1}{2} = \frac{4}{\kappa} + \frac{1}{2}.$$

Let

$$q = r_c - r = 2a + \frac{1}{2} - r > 0.$$

Then the good values of r are those for which $q > 0$. Note that

$$\lambda(r_c) = a + \frac{3}{16a} + 1, \quad \zeta(r_c) = a - \frac{1}{16a}.$$

For $-\infty < r < r_c, -\infty < \lambda < \lambda_c = a + \frac{3}{16a} + 1$, the relationship $r \longleftrightarrow \lambda$ is a bijection and

$$r = 2a + 1 - \sqrt{(2a+1)^2 - 4a\lambda},$$

$$\zeta = \lambda - \frac{r}{2a} = \lambda + \frac{1}{2a}\sqrt{(2a+1)^2 - 4a\lambda} - 1 - \frac{1}{2a}.$$

2.3. Imaginary part parametrization. We assume that $z = i$ and $r < r_c$, that is,

$$q = 2a + \frac{1}{2} - r > 0.$$

We will introduce a time change under which the logarithm of the imaginary part of Z grows linearly. Let

$$\sigma(t) = \inf\{s : Y_s = e^{at}\},$$

and define

$$\hat{Z}_t = Z_{\sigma(t)}, \quad \hat{X}_t = X_{\sigma(t)}, \quad \hat{Y}_t = Y_{\sigma(t)} = e^{at}, \quad \hat{h}_t = h_{\sigma(t)}.$$

We also define

$$K_t = e^{-at}\,\hat{X}_t, \quad \hat{S}_t = S_{\sigma(t)} = \frac{e^{at}}{|\hat{Z}_t|} = \frac{1}{\sqrt{K_t^2+1}}, \quad J_t = \sinh^{-1}(K_t).$$

Under this parametrization, the pair of equations (13) can be written as a single one-variable SDE in K_t or J_t. We will list some computation below, but we summarize the basic idea as follows:

- If $r < r_c$ and we weight by the martingale M_t, then in the weighted measure, J_t is a positive recurrent diffusion.

EXERCISE 17. *Verify the following deterministic relations:*

$$\partial_t \sigma(t) = |\hat{Z}_t|^2,$$

$$\sigma(t) = \int_0^t e^{2as}\,(K_s^2+1)\,ds = \int_0^t e^{2as}\cosh^2 J_s\,ds,$$

$$|\hat{h}_t'(i)| = e^{aL_t},$$

where

$$L_t = \int_0^t \left[1 - \frac{2}{K_s^2+1}\right] ds = t - \int_0^t \frac{2}{\cosh^2 J_s}, ds.$$

$$e^{-at} \leq |\hat{h}_t'(i)| \leq e^{at}.$$

EXERCISE 18. *Show that there exists a standard Brownian motion* $\tilde{B}_t$ *such that*

$$dK_t = -2a\, K_t\, dt + \sqrt{K_t^2 + 1}\, d\tilde{B}_t,$$

$$dJ_t = -(q+r)\, \tanh J_t\, dt + d\tilde{B}_t.$$

Hint: This requires knowing how to handle time changes in SDE*s.*

EXERCISE 19. *If* $N_t = M_{\sigma(t)}$ *where* M_t *is the martingale in* (15), *then*

$$N_t = e^{\nu L_t}\, e^{\xi t}\, [\cosh J_t]^r, \tag{20}$$

where

$$\nu = a\lambda = r\left(\frac{q}{2} + \frac{1}{4}\right) + \frac{r^2}{4},$$

$$\xi = a\zeta = r\left(\frac{q}{2} + \frac{1}{4}\right) - \frac{r^2}{4},$$

Moreover,

$$dN_t = r\, [\tanh J_t]\, N_t\, d\tilde{B}_t.$$

Using the last exercise, we see that we must analyze the SDE

$$dJ_t = -(q+r)\, [\tanh J_t]\, dt + \tilde{B}_t, \quad J_0 = 0.$$

Note that this equation is written in terms of q, r; the parameter a has disappeared. We consider the martingale N_t in (20) which satisfies

$$dN_t = r\, [\tanh J_t]\, N_t\, d\tilde{B}_t, \quad N_0 = 1.$$

Let $\mathbb{P}^*, \mathbb{E}^*$ denote probabilities and expectations with respect to the measure obtained by weighting by the martingale N_t. Then

$$d\tilde{B}_t = r\, [\tanh J_t]\, N_t\, dt + dW_t,$$

where W_t is a standard Brownian motion with respect to $\mathbb{P}^*$. In particular,

$$dJ_t = -q\, [\tanh J_t]\, dt + dW_t.$$

♦Time changes of martingales (under some boundedness conditions) give martingales. Weighting by a time change of a martingale produces the same probability measure (on the σ-algebra $\mathcal{F}_\infty$) on a space as that obtained by weighting by the martingale. (This is subtle — a time changed process is not the same as the original process; it is the underlying measure on the probability space that is the same.) This is why we use the same letter $\mathbb{P}^*, \mathbb{E}^*$ for weighting by N_t as for M_t.

2.4. The one-variable SDE.

♦Many of the one-variable SDEs that arise in studying SLE can be viewed as equations arising from the Girsanov theorem by "weighting Brownian motion locally" by a function. Suppose F is a positive C^2 function on $\mathbb{R}$. Suppose B_t is a standard one-dimensional Brownian motion. Then Itô's formula gives

$$dF(B_t) = F(B_t)\,[A_t\,dt + \Phi_t\,dB_t],$$

where

$$\Phi_t = [\log F(B_t)]' = \frac{F'(B_t)}{F(B_t)}, \qquad A_t = \frac{F''(B_t)}{2\,F(B_t)}.$$

In other words, if

$$M_t = F(B_t)\,\exp\left\{-\int_0^t \frac{F''(B_s)}{2\,F(B_s)}\,ds\right\},$$

then M_t is a local martingale satisfying

$$dM_t = [\log F(B_t)]'\,M_t\,dB_t.$$

If F satisfies some mild restrictions, then M_t is a martingale. If we let $\mathbb{P}^*$ be the measure obtained by weighting by M_t, then the Girsanov theorem implies that

$$dB_t = [\log F(B_t)]'\,dt + dW_t, \tag{21}$$

where W_t is a $\mathbb{P}^*$-Brownian motion. Let $p_t(x,y)$ denote the transition probabilities for Brownian motion and $p_t^*(x,y)$ the transitions for B_t under $\mathbb{P}^*$, that is, the transitions for the equation (21). We know that $p_t(x,y) = p_t(y,x)$. In general, it is hard to give an expression for $p_t^*(x,y)$; however, if we consider a path $\omega(s), 0 \leq s \leq t$, from x to y of time duration t, then the Radon-Nikodym derivative of $\mathbf{P}^*$ with respect to $\mathbb{P}$ on this path is given by

$$\frac{F(y)}{F(x)}\,\exp\left\{-\int_0^t \frac{F''(\omega(s))}{2\,F(\omega(s))}\,ds.\right\}$$

The expression in the exponential may be complicated, but the key fact is that it is the same for the reversed path $\omega^R(s) = \omega(t-s)$ which goes from y to x. Hence we get the reversibility relation

$$p_t^*(x,y) = \frac{F(y)^2}{F(x)^2}\,p_t^*(y,x).$$

This implies that F^2 gives an invariant density for the SDE (21).

EXERCISE 20. *Use the ideas above to find the invariant probability for diffusions satisfying the following:*

$$dX_t = a\,[\cot X_t]\,dt + dW_t, \qquad a \geq 1/2, \qquad 0 < X_t < \pi.$$

$$dX_t = -q\,X_t\,dt + dW_t, \qquad q > 0.$$

In both cases, try to find a function F such that the equations arise by starting with a Brownian motion X_t and then weighting locally by the function F.

Let us consider the SDE

$$dJ_t = -q\,[\tanh J_t]\,dt + dW_t, \qquad J_0 = 0,$$

with $q > 0$. This is the equation obtained by weighting a Brownian motion locally by the function $f(x) = [\cosh x]^{-q}$ and, using this (see note above) or other standard techniques, one can see that this is positive recurrent diffusion with invariant probability density

$$v_q(x) = \frac{C_q}{\cosh^{2q} x}, \quad C_q^{-1} = \int_{-\infty}^{\infty} \frac{dx}{\cosh^{2q} x} = \frac{\Gamma(\frac{1}{2})\,\Gamma(q)}{\Gamma(q+\frac{1}{2})}.$$

Consider the functional L_t that appears in Exercise 17:

$$L_t = t - \int_0^t \frac{2}{\cosh^2 J_s}\, ds.$$

Since J_t is a positive recurrent distribution, at large times t the distribution of J_t is very close to the invariant distribution. If we write J_∞ for a random variable with the invariant distribution, we get

$$\lim_{t\to\infty} \frac{\mathbb{E}[L_t]}{t} = 1 - \mathbb{E}\left[\frac{2}{\cosh^2 J_\infty}\right] = 1 - \int_{-\infty}^{\infty} \frac{2\, C_q\, dx}{\cosh^{2q+2} x} = \beta := \frac{1-2q}{1+2q}.$$

Indeed, one expects more than convergence in expectation. Assuming that an appropriate strong law of large numbers and central limit theorem hold, we would expect

$$L_t = \beta t + O(t^{1/2}).$$

Indeed, one can give exponential estimates using the martingales for values of $\tilde{r}$ near r to show that there exists b such that

$$\mathbb{E}\left[\exp\left\{\frac{b|L_t - \beta t|}{\sqrt{t}}\right\}\right] \le c < \infty.$$

This gives immediate bounds on probabilities

$$\mathbb{P}\left\{|L_t - \beta t| \ge u\sqrt{t}\right\} = \mathbb{P}\left\{\exp\left\{\frac{b|L_t - \beta\, t|}{\sqrt{t}}\right\} \ge e^{bu}\right\} \le c\, e^{-bu}.$$

2.5. Returning to the reverse flow. Here we will not give complete details. We assume $z = i$. Since $N_t = M_{\sigma(t)}$ is a martingale,

$$\mathbb{E}[N_t] = \mathbb{E}[N_0] = 1.$$

Moreover, if E is any event depending on $B_s, 0 \le s \le \sigma(t)$, then

$$\mathbb{E}\left[N_t\, 1_E\right] = \mathbb{E}^*\left[1_E\right] = \mathbb{P}^*(E),$$

where as before $\mathbb{P}^*$ denotes probabilities obtained by weighting by the martingale. To compute $\mathbb{P}^*(E)$, one only needs to consider the one-variable SDE of the previous section.

For example, for some subpower function ψ we might specify the event $E_{t,c}$ such that the following holds for $0 \le s \le t$:

$$J_s \le c\,\psi(s),$$

$$J_s \le c\,\psi(t-s), \tag{22}$$

$$|L_s - \beta s| \le c\,\psi(s)\,\sqrt{s+1},$$

$$|L_s - \beta s| \le c\,\psi(t-s)\,\sqrt{s+1}. \tag{23}$$

Using the one-variable SDE we can show that

$$\lim_{c\to\infty} \liminf_{t\to\infty} \mathbb{P}^*(E_{t,c}) = 1. \tag{24}$$

♦The subpower function ψ needs to grow to infinity sufficiently fast for our estimates. However, for everything we do here, we could choose

$$\psi(s) = 1 \vee \exp\left\{(\log s)^{1/2}\right\},$$

which grows faster than any power of $\log s$. For some of our estimates we do need that

$$J_s \le c, \quad |L_s - \beta s| \le c\sqrt{s},$$

for $s = t - O(1)$. This is why we include the conditions (22) and (23).

We will now establish one direction of Theorem 11. Note that

$$\beta(\lambda) = \frac{1-2q}{1+2q} = \frac{2r-4a}{2+4a-2r} = \frac{1}{\sqrt{(2a+1)^2 - 4a\lambda}} - 1 = -\zeta'(\lambda).$$

(If one wants to understand why $\beta(\lambda)$ should equal $-\zeta'(\lambda)$, see the next subsection.) Choose c sufficiently large so that if $E = E_{t,c}$, then $\mathbb{P}^*(E) \ge 1/2$. Then

$$\frac{1}{2} \le \mathbb{E}\left[M_{\sigma(t)}\, 1_E\right] \le 1.$$

On the event E,

$$\sigma(t) = \int_0^t e^{2as} \cosh^2 J_s \asymp e^{2at}.$$

Here $\asymp$ means up to constants, which is stronger than up to " a subpower function". This uses the conditions (22) and (23). Also recall that

$$|h'_{\sigma(t)}(i)| = e^{aL_t} = \exp\left\{a\beta t + O(t^{1/2}\psi(t))\right\}.$$

With a little argument using distortion theorem ideas and the Loewner equation, we get (letting $T = e^{at}$),

$$\mathbb{E}\left[|h'_{T^2}(i)|^\lambda\, Y_{T^2}^\zeta\, S_{T^2}^{-r}\, 1_E\right] \asymp 1,$$

on an event on which

$$Y_{T^2} \asymp T, \quad S_{T^2} \asymp 1, \quad |h'_{T^2}(i)| \approx T^\beta.$$

In particular,

$$\mathbb{E}\left[|h'_{T^2}(i)|^\lambda\right] \ge c\,T^{-\zeta}.$$

The other direction takes a little more work, which we do not do here although it uses some of the same ideas. For the upper bound, we need to control the terms for which Y_t and S_t are far from their typical values. This was done for many values of r in [**6**] and for all $r < r_c$ in [**4**].

♦The remainder of this section will discuss more advanced topics relating to the reverse Loewner flow. This will not be needed in the later sections, so readers should feel free to skip now to Section 3.

2.6. Multifractal spectra. The basic ideas of multifractal spectrum are the same as those in basic "large deviation" theory and uses a simple idea that sometimes goes under the name of the Legendre transform. Let us explain it heuristically. Suppose that Z_n is a sequence of random variables for which we know the asymptotics of the moment generating function,

$$\mathbb{E}\left[e^{\lambda Z_n}\right] \approx e^{-n\zeta(\lambda)}, \tag{25}$$

for λ in an open interval about the origin. Define $\rho(s)$ roughly by

$$\mathbb{P}\{Z_n \approx sn\} \approx e^{-n\rho(s)}.$$

Then,

$$\mathbb{E}\left[e^{\lambda Z_n}; Z_n \approx sn\right] \gtrsim e^{[\lambda s - \rho(s)]n}.$$

The exponent $\zeta(\lambda)$ can be obtained by maximizing the right-hand side in s,

$$\zeta(\lambda) = \inf_s \left[\lambda s - \rho(s)\right].$$

If ρ is smooth enough, the infimum is obtained at s_λ where

$$\rho'(s_\lambda) = \lambda.$$

Conversely, the Chebyshev inequality gives

$$\mathbb{P}\{Z_n \approx sn\} \lesssim \mathbb{E}[e^{\lambda Z_n}]\, e^{-\lambda sn},$$

and equality is obtained for the optimal s. In other words,

$$\rho(s) = \max_\lambda[-\zeta(\lambda) - s\lambda].$$

The maximizer is obtained at λ_s satisfying

$$\zeta'(\lambda_s) = -s.$$

The "multifractal regime" is the regime where different values of s give different values of λ. In this case we can say roughly:

- The expectation in (25) is carried on the event $Z_n \approx s_\lambda n$. This event has probability about

$$e^{-n\rho(s_\lambda)} \approx e^{-[\zeta(\lambda)+\lambda\, s_\lambda]n}.$$

♦Large deviation theory generally discusses events whose probabilities decay exponentially. In critical phenomena, one generally has events whose probability decays like a power law. It is easy to convert to exponential scales. For example, if we want to study $|h'_{t^2}(i)|$, the corresponding random variable might be

$$Z_n = \log\left|h'_{e^{2n}}(i)\right|.$$

2.7. The tip multifractal spectrum for SLE. Consider the number of times $t \in \mathcal{D}_{2n}$ such that

$$|f_t'(i2^{-n})| \approx 2^{n\beta}.$$

This is imprecise, but I will state some theorems below. If $\beta < 1$, the expected number of such times is

$$2^{n(2-\zeta-\lambda\beta)}, \tag{26}$$

where

$$\beta = \frac{1}{\sqrt{(2a+1)^2 - 4a\lambda}} - 1, \quad \zeta = \lambda + \frac{1}{2a}\sqrt{(2a+1)^2 - 4a\lambda} - 1 - \frac{1}{2a}.$$

The condition $\beta < 1$ corresponds to

$$\lambda < \lambda_c = 1 + a + \frac{3}{16a},$$

and in this range we can solve for λ as a function of β. Since there are 2^{2n} intervals of length 2^{-2n} in $[0,1]$, we can interpret (26) as saying that the "fractal dimension" of the set of times in $[0,1]$ at which $|f_t'(i2^{-n})| \approx 2^{n\beta}$ is $(2-\zeta-\lambda\beta)/2$. This fractal dimension is maximized when $r = \zeta = \lambda = 0$. In this case,

$$\beta = \beta(0) = -\frac{2a}{2a+1} = -\frac{4}{4+\kappa}.$$

♦This says that for small y, the typical value of $|f_t'(iy)|$, *when the curve is parametrized by capacity*, is $y^{\frac{4}{4+\kappa}}$. As one example, consider the limit as $\kappa \to 0$ (let's choose the capacity parametrization at rate 2). In this case, the curve grows deterministically, and by solving the Loewner equation one can check that $|f_t'(iy)| \approx y$.

Let us consider $\beta(0) \leq \beta < 1$. This represents larger than typical values of $|f_t'(iy)|$ and corresponds to $0 \leq \lambda < \lambda_c$. Let

$$K_\beta = \left\{ t \in [0,1] : \lim_{y \to 0+} \frac{\log |f_t'(iy)|}{-\log y} = \beta \right\}.$$

$$\overline{K}_\beta = \left\{ t \in [0,1] : \limsup_{y \to 0+} \frac{\log |f_t'(iy)|}{-\log y} \geq \beta \right\}.$$

THEOREM 21. **[5, 6]** *Suppose $\beta(0) \leq \beta < 1$ and let $\rho = \zeta + \lambda\beta$. The following hold with probability one.*

- *If $\rho > 2$, $\overline{K}_\beta$ is empty.*
- *If $\rho < 2$,*

$$\dim_h \left[\overline{K}_\beta\right] \leq \frac{2-\rho}{2}, \quad \dim_h \left[\gamma(\overline{K}_\beta)\right] \leq \frac{2-\rho}{1-\beta}.$$

- *If $\rho < 2$,*

$$\dim_h \left[K_\beta\right] = \frac{2-\rho}{2}, \quad \dim_h \left[\gamma(K_\beta)\right] = \frac{2-\rho}{1-\beta}.$$

Here $\dim_h$ *denotes Hausdorff dimension.*

- Brownian motion in $\mathbb{R}^d$, $d \geq 1$, has the property that with probability one, if $A \subset [0,1]$ then $\dim_h(B(A)) = 2\dim_h(A)$. SLE_κ in the capacity parametrization does not have the property that $\dim_h(\gamma(A))$ depends only on $\dim_h(A)$.

- The relationship between $\dim_h[K_\beta]$ and $\dim_h[\gamma(K_\beta)]$ can be seen heuristically as follows. If $\dim_h[K_\beta] = \alpha$, then it takes about n^α intervals of length $1/n$ to cover K_β. The image of these intervals is dilated by a factor of about $|f'(1/\sqrt{n})| \approx n^{\beta/2}$. Hence it takes about $n^\alpha = n^{\frac{2\alpha}{1-\beta}\frac{2}{\beta-1}}$ sets of diameter $n^{2/(\beta-1)}$ to cover $\gamma(K_\beta)$. This suggests that the dimension of $\gamma(K_\beta)$ is $\frac{2\alpha}{1-\beta}$.
- The first two parts of the theorem can be deduced from the first moment bound Theorem 11. A similar bound was found in [**3**]. The lower bound is harder, and I will discuss this somewhat below.
- The maximum value of $(2-\rho)/(1-\beta)$ is $d = 1 + \frac{\kappa}{8}$ and occurs when

$$\beta_\# = \frac{\kappa}{\max\{4, \kappa - 4\}} - 1 > \beta(0).$$

A corollary of the last result is that the Hausdorff dimension of the SLE_κ path is d; this result was first proved in [**2**].

EXERCISE 22. *Use the conformal property of SLE_κ to show that for each β, the random variable $\dim_h[K_\beta]$ is constant with probability one.*

To prove the lower bound for a fixed $\beta(0) < \beta < 1$, we construct nontrivial (positive) measures ν, μ carried on $K_\beta, \gamma(K_\beta)$, respectively, such that

$$\int_0^1 \int_0^1 \frac{\nu(ds)\,\nu(dt)}{|s-t|^\alpha} < \infty, \quad \alpha < \frac{2-\rho}{2},$$

and

$$\int_{\mathbb{C}} \int_{\mathbb{C}} \frac{\mu(dz)\,\mu(dw)}{|z-w|^\alpha} < \infty, \quad \alpha < \frac{2-\rho}{\beta}. \tag{27}$$

We will focus on the latter, which is slightly more difficult to handle. By the exercise above, it suffices to show that such a measure exists with positive probability.

We construct μ as a (subsequential) limit of a sequence μ_n of approximating measures. The form of the approximating measures is

$$\mu_n = \sum_{1/2 \le t \le 1, t \in \mathcal{D}_{2n}} \mu(n,t),$$

where $\mu(n,t)$ is a multiple of Lebesgue measure on the disk of radius r_n about $f_t(i2^{-n})$, where the multiple is chosen so that the total mass of $\mu(n,t)$ is

$$2^{n(\zeta-2)}\,|f_t'(i2^{-n})|^\lambda\, J(t,n).$$

Here λ is the exponent associated to β, r_n is a deterministic sequence decreasing to zero, and $J(t,n)$ is the indicator function of a nice event. We will not be precise, but the form of the event is

$$y^{-\beta}\,\psi(1/y)^{-1} \le |f_t'(iy)| \le y^{-\beta}\,\psi(1/y), \quad 2^{-n} \le y \le 1.$$

The subpower function ψ is chosen sufficiently large so that

$$\mathbb{E}\left[|f_t'(i2^{-n})|^\lambda\, J(t,n)\right] \ge c\,\mathbb{E}\left[|f_t'(i2^{-n})|^\lambda\right] \asymp 2^{-n\zeta},$$

which implies $\mathbb{E}[\mu_n] \ge c_1$. If one can show that

$$\mathbb{E}[\mu_n^2] \le c_2,$$

$$\mathbb{E}\left[\int\int \frac{\mu_n(z)\,\mu_n(w)}{|z-w|^\alpha}\right] \le c_\alpha < \infty,$$

then standard techniques (see Section 3.7) imply that with probability $p = p(c_1, c_2) > 0$, we can find a subsequential limit satisfying (27).

We will give only a sketch as to why one would hope to get such an estimate. Let us fix $t, t + s^2 \in \mathcal{D}_n$ with $1/2 \le t < t + s^2 \le 1$. We would like to write this as

$$\mathbb{E}\left[|f_t'(i2^{-n})|^\lambda \, |f_{t+s}'(i2^{-n})|^\lambda J(t,n)\, J(t+s,n)\right].$$

Using the relationship with the reverse Loewner flow as in Section 2.1, we consider an expectation of the form

$$\mathbb{E}\left[|\tilde h_t'(z)|^\lambda \, |h_{t+s^2}'(z)|^\lambda J(t,n)\, J(t+s^2,n)\right], \qquad z = i2^{-n}.$$

(We abuse notation by using J for the corresponding events for the reverse flow.) Recall from Section 2.1 that

$$\tilde h_t'(z)\, h_{t+s^2}'(z) = \tilde h_t'(z)\, \tilde h_t'(Z_{s^2}(z))\, h_{s^2}'(z),$$

and the maps $\tilde h_t$ and h_{s^2} are independent. We also consider the map $\hat h$ which denotes the corresponding map at time $t - s^2$. Then we have

$$\tilde h_t'(z)\, h_{t+s^2}'(z) = \tilde h_{s^2}'(z)\, h_{2s^2}'(z)\, \hat h_{t-s^2}'(\tilde Z_{s^2}(z))\, \hat h_{t-s^2}'(Z_{2s^2}(z)).$$

Here we are writing

$$Z_r(z) = h_r(z) - V_r, \qquad \tilde Z_r(z) = \tilde h_r(z) - \tilde V_r.$$

- The probability that $|h_{s^2}'(z)| \approx [s2^n]^\beta$ is the same as
$$\mathbb{P}\{|h_{s^2 2^{2n}}'(i)| \approx 2^{\beta n}\, s^\beta\} \approx 2^{-n\xi}\, s^{-\xi}.$$
- When we weight paths by $|h_{s^2}'(z)|^\lambda$, and let $Z_{s^2}(z) = X_{s^2} + iY_{s^2}$, then $Y_{s^2} \approx s, |X_{s^2}| \le s\psi(1/s)$. Using the distortion theorem, if f is any conformal transformation of $\mathbb{H}$,
$$|f'(Z_{s^2}(z))| \approx |f'(is)|.$$
- The probability that $|\tilde h_{s^2}'(z)| \approx [s2^n]^\beta$ is about $2^{-n\xi}\, s^{-\xi}$.
- Since $Y_{s^2} \approx s, |X_{s^2}| \le s\psi(1/s)$.
$$|\tilde h_{s^2}'(Z_{s^2}(z))| \approx 1,$$
and hence
$$|h_{2s^2}'(z)| = |h_{s^2}'(z)\, \tilde h_{s^2}'(Z_{s^2}(z))| \approx [s2^n]^\beta,$$
and
$$|\tilde h_{s^2}'(z)\, h_{2s^2}'(z)| \approx [s2^n]^{2\beta}.$$
- When we weight by $|\tilde h_{s^2}'(z)\, h_{2s^2}'(z)|^\lambda$, then the typical path is as above, so that
$$|\hat h_{t-s^2}'(\tilde Z_{s^2}(z))| \approx |\hat h_{t-2s^2}'(Z_{2s^2}(z))| \approx |\hat h_{t-s^2}'(si)|.$$
- The probability that $|\hat h_{t-s^2}'(si)| \approx s^{-\beta}$ is about s^ζ.
- If we carry this argument out carefully, then we can choose appropriate events $J(t,n)$ such that

$$\mathbb{E}\left[|\tilde h_t'(z)|^\lambda \, |h_{t+s^2}'(z)|^\lambda J(t,n)\, J(t+s^2,n)\right] \le 2^{-2n\xi}\, s^{-\zeta}\, \psi(1/s). \tag{28}$$

- Using the Koebe-1/4 theorem, we can see that on this event
$$|\tilde{Z}_{s^2}(z) - Z_{2s^2}(z)| \geq s\,\psi(1/s),$$
and hence
$$|\gamma(t) - \gamma(t+s^2)| \geq s^{1-\beta}\,\psi(1/s).$$

♦This is only a basic sketch of the argument. I am not putting in more details, but let me mention some of the reasons for defining the measure the way that I did.

- One might try to avoid the event $J(t,n)$ and define a measure proportional to $|f_t'(i2^{-n})|^\lambda$. It is probably true that this would concentrate on the correct set. However, the second moment estimates become tricky, because one starts to estimate
$$\mathbb{E}\left[|f_t'(i2^{-n})|^\lambda\,|f_s'(i2^{-n})|^\lambda\right].$$
For s near t this starts looking like the 2λ power and the 2λ moment concentrates on a different event.
- One might try to put on a measure proportional to the indicator function of an event. However, we do not have as sharp estimates for this probability. It is important in getting the second moment estimate that we have an estimate as sharp as (28). In particular, for $1/s$ of order 1, the right-hand side is bounded by a constant times $2^{-n\zeta}$. It is because this is needed that the one-variable analysis leading to (24) was done.

---

3. The forward flow and dimension

3.1. Some intuition. Suppose $z \in \mathbb{H}$ and γ is a chordal SLE_κ path from 0 to ∞ in $\mathbb{H}$. We will ask two related questions:

- Does the path hit z? If not, how close does it get?
- What is the Hausdorff dimension of the path $\gamma(0,\infty)$?

It is known, and we will give a derivation here, that the curve is plane-filling if and only if $\kappa \geq 8$. In other words,
$$\mathbb{P}\{z \in \gamma(0,\infty)\} = \begin{cases} 1 & \kappa \geq 8 \\ 0 & \kappa < 8 \end{cases}.$$
Suppose D is a bounded domain, bounded away from the real line. Suppose $\kappa < 8$. If the fractal dimension of $\gamma \cap D$ is d, then we expect that the number of disks of radius ϵ needed to cover the curve is of order ϵ^{-d}. If we divide D into ϵ^{-2} disks of radius ϵ, then the fraction of these disks needed to cover D is ϵ^{2-d}. In other words, the probability that a particular disk of radius ϵ is needed should be about ϵ^{2-d}. Using this as intuition, we expect as $\epsilon \to 0$,
$$\mathbb{P}\{\mathrm{dist}(\gamma, z) \leq \epsilon\} \approx \epsilon^{2-d}. \tag{29}$$

The goal of this section is to give a precise version of the relation (29). Rohde and Schramm [**15**] first showed that the correct value is
$$d = 1 + \frac{\kappa}{8}.$$
More precise estimates [**2**] are needed to give the result about the Hausdorff dimension, which we state now.

THEOREM 23. [**2**] *If $\kappa < 8$, then with probability one, the Hausdorff dimension of γ_t for $t > 0$ is $1 + \frac{\kappa}{8}$.*

3.2. Basic definitions. Distance to the curve is not a conformal invariant or conformal covariant. A more useful, but similar, notion is conformal radius. Recall that $\mathbb{H}$ denotes the upper half-plane, and let $\mathbb{D} = \{z \in \mathbb{C} : |z| < 1\}$ denote the unit disk.

Definition If D is a (proper) simple connected domain and $z \in D$, we define $\Upsilon_D(z)$ to be one-half times the *conformal radius* of D with respect to z. In other words, if $f : \mathbb{D} \to D$ is a conformal transformation with $f(0) = z$, then

$$\Upsilon_D(z) = \frac{1}{2}\,|f'(0)|.$$

The factor of $1/2$ is a convenience so that

$$\Upsilon_{\mathbb{H}}(x + iy) = \mathrm{Im}(y).$$

It follows from the definition that the conformal radius is conformally *covariant* in the sense that if $F : D \to F(D)$ is a conformal transformation,

$$\Upsilon_{F(D)}(F(z)) = |F'(z)|\,\Upsilon_D(z). \tag{30}$$

By definition, we set $\Upsilon_{\mathbb{C}}(z) = \infty$. The conformal radius is closely related to the *inradius* defined by

$$\mathrm{inrad}_D(z) = \mathrm{dist}(z, \partial D).$$

EXERCISE 24. *Use the Koebe* $1/4$*-theorem to show that for any simply connected domain* D *and* $z \in D$*,*

$$\frac{1}{2}\,\Upsilon_D(z) \le \mathrm{dist}(z, \partial D) \le 2\,\Upsilon_D(z). \tag{31}$$

Definition If D is a simply connected domain and w_1, w_2 are distinct points in ∂D, then

$$S_D(z; w_1, w_2) = \sin[\arg f(z)].$$

where $f : D \to \mathbb{H}$ is a conformal transformation with $f(0) = w_1, f(\infty) = w_2$.

The transformation f is unique up to a dilation, and hence the argument of $f(z)$ does not depend on the choice of f. By definition, S_D is a conformal *invariant,*

$$S_{F(D)}(F(z); F(w_1), F(w_2)) = S_D(z; w_1, w_2).$$

Let $\mathrm{hm}_D(z, \cdot)$ denote harmonic measure, which is defined by saying that the probability that a Brownian motion starting at z exits D at $V \subset \partial D$ is given by $\mathrm{hm}_D(z, V)$.

EXERCISE 25.

- *Consider* $z = re^{i\theta} \in \mathbb{H}$*. Show that the probability that a Brownian motion starting at* z *exits* $\mathbb{H}$ *on* $(-\infty, 0)$ *equals* θ/π*. (Hint: what is the Poisson kernel in* $\mathbb{H}$*?)*
- *Find constants* $0 < c_1 < c_2 < \infty$ *such that the following holds. Suppose* D *is a simply connected domain and* w_1, w_2 *are distinct boundary points of* D*. Write*

  $$\partial D = \{w_1, w_2\} \cup A_1 \cup A_2$$

  where A_1, A_2 *are the two connected subarcs of* $D \setminus \{w_1, w_2\}$*. Then*

$$c_1\, S_D(z; w_1, w_2) \le \min\{\mathrm{hm}_D(z, A_1), \mathrm{hm}_D(z, A_2)\} \le c_2\, S_D(z; w_1, w_2). \tag{32}$$

♦When I discuss boundaries of simply connected domains D, I am using "prime ends". I will not give the precise definition, but the basic idea is that boundary points can be reached in different directions. For example, if $D = \mathbb{D} \setminus [0,1)$ and the boundary points are 0 and -1, then

$$A_1 = \{e^{i\theta} : 0 < \theta < \pi\} \cup (0,1]_+,$$
$$A_2 = \{e^{i\theta} : \pi < \theta < 2\pi\} \cup (0,1]_-,$$

where $(0,1]_+$ (resp., $[0,1]_-$) denotes the points in (0,1] reached from the upper half-plane (lower half-plane).

We now take a chordal SLE_κ path γ parametrized as in Section 0.1. For fixed $z \in \mathbb{H}$, we set $Z_t = Z_t(z) = X_t + iY_t = g_t(z) - U_t$, which satisfies

$$dZ_t = \frac{a}{Z_t}\, dt + dB_t.$$

3.3. Radial parametrization. Chordal SLE_κ uses the half-plane capacity. If $z \in \mathbb{H}$, we can choose a different parametrization such that $\log \Upsilon_{H_t}(z)$ decays linearly (this will be valid at least as long as $z \in H_t$.) Let us fix z, and let

$$\Upsilon_t = \Upsilon_t(z) = \Upsilon_{H_t}(z) = \frac{Y_t}{|g_t'(z)|}.$$

The last equality uses the scaling rule (30) with $F = g_t$. From the (deterministic) Loewner equation (1), we can compute

$$\partial_t |g_t'(z)| = \frac{a\,(Y_t^2 - X_t^2)}{(X_t^2 + Y_t^2)^2}, \qquad \partial_t \Upsilon_t = -\Upsilon_t\, \frac{2aY_t^2}{(X_t^2+Y_t^2)^2}.$$

Since Υ_t decreases with t we can define

$$\Upsilon_\infty = \lim_{t\to\infty} \Upsilon_t.$$

Definition If $z \in \mathbb{H}$, let

$$\sigma(t) = \sigma_z(t) = \inf\left\{s : \Upsilon_s \le e^{-2at}\right\},$$
$$\rho(t) = \rho_z(t) = \inf\left\{s : \operatorname{dist}(\partial H_t, z) \le e^{-2at}\right\}.$$

We call $\sigma(t)$ the *radial parametrization (with rate $2a$)*.

In the radial parametrization $\log \hat{\Upsilon}_t := \log \Upsilon_{\sigma(t)}$ decays linearly. We choose rate $2a$ to make some equations below a little nicer.

EXERCISE 26. *Suppose* $\operatorname{Im}(z) = 1$.

- *Find a constant c such that for all $t \ge 0$*

$$|\log \Upsilon_{\rho(t)} + 2at| \le c.$$

- *Show that there exists a $c > 0$ such that for all $t \ge 0$, $0 \le s \le 1$,*

$$\log \Upsilon_{\rho(t+s)} \le \log \Upsilon_{\rho(t)} + cs. \tag{33}$$

Hint: The first part is straightforward using (31), *but the second part requires more argument.*

In the radial parametrization, the argument behaves in a relatively simple fashion. Let

$$\Theta_t = \arg Z_{\sigma(t)}.$$

The next exercise gives the equation that we will use.

EXERCISE 27. *Show that there is a standard Brownian motion $\hat{W}_t$ such that Θ_t satisfies*

$$d\Theta_t = (2a-1)\,[\cot \Theta_t]\, dt + d\hat{W}_t. \tag{34}$$

Hint: There are two parts to this. First, one uses Itô's formula to give an equation for $\arg Z_t$, and then one converts to the radial parametrization. The rate $2a$ is chosen so that $\hat{W}_t$ is a standard Brownian motion.

♦By comparison with the Bessel equation, one can show that solutions to the "radial Bessel equation"

$$dX_t = \beta\,[\cot X_t]\, dt + dW_t,$$

reach the origin in finite time if and only if $\beta < 1/2$. Hence solutions to the equation (34) reach the origin in finite time if and only if $a > 1/4, \kappa < 8$. Note that finite time in the radial parametrization corresponds to $\Upsilon_\infty > 0$ which corresponds to $z \notin \gamma(0,\infty)$. Hence from this we can see that SLE_κ is plane-filling if and only $\kappa \geq 8$. Another observation is that Θ_t is a martingale if and only if $\kappa = 4$.

♦When considering SLE_κ near an interior point z, the radial parametrization is very useful. However, the parametrization depends on the point z, so it is not as useful for considering two interior points simultaneously.

3.4. Green's function and one-point estimate. If $\kappa < 8$, let

$$d = 1 + \frac{\kappa}{8} = 1 + \frac{1}{4a}.$$

This will be the fractal dimension of the paths, but for the time being, let us consider this only as a notation.

Definition The *Green's function (for chordal SLE_κ from w_1 to w_2 in simply connected D)* is

$$G_D(z; w_1, w_2) = \Upsilon_D(z)^{d-2}\, S_D(z; w_1, w_2)^{4a-1} = \Upsilon_D(z)^{\frac{\kappa}{8}-1}\, S_D(z; w_1, w_2)^{\frac{8}{\kappa}-1}.$$

We set

$$G(z) = G_{\mathbb{H}}(z; 0, \infty) = [\operatorname{Im} z]^{d-2} \sin^{4a-1}[\arg z].$$

The scaling rules for Υ_D and S_D imply the following scaling rule for G_D

$$G_D(z; w_1, w_2) = |F'(z)|^{2-d}\, G_{F(D)}(F(z); F(w_1), F(w_2)). \tag{35}$$

Roughly speaking, we think of $G_D(z; w_1, w_2)$ as representing the probability that the SLE_κ path gets close to z. A precise formulation of this comes in the following theorem which we prove in this section. If γ is an SLE_κ curve in D from w_1 to w_2, we write γ_t for $\gamma(0,t]$, D_t for the unbounded component of $D \setminus \gamma_t$ containing w_2 on its boundary, and if $z \in D$, $\Upsilon_t = \Upsilon_t(z) = \Upsilon_{D_t}(z; w_1, w_2)$ (if $z \notin D_t$, then $\Upsilon_t = 0$), and $\Upsilon_\infty = \lim_{t\to\infty} \Upsilon_t$. The next proposition shows that for all D, z, w_1, w_2,

$$\mathbb{P}\{\Upsilon_t \leq r\} \sim c_*\, G_D(z; w_1, w_2)\, r^{2-d}, \quad r \to 0+. \tag{36}$$

Using the scaling rule (35), we can write this as

$$\mathbb{P}\{\Upsilon_t \leq r\,\Upsilon_0\} \sim c\,(r\Upsilon_0)^{2-d}\, G_D(z; w_1, w_2) = c_*\, S_D(z; w_1, w_2)^{4a-1}\, r^{2-d}.$$

THEOREM 28. *For every $\kappa < 8$ there exists $u > 0$ such that the following holds. Suppose γ is an SLE_κ curve from w_1 to w_2 in D. Then for every $z \in D$,*

$$\mathbb{P}\{\Upsilon_\infty \leq r\,\Upsilon_0\} = c_*\, S_D(z; w_1, w_2)^{4a-1}\, r^{2-d}\, [1 + O(r^u)], \tag{37}$$

where

$$c_*^{-1} = \frac{1}{2}\int_0^\pi \sin^{4a} x\, dx.$$

In particular, if $r_0 < 1$, there exists $0 < c_1 < c_2 < \infty$ such that for all D, z, w_1, w_2 and all $0 < r \leq r_0$,

$$c_1\, S_D(z; w_1, w_2)^{4a-1}\, r^{2-d} \leq \mathbb{P}\{\Upsilon_\infty \leq r\,\Upsilon_0\} \leq c_1\, S_D(z; w_1, w_2)^{4a-1}\, r^{2-d}.$$

The relation (37) means the following. For each $r_0 < 1$, there exists $c < \infty$ which may depend on κ and r_0 but does not depend on r, D, w_1, w_2, z such that if $r \leq r_0$,

$$\left|\mathbb{P}\{\Upsilon_\infty \leq r\,\Upsilon_0\} - c_*\, S_D(z; w_1, w_2)^{4a-1}\, r^{2-d}\right| \leq c\, S_D(z; w_1, w_2)^{4a-1}\, r^{(2-d)+u}.$$

We have not motivated why the function G_D or the value d sould be as given. We do so now. Suppose G_D, d exist satisfying (36). Let $\mathcal{F}_t$ denote the σ-algebra generated by $\{B_s : 0 \leq s \leq t\}$ (or, equivalently, generated by $\{\gamma(s), 0 \leq s \leq t\}$). If $\rho = \inf\{t : \Upsilon_t = r\}$, then

$$N_t = \mathbb{P}\{\Upsilon_\infty \leq r \mid \mathcal{F}_t\}$$

should be a local martingale for $0 \leq t < \rho$. If this is true, then

$$M_t = \mathbb{E}\,[G_D(z; w_1, w_2) \mid \mathcal{F}_t]$$

should be a local martingale. The domain Markov property of SLE_κ and (37) imply that

$$\begin{aligned}
\mathbb{E}\,[G_D(z; w_1, w_2) \mid \mathcal{F}_t] &= G_{H_t}(z; \gamma(t), \infty) \\
&= |g_t'(z)|^{2-d}\, G_{\mathbb{H}}(g_t(z); U_t, \infty) \\
&= |g_t'(z)|^{2-d}\, G(Z_t(z)), \qquad Z_t(z) = g_t(z) - U_t.
\end{aligned}$$

The function G was first computed in [**15**] by essentially doing the following exercise.

EXERCISE 29. *Suppose $\kappa < 8$.*

- *Let*

  $$M_t = M_t(z) = G_{H_t}(z; \gamma(t), \infty) = |g_t'(z)|^{2-d} G(Z_t(z)).$$

  Then M_t is a local martingale satisfying

  $$dM_t = \frac{(1-4a)\, X_t}{X_t^2 + Y_t^2}\, dB_t.$$

- *Suppose $\Phi : \mathbb{H} \to (0, \infty)$ is a C^2 function satisfying $\Phi(rz) = r^{\alpha-2}\Phi(z)$ for some $\alpha > 0$. Suppose also that for each z,*

  $$|g_t'(z)|^{2-\alpha}\, \Phi(Z_t(z))$$

  is a local martingale. Prove that $\alpha = d$ and $\Phi = cG$ for some c.

Unlike the similar local martingale M_t from the reverse Loewner flow in the previous section, this local martingale is not a martingale. It "blows up" on the event of probability zero that $z \in \gamma(0,\infty)$.

In the radial parametrization,

$$\hat{M}_t = M_{\sigma(t)} = e^{2at(2-d)}\,[\sin\Theta_t]^{4a-1}.$$

(If $\Upsilon_\infty > e^{-2at}$, then $\sigma(t) = \infty$ and $\hat{M}_t = 0$.) Recall that in the radial parametrization,

$$d\Theta_t = (1-2a)\,\Theta_t\,dt + d\hat{W}_t.$$

If

$$T = \inf\{t : \sin\Theta_t = 0\},$$

we can write

$$\hat{M}_t = e^{2at(2-d)}\,[\sin\Theta_{t\wedge T}]^{4a-1}.$$

This is a martingale satisfying

$$d\hat{M}_t = (4a-1)\,[\cot\Theta_t]\,\hat{M}_t\,d\hat{W}_t, \qquad t < T.$$

(Itô's formula shows that this is a local martingale, and since $\hat{M}_t$ is uniformly bounded on every compact interval, we can see that it is actually a martingale.) If we weight by the martingale $\hat{M}_t$, then

$$d\hat{W}_t = (4a-1)\,\cot\Theta_t\,dt + dW_t,$$

where W_t is a Brownian motion in the new measure, which we denote by $\mathbb{P}^*$ (with expectations $\mathbb{E}^*$). In particular,

$$d\Theta_t = 2a\,\cot\Theta_t\,dt + dW_t.$$

Since $2a > 1/2$, with probability one with respect to the measure $\mathbb{P}^*$, the process never exits the open interval $(0,\pi)$. (Of course, since the martingale $\hat{M}_t$ equals zero when $\sin\Theta_t = 0$, it is obvious that if we weight by $\hat{M}_t$, the process should never leave $(0,\pi)$!)

♦Consider the SDE

$$dX_t = \beta\,\cot X_t\,dt + dB_t, \qquad 0 < X_0 < \pi,$$

where $\beta > 1/2$. This is the equation obtained by starting with a Brownian motion X_t and weighting locally by the function $f(x) = [\sin x]^\beta$. By comparison with a Bessel equation, it is not hard to show that with probability one $0 < X_t < \pi$ for all times. The invariant probability distribution (see Exercise 20 and the comment above that), is

$$v(x) = C_{2\beta}\,[\sin x]^{2\beta}, \qquad C_{2\beta}^{-1} = \int_0^\pi [\sin y]^{2\beta}\,dy.$$

If $p_t(x,y)$ denotes the density (as a function of y) of X_t given $X_0 = x$, it is standard to show that

$$p_t(x,y) = v(x)\,[1 + O(e^{-\alpha t})], \qquad t \geq 1,$$

for some $\alpha = \alpha_\beta > 0$. In particular, if Φ is a nonnegative function on $(0,\pi)$,

$$\mathbb{E}\,[\Phi(X_t)] = \left[\int_0^\pi \Phi(x)\,v(x)\,dx\right]\,[1 + O(e^{-\alpha t})], \qquad t \geq 1.$$

We can now prove Theorem 28. By conformal invariance, it suffices to consider SLE_κ from 0 to ∞ in $\mathbb{H}$ and $\mathrm{Im}(z) = 1$. Note that $\Upsilon_0 = 1$. Let $r = e^{-2at}$ and $v(x) = C_{4a} \sin^{4a} x$. Then,

$$\begin{aligned}
\mathbb{P}\{\Upsilon_\infty \leq r\} &= \mathbb{E}\,[1\{\Upsilon_\infty \leq r\}] \\
&= \mathbb{E}^*\left[\hat{M}_t^{-1}; T > t\right] \\
&= \mathbb{E}^*\left[\hat{M}_t^{-1}\right] \\
&= r^{2-d}\left[\int_0^\pi [\sin x]^{1-4a}\, v(x)\, dx\right][1 + O(r^u)] \\
&= 2\, C_{4a}\, r^{2-d}\,[1 + O(r^u)].
\end{aligned}$$

Using (33), we get the following corollary in terms of distances.

COROLLARY 30. *If $\kappa < 8$ there exists $c < \infty$ such that for every D, z, w_1, w_2 and every $0 < r \leq 1/2$, if γ is the path of SLE_κ from w_1 to w_2 in D, then*

$$\mathbb{P}\{\mathrm{dist}(\gamma, z) \leq r\,\mathrm{inrad}_D(z)\} \leq c\, S_D(z; w_1, w_2)^{4a-1}\, r^{2-d}. \tag{38}$$

The restriction $r \leq 1/2$ is not required if $D = \mathbb{H}$. This can be seen from the following estimate that we will not prove here. One can prove this similarly to the proofs in this section, but another simple proof can be found in [**1**]. Note that

$$S_{\mathbb{H}}(x + x\epsilon i; 0, \infty) \sim \epsilon, \qquad \epsilon \to 0+\,.$$

PROPOSITION 31. *If $\kappa < 8$, there exists $c < \infty$ such that if γ is the path of SLE_κ from 0 to ∞ in $\mathbb{H}$, $x, \epsilon > 0$, then*

$$\mathbb{P}\{\mathrm{dist}(\gamma, x) < \epsilon\, x\} \leq c\, \epsilon^{4a-1}.$$

3.5. Two-sided radial and radial SLE_κ. Two-sided radial SLE_κ from 0 to ∞ through z can be thought of as chordal SLE_κ from 0 to ∞ conditioned to go through z. This is conditioning on an event of probability zero, so we need to be careful in the definition. A standard way to define events "conditioned on events of measure zero" is to consider a sequence of events of positive probability decreasing to the event, condition with respect to these events of positive probability, and hope to obtain a limit of the measures. It is more convenient to use the Girsanov theorem directly to define two-sided radial SLE_κ but our definition is equivalent to other natural ways of defining the measure (see Exercise 32). The term "two-sided radial" comes from thinking of the path as two (interacting) radial paths from z to 0 and ∞, respectively.

Definition If $\kappa < 8$, then *two-sided radial SLE_κ from 0 to ∞ through z (up to time T_z)* is chordal SLE_κ weighted by the local martingale

$$M_t = G_{H_t}(z; \gamma(t), \infty).$$

If we use the radial parametrization as above, then for two-sided radial SLE,

$$d\Theta_t = 2a\, \sin\Theta_t\, dt + dW_t,$$

where W_t is a Brownian motion. In the half-plane capacity parametrization, two-sided radial SLE_κ satisfies

$$dX_t = \frac{(1 - 3a)X_t}{X_t^2 + Y_t^2}\, dt + dW_t, \qquad \partial_t Y_t = -\frac{aY_t}{X_t^2 + Y_t^2}. \tag{39}$$

EXERCISE 32. *Suppose $\kappa < 8$ and $z \in \mathbb{H}$. For $t < t'$ consider the following probability measures on paths $\gamma(s), 0 \leq s \leq \sigma(t)$. Here we use the half-plane capacity.*

- $\mathbb{P}_1 = \mathbb{P}_{1,t}$ *is chordal SLE_κ conditioned on the event $\{\sigma(t) < \infty\}$ stopped at time $\sigma(t)$.*
- $\mathbb{P}_2 = \mathbb{P}_{2,t}$ *is two-sided radial SLE_κ stopped at time $\sigma(t)$.*
- $\mathbb{P}_{3,t'} = \mathbb{P}_{3,t',t}$ *is chordal SLE_κ conditioned on the event $\{\sigma(t') < \infty\}$ stopped at time $\rho(t)$.*

Then,

- *Show that P_1, P_2 are mutually absolutely continuous and give the Radon-Nikodym derivative.*
- *Show that*
$$\lim_{t' \to \infty} \|P_2 - P_{3,t'}\| = 0,$$
where $\| \cdot \|$ denotes variation distance.

In the discussion on natural parametrization, we will need to consider the time duration in the half-plane capacity of radial $SLE_\kappa, \kappa < 8$. Suppose γ is two-sided radial SLE from 0 to ∞ in z and let $T_z = \inf\{t : \gamma(t) = z\}$. Then with probability one, $T_z < \infty$. We let $\phi(z;t)$ denote the distribution function,

$$\phi(z;t) = \mathbb{P}^*\{T_z \leq t\},$$

where $\mathbb{P}^*$ denotes probabilities using two-sided radial SLE_κ. This is also the distribution time for $\inf\{t : Y_t = 0\}$ where X_t, Y_t satisfy (39) with $X_0 + iY_0 = z$. Let $\phi(z) = \phi(z;1)$; scaling implies

$$\phi(z;t) = \phi(z/t^2).$$

We think of $\phi(z;t)$ as the probability that $z \in \gamma(0,t]$ given that $z \in (0,\infty)$. One can show that

$$\mathbb{E}\left[M_t(z)\right] = \left[1 - \phi(z;t)\right] M_0(z).$$

♦Two-sided radial can be considered as a type of "$SLE(\kappa, \rho)$" process. We will not define these processes here. In fact, most, if not all such processes, can be viewed as processes obtained from the Girsanov theorem by weighting by a local martingale. I find the Girsanov viewpoint more natural, because it generally can be seen as weighting locally by a function. The function may depend on a number of marked points, e.g., the SLE_κ Green's function.

3.5.1. *Radial SLE_κ.* Another example of a process that can be obtained from chordal SLE_κ by a local martingale is radial SLE_κ. If γ is an SLE_κ process, let

$$\Phi_t = H_{\mathbb{H}}(g_t(z), U_t) = |Z_t|^{-1} \arg[Z_t].$$

Here $H_{\mathbb{H}}(z,x)$ denotes π times the Poisson kernel for Brownian motion in $\mathbb{H}$. We will not need radial SLE in these notes, so we leave this relationship as an exercise.

EXERCISE 33. *Let γ denote SLE_κ. Let*

$$b = \frac{3a-1}{2} = \frac{6-\kappa}{2\kappa}.$$

Show that

$$d\Phi_t^b = \Phi_t^b \left[A_t\, dt + \frac{(1-3a)\, X_t}{X_t^2 + Y_t^2}\, dB_t \right],$$

for some A_t. In particular,

$$M_t = \exp\left\{-\int_0^t A_s\,ds\right\}\Phi_t^b,$$

is a local martingale satisfying

$$dM_t = \frac{(1-3a)\,X_t}{X_t^2+Y_t^2}\,M_t\,dB_t.$$

The process that one obtains by weighting by M_t in this exercise is *radial* SLE_κ from 0 to z in $\mathbb{H}$. More precisely, it is radial SLE_κ defined up to the first time that the path separates z from ∞. For this process, in the radial parametrization one gets

$$d\Theta_t = a\,\sin\Theta_t\,dt + dW_t,$$

where W_t is a standard Brownian motion. Radial SLE_κ is usually described as a random curve growing from the boundary of the unit disk to the origin. It is usually parametrized using a radial parametrization (so that the logarithm of the conformal radius decays linearly). Note that this process reaches the origin in finite time if and only if $a < 1/2$ ($\kappa > 4$). If $\kappa > 4$, the path disconnects z from ∞ in finite (radial) time.

3.6. Beffara's two-point estimate. The first proof of Theorem 23 as well as one proof of existence of natural parametrization uses the following estimate.

PROPOSITION 34. *Suppose $\kappa < 8$ and D is a bounded domain bounded away from the real line. Then there exists $c < \infty$ such that for all $\epsilon, \delta > 0$ and $z, w \in D$,*

$$\mathbb{P}\{\Upsilon_\infty(z) \le \epsilon, \Upsilon_\infty(w) \le \delta\} \le c\,\epsilon^{2-d}\,\delta^{2-d}\,|z-w|^{d-2}.$$

This proposition is not easy to prove. The hard work is showing the estimate when $|z-w|$ is of order 1. In Section 4.2 we discuss the proof of the following.

PROPOSITION 35. *Suppose $\kappa < 8$, and $0 < u_1 < u_2 < \infty$ Then there exists $c < \infty$ such that for all $\epsilon, \delta > 0$ and all z, w with*

$$u_1 \le \mathrm{Im}(z), \mathrm{Im}(w) \le u_2,$$

$$u_1 \le |z-w| \le u_2,$$

then

$$\mathbb{P}\{\Upsilon_\infty(z) \le \epsilon, \Upsilon_\infty(w) \le \delta\} \le c\,\epsilon^{2-d}\,\delta^{2-d}.$$

Let us discuss how to get Proposition 34 from Proposition 35. It suffices to consider $\epsilon, \delta, |z-w|$ sufficiently small and without loss of generality assume

$$\delta \le \epsilon \le |z-w|.$$

(If $|z-w| \le \epsilon$, we can use the estimate on $\mathbb{P}\{\mathrm{Im}(w) \le \delta\}$.) Let ρ be the first time t that $\Upsilon_t(z) \le 10|z-w|$. Then

$$5\,|z-w| \le \mathrm{inrad}_{H_\rho}(z) \le 20\,|z-w|.$$

$$\mathbb{P}\{\rho < \infty\} \asymp |z-w|^{2-d}. \tag{40}$$

Applying distortion estimates to g_ρ on the disk of radius $5|z-w|$ about z, we see that

$$|g_\rho(z) - g_\rho(w)| \asymp |g_\rho'(z)|\,|z-w| \asymp |g_\rho'(w)|\,|z-w| \asymp \mathrm{Im}[g_\rho(z)] \asymp \mathrm{Im}[g_\rho(w)].$$

(The reader may wish to verify this. Note that we get both lower and upper bounds on $|g_\rho(z) - g_\rho(w)|$. This uses the univalence of g_ρ.) Therefore $\Upsilon_\rho(w) \asymp |z - w|$. Therefore, for some c,

$$\mathbb{P}\{\Upsilon_\infty(z) \leq \epsilon, \Upsilon_\infty(w) \leq \delta \mid \mathcal{F}_\rho\} \leq \mathbb{P}\left\{\Upsilon_\infty(z) \leq \frac{\epsilon\, \Upsilon_\rho(z)}{|z-w|}, \Upsilon_\infty(w) \leq \frac{c\,\delta\, \Upsilon_\rho(w)}{|z-w|} \mid \mathcal{F}_\rho\right\}.$$

By conformal invariance, the right-hand side equals

$$\mathbb{P}\left\{\Upsilon_\infty(g_\rho(z)) \leq \frac{\epsilon\, \mathrm{Im}(g_\rho(z))}{|z-w|}, \Upsilon_\infty(g_\rho(w)) \leq \frac{c\,\delta\, \mathrm{Im}(g_\rho(w))}{|z-w|}\right\}.$$

If we let $z_1 = g_\rho(z)/\mathrm{Im}(g_\rho(z)), w_1 = g_\rho(w)/\mathrm{Im}(g_\rho(z))$, then by conformal invariance, this equals

$$\mathbb{P}\left\{\Upsilon_\infty(z_1) \leq \frac{\epsilon}{|z-w|}, \Upsilon_\infty(w_1) \leq \frac{\tilde{c}\,\delta}{|z-w|}\right\},$$

where $\tilde{c} = c\,\mathrm{Im}(w)/\mathrm{Im}(z)$. Since $|z_1 - w_1| \asymp \mathrm{Im}(w_1) \asymp \mathrm{Im}(z_1) = 1$, Proposition 35 shows that this probability is bounded above by

$$c\,\epsilon^{2-d}\,\delta^{2-d}\,|z-w|^{2(d-2)}.$$

Combining this with (40), we get Proposition 34.

3.7. Hausdorff dimension. Proposition 34 was the hard step in Beffara's proof of the lower bound of the Hausdorff dimension of SLE_κ curves. In this subsection, we will sketch the basic technique to convert such two-point estimates into estimates on dimension.

THEOREM 36. *If $\kappa < 8$, then with probability one the Hausdorff dimension of $\gamma[0,\infty)$ is $d = 1 + \frac{\kappa}{8}$.*

It is not difficult to see that the value of the dimension is almost surely constant. Giving the upper bound is not difficult using the one-point estimate. We will only discuss the hard direction, the lower bound. It suffices to show that for some domain D and every $\alpha < d$, there is a positive probability that the Hausdorff dimension of $\gamma \cap D$ is at least α. The hard work is the two-point estimate Proposition 34. The rest follows from the proposition below which has appeared a number of places.

PROPOSITION 37. *Suppose $0 < \beta < m$ and A is a random closed subset of $[0,1]^m$. If $\mathbf{j} = (j_1, \ldots, j_m) \in S_n := \{1, \ldots, 2^n\}^m$. Let*

$$V_n(\mathbf{j}) = \left[\frac{j_1 - 1}{2^n}, \frac{j_1}{2^n}\right] \times \cdots \times \left[\frac{j_m - 1}{2^n}, \frac{j_m}{2^n}\right],$$

let $K_n(\mathbf{j})$ denote the indicator function of the event

$$\{A \cap V_m(\mathbf{j}) \neq \emptyset\}.$$

Assume there exist $0 < c_1 < c_2 < \infty$ and a subpower function ψ such that

$$c_1\, 2^{(\beta-m)n} \leq \mathbb{E}\left[K_n(\mathbf{j})\right] \leq c_2\, 2^{(\beta-m)n}, \tag{41}$$

$$\mathbb{E}\left[K_n(\mathbf{j}_1)\, K_n(\mathbf{j}_2)\right] \leq 2^{2(\beta-m)n} \left(\frac{|\mathbf{j}_1 - \mathbf{j}_2|}{2^n}\right)^{(\beta-m)n} \psi\left(\frac{2^n}{|\mathbf{j}_1 - \mathbf{j}_2|}\right). \tag{42}$$

Then, with positive probability, $\dim_h(A) \geq \beta$.

SKETCH OF PROOF. Let

$$A_n = \bigcup_{K_n(\mathbf{j})=1} V_n(\mathbf{j}).$$

Then $A_1 \supset A_2 \supset A_3 \supset \cdots$ and

$$A = \bigcap_{n=1}^{\infty} A_n.$$

Let μ_n denote the (random) measure whose density with respect to Lebesgue measure on $\mathbb{R}^m$ is

$$f_n(x) = 2^{\beta n} 1\{x \in A_n\}.$$

The estimates (41) and (42) imply that there exist c_3, c_4 such that

$$c_3 \leq \mathbb{E}\left[\mu_n(A_n)\right]^2 \leq \mathbb{E}\left[\mu_n(A_n)^2\right] \leq c_4.$$

Moreover, for every $\alpha < \beta$, there exists $C_\alpha < \infty$ such that

$$\mathbb{E}\left[\mathcal{E}_\alpha(\mu_n)\right] \leq C_\alpha,$$

where

$$\mathcal{E}_\alpha(\mu) = \int\int \frac{\mu(dx)\,\mu(dy)}{|x-y|^\alpha}.$$

Using second moment methods and the Markov inequality, we see that there exists $c_5, c_6 > 0$ (independent of α) and $\tilde{C}_\alpha < \infty$ such that

$$\mathbb{P}\left\{c_5 \leq \mu_n(A_n) \leq c_6 \text{ and } \mathcal{E}_\alpha(\mu_n) \leq \tilde{C}_\alpha \text{ for infinitely many } n\right\} \geq c_5.$$

On the event on the left-hand side, we can take a subsequential limit and construct a measure μ supported on A with $\mu(A) \geq c_5$ and $\mathcal{E}_\alpha(A) \leq C_\alpha\}$. On this event, Frostman's lemma implies that $\dim_h(A) \geq \alpha$. □

4. Two-point estimates

In this section we sketch the main idea in the proof of Proposition 35 and discuss the multi-point Green's function for *SLE*.

4.1. Beurling estimate. There is one standard estimate for Brownian motion (harmonic measure) that we will use in the next section. We state it here. A proof can be found in [**7**].

PROPOSITION 38 (Beurling estimate). *There is a $c < \infty$ such that the following is true. Suppose $\eta : [0,1] \to \overline{\mathbb{D}}$ is a curve with $|\eta(0)| = \epsilon, |\eta(1)| = 1$. Then the probability that a Brownian motion starting at the origin reaches the unit circle without hitting η is bounded above by $c\epsilon^{1/2}$.*

♦This estimate is a corollary of a stronger result, the Beurling projection theorem, which implies that for fixed ϵ, the radial line segment $\{re^{i\theta} : \epsilon \leq r \leq 1\}$ maximizes the probability. By finding an appropriate conformal transformation, one can see that the probability in this case is asymptotic to $c\,\epsilon^{1/2}$.

4.2. Proof of Proposition 35. We will only sketch some of the main ideas in the proof of Proposition 35 following [**9**]. For simplicity we will consider $z = -1+i, w = 1+i$ and $\epsilon, \delta < 1/4$ with $n \geq 2$. The main idea is to show that if a curve is going to get very close to z and very close to w, then it does one of two things. Either it gets very close to z without having gotten very close to w and then gets close to w or vice versa. What is unlikely to happen is that the curves gets near z and then near w and then nearer to z and then nearer to w, etc. In order to keep track of this, we note that any time the path goes near z and then near w it must go through the imaginary axis $\mathcal{I} = \{iy : y > 0\}$. In fact, there are crosscuts contained in $\mathcal{I}$ that it must cross.

Some topological issues come up. One is to show that for each t, there exists a unique $\mathcal{I}_t$ with $\mathcal{I}_0 = \mathcal{I}$ such that $\{\mathcal{I}_t : t \geq 0\}$ satisfies the following properties. Recall that H_t is the unbounded component of $\mathbb{H} \setminus \gamma_t$,

- Each $\mathcal{I}_s$ is an open connected subarc of $\mathcal{I}$ that divides H_s into two components, one containing z and the other containing w.
- $s < t$ implies $\mathcal{I}_s \supset \mathcal{I}_t$.
- If $s < t$ and $\gamma[s,t] \cap \mathcal{I} = \emptyset$, then $\mathcal{I}_s = \mathcal{I}_t$.

We define a sequence of stopping times τ_j and radii q_j, r_j as follows.

- $\tau_0 = 0, q_0 = 1, r_0 = 1,$
- If $\tau_j < \infty$, let

$$\gamma^j = \gamma[0,\tau_j], \quad H^j = H_{\tau_j},$$

$$q_j = \text{dist}\left[z, \gamma^j\right], \quad r_j = \text{dist}\left[w, \gamma^j\right].$$

$$\sigma_{j+1} = \inf\left\{t > \tau_j : |\gamma(t) - z| \leq q_j/2 \text{ or } |\gamma(t) - w| \leq r_j/2\right\},$$

$$\tau_{j+1} = \inf\{t > \sigma_{j+1} : \gamma(t) \in \overline{\mathcal{I}_{\sigma_{j+1}}}\}.$$

- Note that if $\tau_{j+1} < \infty$, then either

$$q_{j+1} \leq q_j/2 \text{ and } r_{j+1} > r_j/2$$

or

$$q_{j+1} > q_j/2 \text{ and } r_{j+1} \leq r_j/2.$$

We call the two cases z-excursions and w-excursions, respectively.

What makes the proof tricky is that there are many different combinations of z and w excursions that can occur for which eventually $\text{dist}(z,\gamma) \leq \epsilon$ and $\text{dist}(w,\gamma) \leq \epsilon$. Let us consider the case of z-excursions. What we need to show is something like the following.

- There exists $\alpha > 0$ such that the probability of a successful z-excursion at the $(j+1)$st step with $q_{j+1} \leq \delta q_j$ given γ^j is bounded above a constant times

$$\delta^{2-d}\, q_j^{\alpha}. \tag{43}$$

We will discuss the proof of (43); the proof actually gives a particular value of α but this is not important. With this estimate, one can sum over all possible combinations of z and w excursions and get the result. This last step is fairly straightforward and we will not do this here.

To prove (43), it is important to realize that a successful z-excursion requires two events to occur.

- There exists a first time $T > \tau_j$ such that $|\gamma(T) - z| \leq \delta\, q_j$.
- The SLE_κ path hits the crosscut $\mathcal{I}_T$ after time T.

There are different topological situations to consider but the basic idea is that those configurations for which the first event is relatively likely to occur are those for which the second event is unlikely.

To separate into two cases, consider the domain H^j and let A_1, A_2 be the two arcs in the boundary as in (32). Let

$$\hat{S}_j = S_{H^j}(z; \gamma(\tau_j), \infty).$$

Recall from that equation that

$$\hat{S}_j \asymp \min\left\{\mathrm{hm}_{H^j}(z, A_1), \mathrm{hm}_{H^j}(z, A_2)\right\}.$$

Case 1. Suppose $I = \mathcal{I}_{\tau_j}$ is an unbounded segment. Then the endpoint of this component is $\gamma(\tau_j)$. Since I disconnects z and w, we can see that one of the arcs A_1, A_2, say A_2, defined at time τ_j has the following property: any path from z to A_2 staying in H^j must go through I. This observation and the Beurling estimate (Proposition 38) give

$$h_{H^j}(z, A_2) \leq c\, q_j^{1/2}$$

and hence

$$\hat{S}_j \leq c\, q_j^{1/2}.$$

Using (38), we see that the probability for γ to get within distance δr_j of z is bounded above by a constant times $\delta^{2-d}\, q_j^{(4a-1)/2}$.

Case 2. Suppose $I = \mathcal{I}_{\tau_j}$ is a bounded segment and let A_1, A_2 be the arcs in ∂H^j. There is a radial line of length q_j from z to ∂H^j; it hits one of these arcs, let us assume it is A_1. Let Δ_j denote the infimum of all r such that there exists a curve in H^j from z to A_2 that stays in the disk of radius r about z. The Beurling estimate and (32) as above imply

$$\hat{S}_j \leq c\,(q_j/\Delta_j)^{1/2}.$$

We split into two possibilities: $\Delta_j < \sqrt{q_j}$ and $\Delta_j \geq \sqrt{q_j}$.

- If $\Delta_j \geq \sqrt{q_j}$, then we can use (38) to say that the probability for γ to get within distance δr_j of z is bounded above by a constant times $\delta^{2-d}\, q_j^{(4a-1)/4}$.
- If $\Delta_j \leq \sqrt{q_j}$, there exist curves η_1, η_2 in the intersection of H^j with the disk of radius $q_j^{1/2}$ about z from z to A_1, A_2, respectively. We can concatenate these curves to get a crosscut of H^j from A_1 to A_2 going through z, staying in the disk of radius $q_j^{1/2}$ of z. This curve disconnects I from ∞ in H^j. Let

$$T = \inf\{t \geq \tau_j : |\gamma(t) - z| = \delta\, q_j\}.$$

By (38), the probability that $T < \infty$ is bounded above by a constant times δ^{2-d}. The claim is that the conditional probability that γ intersects $I = \mathcal{I}_T$ after time T given $T < \tau_{j+1}$ is bounded above by a constant times q^{α}. To see this we take a radial line segment from $\gamma(T)$ to z. Combining this with a subset of η, we can see that there is a curve from $\gamma(T)$ to ∂H_T that disconnects I from ∞ in H_T and has diameter less than $2q_j^{1/2}$. Using this and estimates on Brownian excursion measure (details omitted

— the Beurling estimate is used again), we can see that if l is the image of I under g_T, then

$$\mathrm{diam}(l)/\mathrm{dist}(U_T, l) \leq c\, q_j^{1/2}.$$

The probability that SLE_κ in H_T from $\gamma(T)$ to ∞ hits I can now be bounded using Proposition 31.

4.3. Multi-point Green's function for SLE_κ. The techniques to prove the two-point estimate in [9] can be used to prove the following two-point version of (38).

THEOREM 39. [9] *For every $\kappa < 8$, there exists a function $G(z, w)$ such that if $z, w \in \mathbb{H}$, then*

$$\lim_{\epsilon,\delta \to 0} \epsilon^{d-2}\, \delta^{d-2}\, \mathbb{P}\{\Upsilon_\infty(z) \leq \epsilon, \Upsilon_\infty(w) \leq \delta\} = c_*^2\, G(z, w).$$

This theorem does not give an explicit form for G. We can write

$$G(z, w) = \tilde{G}(z, w) + \tilde{G}(w, z),$$

where $\tilde{G}$ represents the "ordered" Green's function. For example $\tilde{G}(z, w)$ represents the probability of visiting z and then later visiting w. We can represent $\tilde{G}(z, w)$ in terms of two-sided radial SLE_κ through z as we now show. The limit in the theorem is independent of how ϵ, δ go to zero. Let us stretch things a bit and let ϵ go to zero fixing δ. Recall that (38) implies

$$\lim_{\epsilon \to 0} \epsilon^{d-2}\, \mathbb{P}\{\Upsilon_\infty(z) \leq \epsilon\} = c_*\, G(z).$$

The limiting distribution on paths is that of two-sided radial SLE_κ going through z. The conditional probability that $\Upsilon_\infty(w) \leq \delta$ "after γ hits z given γ goes through z" should be

$$\mathbb{P}\{\Upsilon_\infty(w) \leq \delta\}$$

for SLE_κ in H_{T_z} from z to ∞. Using (38) again we see that

$$\lim_{\epsilon \to 0} \epsilon^{d-2}\, \mathbb{P}\{\Upsilon_\infty(w) \leq \epsilon\} = c_*\, G_{H_{T_z}}(w; z, \infty).$$

Therefore,

$$\tilde{G}(z, w) = \mathbb{E}^*\left[G_{H_{T_z}}(w; z, \infty)\right] = \mathbb{E}^*\left[|g'_{T_z}(w)|^{2-d}\, G(Z_{T_z}(w))\right],$$

where $\mathbb{E}^*$ denotes expectation with respect to two-sided radial SLE_κ going through z.

The basic idea is that if γ is going to get very close to both z and w, it either first gets very close to z and then gets very close to w or vice versa. It does not keep going back and forth. The estimate (43) is the critical step for making this rigorous.

With the two-point Green's function, we have a two-point local martingale. Let

$$M_t(z, w) = |g'_t(z)|^{2-d}\, |g'_t(w)|^{2-d}\, G(Z_t(z), Z_t(w)). \tag{44}$$

Recall that

$$\sigma_z(s) = \inf\left\{t : \Upsilon_t(z) \leq e^{-2as}\right\}.$$

It follows from Theorem 39 that for all s, $M_{t\wedge(\sigma_z(s)\wedge\sigma_w(s))}$ is a martingale and hence $M_t(z, w)$ is a local martingale.

♦Another possible approach to find the multi-point Green's function is to find a function such that $M_t(z,w)$ in (44) is a local martingale. Using Itô's formula and the product rule, this gives a differential equation in three real variables. (Two complex variables gives four real variables, but a scaling relation reduces the number of variables by one.) This is a possible approach to finding a closed form for this function. However, I suspect that one would need an estimate similar to (43) to prove Theorem 39.

5. Natural parametrization or length

5.1. Motivation and heuristics. The capacity parametrization for SLE is very useful for analyzing the process. In particular, it is the parametrization in which the maps g_t are differentiable in t. However, if one considers scaling limits of discrete processes, there are other parametrizations that one would choose. For example, if one scales self-avoiding walks on the lattice, it is standard to parametrize so that each lattice step is taken in the same amount of time. We can ask if we can find such a parametrization for SLE_κ.

Let us suppose for the moment, that such a parametrization exists. Suppose γ is the curve of SLE_κ parametrized by half-plane capacity; for convenience, let us choose the rate so that the half-plane capacity of $\gamma(0,t]$ is t.

Let Θ_t denote the amount of "natural time" needed to traverse the curve $\gamma(0,t]$. Here, we are starting γ in the half-plane capacity parametrization. We would hope that Θ_t is a continuous, increasing process. We can also view Θ_t as a measure supported on the path where the measure of $\gamma(0,t]$ is Θ_t. In the case $\kappa \geq 8$, one natural choice would be $\Theta_t = \text{area}(\gamma(0,t])$. We will restrict our consideration to $\kappa < 8$. In this case, we might want to define Θ_t some d-dimensional measure of $\gamma(0,t]$. Suppose Θ_t exists, and consider $\Theta_{t+\Delta t} - \Theta_t$ which is the amount of "natural time" needed to traverse $\gamma[t,t+\Delta t]$. Let $\eta = g_t(\gamma[t,t+\Delta t])$. The capacity parametrization is invariant under g_t in the sense that the half-plane capacity of η is Δt. Hence we expect that $\text{diam}\,\eta \approx \sqrt{\Delta t}$. If the natural parametrization is a d-dimensional measure then the natural time needed to traverse η is about

$$(\text{diam}\,\eta)^d \approx (\Delta t)^{d/2}.$$

Also, d-dimensional measures have the property that if one blows up a set by a factor of r, then the measure is multiplied by r^d. If we consider $U_t + i\sqrt{\Delta t}$ as a typical point on η, we might guess that the amount of natural time to traverse $\gamma(t,t+\Delta t]$ is

$$|f_t'(i\sqrt{\Delta t})|^d\,(\Delta t)^{d/2}.$$

Using this as motivation and setting $\Delta t = n$ or 2^{-n}, we might conjecture the following:

$$\Theta_t = \lim_{n\to\infty} n^{-d/2} \sum_{j\leq tn} |f_{j/n}'(i/\sqrt{n})|^d,$$

$$\Theta_t = \lim_{n\to\infty} 2^{-nd/2} \sum_{j\leq t2^n} |f_{j2^{-n}}'(i2^{-n/2})|^d. \tag{45}$$

Much of the more intricate analysis in Section 2 was developed to try to understand the right-hand side. Although we still do not know that the limit exists, there are some things we can say.

In the notation of that section, note that if $r = 1$, then

$$\lambda = d, \quad \zeta = 2 - d, \quad q = 2a - \frac{1}{2}, \quad \frac{\beta}{2} = \frac{1}{4a} - \frac{1}{2} = d - \frac{3}{2}.$$

For $\kappa < 8$, $1 < r_c$, so the approach there works. In particular,

$$\mathbb{E}[|f'_{s/t^2}(i/t)|^d] = \mathbb{E}[|f'_s(i)|^d] \asymp s^{\frac{d}{2}-1}.$$

which gives, for example,

$$\mathbb{E}\left[n^{-d/2} \sum_{j \le n} |f'_{j/n}(i/\sqrt{n})|^d\right] \asymp n^{-d/2} \sum_{j \le n} j^{\frac{d}{2}-1} \asymp 1.$$

(Since we expect $\mathbb{E}[\Theta_1]$ to be positive and finite, this is consistent with our heuristics.) Also, the expectation in

$$\mathbb{E}[|f'_1(i/n)|^d]$$

is carried on an event on which

$$|f'_1(i/\sqrt{n})|^d \approx n^{\beta/2} = n^{d-\frac{3}{2}}.$$

The probability of this event is about $n^{-\alpha}$ where

$$n^{-\alpha}\, n^{d\beta/2} = n^{\frac{d}{2}-1}.$$

Therefore,

$$\alpha = d\beta + 2 - d = d^2 - 2d + 1$$

As we let $n \to \infty$, we get an exceptional set $A \subset [0, 1]$ of times which can be covered by about $n^{1-\alpha}$ intervals of length $1/n$. Hence has fractal dimension

$$1 - \alpha = d(2 - d).$$

However, the images of these exceptional intervals should have diameter of order

$$n^{-1/2} \cdot |f'_1(i/\sqrt{n})| \approx n^{d-2}.$$

Hence the image of A is covered by $n^{1-\alpha}$ intervals of diameter n^{d-2}, which means that the dimension d^* of the image satisfies

$$[n^{d-2}]^{-d^*} = n^{1-\alpha}.$$

This recovers $d^* = d$.

There has been a lot of hand-waving in this argument, but one can make it rigorous [**6**] to show the following.

- With probability one, the Hausdorff dimension of the set of times t such that

$$|f'_t(iy)| \approx y^{-\beta}, \quad y \to 0+, \tag{46}$$

  is $d(2 - d)$.
- With probability one, there is a subset A of times satisfying (46) such that $\gamma(A)$ has Hausdorff dimension d.

In particular, the Hausdorff dimension of the curve is at least d. This gives another proof of Beffara's theorem. In fact, this is one case of the analysis in Section 2. The details were carried out in this case is [**6**] and generalized to the tip multifractal spectrum in [**5**].

♦An important corollary of this multifractal analysis is that the natural parametrization is singular with respect to the capacity parametrization.

♦A comment about terminology: for smooth curves, the term natural parametrization is used in some circles for parametrization by arc length. We think of this as the d-dimensional analogue. It is also the scaling limit of parametrization by arc length and so the phrase natural *length* is also used. Natural length may be a better word, but one must remember that this is a d-dimensional quantity.

5.2. A rigorous definition of natural parametrization. Suppose γ is an SLE_κ curve with $\kappa < 8$. We have seen that if $z \in \mathbb{H}$, then

$$\lim_{\epsilon \to 0+} \epsilon^{d-2}\, \mathbb{P}\{\Upsilon(z) \leq \epsilon\} = c_*\, G(z).$$

We conjecture (and think it might not be too difficult to prove) that there exists a constant c_0 such that the following is true.

Conjecture. *There exist $c_0 > 0$, such that if D is a simply connected domain and γ is SLE_κ from w_1 to w_2, then for $z \in D$,*

$$\lim_{\epsilon \to 0+} \epsilon^{d-2}\ \mathbb{P}\{\mathrm{dist}(\gamma, z) \leq \epsilon\} = c_0\, G_D(z; w_1, w_2).$$

We will assume this conjecture. We list another conjecture, which we believe is more difficult.

Conjecture. *If D is a simply connected domain and γ is SLE_κ from w_1 to w_2, then for $z \in D$, $t > 0$, the limit*

$$\Theta_t = \lim_{\epsilon \to 0+} \epsilon^{d-2}\, c_0^{-1}\, \mathrm{area}\{z : \mathrm{dist}(\gamma_t, z) \leq \epsilon\}$$

exists.

Let us assume this conjecture and see what it implies. For ease, let us assume that D is a bounded domain, and γ is an SLE_κ curve from w_1 to w_2 in D. The choice of parametrization is not really important, but let us assume that it retains the capacity parametrization from the upper half-plane. Then

$$\Theta_\infty = \lim_{\epsilon \to 0+} \epsilon^{d-2}\, c_0^{-1}\, \mathrm{area}\{z : \mathrm{dist}(\gamma, z) \leq \epsilon\}. \tag{47}$$

It seems reasonable from (47) that

$$\mathbb{E}[\Theta_\infty] = \int_D G_D(z; w_1, w_2)\, dA(z),$$

where we write dA for integrals with respect to area. Also,

$$\mathbb{E}[\Theta_\infty \mid \gamma_t] = \mathbb{E}[\Theta_t + (\Theta_\infty - \Theta_t) \mid \gamma_t] = \Theta_t + \mathbb{E}[\Theta_\infty - \Theta_t \mid \gamma_t].$$

But the conformal Markov property and (47) would imply that

$$\mathbb{E}[\Theta_\infty - \Theta_t \mid \gamma_t] = \int_{D_t} G_{D_t}(z; \gamma(t), w_2)\, dA(z),$$

where, as before, D_t is the unbounded component of $D \setminus \gamma_t$ containing w_2 on the boundary. But $\mathbb{E}[\Theta_\infty \mid \gamma_t]$ is a martingale. Hence we get the following characterization.

- Θ_t is the unique increasing, adapted process such that

$$N_t := \Theta_t + \int_{D_t} G_{D_t}(z; \gamma(t), w_2)\, dA(z) \tag{48}$$

  is a martingale.

We will use this to try to construct Θ_t.

Let us return to the upper half-plane. There is a technical issue that we would expect $\mathbb{E}[\Theta_\infty] = \infty$. To avoid this problem, suppose D is a bounded subdomain of $\mathbb{H}$ and let, $\Theta_t = \Theta_t(D)$ be the amount of "natural time" that the process has spent in D up to capacity time t. Then, using the reasoning above, we would expect

$$\mathbb{E}[\Theta_\infty] = \int_D G(z)\, dA(z)$$

$$\mathbb{E}[\Theta_\infty \mid \gamma_t] = \Theta_t + \Psi_t$$

where

$$\Psi_t = \Psi_t(D) = \int_D M_t(z)\, dA(z),$$

and

$$M_t(z) = G_{H_t}(z; \gamma(t), \infty) = |g_t'(z)|^{2-d}\, G(Z_t(z))$$

is the local martingale from Section 3.4.

Definition The *natural parametrization* Θ_t *(restricted to* D*)* is the unique, adapted, continuous, increasing process such that

$$N_t = \Psi_t + \Theta_t$$

is a martingale.

The definition implies the existence of such a process. Uniqueness is not difficult since the difference between any two candidates would be a continuous martingale with paths of bounded variation and hence must be zero. Existence is the issue. Since $M_t(z)$ is a positive local martingale, it is a supermartingale. We recall the function ϕ from Section 3.5 which satisfies

$$\mathbb{E}\left[M_t(z)\right] = \left[1 - \phi(z;t)\right] \mathbb{E}\left[M_t(z)\right], \qquad \phi(z) = \phi(z;1).$$

From this, one can see that Ψ_t is a supermartingale. The Doob-Meyer decomposition implies that there exists Θ_t such that Ψ_t is a local martingale. However, it is not immediate that Θ_t is nontrivial.

♦$M_t(z)$ is a positive supermartingale, but the increasing process Θ_t^* that makes $M_t(z) + \Theta_t^*$ a local martingale is the zero process. This shows that some work is needed to show our Θ_t is not identically zero. In particular, we need to show that Ψ_t is not a local martingale.

It turns out the process Θ_t is nontrivial for all $\kappa < 8$. This was proved for $\kappa < \kappa_0 = 4(7 - \sqrt{33}) = 5.021\cdots$. in [**8**] and for $\kappa < 8$ in [**11**]. The former approach, which should work for all $\kappa < 8$, yields some more information, so we will discuss both methods. Let us consider Θ_1.

The way to construct Θ_1, as in the proof of the Doob-Meyer theorem, is to discretize time. Let $\mathcal{D}_n$ denote the dyadics. Then for $t \in \mathcal{D}_n$, we let

$$\Theta^{(n)}_{t+2^{-n}} - \Theta^{(n)}_t = \mathbb{E}[\Psi_t - \Psi_{t+2^{-n}} \mid \gamma_t].$$

Using the conformal Markov property of SLE and by changing variables, one can show that

$$\mathbb{E}[\Psi_t - \Psi_{t+2^{-n}} \mid \gamma_t] = \int_{\mathbb{H}} |f_t'(z)|^d\, \phi(z; 2^{-n})\, G(z)\, 1\{f_t(z) \in D\}\, dA(z). \tag{49}$$

Hence

$$\Theta_1^{(n)} = \sum_{j=0}^{2^n-1} \int_{\mathbb{H}} |f'_{j2^{-n}}(z)|^d \, \phi(z; 2^{-n}) \, G(z) \, 1\{f'_{j2^{-n}}(z) \in D\} \, dA(z). \tag{50}$$

♦The expression (49) is a little complicated, but let us approximate it. The function $\phi(z; 2^{-n})$ is of order one for $|z| \leq 2^{-n/2}$ and near zero otherwise. For typical z in this disk, $G(z) \asymp (2^{-n/2})^{d-2} = 2^{-n(\frac{d}{2}-1)}$. If we choose $i2^{-n/2}$ as a typical point in the disk, then we can see that the expression is of the same order of magnitude as

$$2^{-n} \, 2^{-n(\frac{d}{2}-1)} \, |f'(i2^{-n/2})|^d = 2^{-nd/2} \, |f'(i2^{-n/2})|^d.$$

Given this we see the similarity between this expression and the one in (45).

It is easy to see that the process

$$N_t^{(n)} = \Psi_t + \Theta_t^{(n)}, \quad t \in \mathcal{D}_n,$$

is a discrete time martingale. The hard step is to take the limit. If the martingales have a uniform L^2 bound, then there one can take the limit easily. In [**8**], this bound was shown directly for $\kappa < \kappa_0$. It was conjectured that this would hold for all $\kappa < 8$. The method gives a bound on the Hólder continuity of Θ_t (with respect to the capacity parametrization).

The Doob-Meyer theorem says that one can take the limit provided that the collection of random variables $\{\Theta_T\}$ is uniformly integrable where T runs over all stopping times with $T \leq 1$. Using this, as in [**11**] adapting an argument from [**16**], we can show existence for all $\kappa < 8$. The argument uses the multi-point Green's function $G(z, w)$. Using Theorem 39, one can see that

$$M_t(z, w) = |g_t'(z)|^{2-d} \, |g_t'(w)|^{2-d} \, G(Z_t(z), Z_t(w))$$

is a positive local martingale and hence a supermartingale. The two-point estimate Proposition 34 combined with Theorem 39 gives

$$G(z, w) \leq c \, |z - w|^{d-2} \qquad z, w \in D.$$

In [**11**], it is shown that there exists $c < \infty$ such that

$$G(z) \, G(w) \leq c \, G(z, w).$$

From this we see that if T is a stopping time,

$$\mathbb{E}\left[M_T(z) \, M_T(w)\right] \leq c \, \mathbb{E}\left[M_T(z, w)\right] \leq c \, \mathbb{E}\left[M_0(z, w)\right] = c \, G(z, w) \leq c \, |z - w|^{d-2}.$$

Therefore,

$$\mathbb{E}\left[\Psi_T^2\right] \leq \int_D \int_D \mathbb{E}\left[M_T(z) M_T(w)\right] \, dA(z) \, dA(w) \leq c \int_D \int_D |z-w|^{d-2} \, dA(z) \, dA(w) \leq c.$$

The uniform bound on $\mathbb{E}\left[\Psi_T^2\right]$ shows the uniform integrability.

5.3. Other domains and other reference measures. We have defined the natural parametrization Θ_t for SLE_κ in $\mathbb{H}$ and given an expression for it as a limit (50). This can be taken as a limit with probability one at least if we take a subsequence. Suppose γ is an SLE_κ curve in $\mathbb{H}$ with the capacity parametrization. Define

$$S_t = \inf\{s : \Theta_s = t\} \quad \eta(t) = \gamma(S_t).$$

Then η is SLE_κ in $\mathbb{H}$ with the natural parametrization.

Suppose

$$F : \mathbb{H} \to D$$

is a conformal transformation with $F(0) = w_1, F(\infty) = w_2$. If γ is an SLE_κ curve in $\mathbb{H}$ with the capacity parametrization, then

$$\tilde{\gamma} = F \circ \gamma(t)$$

is SLE_κ in D from w_1 to w_2 with the capacity parametrization. Define Θ_t^D by

$$\Theta_t^D = \int_0^t |F'(\gamma(s))|^d \, d\Theta_s = \int_0^{\Theta_t} |F'(\eta(s))|^d \, ds.$$

It is not too difficult to show that Θ_t^D satisfies the characteristic property (48) and hence is the correct definition of the natural parametrizaton in D.

If D is a subdomain of $\mathbb{H}$ and $w_1 = 0, w_2 = \infty$, there are two ways to define the natural parametrization for a curve — by considering is as a curve in $\mathbb{H}$ or as a curve in D. If we prove something like (47), it would be obvious that the definitions agree. While we do not have (47), in [**10**] it is shown that the definitions agree. There are many questions which are open. For example, is the natural parametrization for a curve the same as the natural parametrization of the reversal? Dapeng Zhan [**17**] proved that SLE_κ is reversible at least for $0 < \kappa \leq 4$; if we have a result like (47), reversibility of the natural length would be immediate. Unfortunately, it is still an open problem.

Although we used area as our reference measure, we could also define natural parametrization $\Theta_{t,\mu}$ with respect to a different positive measure μ. In this case, we would set

$$\Psi_t = \Psi_{t,\mu} = \int_D G_{D_t}(z; \gamma(t), w_2) \, \mu(dz),$$

and require that $\Psi_t + \Theta_{t,\mu}$ be a martingale. If one follows the proof, one can see that a condition on μ sufficient in order for the $\Theta_{t,\mu}$ to be well defined is

$$\int\int \frac{\mu(dz)\mu(dw)}{|z-w|^{2-d}} < \infty.$$

In other words, if μ is an α-dimensional measure for $\alpha > 2 - d$, then it is well defined.

EXERCISE 40. *Suppose η is SLE_κ in $\mathbb{H}$ from 0 to ∞ in the natural parametrization. Suppose $F : \mathbb{H} \to D$ is a conformal transformation as above, and let $\tilde{\eta}(t) = F(\eta(t))$. Show that $\tilde{\eta}$ has the natural parametrization with respect to the measure μ with*

$$\mu(dz) = |(F^{-1})'(z)|^d \, dA(z).$$

An exercise: Brownian motion in three dimensions

An example which has many of the properties that we have seen in the last three sections is Brownian motion. We will discuss the case of three dimensions and give a sequence of exercises for the reader. This example is somewhat easier than SLE_κ but it contains many of the same ideas. Throughout this subsection, B_t will denote a standard Brownian motion in $\mathbb{R}^3$ and $G(x) = 1/|x|$ will denote the Green's function of three-dimensional Brownian motion. The important properties of G are radial symmetry and the fact that G is harmonic on $\mathbb{R}^3 \setminus \{0\}$. Let

$$M_t(z) = G(B_t - z),$$

$$\rho_z(r) = \inf\{t : |B_t - z| \leq r\},$$

EXERCISE 41. *Show that for $z \neq 0$, $M_t(z)$ is a local martingale satisfying*

$$dM_t(z) = M_t \, \frac{\nabla G(B_t - z)}{G(B_t - z)} \cdot dB_t.$$

Use the local martingle to conclude

$$\mathbb{P}\{\rho_z(r) < \infty\} = \frac{r}{|z|} \wedge 1.$$

For each z, we can consider the measure $\mathbb{P}^*_z$ obtained by weighting by the local martingale $M_t(z)$. Under this measure

$$dB_t = J_t \, dt + dW_t, \qquad J_t = \frac{\nabla G(B_t - z)}{G(B_t - z)},$$

where W_t is a standard three-dimensional Brownian motion with respect to $\mathbb{P}^*_z$. We claim that $M_t(z)$ is not a martingale. This can be derived from either of the next two exercises. Let

$$T_z = \inf\{t : B_t = z\}.$$

If $z \neq 0$, then $\mathbb{P}\{T_z = \infty\} = 1$.

EXERCISE 42. *Show that with probability one with respect to $\mathbb{P}^*_z$, $T_z < \infty$.*

EXERCISE 43. *Show that*

$$\lim_{t \to \infty} \mathbb{E}\left[M_t(z)\right] = 0.$$

The next exercise concerns the two-point Green function. This is the analogue of Theorem (39) although the exercise is significantly easier than the proof of that theorem.

EXERCISE 44. *Show that*

$$\lim_{\epsilon, \delta \to 0+} \epsilon^{-1} \, \delta^{-1} \, \mathbb{P}\left\{\rho_z(\epsilon) < \infty, \rho_w(\delta) < \infty\right\} = G(z, w),$$

where

$$G(z, w) = \left[G(z) + G(w)\right] G(z - w).$$

Show that $M_t(z, w) = G(B_t - z, B_t - w)$ is a local martingale.

EXERCISE 45. *Show there exist c_1, c_2 such that*

$$c_1 G(z) \, G(w) \leq G(z, w) \leq c_2 \, G(z) \, G(w) \, \frac{|z| + |w|}{|z - w|}.$$

EXERCISE 46. *Show there exists $c < \infty$ such that for every stopping time T,*

$$\mathbb{E}\left[M_T(z)\, M_T(w)\right] \leq c\, G(z)\, G(w)\, \frac{|z| + |w|}{|z - w|}.$$

Suppose D is a bounded domain and let

$$\Psi_t = \Psi_t(D) = \int_D M_t(z)\, d^3 z.$$

EXERCISE 47. *Show that Ψ_t is a supermartingale. Let*

$$l_t = \int_0^t 1\{B_s \in D\}\, ds.$$

Show that there exists c such that

$$\Psi_t + c\, l_t$$

is a martingale.

The take-home message from the last exercise is that the usual parametrization of Brownian motion is (up to a multiplicative constant) its natural parametrization.

References

[1] T. Alberts and M.J. Kozdron. Intersection probabilities for a chordal SLE path and a semicircle. *Electron. Commun. Probab.*, 13:448–460, 2008.

[2] V. Beffara. The dimension of the SLE curves. *Ann. Probab.*, 36(4):1421–1452, 2008.

[3] D. Beliaev and S. Smirnov. Harmonic measure and SLE. *Comm. Math. Phys.*, 290(2):577–595, 2009.

[4] F. Johansson Viklund and G. Lawler. Optimal Hölder exponent for the SLE path. *Duke Math J.*, 159(3):351–383, 2011.

[5] F. Johansson Viklund and G. Lawler. Multifractal spectrum for the tip of an SLE curve. *Arxiv preprint arXiv:0911-1.3983*, 2009.

[6] G. Lawler. Multifractal analysis of the reverse flow for the Schramm-Loewner evolution. In *Fractal Geometry and Stochastics IV*, volume 61 of *Progress in Probability*, pages 73–107. Birkhäuser, Basel, 2009.

[7] G. Lawler. *Conformally invariant processes in the plane.* Amer Mathematical Society, 2008.

[8] G. Lawler and S. Sheffield. A natural parameterization for the Schramm-Loewner evolution. *Ann. Probab.* 39(5):1896–1937, 2011

[9] G. Lawler and B. Werness. Multi-point Green's function for *SLE* and an estimate of Beffara. to appear in *Ann. Probab.*

[10] G. Lawler and M. Rezeai. Basic properties of the natural parametrization for the Schramm-Loewner evolution. preprint.

[11] G. Lawler and W. Zhou. SLE curves and natural parametrization. to appear in *Ann. Probab.*

[12] G. Lawler, O. Schramm, and W. Werner. Conformal invariance of planar loop-erased random walks and uniform spanning trees. *Ann. Probab.*, 32(1B):939–995, 2004.

[13] J. Lind. Hölder regularity of the SLE trace. *Trans. Amer. Math. Soc.*, 360(7):3557–3578, 2008.

[14] D. Marshall and S. Rohde. The Loewner differential equation and slit mappings. *J. Amer. Math. Soc.*, 18(4):763–778, 2005.

[15] O. Schramm and S. Rohde. Basic properties of SLE. *Annals of mathematics*, 161(2):883, 2005.

[16] O. Schramm and W. Zhou. Boundary proximity of SLE. *Probab. Theory Related Fields*, 146(3-4):435–450, 2010.

[17] D. Zhan. Reversibility of chordal SLE. *Ann. Probab.*, 36(4):1472–1494, 2008.

DEPARTMENT OF MATHEMATICS, UNIVERSITY OF CHICAGO
E-mail address: `lawler@math.uchicago.edu`

Clay Mathematics Proceedings
Volume 15, 2012

Lectures on Random Polymers

Francesco Caravenna, Frank den Hollander, and Nicolas Pétrélis

ABSTRACT. These lecture notes are a guided tour through the fascinating world of polymer chains interacting with themselves and/or with their environment. The focus is on the mathematical description of a number of physical and chemical phenomena, with particular emphasis on phase transitions and space-time scaling. The topics covered, though only a selection, are typical for the area. Sections 1–3 describe models of polymers without disorder, Sections 4–6 models of polymers with disorder. Appendices A–E contain tutorials in which a number of key techniques are explained in more detail.

CONTENTS

Foreword

These notes are based on six lectures by Frank den Hollander and five tutorials by Francesco Caravenna and Nicolas Pétrélis. The final manuscript was prepared jointly by the three authors. A large part of the material is drawn from the monographs by Giambattista Giacomin [**55**] and Frank den Hollander [**70**]. Links are

2010 *Mathematics Subject Classification.* Primary: 60F10, 60K37, 82B26, 82D60; Secondary: 60F67, 82B27, 82B44, 92D20.

made to some of the lectures presented elsewhere in this volume. In particular, it is argued that in two dimensions the Schramm-Loewner Evolution (SLE) is a natural candidate for the scaling limit of several of the "exotic lattice path" models that are used to describe self-interacting random polymers. Each lecture provides a snapshot of a particular class of models and ends with a formulation of some open problems. The six lectures can be read independently.

Random polymers form an exciting, highly active and challenging field of research that lies at the crossroads between mathematics, physics, chemistry and biology. DNA, arguably the most important polymer of all, is subject to several of the phenomena that are described in these lectures: *folding* (= collapse), *denaturation* (= depinning due to temperature), *unzipping* (= depinning due to force), *adsorption* (= localization on a substrate).

1. Background, model setting, free energy, two basic models

In this section we describe the physical and chemical background of random polymers (Sections 1.1–1.4), formulate the model setting in which we will be working (Section 1.5), discuss the central role of free energy (Section 1.6), describe two basic models of random polymer chains: the simple random walk and the self-avoiding walk (Section 1.7), and formulate a key open problem for the latter (Section 1.8).

1.1. What is a polymer? A polymer is a large molecule consisting of monomers that are tied together by *chemical bonds*. The monomers can be either small units (such as CH_2 in polyethylene; Fig. 1) or larger units with an internal structure (such as the adenine-thymine and cytosine-guanine base pairs in the DNA double helix; Fig. 2). Polymers abound in nature because of the *multivalency* of atoms like carbon, oxygen, nitrogen and sulfur, which are capable of forming long concatenated structures.

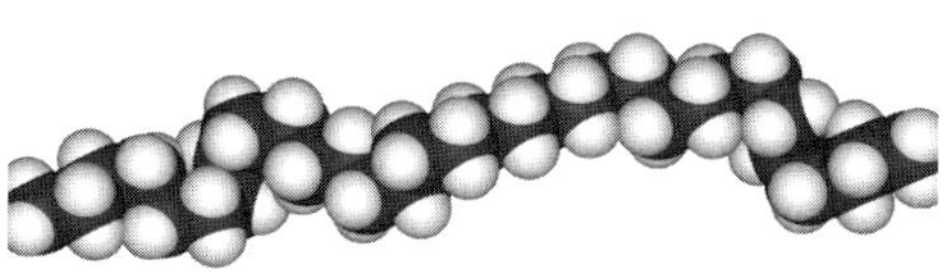

FIGURE 1. Polyethylene.

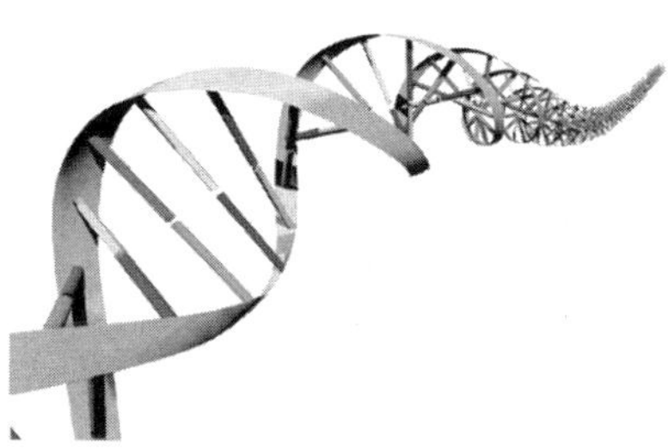

FIGURE 2. DNA.

1.2. What types of polymers occur in nature? Polymers come in two varieties: *homopolymers*, with all their monomers identical (such as polyethylene), and *copolymers*, with two or more different types of monomers (such as DNA). The order of the monomer types in copolymers can be either periodic (e.g. in agar) or random (e.g. in carrageenan).

Another classification is into *synthetic polymers* (like nylon, polyethylene and polystyrene) and *natural polymers* (also called biopolymers). Major subclasses of the latter are: (a) proteins (strings of amino-acids; Fig. 3); (b) nucleic acids (DNA, RNA; Fig. 2); (c) polysaccharides (like agar, alginate, amylopectin, amylose, carrageenan, cellulose); (d) lignin (plant cement); (e) rubber. Apart from (a)–(e), which are organic materials, clays and minerals are inorganic examples of natural polymers. Synthetic polymers typically are homopolymers, while natural polymers typically are copolymers (with notable exceptions). Bacterial polysaccharides tend to be periodic, while plant polysaccharides tend to be random.

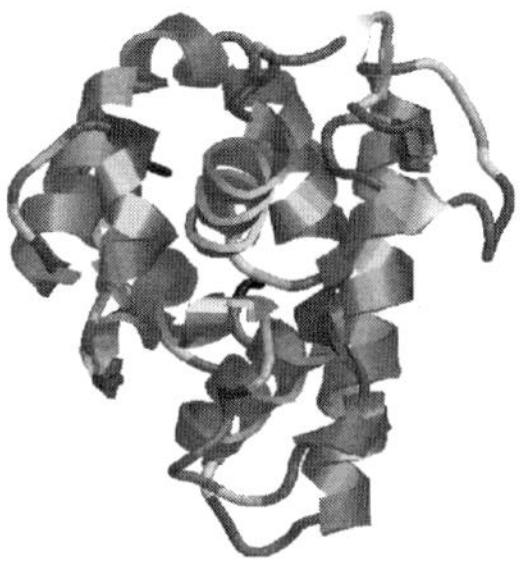

FIGURE 3. A folded-up protein.

Yet another classification is into *linear polymers* and *branched polymers*. In the former, the monomers have one reactive group (such as CH_2), leading to a linear organization as a result of the polymerization process. In the latter, the monomers have two or more reactive groups (such as hydroxy acid), leading to a network organization with multiple cross connections. Most natural polymers are linear, like proteins, DNA, RNA, and the polysaccharides agar, alginate, amylose, carrageenan and cellulose. Some polysaccharides are branched, like amylopectin. Many synthetic polymers are linear, and many are branched. An example of a branched polymer is rubber, both natural and synthetic. The network structure of rubber is what gives it both strength and flexibility!

1.3. What are the size and shape of a polymer? Size and shape are two key properties of a polymer.

Size: The chemical process of building a polymer from monomers is called *polymerization*. The size of a polymer may vary from 10^3 up to 10^{10} (shorter chains do not deserve to be called a polymer, longer chains have not been recorded). Human DNA has $10^9 - 10^{10}$ base pairs, lignin consists of $10^6 - 10^7$ phenyl-propanes, while polysaccharides carry $10^3 - 10^4$ sugar units.

Both in synthetic and in natural polymers, the *size distribution* may either be broad, with numbers varying significantly from polymer to polymer (e.g. nylons, polysaccharides), or be narrow (e.g. proteins, DNA). In synthetic polymers the size distribution can be made narrow through specific polymerization methods.

The length of the monomer units varies from 1.5 Å (for CH_2 in polyethylene) to 20 Å (for the base pairs in DNA), with $1\,\text{Å} = 10^{-10}\,\text{m}$.

Shape: The chemical bonds in a polymer are flexible, so that the polymer can arrange itself in *many different shapes.* The longer the chain, the more involved these shapes tend to be. For instance, the polymer may wind around itself to form a *knot* (Fig. 4), may expand itself to form a *random coil* due to repulsive forces caused by excluded-volume (e.g. when a good solvent surrounds the monomers and prevents them from coming close to each other), or may collapse on itself to form a *compact ball* due to attractive van der Waals forces between the monomers (or repulsive forces between the monomers and a poor solvent causing the polymer to fold itself up).

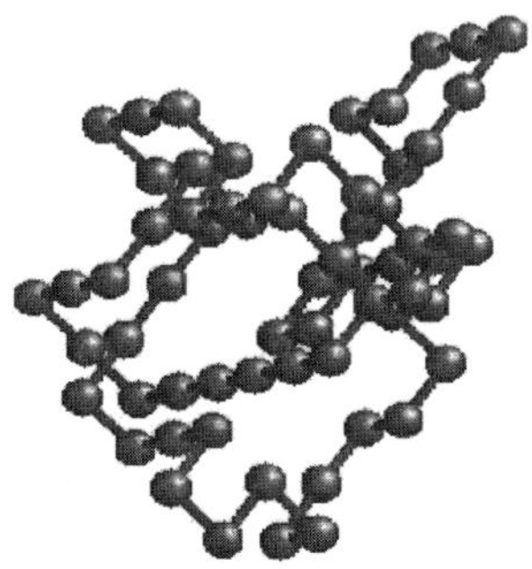

FIGURE 4. A knotted polymer.

In addition, the polymer may interact with a *surface* or with two fluids separated by an *interface*, may interact with a field of *random charges* in which it is immersed, or may be subjected to a *force* applied to one of its endpoints. Many models have been invented to describe such situations. In Sections 2–6 we take a look at some of these models.

1.4. What questions may a mathematician ask and hope to answer? The majority of mathematical research deals with *linear polymers.* Examples of quantities of interest are: number of different spatial configurations, end-to-end distance (subdiffusive/diffusive/superdiffusive), fraction of monomers adsorbed onto a surface, force needed to pull an adsorbed polymer off a surface, effect of randomness in the interactions, all typically in the limit as the polymer gets long (so that techniques from probability theory and statistical physics can be used). In these lectures special attention is given to the *free energy* of the polymer, and to the presence of *phase transitions* as a function of underlying model parameters. Recent surveys are the monographs by Giacomin [**55**] and den Hollander [**70**], and *references therein.*

1.5. What is the model setting? In mathematical models polymers often live on a lattice, like $\mathbb{Z}^d$, $d \geq 1$, and are modelled as random paths, where the monomers are the vertices in the path, and the chemical bonds connecting the monomers are the edges in the path (Fig. 5).

I. Paths and energies: Choosing a polymer model amounts to fixing for each $n \in \mathbb{N}_0 = \mathbb{N} \cup \{0\}$:

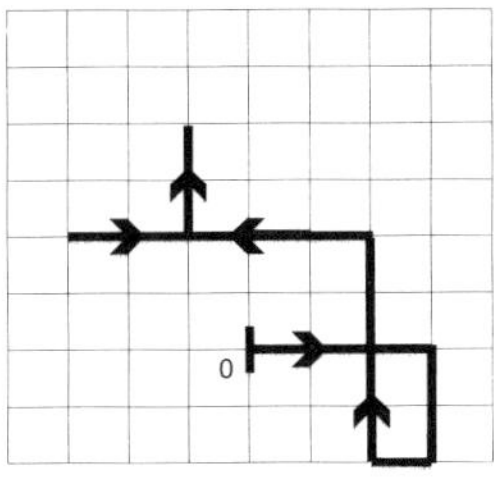

FIGURE 5. A lattice path.

(1) $\mathcal{W}_n$, a set of allowed n-step paths on $\mathbb{Z}^d$,

(2) H_n, a Hamiltonian function that associates an energy to each path in $\mathcal{W}_n$.

The choice of $\mathcal{W}_n$ may allow for directed or undirected paths, possibly with some geometric constraints (see Fig. 6).

FIGURE 6. Three examples of directed paths on $\mathbb{Z}^2$.

The choice of H_n captures the interaction of the polymer with itself and/or its environment. Typically, H_n depends on one or two parameters, including temperature. Sections 2–6 will provide many examples.

II. Path measure: For each $n \in \mathbb{N}_0$, the law of the polymer of length n is defined by assigning to each $w \in \mathcal{W}_n$ a probability given by

$$P_n(w) = \frac{1}{Z_n}\, \mathrm{e}^{-H_n(w)}, \qquad w \in \mathcal{W}_n,$$

where Z_n is the normalizing partition sum. This is called the *Gibbs measure* associated with the pair $(\mathcal{W}_n, H_n)$, and it describes the polymer *in equilibrium* with itself and/or its environment, at a fixed length n. Paths with a low (high) energy have a large (small) probability under the Gibbs measure. *Note:* In the physics and chemistry literature, H_n/kT is put into the exponent instead of H_n, with T the absolute temperature and k the Boltzmann constant. Since kT has the dimension of energy, H_n/kT is a dimensionless quantity. In our notation, however, we absorb kT into H_n.

III. Random environment: In some models H_n also depends on a

$$\text{random environment } \omega$$

describing e.g. a random ordering of the monomer types or a random field of charges in which the polymer is immersed. In this case the Hamiltonian is written as H_n^ω, and the path measure as P_n^ω. The law of ω is denoted by $\mathbb{P}$. (Carefully distinguish between the symbols w and ω.)

Three types of path measures with disorder are of interest:

(1) The *quenched* Gibbs measure

$$P_n^\omega(w) = \frac{1}{Z_n^\omega} \mathrm{e}^{-H_n^\omega(w)}, \qquad w \in \mathcal{W}_n.$$

(2) The *average quenched* Gibbs measure

$$\mathbb{E}(P_n^\omega(w)) = \int P_n^\omega(w)\, \mathbb{P}(\mathrm{d}\omega), \qquad w \in \mathcal{W}_n.$$

(3) The *annealed* Gibbs measure

$$\mathbb{P}_n(w) = \frac{1}{\mathbb{Z}_n} \int \mathrm{e}^{-H_n^\omega(w)}\, \mathbb{P}(\mathrm{d}\omega), \qquad w \in \mathcal{W}_n.$$

These are used to describe a polymer whose random environment is frozen [(1)+(2)], respectively, takes part in the equilibration [(3)]. Note that in (3), unlike in (2), the normalizing partition sum does not (!) appear under the integral.

It is also possible to consider models where the length or the configuration of the polymer changes with time (e.g. due to growing or shrinking), or to consider a Metropolis dynamics associated with the Hamiltonian for an appropriate choice of allowed transitions. These *non-equilibrium* situations are very interesting and challenging, but so far the available mathematics is rather limited. Two recent references are Caputo, Martinelli and Toninelli [**25**], Caputo, Lacoin, Martinelli, Simenhaus and Toninelli [**26**].

1.6. The central role of free energy. The *free energy* of the polymer is defined as

$$f = \lim_{n\to\infty} \frac{1}{n} \log Z_n$$

or, in the presence of a random environment, as

$$f = \lim_{n\to\infty} \frac{1}{n} \log Z_n^\omega \qquad \omega\text{-a.s.}$$

If the limit exists, then it typically is constant ω-a.s., a property referred to as *self-averaging*. We next discuss existence of f and some of its properties.

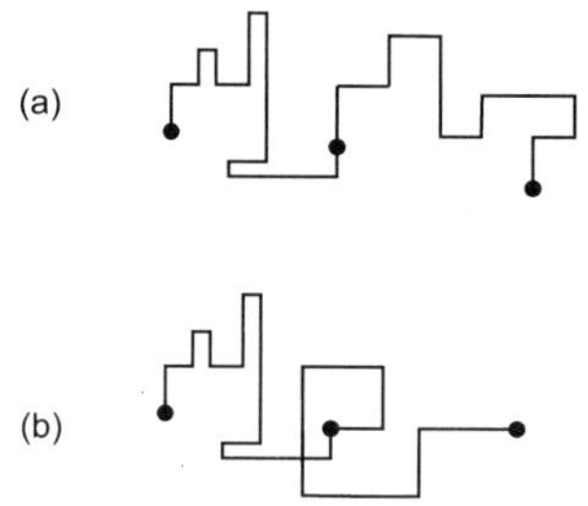

FIGURE 7. Concatenation of two self-avoiding paths: (a) the concatenation is self-avoiding; (b) the concatenation is not self-avoiding.

I. Existence of the free energy: When H_n assigns a repulsive self-interaction to the polymer, the partition sum Z_n satisfies the inequality

$$Z_n \le Z_m\, Z_{n-m} \qquad \forall\, 0 \le m \le n.$$

(See Fig. 7 for an example involving the counting of self-avoiding paths, i.e., $Z_n = |\mathcal{W}_n|$.) Consequently,

$$n \mapsto nf_n = \log Z_n$$

is a *subadditive sequence*, so that

$$f = \lim_{n\to\infty} f_n = \inf_{n\in\mathbb{N}} f_n \in [-\infty, \infty).$$

(See the tutorial in Appendix A.1 of Bauerschmidt, Duminil-Copin, Goodman and Slade [**7**].) If, moreover, $\inf_{w\in\mathcal{W}_n} H_n(w) \le Cn$ for all $n \in \mathbb{N}$ and some $C < \infty$, then $f \neq -\infty$. A similar result holds when H_n assigns an attractive self-interaction to the polymer, in which case the inequalities are reversed, $f \in (-\infty, \infty]$, and $f \neq \infty$ when $|\mathcal{W}_n| \le e^{Cn}$ and $\inf_{w\in\mathcal{W}_n} H_n(w) \ge -Cn$ for all $n \in \mathbb{N}$ and some $C < \infty$.

When H_n assigns both repulsive and attractive interactions to the polymer, then the above argument is generally not available, and the existence of the free energy either remains open or has to be established by other means. Many examples, scenarios and techniques are available. **Tutorial 1 in Appendix A** describes two techniques to prove existence of free energies, in the context of the model of a polymer near a random interface that is the topic of Section 4.

In the presence of a random environment ω, it is often possible to derive a random form of subadditivity. When applicable,

$$n \mapsto nf_n^\omega = \log Z_n^\omega$$

becomes a *subadditive random process*, and Kingman's subadditive ergodic theorem implies the existence of

$$f = \lim_{n\to\infty} f_n^\omega \qquad \omega\text{-a.s.}$$

(as explained in **Tutorial 1 in Appendix A**). This fact is of key importance for polymers with disorder.

II. Convexity of the free energy: Suppose that the Hamiltonian depends linearly on a single parameter $\beta \in \mathbb{R}$, which is pulled out by writing βH_n instead of H_n. Then, by the Hölder inequality, $\beta \mapsto f_n(\beta)$ is convex for all $n \in \mathbb{N}_0$ and hence so is $\beta \mapsto f(\beta)$. Convexity and finiteness imply continuity, and also monotonicity on either side of a minimum. Moreover, at those values of β where $f(\beta)$ is differentiable, convexity implies that

$$f'(\beta) = \lim_{n\to\infty} f_n'(\beta).$$

The latter observation is important because

$$\begin{aligned} f_n'(\beta) &= \left[\frac{1}{n}\log Z_n(\beta)\right]' = \frac{1}{n}\frac{Z_n'(\beta)}{Z_n(\beta)} \\ &= \frac{1}{n}\frac{1}{Z_n(\beta)}\frac{\partial}{\partial\beta}\left(\sum_{w\in\mathcal{W}_n} e^{-\beta H_n(w)}\right) = \frac{1}{n}\sum_{w\in\mathcal{W}_n}[-H_n(w)]\,P_n^\beta(w). \end{aligned}$$

What this says is that $-\beta f'(\beta)$ is the limiting energy per monomer under the Gibbs measure as $n \to \infty$. At those values of β where the free energy fails to be differentiable this quantity is discontinuous, signalling the occurrence of a first-order *phase transition*. (Several examples will be given later on.) Higher-order phase transitions correspond to discontinuity of higher-order derivatives of f.

1.7. Two basic models. The remainder of this section takes a brief look at two basic models for a polymer chain: (1) the *simple random walk*, a polymer without self-interaction; (2) the *self-avoiding walk*, a polymer with excluded-volume self-interaction. In some sense these are the "plain vanilla" and "plain chocolate" versions of a polymer chain. The self-avoiding walk is the topic of the lectures by Bauerschmidt, Duminil-Copin, Goodman and Slade [**7**].

(1) Simple random walk: SRW on $\mathbb{Z}^d$ is the random process $(S_n)_{n\in\mathbb{N}_0}$ defined by

$$S_0 = 0, \qquad S_n = \sum_{i=1}^{n} X_i, \quad n \in \mathbb{N},$$

where $X = (X_i)_{i\in\mathbb{N}}$ is an i.i.d. sequence of random variables taking values in $\mathbb{Z}^d$ with marginal law ($\|\cdot\|$ is the Euclidean norm)

$$P(X_1 = x) = \begin{cases} \frac{1}{2d}, & x \in \mathbb{Z}^d \text{ with } \|x\| = 1, \\ 0, & \text{otherwise.} \end{cases}$$

Think of X_i as the orientation of the chemical bond between the $(i-1)$-th and i-th monomer, and of S_n as the location of the end-point of the polymer of length n. SRW corresponds to choosing

$$\begin{aligned} \mathcal{W}_n &= \big\{w = (w_i)_{i=0}^n \in (\mathbb{Z}^d)^{n+1}\colon \\ &\qquad\qquad w_0 = 0,\ \|w_{i+1} - w_i\| = 1\ \forall\, 0 \le i < n\big\}, \\ H_n &\equiv 0, \end{aligned}$$

so that P_n is the uniform distribution on $\mathcal{W}_n$. In this correspondence, think of $(S_i)_{i=0}^n$ as the realization of $(w_i)_{i=0}^n$ drawn according to P_n.

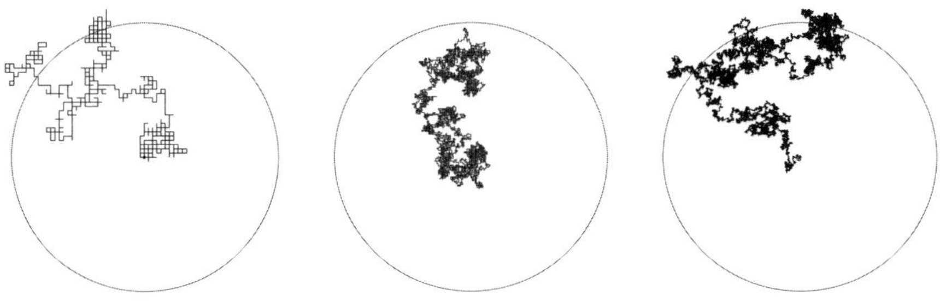

FIGURE 8. Simulation of SRW on $\mathbb{Z}^2$ with $n = 10^3$, 10^4 and 10^5 steps. The circles have radius $n^{1/2}$ in units of the step size. [Courtesy of Bill Casselman and Gordon Slade.]

A distinctive feature of SRW is that it exhibits *diffusive behavior*, i.e.,

$$E_n(S_n) = 0 \quad \text{and} \quad E_n(\|S_n\|^2) = n \qquad \forall n \in \mathbb{N}_0$$

and

$$\left(\frac{1}{n^{1/2}} S_{\lfloor nt \rfloor}\right)_{0\le t\le 1} \quad \Longrightarrow \quad (B_t)_{0\le t\le 1} \qquad \text{as } n \to \infty,$$

where the right-hand side is Brownian motion on $\mathbb{R}^d$, and $\Longrightarrow$ denotes convergence in distribution on the space of càdlàg paths endowed with the Skorohod topology (see Fig. 8).

(2) Self-avoiding walk: SAW corresponds to choosing

$$\begin{aligned} \mathcal{W}_n = \big\{ w = (w_i)_{i=0}^n \in (\mathbb{Z}^d)^{n+1} \colon & \\ & w_0 = 0,\ \|w_{i+1} - w_i\| = 1\ \forall 0 \le i < n, \\ & w_i \neq w_j\ \forall 0 \le i < j \le n \big\}, \\ H_n \equiv 0, & \end{aligned}$$

so that P_n is the uniform distribution on $\mathcal{W}_n$. Again, think of $(S_i)_{i=0}^n$ as the realization of $(w_i)_{i=0}^n$ drawn according to P_n.

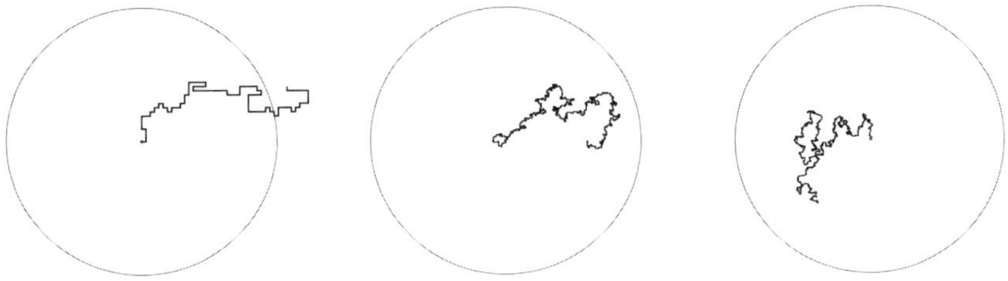

FIGURE 9. Simulation of SAW on $\mathbb{Z}^2$ with $n = 10^2$, 10^3 and 10^4 steps. The circles have radius $n^{3/4}$ in units of the step size. [Courtesy of Bill Casselman and Gordon Slade.]

SAW in $d = 1$ is trivial. In $d \ge 2$ no closed form expression is available for $E_n(\|S_n\|^2)$, but for small and moderate n it can be computed via *exact enumeration methods*. The current record is: $n = 71$ for $d = 2$ (Jensen [**81**]); $n = 36$ for $d = 3$ (Schram, Barkema and Bisseling [**93**]); $n = 24$ for $d \ge 4$ (Clisby, Liang and Slade [**36**]). Larger n can be handled either via numerical simulation (presently up to $n = 2^{25} \approx 3.3 \times 10^7$ in $d = 3$) or with the help of extrapolation techniques.

The mean-square displacement is predicted to scale like

$$E_n(\|S_n\|^2) = \begin{cases} D\, n^{2\nu}\, [1 + o(1)], & d \neq 4, \\ D\, n (\log n)^{\frac{1}{4}}\, [1 + o(1)], & d = 4, \end{cases} \qquad \text{as } n \to \infty,$$

with D a non-universal diffusion constant and ν a universal *critical exponent*. Here, universal refers to the fact that ν is expected to depend only on d, and to be independent of the fine details of the model (like the choice of the underlying lattice or the choice of the allowed increments of the path).

The value of ν is predicted to be

$$\nu = 1\ (d = 1), \quad \tfrac{3}{4}\ (d = 2), \quad 0.588\ldots\ (d = 3), \quad \tfrac{1}{2}\ (d \ge 5).$$

Thus, SAW is *ballistic* in $d = 1$, *subballistic and superdiffusive* in $d = 2, 3, 4$, and *diffusive* in $d \ge 5$.

For $d = 1$ the above scaling is trivial. For $d \ge 5$ a proof has been given by Hara and Slade [**65, 66**]. These two cases correspond to ballistic, respectively, diffusive behavior. The claim for $d = 2, 3, 4$ is open.

- For $d = 2$ the scaling limit is predicted to be $\mathrm{SLE}_{8/3}$ (the Schramm Loewner Evolution with parameter 8/3; see Fig. 9).
- For $d = 4$ a proof is under construction by Brydges and Slade (work in progress).

See the lectures by Bauerschmidt, Duminil-Copin, Goodman and Slade [**7**], Beffara [**8**] and Duminil-Copin and Smirnov [**50**] for more details. SAW in $d \geq 5$ scales to Brownian motion,

$$\left(\frac{1}{Dn^{1/2}} S_{\lfloor nt \rfloor}\right)_{0 \leq t \leq 1} \Longrightarrow (B_t)_{0 \leq t \leq 1} \qquad \text{as } n \to \infty,$$

i.e., SAW is in the *same universality class* as SRW. Correspondingly, $d = 4$ is called the upper critical dimension. The intuitive reason for the crossover at $d = 4$ is that in low dimension long loops are dominant, causing the effect of the self-avoidance constraint in SAW to be long-ranged, whereas in high dimension short loops are dominant, causing it to be short-ranged. Phrased differently, since SRW in dimension $d \geq 2$ has Hausdorff dimension 2, it tends to intersect itself frequently for $d < 4$ and not so frequently for $d > 4$. Consequently, the self-avoidance constraint in SAW changes the qualitative behavior of the path for $d < 4$ but not for $d > 4$.

1.8. Open problems. A version of SAW where self-intersections are not forbidden but are nevertheless discouraged is called the *weakly self-avoiding walk.* Here, $\mathcal{W}_n$ is the same as for SRW, but $H_n(w)$ is chosen to be β times the number of self-intersections of w, with $\beta \in (0, \infty)$ a parameter referred to as the strength of self-repellence. It is predicted that the weakly self-avoiding walk is in the same universality class as SAW (the latter corresponds to $\beta = \infty$). This has been proved for $d = 1$ and $d \geq 5$, but remains open for $d = 2, 3, 4$. The scaling limit of the weakly self-avoiding walk in $d = 2$ is again predicted to be $\mathrm{SLE}_{8/3}$, despite the fact that $\mathrm{SLE}_{8/3}$ does not intersect itself. The reason is that the self-intersections of the weakly self-avoiding walk typically occur close to each other, so that when the scaling limit is taken these self-intersections are lost in the limit. This loss, however, does affect the *time-parametrization* of the limiting $\mathrm{SLE}_{8/3}$, which is predicted to be β-dependent. It is a challenge to prove these predictions. For more details on SLE, we refer to the lectures by Beffara [**8**].

2. Polymer collapse

In this section we consider a polymer that receives a penalty for each *self-intersection* and a reward for each *self-touching.* This serves as a model of a polymer subject to screened van der Waals forces, or a polymer in a poor solvent. It will turn out that there are three phases: *extended, collapsed* and *localized.*

An example is polystyrene dissolved in cyclohexane. At temperatures above 35 degrees Celsius the cyclohexane is a good solvent, at temperatures below 30 it is a poor solvent. When cooling down, the polystyrene collapses from a random coil to a compact ball (see Fig. 10).

In Sections 2.1–2.3 we consider a model with *undirected paths*, in Sections 2.4–2.5 a model with *directed paths.* In Section 2.6 we look at what happens when a force is applied to the endpoint of a collapsed polymer. In Section 2.7 we formulate open problems.

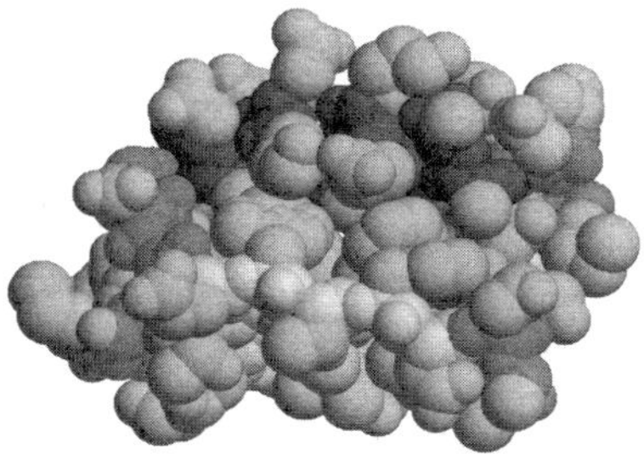

FIGURE 10. A collapsed polymer.

2.1. An undirected polymer in a poor solvent. Our choice for the set of allowed paths and for the interaction Hamiltonian is

$$\mathcal{W}_n = \big\{ w = (w_i)_{i=0}^n \in (\mathbb{Z}^d)^{n+1} : \\ w_0 = 0,\ \|w_{i+1} - w_i\| = 1\ \forall\, 0 \le i < n \big\},$$

$$H_n^{\beta,\gamma}(w) = \beta I_n(w) - \gamma J_n(w),$$

where $\beta, \gamma \in (0, \infty)$, and

$$I_n(w) = \sum_{\substack{i,j=0 \\ i<j}}^{n} 1_{\{\|w_i - w_j\| = 0\}},$$

$$J_n(w) = \tfrac{1}{2d} \sum_{\substack{i,j=0 \\ i<j-1}}^{n} 1_{\{\|w_i - w_j\| = 1\}},$$

count the number of *self-intersections*, respectively, *self-touchings* of w (see Fig. 11). The factor $\frac{1}{2d}$ is added to account for the fact that each site has $2d$ neighboring sites where the polymer can achieve a self-touching. The path measure is

$$P_n^{\beta,\gamma}(w) = \frac{1}{Z_n^{\beta,\gamma}}\, \mathrm{e}^{-H_n^{\beta,\gamma}(w)}\, P_n(w), \qquad w \in \mathcal{W}_n,$$

where P_n is the law of the n-step SRW and $Z_n^{\beta,\gamma}$ is the normalizing partition sum.

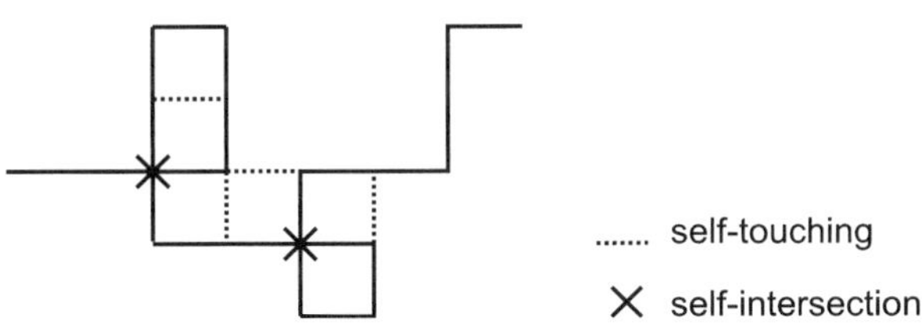

FIGURE 11. A polymer with self-intersections and self-touchings.

Under the law $P_n^{\beta,\gamma}$, self-intersections are penalized while self-touchings are rewarded. The case $\gamma = 0$ corresponds to weakly self-avoiding walk, which falls in the same universality class as SAW as soon as $\beta > 0$ (recall Section 1.8). We expect that for $\beta \gg \gamma$ the polymer is a *random coil*, while for $\gamma \gg \beta$ it is a *compact ball.*

A crossover is expected to occur when β and γ are comparable. In the next two sections we identify two phase transition curves.

2.2. The localization transition. For $L \in \mathbb{N}$, abbreviate $\Lambda(L) = [-L, L]^d \cap \mathbb{Z}^d$.

THEOREM 2.1. [van der Hofstad and Klenke [**67**]] *If $\beta > \gamma$, then the polymer is inflated, i.e., there exists an $\epsilon_0 = \epsilon_0(\beta, \gamma) > 0$ such that for all $0 < \epsilon \leq \epsilon_0$ there exists a $c = c(\beta, \gamma, \epsilon) > 0$ such that*

$$P_n^{\beta,\gamma}\big(S_i \in \Lambda(\epsilon n^{1/d})\ \forall\, 0 \leq i \leq n\big) \leq \mathrm{e}^{-cn} \qquad \forall\, n \in \mathbb{N}.$$

THEOREM 2.2. [van der Hofstad and Klenke [**67**]] *If $\gamma > \beta$, then the polymer is localized, i.e., there exist $c = c(\beta, \gamma) > 0$ and $L_0 = L_0(\beta, \gamma) \in \mathbb{N}$ such that*

$$P_n^{\beta,\gamma}\big(S_i \in \Lambda(L)\ \forall\, 0 \leq i \leq n\big) \geq 1 - \mathrm{e}^{-cLn} \qquad \forall\, n \in \mathbb{N},\ L \geq L_0.$$

Thus, at $\gamma = \beta$ a phase transition takes place, from a phase in which the polymer exits a box of size $n^{1/d}$ to a phase in which it is confined to a finite box. (In Section 2.3 we will see that the inflated phase splits into two subphases: a collapsed phase and an extended phase.)

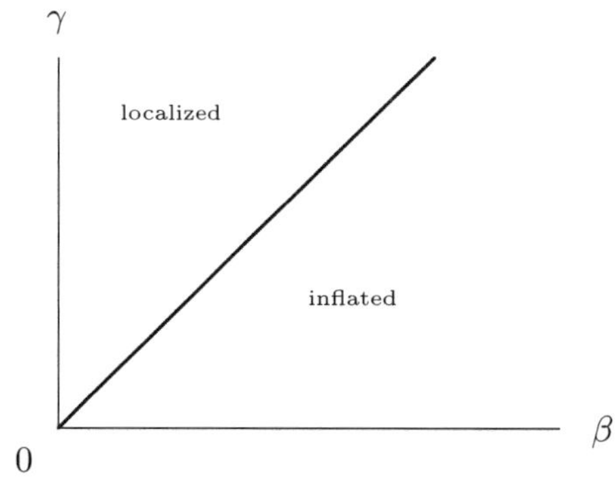

FIGURE 12. Two phases: inflated and localized.

The main ideas behind the proof of Theorems 2.1–2.2 are:

- *Inflated phase:* For ϵ small, most n-step paths that are folded up inside $\Lambda(\epsilon n^{1/d})$ have many self-intersections and many self-touchings. Since $\beta > \gamma$, the former produce more positive energy than the latter produce negative energy, and so the total energy is positive, making such paths unlikely.
- *Localized phase:* Two key ingredients are important:
  - An estimate showing that, since $\gamma > \beta$, the minimum of the Hamiltonian is achieved by a localized path.
  - An estimate showing that, if L is so large that $\Lambda(L)$ contains a minimizing path, then the penalty for leaving $\Lambda(L)$ is severe.

The proof uses a geometric argument based on *folding of paths*, in the spirit of what is done in Section 2.1 of Bauerschmidt, Duminil-Copin, Goodman and Slade [**7**]. It is not known whether or not the minimizing path is unique modulo the symmetries of $\mathbb{Z}^d$.

In terms of the mean-square displacement it is predicted that

$$E_n^{\beta,\gamma}(\|S_n\|^2) \asymp n^{2\nu} \qquad \text{as } n \to \infty,$$

where $\asymp$ stands for "asymptotially the same modulo logarithmic factors" (i.e., $E_n^{\beta,\gamma}(\|S_n\|^2) = n^{2\nu+o(1)}$). Theorems 2.1–2.2 show that $\nu = 0$ in the localized phase and $\nu \geq 1/d$ in the inflated phase. It is conjectured in van der Hofstad and Klenke [**67**] that on the critical line $\gamma = \beta$,

$$\nu = \nu_{\text{loc}} = 1/(d+1).$$

For $d = 1$, this conjecture is proven in van der Hofstad, Klenke and König [**68**]. For $d \geq 2$ it is still open. The key simplification that can be exploited when $\beta = \gamma$ is the relation

$$I_n(w) - J_n(w) = -\frac{n+1}{2} + \frac{1}{8d} \sum_{\{x,y\} \in \mathbb{Z}^d \times \mathbb{Z}^d} |\ell_n(x) - \ell_n(y)|^2,$$

where the sum runs over all unordered pairs of neighboring sites, and $\ell_n(x) = \sum_{i=0}^n 1_{\{w_i = x\}}$ is the local time of w at site x. Since the factor $-\frac{n+1}{2}$ can be absorbed into the partition sum, the model at $\beta = \gamma$ effectively becomes a model where the energy is $\beta/4d$ times the sum of the squares of the gradients of the local times.

2.3. The collapse transition. It is predicted that there is a second phase transition at a critical value $\gamma_c = \gamma_c(\beta) < \beta$ at which the inflated polymer moves from scale $n^{1/d}$ to scale $n^{\nu_{\text{SAW}}}$, with ν_{SAW} the critical exponent for SAW. Thus, it is predicted that the inflated phase splits into two subphases: a *collapsed phase* and an *extended phase*, separated by a second critical curve at which a *collapse transition* takes place. At the second critical curve, the critical exponent is predicted to be

$$\nu = \nu_{\text{coll}} = \begin{cases} \frac{4}{7}, & \text{if } d = 2, \\ \frac{1}{2}, & \text{if } d \geq 3. \end{cases}$$

Thus, the phase diagram for $d \geq 2$ is conjectured to have the shape in Fig. 13. The free energy is known to be ∞ in the localized phase, and is expected to lie in $(-\infty, 0)$ in the two other phases. However, not even the existence of the free energy has been proven in the latter two phases.

Although these predictions are supported by heuristic theories (Duplantier and Saleur [**51**], Seno and Stella [**94**]) and by extensive simulations, a mathematical proof of the existence of the collapse transition and a mathematical verification of the values of the critical exponent have remained open for more than 20 years. For $d = 1$ there is no collapse transition because $\nu_{\text{SAW}} = 1$. Indeed, Theorem 2.1 says that below the critical line $\gamma = \beta$ the polymer is ballistic like SAW.

In $d = 3$, simulations by Tesi, Janse van Rensburg, Orlandini and Whittington [**99**] for SAW with attraction (corresponding to $\beta = \infty$ and $\gamma \in (0, \infty)$) yield $\gamma_c = \gamma_c(\infty) \in [0.274, 0.282]$ and $\nu_{\text{coll}} \in [0.48, 0.50]$, the latter in accordance with the prediction mentioned above.

2.4. A directed polymer in a poor solvent. In order to deal with the collapse transition mathematically, it is necessary to turn to a directed version of the model. The results to be described below are taken from Brak, Guttmann and Whittington [**23**], with refinements carried out in various later papers.

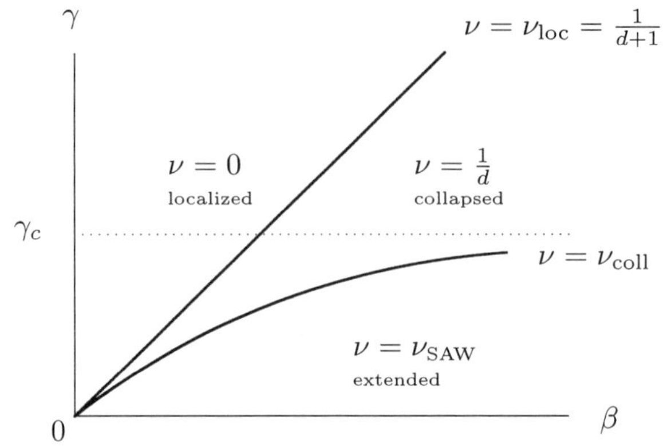

FIGURE 13. Conjectured phase diagram.

Our choice for the set of allowed paths and the interaction Hamiltonian is (see Fig. 14)

$$\begin{aligned}\mathcal{W}_n = \big\{w = (w_i)_{i=0}^n \in (\mathbb{N}_0 \times \mathbb{Z})^{n+1}\colon&\\ w_0 = 0,\ w_1 - w_0 = \rightarrow,&\\ w_{i+1} - w_i \in \{\uparrow, \downarrow, \rightarrow\}\ \forall\, 0 < i < n,&\\ w_i \neq w_j\ \forall\, 0 \leq i < j \leq n\big\},&\end{aligned}$$

$$H_n^\gamma(w) = -\gamma J_n(w),$$

where $\uparrow$, $\downarrow$ and $\rightarrow$ denote steps between neighboring sites in the north, south and east direction, respectively, $\gamma \in \mathbb{R}$ and

$$J_n(w) = \sum_{\substack{i,j=0\\ i<j-1}}^{n} 1_{\{\|w_i - w_j\|=1\}}.$$

The path measure is

$$P_n^\gamma(w) = \frac{1}{Z_n^\gamma}\, \mathrm{e}^{-H_n^\gamma(w)}, \qquad w \in \mathcal{W}_n,$$

with counting measure as the reference law (instead of the uniform measure P_n used in Sections 2.1–2.3) and with normalizing partition sum Z_n^γ. Thus, each self-touching is rewarded when $\gamma > 0$ (= attractive) and penalized when $\gamma < 0$ (= repulsive). Note that, because the path is self-avoiding ($I_n(w) = 0$), the directed model is to be compared with the undirected model at $\beta = \infty$. Also note that the model lives in dimension $1 + 1$ and that no factor $\frac{1}{2}$ is needed in front of the sum defining $J_n(w)$ because the path is directed. The choice that the first step of w must be to the right is made for convenience only. (In the undirected model studied in Sections 2.1–2.3 we did not consider the case $\gamma < 0$ because of the presence of β.)

2.5. Generating functions. The free energy of the directed polymer is given by

$$f(\gamma) = \lim_{n\to\infty} \frac{1}{n} \log Z_n^\gamma,$$

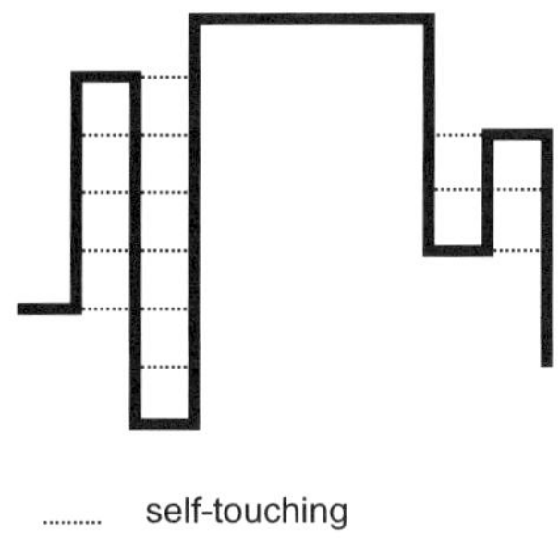

FIGURE 14. A directed SAW with self-touchings.

whenever the limit exists. The following theorem establishes existence and shows that there are two phases: a collapsed phase and an extended phase (see Fig. 15).

THEOREM 2.3. [Brak, Guttmann and Whittington [**23**]] *The free energy exists, is finite, and has a collapse transition at* $\gamma_c = \log x_c$, *with* $x_c \approx 3.382975$ *the unique positive solution of the cubic equation* $x^3 - 3x^2 - x - 1 = 0$. *The collapsed phase corresponds to* $\gamma > \gamma_c$, *the extended phase to* $\gamma < \gamma_c$.

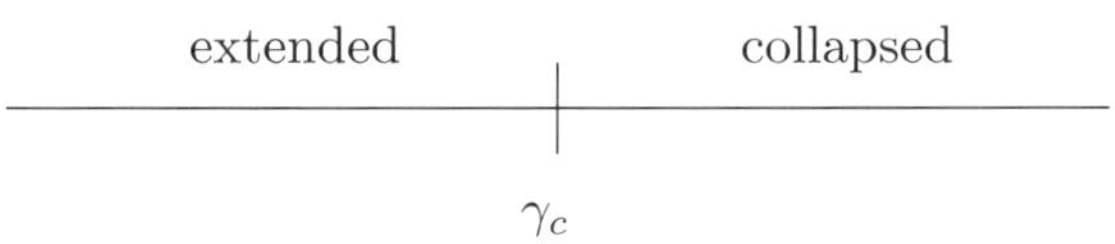

FIGURE 15. Collapse transition for the directed model.

Below we sketch the proof of Theorem 2.3 in 5 Steps. The proof makes use of *generating functions*. The details are worked out in **Tutorial 2 in Appendix B**. In Section 3 we will encounter another model where generating functions lead to a full description of a phase transition.

1. The partition sum $Z_n^\gamma = \sum_{w \in \mathcal{W}_n} \mathrm{e}^{\gamma J_n(w)}$ can be written as $Z_n^\gamma = Z_n(e^\gamma)$ with the power series

$$Z_n(x) = \sum_{m \in \mathbb{N}_0} c_n(m) x^m, \quad x \in [0, \infty),\, n \in \mathbb{N}_0,$$

where

$$\begin{aligned} c_n(m) &= |\{w \in \mathcal{W}_n \colon J_n(w) = m\}| \\ &= \text{the number of } n\text{-step paths with } m \text{ self-touchings.} \end{aligned}$$

2. The existence of the free energy can be proved with the help of a subadditivity argument applied to the coefficients $c_n(m)$, based on concatenation of paths (as in Section 2 in Bauerschmidt, Duminil-Copin, Goodman and Slade [**7**].)

3. The finiteness of the free energy follows from the observation that $c_n(m) = 0$ for $m \geq n$ and $\sum_{m=0}^{\infty} c_n(m) \leq 3^n$, which gives $f(\gamma) \leq \log[3(\mathrm{e}^\gamma \vee 1)] = \log 3 + (\gamma \vee 0)$.

4. The following lemma gives a closed form expression for the generating function ($x = \mathrm{e}^{\gamma}$)

$$G = G(x,y) = \sum_{n\in\mathbb{N}_0} Z_n(x)\, y^n = \sum_{n\in\mathbb{N}_0} \Big[\sum_{m=0}^{n} c_n(m)\, x^m\Big]\, y^n, \qquad x,y \in [0,\infty).$$

LEMMA 2.4. *For $x, y \in [0,\infty)$ the generating function is given by the formal power series*

$$G(x,y) = -\frac{aH(x,y) - 2y^2}{bH(x,y) - 2y^2},$$

where

$$a = y^2(2+y-xy),\ b = y^2(1+x+y-xy),\ H(x,y) = y\,\frac{\bar{g}_0(x,y)}{\bar{g}_1(x,y)},$$

with

$$\bar{g}_r(x,y) = y^r\left(1 + \sum_{k\in\mathbb{N}} \frac{(y-q)^k\, y^{2k}\, q^{\frac{1}{2}k(k+1)}}{\prod_{l=1}^{k}(yq^l - y)(yq^l - q)}\, q^{kr}\right),$$
$$q = xy,\ r = 0,1.$$

The function $H(x,y)$ is a quotient of two q-hypergeometric functions (which are singular at least along the curve $q = xy = 1$). As shown in Brak, Guttmann and Whittington [**23**], the latter can be expressed as continued fractions and therefore can be properly analyzed (as well as computed numerically).

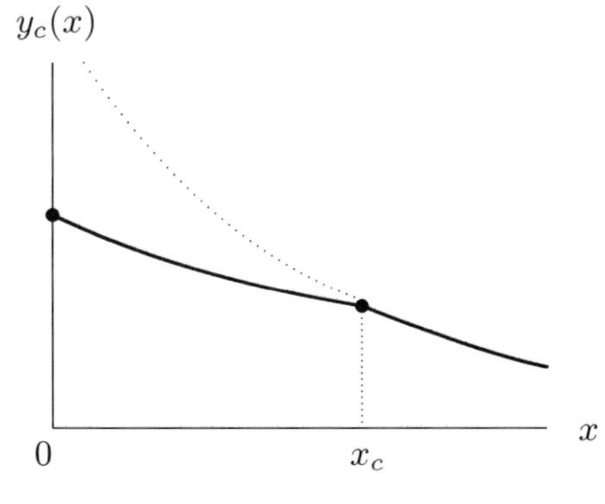

FIGURE 16. The domain of convergence of the generating function $G(x,y)$ lies below the critical curve (= solid curve). The dotted line is the hyperbola $xy = 1$ (corresponding to $q = 1$). The point x_c is identified with the collapse transition, because this is where the free energy is non-analytic.

5. By analyzing the singularity structure of $G(x,y)$ it is possible to compute $f(\gamma)$. Indeed, the task is to identify the critical curve $x \mapsto y_c(x)$ in the (x,y)-plane below which $G(x,y)$ has no singularities and on or above which it does, because this identifies the free energy as

$$f(\gamma) = -\log y_c(\mathrm{e}^{\gamma}), \qquad \gamma \in \mathbb{R}.$$

It turns out that the critical curve has the shape given in Fig. 16, which implies that the free energy has the shape given in Fig. 17.

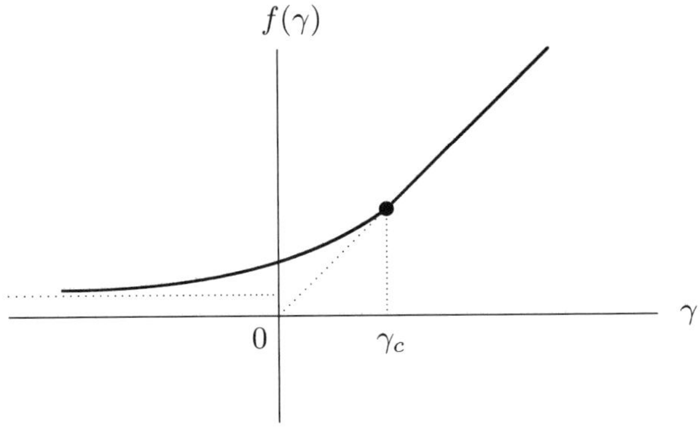

FIGURE 17. Plot of the free energy per monomer. The collapse transition occurs at $\gamma_c = \log x_c$. The limiting value at $\gamma = -\infty$ equals $\log(1/y_c(0))$ with $y_c(0) \approx 0.453397$ the solution of the cubic equation $y^3 + 2y - 1 = 0$, and is the entropy per step of the directed polymer that avoids self-touchings altogether, i.e., $\lim_{n\to\infty} \frac{1}{n} \log c_n(0)$.

The derivative of the free energy is the limiting number of self-touchings per monomer, as plotted in Fig. 18:

$$f'(\gamma) = \lim_{n\to\infty} \frac{1}{n} \sum_{w\in\mathcal{W}_n} J_n(w)\, P_n^\gamma(w).$$

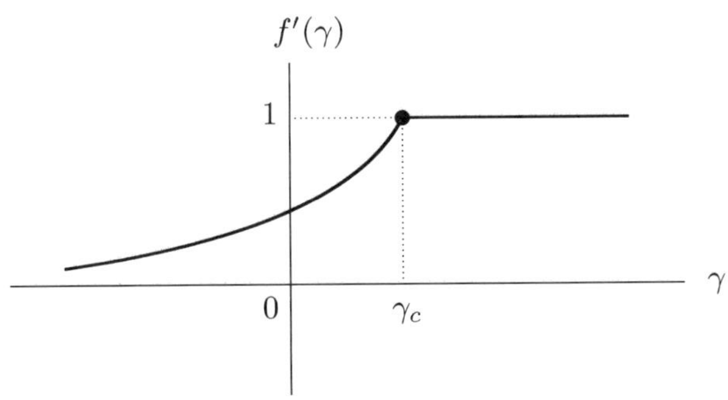

FIGURE 18. Plot of the number of self-touchings per monomer. Since $\gamma \mapsto f'(\gamma)$ is continuous but not differentiable at γ_c, the phase transition is second order.

2.6. Pulling at a collapsed polymer. It is possible to induce a collapse transition by applying a *force* to the endpoint of a polymer rather than changing its interaction strength. The force can be applied, for instance, with the help of optical tweezers. A focused laser beam is used, containing a narrow region – called the beam waist – in which there is a strong electric field gradient. When a dielectric particle, a few nanometers in diameter, is placed in the waist, it feels a strong attraction towards the center of the waist. It is possible to chemically attach such a particle to the end of the polymer and then pull on the particle with the laser beam, thereby effectively exerting a force on the polymer itself. Current experiments allow for forces in the range of $10^{-12} - 10^{-15}$ Newton. With such microscopically small forces the structural, mechanical and elastic properties of polymers can be probed. We refer to Auvray, Duplantier, Echard and Sykes [**6**], Section 5.2, for more details. The force is the result of transversal fluctuations of the dielectric particle, which can be measured with great accuracy.

Ioffe and Velenik [**77, 78, 79, 80**] consider a version of the undirected model in which the Hamiltonian takes the form

$$H_n^{\psi,\phi}(w) = \sum_{x\in\mathbb{Z}^d} \psi\big(\ell_n(x)\big) - (\phi, w_n), \qquad w \in \mathcal{W}_n,$$

where $\mathcal{W}_n$ is the set of allowed n-step paths for the *undirected* model considered in Sections 2.1–2.3, $\ell_n(x) = \sum_{i=0}^n 1_{\{w_i=x\}}$ is the local time of w at site $x \in \mathbb{Z}^d$, $\psi\colon \mathbb{N}_0 \to [0,\infty)$ is non-decreasing with $\psi(0) = 0$, and $\phi \in \mathbb{R}^d$ is a *force* acting on the endpoint of the polymer. Note that (ϕ, w_n) is the *work* exerted by the force ϕ to move the endpoint of the polymer to w_n. The path measure is

$$P_n^{\psi,\phi}(w) = \frac{1}{Z_n^{\psi,\phi}}\, \mathrm{e}^{-H_n^{\psi,\phi}(w)}\, P_n(w), \qquad w \in \mathcal{W}_n,$$

with P_n the law of SRW.

Two cases are considered:

(1) ψ is *superlinear* (= repulsive interaction).
(2) ψ is *sublinear* with $\lim_{\ell\to\infty} \psi(\ell)/\ell = 0$ (= attractive interaction).

Typical examples are:

(1) $\psi(\ell) = \beta\ell^2$ (which corresponds to the weakly self-avoiding walk).
(2) $\psi(\ell) = \sum_{k=1}^{\ell} \beta_k$ with $k \mapsto \beta_k$ non-increasing such that $\lim_{k\to\infty} \beta_k = 0$ (which corresponds to the annealed version of the model of a polymer in a random potential described in Section 6, for the case where the potential is non-negative).

It is shown in Ioffe and Velenik [**77, 78, 79, 80**] (see also references cited therein) that:

(1) The polymer is in an extended phase for all $\phi \in \mathbb{R}^d$.
(2) There is a compact convex set $K = K(\psi) \subset \mathbb{R}^d$, with $\mathrm{int}(K) \ni 0$, such that the polymer is in a collapsed phase (= subballistic) when $\phi \in \mathrm{int}(K)$ and in an extended phase (= ballistic) when $\phi \notin K$.

The proof uses *coarse-graining arguments*, showing that in the extended phase large segments of the polymer can be treated as directed. For $d \geq 2$, the precise shape of the set K is not known. It is known that K has the symmetries of $\mathbb{Z}^d$ and has a locally analytic boundary ∂K with a uniformly positive Gaussian curvature. It is predicted not to be a ball, but this has not been proven. The phase transition at ∂K is first order.

2.7. Open problems. The main challenges are:

- Prove the conjectured phase diagram in Fig. 13 for the undirected (β,γ)-model studied Sections 2.1–2.3 and determine the order of the phase transitions.

- Extend the analysis of the directed γ-model studied in Sections 2.4–2.5 to $1+d$ dimensions with $d \geq 2$.
- Find a closed form expression for the set K of the undirected ψ-model studied in Section 2.6.

For the undirected model in $d = 2$, the scaling limit is predicted to be:

(1) SLE_8 in the collapsed phase (between the two critical curves),
(2) SLE_6 at the collapse transition (on the lower critical curve),
(3) $\mathrm{SLE}_{8/3}$ in the extended phase (below the lower critical curve),

all three with a time parametrization that depends on β and γ (see the lectures by Beffara [**8**] for an explanation of the time parametrization). Case (1) is plausible because SLE_8 is space filling, while we saw in Section 2.2 that the polymer rolls itself up inside a ball with a volume equal to the polymer length. Case (2) is plausible because on the hexagonal lattice the exploration process in critical percolation has a path measure that, apart from higher order terms, is equal to that of the SAW with a critical reward for self-touchings (numerical simulation shows that $\gamma_c \approx \log 2.8$), and this exploration process has been proven to scale to SLE_6 (discussions with Vincent Beffara and Markus Heydenreich). Case (3) is plausible because $\mathrm{SLE}_{8/3}$ is predicted to be the scaling limit of SAW (see Section 1.7).

3. A polymer near a homogeneous interface

This section considers a polymer in the vicinity of a linear interface. Each monomer that touches the interface feels a *binding energy*, resulting in an attractive interaction between the polymer and the interface. The focus is on the occurrence of a phase transition between a *localized phase*, where the polymer stays close to the interface, and a *delocalized phase*, where it wanders away from the interface (see Fig. 19). In Sections 3.1–3.3 we look at the *pinning* version of the model, where the polymer can move on both sides of the interface, and in Section 3.4 at the *wetting* version, where the polymer is constrained to stay on one side of the interface (which acts like a hard wall). In Sections 3.5–3.6 we study how a pinned polymer can be pulled off an interface by applying a force to one of its endpoints. Section 3.7 lists some open problems.

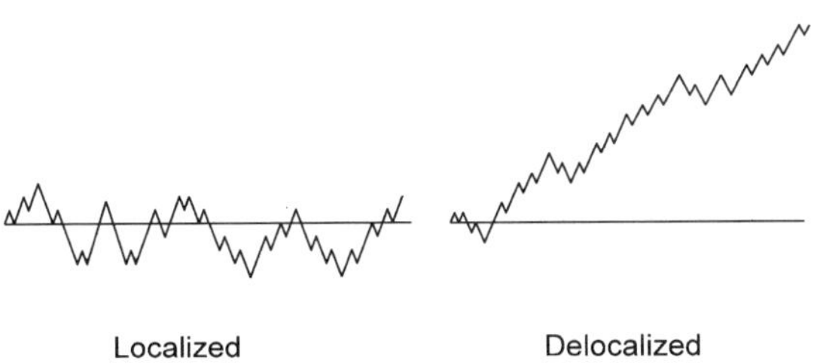

FIGURE 19. Path behavior in the two phases.

Polymers are used as surfactants, foaming and anti-foaming agents, etc. The wetting version of the model considered in the present section can be viewed as describing "paint on a wall".

3.1. Model. Our choices for the set of paths and for the interaction Hamiltonian are

$$\mathcal{W}_n = \{w = (i, w_i)_{i=0}^n : w_0 = 0,\ w_i \in \mathbb{Z}\ \forall 0 \le i \le n\},$$
$$H_n^\zeta(w) = -\zeta L_n(w),$$

with $\zeta \in \mathbb{R}$ and

$$L_n(w) = \sum_{i=1}^n 1_{\{w_i=0\}}, \qquad w \in \mathcal{W}_n,$$

the *local time of* w *at the interface.* The path measure is

$$P_n^\zeta(w) = \frac{1}{Z_n^\zeta}\, e^{-H_n^\zeta(w)}\, P_n(w), \qquad w \in \mathcal{W}_n,$$

where P_n is the projection onto $\mathcal{W}_n$ of the path measure P of an *arbitrary* directed irreducible random walk. This models a $(1+1)$-dimensional directed polymer in $\mathbb{N}_0 \times \mathbb{Z}$ in which each visit to the interface $\mathbb{N} \times \{0\}$ contributes an energy $-\zeta$, which is a reward when $\zeta > 0$ and a penalty when $\zeta < 0$ (see Fig. 20).

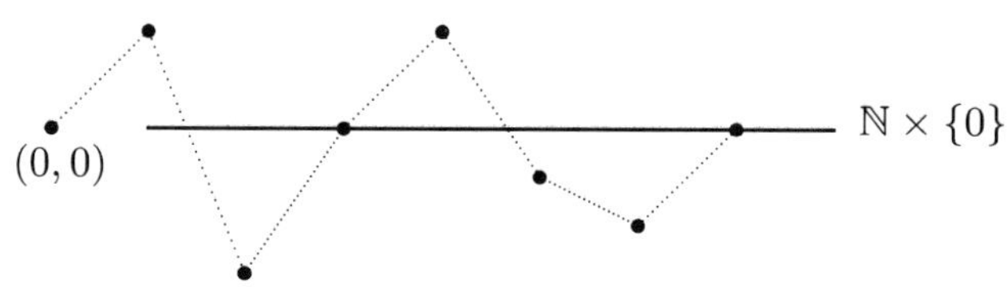

FIGURE 20. A 7-step two-sided path that makes 2 visits to the interface.

Let $S = (S_i)_{i \in \mathbb{N}_0}$ denote the random walk with law P starting from $S_0 = 0$. Let

$$R(n) = P(S_i \neq 0\ \forall 1 \le i < n,\ S_n = 0), \qquad n \in \mathbb{N}.$$

denote the *return time distribution* to the interface. Throughout the sequel it is assumed that $\sum_{n\in\mathbb{N}} R(n) = 1$ and

$$R(n) = n^{-1-a}\,\ell(n), \qquad n \in \mathbb{N},$$

for some $a \in (0,\infty)$ and some $\ell(\cdot)$ slowly varying at infinity (i.e., $\lim_{x\to\infty} \ell(cx)/\ell(x) = 1$ for all $c \in (0,\infty)$). Note that this assumption implies that $R(n) > 0$ for n large enough, i.e., $R(\cdot)$ is aperiodic. It is trivial, however, to extend the analysis below to include the periodic case. SRW corresponds to $a = \frac{1}{2}$ and period 2.

3.2. Free energy. The free energy can be computed explicitly. Let $\phi(x) = \sum_{n\in\mathbb{N}} x^n R(n)$, $x \in [0,\infty)$.

THEOREM 3.1. [Fisher [**53**], Giacomin [**55**], Chapter 2] *The free energy*

$$f(\zeta) = \lim_{n\to\infty} \frac{1}{n} \log Z_n^\zeta$$

exists for all $\zeta \in \mathbb{R}$ *and is given by*

$$f(\zeta) = \begin{cases} 0, & \text{if } \zeta \le 0, \\ r(\zeta), & \text{if } \zeta > 0, \end{cases}$$

where $r(\zeta)$ *is the unique solution of the equation*

$$\phi(e^{-r}) = e^{-\zeta}, \qquad \zeta > 0.$$

PROOF. For $\zeta \leq 0$, estimate

$$\sum_{m>n} R(m) = P(S_i \neq 0 \; \forall \, 1 \leq i \leq n) \leq Z_n^\zeta \leq 1,$$

which implies $f(\zeta) = 0$ because the left-hand side decays polynomially in n.

For $\zeta > 0$, let

$$R^\zeta(n) = \mathrm{e}^{\zeta - r(\zeta) n} \, R(n), \qquad n \in \mathbb{N}.$$

By the definition of $r(\zeta)$, this is a probability distribution on $\mathbb{N}$, with a finite mean $M^\zeta = \sum_{n \in \mathbb{N}} n R^\zeta(n)$ because $r(\zeta) > 0$. The partition sum when the polymer is constrained to end at 0 can be written as

$$Z_n^{*,\zeta} = \sum_{\substack{w \in \mathcal{W}_n \\ w_n = 0}} \mathrm{e}^{\zeta L_n(w)} \, P_n(w) = \mathrm{e}^{r(\zeta) n} \, Q^\zeta(n \in T)$$

with

$$Q^\zeta(n \in T) = \sum_{m=1}^{n} \sum_{\substack{j_1, \ldots, j_m \in \mathbb{N} \\ j_1 + \cdots + j_m = n}} \prod_{k=1}^{m} R^\zeta(j_k),$$

where T is the renewal process whose law Q^ζ is such that the i.i.d. renewals have law R^ζ. Therefore, by the renewal theorem,

$$\lim_{n \to \infty} Q^\zeta(n \in T) = 1/M^\zeta,$$

which yields

$$\lim_{n \to \infty} \frac{1}{n} \log Z_n^{*,\zeta} = r(\zeta).$$

By splitting the partition sum Z_n^ζ according to the last hitting time of 0 (see the end of **Tutorial 1 in Appendix A**), it is straightforward to show that there exists a $C < \infty$ such that

$$Z_n^{*,\zeta} \leq Z_n^\zeta \leq (1 + Cn) Z_n^{*,\zeta} \qquad \forall \, n \in \mathbb{N}_0.$$

It therefore follows that

$$f(\zeta) = \lim_{n \to \infty} \frac{1}{n} \log Z_n^\zeta = r(\zeta).$$

□

For SRW (see Spitzer [**97**], Section 1)

$$\phi(x) = 1 - \sqrt{1 - x^2}, \qquad x \in [0, 1].$$

By Theorem 3.1, this gives

$$f(\zeta) = r(\zeta) = \tfrac{1}{2} \left[\zeta - \log(2 - e^{-\zeta}) \right], \quad f'(\zeta) = \tfrac{1}{2} \left[1 - \tfrac{\mathrm{e}^{-\zeta}}{2 - \mathrm{e}^{-\zeta}} \right], \qquad \zeta > 0,$$

which is plotted in Fig. 21.

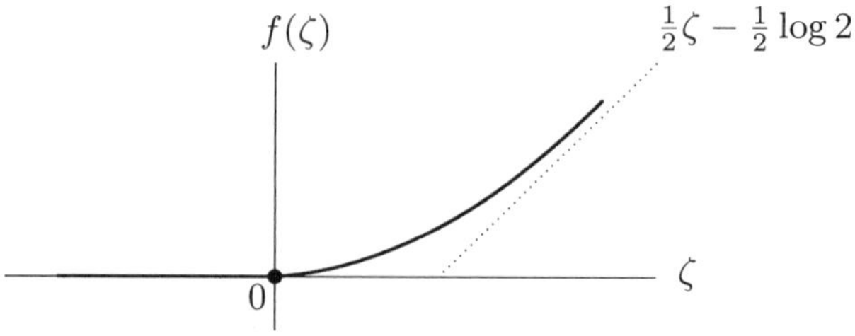

FIGURE 21. Plot of the free energy for pinned SRW.

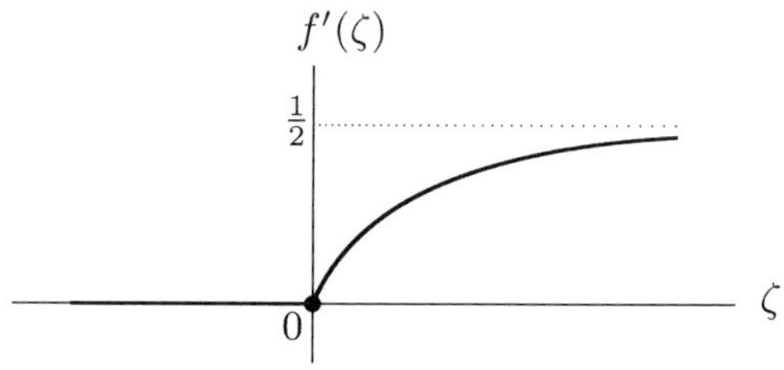

FIGURE 22. Plot of the average fraction of adsorbed monomers for pinned SRW. The phase transition is second order.

3.3. Path properties and order of the phase transition.

THEOREM 3.2. [Deuschel, Giacomin and Zambotti [**49**], Caravenna, Giacomin and Zambotti [**30**], Giacomin [**55**], Chapter 2] *Under the law* P_n^ζ *as* $n \to \infty$*:*
(a) If $\zeta > 0$*, then the path hits the interface with a strictly positive density, while the length and the height of the largest excursion away from the interface up to time* n *are of order* $\log n$*.*
(b) If $\zeta < 0$*, then the path hits the interface finitely often.*
(c) If $\zeta = 0$*, then the number of hits grows like a power of* n*.*

A detailed description of the path measure near the critical value is given in Sohier [**96**].

THEOREM 3.3. [Fisher [**53**], Giacomin [**55**], Chapter 2] *There exists an* $\ell^*(\cdot)$ *slowly varying at infinity such that*

$$f(\zeta) = \zeta^{1/(1\wedge a)}\,\ell^*(1/\zeta)\,[1 + o(1)], \qquad \zeta \downarrow 0.$$

Theorem 3.3 shows that, for all $m \in \mathbb{N}$, the order of the phase transition is m when $a \in [\frac{1}{m}, \frac{1}{m-1})$. For SRW, $a = \frac{1}{2}$ and the phase transition is second order (see Fig. 22).

The proof of Theorem 3.2 depends on fine estimates of the partition sum, beyond the exponential asymptotics found in Theorem 3.1. The proof of Theorem 3.3 is given in **Tutorial 3 in Appendix C**.

3.4. Wetting. What happens when the interface is impenetrable? Then the set of paths is replaced by (see Fig. 23)

$$\mathcal{W}_n^+ = \{w = (i, w_i)_{i=0}^n \colon w_0 = 0,\, w_i \in \mathbb{N}_0 \;\forall\, 0 \le i \le n\}.$$

Accordingly, write $P_n^{\zeta,+}(w)$, $Z_n^{\zeta,+}$ and $f^+(\zeta)$ for the path measure, the partition sum and the free energy. One-sided pinning at an interface is called *wetting*.

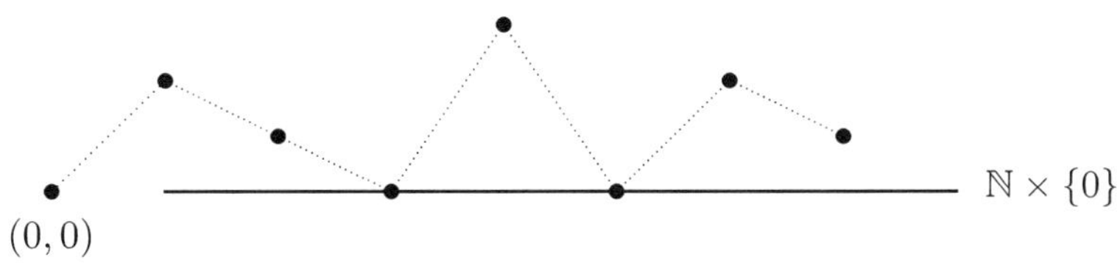

FIGURE 23. A 7-step one-sided path that makes 2 visits to the interface.

Let

$$R^+(n) = P(S_i > 0 \;\forall\, 1 \le i < n,\, S_n = 0), \qquad n \in \mathbb{N}.$$

This is a defective probability distribution. Define

$$\phi^+(x) = \sum_{n \in \mathbb{N}} x^n \, R^+(n), \qquad x \in [0, \infty),$$

and put

$$\widetilde{\phi}(x) = \frac{\phi^+(x)}{\phi^+(1)}, \qquad \zeta_c^+ = \log\left[\frac{1}{\phi^+(1)}\right] > 0.$$

THEOREM 3.4. [Fisher [**53**], Giacomin [**55**], Chapter 2] *The free energy is given by*

$$f^+(\zeta) = \begin{cases} 0, & \text{if } \zeta \le \zeta_c^+, \\ r^+(\zeta), & \text{if } \zeta > \zeta_c^+, \end{cases}$$

where $r^+(\zeta)$ is the unique solution of the equation

$$\widetilde{\phi}(e^{-r}) = e^{-(\zeta - \zeta_c^+)}, \qquad \zeta > \zeta_c^+.$$

The proof is similar to that of the pinned polymer. Localization on an impenetrable interface is harder than on a penetrable interface, because the polymer suffers a larger loss of entropy. This is the reason why $\zeta_c^+ > 0$. For SRW, symmetry gives

$$R^+(n) = \tfrac{1}{2}\, R(n), \qquad n \in \mathbb{N}.$$

Consequently,

$$\zeta_c^+ = \log 2, \quad \widetilde{\phi}(\cdot) = \phi(\cdot),$$

implying that

$$f^+(\zeta) = f(\zeta - \zeta_c^+), \qquad \zeta \in \mathbb{R}.$$

Thus, the free energy suffers a shift (i.e., the curves in Figs. 21–22 move to the right by $\log 2$) and the qualitative behavior is similar to that of pinning.

3.5. Pulling at an adsorbed polymer. A polymer can be pulled off an interface by a force. Replace the pinning Hamiltonian by

$$H_n^{\zeta,\phi}(w) = -\zeta L_n(w) - \phi w_n,$$

where $\phi \in (0,\infty)$ is a force in the upward direction acting on the endpoint of the polymer. Note that ϕw_n is the work exerted by the force to move the endpoint a distance w_n away from the interface. Write $Z_n^{\zeta,\phi}$ to denote the partition sum and

$$f(\zeta,\phi) = \lim_{n\to\infty} \frac{1}{n} \log Z_n^{\zeta,\phi}$$

to denote the free energy. Consider the case where the reference random walk can only make steps of size ≤ 1, i.e., pick $p \in [0,1]$ and put

$$P(S_1 = -1) = P(S_1 = +1) = \tfrac{1}{2}p, \quad P(S_1 = 0) = 1 - p.$$

THEOREM 3.5. [Giacomin and Toninelli [**62**]] *For every $\zeta \in \mathbb{R}$ and $\phi > 0$, the free energy exists and is given by*

$$f(\zeta,\phi) = f(\zeta) \vee g(\phi),$$

with $f(\zeta)$ the free energy of the pinned polymer without force and

$$g(\phi) = \log\big[\,p\cosh(\phi) + (1-p)\,\big].$$

PROOF. Write

$$Z_n^{\zeta,\phi} = Z_n^{*,\zeta} + \sum_{m=1}^{n} Z_{n-m}^{*,\zeta}\, \bar{Z}_m^{\phi},$$

where $Z_n^{*,\zeta}$ is the constrained partition sum without force encountered in Sections 3.1–3.3, and

$$\bar{Z}_m^{\phi} = \sum_{x\in\mathbb{Z}\backslash\{0\}} \mathrm{e}^{\phi x}\, R(m;x), \qquad m \in \mathbb{N},$$

with

$$R(m;x) = P\big(S_i \neq 0\ \forall\, 1 \leq i < m,\ S_m = x\big).$$

It suffices to show that

$$g(\phi) = \lim_{m\to\infty} \frac{1}{m} \log \bar{Z}_m^{\phi},$$

which will yield the claim because

$$f(\zeta) = \lim_{n\to\infty} \frac{1}{n} \log Z_n^{*,\zeta}.$$

The contribution to $\bar{Z}_m^{\phi}$ coming from $x \in \mathbb{Z}\backslash\mathbb{N}_0$ is bounded from above by $1/(1-\mathrm{e}^{-\phi}) < \infty$ and therefore is negligible. (The polymer does not care to stay below the interface because the force is pulling it upwards.) For $x \in \mathbb{N}$ the *reflection principle* gives

$$\begin{aligned} R(m;x) &= \tfrac{1}{2}p\, P\big(S_i > 0\ \forall\, 2 \leq i < m,\ S_m = x \mid S_1 = 1\big) \\ &= \tfrac{1}{2}p\,\big[P(S_m = x \mid S_1 = 1) - P(S_m = x \mid S_1 = -1)\big] \\ &= \tfrac{1}{2}p\,\big[P(S_{m-1} = x-1) - P(S_{m-1} = x+1)\big] \qquad \forall\, m \in \mathbb{N}. \end{aligned}$$

The first equality holds because the path cannot jump over the interface. The second inequality holds because, for any path from 1 to x that hits the interface, the piece of the path until the first hit of the interface can be reflected in the interface to yield a path from -1 to x. Substitution of the above relation into the sum defining $\bar{Z}_m^\phi$ gives

$$\begin{aligned}\bar{Z}_m^\phi &= O(1) + p\sinh(\phi)\sum_{x\in\mathbb{N}} e^{\phi x}\,P(S_{m-1}=x)\\ &= O(1)+O(1)+p\sinh(\phi)\,E\big(e^{\phi S_{m-1}}\big).\end{aligned}$$

But

$$E\big(e^{\phi S_{m-1}}\big) = [p\cosh(\phi)+(1-p)]^{m-1},$$

and so the above claim follows. □

The force either leaves most of the polymer adsorbed, when

$$f(\zeta,\phi) = f(\zeta) > g(\phi),$$

or pulls most of the polymer off, when

$$f(\zeta,\phi) = g(\phi) > f(\zeta).$$

A first-order phase transition occurs at those values of ζ and ϕ where $f(\zeta)=g(\phi)$, i.e., the critical value of the force is given by

$$\phi_c(\zeta) = g^{-1}\big(f(\zeta)\big), \qquad \zeta\in\mathbb{R},$$

with g^{-1} the inverse of g. Think of $g(\phi)$ as the free energy of the polymer with force ϕ not interacting with the interface.

3.6. Re-entrant force-temperature diagram. In order to analyze $\zeta\mapsto\phi_c(\zeta)$, we plot it as a function of temperature, putting

$$\zeta = 1/T, \quad \phi = F/T, \quad F_c(T) = T\phi_c(1/T).$$

It turns out that the curve $T\mapsto F_c(T)$ is increasing when $p\in(0,\frac{2}{3}]$, but has a minimum when $p\in(\frac{2}{3},1)$. The latter behavior is remarkable, since it says that there is a force F such that the polymer is adsorbed both for small T and for large T, but is desorbed for moderate T.

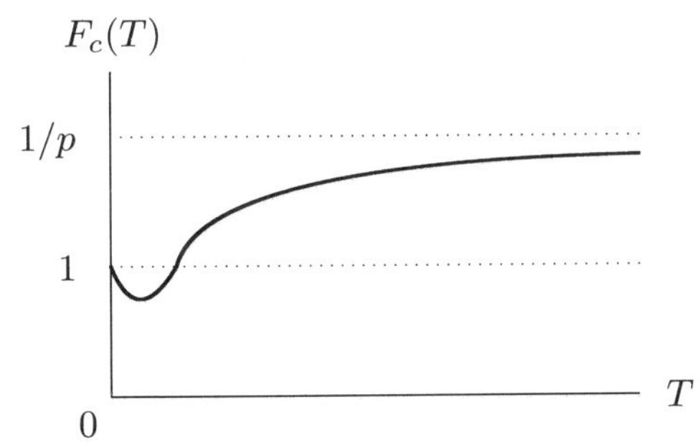

FIGURE 24. Re-entrant force-temperature diagram for $p\in(\frac{2}{3},1)$.

For $p=\frac{2}{3}$ all paths are equally likely, while for $p\in(\frac{2}{3},1)$ paths that move up and down are more likely than paths that stay flat. This leads to the following heuristic explanation of the re-entrant behavior. For every T, the adsorbed polymer makes excursions away from the interface and therefore has a strictly positive entropy. Some of this entropy is lost when a force is applied to the endpoint of the polymer, so that the part of the polymer near the endpoint is pulled away from the interface and is caused to move upwards steeply. There are two cases:

$p = \frac{2}{3}$: As T increases the effect of this entropy loss on the free energy increases, because "free energy = energy − temperature × entropy". This effect must be counterbalanced by a larger force to achieve desorption.

$p \in (\frac{2}{3}, 1)$: Steps in the east direction are favored over steps in the north-east and south-east directions, and this tends to place the adsorbed polymer farther away from the interface. Hence the force decreases for small T (i.e., $F_c(T) < F_c(0)$ for small T, because at $T = 0$ the polymer is fully adsorbed).

3.7. Open problems. Some key challenges are:

- Investigate pinning and wetting of SAW by a linear interface, i.e., study the undirected version of the model in Sections 3.1–3.4. Partial results have been obtained in the works of A.J. Guttmann, J. Hammersley, E.J. Janse van Rensburg, E. Orlandini, A. Owczarek, A. Rechnitzer, C. Soteros, C. Tesi, S.G. Whittington, and others. For references, see den Hollander [**70**], Chapter 7.
- Look at polymers living inside wedges or slabs, with interaction at the boundary. This leads to combinatorial problems of the type described in the lectures by Di Francesco during the summer school, many of which are hard. There is a large literature, with contributions coming from M. Bousquet-Melou, R. Brak, A.J. Guttmann, E.J. Janse van Rensburg, A. Owczarek, A. Rechnitzer, S.G. Whittington, and others. For references, see Guttmann [**64**].
- Caravenna and Pétrélis [**31, 32**] study a directed polymer pinned by a periodic array of interfaces. They identify the rate at which the polymer hops between the interfaces as a function of their mutual distance and determine the scaling limit of the endpoint of the polymer. There are several regimes depending on the sign of the adsorption strength and on how the distance between the interfaces scales with the length of the polymer. Investigate what happens when the interfaces are placed at random distances.
- What happens when the shape of the interface itself is random? Pinning of a polymer by a polymer, both performing directed random walks, can be modelled by the Hamiltonian $H_n^\zeta(w, w') = -\zeta L_n(w, w')$, $\zeta \in \mathbb{R}$, with $L_n(w, w') = \sum_{i=1}^n 1_{\{w_i = w_i'\}}$ the collision local time of $w, w' \in \mathcal{W}_n$, the set of directed paths introduced in Section 3.1. This model was studied by Birkner, Greven and den Hollander [**13**], Birkner and Sun [**14, 15**], Berger and Toninelli [**9**]. A variational formula for the critical adsorption strength is derived in [**13**]. This variational formula turns out to be hard to analyze.

In Sections 1–3 we considered several models of a polymer chain interacting with itself and/or with an interface. In Sections 4–6 we move to models with *disorder*, i.e., there is a random environment with which the polymer chain is interacting. Models with disorder are much harder than models without disorder. In order to advance mathematically, we will restrict ourselves to *directed* paths.

4. A polymer near a random interface

In this section we consider a directed polymer near a linear interface carrying "*random charges*". As in Section 3, the polymer receives an energetic reward or penalty when it hits the interface, but this time the size of the reward or penalty is determined by disorder attached to the interface (see Fig. 25). The goal is to determine under what conditions the disorder is able to pin the polymer to the interface.

In Sections 4.1–4.2 we define the model. In Sections 4.3–4.4 we use large deviation theory to derive a *variational formula* for the critical curve separating a *localized phase* from a *delocalized phase*, both for the quenched and the annealed version of the model (recall part III of Section 1.5). In Section 4.5 we use the two variational formulas to analyze under what conditions the two critical curves are different (= the disorder is relevant) or are the same (= the disorder is irrelevant). In Section 4.6 we explain why denaturation of DNA is described by this model. In Section 4.7 we close by formulating some open problems.

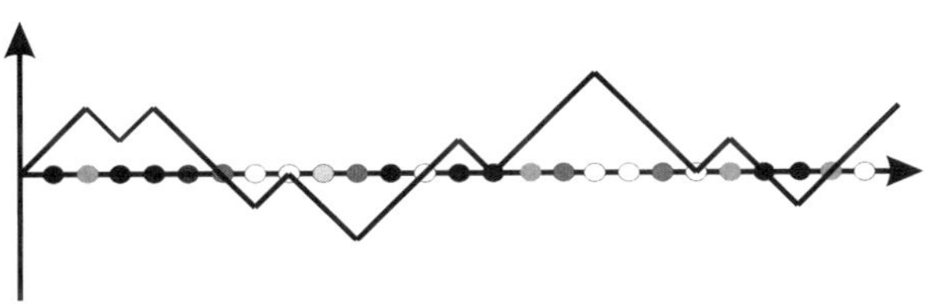

FIGURE 25. Different shades represent different disorder values.

4.1. Model. Let $S = (S_n)_{n\in\mathbb{N}_0}$ be a recurrent Markov chain on a countable state space Υ with a marked point $*$. Write P to denote the law of S given $S_0 = *$. Let

$$R(n) = P(S_i \neq * \;\forall\, 1 \leq i < n,\, S_n = *), \qquad n \in \mathbb{N},$$

denote the *return time distribution* to $*$, and assume that

$$\lim_{n\to\infty} \frac{\log R(n)}{\log n} = -(1+a) \quad \text{for some } a \in [0,\infty).$$

This is a weak version of the regularity condition assumed in Section 3.1 for the homogeneous pinning model.

Let

$$\omega = (\omega_i)_{i\in\mathbb{N}_0}$$

be an i.i.d. sequence of $\mathbb{R}$-valued random variables with marginal law μ_0, playing the role of a random environment. Write $\mathbb{P} = \mu_0^{\otimes\mathbb{N}_0}$ to denote the law of ω. Assume that μ_0 is non-degenerate and satisfies

$$M(\beta) = \mathbb{E}(e^{\beta\omega_0}) = \int_{\mathbb{R}} e^{\beta x} \mu(dx) < \infty \qquad \forall \beta \geq 0.$$

For fixed ω, define a law on the set of directed paths of length $n \in \mathbb{N}_0$ by putting

$$\frac{dP_n^{\beta,h,\omega}}{dP_n}\big((i,S_i)_{i=0}^n\big) = \frac{1}{Z_n^{\beta,h,\omega}} \exp\left[\sum_{i=0}^{n-1}(\beta\omega_i - h)\,1_{\{S_i=*\}}\right],$$

where $\beta \in [0,\infty)$ is the *disorder strength*, $h \in \mathbb{R}$ is the *disorder bias*, P_n is the projection of P onto n-step paths, and $Z_n^{\beta,h,\omega}$ is the normalizing partition sum. Note that the homogeneous pinning model in Section 3 is recovered by putting $\beta = 0$ and $h = -\zeta$ (with the minor difference that now the Hamiltonian includes the term with $i = 0$ but not the term with $i = n$). Without loss of generality we can choose μ_0 to be such that $\mathbb{E}(\omega_0) = 0$, $\mathbb{E}(\omega_0^2) = 1$ (which amounts to a shift of the parameters β, h).

In our standard notation, the above model corresponds to the choice

$$\mathcal{W}_n = \Big\{w = (i,w_i)_{i=0}^n\colon\ w_0 = *,\ w_i \in \Upsilon\ \forall 0 < i \leq n\Big\},$$

$$H_n^{\beta,h,\omega}(w) = -\sum_{i=0}^{n-1}(\beta\omega_i - h)\,1_{\{w_i=*\}}.$$

(As before, we think of $(S_i)_{i=0}^n$ as the realization of $(w_i)_{i=0}^n$ drawn according to $P_n^{\beta,h,\omega}$.) The key example modelling our polymer with pinning is

$$\Upsilon = \mathbb{Z}^d, \quad * = \{0\}, \quad P = \text{ law of directed SRW in } \mathbb{Z}^d, \qquad d = 1,2,$$

for which $a = \frac{1}{2}$ and $a = 0$, respectively. We expect that pinning occurs for large β and/or small h: the polymer gets a large enough energetic reward when it hits the positive charges and does not lose too much in terms of entropy when it avoids the negative charges. For the same reason we expect that no pinning occurs for small β and/or large h. In Sections 4.2–4.6 we identify the phase transition curve and investigate its properties.

4.2. Free energies. The *quenched free energy* is defined as

$$f^{\text{que}}(\beta,h) = \lim_{n\to\infty}\frac{1}{n}\log Z_n^{\beta,h,\omega} \qquad \omega\text{–a.s.}$$

Subadditivity arguments show that ω-a.s. the limit exists and is non-random (see **Tutorial 1 in Appendix A**). Since

$$Z_n^{\beta,h,\omega} = E\left(\exp\left[\sum_{i=0}^{n-1}(\beta\omega_i - h)\,1_{\{S_i=*\}}\right]\right) \geq e^{\beta\omega_0 - h}\sum_{m\geq n} R(m),$$

which decays polynomially in n, it follows that $f^{\text{que}}(\beta,h) \geq 0$. This fact motivates the definition

$$\mathcal{L} = \big\{(\beta,h)\colon\ f^{\text{que}}(\beta,h) > 0\big\},$$
$$\mathcal{D} = \big\{(\beta,h)\colon\ f^{\text{que}}(\beta,h) = 0\big\},$$

which are referred to as the *quenched localized phase*, respectively, the *quenched delocalized phase*. The associated *quenched critical curve* is

$$h_c^{\text{que}}(\beta) = \inf\{h \in \mathbb{R}\colon\ f^{\text{que}}(\beta,h) = 0\}, \qquad \beta \in [0,\infty).$$

Because $h \mapsto f^{\text{que}}(\beta, h)$ is non-increasing, we have $f^{\text{que}}(\beta, h) = 0$ for $h \geq h_c^{\text{que}}(\beta)$. Convexity of $(\beta, h) \mapsto f^{\text{que}}(\beta, h)$ implies that $\beta \mapsto h_c^{\text{que}}(\beta)$ is convex. It is easy to check that both are finite (this uses the bound $f^{\text{que}} \leq f^{\text{ann}}$ with f^{ann} the annealed free energy defined below) and therefore are also continuous. Futhermore, $h_c^{\text{que}}(0) = 0$ (because the critical threshold for the homogeneous pinning model is zero), and $h_c^{\text{que}}(\beta) > 0$ for $\beta > 0$ (see below). Together with convexity the latter imply that $\beta \mapsto h_c^{\text{que}}(\beta)$ is strictly increasing.

Alexander and Sidoravicius [**3**] prove that $h_c^{\text{que}}(\beta) > 0$ for $\beta > 0$ for arbitrary non-degenerate μ_0 (see Fig. 26). This result is important, because it shows that localization occurs even for a *moderately negative* average value of the disorder, contrary to what we found for the homogeneous pinning model in Section 3. Indeed, since $\mathbb{E}(\beta\omega_1 - h) = -h < 0$, even a globally repulsive interface can locally pin the polymer provided the global repulsion is modest: all the polymer has to do is hit the positive charges and avoid the negative charges.

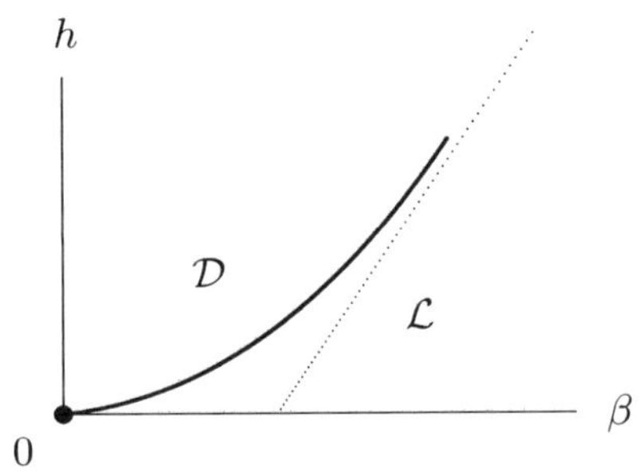

FIGURE 26. Qualitative picture of $\beta \mapsto h_c^{\text{que}}(\beta)$ (the asymptote has finite slope if and only if the support of μ_0 is bounded from above). The details of the curve are known only partially (see below).

The *annealed free energy* is defined by (recall Section 1.5)

$$f^{\text{ann}}(\beta, h) = \lim_{n\to\infty} \frac{1}{n} \log \mathbb{E}\big(Z_n^{\beta,h,\omega}\big).$$

This is the free energy of a homopolymer. Indeed, $\mathbb{E}(Z_n^{\beta,h,\omega}) = Z_n^{h-\log M(\beta)}$, the partition function of the homogeneous pinning model with parameter $h - \log M(\beta)$. The associated *annealed critical curve*

$$h_c^{\text{ann}}(\beta) = \inf\{h \in \mathbb{R} \colon\ f^{\text{ann}}(\beta, h) = 0\}, \qquad \beta \in [0, \infty),$$

can therefore be computed explicitly:

$$h_c^{\text{ann}}(\beta) = \log \mathbb{E}(\mathrm{e}^{\beta\omega_0}) = \log M(\beta).$$

By Jensen's inequality, we have

$$f^{\text{que}} \leq f^{\text{ann}} \quad \longrightarrow \quad h_c^{\text{que}} \leq h_c^{\text{ann}}.$$

In Fig. 28 below we will see how the two critical curves are related.

DEFINITION 4.1. For a given choice of R, μ_0 and β, the disorder is said to be *relevant* when $h_c^{\text{que}}(\beta) < h_c^{\text{ann}}(\beta)$ and *irrelevant* when $h_c^{\text{que}}(\beta) = h_c^{\text{ann}}(\beta)$.

Note: In the physics literature, the notion of relevant disorder is reserved for the situation where the disorder not only changes the critical value but also changes the behavior of the free energy near the critical value. In what follows we adopt the more narrow definition given above. It turns out, however, that for the pinning model considered here a change of critical value entails a change of critical behavior as well.

Some 15 papers have appeared in the past 5 years, containing sufficient conditions for relevant, irrelevant and marginal disorder, based on various types of estimates. Key references are:

- Relevant disorder: Derrida, Giacomin, Lacoin and Toninelli [**48**], Alexander and Zygouras [**4**].
- Irrelevant disorder: Alexander [**2**], Toninelli [**100**], Lacoin [**84**].
- Marginal disorder: Giacomin, Lacoin and Toninelli [**56**].

See also Giacomin and Toninelli [**63**], Alexander and Zygouras [**5**], Giacomin, Lacoin and Toninelli [**57**]. (The word "marginal" stands for "at the border between relevant and irrelevant", and can be either relevant or irrelevant.)

In Sections 4.4–4.6 we derive *variational formulas* for h_c^{que} and h_c^{ann} and provide necessary and sufficient conditions on R, μ_0 and β for relevant disorder. The results are based on Cheliotis and den Hollander [**35**]. In Section 4.3 we give a quick overview of the necessary tools from large deviation theory developed in Birkner, Greven and den Hollander [**12**].

4.3. Preparations. In order to prepare for the large deviation analysis in Section 4.5, we need to place the random pinning problem in a different context.

Think of $\omega = (\omega_i)_{i\in\mathbb{N}_0}$ as a random sequence of *letters* drawn from the alphabet $\mathbb{R}$. Write $\mathcal{P}^{\mathrm{inv}}(\mathbb{R}^{\mathbb{N}_0})$ to denote the set of probability measures on infinite letter sequences that are shift-invariant. The law $\mu_0^{\otimes\mathbb{N}_0}$ of ω is an element of $\mathcal{P}^{\mathrm{inv}}(\mathbb{R}^{\mathbb{N}_0})$. A typical element of $\mathcal{P}^{\mathrm{inv}}(\mathbb{R}^{\mathbb{N}_0})$ is denoted by Ψ.

Let $\widetilde{\mathbb{R}} = \cup_{k\in\mathbb{N}}\mathbb{R}^k$. Think of $\widetilde{\mathbb{R}}$ as the set of *finite words*, and of $\widetilde{\mathbb{R}}^{\mathbb{N}}$ as the set of *infinite sentences*. Write $\mathcal{P}^{\mathrm{inv}}(\widetilde{\mathbb{R}}^{\mathbb{N}})$ to denote the set of probability measures on infinite sentences that are shift-invariant. A typical element of $\mathcal{P}^{\mathrm{inv}}(\widetilde{\mathbb{R}}^{\mathbb{N}})$ is denoted by Q.

The excursions of S away from the interface cut out successive words from the random environment ω, forming an infinite sentence (see Fig. 27). Under the joint law of S and ω, this sentence has law $q_0^{\otimes\mathbb{N}}$ with

$$q_0(\mathrm{d}x_0,\dots,\mathrm{d}x_{k-1}) = R(k)\,\mu_0(\mathrm{d}x_0)\times\cdots\times\mu_0(\mathrm{d}x_{k-1}), \qquad k\in\mathbb{N},\ x_0,\dots,x_{k-1}\in\mathbb{R}.$$

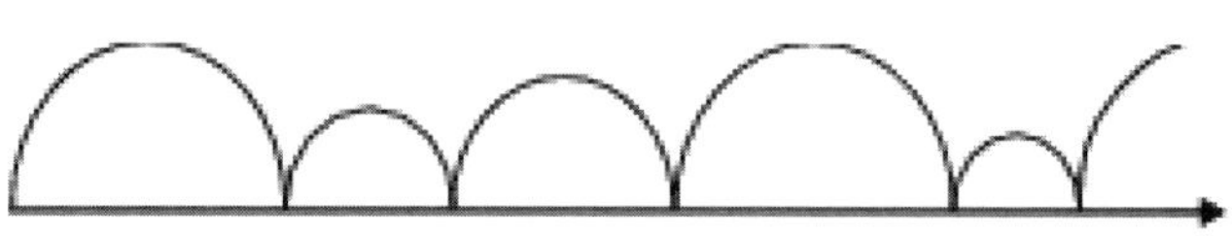

FIGURE 27. Infinite sentence generated by S on ω.

For $Q\in\mathcal{P}^{\mathrm{inv}}(\widetilde{\mathbb{R}}^{\mathbb{N}})$, let

$$I^{\mathrm{que}}(Q) = H\big(Q\,|\,q_0^{\otimes\mathbb{N}}\big) + a\,m_Q\,H\big(\Psi_Q\,|\,\mu_0^{\otimes\mathbb{N}_0}\big),$$

$$I^{\mathrm{ann}}(Q) = H\big(Q\,|\,q_0^{\otimes\mathbb{N}}\big),$$

where

- $\Psi_Q \in \mathcal{P}(\mathbb{R}^{\mathbb{N}_0})$ is the projection of Q via concatenation of words;
- m_Q is the average word length under Q;
- $H(\cdot|\cdot)$ denotes specific relative entropy.

It is shown in Birkner, Greven and den Hollander [**12**] that I^{que} and I^{ann} are the quenched and the annealed rate function in the *large deviation principle* (LDP) for the *empirical process of words*. More precisely,

$$\exp[-NI^{\mathrm{que}}(Q) + o(N)] \quad \text{and} \quad \exp[-NI^{\mathrm{ann}}(Q) + o(N)]$$

are the respective probabilities that the first N words generated by S on ω, periodically extended to form an infinite sentence, have an empirical distribution that is close to $Q \in \mathcal{P}^{\mathrm{inv}}(\widetilde{\mathbb{R}}^{\mathbb{N}})$ in the weak topology. **Tutorial 4 in Appendix D** provides the background of this LDP.

The main message of the formulas for $I^{\mathrm{que}}(Q)$ and $I^{\mathrm{ann}}(Q)$ is that

$$I^{\mathrm{que}}(Q) = I^{\mathrm{ann}}(Q) + \text{ an explicit extra term.}$$

We will see in Section 4.4 that the extra term is crucial for the distinction between relevant and irrelevant disorder.

4.4. Application of the LDP. For $Q \in \mathcal{P}^{\mathrm{inv}}(\widetilde{\mathbb{R}}^{\mathbb{N}})$, let $\pi_{1,1}Q \in \mathcal{P}(\mathbb{R})$ denote the projection of Q onto the first letter of the first word. Define $\Phi(Q)$ to be the average value of the first letter under Q,

$$\Phi(Q) = \int_{\mathbb{R}} x\,(\pi_{1,1}Q)(\mathrm{d}x), \qquad Q \in \mathcal{P}^{\mathrm{inv}}(\widetilde{\mathbb{R}}^{\mathbb{N}}),$$

and $\mathcal{C}$ to be the set

$$\mathcal{C} = \left\{ Q \in \mathcal{P}^{\mathrm{inv}}(\widetilde{\mathbb{R}}^{\mathbb{N}})\colon \int_{\mathbb{R}} |x|\,(\pi_{1,1}Q)(\mathrm{d}x) < \infty \right\}.$$

The following theorem provides variational formulas for the critical curves.

THEOREM 4.2. [Cheliotis and den Hollander [**35**]] *Fix μ_0 and R. For all $\beta \in [0,\infty)$,*

$$h_c^{\mathrm{que}}(\beta) = \sup_{Q \in \mathcal{C}} [\beta\Phi(Q) - I^{\mathrm{que}}(Q)],$$

$$h_c^{\mathrm{ann}}(\beta) = \sup_{Q \in \mathcal{C}} [\beta\Phi(Q) - I^{\mathrm{ann}}(Q)].$$

For $\beta \in [0,\infty)$, let

$$\mu_\beta(\mathrm{d}x) = \frac{1}{M(\beta)}\,\mathrm{e}^{\beta x}\,\mu_0(\mathrm{d}x), \qquad x \in \mathbb{R},$$

and let $Q_\beta = q_\beta^{\otimes\mathbb{N}} \in \mathcal{P}^{\mathrm{inv}}(\widetilde{\mathbb{R}}^{\mathbb{N}})$ be the law of the infinite sentence generated by S on ω when the first letter of each word is drawn from the tilted law μ_β rather than μ_0, i.e.,

$$q_\beta(\mathrm{d}x_0,\dots,\mathrm{d}x_{n-1}) = R(n)\,\mu_\beta(\mathrm{d}x_0)\times\cdots\times\mu_0(\mathrm{d}x_{n-1}), \qquad n \in \mathbb{N},\ x_0,\dots,x_{n-1} \in \mathbb{R}.$$

It turns out that Q_β is the *unique maximizer* of the annealed variational formula. This leads to the following two theorems.

THEOREM 4.3. [Cheliotis and den Hollander [**35**]] *Fix* μ_0 *and* R. *For all* $\beta \in [0,\infty)$,

$$h_c^{\rm que}(\beta) < h_c^{\rm ann}(\beta) \quad \Longleftrightarrow \quad I^{\rm que}(Q_\beta) > I^{\rm ann}(Q_\beta).$$

THEOREM 4.4. [Cheliotis and den Hollander [**35**]] *For all* μ_0 *and* R *there exists a* $\beta_c = \beta_c(\mu_0, R) \in [0,\infty]$ *such that*

$$h_c^{\rm que}(\beta) \begin{cases} = h_c^{\rm ann}(\beta) & \text{if } \beta \in [0,\beta_c], \\ < h_c^{\rm ann}(\beta) & \text{if } \beta \in (\beta_c,\infty). \end{cases}$$

Theorem 4.3 gives a necessary and sufficient condition for relevant disorder, while Theorem 4.4 shows that relevant and irrelevant disorder are separated by a single critical temperature (see Fig. 28).

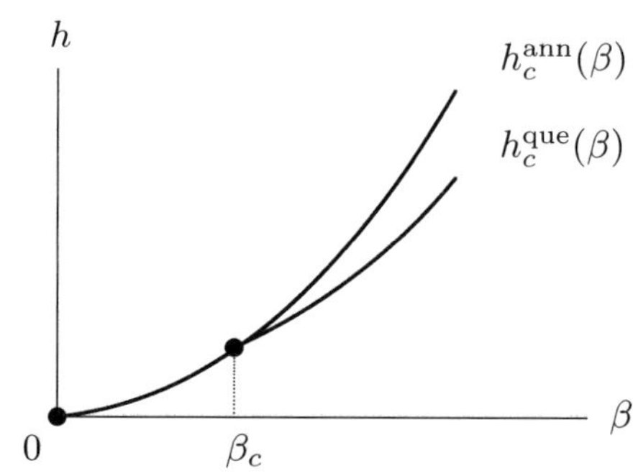

FIGURE 28. Uniqueness of the critical temperature β_c.

4.5. Consequences of the variational characterization. Corollaries 4.5–4.7 give us control over β_c. Abbreviate $\chi = \sum_{n\in\mathbb{N}}[P(S_n = *)]^2$, i.e., the average number of times two independent copies of our Markov chain S meet at $*$.

COROLLARY 4.5. [Cheliotis and den Hollander [**35**]] *(a) If* $a = 0$, *then* $\beta_c = \infty$ *for all* μ_0.
(b) If $a \in (0,\infty)$, *then, for all* μ_0, $\chi < \infty$ *implies that* $\beta_c \in (0,\infty]$.

COROLLARY 4.6. [Cheliotis and den Hollander [**35**]] *(a)* $\beta_c \geq \beta_c^*$ *with*

$$\beta_c^* = \sup\left\{\beta \in [0,\infty)\colon\ M(2\beta)/M(\beta)^2 < 1 + \chi^{-1}\right\}.$$

(b) $\beta_c \leq \beta_c^{**}$ *with*

$$\beta_c^{**} = \inf\left\{\beta \in [0,\infty)\colon\ h(\mu_\beta \mid \mu_0) > h(R)\right\},$$

where $h(\cdot \mid \cdot)$ *is relative entropy and* $h(\cdot)$ *is entropy.*

COROLLARY 4.7. [Cheliotis and den Hollander [**35**]] *If* $a \in (0,\infty)$, *then* $\beta_c \in [0,\infty)$ *for all* μ_0 *with*

$$\mu_0(\{w\}) = 0,$$

where $w = \sup[\mathrm{supp}(\mu_0)]$.

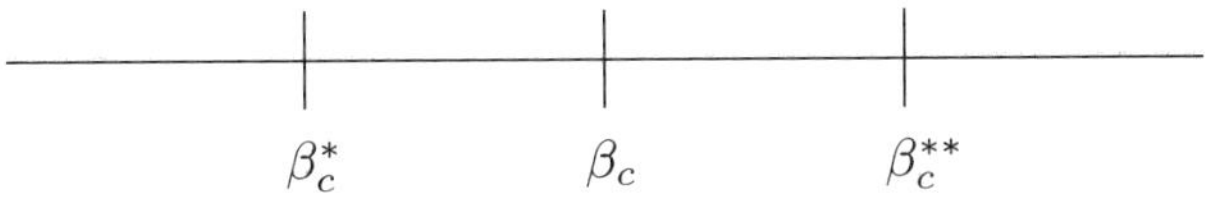

FIGURE 29. Bounds on β_c.

For the case where R is regularly varying at infinity, i.e.,

$$R(n) = n^{-(1+a)}\ell(n), \qquad n \in \mathbb{N},$$

with $\ell(\cdot)$ slowly varying at infinity (which means that $\lim_{x\to\infty} \ell(cx)/\ell(x) = 1$ for all $c \in (0,\infty)$), renewal theory gives

$$P(S_n = *) \sim \begin{cases} \frac{C}{n^{1-a}\ell(n)}, & a \in (0,1), \\ C, & a \in (1,\infty), \\ \ell^*(n), & a = 1, \end{cases} \qquad n \to \infty,$$

for some $C \in (0,\infty)$ and $\ell^*(\cdot)$ slowly varying at infinity. It therefore follows that $\chi < \infty$ if and only if $a \in (0, \frac{1}{2})$ or $a = \frac{1}{2}$, $\sum_{n\in\mathbb{N}} n^{-1}[\ell(n)]^{-2} < \infty$.

A challenging open problem is the following conjecture, which has been proved under more restrictive assumptions on R (see Section 4.7).

CONJECTURE 4.8. [Cheliotis and den Hollander [**35**]] *If $a \in (0,\infty)$, then, for all μ_0, $\chi = \infty$ implies that $\beta_c = 0$.*

Note: The results in Theorem 4.4 and Corollaries 4.5, 4.6 and 4.7 have all been derived in the literature by other means (see the references cited at the end of Section 4.2 and references therein). The point of the above exposition is to show that these results also follow in a natural manner from a *variational analysis* of the random pinning model, based on Theorems 4.2 and 4.3.

The following heuristic criterion, known as the *Harris criterion*, applies to the random pinning model.

- ► "Arbitrary weak disorder modifies the nature of a phase transition when the order of the phase transition in the non-disordered system is < 2."

Since, when R is regularly varying at infinity, the order of the phase transition for the homopolymer is < 2 when $a > \frac{1}{2}$ and ≥ 2 when $a \leq \frac{1}{2}$ (see **Tutorial 3 in Appendix C**), the above results fit with this criterion. It is shown in Giacomin and Toninelli [**60**] that the disorder makes the phase transition smoother: in the random pinning model the order of the phase transition is *at least two*, irrespective of the value of a.

At the critical value $a = \frac{1}{2}$ the disorder can be *marginally relevant* or *marginally irrelevant*, depending on the choice of $\ell(\cdot)$. See Alexander [**2**], Giacomin, Lacoin and Toninelli [**56**].

4.6. Denaturation of DNA. DNA is a string of AT and CG base pairs forming a double helix: A and T share two hydrogen bonds, C and G share three. Think of the two strands as performing random walks in three-dimensional space subject to the restriction that they do not cross each other. Then the distance between the two strands is a random walk conditioned not to return to the origin.

Since three-dimensional random walks are transient, this condition has an effect similar to that of a hard wall.

This view of DNA is called the Poland-Sheraga model (see Fig. 30). The localized phase $\mathcal{L}$ corresponds to the bounded phase of DNA, where the two strands are attached. The delocalized phase $\mathcal{D}$ corresponds to the denaturated phase of DNA, where the two strands are detached.

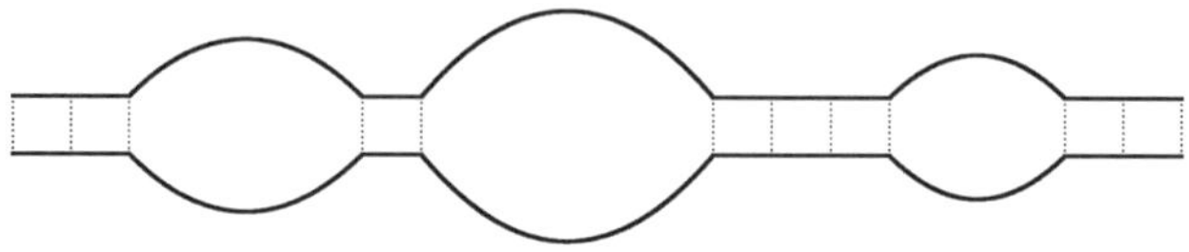

FIGURE 30. Schematic representation of the two strands of DNA in the Poland-Sheraga model. The dotted lines are the interacting base pairs, the loops are the denaturated segments without interaction.

Since the order of the base pairs in DNA is irregular and their binding energies are different, DNA can be thought of as a polymer near an interface with *binary disorder*. Of course, the order of the base pairs will not be i.i.d., but the random pinning model is reasonable at least for a qualitative description. Upon heating, the hydrogen bonds that keep the base pairs together can break and the two strands can separate, either partially or completely. This is called *denaturation*. See Cule and Hwa [**45**], Kafri, Mukamel and Peliti [**82**] for background.

4.7. Open problems. Some key challenges are:

- Provide the proof of Conjecture 4.8. The papers cited at the end of Section 4.2 show that if R is regularly varying at infinity (the condition mentioned below Corollary 4.7), then $\beta_c = 0$ for $a \in (\frac{1}{2}, \infty)$, and also for $a = \frac{1}{2}$ when $\ell(\cdot)$ does not decay too fast.
- Determine whether the phase transition is second order or higher order.
- Find sharp bounds for β_c, in particular, find a necessary and sufficient condition on μ_0 and R under which $\beta_c = \infty$ (i.e., the disorder is irrelevant for all temperatures).
- Bolthausen, Caravenna and de Tilière [**20**] apply a renormalization approach to random pinning. Develop this approach to study the critical curve.

Pétrélis [**89**] studies pinning at an interface with an internal structure. Information on the critical curve is hard to come by.

5. A copolymer interacting with two immiscible fluids

A copolymer is a polymer consisting of different types of monomers. The order of the monomers is determined by the polymerization process through which the copolymer is grown. This section looks at a $(1+1)$-dimensional directed copolymer, consisting of a random concatenation of hydrophobic and hydrophilic monomers, near a linear interface separating two immiscible solvents, oil and water, as depicted in Fig. 31.

The copolymer has a tendency to stay close to the oil-water interface, in order to be able to place as many of its monomers in their preferred fluid. In doing so it lowers energy but loses entropy. A phase transition may be expected between a

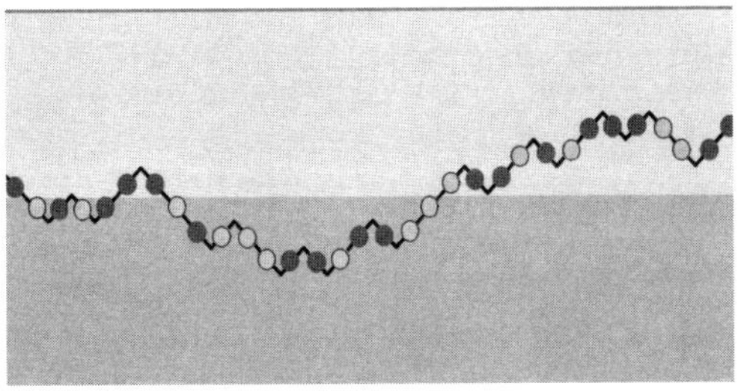

FIGURE 31. A directed copolymer near a linear interface. Oil and hydrophobic monomers are light-shaded, water and hydrophilic monomers are dark-shaded.

localized phase, where the copolymer stays close to the interface, and a *delocalized phase*, where it wanders away. Which of the two phases actually occurs depends on the strengths of the chemical affinities.

Copolymers near liquid-liquid interfaces are of interest due to their extensive application as surfactants, emulsifiers, and foaming or antifoaming agents. Many fats contain stretches of hydrophobic and hydrophilic monomers, arranged in some sort of erratic manner, and therefore are examples of random copolymers. (For the description of such systems, the undirected version of the model depicted in Fig. 32 is of course more appropriate, but we restrict ourselves to the directed version because this is mathematically much more tractable.) The transition between a localized and a delocalized phase has been observed experimentally, e.g. in neutron reflection studies of copolymers consisting of blocks of ethylene oxide and propylene oxide near a hexane-water interface. Here, a thin layer of hexane, approximately 10^{-5} m thick, is spread on water. In the localized phase, the copolymer is found to stretch itself along the interface in a band of width approximately 20 Å.

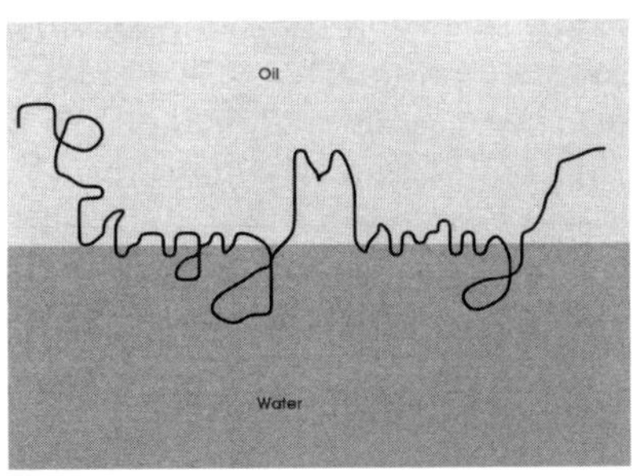

FIGURE 32. An undirected copolymer near a linear interface. The disorder along the copolymer is not indicated.

In Sections 5.1–5.4 we define and study the copolymer model. In Section 5.5 we look at a version of the copolymer model where the linear interface is replaced by a *random interface*, modelling a micro-emulsion. Section 5.6 lists some open problems.

5.1. Model. Let

$$\mathcal{W}_n = \{w = (i, w_i)_{i=0}^n \colon w_0 = 0,\ w_{i+1} - w_i = \pm 1\ \forall\, 0 \le i < n\}$$

denote the set of all n-step directed paths that start from the origin and at each step move either north-east or south-east. Let

$$\omega = (\omega_i)_{i\in\mathbb{N}} \text{ be i.i.d. with } \mathbb{P}(\omega_1 = +1) = \mathbb{P}(\omega_1 = -1) = \tfrac{1}{2}$$

label the order of the monomers along the copolymer. Write $\mathbb{P}$ to denote the law of ω. The Hamiltonian, for fixed ω, is

$$H_n^{\beta,h,\omega}(w) = -\beta \sum_{i=1}^{n} (\omega_i + h)\, \mathrm{sign}(w_{i-1}, w_i), \qquad w \in \mathcal{W}_n,$$

with $\beta, h \in [0,\infty)$ the *disorder strength*, respectively, the *disorder bias* (the meaning of $\mathrm{sign}(w_{i-1}, w_i)$ is explained below). The path measure, for fixed ω, is

$$P_n^{\beta,h,\omega}(w) = \frac{1}{Z_n^{\beta,h,\omega}}\, \mathrm{e}^{-H_n^{\beta,h,\omega}(w)}\, P_n(w), \qquad w \in \mathcal{W}_n,$$

where P_n is the law of the n-step directed random walk, which is the uniform distribution on $\mathcal{W}_n$. Note that P_n is the projection on $\mathcal{W}_n$ of the law P of the infinite directed walk whose vertical steps are SRW.

The interpretation of the above definitions is as follows: $\omega_i = +1$ or -1 stands for monomer i being hydrophobic or hydrophilic; $\mathrm{sign}(w_{i-1}, w_i) = +1$ or -1 stands for monomer i lying in oil or water; $-\beta(\omega_i + h)\mathrm{sign}(w_{i-1}, w_i)$ is the energy of monomer i. For $h = 0$ both monomer types interact equally strongly, while for $h = 1$ the hydrophilic monomers do not interact at all. Thus, only the regime $h \in [0,1]$ is relevant, and for $h > 0$ the copolymer prefers the oil over the water.

Note that the energy of a path is a sum of contributions coming from its *successive excursions away from the interface* (this viewpoint was already exploited in Section 4 for the random pinning model). All that is relevant for the energy of the excursions is what stretch of ω they sample, and whether they are above or below the interface. The copolymer model is harder than the random pinning model, because the energy of an excursion depends on the sum of the values of ω in the stretch that is sampled, not just on the first value. We expect the localized phase to occur for large β and/or small h and the delocalized phase for small β and/or large h. Our goal is to identify the critical curve separating the two phases.

5.2. Free energies. The quenched free energy is defined as

$$f^{\mathrm{que}}(\beta,h) = \lim_{n\to\infty} \frac{1}{n} \log Z_n^{\beta,h,\omega} \qquad \omega\text{−a.s.}$$

Subadditivity arguments show that ω-a.s. the limit exists and is non-random for all $\beta, h \in [0,\infty)$ (see **Tutorial 1 in Appendix A**). The following lower bound holds:

$$f^{\mathrm{que}}(\beta,h) \ge \beta h \qquad \forall\, \beta, h \in [0,\infty).$$

Proof. Abbreviate

$$\Delta_i = \mathrm{sign}(S_{i-1}, S_i)$$

and write

$$\begin{aligned} Z_n^{\beta,h,\omega} &= E\left(\exp\left[\beta\sum_{i=1}^n(\omega_i+h)\Delta_i\right]\right) \\ &\geq E\left(\exp\left[\beta\sum_{i=1}^n(\omega_i+h)\Delta_i\right] 1_{\{\Delta_i=+1\ \forall 1\leq i\leq n\}}\right) \\ &= \exp\left[\beta\sum_{i=1}^n(\omega_i+h)\right] P(\Delta_i=+1\ \forall 1\leq i\leq n) \\ &= \exp[\beta hn + o(n) + O(\log n)] \qquad \omega\text{–a.s.}, \end{aligned}$$

where the last line uses the strong law of large numbers for ω and the fact that $P(\Delta_i = +1\ \forall 1 \leq i \leq n) \geq C/n^{1/2}$ for some $C > 0$. □

Put

$$g^{\text{que}}(\beta,h) = f^{\text{que}}(\beta,h) - \beta h.$$

The above proof shows that $g^{\text{que}}(\beta,h) = 0$ corresponds to the strategy where the copolymer wanders away from the interface in the upward direction. This fact motivates the definition

$$\begin{aligned} \mathcal{L} &= \{(\beta,h)\colon\ g^{\text{que}}(\beta,h) > 0\}, \\ \mathcal{D} &= \{(\beta,h)\colon\ g^{\text{que}}(\beta,h) = 0\}, \end{aligned}$$

referred to as the *localized phase*, respectively, the *delocalized phase*. The associated quenched critical curve is

$$h_c^{\text{que}}(\beta) = \inf\{h \in [0,\infty)\colon\ g^{\text{que}}(\beta,h) = 0\}, \quad \beta \in [0,\infty).$$

Convexity of $(\beta,t) \mapsto g^{\text{que}}(\beta,t/\beta)$ implies that $\beta \mapsto \beta h_c^{\text{que}}(\beta)$ is convex. It is easy to check that both are finite and therefore also continuous. Furthermore, $h_c^{\text{que}}(0) = 0$ and $h_c^{\text{que}}(\beta) > 0$ for $\beta > 0$ (see below). For fixed h, $\beta \mapsto g^{\text{que}}(\beta,h)$ is convex and non-negative, with $g^{\text{que}}(0,h) = 0$, and hence is non-decreasing. Therefore $\beta \mapsto h_c^{\text{que}}(\beta)$ is non-decreasing as well. With the help of the convexity of $\beta \mapsto \beta h_c^{\text{que}}(\beta)$, it is easy to show that $\beta \mapsto \beta h_c^{\text{que}}(\beta)$ is strictly increasing (see Giacomin [**55**], Theorem 6.1). Moreover, $\lim_{\beta\to\infty} h_c^{\text{que}}(\beta) = 1$ (see below). A plot is given in Fig. 33.

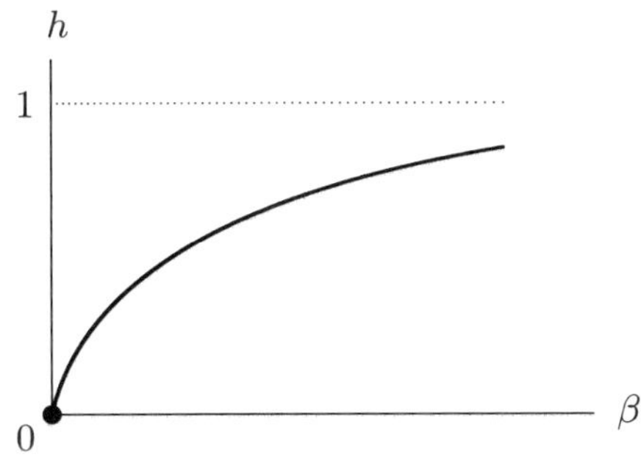

FIGURE 33. Qualitative picture of $\beta \mapsto h_c^{\text{que}}(\beta)$. The details of the curve are known only partially (see below).

The following upper bound on the critical curve comes from an annealed estimate.

THEOREM 5.1. [Bolthausen and den Hollander [**21**]] $h_c^{\rm que}(\beta) \leq \frac{1}{2\beta} \log\cosh(2\beta)$ *for all* $\beta \in (0,\infty)$.

PROOF. Estimate

$$\begin{aligned} g^{\rm que}(\beta,h) &= \lim_{n\to\infty} \frac{1}{n}\, \mathbb{E}\left(\log\left[\mathrm{e}^{-\beta h n} Z_n^{\beta,h,\omega}\right]\right) \\ &= \lim_{n\to\infty} \frac{1}{n}\, \mathbb{E}\left(\log E\left(\exp\left[\beta\sum_{i=1}^n (\omega_i+h)(\Delta_i-1)\right]\right)\right) \\ &\leq \lim_{n\to\infty} \frac{1}{n} \log E\left(\mathbb{E}\left(\exp\left[\beta\sum_{i=1}^n (\omega_i+h)(\Delta_i-1)\right]\right)\right) \\ &= \lim_{n\to\infty} \frac{1}{n} \log E\left(\prod_{i=1}^n \left[\tfrac{1}{2}e^{-2\beta(1+h)} + \tfrac{1}{2}e^{-2\beta(-1+h)}\right]^{1_{\{\Delta_i=-1\}}}\right). \end{aligned}$$

The right-hand side is ≤ 0 as soon as the term between square brackets is ≤ 1. Consequently,

$$(2\beta)^{-1}\log\cosh(2\beta) < h \quad\longrightarrow\quad g^{\rm que}(\beta,h)=0.$$

□

The following lower bound comes from strategies where the copolymer dips below the interface during rare long stretches in ω where the empirical mean is sufficiently biased downwards.

THEOREM 5.2. [Bodineau and Giacomin[**17**]] $h_c^{\rm que}(\beta) \geq (\frac{4}{3}\beta)^{-1}\log\cosh(\frac{4}{3}\beta)$ *for all* $\beta \in (0,\infty)$.

PROOF. See **Tutorial 5 in Appendix E**. □

Theorems 5.1–5.2 are summarized in Fig. 34.

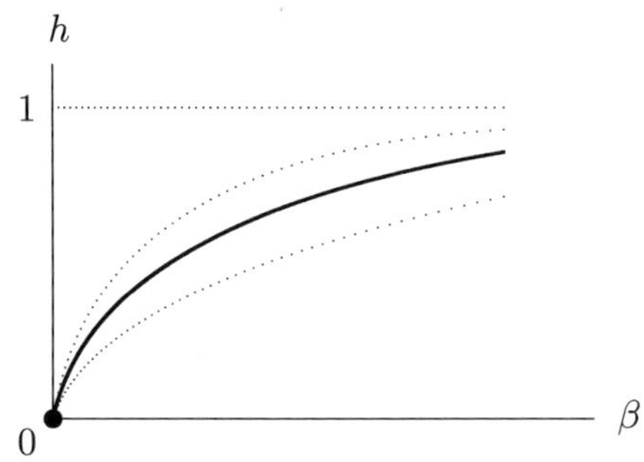

FIGURE 34. Upper and lower bounds on $\beta \mapsto h_c^{\rm que}(\beta)$.

Toninelli [**101**], Toninelli [**102**], Bodineau, Giacomin, Lacoin and Toninelli [**18**] show that the upper and lower bounds on $h_c^{\rm que}(\beta)$ are strict. In fact, the strict inequalities can be extended to the setting considered in Section 4: arbitrary disorder with a finite moment-generating function and excursion length distributions that are regularly varying at infinity). Bolthausen, den Hollander and Opoku [**22**] derive a variational expression for $h_c^{\text{que}}(\beta)$, similar in spirit to what was done in Section 4.4, and extend the strict inequalities to excursion length distributions that are logarithmically equivalent to a power law.

5.3. Weak interaction limit.

THEOREM 5.3. [Boltausen and den Hollander [**21**]] *There exists a $K_c \in (0, \infty)$ such that*

$$\lim_{\beta \downarrow 0} \frac{1}{\beta}\, h_c^{\text{que}}(\beta) = K_c.$$

The idea behind this result is that, as $\beta, h \downarrow 0$, the excursions away from the interface become longer and longer (entropy gradually takes over from energy). As a result, both w and ω can be approximated by Brownian motions. In essence, the weak interaction result follows from the scaling property

$$\lim_{\epsilon \downarrow 0} \epsilon^{-2}\, f^{\text{que}}(\epsilon\beta, \epsilon h) = \widetilde{f}^{\text{que}}(\beta, h), \qquad \beta, h \geq 0,$$

where $\widetilde{f}^{\text{que}}(\beta, h)$ is the quenched free energy of a space-time continuous version of the copolymer model, with Hamiltonian

$$H_t^{\beta,h,b}(B) = -\beta \int_0^t (\mathrm{d}b_s + h\, \mathrm{d}s)\, \mathrm{sign}(B_s)$$

and with path measure given by

$$\frac{\mathrm{d}P_t^{\beta,h,b}}{\mathrm{d}P}(B) = \frac{1}{Z_t^{\beta,h,b}}\, \mathrm{e}^{-H_t^{\beta,h,b}(B)},$$

where $B = (B_s)_{s \geq 0}$ is the polymer path, P is the Wiener measure, and $b = (b_s)_{s \geq 0}$ is a Brownian motion that plays the role of the quenched disorder. The proof is based on a *coarse-graining argument*. Due to the presence of exponential weight factors, the above scaling property is much more delicate than the standard invariance principle relating SRW and Brownian motion.

For the continuum model, a standard scaling argument shows that the quenched critical curve is linear. Its slope K_c is not known and has been the subject of heated debate. The bounds in Theorems 5.1–5.2 imply that $K_c \in [\frac{2}{3}, 1]$. Toninelli [**102**] proved that $K_c < 1$. Caravenna, Giacomin and Gubinelli [**28**] did simulations and found that $K_c \in [0.82, 0.84]$. Moreover, Caravenna, Giacomin and Gubinelli [**28**] and Sohier (private communication) found that

$$h_c^{\text{que}}(\beta) \approx \frac{1}{2K_c\beta} \log\cosh(2K_c\beta)$$

is a good approximation for small and moderate values of β.

The Brownian model describes a continuum copolymer where each infinitesimal element has a random degree of "hydrophobicity" or "hydrophilicity". It turns out that the continuum model is the scaling limit of a whole class of discrete models (see Caravenna and Giacomin [**27**], Caravenna, Giacomin and Toninelli [**29**]), i.e., there is *universality*. This property actually holds for a one-parameter family of continuum models indexed by a tail exponent $a \in (0, 1)$, of which the Brownian copolymer is the special case corresponding to $a = \frac{1}{2}$. It is known that the above approximation of the critical curve is not an equality in general. Bolthausen, den Hollander and Opoku [**22**] obtain sharp upper and lower bounds on K_c.

A related coarse-graining result is proved in Pétrélis [**91**] for a copolymer model with additional random pinning in a finite layer around the interface (of the type considered in Section 4). It is shown that the effect of the disorder in the layer vanishes in the weak interaction limit, i.e., only the disorder along the copolymer is felt in the weak interaction limit.

5.4. Qualitative properties of the phases. We proceed by stating a few path properties in the two phases.

THEOREM 5.4. [Biskup and den Hollander [**16**], Giacomin and Toninelli [**58, 61**]] *(a) If* $(\beta, h) \in \mathcal{L}$, *then the path intersects the interface with a strictly positive density, while the length and the height of the largest excursion away from the interface up to time* n *is order* $\log n$.
(b) If $(\beta, h) \in \mathrm{int}(\mathcal{D})$, *then the path intersects the interface with zero density. The number of intersections is* $O(\log n)$.

For $(\beta, h) \in \mathrm{int}(\mathcal{D})$, the number of intersections is expected to be $O(1)$ under the average quenched path measure (see Part III of Section 1.5). So far this has only been proved for (β, h) above the annealed upper bound.

THEOREM 5.5. [Giacomin and Toninelli [**59, 60**]] *For every* $\beta \in (0, \infty)$,

$$0 \leq g^{\mathrm{que}}(\beta, h) = O\left([h_c^{\mathrm{que}}(\beta) - h]^2\right) \qquad \text{as } h \uparrow h_c^{\mathrm{que}}(\beta).$$

THEOREM 5.6. [Giacomin and Toninelli [**61**]] $(\beta, h) \mapsto f^{\mathrm{que}}(\beta, h)$ *is infinitely differentiable on* $\mathcal{L}$.

Theorem 5.5 says that the phase transition is *at least second order*, while Theorem 5.6 says that the critical curve is the only location where a phase transition of *finite order* occurs. Theorem 5.5 is proved in **Tutorial 5 in Appendix E**.

All of the results in Sections 5.2–5.4 extend to $\omega_i \in \mathbb{R}$ rather than $\omega_i \in \{-1, +1\}$, provided the law of ω_i has a finite moment-generating function, and to more general excursion length distributions, of the type considered in Section 4.1. For an overview, see Caravenna, Giacomin and Toninelli [**29**].

5.5. A copolymer in a micro-emulsion. What happens when the linear interface is replaced by a *random interface*? In particular, what happens when the oil forms droplets that float around in the water, as in Fig. 35? An example is milk, which is a micro-emulsion consisting (among others) of water and tiny fat-droplets. Milk is stabilized by a protein called casein, a copolymer that wraps itself around the droplets and prevents them to coagulate.

A phase transition may be expected between a *localized phase*, where the copolymer spends most of its time near the boundary of the droplets and makes rapid hops from one droplet to the other, and a *delocalized phase*, where it spends most of its time inside and outside of droplets. We will see that the actual behavior is rather more complicated. This is due to the fact that there are three (!) types of randomness in the model: a random polymer path, a random ordering of monomer types, and a random arrangement of droplets in the emulsion.

Here is a quick definition of a model. Split $\mathbb{Z}^2$ into square blocks of size L_n. The copolymer follows a directed self-avoiding path that is allowed to make steps $\uparrow, \downarrow, \rightarrow$ and to enter and exit blocks at diagonally opposite corners (see Fig. 36). Each monomer has probability $\frac{1}{2}$ to be hydrophobic and probability $\frac{1}{2}$ to be hydrophilic,

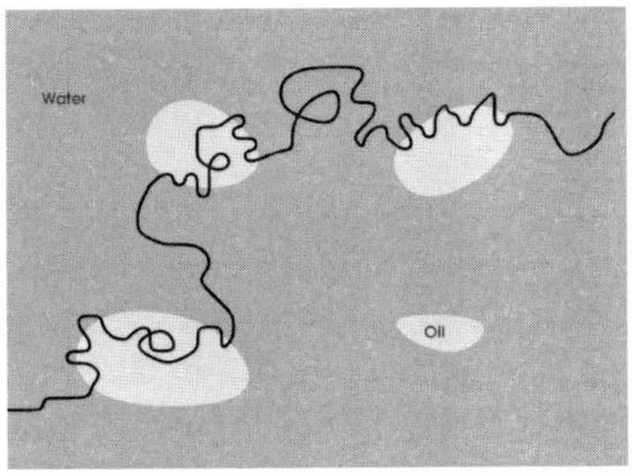

FIGURE 35. An undirected copolymer in an emulsion. The disorder along the copolymer is not indicated.

labeled by ω. Each block has probability p to be filled with oil and probability $1-p$ to be filled with water, labeled by Ω. Assign energies $-\alpha$ and $-\beta$ to the matches hydrophobic/oil, respectively, hydrophilic/water and energy 0 to the mismatches.

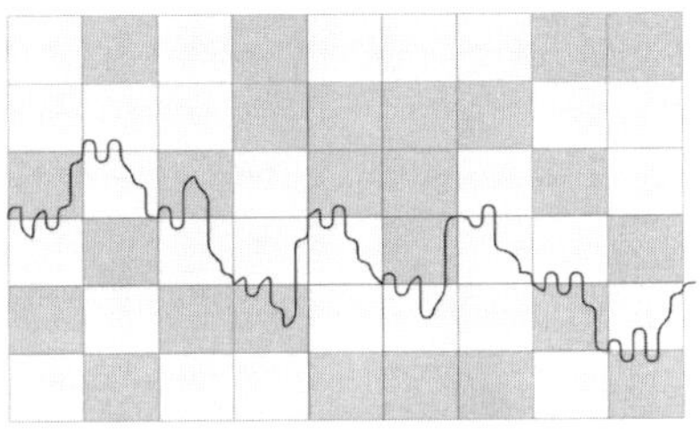

FIGURE 36. A directed self-avoiding path crossing blocks of oil and water, entering and exiting blocks at diagonally opposite corners. The disorder along the copolymer is not indicated.

The above model was studied in den Hollander and Whittington [**75**], den Hollander and Pétrélis [**71, 72, 73**]. The key parameter ranges are $p \in (0,1)$, $\alpha, \beta \in (0,\infty)$, $|\beta| \leq \alpha$. The model is studied in the limit

$$\lim_{n\to\infty} L_n = \infty, \qquad \lim_{n\to\infty} \tfrac{1}{n} L_n = 0.$$

This is a coarse-graining limit in which *the polymer scale and the emulsion scale separate.* In this limit both scales exhibit self-averaging.

Theorems 5.7–5.8 below summarize the main results (in qualitative language), and are illustrated by Figs. 37–40.

THEOREM 5.7. [den Hollander and Whittington [**75**]] *The free energy exists and is non-random ω, Ω-a.s., and is given by a variational formula involving the free energies of the copolymer in each of the four possible pairs of adjacent blocks, the frequencies at which the copolymer visits these pairs on the emulsion scale, and the fractions of time the copolymer spends in these pairs on the polymer scale.*

THEOREM 5.8. [den Hollander and Whittington [**75**], den Hollander and Pétrélis [**71, 72, 73**]] *The analysis of the variational formula reveals that there are two regimes:*
(I) Supercritical: the oil blocks percolate. There are two phases separated by one critical curve.

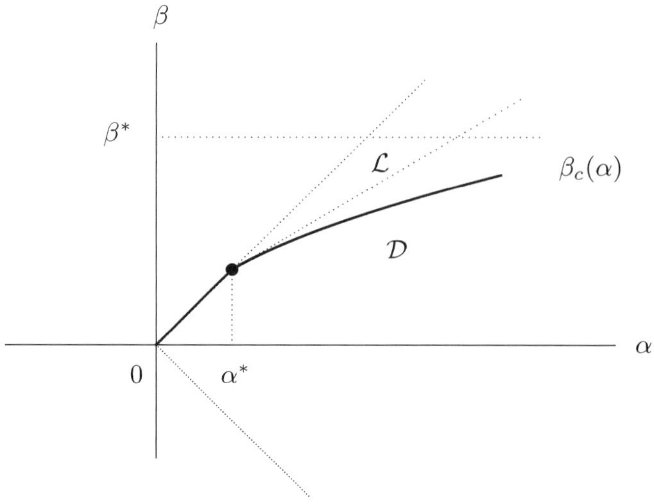

FIGURE 37. Phase diagram in the supercritical regime.

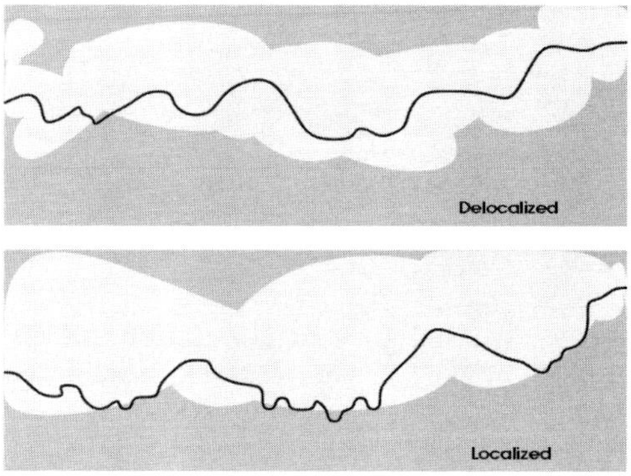

FIGURE 38. Path behavior in the two phases in the supercritical regime.

(II) Subcritical: the oil blocks do not percolate. There are four phases separated by three critical curves meeting in two tricritical points.

As shown in Figs. 37–40, the copolymer-emulsion model shows a remarkably rich phase behavior and associated path behavior. In the supercritical regime the phase diagram shows one critical curve separating two phases. There is a *delocalized* phase $\mathcal{D}$ where the copolymer lies entirely inside the infinite oil cluster, and a *localized phase* $\mathcal{L}$ where part of the copolymer lies near the boundary of the infinite oil cluster. In the subcritical regime the phase diagram shows three critical curves separating four phases meeting at two tricritical points. There are two delocalized phases $\mathcal{D}_1$, $\mathcal{D}_2$ and two localized phases $\mathcal{L}_1$, $\mathcal{L}_2$. For each pair, the distinction comes from the way in which the copolymer behaves near the boundary of the finite oil clusters.

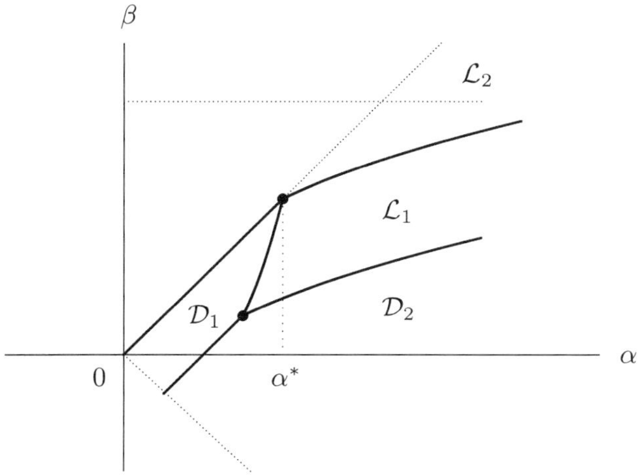

FIGURE 39. Phase diagram in the subcritical regime.

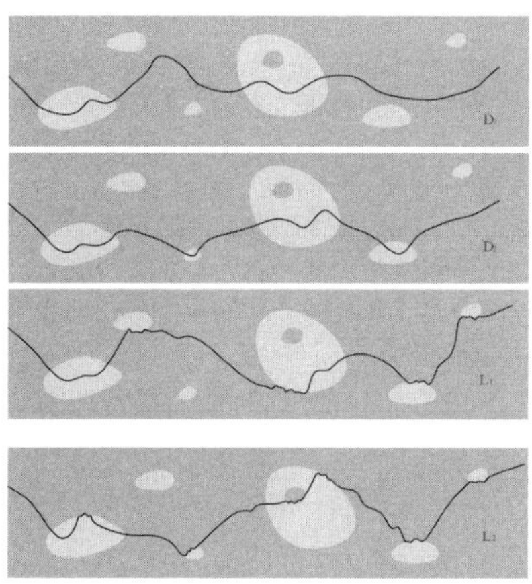

FIGURE 40. Path behavior in the four phases in the subcritical regime.

The corner restriction is unphysical, but makes the model mathematically tractable. In den Hollander and Pétrélis [**74**] this restriction is removed, but the resulting variational formula for the free energy is more complex. The coarse-graining limit is an important simplification: mesoscopic disorder is easier to deal with than microscopic disorder. An example of a model with microscopic disorder in space-time will be the topic of Section 6.

5.6. Open problems. Here are some challenges:

- For the copolymer model in Sections 5.1–5.4, prove that throughout the interior of the delocalized phase the path intersects the interface only finitely often under the average quenched path measure.
- Determine whether the phase transition is second order or higher order.

- Compute the critical slope K_c of the Brownian copolymer.
- For the copolymer/emulsion model in Section 5.5, determine the fine details of the phase diagrams in Figs. 37 and 39, and of the path properties in Figs. 38 and 40.

6. A polymer in a random potential

This section takes a look at a $(1+d)$-dimensional directed polymer in a random potential: the polymer and the potential live on $\mathbb{N} \times \mathbb{Z}^d$, where $\mathbb{N}$ is time and $\mathbb{Z}^d$, $d \geq 1$, is space (see Fig. 41). In Section 6.1 we define the model. In Sections 6.2–6.4 we study the two phases that occur: the *weak disorder phase*, in which the polymer largely ignores the disorder and behaves diffusively, and the *strong disorder phase*, in which the polymer hunts for favorable spots in the disorder and behaves superdiffusively. In Section 6.5 we derive bounds on the critical temperature separating the two phases. Section 6.6 lists a few open problems.

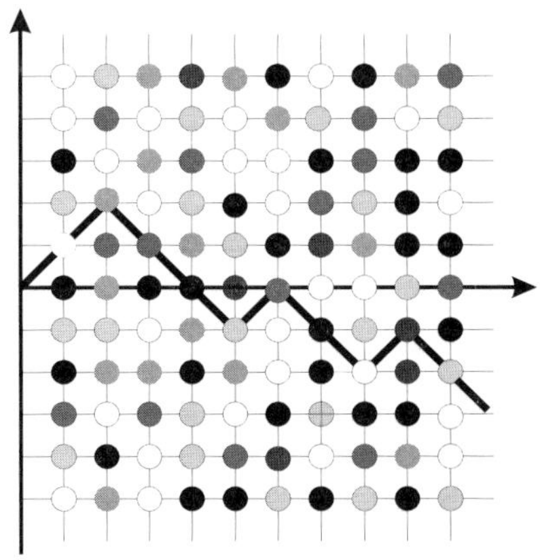

FIGURE 41. A directed polymer in a random potential. Different shades of white, grey and black represent different values of the potential.

6.1. Model.

The set of paths is

$$\mathcal{W}_n = \big\{w = (i, w_i)_{i=0}^n\colon \\ w_0 = 0,\ \|w_{i+1} - w_i\| = 1\ \forall\, 0 \leq i < n\big\}.$$

The random environment

$$\omega = \{\omega(i,x)\colon\ i \in \mathbb{N},\ x \in \mathbb{Z}^d\}$$

consists of an i.i.d. field of $\mathbb{R}$-valued non-degenerate random variables with moment generating function

$$M(\beta) = \mathbb{E}\big(e^{\beta\omega(1,0)}\big) < \infty \qquad \forall\, \beta \in [0,\infty),$$

where $\mathbb{P}$ denotes the law of ω. The Hamiltonian is

$$H_n^{\beta,\omega}(w) = -\beta \sum_{i=1}^n \omega(i, w_i), \qquad w \in \mathcal{W}_n,$$

where β plays the role of the *disorder strength.* The associated quenched path measure is

$$P_n^{\beta,\omega}(w) = \frac{1}{Z_n^{\beta,\omega}}\,\mathrm{e}^{-H_n^{\beta,\omega}(w)}\,P_n(w), \qquad w \in \mathcal{W}_n,$$

where P_n is the projection onto $\mathcal{W}_n$ of the law P of directed SRW on $\mathbb{Z}^d$.

We may think of the model as a version of the "copolymer in emulsion" described in Section 5.5 where the disorder is *microscopic* rather than *mesoscopic.* There are deep relations with several other models in probability theory and statistical physics, including growth and wave-front-propagation models and first-passage percolation. Indeed, for $\beta = \infty$ the polymer follows the path along which the sum of the disorder is largest. This case corresponds to *oriented first-passage percolation,* of which some aspects are discussed in the lectures by Garban and Steif [**54**]. For $\beta < \infty$ the model is sometimes referred to as *oriented first-passage percolation at positive temperature.*

The key object in the analysis of the model is the following quantity:

$$Y_n^{\beta,\omega} = \frac{Z_n^{\beta,\omega}}{\mathbb{E}(Z_n^{\beta,\omega})}, \qquad n \in \mathbb{N}_0.$$

This is the *ratio of the quenched and the annealed partition sum.* The point is that

$$(Y_n^{\beta,\omega})_{n\in\mathbb{N}_0}$$

is a *martingale* w.r.t. the natural filtration generated by ω, i.e., $\mathcal{F} = (\mathcal{F}_n)_{n\in\mathbb{N}_0}$ with $\mathcal{F}_n = \sigma(\omega(i,x)\colon\ 0 \le i \le n,\ x \in \mathbb{Z}^d)$. Indeed, this is seen by writing

$$Y_n^{\beta,\omega} = E\left(\prod_{i=1}^{n}\left[\frac{\mathrm{e}^{\beta\omega(i,S_i)}}{M(\beta)}\right]\right), \qquad Y_0^{\beta,\omega} = 1,$$

from which it is easily deduced that $\mathbb{E}(Y_n^{\beta,\omega}|\mathcal{F}_{n-1}) = Y_{n-1}^{\beta,\omega}$. Note that $\mathbb{E}(Y_n^{\beta,\omega}) = 1$ and $Y_n^{\beta,\omega} > 0$ for all $n \in \mathbb{N}_0$.

6.2. A dichotomy: weak and strong disorder. Since $Y_n^{\beta,\omega} \ge 0$, it follows from the martingale convergence theorem that

$$Y^{\beta,\omega} = \lim_{n\to\infty} Y_n^{\beta,\omega} \quad \text{exists } \omega\text{-a.s.}$$

Moreover, since the event $\{\omega\colon\ Y^{\beta,\omega} > 0\}$ is measurable w.r.t. the tail sigma-algebra of ω, it follows from the Kolmogorov zero-one law that the following *dichotomy* holds:

$$\begin{aligned} &(\mathrm{WD}): && \mathbb{P}(Y^{\beta,\omega} > 0) = 1,\\ &(\mathrm{SD}): && \mathbb{P}(Y^{\beta,\omega} = 0) = 1. \end{aligned}$$

In what follows it will turn out that (WD) characterizes *weak disorder*, for which the behavior of the polymer is diffusive in the $\mathbb{Z}^d$-direction, while (SD) characterizes *strong disorder*, for which the behavior is (expected to be) superdiffusive (see Fig. 42). Note that the nomenclature is appropriate: in phase (WD) the quenched and the annealed partition sum remain comparable in the limit as $n \to \infty$, indicating a weak role for the disorder, while in phase (SD) the annealed partition sum grows faster than the quenched partition sum, indicating a strong role for the disorder.

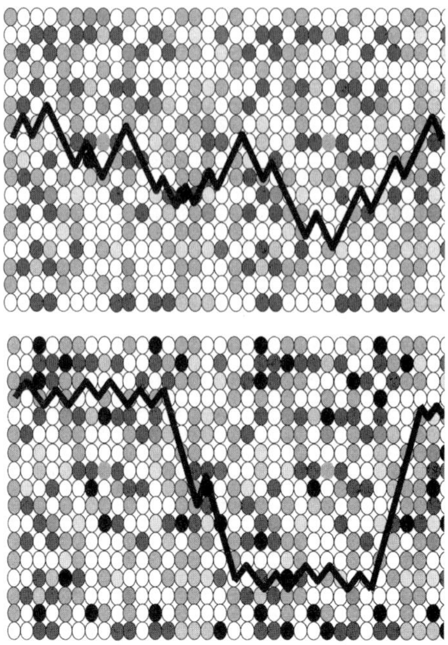

FIGURE 42. Typical path behavior in the two phases.

6.3. Separation of the two phases.

THEOREM 6.1. [Comets and Yoshida [**42**]] *For any choice of the disorder distribution, $\beta \mapsto \mathbb{E}(\sqrt{Y^{\beta,\omega}})$ is non-increasing on $[0,\infty)$. Consequently, there exists a $\beta_c \in [0,\infty]$ such that (see Fig. 43)*

$$\begin{aligned} \beta \in [0,\beta_c) &\quad\longrightarrow\quad (\mathrm{WD}),\\ \beta \in (\beta_c,\infty) &\quad\longrightarrow\quad (\mathrm{SD}). \end{aligned}$$

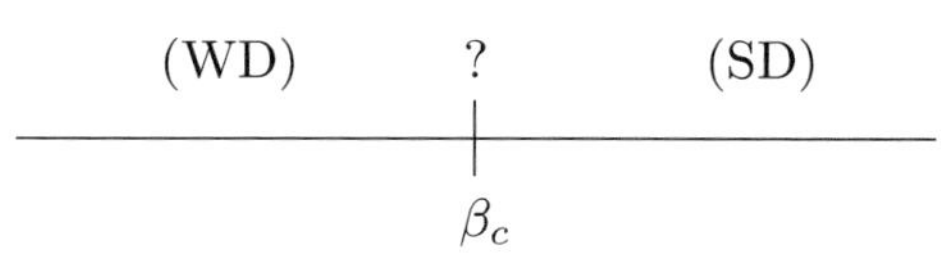

FIGURE 43. Separation of the two phases. It is not known which of the two phases includes β_c.

Since

$$\begin{aligned} f^{\mathrm{que}}(\beta) &= \lim_{n\to\infty} \frac{1}{n} \log Z_n^{\beta,\omega} \qquad \omega\text{-a.s.},\\ f^{\mathrm{ann}}(\beta) &= \lim_{n\to\infty} \frac{1}{n} \log \mathbb{E}(Z_n^{\beta,\omega}), \end{aligned}$$

it follows from the above theorem that

$$f^{\mathrm{que}}(\beta) = f^{\mathrm{ann}}(\beta) \qquad \forall\, \beta \in [0,\beta_c],$$

where the critical value $\beta = \beta_c$ can be added because free energies are continuous. It is expected that (see Fig. 44)

$$f^{\text{que}}(\beta) < f^{\text{ann}}(\beta) \qquad \forall\, \beta \in (\beta_c, \infty),$$

so that for $\beta \in (\beta_c, \infty)$ the quenched and the annealed partition sum have different exponential growth rates, but this remains open. Partial results have been obtained in Comets and Vargas [**39**], Lacoin [**83**].

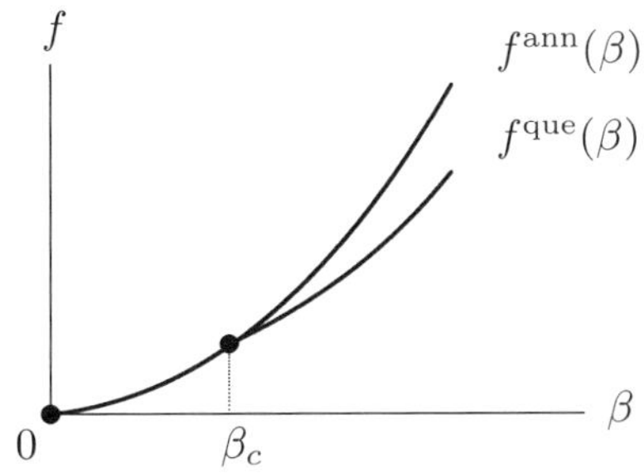

FIGURE 44. Conjectured behavior of the quenched and the annealed free energy.

6.4. Characterization of the two phases. Let

$$\pi_d = (P \otimes P')(\exists\, n \in \mathbb{N}\colon\ S_n = S_n')$$

denote the collision probability of two independent copies of SRW. Note that $\pi_d = 1$ in $d = 1, 2$ and $\pi_d < 1$ in $d \geq 3$. For $\beta \in [0, \infty)$, define

$$\Delta_1(\beta) = \log[M(2\beta)/M(\beta)^2],$$
$$\Delta_2(\beta) = \beta[\log M(\beta)]' - \log M(\beta).$$

Both $\beta \mapsto \Delta_1(\beta)$ and $\beta \mapsto \Delta_2(\beta)$ are strictly increasing on $[0, \infty)$, with $\Delta_1(0) = \Delta_2(0) = 0$ and $\Delta_1(\beta) > \Delta_2(\beta)$ for $\beta \in (0, \infty)$.

Define

$$\max_n^{\beta,\omega} = \max_{x \in \mathbb{Z}^d} P_n^{\beta,\omega}(S_n = x), \qquad n \in \mathbb{N}.$$

This quantity measures how localized the endpoint S_n of the polymer is in the given potential ω: if $\lim_{n\to\infty} \max_n^{\beta,\omega} = 0$, then the path spreads out, while if $\limsup_{n\to\infty} \max_n^{\beta,\omega} > 0$, then the path localizes (at least partially).

THEOREM 6.2. [Imbrie and Spencer [**76**], Bolthausen [**19**], Sinai [**95**], Carmona and Hu [**33**], Comets, Shiga and Yoshida [**37**]] *Suppose that*

(I) $\qquad d \geq 3,\ \Delta_1(\beta) < \log(1/\pi_d).$

Then

$$\lim_{n\to\infty} \frac{1}{n}\, E_n^{\beta,\omega}(\|S_n\|^2) = 1 \qquad \omega\text{-a.s.}$$

and

$$\lim_{n\to\infty} \max_n^{\beta,\omega} = 0 \qquad \omega\text{-a.s.}$$

THEOREM 6.3. [Carmona and Hu [**33**], Comets, Shiga and Yoshida [**37**]] *Suppose that*

$$\text{(II)} \qquad d = 1, 2, \ \beta > 0 \quad \textit{or} \quad d \geq 3, \ \Delta_2(\beta) > \log(2d).$$

Then there exists a $c = c(d, \beta) > 0$ *such that*

$$\limsup_{n\to\infty} \max_n^{\beta,\omega} \geq c \qquad \omega\text{-}a.s.$$

Theorems 6.2–6.3 show that the polymer has qualitatively different behavior in the two regimes. In (I), the scaling is *diffusive*, with the diffusion constant *not renormalized* by the disorder. The reason why the diffusion constant is not renormalized is the directedness of the path: this causes the annealed model to be directed SRW. In (II), there is certainly no scaling to Brownian motion, due to the presence of atoms: the endpoint of the polymer concentrates around one or more most favorable sites whose locations depend on ω. These locations are expected to be at a distance much larger than $\sqrt{n}$, i.e., the scaling is predicted to be *superdiffusive*. This has, however, only been proved in some special cases, in particular, for a one-dimensional model of a directed polymer in a Gaussian random environment (Petermann [**88**]). Further results, also for related models, have been obtained in Piza [**92**], Méjane [**87**], Carmona and Hu [**34**], Bezerra, Tindel and Viens [**10**] and Lacoin [**85**]. The latter reference contains a discussion of the physical conjectures and the mathematical results on this topic.

The proofs of Theorems 6.2–6.3 are based on a series of technical estimates for the martingale $(Y_n^{\beta,\omega})_{n\in\mathbb{N}_0}$. These estimates also show that

$$\text{(I)} \longrightarrow \text{(WD)}, \qquad \text{(II)} \longrightarrow \text{(SD)}.$$

It has been conjectured that, throughout phase (SD),

$$E_n^{\beta,\omega}(\|S_n\|^2) \asymp n^{2\nu} \quad n \to \infty, \ \omega - a.s.$$

($\asymp$ means modulo logarithmic factors), where the exponent ν is predicted not to depend on β and to satisfy

$$\nu = \tfrac{2}{3} \text{ for } d = 1, \qquad \nu \in (\tfrac{1}{2}, \tfrac{2}{3}) \text{ for } d = 2,$$

signalling *superdiffusive behavior.*

6.5. Bounds on the critical temperature. Theorems 6.2–6.3 show that $\beta_c = 0$ for $d = 1, 2$ and $\beta_c \in (0, \infty]$ for $d \geq 3$ (because $\Delta_1(0) = 0$ and $\pi_d < 1$). However, there is a gap between regimes (I) and (II) in $d \geq 3$ (because $\pi_d > 1/2d$ and $\Delta_1(\beta) > \Delta_2(\beta)$ for all $\beta > 0$). Thus, the results do not cover the full parameter regime. In fact, all we know is that

$$\beta_c \in [\beta_c^1, \beta_c^2].$$

with (see Fig. 45)

$$\beta_c^1 = \sup\{\beta \in [0, \infty) \colon \ \Delta_1(\beta) < \log(1/\pi_d)\},$$
$$\beta_c^2 = \inf\{\beta \in [0, \infty) \colon \ \Delta_2(\beta) > \log(2d)\}.$$

Various attempts have been made to sharpen the estimates on β_c: fractional moment estimates on the martingale (Evans and Derrida [**52**], Coyle [**44**], Camanes and Carmona [**24**]); size-biasing of the martingale (Birkner [**11**]). We describe the latter estimate, which involves a critical threshold z^* associated with the collision local time of two independent SRWs.

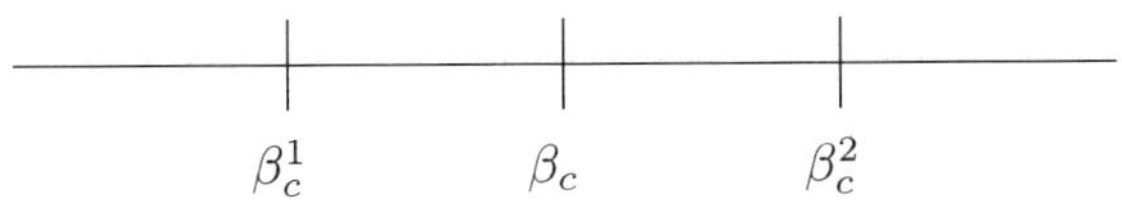

FIGURE 45. For $d \geq 3$ three cases are possible depending on the law $\mathbb{P}$ of the disorder: (1) $0 < \beta_c^1 < \beta_c^2 < \infty$; (2) $0 < \beta_c^1 < \beta_c^2 = \infty$; (3) $\beta_c^1 = \beta_c^2 = \infty$.

THEOREM 6.4. [Birkner [**11**]] *Let*

$$z^* = \sup\{z \geq 1\colon\ E\big(z^{V(S,S')}\big) < \infty\ S' - a.s.\},$$

where

$$V(S,S') = \sum_{n\in\mathbb{N}} 1_{\{S_n = S'_n\}}$$

is the collision local time of two independent SRWs, *and E denotes expectation over S. Define*

$$\beta_c^* = \sup\{\beta \in [0,\infty)\colon\ M(2\beta)/M(\beta)^2 < z^*\}.$$

Then

$$\beta < \beta_c^* \quad \longrightarrow \quad \text{(WD)}$$

and, consequently, $\beta_c \geq \beta_c^*$.

PROOF. Abreviate

$$e = \{e(i,x)\}_{i\in\mathbb{N},x\in\mathbb{Z}^d}$$

with

$$e(i,x) = \mathrm{e}^{\beta\omega(i,x)}/M(\beta).$$

Consider a size-biased version of e, written

$$\hat{e} = \{\hat{e}(i,x)\}_{i\in\mathbb{N},x\in\mathbb{Z}^d},$$

that is i.i.d., is independent of e and has law $\hat{\mathbb{P}}$ given by

$$\hat{\mathbb{P}}(\hat{e}(1,0) \in \cdot\,) = \mathbb{E}\big(e(1,0)\,1_{\{e(1,0)\in\,\cdot\,\}}\big).$$

No normalization is needed because $\mathbb{E}(e(1,0)) = 1$.

Given S', put

$$\hat{e}_{S'} = \{\hat{e}_{S'}(i,x)\}_{i\in\mathbb{N},x\in\mathbb{Z}^d},$$

with

$$\hat{e}_{S'}(i,x) = 1_{\{S'_i \neq x\}}\, e(i,x) + 1_{\{S'_i = x\}}\, \hat{e}(i,x),$$

i.e., size-bias e to $\hat{e}$ everywhere along S', and define

$$\hat{Y}_n^{e,\hat{e},S'} = E\left(\prod_{i=1}^n \hat{e}_{S'}(i,S_i)\right).$$

This is a *size-biased version of the basic martingale*, which in the present notation reads

$$Y_n^e = E\left(\prod_{i=1}^n e(i,S_i)\right).$$

The point of the size-biasing carried out above is that for any bounded function $f\colon [0,\infty) \to \mathbb{R}$,

$$\mathbb{E}\big(Y_n^e\, f(Y_n^e)\big) = (\mathbb{E}\otimes\hat{\mathbb{E}}\otimes E')\Big(f\big(\hat{Y}_n^{e,\hat{e},S'}\big)\Big),$$

where $\mathbb{E}, \hat{\mathbb{E}}, E'$ denote expectation w.r.t. $e, \hat{e}, S'$, respectively. Indeed, the latter follows from the computation

$$\begin{aligned}
\mathbb{E}\big(Y_n^e f(Y_n^e)\big) &= \mathbb{E}\left[E'\left(\prod_{i=1}^n e(i,S_i')\right) f\left(E\left(\prod_{i=1}^n e(i,S_i)\right)\right)\right] \\
&= E'\left(\mathbb{E}\left[\left(\prod_{i=1}^n e(i,S_i')\right) f\left(E\left(\prod_{i=1}^n e(i,S_i)\right)\right)\right]\right) \\
&=^! E'\left((\mathbb{E}\otimes\hat{\mathbb{E}})\left[f\left(E\left(\prod_{i=1}^n \hat{e}_{S'}(i,S_i)\right)\right)\right]\right) \\
&= (\mathbb{E}\otimes\hat{\mathbb{E}}\otimes E')\Big(f\big(\hat{Y}_n^{e,\hat{e},S'}\big)\Big),
\end{aligned}$$

where the third equality uses the definition of $\hat{e}_{S'}$.

The above identity relates the two martingales, and implies that

$$\begin{aligned}
&(Y_n^e)_{n\in\mathbb{N}_0} \text{ is uniformly integrable} \\
&\qquad \Longleftrightarrow \quad (\hat{Y}_n^{e,\hat{e},S'})_{n\in\mathbb{N}_0} \text{ is tight} \qquad (*)
\end{aligned}$$

(as can be seen by picking f such that $\lim_{u\to\infty} f(u) = \infty$). However, an easy computation gives

$$(\mathbb{E}\otimes\hat{\mathbb{E}})\left(\hat{Y}_n^{e,\hat{e},S'}\right) = E\left(z^{\sum_{i=1}^n 1_{\{S_i=S_i'\}}}\right) = E\big(z^{V(S,S')}\big)$$

with $z = M(2\beta)/M(\beta)^2$, where the factor 2β arises because after the size-biasing the intersection sites of S and S' are visited by both paths. Hence

$$E(z^{V(S,S')}) < \infty \quad S'\text{-a.s.}$$

is enough to ensure that the r.h.s. of $(*)$ holds. This completes the proof because the l.h.s. of $(*)$ is equivalent to (WD). Indeed, a.s. convergence plus uniform integrability imply convergence in mean, so that $\mathbb{E}(Y_n^e) = 1$ for all $n \in \mathbb{N}_0$ yields $\mathbb{E}(Y^e) = 1$. □

In Birkner, Greven and den Hollander [**13**] it was proved that $z^* > \pi_d$ in $d \geq 5$, implying that $\beta_c^* > \beta_c^1$. It was conjectured that the same is true in $d = 3, 4$. Part of this conjecture was settled in Birkner and Sun [**14, 15**] and Berger and Toninelli [**9**] (see Fig. 46).

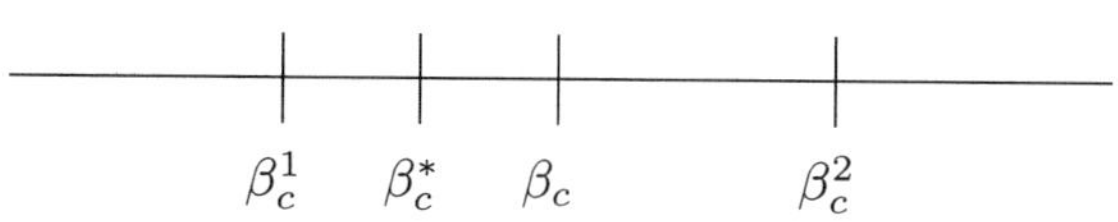

FIGURE 46. Bounds on the critical temperature.

6.6. Open problems.

- Show that in phase (SD) the polymer is concentrated inside a most favorable corridor and identify how this corridor depends on ω.
- Determine whether β_c is part of (WD) or (SD).
- Derive a variational expression for β_c.

- Extend the analysis to undirected random walk. Important progress has been made in Ioffe and Velenik [**77, 78, 79, 80**], Zygouras [**104, 105**], and references cited therein. See also Section 2.7.

Appendix A. Tutorial 1

In this tutorial we describe two methods that can be used to prove the *existence* of the quenched free energy associated with the random pinning model described in Section 4. Section A.1 recalls the model, Sections A.2–A.4 prove existence of the quenched free energy when the endpoint of the polymer is constrained to lie in the interface, while Section A.5 shows how to remove this constraint afterwards. The method of proof is widely applicable, and is not specific to the random pinning model.

A.1. Random pinning of a polymer at an interface. *Configurations of the polymer.* Let $n \in \mathbb{N}$ and consider a polymer made of n monomers. The allowed configurations of this polymer are modeled by the n-step trajectories of a 1-dimensional random walk $S = (S_i)_{i \in \mathbb{N}_0}$. We focus on the case where $S_0 = 0$ and $(S_i - S_{i-1})_{i \in \mathbb{N}}$ is an i.i.d. sequence of random variables satisfying

$$P(S_1 = 1) = P(S_1 = -1) = P(S_1 = 0) = \tfrac{1}{3},$$

although the argument given below applies more generally. We denote by $\mathcal{W}_n$ the set of all n-step trajectories of S.

Disorder at interface. Let $\omega = (\omega_i)_{i \in \mathbb{N}}$ be an i.i.d. sequence of $\mathbb{R}$-valued random variables (which we take bounded for ease of exposition). For $i \in \mathbb{N}$ the interaction intensity between the i-th monomer and the interface takes the value ω_i. Note that ω and S are independent, and write $\mathbb{P}$ for the law of ω. Pick $M > 0$ such that $|\omega_1| \leq M$ $\mathbb{P}$-a.s.

Interaction polymer-interface. The flat interface that interacts with the polymer is located at height 0, so that the polymer hits this interface every time S comes back to 0. Thus, with every $S \in \mathcal{W}_n$ we associate the energy

$$H_n^{\beta,\omega}(S) = -\beta \sum_{i=0}^{n-1} \omega_i \, 1_{\{S_i = 0\}},$$

where $\beta \in (0, \infty)$ stands for the inverse temperature (and for ease of exposition we take zero bias, i.e., we set $h = 0$ in the Hamiltonian in Section 4.1). We think of S as a random realization of the path of the polymer.

Partition function and free energy. For fixed n, the quenched (= frozen disorder) partition function and free energy are defined as

$$Z_n^{\beta,\omega} = E\big(\mathrm{e}^{-H_N^{\beta,\omega}(S)}\big) \quad \text{and} \quad f_n^{\beta,\omega} = \tfrac{1}{n} \log Z_n^{\omega,\beta}.$$

A.2. Convergence of the free energy. Our goal is to prove the following theorem.

THEOREM A.1. *For every $\beta \in \mathbb{R}$ there exists an $f(\beta) \in [0, \beta M]$ such that*

$$\lim_{n \to \infty} \mathbb{E}(f_n^{\beta,\omega}) = f(\beta)$$

and

$$\lim_{n \to \infty} f_n^{\beta,\omega} = f(\beta) \quad \mathbb{P} - a.e.\ \omega.$$

As indicated above, we will prove Theorem A.1 via two different methods. In Section A.3 we will state Kingman's *Subadditive Ergodic Theorem* and see how this can be applied to obtain Theorem A.1. In Section A.4 we will re-prove Theorem A.1 by using a concentration of measure argument. The latter method is more involved, but also more flexible than the former method. For technical reasons, we will first prove Theorem A.1 with the partition function restricted to those trajectories that hit the interface at their right extremity, i.e.,

$$Z_n^{*,\beta,\omega} = E\big(e^{-H_n^{\beta,\omega}(S)}\, 1_{\{S_n=0\}}\big) \quad \text{and} \quad f_n^{*,\beta,\omega} = \tfrac{1}{n}\log Z_n^{*,\beta,\omega}.$$

In Section A.5 we will see that the restriction on the endpoint has no effect on the value of the limiting free energy.

A.3. Method 1: Kingman's theorem.

THEOREM A.2. [Kingman's Subadditive Ergodic Theorem; see Steele [**98**]] *Let* (Ω, A, μ) *be a probability space, let* T *be an ergodic measure-preserving transformation acting on* Ω*, and let* $(g_n)_{n\in\mathbb{N}}$ *be a sequence of random variables in* $L_1(\mu)$ *that satisfy the subadditivity relation*

$$g_{m+n} \geq g_m + g_n(T^m), \qquad m, n \in \mathbb{N}.$$

Then

$$\lim_{n\to\infty} \frac{g_n}{n} = \sup_{k\in\mathbb{N}} E_\mu\left(\frac{g_k}{k}\right) \qquad \mu\text{-a.s.}$$

(1) Let T be the left-shift on $\mathbb{R}^{\mathbb{N}}$. Prove that, for $m, n \in \mathbb{N}$ and $\omega \in \mathbb{R}^{\mathbb{N}}$,

$$\log Z_{m+n}^{*,\beta,\omega} \geq \log Z_m^{*,\beta,\omega} + \log Z_n^{*,\beta,T^m(\omega)}.$$

(2) Apply Theorem A.2 with $(\Omega, A, \mu) = (\mathbb{R}^{\mathbb{N}}, \mathrm{Bor}(\mathbb{R}^{\mathbb{N}}), \mathbb{P})$ and prove Theorem A.1 with the endpoint restriction.

A.4. Method 2: Concentration of measure. This method consists of first proving the first line in Theorem A.1, i.e., the convergence of the *average quenched* free energy, and then using a concentration of measure inequality to show that, with large probability, the quenched free energy is almost equal to its expectation, so that the second line in Theorem A.1 follows. See Giacomin and Toninelli [**58**] for fine details.

(1) Use (A.3) and prove that $(\mathbb{E}(\log Z_n^{*,\beta,\omega}))_{n\in\mathbb{N}}$ is a superadditive sequence, i.e., for $m, n \in \mathbb{N}$,

$$\mathbb{E}(\log Z_{m+n}^{*,\beta,\omega}) \geq \mathbb{E}(\log Z_m^{*,\beta,\omega}) + \mathbb{E}(\log Z_n^{*,\beta,\omega}).$$

(2) Deduce that (see also the tutorial in Appendix A.1 of Bauerschmidt, Duminil-Copin, Goodman and Slade [**7**])

$$\lim_{n\to\infty} \mathbb{E}(f_n^{*,\beta,\omega}) = \sup_{k\in\mathbb{N}} \mathbb{E}(f_k^{*,\beta,\omega}) = f(\beta) \in [0, \beta M].$$

To proceed, we need the following inequality.

THEOREM A.3. [Concentration of measure; see Ledoux [**86**]] *There exist* $C_1, C_2 > 0$ *such that for all* $n \in \mathbb{N}$, $K > 0$, $\varepsilon > 0$ *and* $G_n\colon \mathbb{R}^n \mapsto \mathbb{R}$ *a* K*-Lipschitz (w.r.t. the Euclidean norm) convex function,*

$$\mathbb{P}\Big(\big|G_n(\omega_0, \ldots, \omega_{n-1}) - \mathbb{E}\big(G_n(\omega_0, \ldots, \omega_{n-1})\big)\big| > \varepsilon\Big) \leq C_1 e^{-\frac{C_2\varepsilon^2}{K^2}}.$$

(3) By Hölder's inequality, the function $\omega \in \mathbb{R}^n \mapsto f_n^{*,\beta,\omega} \in \mathbb{R}$ is convex. To prove that it is $(\beta/\sqrt{n})$-Lipschitz, pick $\omega, \omega' \in \mathbb{R}^n$ and compute

$$\begin{aligned}\left|\frac{\partial}{\partial t} f_n^{*,\beta,t\omega+(1-t)\omega'}\right| &= \frac{\beta}{n}\left|\sum_{i=0}^{n-1} P_n^{*,\beta,\omega}(S_i=0)\,(\omega_i-\omega_i')\right| \\ &\le \frac{\beta}{n}\sqrt{\sum_{i=0}^{n-1}\left[P_n^{*,\beta,\omega}(S_i=0)\right]^2}\,\sqrt{\sum_{i=0}^{n-1}(\omega_i-\omega_i')^2} \\ &\le \frac{\beta}{\sqrt{n}}\sqrt{\sum_{i=0}^{n-1}(\omega_i-\omega_i')^2},\end{aligned}$$

where $P_n^{*,\beta,\omega}$ is the path measure with the endpoint restriction.

(4) Apply Theorem A.3 to prove that, for $\varepsilon > 0$,

$$\sum_{n\in\mathbb{N}} \mathbb{P}\big(|f_n^{*,\beta,\omega} - \mathbb{E}(f_n^{*,\beta,\omega})| > \varepsilon\big) < \infty.$$

(5) Combine (2) and (4) to show that, for $\mathbb{P}$-a.e. ω, $f_n^{*,\beta,\omega}$ tends to $f(\beta)$ as $n \to \infty$, which proves Theorem A.1 with the endpoint restriction.

A.5. Removal of the path restriction. The proof of Theorem A.1 will be completed once we show that restricting the partition function to $\{S_n = 0\}$ does not alter the results. To that end, we denote by τ the first time at which the random walk S hits the interface.

(6) Note that there exists a $C_3 > 0$ such that (see Spitzer [**97**], Section 1)

$$P(\tau = n) = \frac{C_3}{n^{3/2}}\,[1+o(1)] \quad \text{and} \quad P(\tau > n) = \frac{2C_3}{n^{1/2}}\,[1+o(1)].$$

(7) Consider the last hit of the interface and show that

$$Z_n^{\beta,\omega} = \sum_{j=0}^{n} Z_j^{*,\beta,\omega}\, P(\tau > n-j).$$

(8) Prove Theorem A.1 by combining (5), (6) and (7).

Appendix B. Tutorial 2

The goal of this tutorial is to provide the combinatorial computation of the free energy for the directed polymer with self-attraction described in Sections 2.4–2.5 leading to Theorem 2.3. This computation is taken from Brak, Guttmann and Whittington [**23**]. Section B.1 recalls the model, Section B.2 proves the existence of the free energy, while Section B.3 derives a formula for the free energy with the help of generating functions.

B.1. Model of a directed polymer in a poor solvent. We begin by recalling some of the notation used in Sections 2.4–2.5.

Configurations of the polymer. For $n \in \mathbb{N}$, the configurations of the polymer are modelled by n-step $(1+1)$-dimensional directed self-avoiding paths $w = (w_i)_{i=0}^n$ that are allowed to move up, down and to the right, i.e.,

$$\begin{aligned}\mathcal{W}_n = \{(w_i)_{i=0}^n \in (\mathbb{N}_0 \times \mathbb{Z})^{n+1} \colon & w_0 = 0, w_1 - w_0 = \rightarrow, \\ & w_i - w_{i-1} = \{\uparrow, \downarrow, \rightarrow\} \ \forall 1 \le i \le n, \\ & w_i \ne w_j \ \forall 0 \le i < j \le n\}.\end{aligned}$$

Self-touchings. The monomers constituting the polymer have an attractive interaction: an energetic reward is given for each *self-touching*, i.e., for each pair (w_i, w_j) with $i < j-1$ and $|w_i - w_j| = 1$. Accordingly, with each $w \in \mathcal{W}_n$ we associate the number of self-touchings

$$J_n(w) = \sum_{0 \le i < j-1 \le n-1} 1_{\{|w_i - w_j| = 1\}},$$

and the energy

$$H_n^\gamma(w) = -\gamma J_n(w),$$

where $\gamma \in \mathbb{R}$ is the interaction parameter.

Partition function, free energy and generating function. For fixed n, the partition function and free energy are defined as

$$Z_n^\gamma = \sum_{w \in \mathcal{W}_n} \mathrm{e}^{-H_n^\gamma(w)}, \qquad f_n(\gamma) = \tfrac{1}{n} \log Z_n^\gamma.$$

For $n \in \mathbb{N}_0$ and $x \in [0, \infty)$, let

$$Z_n(x) = \sum_{m \in \mathbb{N}_0} c_n(m)\, x^m, \quad c_n(m) = |\{w \in \mathcal{W}_n \colon J_n(w) = m\}|.$$

Then $Z_n^\gamma = Z_n(\mathrm{e}^\gamma)$, and the generating function of Z_n^γ can be written as

$$\sum_{n \in \mathbb{N}_0} Z_n^\gamma y^n = G(\mathrm{e}^\gamma, y)$$

with

$$G(x, y) = \sum_{n \in \mathbb{N}_0} \sum_{m \in \mathbb{N}_0} c_n(m) x^m y^n, \qquad x, y \in [0, \infty).$$

B.2. Existence of the free energy. Existence comes in three steps.

(1) Show that for $m, n \in \mathbb{N}_0$ and $x \in [0, \infty)$,

$$Z_{m+n+1}(x) \ge Z_m(x)\, Z_n(x) \quad \text{and} \quad Z_n(x) \le [3(1 \vee x)]^n.$$

(2) Deduce that

$$\lim_{n \to \infty} \frac{1}{n} \log Z_n(x) = \sup_{k \in \mathbb{N}} \frac{1}{k} \log Z_k(x) = \bar{f}(x) \in (0, \log 3 + (0 \vee \log x)].$$

Thus, $f(\gamma) = \bar{f}(\mathrm{e}^\gamma)$, $\gamma \in \mathbb{R}$.

(3) For $x \in [0, \infty)$, let $y_c(x)$ be the radius of convergence of the generating function $G(x, y)$. Show that

$$\bar{f}(x) = -\log y_c(x). \tag{B.1}$$

B.3. Computation of the free energy. To prove Theorem 2.3, we must compute $y_c(x)$, $x \in [0,\infty)$. In what follows we derive the formula for $G(x,y)$ given in Lemma 2.4.

(1) For $n, r, s \in \mathbb{N}_0$, let

$$\begin{aligned}
&\mathcal{W}_{n,r}\\
&\quad = \{w \in \mathcal{W}_n \colon\ w \text{ makes exactly } r \text{ vertical steps after the first step east}\},\\
&\mathcal{W}_{n,r,s}\\
&\quad = \{w \in \mathcal{W}_{n,r} \colon\ w \text{ makes exactly } s \text{ vertical steps after the second step east}\},
\end{aligned}$$

and note that $\mathcal{W}_{n,r} = \emptyset$ if $n < 1+r$ and $\mathcal{W}_{n,r,s} = \emptyset$ if $n < 2+r+s$. Furthermore, for $r, s \in \mathbb{N}$, let

$$\begin{aligned}
&\mathcal{W}^{\uparrow\downarrow}_{n,r,s}\\
&\quad = \{w \in \mathcal{W}_{n,r,s} \colon\ \text{the } r \text{ and } s \text{ vertical steps are made in opposite directions}\},\\
&\mathcal{W}^{\uparrow\uparrow}_{n,r,s}\\
&\quad = \{w \in \mathcal{W}_{n,r,s} \colon\ \text{the } r \text{ and } s \text{ vertical steps are made in the same direction}\},
\end{aligned}$$

so that, for $r \in \mathbb{N}$, $\mathcal{W}_{n,r}$ can be partitioned as

$$\mathcal{W}_{n,r} = \bigcup_{s=0}^{n-r-2} \mathcal{W}_{n,r,s} = \mathcal{W}_{n,r,0} \cup \left[\bigcup_{s=1}^{n-r-2} \left[\mathcal{W}^{\uparrow\downarrow}_{n,r,s} \cup \mathcal{W}^{\uparrow\uparrow}_{n,r,s}\right]\right].$$

For $n, r, m \in \mathbb{N}_0$, let $c_{n,r}(m)$ be the number of n-step paths with m self-touchings making exactly r steps north or south immediately after the first step east, and put

$$g_r(x,y) = \sum_{n\in\mathbb{N}_0} \sum_{m\in\mathbb{N}_0} c_{n,r}(m)\, x^m\, y^n.$$

Clearly, $G(x,y) = \sum_{r\in\mathbb{N}_0} g_r(x,y)$.

(2) Pick $r \in \mathbb{N}$ and use the first equality in the partitioning of $\mathcal{W}_{n,r}$, together with the fact that $c_{n,r}(m) = 0$ when $n < r+1$, to prove that

$$g_r(x,y) = 2y^{r+1} + \sum_{s\in\mathbb{N}_0} \sum_{n=r+2+s}^{\infty} \sum_{w\in\mathcal{W}_{n,r,s}} x^{J_n(w)}\, y^n.$$

For $w \in \mathcal{W}_n$ and $0 \le l < s \le n$, let

$$J_{l,s}(w) = \sum_{l \le i < j-1 \le s-1} 1_{\{|w_i - w_j| = 1\}},$$

which stands for the number of self-touchings made by w between its l-th and s-th step. Clearly, $J_n(w) = J_{0,n}(w)$.

(3) Pick $r, s \in \mathbb{N}$ and $n \ge r+s+2$. Prove that

$$\begin{aligned}
w \in \mathcal{W}^{\uparrow\uparrow}_{n,r,s} \quad &\longrightarrow \quad J_n(w) = J_{r+1,n}(w),\\
w \in \mathcal{W}^{\uparrow\downarrow}_{n,r,s} \quad &\longrightarrow \quad J_n(w) = J_{r+1,n}(w) + \min\{r,s\}.
\end{aligned}$$

(4) Use (2) and (3) to show that

(B.2)
$$g_r(x,y) = y^{r+1}\left[2+\sum_{s=0}^{r}(1+x^s)\,g_s(x,y)+\sum_{s=r+1}^{\infty}(1+x^r)\,g_s(x,y)\right], \qquad r\in\mathbb{N}.$$

In the same spirit show that

(B.3)
$$g_0(x,y) = y + y\,G(x,y).$$

(5) Abbreviate $g_r = g_r(x,y)$. Prove that

(B.4)
$$g_{r+1}-(1+x)yg_r-(1-x)x^ry^{r+2}g_r+xy^2g_{r-1}=0 \qquad r\in\mathbb{N}.$$

To do so, substitute the expressions obtained for g_{r-1}, g_r and g_{r+1} from (B.2) into (B.4), and isolate the terms containing y^{2r+3}. The latter leads to a rewrite of the left-hand side of (B.4) as

(B.5)
$$x^ry^{2r+3}(x-1)\left[2+\sum_{s=0}^{r}(1+x^s)g_s+\sum_{s=r+1}^{\infty}(1+x^r)g_s\right]+x^ry^{r+2}(1-x)g_r.$$

Use (B.2) once more to conclude that (B.5) equals zero.

(6) From (B.4) we see that $(g_r)_{r\in\mathbb{N}_0}$ is determined by g_0 and g_1, while (B.3) constitutes a consistency relation that must be met by the solution of (B.4). Thus, $(g_r)_{r\in\mathbb{N}_0}$ belongs to a two-dimensional vector space generated by any two linearly independent solutions. For this reason, we look for two particular solutions of (B.4) by making an Ansatz. Set $q = xy$, and write g_r in the form

(B.6)
$$g_r = \lambda^r \sum_{l\in\mathbb{N}_0} p_l\, q^{lr}, \qquad r\in\mathbb{N},\ p_0=1,$$

where $\lambda = \lambda(y,q)$ and $p_l = p_l(\lambda,y,q)$, $l\in\mathbb{N}$, are to be determined. Substitute (B.6) into (B.4) to obtain

(B.7)
$$\begin{aligned}&\lambda^2-\lambda(y+q)+yq\\ &+\sum_{l\in\mathbb{N}} q^{l(r-1)}\Big[\big(\lambda^2q^{2l}-\lambda(y+q)q^l+yq\big)p_l+\big(\lambda(q-y)yq^l\big)\,p_{l-1}\Big]=0.\end{aligned}$$

Conclude that (B.7) is satisfied when

(B.8)
$$p_l = \frac{\lambda(y-q)yq^l}{(\lambda q^l-y)(\lambda q^l-q)}\,p_{l-1}, \qquad l\in\mathbb{N},$$

provided λ solves the equation $\lambda^2-\lambda(y+q)+yq=0$, i.e., $\lambda\in\{\lambda_1,\lambda_2\}=\{y,q\}$.

(7) Use (6) to show that $g_r = C_1g_{r,1}+C_2g_{r,2}$, $r\in\mathbb{N}$, where C_1 and C_2 are functions of y,q and

(B.9)
$$g_{r,i}=g_{r,i}(x,y)=(\lambda_i)^r\left(1+\sum_{k\in\mathbb{N}}\frac{(\lambda_i)^k(y-q)^ky^kq^{\frac12 k(k+1)}}{\prod_{l=1}^k(\lambda_iq^l-y)(\lambda_iq^l-q)}\,q^{kr}\right) \qquad i=1,2,\ r\in\mathbb{N}_0.$$

Pick $x>1$ and $0<y<1$ such that $q=xy<1$, and let $r\to\infty$ in (B.9). This gives

$$\lim_{r\to\infty} q^{-r}g_{r,1}(x,y)=0 \quad\text{and}\quad \lim_{r\to\infty} q^{-r}g_{r,2}(x,y)=1.$$

Next, an easy computation shows that $\lim_{r\to\infty} \frac{1}{r} \log |\mathcal{W}_r| = 1+\sqrt{2}$, with $|\mathcal{W}_r| = g_r(1,1)$. Pick $x > 1$ and $0 < y < 1$ such that $q = xy < 1/(1+\sqrt{2})$, and let $r \to \infty$ in (B.2). This gives

$$\lim_{r\to\infty} q^{-r} g_r(x,y) = 0,$$

from which it follows that $C_2 = 0$.

(8) It remains to determine C_1. To that end, note that, by construction, $(g_{r,1})_{r\in\mathbb{N}_0}$ satisfies (B.4) for $r = 0$ as well. Use (B.3) and (B.4) to show that

$$\begin{aligned} \tfrac{1}{2} C_1 g_{0,1} &= g_0 = y + yG, \\ C_1 g_{1,1} &= g_1 = a + bG, \end{aligned} \tag{B.10}$$

with

$$a = y^2(2 + y - xy), \quad b = y^2(1 + x + y - xy).$$

Eliminate C_1 and express G in terms of $g_{0,1}$ and $g_{1,1}$, to obtain

$$G(x,y) = \frac{aH(x,y) - y^2}{bH(x,y) - y^2},$$

where

$$H(x,y) = y\, \frac{g_{0,1}(x,y)}{g_{1,1}(x,y)}.$$

This completes the proof of Lemma 2.4 with $\bar{g}_0 = g_{0,1}$ and $\bar{g}_1 = g_{1,1}$.

(9) Brak, Guttmann and Whittington [**23**] show that the function $H(x,y)$ can be represented as a continued fraction. This representation allows for an analysis of the singularity structure of $G(x,y)$, in particular, for a computation of $y_c(x)$ (the radius of convergence of the power series $y \mapsto G(x,y)$) for fixed x. For instance, from (B.4) it is easily deduced that

$$G(1/y, y) = \sum_{r\in\mathbb{N}_0} g_r(1/y, y) = -1 + \sqrt{\frac{1-y}{1-3y-y^2-y^3}} \qquad (q = 1)$$

and this has a singularity at y_c solving the cubic equation $1 - 3y - y^2 - y^3 = 0$. Fig. 16 gives the plot of $x \mapsto y_c(x)$ that comes out of the singularity analysis. As explained in Section B.3, the free energy is $f(\gamma) = -\log y_c(\mathrm{e}^\gamma)$.

Appendix C. Tutorial 3

The purpose of this tutorial is to take a closer look at the free energy of the homogeneous pinning model described in Section 3. Section C.1 recalls the model, Section C.2 computes the free energy, while Section C.3 identifies the order of the phase transition.

C.1. The model. Let $(S_n)_{n\in\mathbb{N}_0}$ be a random walk on $\mathbb{Z}$, i.e., $S_0 = 0$ and $S_i - S_{i-1}$, $i \in \mathbb{N}$, are i.i.d. Let P denote the law of S. Introducing the first return time to zero $\tau = \inf\{n \in \mathbb{N}\colon\ S_n = 0\}$, we denote by $R(\cdot)$ its distribution:

$$R(n) = P(\tau = n) = P\big(S_i \neq 0\ \forall 1 \le i \le n-1,\ S_n = 0\big), \qquad n \in \mathbb{N}.$$

We require that $\sum_{n\in\mathbb{N}} R(n) = 1$, i.e., the random walk is *recurrent*, and we assume the following *tail asymptotics* for $R(\cdot)$ as $n \to \infty$:

$$R(n) = \frac{c}{n^{1+a}}\,[1 + o(1)], \qquad c > 0,\ a \in (0,1) \cup (1,\infty).$$

The exclusion of $a = 1$ is for simplicity (to avoid logarithmic corrections in later statements). The constant c could be replaced by a *slowly varying function* at the expense of more technicalities, which however we avoid. We recall that, for a nearest-neighbor symmetric random walk, i.e., when $P(S_1 = 1) = P(S_1 = -1) = p$ and $P(S_1 = 0) = 1 - 2p$ with $p \in (0, \frac{1}{2})$, the above tail asymptotics holds with $\alpha = \frac{1}{2}$.

The set of allowed polymer configurations is $\mathcal{W}_n = \{w = (i, w_i)_{i=0}^n : w_0 = 0, w_i \in \mathbb{Z}\, \forall 0 \le i \le n\}$, on which we define the Hamiltonian $H_n^\zeta(w) = -\zeta L_n(w)$, where $\zeta \in \mathbb{R}$ and

$$L_n(w) = \sum_{i=1}^n 1_{\{w_i = 0\}}, \qquad w \in \mathcal{W}_n,$$

is the so-called *local time* of the polymer at the interface (which has height zero). We denote by P_n the projection of P onto $\mathcal{W}_n$, i.e., $P_n(w) = P(S_i = w_i\ \forall 1 \le i \le n)$ for $w \in \mathcal{W}_n$. This is the *a priori* law for the non-interacting polymer. We define our polymer model as the law P_n^ζ on $\mathcal{W}_n$ given by

$$P_n^\zeta(w) = \frac{1}{Z_n^\zeta}\, \mathrm{e}^{-H_n^\zeta(w)}\, P_n(w), \qquad w \in \mathcal{W}_n.$$

The normalizing constant Z_n^ζ, called the *partition function*, is given by

$$Z_n^\zeta = \sum_{w \in \mathcal{W}_n} \mathrm{e}^{-H_n^\zeta(w)}\, P_n(w) = E_n\left(\mathrm{e}^{-H_n^\zeta(w)}\right) = E\left(\mathrm{e}^{\zeta \sum_{i=1}^n 1_{\{S_i = 0\}}}\right).$$

The *free energy* $f(\zeta)$ is defined as the limit

$$f(\zeta) = \lim_{n\to\infty} \frac{1}{n} \log Z_n^\zeta,$$

which has been shown to exist in **Tutorial 1**. From a technical viewpoint it is more convenient to consider the *constrained partition sum* $Z_n^{*,\zeta}$ defined by

$$Z_n^{*,\zeta} = \sum_{\substack{w \in \mathcal{W}_n \\ w_n = 0}} \mathrm{e}^{-H_n^\zeta(w)}\, P_n(w) = E\left(\mathrm{e}^{\zeta \sum_{i=1}^n 1_{\{S_i = 0\}}}\, 1_{\{S_n = 0\}}\right).$$

As shown in **Tutorial 1**, if we replace Z_n^ζ by $Z_n^{*,\zeta}$ in the definition of $f(\zeta)$, then this does not change the value of the limit. Therefore we may focus on $Z_n^{*,\zeta}$.

C.2. Computation of the free energy. We repeat in more detail the derivation of the formula for the free energy $f(\zeta)$ given in Section 3.

(1) Prove that $Z_n^{*,\zeta} \ge \mathrm{e}^\zeta\, P(\tau = n) = \mathrm{e}^\zeta\, R(n)$. Deduce that $f(\zeta) \ge 0$ for every $\zeta \in \mathbb{R}$.
(2) Show that $Z_n^{*,\zeta} \le 1$ for $\zeta \in (-\infty, 0]$. Deduce that $f(\zeta) = 0$ for every $\zeta \in (-\infty, 0]$.
(3) Henceforth we focus on $\zeta \in [0, \infty)$. Define for $x \in [0, 1]$ the generating function $\phi(x) = \sum_{n\in\mathbb{N}} R(n)\, x^n$. Observe that $x \mapsto \phi(x)$ is strictly increasing with $\phi(0) = 0$ and $\phi(1) = 1$. Deduce that for every $\zeta \in [0, \infty)$ there is exactly one value $r = r(\zeta)$ that solves the equation $\phi(\mathrm{e}^{-r}) = \mathrm{e}^{-\zeta}$. Observe that $\tilde{R}_\zeta(n) = \mathrm{e}^\zeta\, R(n)\, \mathrm{e}^{-r(\zeta)n}$ defines a probability distribution on $\mathbb{N}$.

(4) For $n \in \mathbb{N}$ and $1 \le k \le n$, denote by $\Theta_{n,k}$ the set consisting of $k+1$ points drawn from the interval $\{0,\dots,n\}$, including 0 and n. More explicitly, the elements of $\Theta_{n,k}$ are of the form $j = (j_0, j_1, \dots, j_k)$ with $j_0 = 0$, $j_k = n$ and $j_{i-1} < j_i$ for all $1 \le i \le n$. By summing over the locations i at which $S_i = 0$, prove that

$$Z_n^{*,\zeta} = \sum_{k=1}^{n} e^{\zeta k} \sum_{j\in\Theta_{n,k}} \prod_{i=1}^{k} R(j_i - j_{i-1}).$$

Note that this equation can be rewritten as

$$Z_n^{*,\zeta} = e^{r(\zeta)n}\, u_\zeta(n), \qquad u_\zeta(n) = \sum_{k=1}^{n} \sum_{j\in\Theta_{n,k}} \prod_{i=1}^{k} \tilde{R}_\zeta(j_i - j_{i-1}).$$

(5) For fixed $\zeta \in (0,\infty)$, we introduce a *renewal process* $(\tau_n)_{n\in\mathbb{N}_0}$ with law P_ζ, which is a random walk on $\mathbb{N}_0$ with positive increments, i.e., $\tau_0 = 0$ and $\tau_n - \tau_{n-1}$, $n \in \mathbb{N}$, are i.i.d. under P_ζ with law $P_\zeta(\tau_1 = n) = \tilde{R}_\zeta(n)$. Show that the following representation formula holds:

$$u_\zeta(n) = \sum_{k=1}^{n} P_\zeta(\tau_k = n) = P_\zeta\left(\bigcup_{k\in\mathbb{N}} \{\tau_k = n\}\right).$$

In particular, $u_\zeta(n) \le 1$. We will use the following important result known as the *renewal theorem*:

$$\lim_{n\to\infty} u_\zeta(n) = C \in (0,\infty).$$

Here $C = C(\zeta) = [\sum_{m\in\mathbb{N}} m\tilde{R}_\zeta(m)]^{-1} \in (0,\infty)$.

(6) Conclude that $\lim_{n\to\infty} \frac{1}{n}\log Z_n^{*,\zeta} = r(\zeta)$ for every $\zeta \in [0,\infty)$. This means that for $\zeta \in [0,\infty)$ the free energy $f(\zeta)$ coincides with $r(\zeta)$ and therefore satisfies the equation $\phi(e^{-f(\zeta)}) = e^{-\zeta}$.

Note that (4) and (5) give a sharp asymptotics of the constrained partition sum. Also note that the argument only uses the renewal structure of the excursions of the polymer away from the interface, and therefore can be extended to deal with a priori random processes other than random walks.

C.3. Order of the phase transition. From the relation $\phi(e^{-f(\zeta)}) = e^{-\zeta}$ we next derive some interesting properties of the free energy.

(1) Observe that for $x \in (0,1)$ the function $\phi(x) = \sum_{n\in\mathbb{N}} R(n)\, x^n$ is strictly increasing, with non-vanishing first derivative, and is real analytic. Since $\phi(0) = 0$ and $\phi(1) = 1$, its inverse ϕ^{-1}, defined from $(0,1)$ onto $(0,1)$, is real analytic too, by the *Lagrange inversion theorem*. Deduce that the free energy $\zeta \mapsto f(\zeta) = -\log\phi^{-1}(e^{-\zeta})$ restricted to $\zeta \in (0,\infty)$ is real analytic. The same is trivially true for $\zeta \in (-\infty,0)$, since $f(\zeta) = 0$.

(2) Conclude that the free energy $\zeta \mapsto f(\zeta)$ is not analytic at $\zeta = 0$, by the *identity theorem of analytic functions*. Observe that nevertheless the free energy is continuous at $\zeta = 0$.

(3) Introduce the integrated tail probability $\overline{R}(n) = \sum_{k=n+1}^{\infty} R(k)$ for $n \in \mathbb{N}_0$. Deduce from our tail assumption on $R(\cdot)$ that $\overline{R}(n) = \frac{c}{a} n^{-a}[1 + o(1)]$ as $n \to \infty$.

(4) Use summation by parts to show that $1-\phi(x)=(1-x)\sum_{n\in\mathbb{N}_0}\overline{R}(n)x^n$ for $x\in(0,1)$.

PROOF.

$$\begin{aligned}1-\phi(x)&=1-\sum_{n\in\mathbb{N}}R(n)x^n=1-\sum_{n\in\mathbb{N}}(\overline{R}(n-1)-\overline{R}(n))x^n\\&=\left(1+\sum_{n\in\mathbb{N}}\overline{R}(n)x^n\right)-\sum_{n\in\mathbb{N}}\overline{R}(n-1)x^n=\sum_{n\in\mathbb{N}_0}\overline{R}(n)x^n-\sum_{n\in\mathbb{N}_0}\overline{R}(n)x^{n+1}\\&=(1-x)\sum_{n\in\mathbb{N}_0}\overline{R}(n)x^n.\end{aligned}$$

□

(5) Put $\psi(x)=\sum_{n\in\mathbb{N}_0}\overline{R}(n)x^n$, so that $1-\phi(x)=(1-x)\psi(x)$. We first focus on $a\in(1,\infty)$. Show that in that case $\psi(1)=E(\tau)=\sum_{n\in\mathbb{N}}nR(n)\in(0,\infty)$. Deduce from $\phi(\mathrm{e}^{-f(\zeta)})=\mathrm{e}^{-\zeta}$ that, as $\zeta\downarrow 0$,

$$f(\zeta)=\frac{1}{E(\tau)}\,\zeta\,[1+o(1)],\qquad a\in(1,\infty).$$

(6) We next focus on $a\in(0,1)$. Use a Riemann sum approximation to show that, as $r\downarrow 0$,

$$\psi(\mathrm{e}^{-r})=\left(\frac{c\Gamma(1-a)}{a}\right)r^{a-1}\,[1+o(1)],$$

where

$$\Gamma(1-a)=\int_0^\infty\frac{\mathrm{e}^{-t}}{t^a}\,\mathrm{d}t\in(0,\infty).$$

PROOF. Note that, for $a\in(0,1)$, $\psi(\mathrm{e}^{-r})\uparrow\infty$ as $r\downarrow 0$, because $\overline{R}(n)=\frac{c}{a}n^{-a}\,[1+o(1)]$. Therefore, for any fixed $n_0\in\mathbb{N}$, we can safely neglect the first n_0 terms in the sum defining $\psi(\cdot)$, because they give a finite contribution as $r\downarrow 0$. This gives

$$\begin{aligned}\psi(\mathrm{e}^{-r})&\sim\sum_{n=n_0}^\infty\overline{R}(n)\mathrm{e}^{-nr}\sim\frac{c}{a}\sum_{n=n_0}^\infty\frac{\mathrm{e}^{-nr}}{n^a}=\frac{c}{a}r^{a-1}\sum_{n=n_0}^\infty r\,\frac{\mathrm{e}^{-nr}}{(nr)^a}\\&\sim\frac{c}{a}r^{a-1}\left(\int_0^\infty\frac{\mathrm{e}^{-t}}{t^a}\,\mathrm{d}t\right),\end{aligned}$$

where $\sim$ refers to $n_0\to\infty$. □

(7) Deduce from $\phi(\mathrm{e}^{-f(\zeta)})=\mathrm{e}^{-\zeta}$ that, as $\zeta\downarrow 0$,

$$f(\zeta)=\left(\frac{a}{c\Gamma(1-a)}\right)^{1/a}\zeta^{1/a}\,[1+o(1)],\qquad a\in(0,1).$$

Note that the smaller a is, the more regular is the free energy for $\zeta\downarrow 0$, i.e., the higher is the order of the phase transition at $\zeta=0$. For $a\in(1,\infty)$ the derivative of the free energy is discontinuous at $\zeta=0$, which corresponds to a first-order phase transition.

Appendix D. Tutorial 4

The purpose of this (long) tutorial is to provide further detail on the variational approach to the random pinning model described in Section 4. Section D.1 recalls the model, Section D.2 provides the necessary background on large deviation theory, Section D.3 explains the large deviation principles for the empirical process of random words cut out from a random letter sequence according to a renewal process, while Section D.4 shows how the latter are applied to the random pinning model to derive a variational formula for the critical curve.

D.1. The model. Let $S = (S_n)_{n\in\mathbb{N}_0}$, be a Markov chain on a countable space Υ that contains a marked point $*$. Let P denote the law of S, and assume that $S_0 = *$. We introduce the first return time to $*$, namely, $\tau = \inf\{n \in \mathbb{N}\colon\ S_n = *\}$, and we denote by $R(\cdot)$ its distribution:

$$R(n) = P(\tau = n) = P(S_i \neq * \;\forall 1 \leq i \leq n-1,\ S_n = *), \qquad n \in \mathbb{N}.$$

We require that $\sum_{n\in\mathbb{N}} R(n) = 1$, i.e., the Markov chain is *recurrent*, and assume the following *logarithmic tail asymptotics* as $n \to \infty$:

$$\lim_{n\to\infty} \frac{\log R(n)}{\log n} = -(1+a)\,, \qquad \text{with } a \in [0,\infty)\,.$$

For a nearest-neighbor and symmetric random walk on $\mathbb{Z}$, i.e.,

$$P(S_1 = 1) = P(S_1 = -1) = p, \qquad P(S_1 = 0) = 1 - 2p, \qquad p \in (0, \frac{1}{2}),$$

this asympotics holds with $a = \frac{1}{2}$.

The set of allowed polymer configurations is $\mathcal{W}_n = \{w = (i, w_i)_{i=0}^n\colon\ w_0 = *,\ w_i \in \Upsilon\ \forall 0 < i \leq n\}$ on which we define the Hamiltonian

$$H_n^{\beta,h,\omega}(w) = -\sum_{i=0}^{n} (\beta\omega_i - h) 1_{\{w_i = *\}},$$

where $\beta, h \geq 0$ are two parameters that tune the interaction strength and $\omega = (\omega_i)_{i\in\mathbb{N}_0}$ is the *random environment*, a typical realization of a sequence of i.i.d. $\mathbb{R}$-valued random variables with marginal law μ_0. The law of the full sequence ω is therefore $\mathbb{P} = \mu_0^{\otimes\mathbb{N}_0}$. We assume that $M(\beta) = \mathbb{E}(e^{\beta\omega_0}) < \infty$ for all $\beta \in \mathbb{R}$, and w.l.o.g. we assume that $\mathbb{E}(\omega_0) = 0$ and $E(\omega_0^2) = 1$.

We denote by P_n the projection onto $\mathcal{W}_n$ of the law of S, i.e., $P_n(w) = P(S_i = w_i\ \forall 0 \leq i \leq n)$ for $w \in \mathcal{W}_n$. This is the *a priori* law for the non-interacting polymer. We define our polymer model as the law $P_n^{\beta,h,\omega}$ on $\mathcal{W}_n$ given by

$$P_n^{\beta,h,\omega}(w) = \frac{1}{Z_n^{\beta,h,\omega}}\, e^{-H_n^{\beta,h,\omega}(w)}\, P_n(w).$$

The normalizing constant $Z_n^{\beta,h,\omega}$ is the *partition sum* and is given by

$$\begin{aligned} Z_n^{\beta,h,\omega} &= \sum_{w\in\mathcal{W}_n} e^{-H_n^{\beta,h,\omega}(w)}\, P_n(w) \\ &= E_n\left(e^{-H_n^{\beta,h,\omega}(w)}\right) = E\left(e^{\sum_{i=0}^n (\beta\omega_i - h) 1_{\{w_i=*\}}}\right) .D \end{aligned}$$

The *quenched free energy* $f^{\text{que}}(\beta, h)$ is defined as the limit

$$f^{\text{que}}(\beta,h) = \lim_{n\to\infty} \frac{1}{n} \log Z_n^{\beta,h,\omega} \qquad \mathbb{P}\text{-a.s. and in } L^1(\mathbb{P}),$$

which has been shown in Tutorial 1 to exist and to be non-random. It can be easily shown that $f^{\mathrm{que}}(\beta,h)\geq 0$, which motivates the introduction of a *localized phase* $\mathcal{L}$ and a *delocalized phase* $\mathcal{D}$ defined by

$$\mathcal{L}=\{(\beta,h)\colon\ f^{\mathrm{que}}(\beta,h)>0\},\qquad \mathcal{D}=\{(\beta,h)\colon\ f^{\mathrm{que}}(\beta,h)=0\}.$$

It follows from the convexity and the monotonicity of the free energy that these phases are separated by a *quenched critical curve*

$$\beta\mapsto h_c^{\mathrm{que}}(\beta)=\inf\{h\in\mathbb{R}\colon\ f^{\mathrm{que}}(\beta,h)=0\}.$$

In the remainder of this tutorial we develop insight into the *variational formula* for h_c^{que} that was put forward in Section 5.

Note that $f^{\mathrm{que}}(\beta,h)=\lim_{n\to\infty}\frac{1}{n}\mathbb{E}(\log Z_n(\beta,h,\omega))$. Interchanging the expectation $\mathbb{E}$ and the logarithm, we obtain the *annealed free energy*:

$$f^{\mathrm{ann}}(\beta,h)=\lim_{n\to\infty}\frac{1}{n}\log\mathbb{E}\big(Z_n^{\beta,h,\omega}\big)=\lim_{n\to\infty}\frac{1}{n}\log E\big(\mathrm{e}^{(\log M(\beta)-h)\sum_{i=0}^{n}1_{\{S_i=*\}}}\big),$$

which is nothing but the free energy $f(\zeta)$ of a homogeneous pinning model with $\zeta=\log M(\beta)-h$. Recall from Tutorial 3 that $f(\zeta)>0$ for $\zeta>0$ and $f(\zeta)=0$ for $\zeta\leq 0$. Introducing the *annealed critical curve*

$$h_c^{\mathrm{ann}}(\beta)=\inf\{h\in\mathbb{R}\colon\ f^{\mathrm{ann}}(\beta,h)=0\},$$

we find that $h_c^{\mathrm{ann}}(\beta)=\log M(\beta)$. Jensen's inequality yields $f^{\mathrm{que}}(\beta,h)\leq f^{\mathrm{ann}}(\beta,h)$, so that $h_c^{\mathrm{que}}(\beta)\leq h_c^{\mathrm{ann}}(\beta)$. The disorder is said to be *irrelevant* if $h_c^{\mathrm{que}}(\beta)=h_c^{\mathrm{ann}}(\beta)$ and *relevant* if $h_c^{\mathrm{que}}(\beta)<h_c^{\mathrm{ann}}(\beta)$.

D.2. Some background on large deviation theory. Before we proceed with our analysis of the copolymer model we make an intermezzo, namely, we give a brief summary of some basic large deviation results. For more details, see the monographs by Dembo and Zeitouni [**47**] and den Hollander [**69**].

D.2.1. *Relative entropy.* Let ν,ρ be two probabilities on a measurable space $(\Gamma,\mathcal{G})$, i.e., $\nu,\rho\in\mathcal{M}_1(\Gamma)$, the space of probability measures on Γ. For $\nu\ll\rho$ (i.e., ν is absolutely continuous with respect to ρ), we denote by $\frac{\mathrm{d}\nu}{\mathrm{d}\rho}$ the corresponding Radon-Nikodým derivative and we define the *relative entropy* $h(\nu|\rho)$ of ν with respect to ρ by the formula

$$h(\nu|\rho)=\int_\Gamma\log\left(\frac{\mathrm{d}\nu}{\mathrm{d}\rho}\right)\mathrm{d}\nu=\int_\Gamma\left(\frac{\mathrm{d}\nu}{\mathrm{d}\rho}\right)\log\left(\frac{\mathrm{d}\nu}{\mathrm{d}\rho}\right)\mathrm{d}\rho.$$

For $\nu\not\ll\rho$, we simply put $h(\nu|\rho)=\infty$. Note that the function $g(x)=x\log x$ with $g(0)=0$ is convex (hence continuous) and is bounded from below on $[0,\infty)$, so that the integral defining $h(\nu|\rho)$ is well-defined in $\mathbb{R}\cup\{\infty\}$.

- Use Jensen's inequality to show that $h(\nu|\rho)\geq 0$ for all ν,ρ, with $h(\nu|\rho)=0$ if and only if $\nu=\rho$.

For fixed ρ, the function $\nu\mapsto h(\nu|\rho)$ is convex on $\mathcal{M}_1(\Gamma)$. Note that if Γ is a finite set, say $\Gamma=\{1,\dots,r\}$ with $r\in\mathbb{N}$, then we can write

$$h(\nu|\rho)=\sum_{i=1}^{r}\nu_i\log\left(\frac{\nu_i}{\rho_i}\right).$$

D.2.2. *Sanov's Theorem in a finite space.* Let $Y = (Y_n)_{n\in\mathbb{N}}$ be an i.i.d. sequence of random variables taking values in a finite set, which we identify with $\Gamma = \{1, \dots, r\}$ with $r \in \mathbb{N}$. Let $\rho = \{\rho_i\}_{i=1}^r$ with $\rho_i = P(Y_1 = i) > 0$ be the marginal law of this random sequence. Note that $\rho \in \mathcal{M}_1(\Gamma)$. For $n \in \mathbb{N}$ we define the *empirical measure*

$$L_n = \frac{1}{n} \sum_{k=1}^{n} \delta_{Y_k},$$

where δ_x denotes the Dirac mass at x. Note that L_n is a *random element* of $\mathcal{M}_1(\Gamma)$, i.e., a random variable taking values in $\mathcal{M}_1(\Gamma)$, which describes the relative frequency of the "letters" appearing in the sequence $Y_1, \dots, Y_n$.

The space $\mathcal{M}_1(\Gamma)$ can be identified with the simplex $\{x \in (\mathbb{R}^+)^r \colon \sum_{i=1}^r x_i = 1\} \subset (\mathbb{R}^+)^r$, and hence $\mathcal{M}_1(\Gamma)$ can be equipped with the standard Euclidean topology, and we can talk about convergence in $\mathcal{M}_1(\Gamma)$ (which is nothing but convergence of every component). With this identification we have $L_n = \{L_n(i)\}_{i=1}^r$, where $L_n(i)$ is the relative frequency of the symbol i in the sequence $Y_1, \dots, Y_n$, i.e., $L_n(i) = \frac{1}{n}\sum_{k=1}^n 1_{\{Y_k = i\}}$.

- Show that the strong law of large numbers yields the a.s. convergence $\lim_{n\to\infty} L_n = \rho$, where the limit is in $\mathcal{M}_1(\Gamma)$.

The purpose of large deviation theory is to quantify the probability that L_n differs from its limit ρ: given a $\nu \in \mathcal{M}_1(\Gamma)$ different from μ, what is the probability that L_n is close to ν? Take for simplicity $\nu = \{\nu_i\}_{i=1}^r$ of the form $\nu_i = \frac{k_i}{n}$ with $k_i \in \mathbb{N}$ and $\sum_{i=1}^r k_i = n$. (Note that this is the family of laws that can be attained by L_n.)

- Prove that $P(L_n = \nu) = n! \prod_{i=1}^r \frac{{\rho_i}^{k_i}}{k_i!}$.
- Use Stirling's formula $n! = n^n \mathrm{e}^{-n+o(n)}$ to deduce that $P(L_n = \nu) = \mathrm{e}^{-nh(\nu|\rho)+o(n)}$, where $h(\nu|\rho)$ is the relative entropy defined above.

In this sense, the relative entropy $h(\nu|\rho)$ gives the rate of exponential decay for the probability that L_n is close to ν instead of ρ. More generally, it can be shown that if O and C are, respectively, an open and a closed subset of $\mathcal{M}_1(\Gamma)$, then, with the notation $I(\nu) = h(\nu|\rho)$, the following relations hold:

$$\text{(D.1)} \qquad \begin{aligned} \liminf_{n\to\infty} \frac{1}{n} \log P(L_n \in O) &\geq - \inf_{\nu\in O} I(\nu), \\ \limsup_{n\to\infty} \frac{1}{n} \log P(L_n \in C) &\leq - \inf_{\nu\in C} I(\nu). \end{aligned}$$

Whenever the above inequalities hold, we say that the sequence of random variables $(L_n)_{n\in\mathbb{N}}$ satisfies the *large deviation principle* (LDP) with rate n and with rate function $I(\cdot)$.

D.2.3. *Sanov's theorem in a Polish space.* In the previous section we have worked under the assumption that the space Γ is finite. However, everything can be generalized to the case when Γ is *Polish* (a complete separable metric space) equipped with the Borel σ-field. Let $Y = (Y_n)_{n\in\mathbb{N}}$ be an i.i.d. sequence of random variables taking values in Γ and denote by $\rho \in \mathcal{M}_1(\Gamma)$ the law of Y_1. We equip the space $\mathcal{M}_1(\Gamma)$ of probability measures on Γ with the topology of weak convergence (i.e., $\nu_n \to \nu$ in $\mathcal{M}_1(\Gamma)$ if and only if $\int f \mathrm{d}\nu_n \to \int f \mathrm{d}\nu$ for every bounded and continuous $f \colon \Gamma \to \mathbb{R}$). This topology turns $\mathcal{M}_1(\Gamma)$ into a Polish space too, which we equip with the corresponding Borel σ-field. We can therefore speak of convergence in $\mathcal{M}_1(\Gamma)$ as well as random elements of $\mathcal{M}_1(\Gamma)$ (random variables taking values in $\mathcal{M}_1(\Gamma)$).

In particular, the empirical measure L_n introduced above is well defined in this generalized setting as a random element of $\mathcal{M}_1(\Gamma)$. With the help of the ergodic theorem it is possible to show that, in analogy with the case of finite Γ, $\lim_{n\to\infty} L_n = \rho$ a.s. in $\mathcal{M}_1(\Gamma)$. Also, the large deviation inequalities mentioned above continue to hold, again with $I(\nu) = h(\nu|\rho)$ as defined earlier. The formal tool to prove this is the *projective limit LDP* of Dawson and Gärtner [**46**].

D.2.4. *Process level large deviations.* One can take a step further and consider an extended empirical measure, keeping track of "words" instead of single "letters". More precisely, let again $Y = (Y_n)_{n\in\mathbb{N}}$ be an i.i.d. sequence of random variables taking values in a Polish space Γ and denote by $\rho \in \mathcal{M}_1(\Gamma)$ the law of Y_1. For $\ell \in \mathbb{N}$ fixed, one can consider the empirical distribution of ℓ consecutive variables ("words consisting of ℓ letters") appearing in the sequence $Y_1, \ldots, Y_n$:

$$L_n^\ell = \frac{1}{n}\sum_{i=1}^{n} \delta_{(Y_i, Y_{i+1}, \ldots, Y_{i+\ell-1})},$$

where we use for convenience periodic boundary conditions: $Y_{n+i} = Y_i$ for $i = 1, \ldots, \ell-1$. Note that L_n^ℓ is a random element of the space $\mathcal{M}_1(\Gamma^\ell)$ of probability measures on Γ^ℓ. One can show that $\lim_{n\to\infty} L_n^\ell = \rho^{\otimes\ell}$ a.s. and one can obtain the large deviations of L_n^ℓ with an explicit rate function (not pursued here).

One can even go beyond and consider the empirical measure associated with "words of arbitrary length". To do so, it is convenient to denote by $(Y_1, \ldots, Y_n)^{\mathrm{per}}$ the *infinite* sequence obtained by repeating periodically $(Y_1, \ldots, Y_n)$, i.e.,

$$((Y_1, \ldots, Y_n)^{\mathrm{per}})_{mn+j} = Y_j \text{ for } m \in \mathbb{N}_0 \text{ and } j \in \{1, \ldots, n\}.$$

Note that $(Y_1, \ldots, Y_n)^{\mathrm{per}}$ takes values in $\Gamma^{\mathbb{N}}$. Denoting by θ the left shift on $\Gamma^{\mathbb{N}}$, i.e., $(\theta x)_i = x_{i+1}$ for $x = (x_i)_{i\in\mathbb{N}}$, we can therefore introduce the *empirical process*

$$R_n = \frac{1}{n}\sum_{i=0}^{n-1} \delta_{\theta^i (Y_1, \ldots, Y_n)^{\mathrm{per}}},$$

which is by definition a random element of the space $\mathcal{M}_1^{\mathrm{inv}}(\Gamma^{\mathbb{N}})$ of shift-invariant probability measures on the Polish space $\Gamma^{\mathbb{N}}$, which is equipped with the product topology and the product σ-field.

Again, one can show that $\lim_{n\to\infty} R_n = \rho^{\otimes\mathbb{N}}$ a.s. on $\mathcal{M}_1^{\mathrm{inv}}(\Gamma^{\mathbb{N}})$. Furthermore, $(R_n)_{n\in\mathbb{N}}$ satisfies an LDP, namely, for every open set O and closed set C in $\mathcal{M}_1^{\mathrm{inv}}(\Gamma^{\mathbb{N}})$:

$$\liminf_{n\to\infty} \frac{1}{n} \log P(R_n \in O) \geq -\inf_{\nu\in O} I(\nu),$$

$$\limsup_{n\to\infty} \frac{1}{n} \log P(R_n \in C) \leq -\inf_{\nu\in C} I(\nu),$$

where the rate function $I(\nu) = H(\nu|\rho^{\otimes\mathbb{N}})$ is the so-called *specific relative entropy*:

$$H(\nu|\rho^{\otimes\mathbb{N}}) = \lim_{n\to\infty} \frac{1}{n} h(\pi_n\nu|\rho^{\otimes n}),$$

where $h(\cdot\,|\,\cdot)$ is the relative entropy defined earlier and π_n denotes the projection from $\Gamma^{\mathbb{N}}$ to Γ^n onto the first n components. The limit can be shown to be non-decreasing: in particular, $H(\nu|\rho) = 0$ if and only if $\pi_n\nu = \rho^{\otimes n}$ for every $n \in \mathbb{N}$, i.e., $\nu = \rho^{\otimes\mathbb{N}}$.

D.3. Random words cut out from a random letter sequence. Let us apply the large deviation theory sketched in the previous section to study the sequence of random words cut out from a random letter sequence according to an independent renewal process. Our "alphabet" will be $\mathbb{R}$, while $\widetilde{\mathbb{R}} = \bigcup_{n\in\mathbb{N}} \mathbb{R}^k$ will be the set of finite words drawn from $\mathbb{R}$, which can be metrized to become a Polish space.

We recall from D.1 that $\omega = (\omega_i)_{i\in\mathbb{N}_0}$ with law $\mathbb{P}$ is an i.i.d. sequence of $\mathbb{R}$-valued random variables with marginal distribution μ_0, and $S = (S_n)_{n\in\mathbb{N}_0}$ with law P is a recurrent Markov chain on the countable space Υ containing a marked point $*$. The sequences ω and S are independent. From the sequence of letters ω we cut out a sequence of words $Y = (Y_i)_{i\in\mathbb{N}}$ using the successive excursions of S out of $*$. More precisely, we let T_k denote the epoch of the k-th return of S to $*$:

$$T_0 = 0, \qquad T_{k+1} = \inf\{m > T_k\colon\ S_m = *\},$$

and we set $Y_i = (\omega_{T_{i-1}}, \omega_{T_{i-1}+1}, \ldots, \omega_{T_i-1})$. Note that $Y = (Y_i)_{i\in\mathbb{N}} \in \widetilde{\mathbb{R}}^{\mathbb{N}}$.

We next define the empirical process associated with Y:

$$R_n = \frac{1}{n}\sum_{i=0}^{n-1} \delta_{\tilde{\theta}^i(Y_1,\ldots,Y_n)^{\mathrm{per}}},$$

where we denote by $\tilde{\theta}$ the shift acting on $\widetilde{\mathbb{R}}$. By definition, R_n is a random element of the space $\mathcal{M}_1^{\mathrm{inv}}(\widetilde{\mathbb{R}}^{\mathbb{N}})$ of shift-invariant probabilities on $\widetilde{\mathbb{R}}^{\mathbb{N}}$.

We may look at Y and R_n in at least two ways: either under the law $P^* = \mathbb{P}\otimes P$ (= annealed) or under the law P (= quenched). We start with the *annealed viewpoint.*

- Show that under P^* the sequence Y is i.i.d. with marginal law q_0 given by

$$q_0(\mathrm{d}x_1,\ldots,\mathrm{d}x_n) \;=\; R(n)\,\mu_0(\mathrm{d}x_1)\times\cdots\times\mu_0(\mathrm{d}x_n).$$

- Conclude from D.2 that under P^* the sequence $(R_n)_{n\in\mathbb{N}}$ satisfies an LDP on $\mathcal{M}_1^{\mathrm{inv}}(\Gamma^{\mathbb{N}})$ with rate function $I^{\mathrm{ann}}(Q) = H(Q|\mu_0^{\otimes\mathbb{N}})$, the specific relative entropy of Q w.r.t. $\mathbb{P} = \mu_0^{\otimes\mathbb{N}}$.

In words, the probability under P^* that the first n words cuts out of ω by S, periodically extended to an infinite sequence, have an empirical distribution that is close to a law $Q \in \mathcal{M}_1^{\mathrm{inv}}(\Gamma^{\mathbb{N}})$ decays exponentially in n with rate $I^{\mathrm{ann}}(Q)$:

$$P^*(R_n \approx Q) = \exp[-nI^{\mathrm{ann}}(Q) + o(n)].$$

We note that $I^{\mathrm{ann}}(Q) \geq 0$ and $I^{\mathrm{ann}}(Q) = 0$ if and only if $Q = \mu_0^{\otimes\mathbb{N}}$.

We next consider the *quenched viewpoint*, i.e., we fix ω and we write R_n^ω instead of R_n. It is intuitively clear that, when the average is over S only, it is more difficult to observe a large deviation. Therefore, if under P the sequence $(R_n^\omega)_{n\in\mathbb{N}}$ satisfies an LDP on $\mathcal{M}_1^{\mathrm{inv}}(\Gamma^{\mathbb{N}})$ with rate function I^{que}, i.e., if $P(R_n^\omega \approx Q) = \exp[-nI^{\mathrm{que}}(Q) + o(n)]$, then we should have $I^{\mathrm{que}}(Q) \geq I^{\mathrm{ann}}(Q)$. Indeed, this is the case: the difference between $I^{\mathrm{que}}(Q)$ and $I^{\mathrm{ann}}(Q)$ can in fact be explicitly quantified. For details we refer to Birkner, Greven and den Hollander [**12**].

D.4. The empirical process of words and the pinning model. We are finally ready to explore the link between the process of random words Y described in the previous section and our random pinning model. Define for $z \in [0,1]$ the generating function

$$G(z) = \sum_{n \in \mathbb{N}} z^n \, Z_n^{*,\beta,h,\omega},$$

where $Z_n^{*,\beta,h,\omega}$ denotes the *constrained* partition sum

$$Z_n^{*,\beta,h,\omega} = E\left(\mathrm{e}^{\sum_{i=0}^{n-1} (\beta\omega_i - h) 1_{\{S_i = *\}}} \, 1_{\{S_n = *\}} \right).$$

We recall that $Z_n^{*,\beta,h,\omega}$ yields the same free energy as the original partition function $Z_n^{\beta,h,\omega}$, i.e.,

$$f^{\mathrm{que}}(\beta, h) = \lim_{n\to\infty} \frac{1}{n} \log Z_n^{*,\beta,h,\omega} \qquad \mathbb{P}\text{-a.s. and in } L^1(\mathbb{P}).$$

- Prove that the radius of convergence $\overline{z}$ of $G(z)$ equals $\mathrm{e}^{-f^{\mathrm{que}}(\beta,h)}$.
- In analogy with Tutorial 3, show that

$$z^n \, Z_n^{*,\beta,h,\omega} = \sum_{N\in\mathbb{N}} \sum_{0=k_0<k_1<\cdots<k_N=n} \prod_{i=1}^{N} z^{k_i - k_{i-1}} \, R(k_i - k_{i-1}) \, \mathrm{e}^{\beta\omega_{k_{i-1}} - h}.$$

- Deduce that $G(z) = \sum_{N\in\mathbb{N}} F_N^{\beta,h,\omega}(z)$, where

$$\begin{aligned} F_N^{\beta,h,\omega}(z) &= \sum_{0=k_0<k_1<\cdots<k_N<\infty} \prod_{i=1}^{N} z^{k_i - k_{i-1}} \, R(k_i - k_{i-1}) \, \mathrm{e}^{\beta\omega_{k_{i-1}} - h} \\ &= E\left(\prod_{i=1}^{N} z^{T_i - T_{i-1}} \, \mathrm{e}^{\beta\omega_{T_{i-1}} - h} \right) \\ &= \mathrm{e}^{N[S_N^{\beta,\omega}(z) - h]} \end{aligned}$$

  with

$$S_N^{\beta,\omega}(z) = \frac{1}{N} \log E\left(\exp\left[\sum_{i=1}^{N} (T_i - T_{i-1}) \log z + \beta\omega_{T_{i-1}} \right] \right).$$

Given an infinite "sentence" $y = (y_k)_{k\in\mathbb{N}} \in \widetilde{\mathbb{R}}^{\mathbb{N}}$, we denote by $y_1 \in \widetilde{\mathbb{R}}$ its first "word". For a "word" $x \in \widetilde{\mathbb{R}}$, we denote by $\ell(x)$ the length of x and by $c(x)$ the first letter of x.

- Recalling that $Y_i = (\omega_{T_{i-1}}, \omega_{T_{i-1}+1}, \ldots, \omega_{T_i - 1})$, with the T_i's the hitting times of the interface $*$, prove that

$$m(R_N^\omega) = \int_{\widetilde{\mathbb{R}}^{\mathbb{N}}} \ell(y_1) \, R_N^\omega(\mathrm{d}y) = \frac{1}{N} \sum_{i=1}^{N} \ell(Y_i) = \frac{1}{N} \sum_{i=1}^{N} (T_i - T_{i-1}),$$

$$\Phi(R_N^\omega) = \int_{\widetilde{\mathbb{R}}^{\mathbb{N}}} c(y_1) \, R_N^\omega(\mathrm{d}y) = \frac{1}{N} \sum_{i=1}^{N} c(Y_i) = \frac{1}{N} \sum_{i=1}^{N} \omega_{T_{i-1}}.$$

  Hence

$$S_N^{\beta,\omega}(z) = \frac{1}{N} \log E\left(\exp\left[N\left[m(R_N^\omega) \log z + \beta\Phi(R_N^\omega) \right] \right] \right).$$

This shows that $S_N^{\beta,\omega}(z)$ is the expectation of an exponential function of R_N^ω. It is therefore clear that the properties of the generating function $G(z)$, in particular, its radius of convergence $\bar{z}$ (and hence the quenched free energy) can be deduced from the large deviation properties of R_N^ω. Let us therefore set

$$S^{\mathrm{que}}(\beta, z) = \limsup_{N\to\infty} S_N^{\beta,\omega}(z)$$

and $S^{\mathrm{que}}(\beta, 1-) = \lim_{z\uparrow 1} S^{\mathrm{que}}(\beta, z)$.

- Prove that if $h > S^{\mathrm{que}}(\beta, z)$ then $G(z) < \infty$, while if $h < S^{\mathrm{que}}(\beta, z)$ then $G(z) = \infty$.
- Deduce that if $S^{\mathrm{que}}(\beta, 1-) < h$ then $f^{\mathrm{que}}(\beta, h) = 0$, while if $S^{\mathrm{que}}(\beta, 1-) > h$ then $f^{\mathrm{que}}(\beta, h) > 0$. Therefore $h_c^{\mathrm{que}}(\beta) = S^{\mathrm{que}}(\beta, 1-)$.

Finally, with the help of Varadhan's lemma in large deviation theory it can be shown that

$$h_c^{\mathrm{que}}(\beta) = S^{\mathrm{que}}(\beta, 1-) = \sup_{Q\in\mathcal{M}_1^{\mathrm{inv}}(\widetilde{\mathbb{R}}^{\mathbb{N}})} \left[\beta\Phi(Q) - I^{\mathrm{que}}(Q)\right].$$

This gives an explicit variational characterization of the quenched critical curve. An analogous characterization holds for the annealed critical curve too. For details see Cheliotis and den Hollander [**35**].

Appendix E. Tutorial 5

In this tutorial we return to the copolymer model treated in Sections 5.1–5.4 and prove Theorem 5.2 (lower bound on the critical curve) and Theorem 5.5 (order of the phase transition is at least two). Section E.1 recalls the model, Section E.2 proves Theorem 5.2, while Section E.3 proves Theorem 5.5.

E.1. The model. We begin by recalling some of the notation used in Sections 5.1–5.4.

Configurations of the copolymer. For $n \in \mathbb{N}$ the allowed configurations of the copolymer are modelled by the n-step paths of a $(1+1)$-dimensional simple random walk $S = (S_i)_{i\in\mathbb{N}_0}$, i.e., $S_0 = 0$ and $(S_i - S_{i-1})_{i\in\mathbb{N}}$ is an i.i.d. sequence of Bernoulli trials with

$$P(S_1 = +1) = P(S_1 = -1) = \tfrac{1}{2},$$

where we write P for the law of S. The set of n-step paths is denoted by $\mathcal{W}_n$.

Disorder: randomness of the monomer types. The monomers in the copolymer are either hydrophilic or hydrophobic. Their order of appearance is encoded by an i.i.d. sequence $\omega = (\omega_i)_{i\in\mathbb{N}}$ of Bernouilli trials with

$$\mathbb{P}(\omega_1 = +1) = \mathbb{P}(\omega_1 = -1) = \tfrac{1}{2},$$

where we write $\mathbb{P}$ for the law of ω, and we assume that ω and S are independent.

Interaction polymer-interface. The medium is made up of oil and water separated by a flat interface located at height 0, oil being above the interface and water below. The copolymer gets an energetic reward for each monomer it puts in its preferred solvent. Thus, $S \in \mathcal{W}_n$ has energy

$$H_n^{\beta,h,\omega}(S) = -\beta\sum_{i=1}^{n}(\omega_i + h)(\Delta_i - 1), \qquad S \in \mathcal{W}_n,$$

where $\Delta_i = \text{sign}(S_{i-1}, S_i)$ and $\beta \in (0, \infty)$ stands for the inverse temperature. The presence of the -1 in this Hamiltonian is for later convenience and has no effect on the polymer measure. Indeed, by the law of large numbers for ω, we have $\beta \sum_{i=1}^{n} (\omega_i + h) = \beta hn + o(n)$. The term βhn can be moved to the normalizing partition sum, while the term $o(n)$ does not affect the free energy in the limit as $n \to \infty$.

Partition function and free energy. For fixed n, the quenched (= frozen disorder) partition sum and finite-volume free energy are defined as

$$Z_n^{\beta,h,\omega} = E\big(e^{-H_n^{\beta,h,\omega}(S)}\big), \qquad g_n^{\omega}(\beta, h) = \tfrac{1}{n} \log Z_n^{\beta,h,\omega}.$$

Recall that the localized phase $\mathcal{L}$ and the delocalized phase $\mathcal{D}$ are defined by

$$\mathcal{L} = \{(\beta, h) \colon g^{\text{que}}(\beta, h) > 0\}, \qquad \mathcal{D} = \{(\beta, h) \colon g^{\text{que}}(\beta, h) = 0\},$$

where $g^{\text{que}}(\beta, h) = \lim_{n\to\infty} g_n^{\omega}(\beta, h)$ ω-a.s.

E.2. Lower bound on the critical curve. Fix $l \in 2\mathbb{N}$. For $j \in \{1, \ldots, n/l\}$ (for simplicity we pretend that n/l is integer), let

$$I_j = \{(j-1)l + 1, \ldots, jl\}, \qquad \Omega_j = \sum_{i \in I_j} \omega_i.$$

Fix $\delta \in (0, 1]$, and define

$$i_0^{\omega} = 0, \qquad i_{j+1}^{\omega} = \inf\{k \geq i_j^{\omega} + 2 \colon \Omega_k \leq -\delta l\}, \quad j \in \mathbb{N}.$$

These are the stretches of length l where the empirical average of the disorder is $\leq -\delta$, trimmed so that no two stretches occur next to each other, which guarantees that $\tau_j^{\omega} = i_{j+1}^{\omega} - i_j^{\omega} - 1$, $j \in \mathbb{N}$, are ≥ 1. (The copolymer gets a substantial reward when it moves below the interface during these stretches.) Let

$$t_n^{\omega} = \sup\{j \in \mathbb{N}_0 \colon i_j^{\omega} \leq n/l\}.$$

In the estimate below we will need the subset of paths defined by (see Fig. 47)

$$\mathcal{W}_n^{\omega} = \{S \colon S_i < 0 \ \forall\, i \in \cup_{j=1}^{t_n^{\omega}} I_{i_j^{\omega}} \backslash \partial I_{i_j^{\omega}}\} \cap \{S \colon S_i > 0 \ \forall\, i \in \{0, \ldots, n\} \setminus \cup_{j=1}^{t_n^{\omega}} I_{i_j^{\omega}}\}.$$

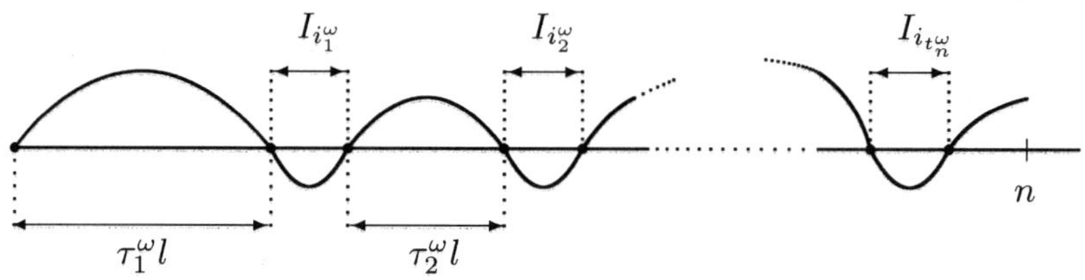

FIGURE 47. A path in the set $\mathcal{W}_n^{\omega}$.

(1) Let

$$R(n) = P(S_i > 0 \ \forall\, 0 < i < n, \ S_n = 0),$$
$$\bar{R}(n) = P(S_i > 0 \ \forall\, 0 < i \leq n).$$

Insert the indicator of the set $\mathcal{W}_n^\omega$ into the definition of the partition sum, to estimate

$$\log Z_n^{\beta,h,\omega} \geq \sum_{j=1}^{t_n^\omega} \log R(\tau_j^\omega l) + t_n^\omega \left[\log R(l) + 2\beta(\delta - h)l\right] + \log \bar{R}(n - i_{t_n^\omega} l).$$

(2) Note that there exists a $C > 0$ such that $R(n) \geq C/n^{3/2}$ for $n \in \mathbb{N}$. Use this to deduce from (1) that

$$\log Z_n^{\beta,h,\omega} \geq t_n^\omega \big[\log C - \tfrac{3}{2} \log(\tfrac{n}{t_n^\omega} - l)\big] + t_n^\omega \left[\log C - \tfrac{3}{2} \log l + 2\beta(\delta - h)l\right] + O(\log n),$$

where the first term arises after we apply Jensen's inequality:

$$\frac{1}{t_n^\omega} \sum_{j=1}^{t_n^\omega} \log \tau_j^\omega \leq \log \left(\frac{1}{t_n^\omega} \sum_{j=1}^{t_n^\omega} \tau_j^\omega \right).$$

(3) Abbreviate

$$q_{l,\delta} = \mathbb{P}\Big(\Omega_1 \leq -\delta l\Big).$$

Use the ergodic theorem to prove that

$$\lim_{n\to\infty} \frac{t_n^\omega}{n} = \frac{1}{l} \frac{q_{l,\delta}}{1 + q_{l,\delta}} = p_{l,\delta} \qquad \omega\text{-a.s..}$$

(Note that $k \in \cup_{j\in\mathbb{N}_0} i_j^\omega$ if and only if $\Omega_k \leq -\delta l$ and $k - 1 \notin \cup_{j\in\mathbb{N}_0} i_j^\omega$.) Since $\sum_{j=1}^{t_n^\omega} \tau_j^\omega l \leq n - t_n^\omega l$, it follows that

$$\limsup_{n\to\infty} \frac{\sum_{j=1}^{t_n^\omega} \tau_i^\omega l}{t_n^\omega} \leq \lim_{n\to\infty} \frac{n - t_n^\omega l}{t_n^\omega} = p_{l,\delta}^{-1} - l \qquad \omega\text{-a.s.}$$

Conclude from (2) that

$$\liminf_{n\to\infty} \frac{1}{n} \log Z_n^{\beta,h,\omega} \geq p_{l,\delta}[-\tfrac{3}{2} \log(p_{l,\delta}^{-1} - l) + 2\beta(\delta - h)l + O(\log l)] \qquad \omega\text{-a.s.}$$

This inequality is valid for all $l \in 2\mathbb{N}$.

(4) Show, with the help of Cramér's theorem of large deviation theory applied to ω, that

$$\lim_{l\to\infty} \tfrac{1}{l} \log q_{l,\delta} = -\sup_{\lambda>0} \left[\lambda\delta - \log M(-\lambda)\right] = -\Sigma(\delta),$$

where $M(\lambda) = \mathbb{E}(\mathrm{e}^{\lambda\omega_1})$, the supremum may be trivially restricted to $\lambda > 0$, and the right-hand side is the Legendre transform of the cumulant generating function $\lambda \mapsto \log M(-\lambda)$. Use the last display and the relation $p_{l,\delta}^{-1} - l = l/q_{l,\delta}$ to show that

$$\lim_{l\to\infty} \tfrac{1}{l} \log(p_{l,\delta}^{-1} - l) = \Sigma(\delta).$$

(5) So far $\delta \in (0, 1]$ is arbitrary. Now combine (3) and (4), optimize over δ, and use that

$$\tfrac{3}{4} \log M(\tfrac{4}{3}\beta) = \sup_{\delta\in(0,1]} \left[-\tfrac{3}{4}\Sigma(\delta) + \beta\delta\right] = \tfrac{3}{4} \sup_{\delta\in(0,1]} \left[\tfrac{4}{3}\beta\delta - \Sigma(\delta)\right],$$

which is the (inverse) Legendre transform of the rate function in Cramér's theorem, to conclude that $g^{\mathrm{que}}(\beta, h) > 0$ as soon as

$$\tfrac{3}{4} \log M(\tfrac{4}{3}\beta) - \beta h > 0.$$

This completes the proof because $M(\frac{4}{3}\beta) = \cosh(\frac{4}{3}\beta)$.

E.3. Order of the phase transition. In the proof below we pretend that ω is an i.i.d. sequence of standard normal random variables, rather than Bernoulli random variables. At the end of the proof we will see how to adapt the argument.

Define the set of trajectories

$$\widetilde{\mathcal{W}}_n^\omega = \{S\colon\ S_i = 0\ \forall j \in \cup_{j=1}^{t_n^\omega} \partial I_{i_j^\omega}\} \cap \{S\colon\ S_i > 0\ \forall i \in \{0,\dots,n\} \setminus \cup_{j=1}^{t_n^\omega} I_{i_j^\omega}\}.$$

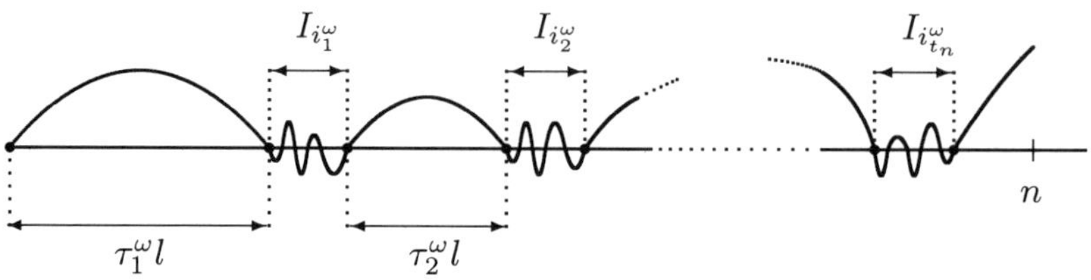

FIGURE 48. A path in the set $\widetilde{\mathcal{W}}_n^\omega$.

(6) Similarly as in (1), insert the indicator of the set $\widetilde{\mathcal{W}}_n^\omega$ into the definition of the partition function to estimate

$$\log Z_n^{\beta,h_c,\omega} \geq \sum_{j=1}^{t_n^\omega} \log R(\tau_j^\omega l) + \sum_{j=1}^{t_n^\omega} \log Z_l^{\beta,h_c,\,\theta^{i_j^\omega l}(\omega)} + \log \bar{R}\big(n - (i_{t_n^\omega}^\omega + 1)l\big),$$

where $\theta^l(\omega) = (\omega_{i+l})_{i\in\mathbb{N}}$.

(7) Take the expectation over $\mathbb{P}$ on both sides of (6), divide by n and use (3), to obtain

$$g^{\mathrm{que}}(\beta,h_c) \geq p_{l,\delta}\big[-\tfrac{3}{2}\log(p_{l,\delta}^{-1} - l) + O(\log l)\big] + \liminf_{n\to\infty} \tfrac{1}{n}\,\mathbb{E}\left(\sum_{j=1}^{t_n^\omega} \log Z_l^{\beta,h_c,\,\theta^{i_j^\omega l}(\omega)}\right).$$

(8) Use a martingale property to prove that

$$\frac{1}{n}\mathbb{E}\left(\sum_{j=1}^{t_n^\omega} \log Z_l^{\beta,h_c,\,\theta^{i_j^\omega l}(\omega)}\right) = E\left(\frac{t_n^\omega}{n}\right)\mathbb{E}\left(\log Z_l^{\beta,h_c,\omega} \mid \Omega_1 \leq -\delta l\right),$$

which gives

$$g^{\mathrm{que}}(\beta,h_c) \geq p_{l,\delta}\left[-\tfrac{3}{2}\log(p_{l,\delta}^{-1} - l) + O(\log l) + \mathbb{E}\left(\log Z_l^{\beta,h_c,\omega} \mid \Omega_1 \leq -\delta l\right)\right].$$

(9) Deduce from (8) and (3) that

$$-\tfrac{3}{2}\Sigma(\delta) + \tfrac{1}{l}\,\mathbb{E}\left(\log Z_l^{\beta,h_c,\omega} \mid \Omega_1 \leq -\delta l\right) + o(1) \leq 0, \qquad \delta > 0,\ l \in 2\mathbb{N},\ l \to \infty.$$

For large l, considering l i.i.d. Gaussian random variables with mean 0 and variance 1 conditioned to have sum $\leq -\delta l$ is equivalent to considering l i.i.d. Gaussian random variables with mean $-\delta$ and variance 1. Therefore we can replace

$\mathbb{E}(\log Z_l^{\beta,h_c,\omega} \mid \Omega_1 \leq -\delta l)$ by $\mathbb{E}(\log Z_l^{\beta,h_c-\delta,\omega}) + o(1)$ and so, after we let $l \to \infty$, the inequality in the last display yields

$$g^{\mathrm{que}}(\beta, h_c - \delta) \leq \tfrac{3}{2}\Sigma_0(\delta).$$

Combine the lower bound on $g^{\mathrm{que}}(\beta, h_c)$ with the upper bound on $g^{\mathrm{que}}(\beta, h_c - \delta)$, and use that $\Sigma_0(\delta) = \frac{1}{2}\delta^2[1 + o(1)]$ as $\delta \downarrow 0$, to obtain that

$$g^{\mathrm{que}}(\beta, h_c - \delta) - g(\beta, h_c) \leq \tfrac{1}{4}\delta^2 \qquad \text{for } \delta \text{ small enough.}$$

This completes the proof for standard Gaussian disorder.

(10) It is easy to extend the proof to binary disorder. All that is needed is to show that the Gaussian approximation in (9) carries through.

References

[1] K.S. Alexander, Ivy on the ceiling: first-order polymer depinning transitions with quenched disorder, Markov Proc. Relat. Fields 13 (2007) 663–680.

[2] K.S. Alexander, The effect of disorder on polymer depinning transitions, Commun. Math. Phys. 279 (2008) 117–146.

[3] K.S. Alexander and V. Sidoravicius, Pinning of polymers and interfaces by random potentials, Ann. Appl. Probab. 16 (2006) 636–669.

[4] K.S. Alexander and N. Zygouras, Quenched and annealed critical points in polymer pinning models, Commun. Math. Phys. 291 (2008) 659–689.

[5] K.S. Alexander and N. Zygouras, Equality of critical points for polymer depinning transitions with loop exponent one, Ann. Appl. Probab. 20 (2010) 356–66.

[6] L. Auvray, B. Duplantier, A. Echard and C. Sykes, Physique des polymères et membranes biologiques, Partie 1, Ecole Polytechnique, Palaiseau, France.

[7] R. Bauerschmidt, H. Duminil-Copin, J. Goodman, G. Slade, Lectures on self-avoiding walks, in this volume.

[8] V. Beffara, Schramm-Loewner evolution and other conformally invariant objects, in this volume.

[9] Q. Berger and F.L. Toninelli, On the critical point of the random walk pinning model in dimension $d = 3$, Electr. J. Probab. 15 (2010) 654–683.

[10] S. Bezerra, S. Tindel and F. Viens, Superdiffusivity for a Brownian polymer in a continuous Gaussian environment, Ann. Probab. 36 (2008) 1642–1675.

[11] M. Birkner, A condition for weak disorder for directed polymers in random environment, Electr. Comm. Prob. 9 (2004) 22–25.

[12] M. Birkner, A. Greven and F. den Hollander, Quenched large deviation principle for words in a letter sequence, Probab. Theory Relat. Fields 148 (2010) 403–456.

[13] M. Birkner, A. Greven and F. den Hollander, Collision local time of transient random walks and intermediate phases in interacting stochastic systems, Electr. J. Probab. 16 (2011) 552–586.

[14] M. Birkner and R. Sun, Annealed vs quenched critical points for a random walk pinning model, Ann. Inst. H. Poincaré Probab. Stat. 46 (2010) 414–441.

[15] M. Birkner and R. Sun, Disorder relevance for the random walk pinning model in dimension 3, Ann. Inst. H. Poincaré Probab. Stat. 47 (2011) 259–293.

[16] M. Biskup and F. den Hollander, A heteropolymer near a linear interface, Ann. Appl. Probab. 9 (1999) 668–687.

[17] T. Bodineau and G. Giacomin, On the localization transition of random copolymers near selective interfaces, J. Stat. Phys. 117 (2004) 801–818.

[18] T. Bodineau, G. Giacomin, H. Lacoin and F.L. Toninelli, Copolymers at selective interfaces: new bounds on the phase diagram, J. Stat. Phys. 132 (2008) 603–626.

[19] E. Bolthausen, A note on diffusion of directed polymers in a random environment, Commun. Math. Phys. 123 (1989) 529–534.

[20] E. Bolthausen, F. Caravenna and B. de Tilière, The quenched critical point of a diluted disordered polymer model, Stoch. Proc. Appl. 119 (2009) 1479–1504.

[21] E. Bolthausen and F. den Hollander, Localization transition for a polymer near an interface, Ann. Probab. 25 (1997) 1334–1366.
[22] E. Bolthausen, F. den Hollander and A. Opoku, A copolymer near a selective interface: variational characterization of the free energy, arXiv.org: 1110.1315 [math.PR], manuscript in preparation.
[23] R. Brak, A.J. Guttmann and S.G. Whittington, A collapse transition in a directed walk model, J. Phys. A: Math. Gen. 25 (1992) 2437–2446.
[24] A. Camanes and P. Carmona, The critical temperature of a directed polymer in a random environment, Markov Proc. Relat. Fields 15 (2009) 105–116.
[25] P. Caputo, F. Martinelli and F.L. Toninelli, On the approach to equilibrium for a polymer with adsorption and repulsion, Electr. J. Probab. 13 (2008) 213–258.
[26] P. Caputo, H. Lacoin, F. Martinelli, F. Simenhaus and F.L. Toninelli, Polymer dynamics in the depinned phase: metastability with logarithmic barriers, arXiv.org: 1007.4470 [math.PR], to appear in Probab. Theory Relat. Fields.
[27] F. Caravenna and G. Giacomin, The weak coupling limit of disordered copolymer models, Ann. Probab. 38 (2010) 2322–2378.
[28] F. Caravenna, G. Giacomin and M. Gubinelli, A numerical approach to copolymers at selective interfaces, J. Stat. Phys. 122 (2006) 799–832.
[29] F. Caravenna, G. Giacomin and F.L. Toninelli, Copolymers at selective interfaces: settled issues and open problems, In: Probability in complex physical systems. In honour of Erwin Bolthausen and Jrgen Grtner. Edited by J.-D. Deuschel, B. Gentz, W. Knig, M. von Renesse, M. Scheutzow, U. Schmock. Springer Proceedings in Mathematics 11 (2012), 289-312.
[30] F. Caravenna, G. Giacomin and L. Zambotti, Sharp asymptotic behavior for wetting models in (1+1)-dimension, Electr. J. Probab. 11 (2006) 345–362.
[31] F. Caravenna and N. Pétrélis, A polymer in a multi-interface medium, Ann. Appl. Probab. 19 (2009) 1803–1839.
[32] F. Caravenna and N. Pétrélis, Depinning of a polymer in a multi-interface medium, Electr. J. Probab. 14 (2009) 2038–2067.
[33] P. Carmona and Y. Hu, On the partition function of a directed polymer in a Gaussian random environment, Probab. Theory Relat. Fields 124 (2002) 431–457.
[34] P. Carmona and Y. Hu, Fluctuation exponents and large deviations for directed polymers in a random environment, Stoch. Proc. Appl. 112 (2004) 285–308.
[35] D. Cheliotis and F. den Hollander, Variational characterization of the critical curve for pinning of random polymers, arXiv.org: 1005.3661v1 [math.PR], to appear in Ann. Probab.
[36] N. Clisby, R. Liang and G. Slade, Self-avoiding walk enumeration via the lace expansion. J. Phys. A: Math. Theor. 40 (2007) 10973–11017.
[37] F. Comets, T. Shiga and N. Yoshida, Directed polymers in random environment: Path localization and strong disorder, Bernoulli 9 (2003) 705–723.
[38] F. Comets, T. Shiga and N. Yoshida, Probabilistic analysis of directed polymers in a random environment: a review, in: *Stochastic Analysis on Large Scale Systems*, Adv. Stud. Pure Math. 39 (2004), pp. 115–142.
[39] F. Comets and V. Vargas, Majorizing multiplicative cascades for directed polymers in random media, Alea 2 (2006) 267–277.
[40] F. Comets and N. Yoshida, Brownian directed polymers in random environment, Commun. Math. Phys. 254 (2004) 257–287.
[41] F. Comets and N. Yoshida, Some new results on Brownian directed polymers in random environment, RIMS Kokyuroku 1386 (2004) 50–66.
[42] F. Comets and N. Yoshida, Directed polymers in random environment are diffusive at weak disorder, Ann. Probab. 34 (2006) 1746–1770.
[43] L.N. Coyle, A continuous time version of random walks in a random potential, Stoch. Proc. Appl. 64 (1996) 209–235.
[44] L.N. Coyle, Infinite moments of the partition function for random walks in a random potential, J. Math. Phys. 39 (1998) 2019–2034.
[45] D. Cule and T. Hwa, Denaturation of heterogeneous DNA, Phys. Rev. Lett. 79 (1997) 2375–2378.
[46] D.A. Dawson and J. Gärtner, Large deviations from the McKean-Vlasov limit for weakly interacting diffusions, Stochastics 20 (1987) 247–308.
[47] A. Dembo and O. Zeitouni, *Large Deviations Techniques and Applications*, Springer, 1998.

[48] B. Derrida, G. Giacomin, H. Lacoin and F.L. Toninelli, Fractional moment bounds and disorder relevance for pinning models, Commun. Math. Phys. 287 (2009) 867–887.
[49] J.-D. Deuschel, G. Giacomin and L. Zambotti, Scaling limits of equilibrium wetting models in (1+1)-dimension, Probab. Theory Relat. Fields 132 (2005) 471–500.
[50] H. Duminil-Copin and S. Smirnov, Conformal invariance of lattice models, in this volume.
[51] B. Duplantier and H. Saleur, Exact tricritical exponents for polymers at the FTHETA point in two dimensions, Phys. Rev. Lett. 59 (1987) 539–542.
[52] M.R. Evans and B. Derrida, Improved bounds for the transition temperature of directed polymers in a finite-dimensional random medium, J. Stat. Phys. 69 (1992) 427–437.
[53] M.E. Fisher, Walks, walls, wetting and melting, J. Stat. Phys. 34 (1984) 667–729.
[54] C. Garban and J.E. Steif, Noise-sensitivity and percolation, in this volume.
[55] G. Giacomin, *Random Polymer Models*, Imperial College Press, London, 2007.
[56] G. Giacomin, H. Lacoin and F.L. Toninelli, Marginal relevance of disorder for pinning models, Commun. Pure Appl. Math. 63 (2010) 233–265.
[57] G. Giacomin, H. Lacoin and F.L. Toninelli, Disorder relevance at marginality and critical point shift, Ann. Inst. H. Poincaré Probab. Stat. 47 (2011) 148–175.
[58] G. Giacomin and F.L. Toninelli, Estimates on path delocalization for copolymers at selective interfaces, Probab. Theory Relat. Fields 133 (2005) 464–482.
[59] G. Giacomin and F.L. Toninelli, Smoothing of depinning transitions for directed polymers with quenched disorder, Phys. Rev. Lett. 96 (2006) 070602.
[60] G. Giacomin and F.L. Toninelli, Smoothing effect of quenched disorder on polymer depinning transitions, Commun. Math. Phys. 266 (2006) 1–16.
[61] G. Giacomin and F.L. Toninelli, The localized phase of disordered copolymers with adsorption, Alea 1 (2006) 149–180.
[62] G. Giacomin and F.L. Toninelli, Force-induced depinning of directed polymers, J. Phys. A: Math. Gen. 40 (2007) 5261–5275.
[63] G. Giacomin and F.L. Toninelli, On the irrelevant disorder regime of pinning models, Ann. Probab. 37 (2009) 1841–1875.
[64] A.J. Guttmann (Ed.), *Polygons, Polyominoes and Polycubes*, Lecture Notes in Physics 775, Springer and Canopus Academic Publishing Ltd., 2009.
[65] T. Hara and G. Slade, Self-avoiding walk in five or more dimensions, I. The critical behaviour, Commun. Math. Phys. 147 (1992) 101–136.
[66] T. Hara and G. Slade, The lace expansion for self-avoiding walk in five or more dimensions, Rev. Math. Phys. 4 (1992) 235–327.
[67] R. van der Hofstad and A. Klenke, Self-attractive random polymers, Ann. Appl. Prob. 11 (2001) 1079–1115.
[68] R. van der Hofstad, A. Klenke and W. König, The critical attractive random polymer in dimension one, J. Stat. Phys. 106 (2002) 477–520.
[69] F. den Hollander, *Large Deviations*, Fields Institute Monographs, AMS, Providence RI, 2000.
[70] F. den Hollander, *Random Polymers*, Lecture Notes in Mathematics 1974, Springer, Berlin, 2009.
[71] F. den Hollander and N. Pétrélis, On the localized phase of a copolymer in an emulsion: supercritical percolation regime, Commun. Math. Phys. 285 (2009) 825–871.
[72] F. den Hollander and N. Pétrélis, On the localized phase of a copolymer in an emulsion: subcritical percolation regime, J. Stat. Phys. 134 (2009) 209–241.
[73] F. den Hollander and N. Pétrélis, A mathematical model for a copolymer in an emulsion, J. Math. Chem. 48 (2010) 83–94.
[74] F. den Hollander and N. Pétrélis, Free energy of a copolymer in a micro-emulsion, arXiv.org: 1204.1234 [math.PR], manuscript in preparation.
[75] F. den Hollander and S.G. Whittington, Localization transition for a copolymer in an emulsion, Theor. Prob. Appl. 51 (2006) 193–240.
[76] J.Z. Imbrie and T. Spencer, Diffusion of directed polymers in a random environment, J. Stat. Phys. 52 (1988) 609–626.
[77] D. Ioffe and Y. Velenik, The statistical mechanics of stretched polymers, Braz. J. Probab. Stat. 24 (2010) 279–299.
[78] D. Ioffe and Y. Velenik, Crossing random walks and stretched polymers at weak disorder, arXiv.org: 1002.4289v2 [math.PR], to appear in Ann. Probab.

[79] D. Ioffe and Y. Velenik, Stretched polymers in random environment, arXiv.org: 1011.0266v1 [math.PR].

[80] D. Ioffe and Y. Velenik, Self-attractive random walks: The case of critical drifts, arXiv.org: 1104.4615v1 [math.PR].

[81] I. Jensen, homepage (`www.ms.unimelb.edu.au/~iwan`).

[82] Y. Kafri, D. Mukamel and L. Peliti, Why is the DNA denaturation transition first order?, Phys. Rev. Lett. 85 (2000) 4988–4991.

[83] H. Lacoin, New bounds for the free energy of directed polymer in dimension $1+1$ and $1+2$, Commun. Math. Phys. 294 (2010) 471–503.

[84] H. Lacoin, The martingale approach to disorder relevance for pinning models, Electr. Comm. Probab. 15 (2010) 418–427.

[85] H. Lacoin, Influence of spatial correlation for directed polymers, Ann. Probab. 39 (2011) 139–175.

[86] M. Ledoux, *The Concentration of Measure Phenomenon*, Mathematical Surveys and Monographs 89, American Mathematical Society, 2001.

[87] O. Mejane, Upper bound of a volume exponent for directed polymers in a random environment, Ann. Inst. H. Poincaré Probab. Stat. 40 (2004) 299–308.

[88] M. Petermann, Superdiffusivity of directed polymers in random environment, Ph.D. thesis, University of Zürich, 2000.

[89] N. Pétrélis, *Localisation d'un Polymère en Interaction avec une Interface*, Ph.D. Thesis, University of Rouen, France, February 2, 2006.

[90] N. Pétrélis, Polymer pinning at an interface, Stoch. Proc. Appl. 116 (2006) 1600–1621.

[91] N. Pétrélis, Copolymer at selective interfaces and pinning potentials: weak coupling limits, Ann. Inst. H. Poincaré Probab. Stat. 45 (2009) 175–200.

[92] M.S.T. Piza, Directed polymers in a random environment: some results on fluctuations, J. Stat. Phys. 89 (1997) 581–603.

[93] R.D. Schram, G.T. Barkema and R.H. Bisseling, Exact enumeration of self-avoiding walks, arXiv.org: 1104.2184 [physics.math-ph].

[94] F. Seno and A.L. Stella, θ point of a linear polymer in 2 dimensions: a renormalization group analysis of Monte Carlo enumerations, J. Physique 49 (1988) 739–748.

[95] Ya.G. Sinai, A remark concerning random walks with random potentials, Fund. Math. 147 (1995) 173–180.

[96] J. Sohier, Finite size scaling for homogene¡ous pinning models, ALEA 6 (2009) 163–177.

[97] F. Spitzer, *Principles of Random Walk* (2nd. ed.), Springer, New York, 1976.

[98] J.M. Steele, Kingman's subadditive ergodic theorem, Ann. Inst. Henri Poincaré 25 (1989) 93–98.

[99] M.C. Tesi, E.J. Janse van Rensburg, E. Orlandini and S.G. Whittington, Monte Carlo study of the interacting self-avoiding walk model in three dimensions, J. Stat. Phys. 82 (1996) 155–181.

[100] F.L. Toninelli, A replica-coupling approach to disordered pinning models, Commun. Math. Phys. 280 (2008) 389–401.

[101] F.L. Toninelli, Disordered pinning models and copolymers: beyond annealed bounds, Ann. Appl. Probab. 18 (2008) 1569–1587.

[102] F.L. Toninelli, Coarse graining, fractional moments and the critical slope of random polymers, Electr. J. Probab. 14 (2009) 531–547.

[103] F.L. Toninelli, Localization transition in disordered pinning models. Effect of randomness on the critical properties, in: Lecture Notes in Mathematics 1970, Springer, Berlin, 2009, pp. 129-176.

[104] N. Zygouras, Lyapounov norms for random walks in low disorder and dimension greater than three, Probab. Theory Relat. Fields 143 (2009) 667–683.

[105] N. Zygouras, Strong disorder in semidirected random polymers, arXiv.org: 1009.2693v2 [math.PR].

Dipartimento di Matematica e Applicazioni, Università degli Studi di Milano - Bicocca, Via Cozzi 53, 20125 Milano, Italy
E-mail address: francesco.caravenna@unimib.it

Mathematical Institute, Leiden University, P.O. Box 9512, 2300 RA Leiden, The Netherlands
E-mail address: denholla@math.leidenuniv.nl

Laboratoire de Mathématiques Jean Leray, 2 rue de la Houssinière - BP 92208, F-44322 Nantes Cedex 3, France
E-mail address: petrelis@univ-nantes.fr

Clay Mathematics Proceedings
Volume 15, 2012

Lectures on Self-Avoiding Walks

Roland Bauerschmidt, Hugo Duminil-Copin, Jesse Goodman, and Gordon Slade

ABSTRACT. These lecture notes provide a rapid introduction to a number of rigorous results on self-avoiding walks, with emphasis on the critical behaviour. Following an introductory overview of the central problems, an account is given of the Hammersley–Welsh bound on the number of self-avoiding walks and its consequences for the growth rates of bridges and self-avoiding polygons. A detailed proof that the connective constant on the hexagonal lattice equals $\sqrt{2+\sqrt{2}}$ is then provided. The lace expansion for self-avoiding walks is described, and its use in understanding the critical behaviour in dimensions $d > 4$ is discussed. Functional integral representations of the self-avoiding walk model are discussed and developed, and their use in a renormalisation group analysis in dimension 4 is sketched. Problems and solutions from tutorials are included.

CONTENTS

Foreword

These notes are based on a course on Self-Avoiding Walks given in Búzios, Brazil, in August 2010, as part of the Clay Mathematics Institute Summer School and the XIV Brazilian Probability School. The course consisted of six lectures by

2010 *Mathematics Subject Classification.* Primary 82B41; Secondary 60K35.

Gordon Slade, a lecture by Hugo Duminil-Copin based on recent joint work with Stanislav Smirnov (see Section 3), and tutorials by Roland Bauerschmidt and Jesse Goodman. The written version of Slade's lectures was drafted by Bauerschmidt and Goodman, and the written version of Duminil-Copin's lecture was drafted by himself. The final manuscript was integrated and prepared jointly by the four authors.

1. Introduction and overview of the critical behaviour

These lecture notes focus on a number of rigorous results for self-avoiding walks on the d-dimensional integer lattice $\mathbb{Z}^d$. The model is defined by assigning equal probability to all paths of length n starting from the origin and without self-intersections. This family of probability measures is not consistent as n is varied, and thus does not define a stochastic process; the model is combinatorial in nature. The natural questions about self-avoiding walks concern the asymptotic behaviour as the length of the paths tends to infinity. Despite its simple definition, the self-avoiding walk is difficult to study in a mathematically rigorous manner. Many of the important problems remain unsolved, and the basic problems encompass many of the features and challenges of critical phenomena. This section gives the basic definitions and an overview of the critical behaviour.

1.1. Simple random walks. The basic reference model is *simple random walk* (SRW). Let $\Omega \subset \mathbb{Z}^d$ be the set of possible steps. The primary examples considered in these lectures are

$$\begin{aligned} &\text{the nearest-neighbour model:} && \Omega = \left\{x \in \mathbb{Z}^d : \|x\|_1 = 1\right\}, \\ &\text{the spread-out model:} && \Omega = \left\{x \in \mathbb{Z}^d : 0 < \|x\|_\infty \le L\right\}, \end{aligned} \tag{1.1}$$

where L is a fixed integer, usually large. An *n-step walk* is a sequence $\omega = (\omega(0), \omega(1), \ldots, \omega(n))$ with $\omega(j) - \omega(j-1) \in \Omega$ for $j = 1, \ldots, n$. The *n-step simple random walk* is the uniform measure on n-step walks. We define the sets

$$\mathcal{W}_n(0,x) = \{\omega : \omega \text{ is an } n\text{-step walk with } \omega(0) = 0 \text{ and } \omega(n) = x\} \tag{1.2}$$

and

$$\mathcal{W}_n = \bigcup_{x \in \mathbb{Z}^d} \mathcal{W}_n(0,x). \tag{1.3}$$

1.2. Self-avoiding walks. The *weakly self-avoiding walk* and the *strictly self-avoiding walk* (the latter also called simply *self-avoiding walk*) are the main subjects of these notes. These are random paths on $\mathbb{Z}^d$, defined as follows. Given an n-step walk $\omega \in \mathcal{W}_n$, and integers s, t with $0 \le s < t \le n$, let

$$U_{st} = U_{st}(\omega) = -1_{\{\omega(s) = \omega(t)\}} = \begin{cases} -1 & \text{if } \omega(s) = \omega(t), \\ 0 & \text{if } \omega(s) \ne \omega(t). \end{cases} \tag{1.4}$$

Fix $\lambda \in [0,1]$. We assign to each path $\omega \in \mathcal{W}_n$ the weighting factor

$$\prod_{0 \le s < t \le n} (1 + \lambda U_{st}(\omega)). \tag{1.5}$$

The weights can also be expressed as Boltzmann weights:

$$\prod_{0 \le s < t \le n} (1 + \lambda U_{st}(\omega)) = \exp\Big(-g \sum_{0 \le s < t \le n} 1_{\{\omega(s) = \omega(t)\}}\Big) \tag{1.6}$$

with $g = -\log(1-\lambda) \in [0,\infty)$ for $\lambda \in [0,1)$. Making the convention $\infty \cdot 0 = 0$, the case $\lambda = 1$ corresponds to $g = \infty$.

The choice $\lambda = 0$ assigns equal weight to all walks in $\mathcal{W}_n$; this is the case of the simple random walk. For $\lambda \in (0,1)$, self-intersections are penalised but not forbidden, and the model is called the *weakly self-avoiding walk*. The choice $\lambda = 1$ prevents any return to a previously visited site, and defines the *self-avoiding walk* (SAW). More precisely, an n-step walk ω is a self-avoiding walk if and only if the expression (1.5) is non-zero for $\lambda = 1$, which happens if and only if ω visits each site at most once, and for such walks the weight equals 1.

These weights give rise to associated partition sums $c_n^{(\lambda)}(x)$ and $c_n^{(\lambda)}$ for walks in $\mathcal{W}_n(0,x)$ and $\mathcal{W}_n$, respectively:

$$c_n^{(\lambda)}(x) = \sum_{\omega \in \mathcal{W}_n(0,x)} \prod_{0 \le s < t \le n} (1 + \lambda U_{st}(\omega)), \quad c_n^{(\lambda)} = \sum_{x \in \mathbb{Z}^d} c_n^{(\lambda)}(x). \tag{1.7}$$

In the case $\lambda = 1$, $c_n^{(1)}(x)$ counts the number of self-avoiding walks of length n ending at x, and $c_n^{(1)}$ counts all n-step self-avoiding walks. The case $\lambda = 0$ reverts to simple random walk, for which $c_n^{(0)} = |\Omega|^n$. When $\lambda = 1$ we will often drop the superscript (1) and write simple c_n instead of $c_n^{(1)}$.

We also define probability measures $\mathbb{Q}_n^{(\lambda)}$ on $\mathcal{W}_n$ with expectations $\mathbb{E}_n^{(\lambda)}$:

$$\mathbb{Q}_n^{(\lambda)}(A) = \frac{1}{c_n^{(\lambda)}} \sum_{\omega \in A} \prod_{0 \le s < t \le n} (1 + \lambda U_{st}(\omega)) \quad (A \subset \mathcal{W}_n), \tag{1.8}$$

$$\mathbb{E}_n^{(\lambda)}(X) = \frac{1}{c_n^{(\lambda)}} \sum_{\omega \in \mathcal{W}_n} X(\omega) \prod_{0 \le s < t \le n} (1 + \lambda U_{st}(\omega)) \quad (X : \mathcal{W}_n \to \mathbb{R}). \tag{1.9}$$

The measures $\mathbb{Q}_n^{(\lambda)}$ define the weakly self-avoiding walk when $\lambda \in (0,1)$ and the strictly self-avoiding walk when $\lambda = 1$. Occasionally we will also consider self-avoiding walks that do not begin at the origin.

1.3. Subadditivity and the connective constant. The sequence $c_n^{(\lambda)}$ has the following submultiplicativity property:

$$c_{n+m}^{(\lambda)} \le \sum_{\omega \in \mathcal{W}_{n+m}} \prod_{0 \le s < t \le n} (1 + \lambda U_{st}) \prod_{n \le s' < t' \le n+m} (1 + \lambda U_{s't'}) \le c_n^{(\lambda)} c_m^{(\lambda)}. \tag{1.10}$$

Therefore, $\log c_n^{(\lambda)}$ is a *subadditive* sequence: $\log c_{n+m}^{(\lambda)} \le \log c_n^{(\lambda)} + \log c_m^{(\lambda)}$.

LEMMA 1.1. *If $a_1, a_2, \ldots \in \mathbb{R}$ obey $a_{n+m} \le a_n + a_m$ for every n, m, then*

$$\lim_{n \to \infty} \frac{a_n}{n} = \inf_{n \ge 1} \frac{a_n}{n} \in [-\infty, \infty). \tag{1.11}$$

PROOF. See Problem 1.1. The value $-\infty$ is possible, e.g., for the sequence $a_n = -n^2$. □

Applying Lemma 1.1 to $c_n^{(\lambda)}$ gives the existence of μ_λ such that $\lim \frac{1}{n} \log c_n^{(\lambda)} = \log \mu_\lambda \le \frac{1}{n} \log c_n^{(\lambda)}$ for all n, i.e.,

$$\mu_\lambda = \lim_{n \to \infty} (c_n^{(\lambda)})^{1/n} \text{ exists, and} \quad c_n^{(\lambda)} \ge \mu_\lambda^n \quad \text{for all } n. \tag{1.12}$$

In the special case $\lambda = 1$, we write simply $\mu = \mu_1$. This μ, which depends on d (and also on L for the spread-out model), is called the *connective constant*. For the nearest-neighbour model, by counting only walks that move in positive coordinate directions, and by counting walks that are restricted only to prevent immediate reversals of steps, we obtain

$$d^n \leq c_n \leq 2d(2d-1)^{n-1} \quad \text{which implies} \quad d \leq \mu \leq 2d-1. \tag{1.13}$$

For $d = 2$, the following rigorous bounds are known:

$$\mu \in [2.625\,622, 2.679\,193]. \tag{1.14}$$

The lower bound is due to Jensen [**47**] via bridge enumeration (bridges are defined in Section 2.1 below), and the upper bound is due to Pönitz and Tittmann [**64**] by comparison with finite-memory walks. The estimate

$$\mu = 2.638\,158\,530\,31(3) \tag{1.15}$$

is given in [**45**]; here the 3 in parentheses represents the subjective error in the last digit. It has been observed that $1/\mu$ is well approximated by the smallest positive root of $581x^4 + 7x^2 - 13 = 0$ [**23, 48**], though no derivation or explanation of this quartic polynomial is known, and later evidence has raised doubts about its validity [**45**].

Even though the definition of self-avoiding walks has been restricted to the graph $\mathbb{Z}^d$ thus far, it applies more generally. In 1982, arguments based on a Coulomb gas formalism led Nienhuis [**61**] to predict that on the hexagonal lattice the connective constant is equal to $\sqrt{2+\sqrt{2}}$. This was very recently proved by Duminil-Copin and Smirnov [**24**], whose theorem is the following.

THEOREM 1.2. *The connective constant for the hexagonal lattice is*

$$\mu = \sqrt{2+\sqrt{2}}. \tag{1.16}$$

The proof of Theorem 1.2 is presented in Section 3 below. Except for trivial cases, this is the only lattice for which the connective constant is known explicitly.

Returning to $\mathbb{Z}^d$, in 1963, Kesten [**50**] proved that

$$\lim_{n\to\infty} \frac{c_{n+2}}{c_n} = \mu^2, \tag{1.17}$$

but it remains an open problem (for $d = 2, 3, 4$) to prove that

$$\lim_{n\to\infty} \frac{c_{n+1}}{c_n} = \mu. \tag{1.18}$$

Even the proof of $c_{n+1} \geq c_n$ is a non-trivial result, proved by O'Brien [**62**], though it is not hard to show that $c_{n+2} \geq c_n$.

1.4. $1/d$ **expansion.** It was proved by Hara and Slade [**35**] that the connective constant $\mu(d)$ for $\mathbb{Z}^d$ (with nearest-neighbour steps) has an asymptotic expansion in powers of $1/2d$ as $d \to \infty$: There exist integers $a_i \in \mathbb{Z}$, $i = -1, 0, 1, \ldots$ such that

$$\mu(d) \sim \sum_{i=-1}^{\infty} \frac{a_i}{(2d)^i} \tag{1.19}$$

in the sense that $\mu(d) = a_{-1}(2d) + a_0 + \cdots + a_{M-1}(2d)^{-(M-1)} + O(d^{-M})$, for each fixed M. In Problem 5.1 below, the first three terms are computed. The constant in the $O(d^{-M})$ term may depend on M. It is expected, though not proved, that the asymptotic series in (1.19) has radius of convergence 0, so that the right-hand side of (1.19) diverges for each fixed d. The values of a_i are known for $i = -1, 0, \ldots, 11$ and grow rapidly in magnitude; see Clisby, Liang, and Slade [**21**].

Graham [**26**] has proved Borel-type error bounds for the asymptotic expansion of $z_c = z_c(d) = \mu^{-1}$. Namely, writing the asymptotic expansion of z_c as $\sum_{i=1}^{\infty} \alpha_i (2d)^{-i}$, there is a constant C, independent of d and M, such that for each M and for all $d \geq 1$,

$$\left| z_c - \sum_{i=1}^{M-1} \frac{\alpha_i}{(2d)^i} \right| \leq \frac{C^M M!}{(2d)^M}. \tag{1.20}$$

An extension of (1.20) to *complex* values of the dimension d would be needed in order to apply the method of Borel summation to recover the value of z_c, and hence of $\mu(d)$, from the asymptotic series.

1.5. Critical exponents. It is a characteristic feature of models of statistical mechanics at the critical point that there exist *critical exponents* which describe the asymptotic behaviour on the large scale. It is a deep conjecture, not yet properly understood mathematically, that these critical exponents are *universal*, meaning that they depend only on the spatial dimension of the system, but not on details such as the specific lattice in $\mathbb{R}^d$. For the case of the self-avoiding walk, this conjecture of universality extends to lack of dependence on the constant λ, as soon as $\lambda > 0$. We now introduce the critical exponents, and in Section 1.6 we will discuss what is known about them in more detail.

1.5.1. *Number of self-avoiding walks.* It is predicted that for each d there is a constant γ such that for all $\lambda \in (0, 1]$, and for both the nearest-neighbour and spread-out models,

$$c_n^{(\lambda)} \sim A_\lambda \mu_\lambda^n n^{\gamma - 1}. \tag{1.21}$$

Here $f(n) \sim g(n)$ means $\lim_{n\to\infty} f(n)/g(n) = 1$. The predicted values of the critical exponent γ are:

$$\gamma = \begin{cases} 1 & d = 1, \\ \frac{43}{32} & d = 2, \\ 1.16\ldots & d = 3, \\ 1 & d = 4, \\ 1 & d \geq 5. \end{cases} \tag{1.22}$$

In fact, for $d = 4$, the prediction involves a logarithmic correction:

$$c_n^{(\lambda)} \sim A_\lambda \mu_\lambda^n (\log n)^{1/4}. \tag{1.23}$$

This situation should be compared with simple random walk, for which $c_n^{(0)} = |\Omega|^n$, so that μ_0 is equal to the degree $|\Omega|$ of the lattice, and $\gamma = 1$.

In the case of the self-avoiding walk (i.e., $\lambda = 1$), γ has a probabilistic interpretation. Sampling independently from two n-step self-avoiding walks uniformly,

$$\mathbb{P}(\omega_1 \cap \omega_2 = \{0\}) = \frac{c_{2n}}{c_n^2} \sim \text{const} \frac{1}{n^{\gamma-1}}, \tag{1.24}$$

so γ is a measure of how likely it is for two self-avoiding walks to avoid each other. The analogous question for SRW is discussed in [**53**].

Despite the precision of the prediction (1.21), the best rigorously known bounds in dimension $d = 2, 3, 4$ are very far from tight and almost 50 years old. In [**29**], Hammersley and Welsh proved that, for all $d \geq 2$,

$$\mu^n \leq c_n \leq \mu^n e^{\kappa\sqrt{n}} \tag{1.25}$$

(the lower bound is just subadditivity, the upper bound is nontrivial). This was improved slightly by Kesten [**50**], who showed that for $d = 3, 4, \ldots$,

$$\mu^n \leq c_n \leq \mu^n \exp\left(\kappa n^{2/(d+2)} \log n\right). \tag{1.26}$$

The proof of the Hammersley–Welsh bound is the subject of Section 2.1.

1.5.2. *Mean-square displacement.* Let $|x|$ denote the Euclidean norm of $x \in \mathbb{R}^d$. It is predicted that for $\lambda \in (0, 1]$, and for both the nearest-neighbour and spread-out models,

$$\mathbb{E}_n^{(\lambda)} |\omega(n)|^2 \sim D_\lambda n^{2\nu}, \tag{1.27}$$

with

$$\nu = \begin{cases} 1 & d = 1, \\ \frac{3}{4} & d = 2, \\ 0.588\ldots & d = 3, \\ \frac{1}{2} & d = 4, \\ \frac{1}{2} & d \geq 5. \end{cases} \tag{1.28}$$

Again, a logarithmic correction is predicted for $d = 4$:

$$\mathbb{E}_n^{(\lambda)} |\omega(n)|^2 \sim D_\lambda n (\log n)^{1/4}. \tag{1.29}$$

This should be compared with the SRW, for which $\nu = \frac{1}{2}$ in all dimensions.

Almost nothing is known rigorously about ν in dimensions 2, 3, 4. It is an open problem to show that the mean-square displacement grows at least as rapidly as simple random walk, and grows more slowly than ballistically, i.e., it has not been proved that

$$cn \leq \mathbb{E}_n^{(1)} |\omega(n)|^2 \leq Cn^{2-\epsilon}, \tag{1.30}$$

or even that the endpoint is typically as far away as the surface of a ball of volume n, i.e., $cn^{2/d} \leq \mathbb{E}_n^{(1)} |\omega(n)|^2$. Madras (unpublished) has shown $\mathbb{E}_n^{(1)} |\omega(n)|^2 \geq cn^{4/3d}$.

1.5.3. *Two-point function and susceptibility.* The two-point function is defined by

$$G_z^{(\lambda)}(x) = \sum_{n=0}^{\infty} c_n^{(\lambda)}(x) z^n, \tag{1.31}$$

and the susceptibility by

$$\chi^{(\lambda)}(z) = \sum_{x \in \mathbb{Z}^d} G_z^{(\lambda)}(x) = \sum_{n=0}^{\infty} c_n^{(\lambda)} z^n. \tag{1.32}$$

Since $\chi^{(\lambda)}$ is a power series whose coefficients satisfy (1.12), its radius of convergence $z_c^{(\lambda)}$ is given by $z_c^{(\lambda)} = \mu_\lambda^{-1}$. The value $z_c^{(\lambda)}$ is referred to as the *critical point.*

PROPOSITION 1.3. *Fix $\lambda \in [0, 1], z \in (0, z_c^{(\lambda)})$. Then $G_z^{(\lambda)}(x)$ decays exponentially in x.*

PROOF. For simplicity, we consider only the nearest-neighbour model, and we omit λ from the notation. Since $c_n(x) = 0$ if $n < \|x\|_1$,

$$G_z(x) = \sum_{n=\|x\|_1}^{\infty} c_n(x) z^n \le \sum_{n=\|x\|_1}^{\infty} c_n z^n. \tag{1.33}$$

Fix $z < z_c = 1/\mu$ and choose $\epsilon > 0$ such that $z(\mu + \epsilon) < 1$. Since $c_n^{1/n} \to \mu$, there exists $K = K(\epsilon)$ such that $c_n \le K(\mu+\epsilon)^n$ for all n. Hence

$$G_z(x) \le K \sum_{n=\|x\|_1}^{\infty} (z(\mu+\epsilon))^n \le K'(z(\mu+\epsilon))^{\|x\|_1}, \tag{1.34}$$

as claimed. □

We restrict temporarily to $\lambda = 1$. Much is known about $G_z(x)$ for $z < z_c$: there is a norm $|\cdot|_z$ on $\mathbb{R}^d$, satisfying $\|u\|_\infty \le |u|_z \le \|u\|_1$ for all $u \in \mathbb{R}^d$, such that $m(z) = \lim_{|x|_z \to \infty} -\frac{\log G_z(x)}{|x|_z}$ exists and is finite. The *correlation length* is defined by $\xi(z) = 1/m(z)$, and hence approximately

$$G_z(x) \approx e^{-|x|_z/\xi(z)}. \tag{1.35}$$

Indeed, more precise asymptotics (Ornstein–Zernike decay) are known [**17, 57, 15**]:

$$G_z(x) \sim \frac{c}{|x|_z^{(d-1)/2}} e^{-|x|_z/\xi(z)} \quad \text{as } x \to \infty, \tag{1.36}$$

and the arguments leading to this also prove that

$$\lim_{z \nearrow z_c} \xi(z) = \infty. \tag{1.37}$$

As a refinement of (1.37), it is predicted that as $z \nearrow z_c$,

$$\xi(z) \sim \text{const} \left(1 - \frac{z}{z_c}\right)^{-\nu}, \tag{1.38}$$

and that, in addition, as $|x| \to \infty$ (for $d \ge 2$),

$$G_{z_c}(x) \sim \frac{\text{const}}{|x|^{d-2+\eta}}. \tag{1.39}$$

The exponents γ, η and ν are predicted to be related to each other via *Fisher's relation* (see, e.g., [**57**]):

$$\gamma = (2-\eta)\nu. \tag{1.40}$$

There is typically a correspondence between the asymptotic growth of the coefficients in a generating function and the behaviour of the generating function near its dominant singularity. For our purpose we note that, under suitable hypotheses,

$$a_n \sim \frac{n^{\gamma-1}}{R^n} \text{ as } n \to \infty \quad \overset{\approx}{\Longleftrightarrow} \quad \sum_n a_n z^n \sim \frac{C}{(1-z/R)^\gamma} \text{ as } z \nearrow R. \tag{1.41}$$

The easier $\Longrightarrow$ direction is known as an Abelian theorem, and the more delicate $\Longleftarrow$ direction is known as a Tauberian theorem [**36**]. With this in mind, our earlier prediction for $c_n^{(\lambda)}$ for $\lambda \in (0,1]$ corresponds to:

$$\chi^{(\lambda)}(z) \sim \frac{\text{const}_\lambda}{(1-z/z_c)^\gamma} \tag{1.42}$$

as $z \nearrow z_c$, with an additional factor $|\log(1 - z/z_c)|^{1/4}$ on the right-hand side when $d = 4$.

1.6. Effect of the dimension. Universality asserts that self-avoiding walks on different lattices in a fixed dimension d should behave in the same way, independently of the fine details of how the model is defined. However, the behaviour does depend very strongly on the dimension.

1.6.1. $d = 1$. For the nearest-neighbour model with $\lambda = 1$ it is a triviality that $c_n^{(1)} = 2$ for all $n \geq 1$ and $|\omega(n)| = n$ for all ω, since a self-avoiding walk must continue either in the negative or in the positive direction. Any configuration $\omega \in \mathcal{W}_n$ is possible when $\lambda \in (0, 1)$, however, and it is by no means trivial to prove that the critical behaviour when $\lambda \in (0, 1)$ is similar to the case of $\lambda = 1$. The following theorem of König [**52**] (extending a result of Greven and den Hollander [**27**]) proves that the weakly self-avoiding walk measure (1.8) does have ballistic behaviour for all $\lambda \in (0, 1)$.

THEOREM 1.4. *Let $d = 1$. For each $\lambda \in (0, 1)$, there exist $\theta(\lambda) \in (0, 1)$ and $\sigma(\lambda) \in (0, \infty)$ such that for all $u \in \mathbb{R}$,*

$$\lim_{n\to\infty} \mathbb{Q}_n^{(\lambda)}\left(\frac{|\omega(n)| - n\theta}{\sigma\sqrt{n}} \leq u\right) = \int_{-\infty}^{u} \frac{e^{-t^2/2}}{\sqrt{2\pi}}\, dt. \tag{1.43}$$

A similar result is proved in [**52**] for the 1-dimensional spread-out strictly self-avoiding walk. The result of Theorem 1.4 should be contrasted to the case $\lambda = 0$, which has diffusive rather than ballistic behaviour. It remains an open problem to prove the intuitively appealing statement that θ should be an increasing function of λ. A review of results for $d = 1$ is given in [**40**].

1.6.2. $d = 2$. Based on non-rigorous Coulomb gas methods, Nienhuis [**61**] predicted that $\gamma = \frac{43}{32}$, $\nu = \frac{3}{4}$. These predicted values have been confirmed numerically by Monte Carlo simulation, e.g., [**55**], and exact enumeration of self-avoiding walks up to length $n = 71$ [**46**].

Lawler, Schramm, and Werner [**54**] have given major mathematical support to these predictions. Roughly speaking, they show that if self-avoiding walk has a scaling limit, and if this scaling limit has a certain conformal invariance property, then the scaling limit must be $\mathrm{SLE}_{8/3}$ (the Schramm–Loewner evolution with parameter $\kappa = \frac{8}{3}$). The values of γ and ν are then recovered from an $\mathrm{SLE}_{8/3}$ computation. Numerical evidence supporting the statement that the scaling limit is $\mathrm{SLE}_{8/3}$ is given in [**49**]. However, until now, it remains an open problem to prove the required existence and conformal invariance of the scaling limit.

The result of [**54**] is discussed in greater detail in the course of Vincent Beffara [**1**]. Here, we describe it only briefly, as follows. Consider a simply connected domain Ω in the complex plane $\mathbb{C}$ with two points a and b on the boundary. Fix $\delta > 0$, and let $(\Omega_\delta, a_\delta, b_\delta)$ be a discrete approximation of (Ω, a, b) in the following sense: Ω_δ is the largest finite domain of $\delta\mathbb{Z}^2$ included in Ω, a_δ and b_δ are the closest vertices of $\delta\mathbb{Z}^2$ to a and b respectively. When δ goes to 0, this provides an approximation of the domain.

For fixed $z, \delta > 0$, there is a probability measure on the set of self-avoiding walks ω between a_δ and b_δ that remain in Ω_δ by assigning to ω a *Boltzmann weight* proportional to $z^{\ell(\omega)}$, where $\ell(\omega)$ denotes the length of ω. We obtain a random piecewise linear curve, denoted by ω_δ.

It is possible to prove that when $z < z_c = 1/\mu$, walks are penalised so much with respect to their length that ω_δ becomes straight when δ goes to 0; this is closely related to the Ornstein–Zernike decay results. On the other hand, it is expected that, when $z > z_c$, the entropy wins against the penalisation and ω_δ becomes space filling when δ tends to 0. Finally, when $z = z_c$, the sequence of measures conjecturally converges to a random continuous curve. It is for this case that we have the following conjecture of Lawler, Schramm and Werner [**54**].

CONJECTURE 1.5. *For $z = z_c$, the random curve ω_δ converges to* $\mathrm{SLE}_{8/3}$ *from a and b in the domain Ω.*

It remains a major open problem in 2-dimensional statistical mechanics to prove the conjecture.

1.6.3. $d = 3$. For $d = 3$, there are no rigorous results for critical exponents, and no mathematically well-defined candidate has been proposed for the scaling limit. An early prediction for the values of ν, referred to as the Flory values [**25**], was $\nu = \frac{3}{d+2}$ for $1 \leq d \leq 4$. This does give the correct answer for $d = 1, 2, 4$, but it is not quite accurate for $d = 3$—the Flory argument is very remote from a rigorous mathematical proof. Flory's interest in the problem was motivated by the use of SAWs to model polymer molecules; this application is discussed in detail in the course of Frank den Hollander [**42**] (see also [**43**]).

For $d = 3$, there are three methods to compute the exponents approximately. In one method, non-rigorous field theory computations in theoretical physics [**28**] combine the $n \to 0$ limit for the $O(n)$ model with an expansion in $\epsilon = 4 - d$ about dimension $d = 4$, with $\epsilon = 1$. Secondly, Monte Carlo studies have been carried out with walks of length 33,000,000 [**20**], using the pivot algorithm [**58, 44**]. Finally, exact enumeration plus series analysis has been used; currently the most extensive enumerations in dimensions $d \geq 3$ use the lace expansion [**21**], and for $d = 3$ walks have been enumerated to length $n = 30$. The exact enumeration estimates for $d = 3$ are $\mu = 4.684043(12)$, $\gamma = 1.1568(8)$, $\nu = 0.5876(5)$ [**21**]. Monte Carlo estimates are consistent with these values: $\gamma = 1.1575(6)$ [**16**] and $\nu = 0.587597(7)$ [**20**].

1.6.4. $d = 4$. Four dimensions is the *upper critical dimension* for the self-avoiding walk. This term encapsulates the notion that for $d > 4$ self-avoiding walk has the same critical behaviour as simple random walk, while for $d < 4$ it does not. The dimension 4 can be guessed by considering the fractal properties of the simple random walk: for $d \geq 2$, the path of a simple random walk is two-dimensional. If $d > 4$, two independent two-dimensional objects should generically not intersect, so that the effect of self-interaction between the past and the future of a simple random walk should be negligible. In $d = 4$, the expected number of intersections between two independent random walks tends to infinity, but only logarithmically in the length. Such considerations are related to the logarithmic corrections that appear in (1.23) and (1.29).

The existence of logarithmic corrections to scaling has been proved for models of weakly self-avoiding walk on a 4-dimensional *hierarchical* lattice, using rigorous renormalisation group methods [**5, 9, 10, 32**]. The hierarchical lattice is a simplification of the hypercubic lattice $\mathbb{Z}^4$ which is particularly amenable to the renormalisation group approach. Recently there has been progress in the application of renormalisation group methods to a continuous-time weakly self-avoiding walk model on $\mathbb{Z}^4$ itself, and in particular it has been proved in this context that the critical two-point function has $|x|^{-2}$ decay [**12**], which is a statement that the critical exponent η is equal to 0. This is the topic of Section 7 below.

1.6.5. $d \geq 5$. Using the lace expansion, it has been proved that for the nearest-neighbour model in dimensions $d \geq 5$ the critical exponents exist and take their so-called *mean field* values $\gamma = 1$, $\nu = \frac{1}{2}$ [**34, 33**] and $\eta = 0$ [**30**], and that the scaling limit is Brownian motion [**33**]. The lace expansion for self-avoiding walks is discussed in Section 4, and its application to prove simple random walk behaviour in dimensions $d \geq 5$ is discussed in Section 5.

1.7. Tutorial.

PROBLEM 1.1. Let (a_n) be a real-valued sequence that is subadditive, that is, $a_{n+m} \leq a_n + a_m$ holds for all n, m. Prove that $\lim_{n\to\infty} n^{-1}a_n$ exists in $[-\infty, \infty)$ and equals $\inf_n n^{-1}a_n$.

PROBLEM 1.2. Prove that the connective constant μ for the nearest-neighbour model on the square lattice $\mathbb{Z}^2$ obeys the strict inequalities $2 < \mu < 3$.

PROBLEM 1.3. A family of probability measures $(\mathbb{P}_n)$ on $\mathcal{W}_n$ is called consistent if $\mathbb{P}_n(\omega) = \sum_{\rho>\omega} \mathbb{P}_m(\rho)$ for all $m > n$ and for all $\omega \in \mathcal{W}_n$, where the sum is over all $\rho \in \mathcal{W}_m$ whose first n steps agree with ω. Show that $\mathbb{Q}_n^{(1)}$, the uniform measure on SAWs, does not provide a consistent family.

PROBLEM 1.4. Show that the Fourier transform of the two-point function of the 1-dimensional strictly self-avoiding walk is given by

$$\hat{G}_z(k) = \frac{1 - z^2}{1 + z^2 - 2z\cos k}. \tag{1.44}$$

Here $\hat{f}(k) = \sum_{x\in\mathbb{Z}^d} f(x)e^{ik\cdot x}$.

PROBLEM 1.5. Suppose that $f(z) = \sum_{n=0}^{\infty} a_n z^n$ has radius of convergence 1. Suppose that $|f(z)| \leq c|1-z|^{-b}$ uniformly in $|z| < 1$, with $b \geq 1$. Prove that, for some constant C, $|a_n| \leq Cn^{b-1}$ if $b > 1$, and that $|a_n| \leq C\log n$ if $b = 1$. Hint:

$$a_n = \frac{1}{2\pi i}\oint_{\Gamma_n} \frac{f(z)}{z^{n+1}}\, dz, \tag{1.45}$$

where $\Gamma_n = \{z \in \mathbb{C} : |z| = 1 - \frac{1}{n}\}$.

PROBLEM 1.6. Consider the nearest-neighbour simple random walk $(X_n)_{n\geq 0}$ on $\mathbb{Z}^d$ started at the origin. Let $D(x) = (2d)^{-1}1_{\{\|x\|_1=1\}}$ denote its step distribution. The two-point function for simple random walk is defined by

$$C_z(x) = \sum_{n\geq 0} c_n^{(0)}(x) z^n = \sum_{n\geq 0} D^{*n}(x)(2dz)^n, \tag{1.46}$$

where D^{*n} denotes the n-fold convolution of D with itself.

(a) Let u denote the probability that the walk ever returns to the origin. The walk is recurrent if $u = 1$ and transient if $u < 1$. Let N denote the random number of visits to the origin, including the initial visit at time 0, and let $m = \mathbb{E}(N)$. Show that $m = (1-u)^{-1}$; so the walk is recurrent if and only if $m = \infty$.

(b) Show that

$$m = \sum_{n \geq 0} \mathbb{P}(X_n = 0) = \int_{[-\pi,\pi]^d} \frac{1}{1 - \hat{D}(k)} \frac{d^d k}{(2\pi)^d}. \tag{1.47}$$

Thus transience is characterised by the integrability of $\hat{C}_{z_0}(k)$, where $z_0 = (2d)^{-1}$.

(c) Show that the walk is recurrent in dimensions $d \leq 2$ and transient for $d > 2$.

PROBLEM 1.7. Let $X^1 = (X^1_i)_{i \geq 0}$ and $X^2 = (X^2_i)_{i \geq 0}$ be two independent nearest-neighbour simple random walks on $\mathbb{Z}^d$ started at the origin, and let

$$I = \sum_{i \geq 0} \sum_{j \geq 0} 1_{\{X^1_i = X^2_j\}} \tag{1.48}$$

be the random number of intersections of the two walks. Show that

$$\mathbb{E}(I) = \int_{[-\pi,\pi]^d} \frac{1}{[1 - \hat{D}(k)]^2} \frac{d^d k}{(2\pi)^d}. \tag{1.49}$$

Thus $\mathbb{E}(I)$ is finite if and only if $\hat{C}_{z_0}$ is square integrable. Conclude that the expected number of intersections is finite if $d > 4$ and infinite if $d \leq 4$.

2. Bridges and polygons

Throughout this section, we consider only the nearest-neighbour strictly self-avoiding walk on $\mathbb{Z}^d$. We will introduce a class of self-avoiding walks called bridges, and will show that the number of bridges grows with the same exponential rate as the number of self-avoiding walks, namely as μ^n. The analogous fact for the hexagonal lattice $\mathbb{H}$ will be used in Section 3 as an ingredient in the proof that the connective constant for $\mathbb{H}$ is $\sqrt{2 + \sqrt{2}}$. The study of bridges will also lead to the proof of the Hammersley–Welsh bound (1.25) on c_n. Finally, we will study self-avoiding polygons, and show that they too grow in number as μ^n.

2.1. Bridges and the Hammersley–Welsh bound.

For a self-avoiding walk ω, denote by $\omega_1(i)$ the first spatial coordinate of $\omega(i)$.

DEFINITION 2.1. An n-step *bridge* is an n-step SAW ω such that

$$\omega_1(0) < \omega_1(i) \leq \omega_1(n) \qquad \text{for } i = 1, 2, \ldots, n. \tag{2.1}$$

Let b_n be the number of n-step bridges with $\omega(0) = 0$ for $n > 1$, and $b_0 = 1$.

While the number of self-avoiding walks is a *sub*multiplicative sequence, the number of bridges is *super*multiplicative:

$$b_{n+m} \geq b_n b_m. \tag{2.2}$$

Thus, applying Lemma 1.1 to $-\log b_n$, we obtain the existence of the bridge growth constant μ_{Bridge} defined by

$$\mu_{\text{Bridge}} = \lim_{n\to\infty} b_n^{1/n} = \sup_{n \geq 1} b_n^{1/n}. \tag{2.3}$$

Using the trivial inequality $\mu_{\text{Bridge}} \leq \mu$ we conclude that

$$b_n \leq \mu^n_{\text{Bridge}} \leq \mu^n. \tag{2.4}$$

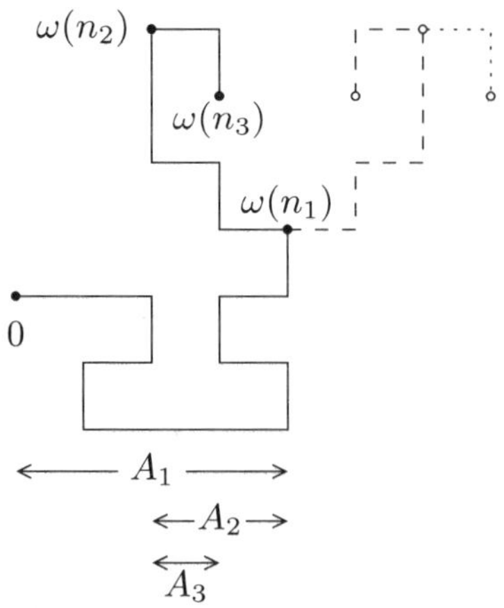

FIGURE 1. A half-space walk is decomposed into bridges, which are reflected to form a single bridge.

DEFINITION 2.2. An n-step *half-space walk* is an n-step SAW ω with

$$\omega_1(0) < \omega_1(i) \qquad \text{for } i = 1, 2, \ldots, n. \tag{2.5}$$

Let $h_0 = 1$, and for $n \geq 1$, let h_n denote the number of n-step half-space walks with $\omega(0) = 0$.

DEFINITION 2.3. The *span* of an n-step SAW ω is

$$\max_{0 \leq i \leq n} \omega_1(i) - \min_{0 \leq i \leq n} \omega_1(i). \tag{2.6}$$

Let $b_{n,A}$ be the number of n-step bridges with span A.

We will use the following result on integer partitions which dates back to 1917, due to Hardy and Ramanujan [**37**].

THEOREM 2.4. *For an integer $A \geq 1$, let $P_D(A)$ denote the number of ways of writing $A = A_1 + \cdots + A_k$ with $A_1 > \cdots > A_k \geq 1$, for any $k \geq 1$. Then*

$$\log P_D(A) \sim \pi \left(\frac{A}{3}\right)^{1/2} \tag{2.7}$$

as $A \to \infty$.

PROPOSITION 2.5. *$h_n \leq P_D(n) b_n$ for all $n \geq 1$.*

PROOF. Set $n_0 = 0$ and inductively define

$$A_{i+1} = \max_{j > n_i} (-1)^i (\omega_1(j) - \omega_1(n_i)) \tag{2.8}$$

and

$$n_{i+1} = \max \left\{ j > n_i : (-1)^i (\omega_1(j) - \omega_1(n_i)) = A_{i+1} \right\}. \tag{2.9}$$

In words, $j = n_1$ maximises $\omega_1(j)$, $j = n_2$ minimises $\omega_1(j)$ for $j > n_1$, n_3 maximises $\omega_1(j)$ for $j > n_2$, and so on in an alternating pattern. In addition $A_1 = \omega_1(n_1) - \omega_1(n_0)$, $A_2 = \omega_1(n_1) - \omega_1(n_2)$ and so on. Moreover, the n_i are chosen to be the last times these extrema are attained.

This procedure stops at some step $K \geq 1$ when $n_K = n$. Since the n_i are chosen maximal, it follows that $A_{i+1} < A_i$. Note that $K = 1$ if and only if ω is a bridge, and in that case A_1 is the span of ω. Let $h_n[a_1, \ldots, a_k]$ denote the number of n-step half-space walks with $K = k$, $A_i = a_i$ for $i = 1, \ldots, k$. We observe that

$$h_n[a_1, a_2, a_3, \ldots, a_k] \leq h_n[a_1 + a_2, a_3, \ldots, a_n]. \tag{2.10}$$

To obtain this, reflect the part of the walk $(\omega(j))_{j \geq n_1}$ across the line $\omega_1 = A_1$; see Figure 1. Repeating this inequality gives

$$h_n[a_1, \ldots, a_k] \leq h_n[a_1 + \cdots + a_k] = b_{n, a_1 + \cdots + a_k}. \tag{2.11}$$

So we can bound

$$\begin{aligned} h_n &= \sum_{k \geq 1} \sum_{a_1 > \cdots > a_k > 0} h_n[a_1, \ldots, a_k] \\ &\leq \sum_{k \geq 1} \sum_{a_1 > \cdots > a_k > 0} b_{n, a_1 + \cdots + a_k} \\ &= \sum_{A=1}^{n} P_D(A) b_{n,A}. \end{aligned} \tag{2.12}$$

Bounding $P_D(A)$ by $P_D(n)$, we obtain $h_n \leq P_D(n) \sum_{A=1}^{n} b_{n,A} = P_D(n) b_n$ as claimed. □

We can now prove the Hammersley–Welsh bound (1.25), from [**29**].

THEOREM 2.6. *Fix $B > \pi(\frac{2}{3})^{1/2}$. Then there is $n_0 = n_0(B)$ independent of the dimension $d \geq 2$ such that*

$$c_n \leq b_{n+1} e^{B\sqrt{n}} \leq \mu^{n+1} e^{B\sqrt{n}} \qquad \text{for } n \geq n_0. \tag{2.13}$$

Note that (2.13), though an improvement over $c_n \leq \mu^n e^{o(n)}$ which follows from the definition (1.12) of μ, is still much larger than the predicted growth $c_n \sim A\mu^n n^{\gamma-1}$ from (1.21). It is an open problem to improve Theorem 2.6 in $d = 2, 3, 4$ beyond the result of Kesten [**50**] shown in (1.26).

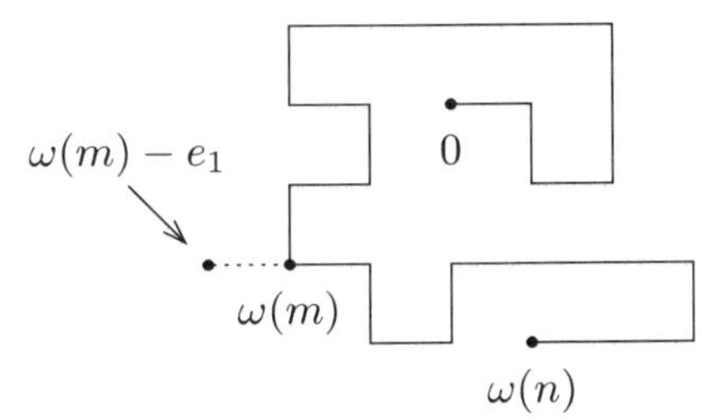

FIGURE 2. The decomposition of a self-avoiding walk into two half-space walks.

PROOF OF THEOREM 2.6. We first prove

$$c_n \leq \sum_{m=0}^{n} h_{n-m} h_{m+1}, \tag{2.14}$$

using the decomposition depicted in Figure 2, as follows. Given an n-step SAW ω, let

$$x_1 = \min_{0\le i\le n} \omega_1(i), \qquad m = \max\{i : \omega_1(i) = x_1\}. \tag{2.15}$$

Write e_1 for the unit vector in the first coordinate direction of $\mathbb{Z}^d$. Then (after translating by $\omega(m)$) the walk $(\omega(m), \omega(m+1), \dots, \omega(n))$ is an $(n-m)$-step half-space walk, and (after translating by $\omega(m)-e_1$) the walk $(\omega(m)-e_1, \omega(m), \omega(m-1), \dots, \omega(1), \omega(0))$ is an $(m+1)$-step half-space walk. This proves (2.14).

Next, we apply Proposition 2.5 in (2.14) and use (2.2) to get

$$\begin{aligned} c_n &\le \sum_{m=0}^{n} P_D(n-m)P_D(m+1)b_{n-m}b_{m+1} \\ &\le b_{n+1}\sum_{m=0}^{n} P_D(n-m)P_D(m+1). \end{aligned} \tag{2.16}$$

Fix $B > B' > \pi(\frac{2}{3})^{1/2}$. By Theorem 2.4, there is $K > 0$ such that $P_D(A) \le K\exp\left(B'(A/2)^{1/2}\right)$ and consequently

$$P_D(n-m)P_D(m+1) \le K^2 \exp\left[B'\left(\sqrt{\frac{n-m}{2}} + \sqrt{\frac{m+1}{2}}\right)\right]. \tag{2.17}$$

The bound $x^{1/2} + y^{1/2} \le (2x+2y)^{1/2}$ now gives

$$c_n \le (n+1)K^2 e^{B'\sqrt{n+1}} b_{n+1} \le e^{B\sqrt{n}} b_{n+1} \tag{2.18}$$

if $n \ge n_0(B)$. By (2.4), the result follows. □

COROLLARY 2.7. *For $n \ge n_0(B)$,*

$$b_n \ge c_{n-1}e^{-B\sqrt{n-1}} \ge \mu^{n-1}e^{-B\sqrt{n-1}}. \tag{2.19}$$

In particular, $b_n^{1/n} \to \mu$ and so $\mu_{Bridge} = \mu$.

COROLLARY 2.8. *Define the bridge generating function $B(z) = \sum_{n=0}^{\infty} b_n z^n$. Then*

$$\chi(z) \le \frac{1}{z} e^{2(B(z)-1)} \tag{2.20}$$

and in particular $B(1/\mu) = \infty$.

PROOF. In the proof of Proposition 2.5, we decomposed a half-space walk into subwalks on $[n_{i-1}, n_i]$ for $i = 1, \dots, K$. Note that each such subwalk was in fact a bridge of span A_i. With this observation, we conclude that

$$h_n \le \sum_{k=1}^{\infty} \sum_{A_1 > \cdots > A_k} \sum_{0=n_0<n_1<\cdots<n_k=n} \prod_{i=1}^{k} b_{n_i-n_{i-1},A_i} \tag{2.21}$$

(the second sum is over A_1 when $k = 1$). The choice of a descending sequence $A_1 > \cdots > A_k$ of arbitrary length is equivalent to the choice of a subset of $\mathbb{N}$, so that taking generating functions gives

$$\sum_{n=0}^{\infty} h_n z^n \le \prod_{A=1}^{\infty}\left(1 + \sum_{m=1}^{\infty} b_{m,A} z^m\right). \tag{2.22}$$

Using the inequality $1 + x \le e^x$, we obtain

$$\sum_{n=0}^{\infty} h_n z^n \le \exp\left(\sum_{A=1}^{\infty}\sum_{m=1}^{\infty} b_{m,A} z^m\right) = e^{B(z)-1}. \tag{2.23}$$

Now using (2.14) gives

$$\begin{aligned}\chi(z) = \sum_{n=0}^{\infty} c_n z^n &\le \frac{1}{z}\sum_{n=0}^{\infty}\sum_{m=0}^{n} h_{n-m} z^{n-m} h_{m+1} z^{m+1} \\ &= \frac{1}{z}\left(\sum_{n=0}^{\infty} h_n z^n\right)\left(\sum_{n=1}^{\infty} h_n z^n\right) \\ &\le \frac{1}{z} e^{2(B(z)-1)}, \end{aligned} \tag{2.24}$$

as required. □

2.2. Self-avoiding polygons. A $2n$-step *self-avoiding return* is a walk $\omega \in \mathcal{W}_{2n}$ with $\omega(2n) = \omega(0) = 0$ and with $\omega(i) \neq \omega(j)$ for distinct pairs i, j other than the pair $0, 2n$. A *self-avoiding polygon* is a self-avoiding return with both the orientation and the location of the origin forgotten. Thus we can count self-avoiding polygons by counting self-avoiding returns up to orientation and translation invariance, and their number is

$$q_{2n} = \frac{2dc_{2n-1}(e_1)}{2 \cdot 2n}, \qquad n \ge 2, \tag{2.25}$$

where $e_1 = (1, 0, \ldots, 0)$ is the first standard basis vector. Here, the 2 in the denominator cancels the choice of orientation, and the $2n$ cancels the choice of origin in the polygon.

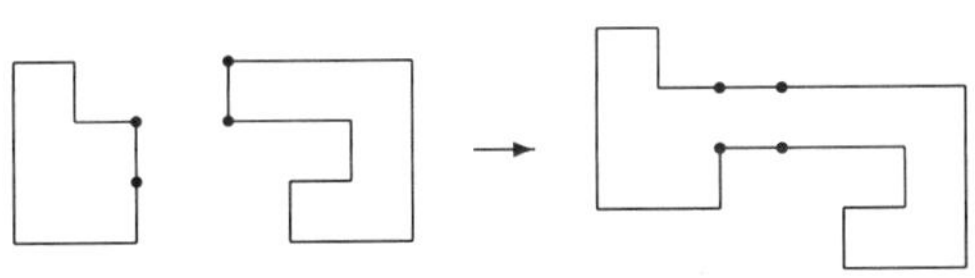

FIGURE 3. Concatenation of a 10-step polygon and a 14-step polygon to produce a 24-step polygon in $\mathbb{Z}^2$.

We first observe that two self-avoiding polygons can be concatenated to form a larger self-avoiding polygon. Consider first the case of $d = 2$. The procedure is as in Figure 3, namely we join a "rightmost" bond of one polygon to a "leftmost" bond of the other. This shows that for even integers $m, n \ge 4$, and for $d = 2$, $q_m q_n \le q_{m+n}$. With a little thought (see [**57**] for details), in general dimensions $d \ge 2$ one obtains

$$\frac{q_m q_n}{d-1} \le q_{m+n}, \tag{2.26}$$

and if we set $q_2 = 1$ and make the easy observation that $q_n \le q_{n+2}$, then (2.26) holds for all even $m, n \ge 2$. It follows from (2.26) that

$$q_{2n}^{1/2n} \to \mu_{\text{Polygon}} \le \mu, \qquad q_{2n} \le \mu_{\text{Polygon}}^{2n} \le \mu^{2n} \quad \text{for all } n \ge 2. \tag{2.27}$$

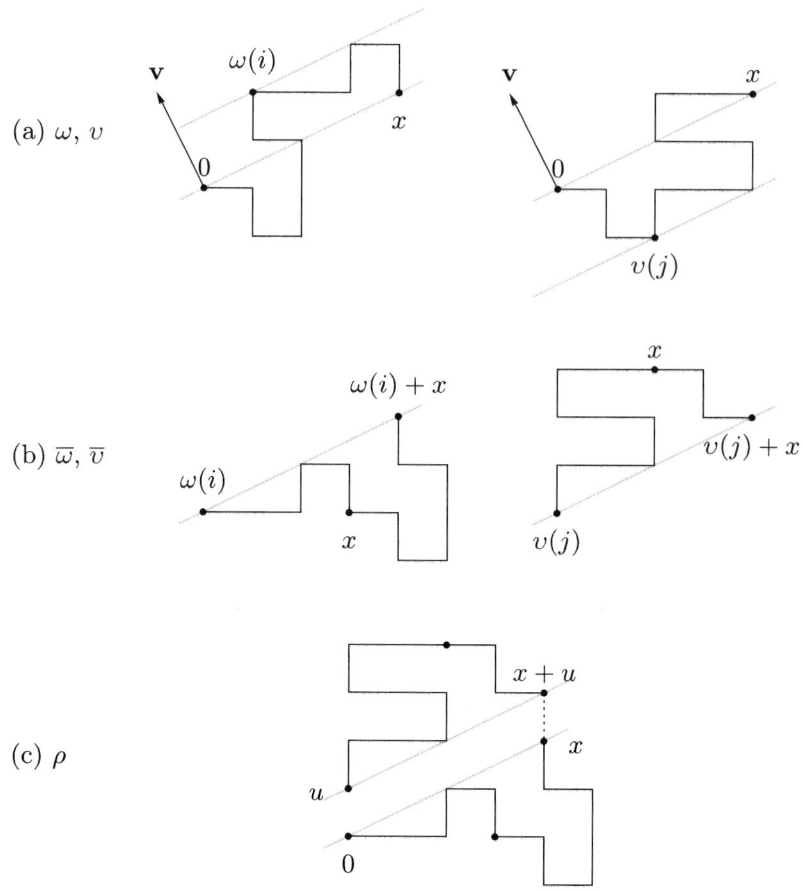

FIGURE 4. Proof of Theorem 2.9. Here $n = 12$. (a) The n-step bridges ω and υ, and the vector $\mathbf{v}$. (b) The derived walks $\overline{\omega}$ and $\overline{\upsilon}$. (c) The $(2n+1)$-step walk ρ; here $u = (1,0)$. The shaded lines are the hyperplanes orthogonal to $\mathbf{v}$.

THEOREM 2.9. *There is a constant $K = K(d)$ such that, for all $n \geq 1$,*

$$c_{2n+1}(e_1) \geq \frac{K}{n^{d+2}} b_n^2. \tag{2.28}$$

PROOF. We first show the inequality

$$\sum_{x \in \mathbb{Z}^d} b_n(x)^2 \leq 2d(n+1)^2 c_{2n+1}(e_1) \tag{2.29}$$

where $b_n(x)$ denotes the number of n-step bridges ending at x. The proof is illustrated in Figure 4. Namely, given n-step bridges ω and υ with $\omega(n) = \upsilon(n) = x \in \mathbb{Z}^d$, let $\mathbf{v} \in \mathbb{R}^d$ be some non-zero vector orthogonal to x, and fix some unit direction $u \in \mathbb{Z}^d$ with $u \cdot \mathbf{v} > 0$. Let $i \in \{0, 1, \ldots, n\}$ be the smallest index maximising $\omega(i) \cdot \mathbf{v}$ and $j \in \{0, 1, \ldots, n\}$ the smallest index minimising $\upsilon(j) \cdot \mathbf{v}$. Split ω into the pieces before and after i and interchange them to produce a walk $\overline{\omega}$, as in Figure 4(b). Do the same for υ and j. Finally combine $\overline{\omega}$ and $\overline{\upsilon}$ with an inserted step u to produce a SAW ρ with $\rho(2n+1) = u$, as in Figure 4(c). The resulting map $(\omega, \upsilon) \mapsto (\rho, i, j)$ is one-to-one, which proves (2.29).

Now, applying the Cauchy-Schwarz inequality to (2.29) gives

$$b_n^2 = \left(\sum_{x \in \mathbb{Z}^d} b_n(x) 1_{\{b_n(x) \neq 0\}} \right)^2 \leq \sum_{x \in \mathbb{Z}^d} b_n(x)^2 \sum_{x \in \mathbb{Z}^d} 1_{\{b_n(x) \neq 0\}}$$

$$\leq n(2n+1)^{d-1} \sum_{x \in \mathbb{Z}^d} b_n(x)^2. \tag{2.30}$$

Thus $2dc_{2n+1}(e_1) \geq \frac{b_n^2}{n(n+1)^2(2n+1)^{d-1}}$, which completes the proof. □

COROLLARY 2.10. *There is a $C > 0$ such that*

$$\mu^{2n} e^{-C\sqrt{n}} \leq c_{2n+1}(e_1) \leq (n+1)\mu^{2n+2}. \tag{2.31}$$

In particular, $\mu_{\text{Polygon}} = \mu$.

PROOF. The lower bound follows from Theorem 2.9 and Corollary 2.7. The upper bound follows from (2.25) and (2.27) (using $d \geq 2$). □

With a little more work, it can be shown that for any fixed $x \neq 0$, $c_n(x)^{1/n} \to \mu$ as $n \to \infty$ along the subsequence of integers whose parity agrees with $\|x\|_1$. The details can be found in [**57**]. Thus the radius of convergence of the two-point function $G_z(x) = \sum_{n=0}^{\infty} c_n(x) z^n$ is equal to $z_c = 1/\mu$ for all x.

3. The connective constant on the hexagonal lattice

Throughout this section, we consider self-avoiding walks on the hexagonal lattice $\mathbb{H}$. Our first and primary goal is to prove the following theorem from [**24**]. The proof makes use of a certain observable of broader significance, and following the proof we discuss this in the context of the $O(n)$ models.

THEOREM 3.1. *For the hexagonal lattice $\mathbb{H}$,*

$$\mu = \sqrt{2 + \sqrt{2}}. \tag{3.1}$$

As a matter of convenience, we extend walks at their extremities by two half-edges in such a way that they start and end at *mid-edges*, i.e., centres of edges of $\mathbb{H}$. The set of mid-edges will be called H. We position the hexagonal lattice $\mathbb{H}$ of mesh size 1 in $\mathbb{C}$ so that there exists a horizontal edge e with mid-edge a being 0. We now write c_n for the number of n-step SAWs on the hexagonal lattice $\mathbb{H}$ which start at 0, and $\chi(z) = \sum_{n=0}^{\infty} c_n z^n$ for the susceptibility.

We first point out that it suffices to count bridges. On the hexagonal lattice, a bridge is defined by the following adaptation of Definition 2.1: a *bridge* on $\mathbb{H}$ is a SAW which never revisits the vertical line through its starting point, never visits a vertical line to the right of the vertical line through its endpoint, and moreover starts and ends at the midpoint of a horizontal edge. We now use b_n to denote the number of n-step bridges on $\mathbb{H}$ which start at 0. It is straightforward to adapt the arguments used to prove Corollary 2.7 to the hexagonal lattice, leading to the conclusion that $\mu_{\text{Bridge}} = \mu$ also on $\mathbb{H}$. Thus it suffices to show that

$$\mu_{\text{Bridge}} = \sqrt{2 + \sqrt{2}}. \tag{3.2}$$

Using notation which anticipates our conclusion but which should not create confusion, we will write

$$z_c = \frac{1}{\sqrt{2+\sqrt{2}}}. \tag{3.3}$$

We also write $B(z) = \sum_{n=0}^{\infty} b_n z^n$ for $z > 0$. To prove (3.2), it suffices to prove that $B(z_c) = \infty$ or $\chi(z_c) = \infty$, and that $B(z) < \infty$ whenever $z < z_c$. This is what we will prove.

3.1. The holomorphic observable. The proof is based on a generalisation of the two-point function that we call the *holomorphic observable*. In this section, we introduce the holomorphic observable and prove its discrete analyticity. Some preliminary definitions are required.

A *domain* $\Omega \subset H$ is a union of all mid-edges emanating from a given connected collection of vertices $V(\Omega)$; see Figure 5. In other words, a mid-edge x belongs to Ω if at least one end-point of its associated edge is in $V(\Omega)$. The boundary $\partial\Omega$ consists of mid-edges whose associated edge has exactly one endpoint in Ω. We further assume Ω to be simply connected, i.e., having a connected complement.

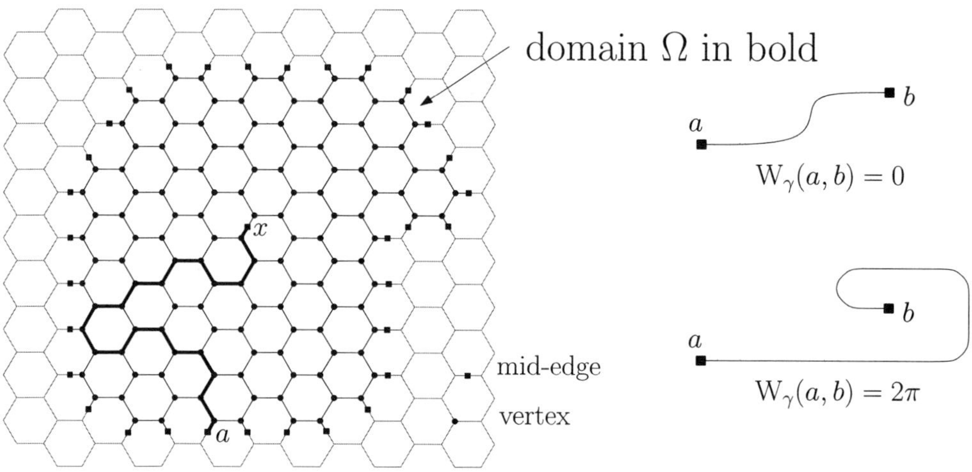

FIGURE 5. Left: A domain Ω whose boundary mid-edges are pictured by small black squares. Vertices of $V(\Omega)$ correspond to circles. Right: Winding of a SAW ω.

DEFINITION 3.2. The winding $\mathrm{W}_\omega(a,b)$ of a SAW ω between mid-edges a and b (not necessarily the start and end of ω) is the total rotation in radians when ω is traversed from a to b; see Figure 5.

We write $\omega : a \to E$ if a walk ω starts at mid-edge a and ends at some mid-edge of $E \subset H$. In the case where $E = \{b\}$, we simply write $\omega : a \to b$. The *length* $\ell(\omega)$ of the walk is the number of vertices belonging to ω. The following definition provides a generalisation of the two-point function $G_z(x)$.

DEFINITION 3.3. Fix $a \in \partial\Omega$ and $\sigma \in \mathbb{R}$. For $x \in \Omega$ and $z \geq 0$, the *holomorphic observable* is defined to be

$$F_z(x) = \sum_{\omega \subset \Omega:\ a \to x} e^{-i\sigma \mathrm{W}_\omega(a,x)} z^{\ell(\omega)}. \tag{3.4}$$

In contrast to the two-point function, the weights in the holomorphic observable need not be positive. For the special case $z = z_c$ and $\sigma = \frac{5}{8}$, F_{z_c} satisfies the relation in the following lemma, a relation which can be regarded as a weak form of discrete analyticity, and which will be crucial in the rest of the proof.

LEMMA 3.4. *If $z = z_c$ and $\sigma = \frac{5}{8}$, then, for every vertex $v \in V(\Omega)$,*

$$(p-v)F_{z_c}(p) + (q-v)F_{z_c}(q) + (r-v)F_{z_c}(r) = 0, \tag{3.5}$$

where p, q, r are the mid-edges of the three edges adjacent to v.

PROOF. Let $z \geq 0$ and $\sigma \in \mathbb{R}$. We will specialise later to $z = z_c$ and $\sigma = \frac{5}{8}$. We assume without loss of generality that p, q and r are oriented counter-clockwise around v. By definition, $(p-v)F_z(p) + (q-v)F_z(q) + (r-v)F_z(r)$ is a sum of contributions $c(\omega)$ over all possible SAWs ω ending at p, q or r. For instance, if ω ends at the mid-edge p, then its contribution will be

$$c(\omega) = (p-v)e^{-i\sigma \mathrm{W}_\omega(a,p)} z^{\ell(\omega)}. \tag{3.6}$$

The set of walks ω finishing at p, q or r can be partitioned into pairs and triplets of walks as depicted in Figure 6, in the following way:

- If a SAW ω_1 visits all three mid-edges p, q, r, then the edges belonging to ω_1 form a SAW plus (up to a half-edge) a self-avoiding return from v to v. One can associate to ω_1 the walk ω_2 passing through the same edges, but traversing the return from v to v in the opposite direction. Thus, walks visiting the three mid-edges can be grouped in pairs.
- If a walk ω_1 visits only one mid-edge, it can be associated to two walks ω_2 and ω_3 that visit exactly two mid-edges by prolonging the walk one step further (there are two possible choices). The reverse is true: a walk visiting exactly two mid-edges is naturally associated to a walk visiting only one mid-edge by erasing the last step. Thus, walks visiting one or two mid-edges can be grouped in triplets.

We will prove that when $\sigma = \frac{5}{8}$ and $z = z_c$ the sum of contributions for each pair and each triplet vanishes, and therefore the total sum is zero.

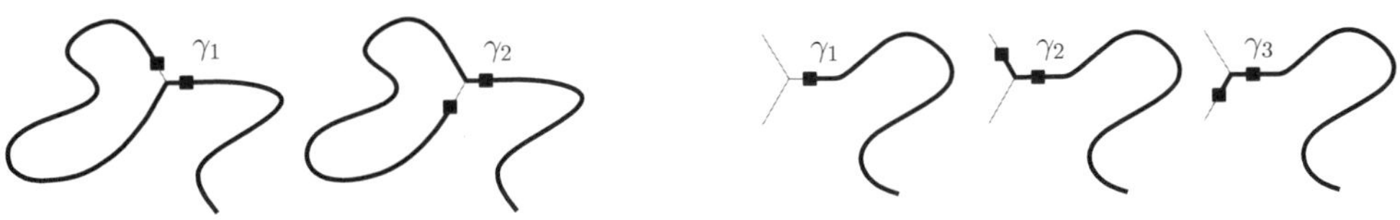

FIGURE 6. Left: a pair of walks visiting the three mid-edges and matched together. Right: a triplet of walks, one visiting one mid-edge, the two others visiting two mid-edges, which are matched together.

Let ω_1 and ω_2 be two walks that are grouped as in the first case. Without loss of generality, we assume that ω_1 ends at q and ω_2 ends at r. Note that ω_1 and ω_2 coincide up to the mid-edge p since (ω_1, ω_2) are matched together. Then

$$\ell(\omega_1) = \ell(\omega_2) \quad \text{and} \quad \begin{cases} \mathrm{W}_{\omega_1}(a,q) = \mathrm{W}_{\omega_1}(a,p) + \mathrm{W}_{\omega_1}(p,q) = \mathrm{W}_{\omega_1}(a,p) - \frac{4\pi}{3} \\ \mathrm{W}_{\omega_2}(a,r) = \mathrm{W}_{\omega_2}(a,p) + \mathrm{W}_{\omega_2}(p,r) = \mathrm{W}_{\omega_1}(a,p) + \frac{4\pi}{3}. \end{cases} \tag{3.7}$$

In evaluating the winding of ω_1 between p and q, we used the fact that $a \in \partial\Omega$ and Ω is simply connected. The term $e^{-i\sigma \mathrm{W}_\omega(a,x)}$ gives a weight λ or $\bar\lambda$ per left or right turn of ω, where

$$\lambda = \exp\left(-i\sigma\frac{\pi}{3}\right). \tag{3.8}$$

Writing $j = e^{i2\pi/3}$, we obtain

$$\begin{aligned} c(\omega_1) + c(\omega_2) &= (q-v)e^{-i\sigma \mathrm{W}_{\omega_1}(a,q)} z^{\ell(\omega_1)} + (r-v)e^{-i\sigma \mathrm{W}_{\omega_2}(a,r)} z^{\ell(\omega_2)} \\ &= (p-v)e^{-i\sigma \mathrm{W}_{\omega_1}(a,p)} z^{\ell(\omega_1)} \left(j\bar\lambda^4 + \bar j \lambda^4\right). \end{aligned} \tag{3.9}$$

Now we set $\sigma = \frac{5}{8}$ so that $j\bar\lambda^4 + \bar j\lambda^4 = 2\cos(\frac{3\pi}{2}) = 0$, and hence

$$c(\omega_1) + c(\omega_2) = 0. \tag{3.10}$$

Let $\omega_1, \omega_2, \omega_3$ be three walks matched as in the second case. Without loss of generality, we assume that ω_1 ends at p and that ω_2 and ω_3 extend ω_1 to q and r respectively. As before, we easily find that

$$\ell(\omega_2) = \ell(\omega_3) = \ell(\omega_1) + 1 \quad \text{and} \quad \begin{cases} \mathrm{W}_{\omega_2}(a,r) = \mathrm{W}_{\omega_2}(a,p) + \mathrm{W}_{\omega_2}(p,q) = \mathrm{W}_{\omega_1}(a,p) - \frac{\pi}{3} \\ \mathrm{W}_{\omega_3}(a,r) = \mathrm{W}_{\omega_3}(a,p) + \mathrm{W}_{\omega_3}(p,r) = \mathrm{W}_{\omega_1}(a,p) + \frac{\pi}{3}, \end{cases} \tag{3.11}$$

and thus

$$c(\omega_1) + c(\omega_2) + c(\omega_3) = (p-v)e^{-i\sigma \mathrm{W}_{\omega_1}(a,p)} z^{\ell(\omega_1)} \left(1 + zj\bar\lambda + z\bar j\lambda\right). \tag{3.12}$$

Now we choose z such that $1 + zj\bar\lambda + z\bar j\lambda = 0$. Due to our choice $\sigma = \frac{5}{8}$, we have $\lambda = \exp(-i\frac{5\pi}{24})$. Thus we choose $z_c^{-1} = 2\cos\frac{\pi}{8} = \sqrt{2+\sqrt{2}}$.

Now the desired identity (3.5) follows immediately by summing over all the pairs and triplets of walks. □

The last step of the proof of Lemma 3.4 is the *only* place where the choice $z = z_c = 1/\sqrt{2+\sqrt{2}}$ is used in the proof of Theorem 3.1.

3.2. Proof of Theorem 3.1 completed. Now we will apply Lemma 3.4 to prove Theorem 3.1.

We consider a vertical strip domain S_T composed of the vertices of T strips of hexagons, and its finite version $S_{T,L}$ cut at height L at an angle of $\frac{\pi}{3}$; see Figure 7. We denote the left and right boundaries of S_T by α and β, respectively, and the top and bottom boundaries of $S_{T,L}$ by ϵ and $\bar\epsilon$, respectively. We also introduce the positive quantities:

$$A_{T,L}(z) = \sum_{\omega \subset S_{T,L}:\ a \to \alpha\setminus\{a\}} z^{\ell(\omega)}, \tag{3.13}$$

$$B_{T,L}(z) = \sum_{\omega \subset S_{T,L}:\ a \to \beta} z^{\ell(\omega)}, \tag{3.14}$$

$$E_{T,L}(z) = \sum_{\omega \subset S_{T,L}:\ a \to \epsilon \cup \bar\epsilon} z^{\ell(\omega)}. \tag{3.15}$$

LEMMA 3.5. *For $z = z_c$,*

$$1 = c_\alpha A_{T,L}(z_c) + B_{T,L}(z_c) + c_\epsilon E_{T,L}(z_c), \tag{3.16}$$

where $c_\alpha = \cos\left(\frac{3\pi}{8}\right)$ and $c_\epsilon = \cos\left(\frac{\pi}{4}\right)$.

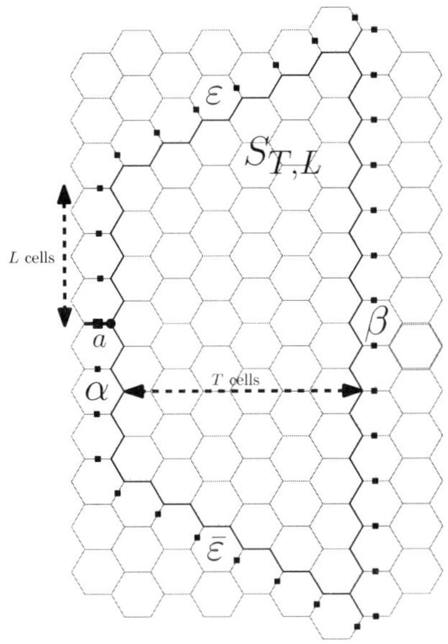

FIGURE 7. Domain $S_{T,L}$ and boundary parts α, β, ϵ and $\bar{\epsilon}$.

PROOF. We fix $z = z_c$ and drop it from the notation. We sum the relation (3.5) over all vertices in $V(S_{T,L})$. Contributions at interior mid-edges vanish and we arrive at

$$-\sum_{x\in\alpha} F(x) + \sum_{x\in\beta} F(x) + j\sum_{x\in\epsilon} F(x) + \bar{j}\sum_{x\in\bar{\epsilon}} F(x) = 0. \tag{3.17}$$

The winding of any SAW from a to the bottom part of α is $-\pi$, while the winding to the top part is π. Using this and symmetry, together with the fact that the only SAW from a to a has length 0, we conclude that

$$\sum_{x\in\alpha} F(x) = F(a) + \sum_{x\in\alpha\setminus\{a\}} F(x) = 1 + \frac{e^{-i\sigma\pi} + e^{i\sigma\pi}}{2} A_{T,L} = 1 - c_\alpha A_{T,L}. \tag{3.18}$$

Similarly, the winding from a to any half-edge in β, ϵ or $\bar{\epsilon}$ is respectively 0, $\frac{2\pi}{3}$ or $-\frac{2\pi}{3}$. Therefore, again using symmetry,

$$\sum_{x\in\beta} F(x) = B_{T,L}, \qquad j\sum_{x\in\epsilon} F(x) + \bar{j}\sum_{x\in\bar{\epsilon}} F(x) = c_\epsilon E_{T,L}. \tag{3.19}$$

The proof is completed by inserting (3.18)–(3.19) into (3.17). □

The sequences $(A_{T,L}(z))_{L>0}$ and $(B_{T,L}(z))_{L>0}$ are increasing in L and are bounded for $z \leq z_c$, thanks to (3.16) and the monotonicity in z. Thus they have limits

$$A_T(z) = \lim_{L\to\infty} A_{T,L}(z) = \sum_{\omega\subset S_T:\ a\to\alpha\setminus\{a\}} z^{\ell(\omega)}, \tag{3.20}$$

$$B_T(z) = \lim_{L\to\infty} B_{T,L}(z) = \sum_{\omega\subset S_T:\ a\to\beta} z^{\ell(\omega)}. \tag{3.21}$$

When $z = z_c$, via (3.16) again, we conclude that $(E_{T,L}(z_c))_{L>0}$ is decreasing and converges to a limit $E_T(z_c) = \lim_{L\to\infty} E_{T,L}(z_c)$. Thus, by (3.16),

$$1 = c_\alpha A_T(z_c) + B_T(z_c) + c_\epsilon E_T(z_c). \tag{3.22}$$

PROOF OF THEOREM 3.1. The bridge generating function is given by $B(z) = \sum_{T=0}^{\infty} B_T(z)$. Recall that it suffices to show that $B(z) < \infty$ for $z < z_c$, and that $B(z_c) = \infty$ or $\chi(z_c) = \infty$.

We first assume $z < z_c$. Since $B_T(z)$ involves only bridges of length at least T, it follows from (3.22) that

$$B_T(z) \le \left(\frac{z}{z_c}\right)^T B_T(z_c) \le \left(\frac{z}{z_c}\right)^T, \tag{3.23}$$

and hence $B(z)$ is finite since the right-hand side is summable.

It remains to prove that $B(z_c) = \infty$ or $\chi(z_c) = \infty$. We do this by considering two separate cases. Suppose first that, for some T, $E_T(z_c) > 0$. As noted previously, $E_{T,L}(z_c)$ is decreasing in L. Therefore, as required,

$$\chi(z_c) \ge \sum_{L=1}^{\infty} E_{T,L}(z_c) \ge \sum_{L=1}^{\infty} E_T(z_c) = \infty. \tag{3.24}$$

It remains to consider the case that $E_T^{z_c} = 0$ for every T. In this case, (3.22) simplifies to

$$1 = c_\alpha A_T(z_c) + B_T(z_c). \tag{3.25}$$

Observe that walks contributing to $A_{T+1}(z_c)$ but not to $A_T(z_c)$ must visit some vertex adjacent to the right edge of S_{T+1}. Cutting such a walk at the first such point (and adding half-edges to the two halves), we obtain two bridges of span $T+1$ in S_{T+1}. We conclude from this that

$$A_{T+1}(z_c) - A_T(z_c) \le z_c \left(B_{T+1}(z_c)\right)^2. \tag{3.26}$$

Combining (3.25) for T and $T+1$ with (3.26), we can write

$$\begin{aligned} 0 &= [c_\alpha A_{T+1}(z_c) + B_{T+1}(z_c)] - [c_\alpha A_T(z_c) + B_T(z_c)] \\ &\le c_\alpha z_c \left(B_{T+1}(z_c)\right)^2 + B_{T+1}(z_c) - B_T(z_c), \end{aligned} \tag{3.27}$$

so

$$c_\alpha z_c \left(B_{T+1}(z_c)\right)^2 + B_{T+1}(z_c) \ge B_T(z_c). \tag{3.28}$$

It is an easy exercise to verify by induction that

$$B_T(z_c) \ge \min\{B_1(z_c), 1/(c_\alpha z_c)\}\frac{1}{T} \tag{3.29}$$

for every $T \ge 1$. This implies, as required, that

$$B(z_c) \ge \sum_{T=1}^{\infty} B_T(z_c) = \infty. \tag{3.30}$$

This completes the proof. ∎

3.3. Conjecture 1.5 and the holomorphic observable. Recall the statement of Conjecture 1.5. When formulated on $\mathbb{H}$, this conjecture concerns a simply connected domain Ω in the complex plane $\mathbb{C}$ with two points a and b on the boundary, with a discrete approximation given by the largest finite domain Ω_δ of $\delta\mathbb{H}$ included in Ω, and with a_δ and b_δ the closest vertices of $\delta\mathbb{H}$ to a and b respectively. A probability measure $\mathbb{P}_{z,\delta}$ is defined on the set of SAWs ω between a_δ and b_δ that remain in Ω_δ by assigning to ω a weight proportional to $z_c^{\ell(\omega)}$. We obtain a random curve denoted ω_δ. We can also define the observable in this context, and we denote it by F_δ. Conjecture 1.5 then asserts that the random curve ω_δ converges to $\mathrm{SLE}_{8/3}$ from a and b in the domain Ω.

A possible approach to proving Conjecture 1.5 might be the following. First, prove a precompactness result for self-avoiding walks. Then, by taking a subsequence, we could assume that the curve γ_δ converges to a continuous curve (in fact, the limiting object would need to be a Loewner chain, see [**1**]). The second step would consist in identifying the possible limits. The holomorphic observable should play a crucial role in this step. Indeed, if F_δ converges when rescaled to an explicit function, one could use the *martingale technique* introduced in [**70**] to verify that the only possible limit is $\mathrm{SLE}_{8/3}$.

Regarding the convergence of F_δ, we first recall that in the discrete setting contour integrals should be performed along dual edges. For $\mathbb{H}$, the dual edges form a triangular lattice, and Lemma 3.4 has the enlightening interpretation that the contour integral vanishes along any elementary dual triangle. Any area enclosed by a discrete closed dual contour is a union of elementary triangles, and hence the integral along any discrete closed contour also vanishes. This is a discrete analogue of Morera's theorem. It implies that if the limit of F_δ (properly rescaled) exists and is continuous, then it is automatically holomorphic. By studying the boundary conditions, it is even possible to identify the limit. This leads to the following conjecture, which is based on ideas in [**70**].

CONJECTURE 3.6. *Let Ω be a simply connected domain (not equal to $\mathbb{C}$), let $z \in \Omega$, and let a, b be two distinct points on the boundary of Ω. We assume that the boundary of Ω is smooth near b. For $\delta > 0$, let F_δ be the holomorphic observable in the domain $(\Omega_\delta, a_\delta, b_\delta)$ approximating (Ω, a, b), and let z_δ be the closest point in Ω_δ to z. Then*

$$\lim_{\delta\to 0} \frac{F_\delta(a_\delta, z_\delta)}{F_\delta(a_\delta, b_\delta)} = \left(\frac{\Phi'(z)}{\Phi'(b)}\right)^{5/8}, \tag{3.31}$$

where Φ is a conformal map from Ω to the upper half-plane mapping a to ∞ and b to 0.

The right-hand side of (3.31) is well-defined, since the conformal map Φ is unique up to multiplication by a real factor.

3.4. Loop models and holomorphic observables. The original motivation for the introduction of the holomorphic observable stems from a more general context, which we now discuss. The *loop $O(n)$ model* is a lattice model on a domain Ω. We restrict attention in this discussion to the hexagonal lattice $\mathbb{H}$. A configuration ω is a family of self-avoiding loops, and its probability is proportional to $z^{\#\mathrm{edges}}n^{\#\mathrm{loops}}$. The *loop parameter n* is taken in $[0, 2]$. There are other variants of the model; for instance, one can introduce an interface going from one point a on the boundary to the inside, or one interface between two points of the boundary. The case $n = 1$ corresponds to the Ising model, while the case $n = 0$ corresponds to the self-avoiding walk (when allowing one interface).

Fix $n \in [0, 2]$. It is a non-rigorous prediction of [**61**] that the model has the following three phases distinguished by the value of z:

- If $z < 1/\sqrt{2 + \sqrt{2 - n}}$, the loops are sparse (typically of logarithmic size in the size of the domain). This phase is subcritical.
- If $z = 1/\sqrt{2 + \sqrt{2 - n}}$, the loops are dilute (there are loops of the size of the domain which are typically separated be a distance of the size of the domain). This phase is critical.
- If $z > 1/\sqrt{2 + \sqrt{2 - n}}$, the loops are dense (there are loops of the size of the domain which are typically separated be a distance much smaller than the size of the domain). This phase is critical as well.

Consider the special case of the Ising model at its critical value $z_c = 1/\sqrt{3}$. Let E denote the set of configurations consisting only of self-avoiding loops, and let $E(a, x)$ denote the set of configurations with self-avoiding loops plus an interface γ from a to x. Then, ignoring the issue of boundary conditions, the Ising spin-spin correlation is given in terms of the loop model by

$$\langle \sigma(a)\sigma(x) \rangle = \frac{\sum_{\omega \in E(a,x)} z_c^{\#\text{edges}}}{\sum_{\omega \in E} z_c^{\#\text{edges}}}. \tag{3.32}$$

A natural operation in physics consists in flipping the sign of the coupling constant of the Ising model along a path from a to x, in such a way that a monodromy is introduced: if we follow a path turning around x, spins are reversed after one whole turn. See, e.g., [**65**]. In terms of the loop representation, the spin-spin correlation $\langle \sigma(a)\sigma(x) \rangle_{\text{monodromy}}$ in this new Ising model is

$$\langle \sigma(a)\sigma(x) \rangle_{\text{monodromy}} = \frac{\sum_{\omega \in E(a,x)} (-1)^{\#\text{turns of } \gamma \text{ around } x} z_c^{\#\text{edges}}}{\sum_{\omega \in E} z_c^{\#\text{edges}}} \tag{3.33}$$

where γ is the interface between a and x.

The numerator of the right-hand side of (3.33) can be rewritten as

$$\sum_{\omega \in E(a,x)} e^{-i\frac{1}{2}\mathrm{W}_\gamma(a,x)} z^{\#\text{edges}} n^{\#\text{loops}} \tag{3.34}$$

with $n = 1$. This is of the same form as the holomorphic observable (3.4). With general values of n, and with the freedom to choose the value of $\sigma \in [0, 1]$, we obtain the observable

$$F_z(x) = \sum_{\omega \in E(a,x)} e^{-i\sigma\mathrm{W}_\gamma(a,x)} z^{\#\text{edges}} n^{\#\text{loops}}. \tag{3.35}$$

The values of σ and z need to be chosen according to the value of n. If $\sigma = \sigma(n)$ satisfies $2\cos[(1 + 2\sigma)2\pi/3] = -n$ and $z = z(n) = 1/\sqrt{2 + \sqrt{2 - n}}$, then the proof of Lemma 3.4 can be modified to yield its conclusion in this more general context.

To conclude this discussion, consider the loop $O(n)$ model with a family of self-avoiding loops and a single interface between two boundary points a and b. For $n = 1$ and $z = 1/\sqrt{3}$, it has been proved that the interface converges to SLE_3 [**18**]. For other values of z and n, the following behaviour is conjectured [**70**].

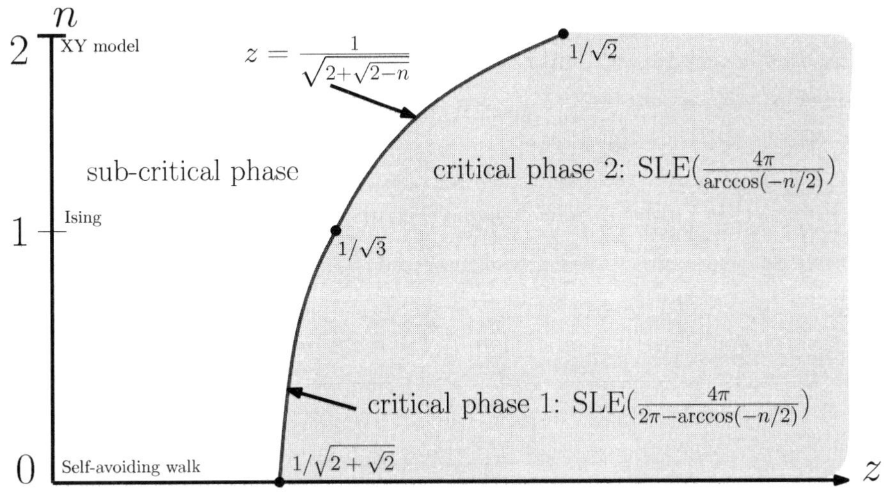

FIGURE 8. Phase diagram for $O(n)$ models.

CONJECTURE 3.7. *Fix $n \in [0,2]$. For $z = 1/\sqrt{2+\sqrt{2-n}}$, the interface between a and b converges, as the lattice spacing goes to zero, to*

$$\text{SLE}_\kappa \quad \textit{with} \quad \kappa = \frac{4\pi}{2\pi - \arccos(-n/2)}. \tag{3.36}$$

For $z > 1/\sqrt{2+\sqrt{2-n}}$, the interface between a and b converges, as the lattice spacing goes to zero, to

$$\text{SLE}_\kappa \quad \textit{with} \quad \kappa = \frac{4\pi}{\arccos(-n/2)}. \tag{3.37}$$

Conjecture 3.7 is summarised in Figure 8. The value of arccos is in $[0, \pi]$, so the first regime corresponds to $\kappa \in [\frac{8}{3}, 4]$ and the second to $\kappa \in [4, 8]$. These two critical regimes do not belong to the same universality class, in the sense that the scaling limit of the interface is not the same. In particular, since SLE_κ curves are simple for $\kappa \le 4$ but not for $\kappa > 4$ (see [**1**]), in the dilute phase the interface is conjectured to be simple in the scaling limit, but not in the dense phase. In addition, all the SLE_κ models for $\frac{8}{3} \le \kappa \le 8$ arise in these $O(n)$ models. This rich behaviour is at the heart of the mathematical interest in $O(n)$ models. To prove the conjecture remains a major challenge in 2-dimensional statistical mechanics.

4. The lace expansion

4.1. Main results. In dimensions $d \ge 5$, it has been proved that SAW has the same scaling behaviour as SRW. The following two theorems, due to Hara and Slade [**33, 34**] and to Hara [**30**], respectively, show that the critical exponents γ, ν, η exist and take the values $\gamma = 1$, $\nu = \frac{1}{2}$, $\eta = 0$, and that the scaling limit is Brownian motion.

THEOREM 4.1. *Fix $d \geq 5$, and consider the nearest-neighbour SAW on $\mathbb{Z}^d$. There exist constants $A, D, \epsilon > 0$ such that, as $n \to \infty$,*

$$c_n = A\mu^n[1 + O(n^{-\epsilon})], \tag{4.1}$$

$$\mathbb{E}_n |\omega(n)|^2 = Dn[1 + O(n^{-\epsilon})]. \tag{4.2}$$

Also,

$$\left(\frac{\omega(\lfloor nt \rfloor)}{\sqrt{Dn}}\right)_{t \geq 0} \to (B_t)_{t \geq 0}, \tag{4.3}$$

where B_t denotes Brownian motion and the convergence is in distribution.

THEOREM 4.2. *Fix $d \geq 5$, and consider the nearest-neighbour SAW on $\mathbb{Z}^d$. There are constants $c, \epsilon > 0$ such that, as $x \to \infty$,*

$$G_{z_c}(x) = \frac{c}{|x|^{d-2}} \left[1 + O\left(|x|^{-\epsilon}\right)\right]. \tag{4.4}$$

The proofs are based on the lace expansion, a technique that was introduced by Brydges and Spencer [**14**] to study the weakly SAW in dimensions $d > 4$. Since 1985, the method of lace expansion has been highly developed and extended to several other models: percolation ($d > 6$), oriented percolation ($d > 4$ *spatial* dimensions), the contact process ($d > 4$), lattice trees and lattice animals ($d > 8$), the Ising model ($d > 4$), and to random subgraphs of high-dimensional transitive graphs such as the Boolean cube. For a review and references, see [**69**].

Versions of Theorems 4.1–4.2 have been proved also for spread-out models; see [**57, 31**]. More recently, the above two theorems have been extended also to study long-range SAWs based on simple random walks which take steps of length r with probability proportional to $r^{-d-\alpha}$ for some α. For $\alpha \in (0, 2)$, the upper critical dimension (recall Section 1.6.4) is reduced from 4 to 2α, and the Brownian limit is replaced by a stable law in dimensions $d > 2\alpha$ [**38**]. Further results in this direction can be found in [**39, 19**].

Our goal now is modest. In this section, we will derive the lace expansion. In Section 5, we will sketch a proof of how it can be used to prove that $\gamma = 1$, in the sense that

$$\chi(z) \asymp (1 - z/z_c)^{-1} \qquad \text{as } z \nearrow z_c, \tag{4.5}$$

both for the nearest-neighbour model with $d \geq d_0 \gg 4$, and for the spread-out model with $L \geq L_0(d) \gg 1$ and any $d > 4$. Here, the notation $f(z) \asymp g(z)$ means that there exist positive c_1, c_2 such that $c_1 g(z) \leq f(z) \leq c_2 g(z)$ holds uniformly in z. The lower bound in (4.5) holds in all dimensions and follows immediately from the elementary observation in (1.12) that $c_n \geq \mu^n = z_c^{-n}$, since

$$\chi(z) = \sum_{n=0}^{\infty} c_n z^n \geq \sum_{n=0}^{\infty} (\mu z)^n = \frac{1}{1 - z/z_c} \tag{4.6}$$

for $z < z_c$. It therefore suffices to prove that in high dimensions we have the complementary upper bound

$$\chi(z) \leq \frac{C}{1 - z/z_c} \tag{4.7}$$

for some finite constant C.

4.2. The differential inequality for $\chi(z)$. We prove (4.7) by means of a *differential inequality*—an inequality relating $\frac{d}{dz}\chi(z)$ to $\chi(z)$. The derivation of the differential inequality and its implication for (4.7) first appeared in [**3**].

The differential inequality is expressed in terms of the quantity

$$\mathsf{B}(z) = \sum_{x \in \mathbb{Z}^d} G_z(x)^2 \tag{4.8}$$

for $z \le z_c$. Proposition 1.3 ensures that $\mathsf{B}(z)$ is finite for $z < z_c$. If we assume, as usual, that $G_{z_c} \sim c|x|^{-(d-2+\eta)}$, then $\mathsf{B}(z_c)$ will be finite precisely when $d > 4 - 2\eta$. With Fisher's relation (1.40) and the predicted values of γ and ν from (1.22) and (1.28), this inequality can be expected to hold, and correspondingly $\mathsf{B}(z_c) < \infty$, only for $d > 4$ (this is a prediction, not a theorem). We refer to $\mathsf{B}(z)$ as the *bubble diagram* because we express (4.8) diagramatically as

(4.9) $\mathsf{B}(z) =$ 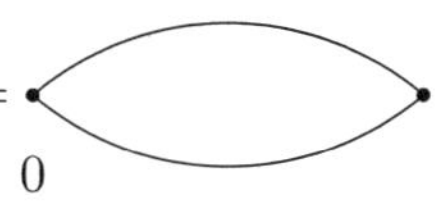 .

0

In this diagram, each line represents a factor $G_z(x)$ and the unlabelled vertex is summed over $x \in \mathbb{Z}^d$. The condition that $\mathsf{B}(z_c) < \infty$ will be referred to as the *bubble condition.*

We now derive the differential inequality

$$\frac{d}{dz}(z\chi(z)) \ge \frac{\chi(z)^2}{\mathsf{B}(z)}. \tag{4.10}$$

Assuming (4.10), we obtain (4.7) as if we were solving a differential equation. Namely, using the monotonicity of B, we first replace $\mathsf{B}(z)$ by $\mathsf{B}(z_c)$ in (4.10). We then rearrange and integrate from z to z_c, using the terminal value $\chi(z_c) = \infty$ from (4.6), to obtain

$$\begin{aligned}
\frac{1}{z^2\chi(z)^2}\frac{d}{dz}(z\chi(z)) &\ge \frac{1}{z^2\mathsf{B}(z_c)}\\
-\frac{d}{dz}\left(\frac{1}{z\chi(z)}\right) &\ge \frac{d}{dz}\left(\frac{-1}{z\mathsf{B}(z_c)}\right)\\
-0 + \frac{1}{z\chi(z)} &\ge \frac{1}{\mathsf{B}(z_c)}\left(-\frac{1}{z_c} + \frac{1}{z}\right)\\
\frac{\mathsf{B}(z_c)}{1 - z/z_c} &\ge \chi(z). \qquad (4.11)
\end{aligned}$$

Thus we have reduced the proof of (4.5) to verifying (4.10) and showing that $\mathsf{B}(z_c) < \infty$ in high dimensions. We will prove (4.10) now, and in Section 5 we will sketch the proof of the bubble condition in high dimensions.

We will use diagrams to derive (4.10). A proof using more conventional mathematical notation can be found, e.g., in [**69**]. In the diagrams in the next two paragraphs, each dot denotes a point in $\mathbb{Z}^d$, and if a dot is unlabelled then it is summed over all points in $\mathbb{Z}^d$. Each arc (or line) in a diagram represents a generating function for a SAW connecting the endpoints. At times SAWs corresponding to distinct lines must be mutually-avoiding. We will indicate this condition by labelling diagram lines and listing in groups those that mutually avoid.

With these conventions, we can describe the two-point function and the susceptibility succinctly by

(4.12) $G_z(x) =$ 

, $\chi(z) =$ 

.

In order to obtain (4.10), let us consider $Q(z) = \frac{d}{dz}(z\chi(z))$. Note that $Q(z)$ can be regarded as the generating function for SAWs weighted by the number of vertices visited in the walk. We represent this diagrammatically as:

(4.13) $Q(z) = \sum_{n=0}^{\infty}(n+1)c_n z^n =$ 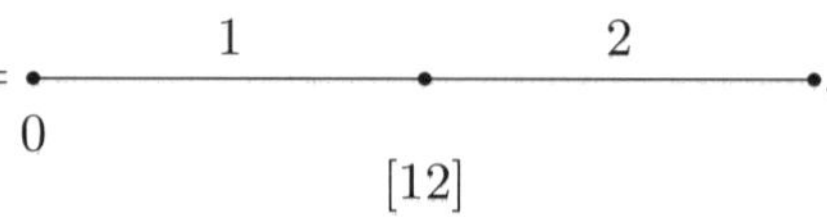

.

In (4.13), each segment represents a SAW path, and the notation [12] indicates that SAWs 1 and 2 must be mutually avoiding, apart from one shared vertex.

We apply inclusion-exclusion to (4.13), first summing over all pairs of SAWs, mutually avoiding or not, and then subtracting configurations where SAWs 1 and 2 intersect. We parametrise the subtracted term according to the *last* intersection point along the second walk. Renumbering the subwalks, we have

(4.14) $Q(z) =$ 

$-$ 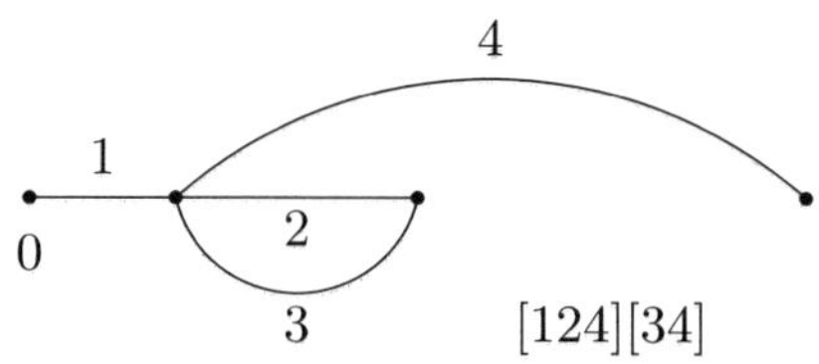

where the notation [124][34] means that walks 1, 2 and 4 must be mutually avoiding except at the endpoints, whereas walk 3 must avoid walk 4 but is allowed to intersect walks 1 and 2. Also, SAWs 2 and 3 must each take at least one step. We obtain an inequality by relaxing the avoidance pattern to [14], keeping the requirement that the walk 23 should be non-empty:

$Q(z) \geq$ 

$-$ 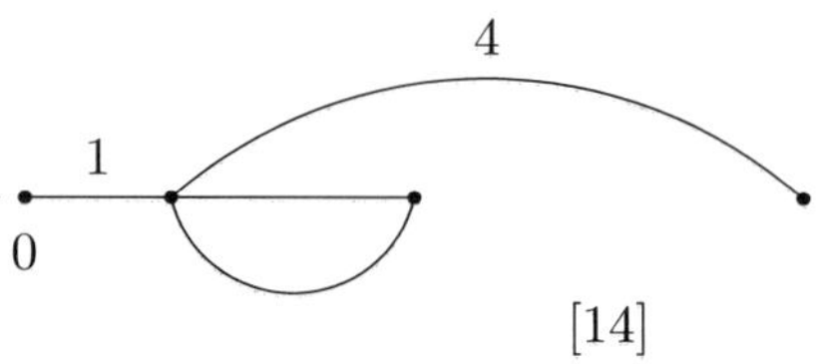

(4.15) $= \chi(z)^2 - Q(z)(\mathsf{B}(z) - 1)$.

Rearranging gives the inequality (4.10).

4.3. The lace expansion by inclusion-exclusion. The proof of the bubble condition is based on the lace expansion. The original derivation of the lace expansion by Brydges and Spencer [**14**] made use of a certain graphical construction called a *lace*. Later, it was realised that repeated inclusion-exclusion leads to the same expansion [**68**]. We present the inclusion-exclusion approach now; the approach via laces is treated in the problems of Section 4.4. The underlying graph plays little role in the derivation, and the following discussion pertains to either nearest-neighbour or spread-out SAWs. Indeed, with minor modifications, the discussion also applies on general graphs [**22**].

We use the convolution $(f * g)(x) = \sum_{y\in\mathbb{Z}^d} f(y)g(x-y)$ of two functions f, g on $\mathbb{Z}^d$. The lace expansion gives rise to a formula for $c_n(x)$, for $n \geq 1$, of the form

$$
\begin{aligned}
c_n(x) &= (c_1 * c_{n-1})(x) + \sum_{m=2}^{n} (\pi_m * c_{n-m})(x) \\
&= \sum_{y\in\mathbb{Z}^d} c_1(y)c_{n-1}(x-y) + \sum_{m=2}^{n} \sum_{y\in\mathbb{Z}^d} \pi_m(y)c_{n-m}(x-y),
\end{aligned}
\tag{4.16}
$$

in which the coefficients $\pi_m(y)$ are certain combinatorial integers that we will define below. Note that the identity (4.16) would hold for SRW with $\pi \equiv 0$. The quantity $\pi_m(y)$ can therefore be understood as a correction factor determining to what degree SAWs fail to behave like SRWs. In this sense, the lace expansion studies the SAW as a perturbation of the SRW.

Our starting point is similar to that of the derivation of the differential inequality (4.10), but now we will work with identities rather than inequalities. Also, rather than working with generating functions, we will work instead with walks with a fixed number of steps and without factors z: diagrams now arise from walks of fixed length. We begin by dividing an n-step SAW ($n \geq 1$) into its first step and the remainder of the walk. Because of self-avoidance, these two parts must be mutually avoiding, and we perform inclusion-exclusion on this condition:

(4.17)

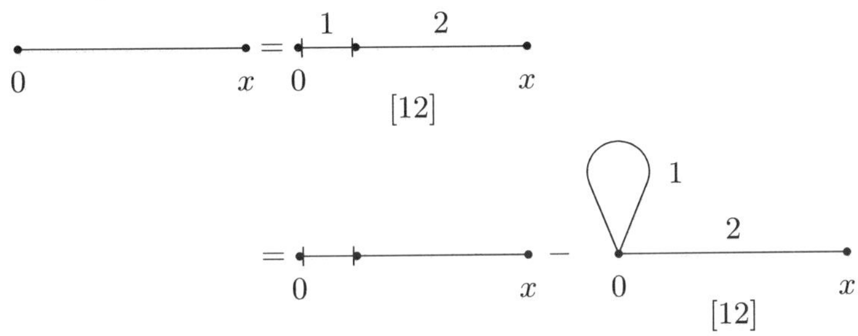

where ↤—↦ indicates a single step. In more detail, the first term on the right-hand side represents $(c_1 * c_{n-1})(x)$, and the subtracted term represents the number of n-step walks from 0 to x which are self-avoiding apart from a single required return to 0. We again perform inclusion-exclusion, first on the avoidance [12] in the second term of (4.17) (noting now the *first* time along walk 2 that walk 1 is hit):

(4.18)

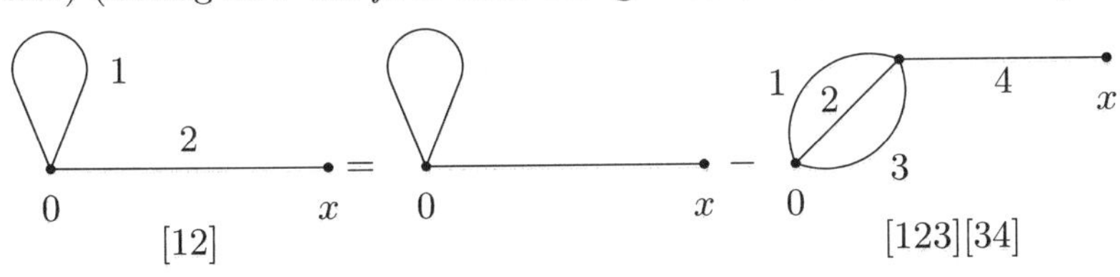

and then on the avoidance [34] in the second term of (4.18) (noting the *first* time along walk 4 that walk 3 is hit):

(4.19)

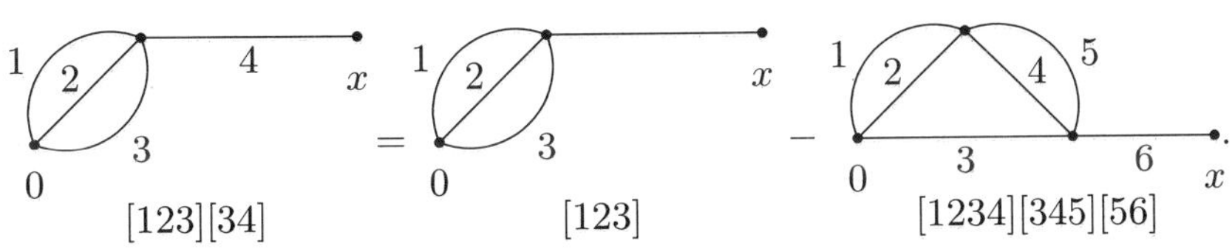

The process is continued recursively. Since the total number n of steps is finite, the above process terminates after a finite number of applications of inclusion-exclusion, because each application uses at least one step. The result is

(4.20)

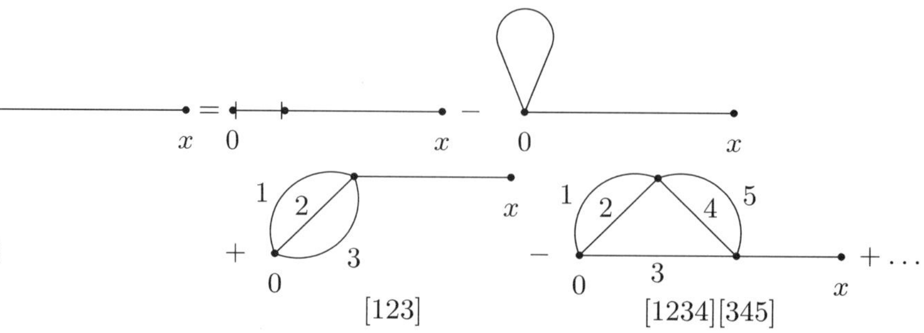

The first term on the right-hand side is just $(c_1 * c_{n-1})(x)$. In the remaining terms on the right-hand side, we regard the line ending at x as having length $n-m$, so that m steps are used by the other lines. We also regard the line ending at x as starting at y. A crucial fact is that the line ending at x has no dependence on the other lines, so it represents $c_{n-m}(x-y)$. Thus, if we define the coefficients $\pi_m(y)$ as

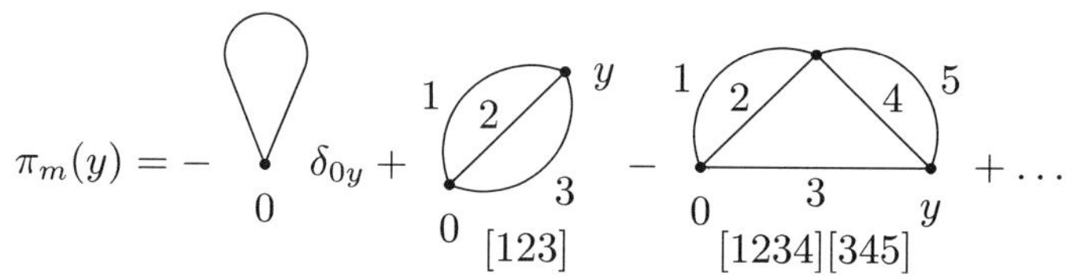

$$= \sum_{N=1}^{\infty} (-1)^N \pi_m^{(N)}(y), \tag{4.21}$$

where $\delta_{xy} = 1_{\{x=y\}}$ denotes the Kronecker delta, then (4.20) becomes (4.16), namely

$$c_n(x) = \sum_{y \in \mathbb{Z}^d} c_1(y) c_{n-1}(x-y) + \sum_{m=2}^{n} \sum_{y \in \mathbb{Z}^d} \pi_m(y) c_{n-m}(x-y). \tag{4.22}$$

By definition, $\pi_m^{(1)}(y)$ counts the number of m-step self-avoiding returns if $y = 0$, and is otherwise 0. Also, $\pi_m^{(2)}(y)$ counts the number of m-step "θ-diagrams" with vertices 0 and y, i.e., the number of m-step walks which start at zero, end at y, and are self-avoiding apart from a required return to 0 and a visit to y before terminating at y. With more attention to the inclusion-exclusion procedure, it can be seen that in the three diagrams on the right-hand side of (4.21) all the individual subwalks must have length at least 1 except for subwalk 3 of the third term which may have length 0. As noted above, the inclusion-exclusion procedure terminates after a finite number of steps, so the terms in the series (4.21) are eventually all zero, but as m increases more and more terms are non-zero. If the diagrams make you uncomfortable, formulas for $\pi_m(y)$ are given in Section 4.4. This completes the derivation of the lace expansion.

Our next task is to relate $\pi_m(y)$ to our goal of proving the bubble condition. Equation (4.16) contains two convolutions: a convolution in space given by the sum over y, and a convolution in time given by the sum over m. To eliminate these and facilitate analysis, we pass to generating functions and Fourier transforms. By definition of the two-point function,

$$G_z(x) = \sum_{n=0}^{\infty} c_n(x) z^n = \delta_{0x} + \sum_{n=1}^{\infty} c_n(x) z^n, \tag{4.23}$$

and we define

$$\Pi_z(x) = \sum_{m=2}^{\infty} \pi_m(x) z^m. \tag{4.24}$$

From (4.16), we obtain

$$\begin{aligned} G_z(x) &= \delta_{0x} + \sum_{y\in\mathbb{Z}^d} z c_1(y) G_z(x-y) + \sum_{y\in\mathbb{Z}^d} \Pi_z(y) G_z(x-y) \\ &= \delta_{0x} + z(c_1 * G_z)(x) + (\Pi_z * G_z)(x). \end{aligned} \tag{4.25}$$

Given an absolutely summable function $f : \mathbb{Z}^d \to \mathbb{C}$, we write its Fourier transform as

$$\hat{f}(k) = \sum_{x\in\mathbb{Z}^d} f(x) e^{ik\cdot x}, \tag{4.26}$$

with $k = (k_1, \ldots, k_d) \in [-\pi, \pi]^d$. Then (4.25) gives

$$\hat{G}_z(k) = 1 + z\hat{c}_1(k)\hat{G}_z(k) + \hat{\Pi}_z(k)\hat{G}_z(k). \tag{4.27}$$

We solve for $\hat{G}_z(k)$ to obtain

$$\hat{G}_z(k) = \frac{1}{1 - z\hat{c}_1(k) - \hat{\Pi}_z(k)}. \tag{4.28}$$

It is convenient to express $c_1(y)$ in terms of the probability distribution for the steps of the corresponding SRW model:

$$D(y) = \frac{c_1(y)}{|\Omega|}, \qquad \hat{c}_1(k) = |\Omega|\, \hat{D}(k), \tag{4.29}$$

where $|\Omega|$ denotes the cardinality of either option for the set Ω defined in (1.1). For the nearest-neighbour model, $|\Omega| = 2d$ and

$$\hat{D}(k) = \frac{1}{2d} \sum_{j=1}^{d} \left(e^{ik_j} + e^{-ik_j}\right) = \frac{1}{d} \sum_{j=1}^{d} \cos k_j. \tag{4.30}$$

To simplify the notation, we define $\hat{F}_z(k)$ by

$$\hat{G}_z(k) = \frac{1}{1 - z\,|\Omega|\, \hat{D}(k) - \hat{\Pi}_z(k)} = \frac{1}{\hat{F}_z(k)}. \tag{4.31}$$

Notice that $\hat{G}_z(0) = \sum_{x\in\mathbb{Z}^d} \sum_{n=0}^{\infty} c_n(x) z^n = \chi(z)$, so that $\hat{G}_z(0)$ will have a singularity at $z = z_c$. To emphasise this, we will write

$$\begin{aligned} \hat{F}_z(k) &= \hat{F}_z(0) + \left(\hat{F}_z(k) - \hat{F}_z(0)\right) \\ &= \chi(z)^{-1} + z\,|\Omega| \left(1 - \hat{D}(k)\right) + \left(\hat{\Pi}_z(0) - \hat{\Pi}_z(k)\right). \end{aligned} \tag{4.32}$$

Now we can make contact with our goal of proving the bubble condition. By Parseval's relation,

$$\mathsf{B}(z) = \sum_{x \in \mathbb{Z}^d} G_z(x)^2 = \int_{[-\pi,\pi]^d} |\hat{G}_z(k)|^2 \frac{d^d k}{(2\pi)^d} \tag{4.33}$$

(this includes the case where one side of the equality, and hence both, are infinite). The issue of whether $\mathsf{B}(z_c) < \infty$ or not boils down to the question of whether the singularity of the integrand is integrable or not, so we will need to understand the asymptotics of the terms in (4.32) as $k \to 0$ and $z \nearrow z_c$. In principle there could be other singularities when $z = z_c$, but for the nearest-neighbour and spread-out models $1 - \hat{D}(k) > 0$ for non-zero k, and one of the goals of the analysis will be to prove that the term $\hat{\Pi}_z(0) - \hat{\Pi}_z(k)$ cannot create a cancellation.

The term $1 - \hat{D}(k)$ is explicit, and for the nearest-neighbour model has asymptotic behaviour

$$1 - \hat{D}(k) = \frac{1}{d} \sum_{j=1}^{d} (1 - \cos k_j) \sim \frac{|k|^2}{2d} \tag{4.34}$$

as $k \to 0$. We need to see that the term $\hat{\Pi}_z(0) - \hat{\Pi}_z(k)$ is relatively small in high dimensions. By symmetry, we can write this term as

$$\hat{\Pi}_z(0) - \hat{\Pi}_z(k) = \sum_{x \in \mathbb{Z}^d} (1 - e^{ik\cdot x}) \Pi_z(x) = \sum_{x \in \mathbb{Z}^d} (1 - \cos k \cdot x) \Pi_z(x). \tag{4.35}$$

Finally, we note that the equation $\chi(z_c) = \infty$ can be rewritten as $0 = \chi(z_c)^{-1} = 1 - z_c |\Omega| - \hat{\Pi}_{z_c}(0)$, from which we see that the critical point z_c is given implicitly by

$$z_c = \frac{1}{|\Omega|} (1 - \hat{\Pi}_{z_c}(0)). \tag{4.36}$$

This equation has been the starting point for the study of z_c, in particular for the derivation of the $1/d$ expansion for the connective constant discussed in Section 1.4. Problem 5.1 below indicates how the first terms are obtained.

4.4. Tutorial. These problems develop the original derivation of the lace expansion by Brydges and Spencer [**14**]. All this material can also be found in [**69**].

We require a notion of graphs on integer intervals, and connectivity of these graphs. We emphasise in advance that the notion of connectivity is *not* the usual graph theoretic one, but that it is the right notion in this context.

DEFINITION 4.3. (i) Let $I = [a, b]$ be an interval of non-negative integers. An *edge* is a pair $st = \{s, t\}$ with $s, t \in \mathbb{Z}$ and $a \le s < t \le b$. A *graph* on $[a, b]$ is a set of edges. We denote the set of all graphs on $[a, b]$ by $\mathcal{B}[a, b]$.

(ii) A graph $\Gamma \in \mathcal{B}[a, b]$ is *connected* if a, b are endpoints of edges, and if for any $c \in (a, b)$, there are $s, t \in [a, b]$ such that $c \in (s, t)$ and $st \in \Gamma$. Equivalently, Γ is connected if $(a, b) = \cup_{st \in \Gamma} (s, t)$. The set of all connected graphs on $[a, b]$ is denoted by $\mathcal{G}[a, b]$.

PROBLEM 4.1. Give an example of a graph which is connected in the above sense, but not path-connected in the usual graph theoretic sense, and give an example which is path-connected, but not connected in the above sense.

Let $U_{st}(\omega) = -1_{\{\omega(s)\neq\omega(t)\}}$, and for $a < b$ define

$$K[a,b](\omega) = \prod_{a\leq s<t\leq b} (1 + U_{st}(\omega)), \quad K[a,a](\omega) = 1, \tag{4.37}$$

so that

$$c_n(x) = \sum_{\omega\in\mathcal{W}_n(0,x)} K[0,n](\omega). \tag{4.38}$$

PROBLEM 4.2. Show that

$$K[a,b](\omega) = \sum_{\Gamma\in\mathcal{B}[a,b]} \prod_{st\in\Gamma} U_{st}(\omega). \tag{4.39}$$

PROBLEM 4.3. For $a < b$, let

$$J[a,b](\omega) = \sum_{\Gamma\in\mathcal{G}[a,b]} \prod_{st\in\Gamma} U_{st}(\omega). \tag{4.40}$$

Show that

$$K[a,b] = K[a+1,b] + \sum_{j=a+1}^{b} J[a,j]K[j,b]. \tag{4.41}$$

PROBLEM 4.4. Define

$$\pi_m(x) = \sum_{\omega\in\mathcal{W}_m(0,x)} J[0,m](\omega) \tag{4.42}$$

for $m \geq 1$. Use Problem 4.3 to show that, for $n \geq 1$,

$$c_n(x) = (c_1 * c_{n-1})(x) + \sum_{m=1}^{n} (\pi_m * c_{n-m})(x). \tag{4.43}$$

(Compared to (4.16), the sum here starts at $m = 1$ instead of $m = 2$. In fact, we will see that $\pi_1(x) = 0$ for the self-avoiding walk, since walks cannot self-intersect in 1 step.)

DEFINITION 4.4. A *lace* is a minimally connected graph, that is, a connected graph for which the removal of any edge would result in a disconnected graph. The set of laces on $[a,b]$ is denoted $\mathcal{L}[a,b]$.

PROBLEM 4.5. Let $L = \{s_1t_1, \ldots, s_Nt_N\}$, where $s_l < t_l$ and $s_l \leq s_{l+1}$ for all l (and all the edges are different). Show that L is a lace if and only if

$$a = s_1 < s_2, \quad s_N < t_{N-1} < t_N = b, \quad s_{l+1} < t_l \leq s_{l+2} \quad (1 \leq l \leq N-2), \tag{4.44}$$

or $L = \{ab\}$ if $N = 1$. In particular, for $N > 1$, L divides $[a,b]$ into $2N - 1$ subintervals,

$$[s_1,s_2],[s_2,t_1],[t_1,s_3][s_3,t_2],\ldots,[t_{N-2},s_N][s_N,t_{N-1}],[t_{N-1},t_N]. \tag{4.45}$$

Determine which of these intervals must have length at least 1, and which can have length 0.

Let $\Gamma \in \mathcal{G}[a,b]$ be a connected graph. We associate a unique lace L_Γ to Γ as follows: Let

$$\begin{aligned} t_1 &= \max\{t : at \in \Gamma\}, \quad s_1 = a, \\ t_{i+1} &= \max\{t : \exists s < t_i \text{ such that } st \in \Gamma\}, \quad s_{i+1} = \min\{s : st_{i+1} \in \Gamma\}. \end{aligned} \tag{4.46}$$

The procedure terminates when $t_N = b$ for some N, and we then define $\mathsf{L}_\Gamma = \{s_1t_1, \dots, s_Nt_N\}$. We define the set of edges *compatible* with a lace $L \in \mathcal{L}[a,b]$ to be

$$\mathcal{C}(L) = \{st : \mathsf{L}_{L\cup\{st\}} = L, st \notin L\}. \tag{4.47}$$

PROBLEM 4.6. Show that $\mathsf{L}_\Gamma = L$ if and only if $L \subset \Gamma$ and $\Gamma \setminus L \subset \mathcal{C}(L)$.

PROBLEM 4.7. Show that

$$J[a,b](\omega) = \sum_{L\in\mathcal{L}[a,b]} \prod_{st\in L} U_{st}(\omega) \sum_{\Gamma:\mathsf{L}_\Gamma = L} \prod_{s't'\in\Gamma\setminus L} U_{s't'}(\omega). \tag{4.48}$$

Conclude from the previous exercise that

$$\sum_{\Gamma:\mathsf{L}_\Gamma=L} \prod_{s't'\in\Gamma\setminus L} U_{s't'}(\omega) = \prod_{s't'\in\mathcal{C}(L)} (1 + U_{s't'}(\omega)), \tag{4.49}$$

and thus

$$J[a,b](\omega) = \sum_{L\in\mathcal{L}[a,b]} \prod_{st\in L} U_{st}(\omega) \prod_{s't'\in\mathcal{C}(L)} (1 + U_{s't'}(\omega)). \tag{4.50}$$

PROBLEM 4.8. Let $\mathcal{L}^{(N)}[a,b]$ denote the set of laces on $[a,b]$ which consist of exactly N edges. Define

$$J^{(N)}[a,b](\omega) = \sum_{L\in\mathcal{L}^{(N)}[a,b]} \prod_{st\in L} U_{st}(\omega) \prod_{s't'\in\mathcal{C}(L)} (1 + U_{s't'}(\omega)) \tag{4.51}$$

and

$$\pi_m^{(N)}(x) = (-1)^N \sum_{\omega\in\mathcal{W}_m(0,x)} J^{(N)}[0,m](\omega). \tag{4.52}$$

(a) Prove that

$$\pi_m(x) = \sum_{N=1}^{\infty} (-1)^N \pi_m^{(N)}(x) \tag{4.53}$$

with $\pi_m^{(N)}(x) \geq 0$.

(b) Describe the walk configurations that correspond to non-zero terms in $\pi_m^{(N)}(x)$, for $N = 1,2,3,4$. What parts of the walk must be mutually avoiding?

(c) What is the interpretation of the possibly empty intervals in Problem 4.5?

5. Lace expansion analysis in dimensions $d > 4$

In this section, we outline a proof that the bubble condition holds for the nearest-neighbour model in sufficiently high dimensions, and for the spread-out model in dimensions $d > 4$ provided L is large enough. As noted above, the bubble condition implies that $\gamma = 1$ in the sense that the susceptibility diverges linearly at the critical point as in (4.5). Proving the bubble condition will require control of the generating function $\hat{\Pi}_z(k)$ at the critical value $z = z_c$. According to (4.21) (see also Problem 4.8), Π_z is given by an infinite series

$$\Pi_z(x) = \sum_{N=1}^{\infty} (-1)^N \Pi_z^{(N)}(x), \qquad \Pi_z^{(N)}(x) = \sum_{m=2}^{\infty} \pi_m^{(N)}(x) z^m. \tag{5.1}$$

The lace expansion is said to converge if $\Pi_z(x)$ is absolutely summable when $z = z_c$, in the strong sense that

$$\sum_{x\in\mathbb{Z}^d}\sum_{N=1}^{\infty}\Pi_{z_c}^{(N)}(x) < \infty. \tag{5.2}$$

There are now several different approaches to proving convergence of the lace expansion. In particular, a powerful but technically demanding method involves the study of (4.16) by induction on n [**41**]. Here we will follow the relatively simple approach of [**69**], which was inspired by a similar argument for percolation in [**2**]. Some details are omitted below; these can all be found in [**69**].

We will make use of the usual ℓ^p norms on functions on $\mathbb{Z}^d$, for $p = 1, 2, \infty$. In addition, when dealing with functions on the torus $[-\pi,\pi]^d$, we will use the usual L^p norms with respect to the probability measure $(2\pi)^{-d}d^dk$ on the torus, for $p = 1, 2$. To simplify the notation, we will sometimes omit the measure, and write, e.g., $\mathsf{B}(z) = \int \hat{G}_z^2 = \|\hat{G}_z\|_2^2$.

5.1. Diagrammatic estimates. We will obtain bounds on $\Pi_z(x)$ in terms of $G_z(x)$ and the closely related quantity $H_z(x)$ defined by

$$H_z(x) = G_z(x) - \delta_{0x} = \sum_{n=1}^{\infty} c_n(x)z^n. \tag{5.3}$$

The trivial term $c_0(x) = \delta_{0x}$ in $G_z(x)$ gives rise to a contribution 1 in the bubble diagram, and it will be important in the following that this contribution sometimes be omitted. It is for this reason that we use H_z as well as G_z.

The following *diagrammatic estimates* bound Π_z in terms of H_z and G_z. Once this theorem has been proved, the details of the definition of Π_z are no longer needed—the rest of the argument is analysis that uses the diagrammatic estimates.

THEOREM 5.1. *For any* $z \geq 0$,

$$\sum_{x\in\mathbb{Z}^d}\Pi_z^{(1)}(x) \leq z\,|\Omega|\,\|H_z\|_\infty, \tag{5.4}$$

$$\sum_{x\in\mathbb{Z}^d}(1-\cos k\cdot x)\Pi_z^{(1)}(x) = 0, \tag{5.5}$$

and for $N \geq 2$,

$$\sum_{x\in\mathbb{Z}^d}\Pi_z^{(N)}(x) \leq \|H_z\|_\infty\,\|G_z * H_z\|_\infty^{N-1}, \tag{5.6}$$

$$\sum_{x\in\mathbb{Z}^d}(1-\cos k\cdot x)\Pi_z^{(N)}(x) \leq N^2\,\|(1-\cos k\cdot x)H_z\|_\infty\,\|G_z * H_z\|_\infty^{N-1}. \tag{5.7}$$

PROOF. We prove just the cases $N = 1, 2$ here; the complete proof can be found in [**69**, Theorem 4.1].

For $N = 1$, since $\pi_m^{(1)}(x)$ is equal to δ_{0x} times the number $\sum_{y\in\Omega} c_{m-1}(y)$ of self-avoiding returns, we have

$$\sum_{x\in\mathbb{Z}^d}\Pi_z^{(1)}(x) = \sum_{y\in\Omega}\sum_{m=2}^{\infty} c_{m-1}(y)z^m = \sum_{y\in\Omega} zH_z(y), \tag{5.8}$$

which implies (5.4). Also, (5.5) follows from

$$\sum_{x\in\mathbb{Z}^d}(1-\cos k\cdot x)\Pi_z^{(1)}(x)=(1-\cos k\cdot 0)\Pi_z^{(1)}(0)=0. \tag{5.9}$$

For $N=2$, dropping the mutual avoidance constraint between the three lines in $\pi_m^{(2)}(x)$ in (4.21) gives

$$\begin{aligned}\sum_{x\in\mathbb{Z}^d}\Pi_z^{(2)}(x)&\le\sum_{x\in\mathbb{Z}^d}H_z(x)^3\le\|H_z\|_\infty(H_z*H_z)(0)\\&\le\|H_z\|_\infty\|H_z*H_z\|_\infty\end{aligned} \tag{5.10}$$

and

$$\sum_{x\in\mathbb{Z}^d}(1-\cos k\cdot x)\Pi_z^{(2)}(x)\le\|(1-\cos k\cdot x)H_z\|_\infty\|H_z*H_z\|_\infty. \tag{5.11}$$

Since $0\le H_z(x)\le G_z(x)$, this is stronger than (5.6) and (5.7). □

5.2. The small parameter. Theorem 5.1 shows that the sum over N in (5.1) can be dominated by the sum of a geometric series with ratio $\|G_z*H_z\|_\infty$. Ideally, we would like this ratio to be small. A Cauchy–Schwarz estimate gives

$$\begin{aligned}\|H_z*G_z\|_\infty&\le\|H_z\|_\infty+\|H_z*H_z\|_\infty\le\|H_z\|_\infty+\|H_z\|_2^2\\&\le\|H_z\|_\infty+\|G_z\|_2^2=\|H_z\|_\infty+\mathsf{B}(z),\end{aligned} \tag{5.12}$$

but this looks problematic because the upper bound involves the bubble diagram —the very quantity we are trying to prove is finite at the critical point! So we will need some insight to make good use of the diagrammatic estimates.

An important idea will be to use not just the finiteness, but also the smallness of H_z. Specifically, we might hope that $\|H_{z_c}\|_2^2=\|\hat{H}_{z_c}\|_2^2=\|\hat{G}_{z_c}-1\|_2^2$ should be small when the corresponding quantity for SRW is small.

Let $C_z(x)=\sum_{n=0}c_n^{(0)}(x)z^n$ be the analogue of $G_z(x)$ for the SRW model. Its critical value is $z_0=|\Omega|^{-1}$, and

$$\hat{C}_z(k)=\frac{1}{1-z|\Omega|\hat{D}(k)},\qquad \hat{C}_{z_0}(k)=\frac{1}{1-\hat{D}(k)}. \tag{5.13}$$

The SRW analogue of $\|\hat{G}_{z_c}-1\|_2^2$ is

$$\|\hat{C}_{z_0}-1\|_2^2=\int\left(\frac{1}{1-\hat{D}}-1\right)^2=\int\frac{\hat{D}^2}{(1-\hat{D})^2}. \tag{5.14}$$

The following elementary proposition shows that the above integral is small for the models we are studying. The hypothesis $d>4$ is needed for convergence, due to the $(|k|^{-2})^2$ singularity at the origin.

PROPOSITION 5.2. *Let $d>4$. Then*

$$\int\frac{\hat{D}^2}{(1-\hat{D})^2}\le\beta \tag{5.15}$$

where, for some constant K,

$$\beta=\begin{cases}\dfrac{K}{d-4} & \text{for the nearest-neighbour model,}\\[2mm] \dfrac{K}{L^d} & \text{for the spread-out model.}\end{cases} \tag{5.16}$$

PROOF. This is a calculus problem. For the nearest-neighbour model, see [**57**, Lemma A.3], and for the spread-out model see [**69**, Proposition 5.3]. □

We will prove the following theorem.

THEOREM 5.3. *There are constants β_0 and C, independent of d and L, such that when* (5.15) *holds with $\beta \le \beta_0$ we have $\mathsf{B}(z_c) \le 1 + C\beta$.*

Theorem 5.3 achieves our goal of proving the bubble condition for the nearest-neighbour model in sufficiently high dimensions, and for the spread-out model with L sufficiently large in dimensions $d > 4$. As noted previously, this gives the following corollary that $\gamma = 1$ in high dimensions.

COROLLARY 5.4. *When* (5.15) *holds with $\beta \le \beta_0$, then as $z \nearrow z_c$,*

$$\chi(z) \asymp \frac{1}{1 - z/z_c}. \tag{5.17}$$

5.3. Proof of Theorem 5.3. We begin with the following elementary lemma, which will be a principal ingredient in the proof.

LEMMA 5.5. *Let $a < b$ be real numbers and let f be a continuous real-valued function on $[z_1, z_2)$ such that $f(z_1) \le a$. Suppose that, for each $z \in (z_1, z_2)$, we have the implication*

$$f(z) \le b \qquad \Longrightarrow \qquad f(z) \le a. \tag{5.18}$$

Then $f(z) \le a$ for all $z \in [z_1, z_2)$.

PROOF. The result is a straightforward application of the Intermediate Value Theorem. □

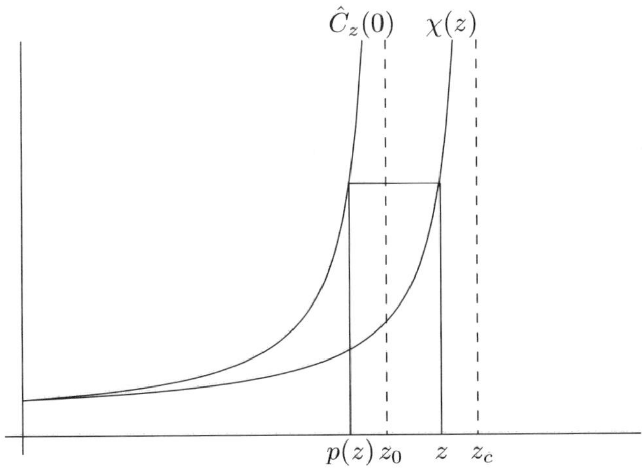

FIGURE 9. The definition of $p(z)$.

We will apply Lemma 5.5 to a carefully chosen function f, based on a *coupling* between $\hat{G}$ on the parameter range $[0, z_c)$, and the SRW analogue $\hat{C}$ on the parameter range $[0, z_0)$. To define the coupling, let $z \in [0, z_c)$ and define $p(z) \in [0, z_0)$

by

$$\hat{G}_z(0) = \chi(z) = \hat{C}_{p(z)}(0) = \frac{1}{1 - p(z)\,|\Omega|}, \tag{5.19}$$

i.e.,

$$p(z)\,|\Omega| = 1 - \chi(z)^{-1} = z\,|\Omega| + \hat{\Pi}_z(0). \tag{5.20}$$

See Figure 9. We expect (or hope!) that $\hat{G}_z(k) \approx \hat{C}_{p(z)}(k)$ for *all* k, not just for $k = 0$, as well as an additional condition that expresses another form of similarity between $\hat{G}_z(k)$ and $\hat{C}_{p(z)}(k)$. For the latter, we define

$$-\tfrac{1}{2}\Delta_k \hat{G}_z(l) = \hat{G}_z(l) - \tfrac{1}{2}\big(\hat{G}_z(l+k) + \hat{G}_z(l-k)\big); \tag{5.21}$$

this is the Fourier transform of $(1 - \cos k \cdot x)G_z(x)$ with l as the dual variable. We aim to apply Lemma 5.5 with $z_1 = 0$, $z_2 = z_c$, $a = 1 + \text{const} \cdot \beta$ (with a constant whose value is determined in (5.27) below), $b = 4$ (in fact, any fixed $b > 1$ will do here), and

$$f(z) = \max\{f_1(z), f_2(z), f_3(z)\} \tag{5.22}$$

where

$$f_1(z) = z\,|\Omega|, \qquad f_2(z) = \sup_{k \in [-\pi,\pi]^d} \frac{|\hat{G}_z(k)|}{|\hat{C}_{p(z)}(k)|}, \tag{5.23}$$

and

$$f_3(z) = \sup_{k,l \in [-\pi,\pi]^d} \frac{\frac{1}{2}|\Delta_k \hat{G}_z(l)|}{|U_{p(z)}(k,l)|}, \tag{5.24}$$

with

$$\begin{aligned} U_{p(z)}(k,l) = 16\hat{C}_{p(z)}(k)^{-1}\Big(&\hat{C}_{p(z)}(l-k)\hat{C}_{p(z)}(l) + \hat{C}_{p(z)}(l+k)\hat{C}_{p(z)}(l) \\ &+ \hat{C}_{p(z)}(l-k)\hat{C}_{p(z)}(l+k)\Big). \end{aligned} \tag{5.25}$$

The choice of $U_{p(z)}(k,l)$ is made for technical reasons not explained here, and should be regarded as a useful replacement for the more natural choice $\frac{1}{2}|\Delta_k \hat{C}_{p(z)}(l)|$.

The conclusion from Lemma 5.5 would be that $f(z) \le a$ for all $z \in [0, z_c)$. The inequality (5.15) can be used to show that $\|(1-\hat{D})^{-1}\|_2^2 \le 1 + 3\beta$ (see [**69**, (5.10)]), and hence we may assume that $\|(1-\hat{D})^{-1}\|_2^2 \le 2$. Using $f_2(z) \le a$ we therefore conclude that

$$\begin{aligned} B(z_c) = \lim_{z \nearrow z_c} B(z) &= \lim_{z \nearrow z_c} \|\hat{G}_z\|_2^2 \\ &\le a^2 \lim_{z \nearrow z_c} \|\hat{C}_{p(z)}\|_2^2 = a^2 \big\|(1-\hat{D})^{-1}\big\|_2^2 \\ &\le 2a^2 < \infty \end{aligned} \tag{5.26}$$

which is our goal. Thus it suffices to verify the hypotheses on $f(z)$ in Lemma 5.5. This is the content of the following lemma.

LEMMA 5.6. *The function $f(z)$ defined by* (5.22)–(5.24) *is continuous on* $[0, z_c)$, *with $f(0) = 1$, and for each $z \in (0, z_c)$,*

$$f(z) \le 4 \quad \Longrightarrow \quad f(z) \le 1 + O(\beta). \tag{5.27}$$

PROOF. It is relatively easy to verify the continuity of f, and we omit the details. To see that $f(0) = 1 \le a$, we observe that $f_1(0) = 0$, $p(0) = 0$ and hence $f_2(0) = 1/1 = 1$, and $f_3(0) = 0$. The difficult step is to prove the implication (5.27), and the remainder of the proof concerns this step. We assume throughout that $f(z) \le 4$.

We consider first $f_1(z) = z\,|\Omega|$. Our goal is to prove that $f_1(z) \le 1 + O(\beta)$, and for this we will only use the assumptions $f_1(z) \le 4$ and $f_2(z) \le 4$; we do not yet need f_3. Since $0 < \chi(z) < \infty$, we have $\chi(z)^{-1} = 1 - z\,|\Omega| - \hat{\Pi}_z(0) > 0$, i.e.,

$$f_1(z) = z\,|\Omega| < 1 - \hat{\Pi}_z(0) \le 1 + |\hat{\Pi}_z(0)|. \tag{5.28}$$

The required bound for $f_1(z)$ will follow once we show that for all $z \in (0, z_c)$ and for all $k \in [-\pi, \pi]^d$,

$$|\hat{\Pi}_z(k)| \le O(\beta). \tag{5.29}$$

To prove (5.29) we use Theorem 5.1 (more precisely, (5.4) and (5.6)), to obtain

$$\begin{aligned} |\hat{\Pi}_z(k)| &\le \sum_{N=1}^{\infty} \sum_{x\in\mathbb{Z}^d} \Pi_z^{(N)}(x) \\ &\le \|H_z\|_\infty \left(f_1(z) + \sum_{N=2}^{\infty} \|G_z * H_z\|_\infty^{N-1} \right). \end{aligned} \tag{5.30}$$

For the first term, we use $f_1(z) \le 4$. For the second term, we need a bound on $\|G_z * H_z\|_\infty$ in order to bound the sum. By definition,

$$\|G_z * H_z\|_\infty \le \|H_z\|_\infty + \|H_z * H_z\|_\infty \le \|H_z\|_\infty + \|H_z\|_2^2 . \tag{5.31}$$

Now H_z is the generating function for SAWs which take at least one step. By omitting the avoidance constraint between the first step and subsequent steps, we obtain

$$H_z(x) \le z\,|\Omega|\,(D * G_z)(x) \le 4(D * G_z)(x). \tag{5.32}$$

Thus we can bound the second term in (5.31), using $f_2(z) \le 4$ and Proposition 5.2, as

$$\begin{aligned} \|H_z\|_2^2 &\le 4^2\,\|D * G_z\|_2^2 = 4^2 \big\|\hat{D}\hat{G}_z\big\|_2^2 \\ &\le 4^4 \big\|\hat{D}\hat{C}_{p(z)}\big\|_2^2 = 4^4\,\big\|D * C_{p(z)}\big\|_2^2 \\ &\le 4^4\,\|D * C_{z_0}\|_2^2 = 4^4 \big\|\hat{D}(1-\hat{D})^{-1}\big\|_2^2 \\ &\le 4^4\beta. \end{aligned} \tag{5.33}$$

Similar estimates show $\|H_z\|_\infty \le O(\beta)$. If we substitute these estimates into (5.30), we obtain

$$|\hat{\Pi}_z(k)| \le C\beta \left(4 + \sum_{N=2}^{\infty} (C\beta)^{N-1} \right) \tag{5.34}$$

for some constant C, so that (5.29) will hold for β sufficiently small. This completes the proof for $f_1(z)$.

We next sketch the proof that $f_2(z) \le 1 + O(\beta)$. Recalling the notation $\hat{F}_z(k) = \hat{G}_z(k)^{-1}$ introduced in (4.31), and using the formulas (5.19) and (5.20) for $p(z)$, we obtain

$$\begin{aligned}
\frac{\hat{G}_z(k)}{\hat{C}_{p(z)}(k)} - 1 &= \frac{1 - p(z)\,|\Omega|\,\hat{D}(k)}{\hat{F}_z(k)} - 1 \\
&= \frac{1 - \big(z\,|\Omega| + \hat{\Pi}_z(0)\big)\hat{D}(k) - \hat{F}_z(k)}{\hat{F}_z(k)} \\
&= \frac{-\hat{\Pi}_z(0)\hat{D}(k) + \hat{\Pi}_z(k)}{\hat{F}_z(k)} \\
&= \frac{\hat{\Pi}_z(0)\big(1 - \hat{D}(k)\big) - \big(\hat{\Pi}_z(0) - \hat{\Pi}_z(k)\big)}{\hat{F}_z(k)}. \qquad (5.35)
\end{aligned}$$

The bound $f_3(z) \leq 4$ and (5.7) can be used to show that $|\hat{\Pi}_z(0) - \hat{\Pi}_z(k)| \leq O(\beta)\big(1 - \hat{D}(k)\big)$ (see **[69]** for details); it is precisely at this point that the need to include f_3 in the definition of f arises. Together with (5.29), this shows that the numerator of (5.35) is $O(\beta)\big(1 - \hat{D}(k)\big)$.

For the denominator, we recall the formula (4.32):

$$\hat{F}_z(k) = \chi(z)^{-1} + z\,|\Omega|\,\big(1 - \hat{D}(k)\big) + \big(\hat{\Pi}_z(0) - \hat{\Pi}_z(k)\big). \qquad (5.36)$$

To bound $\hat{F}_z(k)$ from below, we consider two parameter ranges for z. If $z \leq \frac{1}{2}\,|\Omega|^{-1}$, we can make the trivial estimate $\chi(z)^{-1} \geq \hat{C}_z(0)^{-1} = 1 - z\,|\Omega| \geq \frac{1}{2}$, so that $\hat{F}_z(k) \geq \frac{1}{2} + 0 - O(\beta) \geq \frac{1}{4}$ for small β. Since the numerator of (5.35) is itself $O(\beta)$, this proves that $f_2(z) \leq 1 + O(\beta)$ for this range of z.

It remains to consider $\frac{1}{2}\,|\Omega|^{-1} \leq z \leq z_c$. Now we estimate

$$\hat{F}_z(k) \geq 0 + \tfrac{1}{2}\big(1 - \hat{D}(k)\big) - O(\beta)\big(1 - \hat{D}(k)\big) \geq \tfrac{1}{4}\big(1 - \hat{D}(k)\big). \qquad (5.37)$$

The factors $1 - \hat{D}(k)$ in the numerator and denominator of (5.35) cancel, leaving $O(\beta)$ as desired.

Finally the proof for $f_3(z)$ is similar to the proof for $f_2(z)$, and we refer to **[69]** for the details. □

5.4. Tutorial. For simplicity, we restrict our attention now to the nearest-neighbour model of SAWs in dimensions sufficiently high that the preceding arguments and conclusions apply. In Lemma 5.6, we found that $f_2(z) \leq a = 1 + O(d^{-1})$, since $\beta \leq O((d-4)^{-1}) = O(d^{-1})$. This estimate, which states that

$$\hat{G}_z(k) \leq a\hat{C}_{p(z)}(k) \qquad k \in [\pi, \pi]^d, \quad z \in (0, z_c), \qquad (5.38)$$

is most important for $k \approx 0$, the *small frequencies*, and it is referred to as the *infrared bound.* Other bounds obtained in Lemma 5.6 can be framed as follows: there is a constant c, independent of $z \leq z_c$, such that

$$\|H_z\|_2^2 \leq cd^{-1}, \quad \|H_z\|_\infty \leq cd^{-1}, \quad \|\Pi_z\|_1 \leq cd^{-1}, \qquad (5.39)$$

and

$$\|\Pi_z^{(N)}\|_1 \leq (cd^{-1})^N, \quad \sum_{N=M}^{\infty} \|\Pi_z^{(N)}\|_1 \leq cd^{-M}. \qquad (5.40)$$

We also recall that the Fourier transform of the two-point function can be written as

$$\hat{G}_z(k) = \frac{1}{1 - z|\Omega|\hat{D}(k) - \hat{\Pi}_z(k)}. \tag{5.41}$$

Since $\hat{G}_z(0) \to \infty$ as $z \to z_c$, we obtain the equation

$$1 - z_c|\Omega| - \hat{\Pi}_{z_c}(0) = 0. \tag{5.42}$$

This equation provides a starting point to study the connective constant $\mu = z_c^{-1}$.

PROBLEM 5.1. In this problem, we show that the connective constant obeys

$$\mu = 2d - 1 - (2d)^{-1} + O((2d)^{-2}) \qquad \text{as } d \to \infty. \tag{5.43}$$

This special case of the results discussed in Section 1.4 was first proved by Kesten [**51**], by very different means.

(a) Let $m \geq 1$ be an integer. Show that $\|(1 - \hat{D})^{-m}\|_1$ is non-increasing in $d > 2m$. In particular, it follows that $\|\hat{C}_{z_0}\|_2$ is bounded uniformly in $d > 4$.
Hint: $A^{-m} = \Gamma(m)^{-1} \int_0^\infty u^{m-1} e^{-uA}\, du$.

(b) Let $H_z^{(j)}(x) = \sum_{m=j}^\infty c_m(x) z^m$ be the generating function for SAWs that take at least j steps. By relaxing the condition of mutual self-avoidance for the first j steps, show that

$$\|H_{z_c}^{(j)}\|_\infty \leq O((2d)^{-j/2}), \quad j > 1. \tag{5.44}$$

Hint: Use the infrared bound for the two-point function (5.38), and that the probability that a $2j$-step simple random walk which starts at 0 also ends at 0 is

$$\|\hat{D}^{2j}\|_1 \leq O((2d)^{-j}). \tag{5.45}$$

(c) Recall that $\pi_n^{(1)}(x) = 0$ if $x \neq 0$, so that $\hat{\Pi}_z^{(1)}(0) = \sum_{x \in \mathbb{Z}^d} \Pi_z^{(1)}(x) = \Pi_z^{(1)}(0)$ is the generating function for all self-avoiding returns. Prove that

$$\hat{\Pi}_{z_c}^{(1)}(0) = (2d)^{-1} + 3(2d)^{-2} + O((2d)^{-3}) \tag{5.46}$$

(d) Note that $\hat{\Pi}_z^{(2)}(0)$ is the generating function for all *θ-walks*: paths that visit their eventual endpoint, return to the origin, then return to their endpoint, and are otherwise self-avoiding. Prove that

$$\hat{\Pi}_z^{(2)}(0) = (2d)^{-2} + O((2d)^{-3}). \tag{5.47}$$

(e) Conclude from (c) and (d) that

$$\hat{\Pi}_z(0) = -(2d)^{-1} - 2(2d)^{-2} + O((2d)^{-3}), \tag{5.48}$$

and use this to show

$$\mu = 2d - 1 - (2d)^{-1} + O((2d)^{-2}). \tag{5.49}$$

We have seen in Section 4.2 that $\chi(z) \asymp (1 - z/z_c)^{-1}$ in high dimensions, assuming the bubble condition. The next problem shows that this bound can be improved to an asymptotic formula.

PROBLEM 5.2. (a) Show that

$$\frac{d[z\chi(z)]}{dz} = V(z)\chi(z)^2, \quad \text{where } V(z) = 1 - \hat{\Pi}_z(0) + z\frac{d\hat{\Pi}_z(0)}{dz}. \tag{5.50}$$

Hint: Let $\hat{F}_z(0) = \chi(z)^{-1} = 1 - z|\Omega| - \hat{\Pi}_z(0)$ and express the left-hand side in terms of $\hat{F}_z(0)$.

(b) Show that $\hat{\Pi}_{z_c}(0)$, $\frac{d}{dz}\hat{\Pi}_{z_c}(0)$ and thus $V(z_c)$ are finite. It follows that

$$\frac{d[z\chi(z)]}{dz} = V(z)\chi(z)^2 \sim V(z_c)\chi(z)^2 \quad \text{as } z \nearrow z_c, \tag{5.51}$$

where $f(z) \sim g(z)$ means $\lim_{z\nearrow z_c} f(z)/g(z) = 1$.

(c) Prove that $\chi(z) \sim A(1 - z/z_c)^{-1}$ as $z \nearrow z_c$, where the constant A is given by $A = z_c^{-1}[2d + \frac{d}{dz}|_{z=z_c}\hat{\Pi}_z(0)]^{-1}$.

6. Integral representation for walk models

It has long been understood by physicists that it is sometimes possible to represent random fields by random walks. Ideas in this direction due to Symanzik [**71**] were influential among mathematicians, and inspired, e.g., the analysis of [**6, 7**] who showed how to use random walks to represent and analyse ferromagnetic lattice spin systems. In this section, we develop representations of two random walk models in terms of random fields, via functional integrals. Our ultimate goal is rather the opposite to that of [**6, 7**], namely we wish to study models of random walks via studying their integral representations. This will be the topic of Section 7.

We begin in Section 6.1 with some background material about Gaussian integrals. In Section 6.2, we use these Gaussian integrals to represent a model of SAWs in a background of self-avoiding loops, a model closely related to the $O(n)$ loop model discussed in Section 3.4. The random field in these Gaussian integrals is called a *boson* field in physics. It was realised in the physics literature [**59, 63**] that the loops in the loop model could be eliminated by the use of anti-commuting variables, referred to as a *fermion* field, thereby providing a representation for models of SAWs. The anti-commuting variables can be understood in terms of differential forms with their anti-commuting wedge product, and in Sections 6.3–6.4 we provide the relevant background on differential forms and their integration. Finally, in Section 6.5, we obtain an integral representation for SAWs. The ideas in this section are developed in further detail in [**11**].

6.1. Gaussian integrals. Fix a positive integer M. Later, we identify the set $\{1,\ldots,M\}$ with a finite set Λ on which the walks related to the fields take place, e.g., $\Lambda \subset \mathbb{Z}^d$. Consider a two-component real field

$$(u,v) = (u_x, v_x)_{x\in\{1,\ldots,M\}} \in \mathbb{R}^M \times \mathbb{R}^M. \tag{6.1}$$

From this, we obtain the associated complex field $(\varphi,\bar{\varphi}) = (\varphi_x,\bar{\varphi}_x)_{x\in\{1,\ldots,M\}}$, where

$$\varphi_x = u_x + iv_x, \quad \bar{\varphi}_x = u_x - iv_x; \tag{6.2}$$

this is the so-called *boson field.* We wish to integrate with respect to the variables $(\varphi_x,\bar{\varphi}_x)$, and for this we will use the differentials $d\varphi_x = du_x + i\,dv_x$ and $d\bar{\varphi}_x = du_x - i\,dv_x$. As we will discuss in more detail in Section 6.3, differentials are multiplied using an anti-commuting product, so in particular $du_x\,dv_x = -dv_x\,du_x$, $du_x\,du_x = dv_x\,dv_x = 0$, and $d\bar{\varphi}_x\,d\varphi_x = 2i\,du_x\,dv_x$.

Let $C = (C_{xy})_{x,y\in\{1,\dots,M\}}$ be an $M \times M$ complex matrix with positive Hermitian part, meaning that

$$\sum_{x,y=1}^{M} \varphi_x (C_{xy} + \bar{C}_{yx}) \bar{\varphi}_y > 0 \quad \text{for all } \varphi \neq 0 \text{ in } \mathbb{C}^M. \tag{6.3}$$

It is not difficult to see that this implies that $A = C^{-1}$ exists. The (complex) Gaussian measure with covariance C is defined by

$$d\mu_C(\varphi, \bar{\varphi}) = \frac{1}{Z_C} e^{-\varphi A \bar{\varphi}} \, d\bar{\varphi} \, d\varphi, \tag{6.4}$$

where $\varphi A \bar{\varphi} = \sum_{x,y=1}^{M} \varphi_x A_{xy} \bar{\varphi}_y$, and

$$d\bar{\varphi} \, d\varphi = d\bar{\varphi}_1 \, d\varphi_1 \cdots d\bar{\varphi}_M \, d\varphi_M = (2i)^M du_1 \, dv_1 \cdots du_M \, dv_M \tag{6.5}$$

is a multiple of the Lebesgue measure on $\mathbb{R}^{2M}$. The normalisation constant

$$Z_C = \int_{\mathbb{R}^{2M}} e^{-\varphi A \bar{\varphi}} \, d\bar{\varphi} \, d\varphi \tag{6.6}$$

can be computed explicitly.

Lemma 6.1. *For C with positive Hermitian part, the normalisation of the Gaussian integral is given by*

$$Z_C = \frac{(2\pi i)^M}{\det A}. \tag{6.7}$$

Proof. In this proof, we make the simplifying assumption that C and thus also A are Hermitian, though the result holds more generally; see [**11**]. By the spectral theorem for Hermitian matrices, there is a positive diagonal matrix $D = \operatorname{diag}(d_x)$ and a unitary matrix U such that $A = U^{-1} D U$. Then, $\varphi A \bar{\varphi} = \rho D \bar{\rho}$ where $\rho = \bar{U} \varphi$ ($\bar{U}$ is the complex conjugate of U). By a change of variables in the integral and explicit computation of the resulting 1-dimensional integral,

$$Z_C = \prod_{x=1}^{M} \int_{\mathbb{R}^2} e^{-d_x (u_x^2 + v_x^2)} \, 2i \, du_x \, dv_x = \frac{(2\pi i)^M}{\prod_{x=1}^{M} d_x} = \frac{(2\pi i)^M}{\det A}. \tag{6.8}$$

□

We define the differential operators

$$\frac{\partial}{\partial \varphi_x} = \frac{1}{2} \left(\frac{\partial}{\partial u_x} - i \frac{\partial}{\partial v_x} \right), \quad \frac{\partial}{\partial \bar{\varphi}_x} = \frac{1}{2} \left(\frac{\partial}{\partial u_x} + i \frac{\partial}{\partial v_x} \right). \tag{6.9}$$

It is easy to check that

$$\frac{\partial \varphi_y}{\partial \varphi_x} = \frac{\partial \bar{\varphi}_y}{\partial \bar{\varphi}_x} = \delta_{xy}, \quad \frac{\partial \bar{\varphi}_y}{\partial \varphi_x} = \frac{\partial \varphi_y}{\partial \bar{\varphi}_x} = 0. \tag{6.10}$$

The following integration by parts formula will be useful.

Lemma 6.2. *For C with positive Hermitian part, and for nice functions F,*

$$\int \bar{\varphi}_a F \, d\mu_C(\varphi, \bar{\varphi}) = \sum_{x=1}^{M} C_{ax} \int \frac{\partial F}{\partial \varphi_x} \, d\mu_C(\varphi, \bar{\varphi}). \tag{6.11}$$

PROOF. Integrating by parts, we obtain

$$\int \frac{\partial F}{\partial \varphi_x} e^{-\varphi A\bar\varphi}\, d\bar\varphi\, d\varphi = -\int F \frac{\partial}{\partial \varphi_x} e^{-\varphi A\bar\varphi}\, d\bar\varphi\, d\varphi$$

$$= \int F \sum_y A_{xy}\bar\varphi_y e^{-\varphi A\bar\varphi}\, d\bar\varphi\, d\varphi. \tag{6.12}$$

It follows from the fact that $C = A^{-1}$ that

$$\sum_{x=1}^{M} C_{ax} \int \frac{\partial F}{\partial \varphi_x}\, d\mu_C = \int \sum_{x,y} C_{ax}A_{xy}\bar\varphi_y F\, d\mu_C = \int \bar\varphi_a F\, d\mu_C. \tag{6.13}$$ □

The following application of Lemma 6.2 is a special case of Wick's Theorem. The quantity appearing on the right-hand side of (6.14) is the *permanent* of the submatrix of C indexed by $(x_i, y_j)_{i,j=1}^k$.

LEMMA 6.3. *Let $\{x_1, \ldots, x_k\}$ and $\{y_1, \ldots, y_k\}$ each be sets with k distinct elements from $\{1, \ldots, M\}$. Then*

$$\int \prod_{l=1}^{k} \bar\varphi_{x_l}\varphi_{y_l}\, d\mu_C = \sum_{\sigma\in S_k}\prod_{l=1}^{k} C_{x_l, y_{\sigma(l)}}, \tag{6.14}$$

where the sum is over the set S_k of permutations of $\{1, \ldots, k\}$.

PROOF. This follows by repeated application of the integration by parts formula in Lemma 6.2. Each time the formula is applied, one factor of $\bar\varphi$ disappears on the right-hand side of (6.11), and the partial differentiation eliminates one factor φ as well. □

6.2. Integral representation for a loop model. Let Λ be a finite set of cardinality M. Fix $a, b \in \Lambda$ and a subset $X \subset \Lambda \setminus \{a, b\}$. An example we have in mind is $\Lambda \subset \mathbb{Z}^d$ and $X = \Lambda \setminus \{a, b\}$. We define the integral

$$G_{ab,X} = \int \bar\varphi_a \varphi_b \prod_{x\in X} (1 + \varphi_x\bar\varphi_x)\, d\mu_C. \tag{6.15}$$

As we now explain, this can be interpreted as a loop model whose configurations consist of a self-avoiding walk from a to b whose intermediate steps lie in X, together with a background of closed loops in X. We denote by $\mathcal{S}_{ab}(X)$ the set of sequences $(a, x_1, \ldots, x_{n-1}, b)$ with $n \geq 1$ arbitrary and the $x_i \in X$ distinct—these are SAWs with rather general steps.

Repeated integration by parts gives

$$G_{ab,X} = \sum_{\omega\in\mathcal{S}_{ab}(X)} C^{\omega} \int \prod_{x\in X\setminus\omega} (1 + \varphi_x\bar\varphi_x)\, d\mu_C, \tag{6.16}$$

where $C^{\omega} = \prod_{i=1}^{\ell(\omega)} C_{w(i-1),w(i)}$. Also, by expanding the product and applying Lemma 6.3, we obtain

$$\int \prod_{x\in X\setminus\omega} (1 + \varphi_x\bar\varphi_x)\, d\mu_C = \sum_{Z\subset X\setminus\omega} \int \prod_{x\in Z} \varphi_x\bar\varphi_x\, d\mu_C$$

$$= \sum_{Z\subset X\setminus\omega} \sum_{\sigma\in S(Z)} \prod_{z\in Z} C_{z,\sigma(z)}, \tag{6.17}$$

with $S(Z)$ is the set of permutations of the set Z. Altogether, this gives

$$G_{ab,X} = \sum_{\omega \in \mathcal{S}_{ab}(X)} C^{\omega} \sum_{Z \subset X \setminus \omega} \sum_{\sigma \in S(Z)} \prod_{z \in Z} C_{z,\sigma(z)}. \tag{6.18}$$

Thus, by decomposing the permutation σ into cycles, we can interpret (6.15) as the generating function for self-avoiding walks from a to b in a background of loops with weight C_{xy} for every step between x and y (with each loop corresponding to a cycle of σ). See Figure 10.

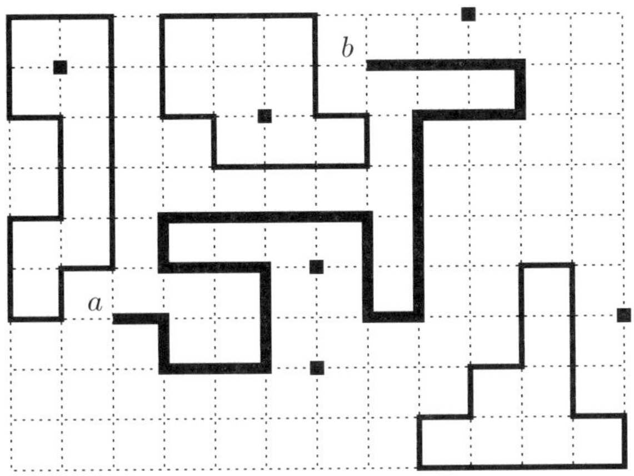

FIGURE 10. Self-avoiding walk from a to b with loop background. Loops can have length zero. The loops will be eliminated by the use of differential forms.

6.3. Differential forms. Our next goal is to modify the example of Section 6.2 with the help of differential forms, which are versions of what physicists call *fermions*, to obtain an integral representation for the generating function for self-avoiding walks *without* the loop background. A gentle introduction to differential forms can be found in [**66**].

The *Grassmann algebra* $\mathcal{N}$ of differential forms is generated by the one-forms $du_1, dv_1, \ldots, du_M, dv_M$, with anticommutative product $\wedge$. A p-form (a differential form of degree p) is a function of the variables (u, v) times a product of p differentials or sum of these. Because of anticommutativity, $du_x \wedge du_x = dv_x \wedge dv_x = 0$, and any p-form with $p > 2M$ must be zero. A form of maximal degree can thus be written uniquely as

$$K = f(u, v)\, du_1 \wedge dv_1 \wedge \cdots \wedge du_M \wedge dv_M, \tag{6.19}$$

where $du_1 \wedge dv_1 \wedge \cdots \wedge du_M \wedge dv_M$ is the standard volume form on $\mathbb{R}^{2M}$. A general differential form is a linear combination of p-forms, where different terms in the sum can have different values of p. Together, the differential forms constitute the algebra $\mathcal{N}$.

We will omit the wedge $\wedge$ from the notation from now on, and write simply $du_1\, dv_1$ for $du_1 \wedge dv_1$, but it should be borne in mind that order is significant in such an expression: $du_x\, dv_y = -dv_y\, du_x$. On the other hand, two forms of *even* degree commute.

We again use complex variables, and write

$$\begin{aligned} \varphi_x &= u_x + iv_x, & \bar{\varphi}_x &= u_x - iv_x, \\ d\varphi_x &= du_x + i\,dv_x, & d\bar{\varphi}_x &= du_x - i\,dv_x. \end{aligned} \tag{6.20}$$

Then

$$d\bar{\varphi}_x \, d\varphi_x = 2i \, du_x \, dv_x. \tag{6.21}$$

Given any fixed choice of the complex square root, we introduce the notation

$$\psi_x = \frac{1}{\sqrt{2\pi i}} d\varphi_x, \quad \bar{\psi}_x = \frac{1}{\sqrt{2\pi i}} d\bar{\varphi}_x. \tag{6.22}$$

The collection of differential forms

$$(\psi, \bar{\psi}) = (\psi_x, \bar{\psi}_x)_{x \in \{1,\dots,M\}} \tag{6.23}$$

is called the *fermion* field. It follows that

$$\bar{\psi}_x \psi_x = \frac{1}{\pi} du_x \, dv_x. \tag{6.24}$$

Let $\Lambda = \{1, \dots, M\}$. Given an $M \times M$ matrix A, we define the differential form

$$S_A = \varphi A \bar{\varphi} + \psi A \bar{\psi} = \sum_{x,y \in \Lambda} \varphi_x A_{xy} \bar{\varphi}_y + \sum_{x,y \in \Lambda} \psi_x A_{xy} \bar{\psi}_y. \tag{6.25}$$

An example of special interest is the case where $A_{uv} = \delta_{ux}\delta_{vx}$ for some fixed $x \in \Lambda$. In this case, we write τ_x in place of S_A, i.e.,

$$\tau_x = \varphi_x \bar{\varphi}_x + \psi_x \bar{\psi}_x. \tag{6.26}$$

6.4. Functions of forms and integrals of forms. The following definition tells us how to integrate a differential form.

DEFINITION 6.4. Let F be a differential form whose term K of maximal degree is as in (6.19). The *integral of* F is then defined to be

$$\int F = \int K = \int_{\mathbb{R}^{2M}} f(u,v) \, du_1 \, dv_1 \cdots du_M \, dv_M. \tag{6.27}$$

In particular, if F contains no term of degree $2M$ then its integral is zero.

We also need to define functions of even differential forms.

DEFINITION 6.5. Let $K = (K_j)_{j \in J}$ be a finite collection of differential forms, with each K_j even (a sum of forms of even degrees). Let $K_j^{(0)}$ be the degree zero part of K_j. Given a C^∞ function $F : \mathbb{R}^J \to \mathbb{C}$, we define $F(K)$ to be the form given by the Taylor polynomial (a polynomial in ψ and $\bar{\psi}$)

$$F(K) = \sum_\alpha \frac{1}{\alpha!} F^{(\alpha)}(K^{(0)})(K - K^{(0)})^\alpha \tag{6.28}$$

where $\alpha = (\alpha_1, \dots, \alpha_j)$ is a multi-index and

$$\alpha! = \prod_{j \in J} \alpha_j!, \quad (K - K^{(0)})^\alpha = \prod_{j \in J} (K_j - K_j^{(0)})^{\alpha_j}. \tag{6.29}$$

The sum in (6.28) is finite due to anticommutativity, and the product in (6.29) is well-defined because all factors are even and thus commute.

EXAMPLE 6.6. A simple but important example is $J = 1$ and $F(t) = e^{-t}$, for which we obtain, e.g.,

$$e^{-\tau_x} = e^{-\varphi_x\bar{\varphi}_x - \psi_x\bar{\psi}_x} = e^{-\varphi_x\bar{\varphi}_x}(1 - \psi_x\bar{\psi}_x), \tag{6.30}$$

$$e^{-S_A} = e^{-\varphi A\bar{\varphi} - \psi A\bar{\psi}} = e^{-\varphi A\bar{\varphi}} \sum_{n=0}^{M} \frac{(-1)^n}{n!}(\psi A\bar{\psi})^n. \tag{6.31}$$

The following lemma displays a remarkable self-normalisation property of these integrals.

LEMMA 6.7. *If A is a complex $M \times M$ matrix with positive Hermitian part, then*

$$\int e^{-S_A} = 1. \tag{6.32}$$

PROOF. Using (6.31) and Definition 6.4,

$$\begin{aligned}\int e^{-S_A} &= \int_{\mathbb{R}^{2M}} e^{-\varphi A\bar{\varphi}} \frac{1}{M!}(-1)^M(\psi A\bar{\psi})^M \\ &= \frac{1}{M!}\left(\frac{-1}{2\pi i}\right)^M \int_{\mathbb{R}^{2M}} e^{-\varphi A\bar{\varphi}}(d\varphi A d\bar{\varphi})^M. \end{aligned} \tag{6.33}$$

By definition,

$$(d\varphi A d\bar{\varphi})^M = \sum_{x_1,y_1} \cdots \sum_{x_M,y_M} A_{x_1y_1} \cdots A_{x_My_M}\, d\varphi_{x_1}\, d\bar{\varphi}_{y_1} \cdots d\varphi_{x_M}\, d\bar{\varphi}_{y_M}. \tag{6.34}$$

Due to the antisymmetry, non-zero contributions to the above sum require that $x_1, \ldots, x_M$ and $y_1, \ldots, y_M$ each be a permutation of $\{1, \ldots, M\}$. Thus, by interchanging the (commuting) pairs $d\varphi_{x_i} d\bar{\varphi}_{y_i}$ so as to place the x_i in the order $1, \ldots, M$, and then relabelling the y_i, we obtain

$$\begin{aligned}(d\varphi A d\bar{\varphi})^M &= M! \sum_{y_1,\ldots,y_M} A_{1y_1} \cdots A_{My_M}\, d\varphi_1\, d\bar{\varphi}_{y_1} \cdots d\varphi_M\, d\bar{\varphi}_{y_M} \\ &= M! \sum_{y_1,\ldots,y_M} \epsilon_{y_1,\ldots,y_M} A_{1y_1} \cdots A_{My_M}\, d\varphi_1\, d\bar{\varphi}_1 \cdots d\varphi_M\, d\bar{\varphi}_M \\ &= M!\,(-1)^M(\det A)\, d\bar{\varphi}\, d\varphi, \end{aligned} \tag{6.35}$$

where $\epsilon_{y_1,\ldots,y_M}$ is the sign of the permutation $(y_1, \ldots, y_M)$ of $\{1, \ldots, M\}$. With Lemma 6.1, it follows that

$$\int e^{-S_A} = \frac{\det A}{(2\pi i)^M} \int_{\mathbb{R}^{2M}} e^{-\varphi A\bar{\varphi}}\, d\bar{\varphi}\, d\varphi = 1. \tag{6.36}$$

□

REMARK 6.8. More generally, the calculation in the previous proof also shows that for a function $f = f(\varphi, \bar{\varphi})$, a form of degree zero,

$$\int e^{-S_A} f = \int f\, d\mu_C \qquad (C = A^{-1}), \tag{6.37}$$

provided f is such that the integral on the right-hand side converges. In our present setup, we have defined $\int e^{-S_A}F$ for more general forms F, so this provides an extension of the Gaussian integral of Section 6.1.

The self-normalisation property of Lemma 6.7 has the following beautiful extension. The precise hypotheses needed on F can be found in [**11**, Proposition 4.4].

LEMMA 6.9. *If A is a complex $M \times M$ matrix with positive Hermitian part, and $F : \mathbb{R}^M \to \mathbb{C}$ is a nice function (exponential growth at infinity is permitted), then*

$$\int e^{-S_A} F(\tau) = F(0), \tag{6.38}$$

where we regard τ as the vector $(\tau_1, \ldots, \tau_M)$.

PROOF (SKETCH). If F is Schwartz class, e.g., then it can be expressed in terms of its Fourier transform as

$$F(t) = \frac{1}{(2\pi)^M} \int_{\mathbb{R}^M} \hat{F}(k) e^{-ik\cdot t}\, dk_1 \cdots dk_M. \tag{6.39}$$

It then follows that

$$\int e^{-S_A} F(\tau) = \frac{1}{(2\pi)^M} \int \hat{F}(k) \left(\int e^{-S_A - ik\cdot\tau} \right) dk = F(0) \tag{6.40}$$

because $S_A + ik \cdot \tau = S_{A+iK}$ with $K = \mathrm{diag}(k_x)_{x=1}^M$, and thus $\int e^{-S_{A+iK}} = 1$ by Lemma 6.7. □

It is not difficult to extend the integration by parts formula for Gaussian measures, Lemma 6.2, to the present more general setting; see [**11**] for details. The result is the following.

LEMMA 6.10. *For $a \in \Lambda$, for $C = A^{-1}$ with positive Hermitian part, and for forms F for which the integrals exist,*

$$\int e^{-S_A} \bar{\varphi}_a F = \sum_{x\in\Lambda} C_{ax} \int e^{-S_A} \frac{\partial F}{\partial \varphi_x}. \tag{6.41}$$

6.5. Integral representation for self-avoiding walk. Let Λ be a finite set and let $a, b \in \Lambda$. In Section 6.2, we showed that the integral

$$\int \bar{\varphi}_a \varphi_b \prod_{x\neq a,b} (1 + \varphi_x \bar{\varphi}_x)\, d\mu_C \tag{6.42}$$

is the generating function for SAWs in a background of self-avoiding loops. The following theorem shows that the loops are eliminated if we replace the factors $(1+\varphi_x\bar{\varphi}_x)$ by $(1+\tau_x) = (1+\varphi_x\bar{\varphi}_x + \psi_x\bar{\psi}_x)$ and replace the Gaussian measure $d\mu_C$ by e^{-S_A} with $A = C^{-1}$.

THEOREM 6.11. *For $C = A^{-1}$ with positive Hermitian part, and for $a, b \in \Lambda$,*

$$\sum_{\omega\in\mathcal{S}_{a,b}(\Lambda)} C^\omega = \int e^{-S_A} \bar{\varphi}_a \varphi_b \prod_{x\neq a,b} (1 + \tau_x). \tag{6.43}$$

PROOF. Exactly as in Section 6.2, but now using the integration by parts formula of Lemma 6.10, we obtain

$$\int e^{-S_A} \bar{\varphi}_a \varphi_b \prod_{x\neq a,b} (1 + \tau_x) = \sum_{\omega\in\mathcal{S}_{a,b}(\Lambda)} C^\omega \int e^{-S_A} \prod_{x\in\Lambda\setminus\omega} (1 + \tau_x). \tag{6.44}$$

However, the integral on the right-hand side, which formerly generated loops, is now equal to 1 by Lemma 6.9. □

7. Renormalisation group analysis in dimension 4

The integral representation of Theorem 6.11 opens up the following possibility for studying SAWs on $\mathbb{Z}^d$: approximate $\mathbb{Z}^d$ by a large finite set Λ, rewrite the SAW two-point function as an integral as in (6.43), and apply methods of analysis to compute the asymptotic behaviour of the integral uniformly in the limit $\Lambda \nearrow \mathbb{Z}^d$. In this section, we sketch how such a program can be carried out for a particular model of continuous-time weakly SAW on the 4-dimensional lattice $\mathbb{Z}^4$, using a variant of Theorem 6.11. In this approach, once the integral representation has been invoked, the original SAWs no longer appear and play no further role in the analysis. The method of proof is a rigorous renormalisation group method [**12, 13**]. There is work in progress, not discussed further here, to attempt to extend this program to a particular spread-out version of the discrete-time strictly SAW model on $\mathbb{Z}^4$ using Theorem 6.11.

We begin in Section 7.1 with the definition of the continuous-time weakly SAW and a statement of the main result for its two-point function, followed by some commentary on related results. The approximation of the two-point function on $\mathbb{Z}^d$ by a two-point function on a d-dimensional finite torus Λ is discussed in Section 7.2, and the integral representation of the two-point function on Λ is explained in Section 7.3. The discussion of integration of differential forms from Section 6.4 is developed further in Section 7.4. At this point, the stage is set for the application of the renormalisation group method, and this is described briefly in Sections 7.5–7.7. A more extensive account of all this can be found in [**12, 13**].

7.1. Continuous-time weakly self-avoiding walk. The definition of the discrete-time weakly self-avoiding walk was given in Section 1.2. With an unimportant change in our conventions, and writing $z = e^{-\nu}$ and using the parameter $g > 0$ of (1.6) rather than λ, the two-point function (1.31) can be rewritten as

$$G_\nu^{(g),\mathrm{DT}}(x) = \sum_{n=0}^{\infty} \sum_{\omega \in \mathcal{W}_n(0,x)} \exp\Big(-g \sum_{i,j=0}^{n} 1_{\{\omega(i)=\omega(j)\}}\Big) e^{-\nu n}, \tag{7.1}$$

where "DT" emphasises the fact that the walks are in discrete time. The *local time* at $v \in \mathbb{Z}^d$ is defined as the number of visits to v up to time n, i.e.,

$$L_{v,n} = L_{v,n}(\omega) = \sum_{i=0}^{n} 1_{\{\omega(i)=v\}}. \tag{7.2}$$

Note that $\sum_{v \in \mathbb{Z}^d} L_{v,n} = n$ is independent of the walk ω, and that

$$\sum_{v \in \mathbb{Z}^d} L_{v,n}^2 = \sum_{v \in \mathbb{Z}^d} \sum_{i,j=0}^{n} 1_{\{\omega(i)=v\}} 1_{\{\omega(j)=v\}} = \sum_{i,j=0}^{n} 1_{\{\omega(i)=\omega(j)\}}. \tag{7.3}$$

Thus, writing $z = e^{-\nu}$, the two-point function can be rewritten as

$$G_\nu^{(g),\mathrm{DT}}(x) = \sum_{n=0}^{\infty} \sum_{\omega \in \mathcal{W}_n(0,x)} e^{-g \sum_{v \in \mathbb{Z}^d} L_{v,n}^2} e^{-\nu n}. \tag{7.4}$$

The two-point function of the continuous-time weakly SAW is a modification of (7.4) in which the underlying random walk model has continuous, rather than discrete, time. To define the modification, we consider the *continuous-time* random walk X which takes nearest-neighbour steps like the usual SRW, but whose jumps occur after independent Exp($2d$) holding times at each vertex. In other words, the steps occur at the events of a rate-$2d$ Poisson process, rather than at integer times. We write $\mathbb{E}_0$ for the expectation associated to the process X started at $X(0) = 0 \in \mathbb{Z}^d$. The local time of X at v up to time T is now defined by

$$L_{v,T} = \int_0^T 1_{\{X(s)=v\}}\, ds. \tag{7.5}$$

The probabilistic structure of (1.7)–(1.9) extends naturally to the continuous-time setting. With this in mind, we define the *two-point function* of continuous-time weakly SAW by

$$G_\nu^{(g)}(x) = \int_0^\infty \mathbb{E}_0(e^{-g\sum_v L_{v,T}^2} 1_{\{X(T)=x\}})e^{-\nu T}\, dT; \tag{7.6}$$

this is a natural modification of (7.1). The continuous-time SAW is predicted to lie in the same universality class as the discrete-time SAW.

Using a subadditivity argument as in Section 1.3, it is not difficult to see that the limit

$$\lim_{T\to\infty} \left(\mathbb{E}_0(e^{-g\sum_v L_{v,T}^2})\right)^{1/T} = e^{\nu_c(g)} \tag{7.7}$$

exists, for some $\nu_c(g) \le 0$. We leave it as an exercise to show that $\nu_c(g) > -\infty$. In particular, $G_\nu^{(g)}(x)$ is well-defined for $\nu > \nu_c(g)$. The following theorem of Brydges and Slade [**12, 13**] shows that the critical exponent η is equal to 0 for this model, in dimensions $d \ge 4$.

THEOREM 7.1. *Let $d \ge 4$. For $g \ge 0$ sufficiently small, there exists $c_g > 0$ such that*

$$G_{\nu_c(g)}^{(g)}(x) = \frac{c_g}{|x|^{d-2}}(1 + o(1)) \quad \text{as } |x| \to \infty. \tag{7.8}$$

Theorem 7.1 should be compared with the result of Theorem 4.2 for $d \ge 5$. The main point in Theorem 7.1 is the inclusion of the upper critical dimension $d = 4$. In particular, there is no logarithmic correction to the leading asymptotic behaviour of the critical two-point function when $d = 4$. The case $g = 0$ is the classical result that the SRW Green function obeys $G_0^{(0)}(x) \sim c_0|x|^{-(d-2)}$, which in fact holds in all dimensions $d > 2$.

The proof of Theorem 7.1 is based on an integral representation combined with a rigorous renormalisation group method, and is inspired by the methods used in [**5, 9, 10**] for the continuous-time weakly self-avoiding walk on the 4-dimensional *hierarchical lattice*. The hierarchical lattice is a modification of the lattice $\mathbb{Z}^d$ that is particularly amenable to a renormalisation group approach. It is predicted that the models on the hierarchical lattice and $\mathbb{Z}^d$ lie in the same universality class. Strong evidence for this is the result of Brydges and Imbrie [**9**] that on the 4-dimensional hierarchical lattice the typical end-to-end distance after time T is given, for small $g > 0$ and as $T \to \infty$, by

$$\frac{\mathbb{E}_0(|\omega(T)|\, e^{-g\sum_v L_{v,T}^2})}{\mathbb{E}_0(e^{-g\sum_v L_{v,T}^2})} = cT^{1/2}(\log T)^{1/8}\left[1 + \frac{\log\log T}{32\log T} + O\left(\frac{1}{\log T}\right)\right]. \tag{7.9}$$

This matches the prediction (1.29) for $\mathbb{Z}^4$. There are related results by Hara and Ohno [**32**], proved with a completely different renormalisation group approach, for the critical two-point function, susceptibility and correlation length of the *discrete-time* weakly self-avoiding walk on the d-dimensional hierarchical lattice for $d \geq 4$.

Recently, Mitter and Scoppola [**60**] used the integral representation and renormalisation group analysis to study a continuous-time weakly self-avoiding walk with long-range steps. In the model of [**60**], each step of length r has a weight decaying like $r^{-d-\alpha}$, with $\alpha = \frac{1}{2}(3+\epsilon)$ for small $\epsilon > 0$, in dimension $d = 3$. This is *below* the upper critical dimension $2\alpha = 3 + \epsilon$ (recall the discussion below Theorem 4.2). The main result is a control of the renormalisation group trajectory, a first step towards the computation of the asymptotic behaviour of the critical two-point function below the upper critical dimension. This is a rigorous version, for the weakly self-avoiding walk, of the expansion in $\epsilon = 4 - d$ discussed in [**72**].

7.2. Finite-volume approximation. Integral representations of the type discussed in Section 6.5 are for walks on a finite set. In preparation for the integral representation, we first discuss the approximation of the two-point function $G_{\nu_c}^{(g)}(x)$ on $\mathbb{Z}^d$ by a two-point function on the finite torus $\Lambda = \mathbb{Z}^d/R\mathbb{Z}^d$ with side length $R \in \mathbb{Z}_+$. For later convenience, we will always take $R = L^N$ with L a large dyadic integer. The parameter g is regarded as a fixed positive number and will sometimes be omitted in what follows, to simplify the notation. We denote by G^Λ the natural modification of (7.6) in which the random walk on $\mathbb{Z}^d$ is replaced by the random walk on Λ.

THEOREM 7.2. *Let $d \geq 1$, $g > 0$, and $x \in \mathbb{Z}^d$. Then for all $\nu \geq \nu_c$,*

$$G_\nu(x) = \lim_{\nu' \searrow \nu} \lim_{N \to \infty} G_{\nu'}^\Lambda(x), \tag{7.10}$$

where, on the right-hand side, x is the canonical representative of x in Λ for L^N large compared to x.

PROOF. This follows from a version of the Simon–Lieb inequality [**67, 56**] for the continuous-time weakly self-avoiding walk. In the problems of Section 7.8 below, we develop the corresponding argument in the discrete-time setting. With a little more work, the same approach can be adapted to continuous time. □

We are most interested in the case $\nu = \nu_c$ in Theorem 7.2. The theorem allows for the study of the critical two-point function on $\mathbb{Z}^d$ via the subcritical two-point function in finite volume, provided sufficient control is maintained to take the limits. Since SRW is recurrent in finite volume, its Green function is infinite, and the flexibility of taking ν slightly larger than ν_c helps bypass this concern.

7.3. Integral representation. We recall the introduction of the boson field $(\varphi_x, \bar\varphi_x)$ in (6.20) and the fermion field $(\psi_x, \bar\psi_x)$ in (6.22), and now index these fields with x in the torus $\Lambda = \mathbb{Z}^d/L^N\mathbb{Z}^d$. We also recall from (6.26) the definition, for $x \in \Lambda$, of the differential form

$$\tau_x = \varphi_x\bar\varphi_x + \psi_x\bar\psi_x. \tag{7.11}$$

The Laplacian Δ applies to the boson and fermion fields according to

$$(\Delta\varphi)_x = \sum_{y:y\sim x}(\varphi_y - \varphi_x), \quad (\Delta\psi)_x = \sum_{y:y\sim x}(\psi_y - \psi_x), \tag{7.12}$$

where the sum is over the neighbours y of x in the torus Λ. We also define the differential forms

$$\tau_{\Delta,x} = \frac{1}{2}(\varphi_x(-\Delta\bar{\varphi})_x + (-\Delta\varphi)_x\bar{\varphi}_x + \psi_x(-\Delta\bar{\psi})_x + (-\Delta\psi)_x\bar{\psi}_x). \tag{7.13}$$

The following theorem is proved in [**9**]; see also [**11**, Theorem 5.1] for a self-contained proof. Its requirement that $G_\nu^\Lambda(x) < \infty$ for large Λ is a consequence of Theorem 7.2.

THEOREM 7.3. *For $\nu > \nu_c$ and $0, x \in \Lambda$, and for Λ large enough that $G_\nu^\Lambda(x) < \infty$, the finite-volume two-point function has the integral representation*

$$G_\nu^\Lambda(x) = \int e^{-\sum_{v\in\Lambda}(\tau_{\Delta,v}+g\tau_v^2+\nu\tau_v)}\bar{\varphi}_0\varphi_x. \tag{7.14}$$

It is the goal of the method to show that the infinite-volume critical two-point function is asymptotically equal to a multiple of the inverse Laplacian on $\mathbb{Z}^d$, for $d \geq 4$. To exhibit an explicit factor to account for this multiple, we introduce a parameter $z_0 > -1$ by making the change of variables $\varphi_x \mapsto (1+z_0)^{1/2}\varphi_x$. With this change of variables, the integral representation (7.14) becomes

$$G_\nu^\Lambda(x) = (1+z_0)\int e^{-S(\Lambda)}e^{-\widetilde{V}_0(\Lambda)}\bar{\varphi}_0\varphi_x, \tag{7.15}$$

where

$$S(\Lambda) = \sum_{v\in\Lambda}(\tau_{\Delta,v} + m^2\tau_v), \tag{7.16}$$

$$\widetilde{V}_0(\Lambda) = \sum_{v\in\Lambda}(g_0\tau_v^2 + \nu_0\tau_v + z_0\tau_{\Delta,v}), \tag{7.17}$$

with

$$g_0 = (1+z_0)^2 g, \quad \nu_0 = (1+z_0)\nu_c, \quad m^2 = (1+z_0)(\nu - \nu_c). \tag{7.18}$$

In particular, the limit $\nu \searrow \nu_c$ corresponds to $m^2 \searrow 0$.

It is often convenient in statistical mechanics to obtain a correlation function by differentiation of a partition function with respect to an external field, and we will follow this approach here. Introducing an *external field* $\sigma \in \mathbb{C}$, we define

$$V_0(\Lambda) = \widetilde{V}_0(\Lambda) + \sigma\bar{\varphi}_0 + \bar{\sigma}\varphi_x. \tag{7.19}$$

Then the two-point function is given by

$$G_\nu^\Lambda(x) = (1+z_0)\frac{\partial^2}{\partial\sigma\partial\bar{\sigma}}\Big|_{\sigma=\bar{\sigma}=0}\int_{\mathbb{C}^\Lambda} e^{-S(\Lambda)-V_0(\Lambda)}. \tag{7.20}$$

Our goal now is the evaluation of the large-x asymptotic behaviour of

$$G_{\nu_c}(x) = \lim_{m^2\searrow 0}\lim_{N\to\infty}(1+z_0)\frac{\partial^2}{\partial\sigma\partial\bar{\sigma}}\Big|_{\sigma=\bar{\sigma}=0}\int_{\mathbb{C}^\Lambda} e^{-S(\Lambda)-V_0(\Lambda)}. \tag{7.21}$$

For the case $\widetilde{V}_0 = 0$ (so in particular $z_0 = 0$), in view of Remark 6.8 the right-hand side becomes

$$\lim_{m^2\searrow 0}\lim_{N\to\infty}\int_{\mathbb{C}^\Lambda} e^{-S(\Lambda)}\bar{\varphi}_0\varphi_x = \lim_{m^2\searrow 0}\lim_{\Lambda\nearrow\mathbb{Z}^d}\int \bar{\varphi}_0\varphi_x d\mu_{(-\Delta_\Lambda+m^2)^{-1}}, \tag{7.22}$$

and by Lemma 6.3 this is equal to

$$\lim_{m^2\searrow 0}\lim_{\Lambda\nearrow\mathbb{Z}^d}(-\Delta_\Lambda + m^2)_{0x}^{-1} = (-\Delta_{\mathbb{Z}^d})_{0x}^{-1} \sim c_0|x|^{-(d-2)} \tag{7.23}$$

(we have added subscripts to the Laplacians to emphasise where they act). The goal of the forthcoming analysis is to show that for small $g > 0$, and with the correct choice of z_0, the effect of $\widetilde{V}_0$ is a small perturbation in the sense that its presence does not change the power in this $|x|^{-(d-2)}$ decay.

7.4. Superexpectation. We will need some further development of the theory of integration of differential forms discussed in Section 6.4. As before, we denote the algebra of differential forms, now with index set Λ, by $\mathcal{N}$. Let C be a $\Lambda \times \Lambda$ matrix, with positive-definite Hermitian part, and with inverse $A = C^{-1}$. The *Gaussian superexpectation* with covariance matrix C is defined by

$$\mathbb{E}_C F = \int e^{-S_A} F \quad \text{for } F \in \mathcal{N}. \tag{7.24}$$

The name "superexpectation" comes from the fact that the integral representation for the two-point function is actually a *supersymmetric* field theory; supersymmetry is discussed in [**11**].

Note that, by Lemma 6.7 and Remark 6.8, $\mathbb{E}_C 1 = 1$, and more generally $\mathbb{E}_C f = \int f d\mu_C$ if f is a zero-form. The latter property shows that the Gaussian superexpectation extends the ordinary Gaussian expectation, and we wish to take this further. Recall the elementary fact that if $X_1 \sim N(0, \sigma_1^2)$ and $X_2 \sim N(0, \sigma_2^2)$ are independent normal random variables, then $X_1 + X_2 \sim N(0, \sigma_1^2 + \sigma_2^2)$. In particular, if $X \sim N(0, \sigma_1^2 + \sigma_2^2)$ then we can evaluate $\mathbb{E}(f(X))$ in stages as

$$\mathbb{E}(f(X)) = \mathbb{E}(\mathbb{E}(f(X_1 + X_2) \,|\, X_2)). \tag{7.25}$$

It will be a crucial ingredient of the following analysis that this has an extension to the superexpectation, as we describe next.

By definition, any form $F \in \mathcal{N}$ is a linear combination of products of factors ψ_{x_i} and $\bar{\psi}_{\bar{x}_i}$, with $x_i, \bar{x}_i \in \Lambda$ and with coefficients given by functions of φ and $\bar{\varphi}$. The coefficients may also depend on the external field $(\sigma, \bar{\sigma})$, but we leave the dependence on $\sigma, \bar{\sigma}$ implicit in the notation. We also define an algebra $\mathcal{N}^\times$ with twice as many fields as $\mathcal{N}$, namely with boson fields (ϕ, ξ) and fermion fields (ψ, η), where $\phi = (\varphi, \bar{\varphi})$, $\xi = (\zeta, \bar{\zeta})$, $\psi = \frac{1}{\sqrt{2\pi i}}(d\varphi, d\bar{\varphi})$, $\eta = \frac{1}{\sqrt{2\pi i}}(d\zeta, d\bar{\zeta})$. For a form $F = f(\varphi, \bar{\varphi})\psi^x \bar{\psi}^y$ (where ψ^x denotes a product $\psi_{x_1} \cdots \psi_{x_j}$), we define

$$\theta F = f(\varphi + \xi, \bar{\varphi} + \bar{\xi})(\psi + \eta)^x (\bar{\psi} + \bar{\eta})^y, \tag{7.26}$$

and we extend this to a map $\theta : \mathcal{N} \to \mathcal{N}^\times$ by linearity. Then we understand the map $\mathbb{E}_C \circ \theta : \mathcal{N} \to \mathcal{N}$ as the integration with respect to the *fluctuation fields* ξ and η, with the fields ϕ and ψ left fixed. This is like a conditional expectation. However, this is not standard probability theory, since $\mathbb{E}_C$ does not arise from a probability measure and takes values in the (non-commutative) algebra of forms.

The superexpectation has the following important convolution property, analogous to (7.25) (see [**9, 13**]).

PROPOSITION 7.4. *Let $F \in \mathcal{N}$, and suppose that C_1 and C' have positive-definite Hermitian parts. Then*

$$\mathbb{E}_{C'+C_1} F = \mathbb{E}_{C'}(\mathbb{E}_{C_1} \theta F). \tag{7.27}$$

Suppose C and C_j, $j = 1, \dots, N$, are $\Lambda \times \Lambda$ matrices with positive-definite Hermitian parts, such that

$$C = \sum_{j=1}^{N} C_j. \tag{7.28}$$

Then, by the above proposition,

$$\mathbb{E}_C F = \big(\mathbb{E}_{C_N} \circ \mathbb{E}_{C_{N-1}}\theta \circ \cdots \circ \mathbb{E}_{C_1}\theta\big) F. \tag{7.29}$$

In the next section, we describe a particular choice of the decomposition (7.28), which will allow us to control the progressive integration in (7.29).

7.5. Decomposition of the covariance. Our goal is to compute the large-x asymptotic behaviour of the two-point function using (7.20), which we can now rewrite as

$$G_\nu^\Lambda(x) = (1 + z_0) \frac{\partial^2}{\partial\sigma\partial\bar\sigma}\bigg|_{\sigma=\bar\sigma=0} \mathbb{E}_C e^{-V_0(\Lambda)}, \tag{7.30}$$

with $C = (-\Delta + m^2)^{-1}$. The Laplacian is on the torus Λ, and we must take the limits as Λ approaches $\mathbb{Z}^d$ and m^2 approaches zero, so C is an approximation to $(-\Delta_{\mathbb{Z}^d})^{-1}$. The operator $(-\Delta_{\mathbb{Z}^d})^{-1}$ decays as $|x|^{-2}$ in dimension $d = 4$, and such long-range correlations make the analysis difficult. The renormalisation group approach takes the long-range correlations into account progressively, by making a good decomposition of the covariance C into a sum of terms with *finite* range, together with progressive integration as in (7.29). The particular decomposition used is given in the following theorem, which extends a result of Brydges, Guadagni and Mitter [**8**]; see also [**4, 13**]. In its statement, $\nabla_x^\alpha = \nabla_{x_1}^{\alpha_1} \cdots \nabla_{x_d}^{\alpha_d}$ for a multi-index $\alpha = (\alpha_1, \dots, \alpha_d)$, where ∇_{x_k} denotes the finite-difference operator $\nabla_{x_k} f(x, y) = f(x + e_k, y) - f(x, y)$.

THEOREM 7.5. *Let $d > 2$ and $N \in \mathbb{Z}_+$, and let Λ be the torus $\mathbb{Z}^d / L^N \mathbb{Z}^d$, with L a sufficiently large dyadic integer. Let $m^2 > 0$ and let $C = (-\Delta + m^2)^{-1}$ on Λ. There exist positive-definite $\Lambda \times \Lambda$ matrices $C_1, \dots, C_N$ such that:*

(a) $C = \sum_{j=1}^N C_j$,

(b) $C_j(x, y) = 0$ *if* $|x - y| \geq \frac{1}{2} L^j$,

(c) *for multi-indices α, β with ℓ^1 norms $|\alpha|_1, |\beta|_1$ at most some fixed value p, and for $j < N$,*

$$|\nabla_x^\alpha \nabla_y^\beta C_j(x, y)| \leq c L^{-(j-1)(2[\phi] - (|\alpha|_1 + |\beta|_1))}, \tag{7.31}$$

where $[\phi] = \frac{1}{2}(d - 2)$, and c is independent of j and N.

The decomposition in Theorem 7.5(a) is called a *finite-range* decomposition because of item (b): the covariance C_j has range $\frac{1}{2} L^j$, and fields at points separated beyond that range are uncorrelated under $\mathbb{E}_{C_j}$.

To compute the important expectation $\mathbb{E}_C e^{-V_0(\Lambda)}$ in (7.30), we use Theorem 7.5 and Proposition 7.4 to evaluate it progressively. Namely, if we define

$$Z_0 = e^{-V_0(\Lambda)}, \quad Z_{j+1} = \mathbb{E}_{C_{j+1}} \theta Z_j \quad (j + 1 < N), \quad Z_N = \mathbb{E}_{C_N} Z_{N-1}, \tag{7.32}$$

then the desired expectation is equal to $Z_N = \mathbb{E}_C e^{-V_0(\Lambda)}$. Thus we are led to study the recursion $Z_j \mapsto Z_{j+1}$.

In the expectation $Z_{j+1} = \mathbb{E}_{C_{j+1}}\theta Z_j$, on the right-hand side we write $\varphi_j = \varphi_{j+1} + \zeta_{j+1}$, as in (7.26), and similarly for $\bar{\varphi}_j$, $d\varphi_j$, $d\bar{\varphi}_j$. The expectation $\mathbb{E}_{C_{j+1}}\theta$ integrates out ζ_{j+1}, $\bar{\zeta}_{j+1}$, $d\zeta_{j+1}$, $d\bar{\zeta}_{j+1}$ leaving dependence of Z_{j+1} on φ_{j+1}, $\bar{\varphi}_{j+1}$, $d\varphi_{j+1}$, $d\bar{\varphi}_{j+1}$. This process is repeated. The ζ_j fields that are integrated out are the fluctuation fields.

It follows from Remark 6.8 and Lemma 6.3 that $\mathbb{E}_{C_{j+1}}|\zeta_{j,x}|^2 = C_{j+1}(x,x)$. With Theorem 7.5(c), this indicates that the typical size of the fluctuation field ζ_j is of order $L^{-j[\phi]}$; the number $[\phi] = \frac{1}{2}(d-2)$ is referred to as the *scaling dimension* or *engineering dimension* of the field. Moreover, Theorem 7.5(c) also indicates that the derivative of $\zeta_{j,x}$ is typically smaller than the field itself by a factor L^{-j}, so that the fluctuation field remains approximately constant over a distance L^j.

To make systematic use of this behaviour of the fields, we introduce nested pavings of Λ by sets of *blocks* $\mathcal{B}_j$ on scales $j = 0, \ldots, N$. The blocks in $\mathcal{B}_0$ are simply the points in Λ. The blocks in $\mathcal{B}_1$ form a disjoint paving of Λ by boxes of side L. More generally, each block in $\mathcal{B}_j$ has side L^j and consists of L^d disjoint blocks in $\mathcal{B}_{j-1}$. A *polymer* on scale j is any union of blocks in $\mathcal{B}_j$, and we denote the set of scale-j polymers by $\mathcal{P}_j$. (This terminology is standard but these polymers have nothing to do with physical polymers or random walks, they merely provide a means of organising subsets in the pavings of the torus.)

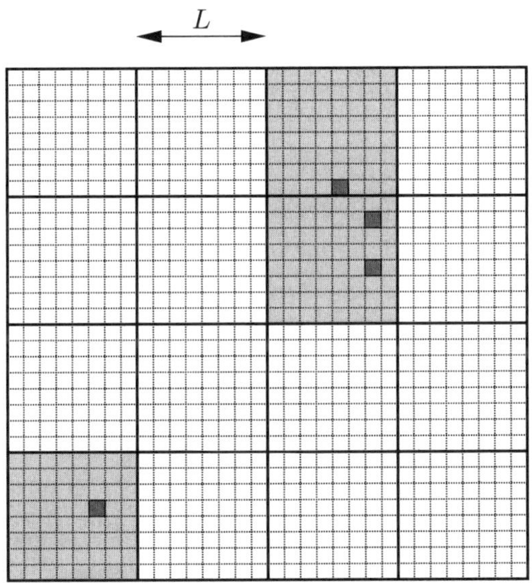

FIGURE 11. The four small shaded squares represent a polymer in $\mathcal{P}_0$, and the three larger shaded squares represent its closure in $\mathcal{P}_1$.

For a block $B \in \mathcal{B}_j$, the above considerations concerning the typical size of the fluctuation field suggest that, at each of the L^{dj} points $x \in B$, $\zeta_{j,x}$ has typical size $L^{-j[\phi]}$, and hence

$$\sum_{x\in B} \zeta_{j,x}^p \approx L^{dj} L^{-pj[\phi]} = L^{(d-p[\phi])j}. \tag{7.33}$$

The above sum is *relevant* (growing exponentially in j) for $p[\phi] < d$, *irrelevant* (decaying exponentially in j) for $p[\phi] > d$, and *marginal* (neither growing or decaying) for $p[\phi] = d$. Since $\tau_x = \varphi_x\bar{\varphi}_x + \psi_x\bar{\psi}_x$ is quadratic in the fields, it corresponds to $p = 2$. Thus $p[\phi] = 2[\phi] = d - 2 < d$ and τ_x is relevant in all dimensions. Similarly, τ_x^2 corresponds to $p = 4$ with $p[\phi] = 4[\phi] = 2d - 4$, so that τ_x^2 is irrelevant for $d > 4$, marginal for $d = 4$, and relevant for $d < 4$. The monomial $\tau_{\Delta,x}$ is marginal in all dimensions. In fact, the three monomials τ_x^2, τ_x and $\tau_{\Delta,x}$, which constitute the initial potential $\widetilde{V}_0$, are precisely the marginal and relevant local monomials that are Euclidean invariant and obey an additional symmetry between bosons and fermions called *supersymmetry* (see [**11**]).

7.6. The map $Z_0 \mapsto Z_1$. For an idea of how the recursion $Z_j \mapsto Z_{j+1}$ might be studied, let us take $j = 0$ and consider the map $Z_0 \mapsto Z_1 = \mathbb{E}_{C_1}\theta Z_0$.

For simplicity, we set $\sigma = \bar{\sigma} = 0$, so that $V_0 = g_0\tau^2 + \nu_0\tau + z_0\tau_\Delta$ is translation invariant. As usual, the monomials in V_0 depend on the fields $\varphi, \bar{\varphi}, \psi, \bar{\psi}$. As discussed above, we decompose the field φ as $\varphi = \varphi_1 + \zeta_1$, and similarly for $\bar{\varphi}, \psi, \bar{\psi}$. The operation $\mathbb{E}_{C_1}\theta$ integrates out the fields $\zeta_1, \bar{\zeta}_1, d\zeta_1, d\bar{\zeta}_1$. Recall that, by definition, $\mathcal{P}_0$ is the set of subsets of Λ. We write $I_0(x) = e^{-V_0(x)}$, and, for $X \in \mathcal{P}_0$, write $I_0^X = \prod_{x\in X} I_0(x) = e^{-V_0(X)}$ where $V_0(X) = \sum_{x\in X} V_0(x)$. In this notation, the dependence on the fields is left implicit. Let

$$V_1 = g_1\tau^2 + \nu_1\tau + z_1\tau_\Delta \tag{7.34}$$

denote a modification of V_0 in which the coupling constants in V_0 have been adjusted, or *renormalised*, to some new values g_1, ν_1, z_1. This is the origin of the term "renormalisation" in the renormalisation group. We set $I_1^X = e^{-V_1(X)}$, but with the fields in V_1 given by $\varphi_1, \bar{\varphi}_1, d\varphi_1, d\bar{\varphi}_1$. Let $\delta I_1^X = \prod_{x\in X}(I_1(x) - \theta I_0(x))$; this is an element of $\mathcal{N}^\times$ since I_1 depends on the fields φ_1 and so on, while θI_0 depends on $\varphi_1 + \zeta_1$ and so on.

Then we obtain

$$\begin{aligned} Z_1(\Lambda) &= \mathbb{E}_{C_1}\theta I_0(\Lambda) = E_{C_1}\prod_{x\in\Lambda}(I_1(x) + \delta I_1(x)) \\ &= \mathbb{E}_{C_1}\sum_{X\in\mathcal{P}_0} I_1^{\Lambda\setminus X}\delta I_1^X = \sum_{X\in\mathcal{P}_0} I_1^{\Lambda\setminus X}\mathbb{E}_{C_1}\delta I_1^X. \end{aligned} \tag{7.35}$$

Here we have expressed Z_1 as a sum over a polymer on scale 0; we wish to express it as a sum over a polymer on scale 1. To this end, for a polymer X on scale 0, we define the *closure* $\overline{X}$ to be the smallest polymer on scale 1 containing X: see Figure 11. We can now write

$$Z_1(\Lambda) = \sum_{U\in\mathcal{P}_1} I_1^{\Lambda\setminus U} K_1(U), \tag{7.36}$$

where

$$K_1(U) = \sum_{X\in\mathcal{P}_0:\overline{X}=U} I_1^{U\setminus X}\mathbb{E}_{C_1}\delta I_1^X. \tag{7.37}$$

DEFINITION 7.6. For $j = 0, 1, 2, \ldots, N$, and for $F, G : \mathcal{P}_j \to \mathcal{N}_{\text{even}}$, where $\mathcal{N}_{\text{even}}$ denotes the forms of even degree, the *circle product* of F, G is

$$(F \circ G)(\Lambda) = \sum_{U\in\mathcal{P}_j(\Lambda)} F(\Lambda \setminus U)G(U). \tag{7.38}$$

Note that the circle product depends on the scale j.

The circle product is associative and commutative (the latter due to the restriction to forms of even degree). With the circle product, we can encode the formula (7.36) compactly as $Z_1(\Lambda) = (I_1 \circ K_1)(\Lambda)$, with the convention that $I_1(U) = I_1^U$. The identity element for the circle product is $1_{\{U=\varnothing\}}$. Thus, if we define $K_0(X) = 1_{\{X=\varnothing\}}$, then $Z_0(\Lambda) = I_0(\Lambda) = (I_0 \circ K_0)(\Lambda)$.

All later stages of the recursion proceed inductively from $Z_j = (I_j \circ K_j)(\Lambda)$. The interaction I_j continues to be defined by a potential V_j, but the form of the dependence will not, in general, be as simple as $I = e^{-V}$. The interaction does, however, obey $I_j(X) = \prod_{B\in\mathcal{B}_j(X)} I_j(B)$, for all $X \in \mathcal{P}_j$ and for all j. The following factorisation property of K_1, which can be verified from (7.37), allows the induction to proceed. If $U \in \mathcal{P}_1$ has connected components $U_1, \ldots, U_k$, then $K_1(U) = \prod_{i=1}^k K_1(U_i)$; the notion of connectivity here includes blocks touching at a corner. The induction will preserve this key property for K_j and $\mathcal{P}_j$, for all j.

7.7. Remaining steps in the proof. Our goal is to prove Theorem 7.1. According to (7.21), we need to show that there is a choice of z_0 such that, for g small and positive,

$$G_{\nu_c}(x) = \lim_{m^2\searrow 0}\lim_{N\to\infty}(1+z_0)\frac{\partial^2}{\partial\sigma\partial\bar\sigma}\Big|_{\sigma=\bar\sigma=0} Z_N(\Lambda) \sim c_g|x|^{-(d-2)}. \tag{7.39}$$

In particular, we see from this that the correct choice of z_0 will appear in the value of the constant c_g. The remaining steps in the proof of (7.39) are summarised, imprecisely, as follows. Much is left unsaid here, and details can be found in [**13**].

THEOREM 7.7. *Let $d \geq 4$, and let $g > 0$ be sufficiently small. There is a choice of $V_1, \ldots, V_N$ given, for $X \subset \Lambda$, by*

$$V_j(X) = \sum_{v\in X}(g_j\tau_v^2 + \nu_j\tau_v + z_j\tau_{\Delta,v}) + \lambda_j(\sigma\bar\varphi_0 + \bar\sigma\varphi_x) + q_j^2\sigma\bar\sigma, \tag{7.40}$$

with V_j determining I_j, and a choice of $K_1, \ldots, K_N$ with $K_j : \mathcal{P}_j \to \mathcal{N}$ obeying the key factorisation property mentioned above, such that

$$Z_j(\Lambda) = (I_j \circ K_j)(\Lambda) \tag{7.41}$$

obeys the recursion $Z_{j+1} = \mathbb{E}_{C_{j+1}}\theta Z_j$. Moreover, $(V_j, K_j)_{0\leq j\leq N}$ obeys the flow equations

$$g_{j+1} = g_j - cg_j^2 + r_{g,j} \tag{7.42}$$

$$\nu_{j+1} = \nu_j + 2g_jC_{j+1}(0,0) + r_{\nu,j} \tag{7.43}$$

$$z_{j+1} = z_j + r_{z,j} \tag{7.44}$$

$$K_{j+1} = r_{K,j} \tag{7.45}$$

where the r terms represent error terms. Further equations define the evolution of λ_j and q_j.

The previous theorem represents the recursion $Z_j \mapsto Z_{j+1}$ as a dynamical system. A fixed-point theorem is used to make the correct choice of the initial value z_0 so that the r terms remain small on all scales, and so that (g_j, ν_j, z_j, K_j) flows to $(0, 0, 0, 0)$. The latter is referred to as *infrared asymptotic freedom*, and is the effect anticipated below (7.23). This final ingredient is summarised in the following theorem.

THEOREM 7.8. *If $g > 0$ is sufficiently small (independent of N and m^2), there exists z_0 such that*

$$\lim_{m^2 \searrow 0} \lim_{N\to\infty} V_N = \lambda_\infty(\sigma\bar{\varphi}_0 + \bar{\sigma}\varphi_x) + q_\infty \sigma\bar{\sigma}, \tag{7.46}$$

with $\lambda_\infty > 0$ and, as $x \to \infty$, $q_\infty \sim \lambda_\infty^2(-\Delta_{\mathbb{Z}^d})^{-1}_{0x}$. Moreover, in an appropriately defined Banach space,

$$\lim_{m^2 \searrow 0} \lim_{N\to\infty} K_N(\Lambda) = 0. \tag{7.47}$$

At scale N there are only two polymers, namely the single block Λ and the empty set $\varnothing$. By definition, $I_N(\varnothing) = K_N(\varnothing) = 1$. Also, the field has been entirely integrated out at scale N, and from Theorem 7.8 and the definition of the circle product, we obtain

$$Z_N(\Lambda) = I_N(\Lambda) + K_N(\Lambda) \approx I_N(\Lambda) \approx e^{-q_N\sigma\bar{\sigma}}. \tag{7.48}$$

Let $z_0^* = \lim_{m^2\searrow 0} z_0$. With (7.39) and $q_N \to q_\infty$, this gives

$$G_{\nu_c}(x) = (1+z_0^*)q_\infty \sim (1+z_0^*)\lambda_\infty^2(-\Delta_{\mathbb{Z}^d})^{-1}_{0x} \sim (1+z_0^*)\lambda_\infty^2 c_0|x|^{-(d-2)}. \tag{7.49}$$

This is the desired conclusion of Theorem 7.1.

7.8. Tutorial. These problems develop a proof of the discrete-time version of Theorem 7.2. The proof makes use of a Simon–Lieb inequality—this is now a generic term for inequalities of the sort introduced in [**67, 56**] for the Ising model. The approach developed here can be adapted to prove Theorem 7.2.

Let Γ represent either $\Gamma = \mathbb{Z}^d$ or the discrete torus $\Gamma = \mathbb{Z}^d/R\mathbb{Z}^d$. Let $\mathbb{E}_x$ denote the expectation for the usual discrete-time SRW on Γ, which we denote now by $(X_n)_{n\geq 0}$, starting at x. Let $I_{m,n}$ denote the number of self-intersections of X between times m and n:

$$I_{m,n} = \sum_{m\leq i<j\leq n} 1_{\{X_i=X_j\}}, \qquad I_n = I_{0,n}. \tag{7.50}$$

We define the two-point function of the weakly SAW in the domain $D \subset \Gamma$ by

$$G_{\nu,D}(x,y) = \sum_{n\geq 0} \mathbb{E}_x(e^{-gI_n}1_{\{X_n=y,n<T_D\}})e^{-\nu n}, \quad x,y\in\Gamma,\ \nu\in\mathbb{R}, \tag{7.51}$$

where $T_D = \inf\{n \geq 0 : X_n \notin D\}$ is the exit time of D. We define the boundary $\partial D = \{x \notin D : \exists y \in D \text{ s.t. } x \sim y\}$, and the closure $\bar{D} = D \cup \partial D$. The two-point function on the entire graph is written as G_ν rather than $G_{\nu,\Gamma}$. Let $c_n(x,y) = E_x(e^{-gI_n}1_{\{X_n=y\}})$, let $c_n = \sum_{y\in\Gamma} c_n(0,y)$, and define the susceptibility by

$$\chi(\nu) = \sum_{y\in\Gamma} G_\nu(0,y) = \sum_{n\geq 0} c_n e^{-\nu n}. \tag{7.52}$$

PROBLEM 7.1. Verify that $(c_n)_{n\geq 0}$ is a submultiplicative sequence, i.e. $c_{n+m} \leq c_n c_m$, and conclude that $\frac{1}{n}\log(c_n)$ converges to its infimum, which is ν_c by definition. In particular, notice that for $\nu < \nu_c$, $\chi(\nu) = \infty$ and for $\nu > \nu_c$, $\chi(\nu) < \infty$.

PROBLEM 7.2. Let $\chi^R(\nu)$ be the susceptibility for $\mathbb{Z}^d/R'\mathbb{Z}^d$ where $R' = 2R+1$, and let $\chi(\nu)$ be the susceptibility for $\mathbb{Z}^d$. Prove that $\chi^R(\nu) \leq \chi(\nu)$ for $R' \geq 3$, and, in particular, that $\nu_c(\mathbb{Z}^d) \geq \nu_c(\mathbb{Z}^d/R\mathbb{Z}^d)$. Here, $\nu_c(\Gamma)$ denotes the critical point of the weakly SAW on Γ.

PROBLEM 7.3. Prove the following version of the Simon-Lieb inequality for the discrete-time weakly SAW on Γ. Given $D \subset \Gamma$, show that

$$G_\nu(x, y) - G_{\nu,D}(x, y) \leq \sum_{z \in \partial D} G_{\nu,\bar{D}}(x, z) G_\nu(z, y). \tag{7.53}$$

Note that if $x \in D$ and $y \in D^c$, then $G_{\nu,D}(x, y) = 0$.

The following problem provides an approach to proving exponential decay of a subcritical two-point function which, unlike Proposition 1.3, adapts well to the continuous-time setting.

PROBLEM 7.4. Let $\Lambda_R = \{-R+1, \ldots, R\}^d \subset \mathbb{Z}^d$. For $\nu > \nu_c$, $\sum_{y \in \mathbb{Z}^d} G_\nu(0, y)$ is finite, and thus $\theta = \sum_{y \in \partial \Lambda_R} G_\nu(0, y) < 1$ for R sufficiently large. Conclude from Problem 7.3 with $D = \Lambda_R$ that for $y \notin \Lambda_R$,

$$G_\nu(0, y) \leq \theta^{\lfloor |y|_\infty/(R+1) \rfloor} \sup_{x \in \mathbb{Z}^d} G_\nu(0, x). \tag{7.54}$$

PROBLEM 7.5. Let $(T_R)_{R \in \mathbb{N}}$ be a sequence of discrete tori with the vertex sets V_R embedded in $\mathbb{Z}^d$ by $V_R = \Lambda_R$ where Λ_R is as in Problem 7.4; in particular, $V_R \subset V_{R+1}$. Let G_ν^R be the two-point function on T_R, and G_ν be the two-point function on $\mathbb{Z}^d$. Use Problem 7.2 and Problem 7.4 to prove that for all $\nu > \nu_c = \nu_c(\mathbb{Z}^d)$, $x, y \in \mathbb{Z}^d$,

$$G_\nu^R(x, y) \to G_\nu(x, y) \quad \text{as } R \to \infty. \tag{7.55}$$

Conclude that

$$G_{\nu_c}(x, y) = \lim_{\nu \searrow \nu_c} \lim_{R \to \infty} G_\nu^R(x, y). \tag{7.56}$$

Appendix A. Solutions to the problems

A.1. Solutions for Tutorial 1.7.

PROBLEM 1.1. Let M be an integer, and for every $n \in \mathbb{N}$, write $n = Mk + r$ with $0 \leq r < M$. Then,

$$\frac{1}{n} a_n \leq \frac{k}{n} a_M + \frac{1}{n} a_r, \quad \text{and, thus,} \quad \limsup_{n \to \infty} \frac{1}{n} a_n \leq \frac{1}{M} a_M. \tag{A.1}$$

In particular,

$$\limsup_{n \to \infty} \frac{1}{n} a_n \leq \inf_{M \in \mathbb{N}} \frac{1}{M} a_M \leq \liminf_{M \to \infty} \frac{1}{n} a_n, \tag{A.2}$$

which implies both statements of the claim. □

PROBLEM 1.2. The number of n-step walks with steps only in positive coordinate directions is d^n. The number of walks which do not reverse direction is $2d(2d-1)^{n-1}$. Thus,

$$d^n \leq c_n \leq 2d(2d-1)^{n-1} \quad \text{and therefore} \quad d \leq \mu \leq 2d - 1. \tag{A.3}$$

The upper bound can easily be improved by excluding more patterns that lead to self-intersecting walks than merely reversals of steps. For example, by considering walks which do not contain anti-clockwise "unit squares" (see Figure 12), we obtain

$$c_{3n+1} \leq 2d((2d-1)^3 - 1)^n = 4(26^{1/3})^{3n}, \tag{A.4}$$

giving $\mu \le 26^{1/3} < 3$. Similarly, the lower bound can be improved by considering walks that take steps either in positive coordinate directions, i.e., north or east, or in an east-north-west-north pattern: see Figure 12. It follows that

$$c_{4n} \ge (d^4 + 1)^n = (17^{1/4})^{4n}, \tag{A.5}$$

where $17^{1/4} > 2$. In particular, $2 < 17^{1/4} \le \mu \le 26^{1/3} < 3$. □

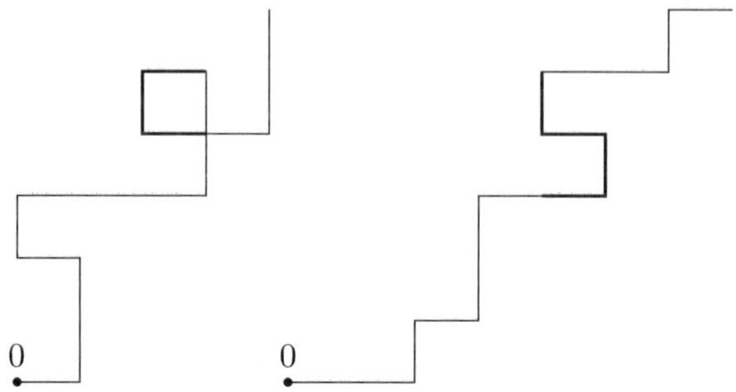

FIGURE 12. Left: The walk does contain a unit square. Right: The walk only takes steps east, north, or in east-north-west-north patterns (thick line).

PROBLEM 1.3. SAWs can get trapped: see Figure 13. A trapped walk ω of length n does not arise as the restriction of a walk ρ of length $m > n$ to the first n steps. Thus, under $\mathbb{Q}_n^{(1)}$, ω has positive probability, while $\sum_{\rho > \omega} \mathbb{Q}_m^{(1)}(\rho) = 0$. □

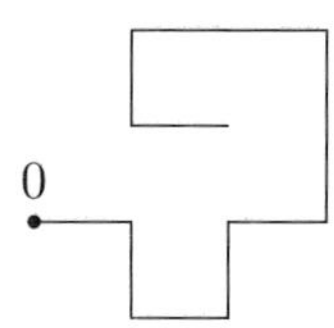

FIGURE 13. Trapped walk.

PROBLEM 1.4. $c_n(x) = 1_{\{|x|=n\}}$, so $G_z(x) = \sum_{n\ge 0} c_n(x) z^n = z^{|x|}$, and

$$\hat{G}_z(k) = \sum_{x\in\mathbb{Z}} z^{|x|} e^{ikx} = -1 + \sum_{n\ge 0} z^n (e^{ikn} + e^{-ikn})$$

$$= -1 + (1 - ze^{ik})^{-1} + (1 - ze^{-ik})^{-1} = \frac{1 - z^2}{1 - 2z\cos k + z^2}, \tag{A.6}$$

as claimed. □

PROBLEM 1.5. The assumption implies

$$|f((1 - 1/n)e^{i\varphi})| \le c|1 - (1 - 1/n)e^{i\varphi}|^{-b}. \tag{A.7}$$

Note that for $\varphi \in [0, \pi/2]$,

$$|\mathrm{Re}(1 - (1 - 1/n)e^{i\varphi})| = 1 - (1 - 1/n)\cos\varphi \ge 1/n, \tag{A.8}$$

$$\text{(A.9)} \qquad |\mathrm{Im}(1-(1-1/n)e^{i\varphi})| = |(1-1/n)\sin\varphi| \geq (1-1/n)\frac{2\varphi}{\pi}.$$

Suppose $b > 1$. The integral is estimated using $|z|^n \geq Ce^{-1}$ for $|z| = 1 - 1/n$,

$$\text{(A.10)} \qquad \frac{1}{2\pi}\int_0^{\pi/2} |f((1-1/n)e^{i\varphi})|\, d\varphi \leq c\int_0^{\pi/2}\left(\frac{1}{n} + (1-1/n)\frac{2\varphi}{\pi}\right)^{-b} d\varphi$$
$$= (1-1/n)^{-1}c\int_{1/n}^1 t^{-b}\, dt = (1-1/n)^{-1}cb(n^{b-1}-1) \leq cn^{b-1},$$

and, since $|f(z)|$ is bounded for z bounded away from 1,

$$\text{(A.11)} \qquad \frac{1}{2\pi}\int_{\pi/2}^{\pi} |f((1-1/n)e^{i\varphi})|\, d\varphi \leq c.$$

Likewise, the contributions for the interval $[\pi, 2\pi]$ are estimated and we obtain

$$\text{(A.12)} \qquad |a_n| \leq cn^{b-1}.$$

The above assumed $b > 1$ but the extension to $b = 1$ is easy. □

PROBLEM 1.6. (a) Let $T_0 = 0$ and $T_k = \inf\{n > T_{k-1} : X_n = 0\}$. Then $u = P(T_1 < \infty)$, and by induction and the strong Markov property, $P(T_k < \infty) = u^k$. It follows that

$$\text{(A.13)} \qquad m = \mathbb{E}(N) = \sum_{k\geq 0}\mathbb{P}(T_k < \infty) = (1-u)^{-1}.$$

(b) The solution relies on the formula

$$\text{(A.14)} \qquad \mathbb{P}(X_n = 0) = \int_{[-\pi,\pi]^d} \hat{D}(k)^n \frac{d^dk}{(2\pi)^d}.$$

Some care is required when performing the sum over n since the best uniform bound on $\hat{D}^n$ is 1 which is not summable. A solution is to make use of monotone convergence first, and then apply the dominated convergence theorem, as follows,

$$\text{(A.15)} \qquad m = \lim_{t\nearrow 1}\sum_{n\geq 0}\mathbb{P}(X_n = 0)t^n = \lim_{t\nearrow 1}\int_{[-\pi,\pi]^d}\frac{1}{1-t\hat{D}(k)}\frac{d^dk}{(2\pi)^d}.$$

Note that $\hat{D}$ is a real-valued function and that

$$\text{(A.16)} \qquad \frac{1}{1-t\hat{D}(k)} \leq \frac{2}{1-\hat{D}(k)} \quad \text{for } t \in [1/2, 1],$$

so that if $(1-\hat{D})^{-1} \in L^1$, then the claim follows by dominated convergence. In the case that $(1-\hat{D})^{-1} \notin L^1$, the claim follows from Fatou's lemma.

(c) $\hat{D}(k) = \sum_{j=1}^d (e^{ik_j} + e^{-ik_j}) = 2\sum_{j=1}^d \cos(k_j)$ and thus $1 - \hat{D}(k) = O(1)|k|^2$ as $k \to 0$. Note further that

$$\text{(A.17)} \qquad \int_{\mathbb{R}^d} f(|k|)\, dk = V_{d-1}\int_0^\infty f(r)\, r^{d-1}\, dr,$$

where V_{d-1} is the volume of the $(d-1)$-dimensional sphere, and in particular,

$$\text{(A.18)} \qquad \int_{[-\epsilon,\epsilon]^d} |k|^{-p}\, dk \text{ is integrable} \quad \text{if and only if} \quad d > p.$$

□

PROBLEM 1.7. Note that

$$I = \sum_{x\in\mathbb{Z}^d} \Big(\sum_{i\geq 0} 1_{\{X_i^1=x\}}\Big)\Big(\sum_{j\geq 0} 1_{\{X_j^2=x\}}\Big), \tag{A.19}$$

and thus, by Parseval's theorem, if $f \in L^2(\mathbb{Z}^d)$,

$$\mathbb{E}(I) = \sum_{x\in\mathbb{Z}^d} f(x)^2 = \int_{[-\pi,\pi]^d} |\hat{f}(k)|^2 \frac{d^dk}{(2\pi)^d}, \tag{A.20}$$

where

$$f(x) = \sum_{j\geq 0} \mathbb{P}\{X_j^1 = x\} = \sum_{j\geq 0} D^{*j}(x), \quad \hat{f}(k) = \sum_{j\geq 0} \hat{D}(k)^j = \frac{1}{1-\hat{D}(k)}. \tag{A.21}$$

If $f \notin L^2(\mathbb{Z}^d)$, then both sides must be infinite. □

A.2. Solutions for Tutorial 4.4.

PROBLEM 4.1. The graph $\{0n\}$ is connected on $[0,n]$ in the above sense but not path-connected. Also, $\{01, 12, \ldots, (n-1)n\}$ is path-connected but not connected in the above sense since the open intervals $(i-1,i)$ do not overlap. □

PROBLEM 4.2. This is an application of the identity

$$\prod_{i\in I}(1+u_i) = \sum_{S\subset I}\prod_{i\in S} u_i \tag{A.22}$$

with I being the set of edges on $[a,b]$. □

PROBLEM 4.3. The identity corresponds to a decomposition of $\mathcal{B}[a,b]$ by connected components. The term $K[a+1,b]$ corresponds to graphs Γ for which $a \notin \Gamma$.

So assume $a \in \Gamma$. We shall show that Γ can be written uniquely as $\Gamma = \Gamma' \cup \Gamma''$ where $\Gamma' \in \mathcal{G}[a,j]$ and $\Gamma'' \in \mathcal{B}[j,b]$ for some $j \in (a,b]$. Informally, Γ' is the connected component of Γ containing a, though we must verify that this notion is well-defined. Conversely it is clear that if $\Gamma' \in \mathcal{G}[a,j]$, $\Gamma'' \in \mathcal{B}[j,b]$ for some $j \in (a,b]$, then $\Gamma = \Gamma' \cup \Gamma'' \in \mathcal{B}[a,b]$ with $a \in \Gamma$. Then the result will follow since

$$\prod_{st\in\Gamma'\cup\Gamma''} U_{st} = \prod_{st\in\Gamma'} U_{st} \prod_{st\in\Gamma''} U_{st}. \tag{A.23}$$

Let

$$j = \min\{i \in (a,b] : i \notin (s,t) \text{ for some } st \in \Gamma\}. \tag{A.24}$$

The minimum is well defined since there can be no $st \in \Gamma$ for which $b \in (s,t)$. By construction, every edge $st \in \Gamma$ satisfies $t \leq j$ or $s \geq j$, so that we can write $\Gamma = \Gamma' \cup \Gamma''$ where $\Gamma' \in \mathcal{B}[a,j]$, $\Gamma'' \in \mathcal{B}[j,b]$. We must show that $\Gamma' \in \mathcal{G}[a,j]$, i.e., that Γ' is connected. But $\cup_{st\in\Gamma'}(s,t) = (a,j) \cap \cup_{st\in\Gamma}(s,t) = (a,j)$ by the minimality of j.

Finally we check that the decomposition $\Gamma = \Gamma' \cup \Gamma''$ is unique: this follows because if $\Gamma' \in \mathcal{G}[a,j']$ and $\Gamma'' \in \mathcal{B}[j',b]$ then the formula (A.24) recovers $j = j'$. □

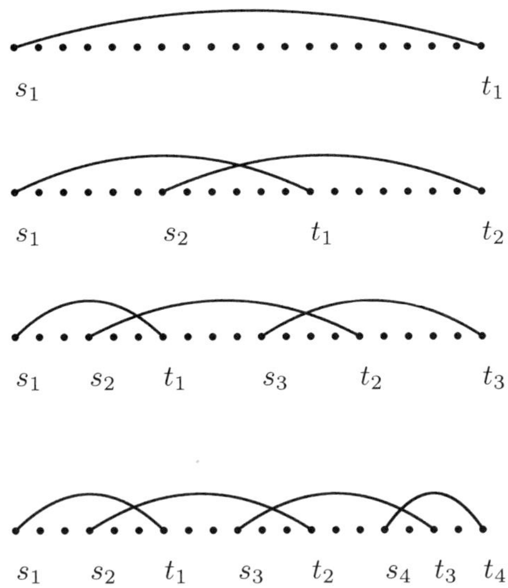

FIGURE 14. Laces in $\mathcal{L}^{(N)}[a,b]$ for $N = 1,2,3,4$, with $s_1 = a$ and $t_N = b$.

PROBLEM 4.4. The convolutions correspond to summing over the values of $\omega(1)$ and $\omega(m)$. Namely, noting that a walk ω on $[0,n]$ is equivalent to a pair of walks ω_0 on $[0,m]$ and ω_1 on $[m,n]$ with $\omega_0(m) = \omega_1(m)$, we have

$$\begin{aligned} c_n(x) &= \sum_{\omega\in\mathcal{W}_n(0,x)} K[1,n](\omega) + \sum_{m=1}^{n} \sum_{\omega\in\mathcal{W}_n(0,x)} J[0,m](\omega)K[m,n](\omega) \\ &= \sum_{y\in\mathbb{Z}^d} \sum_{\omega_0\in\mathcal{W}_1(0,y)} \sum_{\substack{\omega_1:[1,n]\to\mathbb{Z}^d,\\ \omega_1(1)=y,\,\omega_1(n)=x}} K[1,n](\omega_1) \\ &\quad + \sum_{m=1}^{n} \sum_{y\in\mathbb{Z}^d} \sum_{\omega_0\in\mathcal{W}_m(0,y)} \sum_{\substack{\omega_1:[m,n]\to\mathbb{Z}^d,\\ \omega_1(m)=y,\,\omega_1(n)=x}} J[0,m](\omega_0)K[m,n](\omega_1) \\ &= \sum_{y\in\mathbb{Z}^d} 1_{\{y\in\Omega\}} c_{n-1}(x-y) + \sum_{m=1}^{n}\sum_{y\in\mathbb{Z}^d} \pi_m(y) c_{n-m}(x-y) \qquad \text{(A.25)} \end{aligned}$$

where we use the translation invariance (in time and space) of K. Since $c_1(y) = 1_{\{y\in\Omega\}}$, this is the desired equation. □

PROBLEM 4.5. Figure 14 is helpful. Note first that if L is a lace, then $s_l < s_{l+1}$ for each l. Indeed, if $s_l = s_{l+1}$, we may assume that $t_l < t_{l+1}$. But then $(s_l, t_l) \subset (s_{l+1}, t_{l+1})$ so that $L \setminus \{s_l t_l\}$ is still connected. A similar argument gives $t_l < t_{l+1}$. The requirement that L is connected implies that $a = s_1$ and $b = t_N$.

Suppose to the contrary that (1) $s_{l+1} \geq t_l$ $(1 \leq l \leq N-1)$ or (2) $s_{l+2} < t_l$ $(1 \leq l \leq N-2)$. In case (1), L is not connected, since $s_i \geq t_l$ for $i \geq l+1$ while

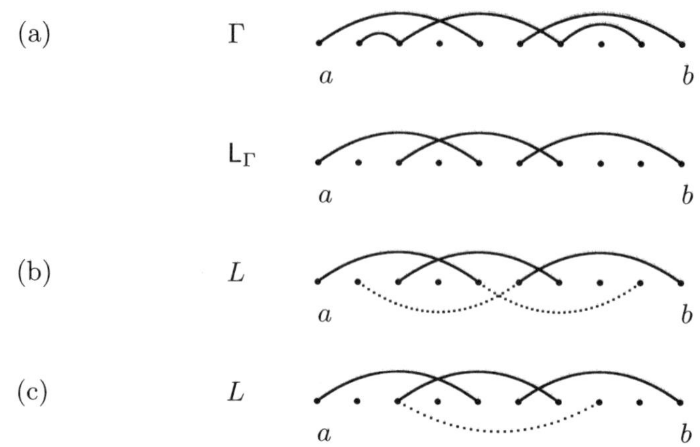

FIGURE 15. (a) A connected graph Γ and its associated lace $L = \mathsf{L}_\Gamma$. (b) The dotted edges are compatible with the lace L. (c) The dotted edge is not compatible with the lace L.

$t_i \le t_l$ for $i \le l$. In case (2), the edge $s_{l+1}t_{l+1}$ is redundant since $(s_{l+1}, t_{l+1}) \subset (s_l, t_l) \cup (s_{l+2}, t_{l+2}) = (s_l, t_{l+2})$.

For the converse, the hypotheses imply that $\cup_{st\in L}(s,t) = (a,b)$, so L is connected. Neither s_1t_1 nor s_Nt_N can be removed from L since they are the only edges containing the endpoints. If s_lt_l is removed, $2 \le l \le N-1$, then $t_{l-1} \le s_{l+1}$ implies that $\cup_{st\in L}(s,t) = (a, t_{l-1}) \cup (s_{l+1}, b) \neq (a,b)$. So $L \setminus \{st\}$ is not connected. Since connectedness is a monotone property, no strict subset of L can be connected, so L is minimally connected, i.e., a lace.

Finally the intervals are as follows: the first and last intervals are $[s_1, s_2]$ and $[t_{N-1}, t_N]$; the $2i^{\text{th}}$ interval is $[s_{i+1}, t_i]$ $(1 \le i \le N-1)$; and the $(2i+1)^{\text{st}}$ interval is $[t_i, s_{i+2}]$, $1 \le i \le N-2$. The inequalities above show that the points $\{s_i, t_i\}$ do indeed form the intervals claimed, and that the intervals $[t_i, s_{i+2}]$ can be empty while the other intervals must be non-empty. □

PROBLEM 4.6. Figure 15 is helpful. First, since necessarily $\mathsf{L}_\Gamma \subset \Gamma$, we may assume that $L \subset \Gamma$, and we write $\Gamma = L \cup A$ with $A \cap L = \emptyset$.

Next, we reformulate the inductive procedure for selecting the edges of L_Γ. At each step, the edge $s_{i+1}t_{i+1}$ is, among all edges $st \in \Gamma$ satisfying $s < t_i$, the one that is *maximal* with respect to the following order relation: $st \succ s't'$ if and only if $t > t'$ or $t = t'$ and st is longer than $s't'$ (i.e., $t - s > t' - s'$).

The result follows at once from this observation. Indeed, $\mathsf{L}_\Gamma = L$ means that at each inductive step, $s_{i+1}t_{i+1} \in L$ is the maximal edge st satisfying $s < t_i$, among all edges of $L \cup A$. This is equivalent to saying that for each $s't' \in A$, at each inductive step, $s_{i+1}t_{i+1}$ is the maximal edge among all edges of $L \cup \{s't'\}$. But this is precisely the condition that $A \subset \mathcal{C}(L)$. □

PROBLEM 4.7. The first equation is simply a decomposition of $\Gamma \in \mathcal{G}[a,b]$ according to the value of L_Γ. The second equation follows using (A.22) because Problem 4.6 shows that Γ for which $\mathsf{L}_\Gamma = L$ can be identified as L together with an arbitrary subset of edges from $\mathcal{C}(L)$. The last equation is immediate from the preceding ones. □

PROBLEM 4.8. (a) By definition,

$$\begin{aligned}\sum_{N=1}^{\infty}(-1)^N\pi_m^{(N)}(x) &= \sum_{\omega\in\mathcal{W}_m(0,x)}\sum_{N=1}^{\infty}J^{(N)}[0,m](\omega)\\ &= \sum_{\omega\in\mathcal{W}_m(0,x)}J[0,m](\omega)=\pi_m(x).\end{aligned}\tag{A.26}$$

Each of the N factors U_{st}, $st \in L$, contributes -1, so $\pi_m^{(N)}(x) \geq 0$.

(b) $N = 1$: The only lace with 1 edge is $L = \{0m\}$, and every edge except $0m$ is compatible with L. So $J^{(1)}[0,m]$ contains the single factor U_{0m}, and the factor $1 + U_{s't'}$ for each $s't' \neq 0m$. So a contributing ω must have $\omega(s') \neq \omega(t')$ whenever $s't' \neq 0m$, as well as $0 = \omega(0) = \omega(m) = x$. Hence $\pi_m^{(1)}(x) = 0$ for $x \neq 0$, and $\pi_m^{(1)}(0)$ is the number of m-step self-avoiding returns.

$N = 2$: For $L = \{0t_1, s_1m\}$, the factors U_{st}, $st \in L$, require that ω should start at 0, visit x (at step s_1), return to 0 (at step t_1), then return to x. The compatible edges consist of every edge except the edges of L and the edges $0t$, $t > t_1$ and sm, $s < s_1$. This implies that each of the three intervals in ω must be self-avoiding and mutually avoiding, except for the intersections required above. (In particular, $x \neq 0$.) (Intersections of the form $\omega(0) = \omega(t)$, $t > t_1$, might not appear to be forbidden, but actually they are impossible since we require $\omega(t) \neq \omega(t_1) = \omega(0)$.)

$N = 3, 4, \ldots$: As for $N = 2$, ω must have self-intersections corresponding to the edges of the lace and self-avoidance corresponding to each compatible edge. It is convenient to recall the $2N - 1$ intervals from Problem 4.5. Because of compatible edges, ω is required to be self-avoiding on each of these intervals. In addition, certain of these intervals are required to be mutually avoiding, but not all of them need be, corresponding to the fact that an edge spanning too many intervals cannot be compatible. The pattern of mutual avoidance is described as follows: for $N = 3$,

$$[1234][345] \tag{A.27}$$

and for $N = 4$,

$$[1234][3456][567] \tag{A.28}$$

where, for instance [3456] indicates that the third to sixth interval must be mutually self-avoiding, except for the required intersections. These intersections require that at the endpoints of the intervals, ω must visit the following points (for the case $N = 4$):

$$0, x_1, 0, x_2, x_1, x_3, x_2, x_3 \tag{A.29}$$

where $x_3 = x$, corresponding to the intervals $[s_1, s_2]$, $[s_2, t_1]$, $[t_1, s_3]$, $[s_3, t_2]$, $[t_2, s_4]$, $[s_4, t_3]$, $[t_3, t_4]$.

To prove the avoidance patterns amounts to analysing exactly which edges are compatible. For instance, it is easy to verify that if $s_{i+1} \leq s < t_i$ for $1 \leq i \leq N-1$, then $st \in \mathcal{C}(L)$ if and only if $t \leq t_{i+1}$ (assuming $st \notin L$).

(c) The possibly empty intervals indicate that the $3^{\text{rd}}, 5^{\text{th}}, \ldots, (2N-3)^{\text{rd}}$ segments of the diagrams above can be empty, whereas all other segments must have non-zero length. The picture for $N = 11$ is

where the lines that are slashed are exactly the lines that are permitted to have length zero. □

A.3. Solutions for Tutorial 5.4.

PROBLEM 5.1. (a) Using the hint and Fubini's theorem (which is applicable because $d > 2m$),

$$\int_{[-\pi,\pi]^d} \frac{1}{[1-\hat{D}(k)]^m} \frac{d^d k}{(2\pi)^d} = \Gamma(m)^{-1} \int_0^\infty \left(\int_{[-\pi,\pi]^d} e^{-u[1-\hat{D}(k)]} \frac{d^d k}{(2\pi)^d} \right) u^{m-1}\, du$$

$$= \Gamma(m)^{-1} \int_0^\infty \left(\int_{-\pi}^{\pi} e^{-u(1-\cos k)/d} \frac{dk}{2\pi} \right)^d u^{m-1}\, du. \tag{A.30}$$

The inner integral is decreasing as a function of d by Hölder's inequality.

(b) Relaxing the self-avoidance for the first j steps gives the inequality

$$H_z^{(j)}(x) \le (z|\Omega|D)^{*j} * G_z(x), \tag{A.31}$$

and thus, by Cauchy-Schwarz,

$$\|H_z^{(j)}\|_\infty \le \|\hat{H}_z^{(j)}\|_1 \le (z|\Omega|)^j \|\hat{D}^j\|_2 \|\hat{G}_z\|_2. \tag{A.32}$$

The claim now follows from $z_c|\Omega| \le a$,

$$\|\hat{D}^j\|_2 = \|\hat{D}^{2j}\|_1^{1/2} \le O((2d)^{-j/2}), \tag{A.33}$$

and

$$\|\hat{G}_z\|_2 \le a\|\hat{C}_{p(z)}\|_2 \le a\|\hat{C}_{1/|\Omega|}\|_2 \le O(1) \tag{A.34}$$

by the infrared bound (5.38) and (a).

(c) Calculating the first two terms explicitly, we obtain

$$\hat{\Pi}_z^{(1)}(0) = (2d)z^2 + (2d)(2d-2)z^4 + \sum_{m\ge 6} \hat{\pi}_m^{(1)}(0) z^m$$

$$= (2d)z^2 + (2d)(2d-2)z^4 + O((2d)^{-3}), \tag{A.35}$$

where the remainder was estimated as in (b), and using the symmetry of D:

$$\sum_{m\ge j} \hat{\pi}_m^{(1)}(0) z^m = (H_z^{(j-1)} * z|\Omega|D)(0)$$

$$\le (z|\Omega|)^j \, \|D^{*j} * G_z\|_\infty \le O((2d)^{-j/2}). \tag{A.36}$$

Using (5.40), we obtain

$$\hat{\Pi}_z(0) = -\hat{\Pi}_z^{(1)}(0) + O((2d)^{-2}). \tag{A.37}$$

Equation (5.42) gives $z_c = (2d)^{-1} - (2d)^{-1}\hat{\Pi}_{z_c}(0) = (2d)^{-1} + O((2d)^{-2})$, from which we obtain

$$\begin{aligned} z_c &= (2d)^{-1} + (2d)^{-1}\hat{\Pi}^{(1)}_{z_c}(0) + O((2d)^{-3}) \\ &= (2d)^{-1} + (2d)^{-2} + O((2d)^{-3}). \end{aligned} \tag{A.38}$$

Finally, we obtain

$$\hat{\Pi}^{(1)}_{z_c}(0) = [(2d)^{-1} + 2(2d)^{-2}] + (2d)^{-2} + O((2d)^{-3}). \tag{A.39}$$

(d) The generating function for θ-walks from 0 to x can be written as

$$\hat{\Pi}^{(2)}_z(0) = (2d)z^3 + 3(2d)(2d-2)z^5 + \sum_{m\geq 7} \hat{\pi}^{(2)}_m(0) z^m. \tag{A.40}$$

Using $z_c = (2d)^{-1} + (2d)^{-2} + O((2d)^{-3})$ from part (c), we obtain

$$(2d)z_c^3 + 3(2d)(2d-2)z_c^5 = (2d)^{-2} + O((2d)^{-3}). \tag{A.41}$$

The remainder is estimated using the fact that in a θ-walks from 0 to x of length $m \geq 7$, either two of the subwalks take just one step and the other takes at least 5 steps, or at least two of the subwalks take at least 3 steps. Thus there is a combinatorial constant K such that

$$\begin{aligned} \sum_{m\geq 7} \hat{\pi}^{(2)}_m(0) z^m \leq K \sum_e H^{(5)}_z(e) z|\Omega|D(e) z|\Omega|D(e) \\ + K \sum_x H^{(1)}_z(x) H^{(3)}_z(x) H^{(3)}_z(x). \end{aligned} \tag{A.42}$$

The first term can be estimated by an L^∞ bound (use $z|\Omega| \leq a$ and $|\Omega| = 2d$):

$$\begin{aligned} \sum_e H^{(5)}_z(e) z|\Omega|D(-e) z|\Omega|D(e) &\leq \|H^{(5)}_z\|_\infty a^2 (2d)^{-1} \\ &\leq O((2d)^{-7/2}) \leq O((2d)^{-3}). \end{aligned} \tag{A.43}$$

The second term is estimated in the spirit of (b):

$$\begin{aligned} \sum_x H^{(1)}_z(x) H^{(3)}_z(-x) H^{(3)}_z(x) &\leq O((2d)^{-1/2}) (H^{(3)}_z * H^{(3)}_z)(0) \\ &\leq O((2d)^{-1/2}) ((z|\Omega|D)^{*6} * G_z^{*2})(0) \\ &\leq O((2d)^{-1/2}) \|\hat{D}^6 \hat{G}_z^2\|_1 \\ &\leq O((2d)^{-1/2}) \|\hat{D}^6\|_2 \|\hat{G}_z^2\|_2 \leq O((2d)^{-7/2}). \end{aligned} \tag{A.44}$$

(e) Using (5.40), we obtain

$$\hat{\Pi}_{z_c}(0) = -(2d)^{-1} - 2(2d)^{-2} + O((2d)^{-3}). \tag{A.45}$$

From (5.42), it then follows that

$$z_c = (2d)^{-1} + (2d)^{-2} + 2(2d)^{-3} + O((2d)^{-4}). \tag{A.46}$$

Inverting this finally yields

$$\mu = 2d - 1 - (2d)^{-1} + O((2d)^{-2}). \tag{A.47}$$

□

PROBLEM 5.2. (a) This is a straightforward calculation.

(b) This requires an extension of the diagrammatic estimates. The argument is sketched in [**69**, Section 5.4].

(c) Note that

$$\frac{d[z\chi(z)]^{-1}}{dz} = -[z\chi(z)]^{-2}\frac{d[z\chi(z)]}{dz} \sim -\frac{V(z_c)}{z^2} \sim -\frac{V(z_c)}{z_c^2}. \tag{A.48}$$

Integrating this asymptotic relation, we obtain

$$\lim_{z\to z_c}[z\chi(z)]^{-1} - [z\chi(z)]^{-1} \sim -\frac{V(z_c)}{z_c^2}(z_c - z). \tag{A.49}$$

The limit vanishes and thus

$$\chi(z)^{-1} \sim V(z_c)^{-1}(1 - z/z_c)^{-1}. \tag{A.50}$$

The claim then follows from the definition of $V(z_c)$ and (5.42). □

A.4. Solutions for Tutorial 7.8.

PROBLEM 7.1. For $m, n \geq 0$,

$$\begin{aligned} I_{0,n+m} &= \sum_{0\leq i<j\leq n+m} 1_{\{X_i=X_j\}} \\ &\geq \sum_{0\leq i<j\leq m} 1_{\{X_i=X_j\}} + \sum_{m\leq i<j\leq n+m} 1_{\{X_i=X_j\}} = I_{0,m} + I_{m,n+m}. \end{aligned} \tag{A.51}$$

By translation invariance and the Markov property, $I_{m,n+m}$ is independent of $I_{0,m}$ and has the same law as $I_{0,n}$. Therefore

$$c_{n+m} \leq \mathbb{E}_0(e^{-gI_{0,m}}e^{-gI_{m,m+n}}) = \mathbb{E}_0(e^{-gI_{0,m}})\mathbb{E}_0(e^{-gI_{0,n}}) = c_m c_n \tag{A.52}$$

as claimed. The remaining statements follow since (A.52) implies that $(\log c_n)_{n\geq 0}$ is a subadditive sequence, and Lemma 1.1 can be applied. □

PROBLEM 7.2. Note that there is a one-to-one correspondence between nearest-neighbour walks on $\mathbb{Z}^d$ and such walks on the torus $\mathbb{Z}^d/R\mathbb{Z}^d$, $R \geq 3$, by folding a walk on $\mathbb{Z}^d$ (the image under the canonical projection $\mathbb{Z}^d \twoheadrightarrow \mathbb{Z}^d/R\mathbb{Z}^d$), and corresponding unfolding of walks on $\mathbb{Z}^d/R\mathbb{Z}^d$ (unique for the nearest-neighbour step distribution provided $R \geq 3$). Given a walk $X = (X_n)_{n\geq 0}$ on $\mathbb{Z}^d$ starting at 0, we denote the folded (or projected) walk by X'. Write $\Lambda_R = \{-R+1, \ldots, R\}^d$ and $R' = 2R + 1$; then

$$\begin{aligned} I_n(X) &= \sum_{0\leq i<j\leq n}\sum_{x\in\mathbb{Z}^d} 1_{\{X_i=X_j=x\}} = \sum_{0\leq i<j\leq n}\sum_{x\in\Lambda_R}\sum_{y\in\mathbb{Z}^d} 1_{\{X_i=X_j=x+yR'\}} \\ &\leq \sum_{0\leq i<j\leq n}\sum_{x\in\Lambda_R}\sum_{y_1,y_2\in\mathbb{Z}^d} 1_{\{X_i=x+y_1R'\}}1_{\{X_j=x+y_2R'\}} \\ &= I_n(X'), \end{aligned} \tag{A.53}$$

and thus

$$\mathbb{E}(e^{-gI_n}) \geq \mathbb{E}^R(e^{-gI_n}). \tag{A.54}$$

The desired inequalities both follow from this one. □

PROBLEM 7.3. Note that

$$G_\nu(x,y) - G_{\nu,D}(x,y) = \sum_{n=0}^{\infty} \mathbb{E}_x(e^{-gI_n} 1_{\{X_n=y,\, n\geq T_D\}})e^{-\nu n}, \tag{A.55}$$

and, by partitioning in T_D and X_{T_D}, we obtain

$$\mathbb{E}_x(e^{-gI_n} 1_{\{X_n=y, n\geq T_D\}}) = \sum_{z\in\partial D}\sum_{m=0}^{n} \mathbb{E}_x(e^{-gI_n} 1_{\{X_n=y\}} 1_{\{X_{T_D}=z\}} 1_{\{T_D=m\}}). \tag{A.56}$$

Using $I_n \geq I_m + I_{m,n}$ and the Markov property, it follows that

$$\begin{aligned} &\mathbb{E}_x(e^{-gI_n} 1_{\{X_n=y\}} 1_{\{X_{T_D}=z\}} 1_{\{T_D=m\}}) \\ &\quad\leq \mathbb{E}_x(e^{-gI_m} 1_{\{X_m=z\}} 1_{\{T_D=m\}} e^{-gI_{m,n}} 1_{\{X_n=y\}}) \\ &\quad= \mathbb{E}_x(e^{-gI_m} 1_{\{X_m=z\}} 1_{\{T_D=m\}}) \mathbb{E}_z(e^{-gI_{n-m}} 1_{\{X_{n-m}=y\}}). \end{aligned} \tag{A.57}$$

Thus, because $\{T_D = m, X_m = z\} = \{m \leq T_D, X_m = z\}$ for $z \in \partial D$,

$$\begin{aligned} G_\nu(x,y) - G_{\nu,D}(x,y) &\leq \sum_{z\in\partial D}\sum_{n=0}^{\infty}\sum_{m=0}^{n} \mathbb{E}_x(e^{-gI_m} 1_{\{X_m=z\}} 1_{\{T_D=m\}}) \\ &\qquad\qquad \cdot \mathbb{E}_z(e^{-gI_{n-m}} 1_{\{X_{n-m}=y\}}) e^{-\nu n} \\ &\leq \sum_{z\in\partial D} G_{\nu,\bar{D}}(x,z) G_\nu(z,y), \end{aligned} \tag{A.58}$$

as claimed. □

PROBLEM 7.4. Let $m = \lfloor |y|_\infty/(R+1)\rfloor$. By the Simon-Lieb inequality (7.53), translation invariance, and the bound $G_{\nu,\bar{D}}(x,z) \leq G_\nu(x,z)$, we have

$$\begin{aligned} G_\nu(x,y) &\leq \sum_{z_1\in x+\partial\Lambda_R} G_\nu(x,z_1) G_\nu(z_1,y) \\ &\leq \cdots \leq \sum_{z_1\in z_0+\partial\Lambda_R} \cdots \sum_{z_m\in z_{m-1}+\partial\Lambda_R} G_\nu(x,z_1)G_\nu(z_1,z_2)\cdots G_\nu(z_m,y) \\ &\leq \theta^m \sup_{x\in\mathbb{Z}^d} G_\nu(0,x). \end{aligned} \tag{A.59}$$

Note that we applied (7.53) in such a manner that the term $G_{\nu,D}(x,y)$ vanishes. □

PROBLEM 7.5. Fix $\nu > \nu_c$, and let $D_R = \{-R+2,\dots,R-1\}^d$ be the interior of Λ_R. By monotone convergence,

$$G_\nu(x,y) = \lim_{R\to\infty} G_{\nu,D_R}(x,y). \tag{A.60}$$

Hence, to prove (7.55), it suffices to show that $\lim_{R\to\infty} G_\nu^R(x,y) - G_{\nu,D_R}(x,y) = 0$. Now,

$$G_\nu^R(x,y) - G_{\nu,D_R}(x,y) = G_\nu^R(x,y) - G_{\nu,D_R}^R(x,y), \tag{A.61}$$

and thus, from the Simon-Lieb inequality (Problem 7.3), it follows that

$$\begin{aligned} G_\nu^R(x,y) - G_{\nu,D_R}(x,y) &\leq \sum_{z\in\partial D_R} G_{\nu,\bar{D}_R}^R(x,z) G_\nu^R(z,y) \\ &\leq \Big(\sup_{z\in\partial D_R} G_{\nu,\bar{D}_R}^R(x,z)\Big)\Big(\sum_{z\in\partial D_R} G_\nu^R(z,y)\Big). \end{aligned} \tag{A.62}$$

By Problem 7.2,

$$\sum_{z\in\partial D_R} G_\nu^R(z,y) \le \sum_{z\in\Lambda_R} G_\nu^R(z,y) = \chi^R(\nu) \le \chi(\nu) < \infty, \tag{A.63}$$

and, by Problem 7.4 and the fact that $G_\nu(0,x)$ is uniformly bounded since the susceptibility is finite,

$$\begin{aligned} \sup_{z\in\partial D_R} G^R_{\nu,\bar{D}_R}(x,z) &= \sup_{z\in\partial D_R} G_{\nu,\bar{D}_R}(x,z) \le \sup_{z\in\partial D_R} G_\nu(x,z) \\ &\le \sup_{z\in\partial D_R} Ce^{-\gamma|z-x|} \le Ce^{-\gamma(R-|x|)} \to 0 \end{aligned} \tag{A.64}$$

as $R\to\infty$. Therefore $\lim_{R\to\infty} G_\nu^R(x,y) - G_{\nu,\bar{D}_R}(x,y) = 0$, proving (7.55). Finally (7.56) follows since $G_{\nu_c}(x,y) = \lim_{\nu\searrow\nu_c} G_\nu(x,y)$ by monotone convergence. □

Acknowledgements

The authors are grateful to the organisers of both the Clay Mathematics Institute Summer School and the XIV Brazilian Probability School for having delivered such a stimulating and energetic event. We are particularly grateful to Vladas Sidoravicius for his tireless efforts on behalf of the School and its participants.

Hugo Duminil-Copin thanks Stanislav Smirnov for very stimulating discussions and encouragement related to their joint work presented in Section 3. Gordon Slade thanks David Brydges for many discussions related to their joint work presented in Section 7.

The work of Hugo Duminil-Copin was supported by the ANR grant BLAN06-3-134462, the EU Marie-Curie RTN CODY, the ERC AG CONFRA, and the Swiss FNS. The work of Gordon Slade was supported in part by NSERC of Canada.

References

[1] V. Beffara, SLE *and other conformally invariant objects*, Lectures in this Summer School.

[2] C. Borgs, J.T. Chayes, R. van der Hofstad, G. Slade, and J. Spencer, *Random subgraphs of finite graphs: II. The lace expansion and the triangle condition*, Ann. Probab. **33** (2005), 1886–1944.

[3] A. Bovier, G. Felder, and J. Fröhlich, *On the critical properties of the Edwards and the self-avoiding walk model of polymer chains*, Nucl. Phys. B **230** **[FS10]** (1984), 119–147.

[4] D.C. Brydges, *Lectures on the renormalisation group*, Statistical Mechanics (S. Sheffield and T. Spencer, eds.), American Mathematical Society, Providence, (2009), IAS/Park City Mathematics Series, Volume 16, pp. 7–93.

[5] D. Brydges, S.N. Evans, and J.Z. Imbrie, *Self-avoiding walk on a hierarchical lattice in four dimensions*, Ann. Probab. **20** (1992), 82–124.

[6] D.C. Brydges, J. Fröhlich, and A.D. Sokal, *The random walk representation of classical spin systems and correlation inequalities. II. The skeleton inequalities*, Commun. Math. Phys. **91** (1983), 117–139.

[7] D.C. Brydges, J. Fröhlich, and T. Spencer, *The random walk representation of classical spin systems and correlation inequalities*, Commun. Math. Phys. **83** (1982), 123–150.

[8] D.C. Brydges, G. Guadagni, and P.K. Mitter, *Finite range decomposition of Gaussian processes*, J. Stat. Phys. **115** (2004), 415–449.

[9] D.C. Brydges and J.Z. Imbrie, *End-to-end distance from the Green's function for a hierarchical self-avoiding walk in four dimensions*, Commun. Math. Phys. **239** (2003), 523–547.

[10] ———, *Green's function for a hierarchical self-avoiding walk in four dimensions*, Commun. Math. Phys. **239** (2003), 549–584.

[11] D.C. Brydges, J.Z. Imbrie, and G. Slade, *Functional integral representations for self-avoiding walk*, Probab. Surveys **6** (2009), 34–61.

[12] D. Brydges and G. Slade, *Renormalisation group analysis of weakly self-avoiding walk in dimensions four and higher*, To appear in *Proceedings of the International Congress of Mathematicians*, Hyderabad, 2010, ed. R. Bhatia, Hindustan Book Agency, Delhi.

[13] ———, *Weakly self-avoiding walk in dimensions four and higher: a renormalisation group approach*, In preparation.

[14] D.C. Brydges and T. Spencer, *Self-avoiding walk in 5 or more dimensions*, Commun. Math. Phys. **97** (1985), 125–148.

[15] M. Campanino, D. Ioffe, and Y. Velenik, *Random path representation and sharp correlations asymptotics at high-temperatures*, Adv. Stud. Pure Math. **39** (2004), 29–52.

[16] S. Caracciolo, M.S. Causo, and A. Pelissetto, *High-precision determination of the critical exponent γ for self-avoiding walks*, Phys. Rev. E **57** (1998), 1215–1218.

[17] J.T. Chayes and L. Chayes, *Ornstein-Zernike behavior for self-avoiding walks at all non-critical temperatures*, Commun. Math. Phys. **105** (1986), 221–238.

[18] D. Chelkak and S. Smirnov, *Universality in the 2D Ising model and conformal invariance of fermionic observables*, to appear in Inv. Math., (2009).

[19] L.-C. Chen and A. Sakai, *Asymptotic behavior of the gyration radius for long-range self-avoiding walk and long-range oriented percolation*, to appear in *Ann. Probab.* Preprint, (2010).

[20] N. Clisby, *Accurate estimate of the critical exponent ν for self-avoiding walks via a fast implementation of the pivot algorithm*, Phys. Rev. Lett. **104** (2010), 055702.

[21] N. Clisby, R. Liang, and G. Slade, *Self-avoiding walk enumeration via the lace expansion*, J. Phys. A: Math. Theor. **40** (2007), 10973–11017.

[22] N. Clisby and G. Slade, *Polygons and the lace expansion*, Polygons, Polyominoes and Polycubes (A.J. Guttmann, ed.), Springer, Dordrecht, (2009), Lecture Notes in Physics Vol. 775, pp. 117–142.

[23] A.R. Conway, I.G. Enting, and A.J. Guttmann, *Algebraic techniques for enumerating self-avoiding walks on the square lattice*, J. Phys. A: Math. Gen. **26** (1993), 1519–1534.

[24] H. Duminil-Copin and S. Smirnov, *The connective constant of the honeycomb lattice equals* $\sqrt{2+\sqrt{2}}$, Annals of Math., **175(3)**, 1653–1665 (2012).

[25] P.J. Flory, *The configuration of a real polymer chain*, J. Chem. Phys. **17** (1949), 303–310.

[26] B.T. Graham, *Borel-type bounds for the self-avoiding walk connective constant*, J. Phys. A: Math. Theor. **43** (2010), 235001.

[27] A. Greven and F. den Hollander, *A variational characterization of the speed of a one-dimensional self-repellent random walk*, Ann. Appl. Probab. **3** (1993), 1067–1099.

[28] R. Guida and J. Zinn-Justin, *Critical exponents of the N-vector model*, J. Phys. A: Math. Gen. **31** (1998), 8103–8121.

[29] J.M. Hammersley and D.J.A. Welsh, *Further results on the rate of convergence to the connective constant of the hypercubical lattice*, Quart. J. Math. Oxford **(2), 13** (1962), 108–110.

[30] T. Hara, *Decay of correlations in nearest-neighbor self-avoiding walk, percolation, lattice trees and animals*, Ann. Probab. **36** (2008), 530–593.

[31] T. Hara, R. van der Hofstad, and G. Slade, *Critical two-point functions and the lace expansion for spread-out high-dimensional percolation and related models*, Ann. Probab. **31** (2003), 349–408.

[32] T. Hara and M. Ohno, *Renormalization group analysis of hierarchical weakly self-avoiding walk in four dimensions*, In preparation.

[33] T. Hara and G. Slade, *Self-avoiding walk in five or more dimensions. I. The critical behaviour*, Commun. Math. Phys. **147** (1992), 101–136.

[34] ———, *The lace expansion for self-avoiding walk in five or more dimensions.*, Reviews in Math. Phys. **4** (1992), 235–327.

[35] ———, *The self-avoiding-walk and percolation critical points in high dimensions*, Combin. Probab. Comput. **4** (1995), 197–215.

[36] G.H. Hardy, *Divergent series*, Oxford University Press, Oxford, (1949).

[37] G.H. Hardy and S. Ramanujan, *Asymptotic formulae for the distribution of integers of various types*, Proc. Lond. Math. Soc. **(2) 16** (1917), 112–132.

[38] M. Heydenreich, *Long-range self-avoiding walk converges to alpha-stable processes*, To appear in *Ann. I. Henri Poincaré Probab. Statist.* Preprint, (2008).

[39] M. Heydenreich, R. van der Hofstad, and A. Sakai, *Mean-field behavior for long- and finite range Ising model, percolation and self-avoiding walk*, J. Stat. Phys. **132** (2008), 1001–1049.

[40] R. van der Hofstad and W. König, *A survey of one-dimensional random polymers*, J. Stat. Phys. **103** (2001), 915–944.

[41] R. van der Hofstad and G. Slade, *A generalised inductive approach to the lace expansion*, Probab. Theory Related Fields **122** (2002), 389–430.

[42] F. den Hollander, *Random polymers*, Lectures in this Summer School.

[43] ———, *Random polymers*, Springer, Berlin, (2009), Lecture Notes in Mathematics Vol. 1974. Ecole d'Eté de Probabilités de Saint–Flour XXXVII–2007.

[44] E. J. Janse van Rensburg, *Monte Carlo methods for the self-avoiding walk*, J. Phys. A: Math. Theor. **42** (2009), 323001.

[45] I. Jensen, *A parallel algorithm for the enumeration of self-avoiding polygons on the square lattice*, J. Phys. A: Math. Gen. **36** (2003), 5731–5745.

[46] ———, *Enumeration of self-avoiding walks on the square lattice*, J. Phys. A: Math. Gen. **37** (2004), 5503–5524.

[47] ———, *Improved lower bounds on the connective constants for two-dimensional self-avoiding walks*, J. Phys. A: Math. Gen. **37** (2004), 11521–11529.

[48] I. Jensen and A.J. Guttmann, *Self-avoiding polygons on the square lattice*, J. Phys. A: Math. Gen. **32** (1999), 4867–4876.

[49] T. Kennedy, *Conformal invariance and stochastic Loewner evolution predictions for the 2D self-avoiding walk—Monte Carlo tests*, J. Stat. Phys. **114** (2004), 51–78.

[50] H. Kesten, *On the number of self-avoiding walks*, J. Math. Phys. **4** (1963), 960–969.

[51] ———, *On the number of self-avoiding walks. II*, J. Math. Phys. **5** (1964), 1128–1137.

[52] W. König, *A central limit theorem for a one-dimensional polymer measure*, Ann. Probab **24** (1996), 1012–1035.

[53] G.F. Lawler, *Intersections of random walks*, Birkhäuser, Boston, (1991).

[54] G.F. Lawler, O. Schramm, and W. Werner, *On the scaling limit of planar self-avoiding walk*, Proc. Symposia Pure Math. **72** (2004), 339–364.

[55] B. Li, N. Madras, and A.D. Sokal, *Critical exponents, hyperscaling, and universal amplitude ratios for two- and three-dimensional self-avoiding walks*, J. Stat. Phys. **80** (1995), 661–754.

[56] E.H. Lieb, *A refinement of Simon's correlation inequality*, Commun. Math. Phys. **77** (1980), 127–136.

[57] N. Madras and G. Slade, *The self-avoiding walk*, Birkhäuser, Boston, (1993).

[58] N. Madras and A.D. Sokal, *The pivot algorithm: A highly efficient Monte Carlo method for the self-avoiding walk*, J. Stat. Phys. **50** (1988), 109–186.

[59] A.J. McKane, *Reformulation of $n \to 0$ models using anticommuting scalar fields*, Phys. Lett. A **76** (1980), 22–24.

[60] P.K. Mitter and B. Scoppola, *The global renormalization group trajectory in a critical supersymmetric field theory on the lattice $\mathbb{Z}^3$*, J. Stat. Phys. **133** (2008), 921–1011.

[61] B. Nienhuis, *Exact critical exponents of the $O(n)$ models in two dimensions*, Phys. Rev. Lett. **49** (1982), 1062–1065.

[62] G.L. O'Brien, *Monotonicity of the number of self-avoiding walks*, J. Stat. Phys. **59** (1990), 969–979.

[63] G. Parisi and N. Sourlas, *Self-avoiding walk and supersymmetry*, J. Phys. Lett. **41** (1980), L403–L406.

[64] A. Pönitz and P. Tittmann, *Improved upper bounds for self-avoiding walks in Z^d*, Electron. J. Combin. **7** (2000), Paper R21.

[65] V. Riva and J. Cardy, *Holomorphic parafermions in the Potts model and stochastic Loewner evolution*, J. Stat. Mech.: Theory Exp. (2006), P12001.

[66] W. Rudin, *Principles of mathematical analysis*, 3rd ed., McGraw–Hill, New York, (1976).

[67] B. Simon, *Correlation inequalities and the decay of correlations in ferromagnets*, Commun. Math. Phys. **77** (1980), 111–126.

[68] G. Slade, *The lace expansion and the upper critical dimension for percolation*, Lectures in Applied Mathematics **27** (1991), 53–63, (Mathematics of Random Media, eds. W.E. Kohler and B.S. White, A.M.S., Providence).

[69] ———, *The lace expansion and its applications.*, Springer, Berlin, (2006), Lecture Notes in Mathematics Vol. 1879. Ecole d'Eté de Probabilités de Saint–Flour XXXIV–2004.

[70] S. Smirnov, *Towards conformal invariance of 2D lattice models*, International Congress of Mathematicians (Zürich), vol. II, Eur. Math. Soc., (2006), pp. 1421–1451.

[71] K. Symanzik, *Euclidean quantum field theory*, Local Quantum Field Theory (New York) (R. Jost, ed.), Academic Press, (1969).

[72] K. Wilson and J. Kogut, *The renormalization group and the ϵ expansion*, Phys. Rep. **12** (1974), 75–200.

Department of Mathematics, University of British Columbia, 1984 Mathematics Road, Vancouver, BC, Canada V6T 1Z2

E-mail address: brt@math.ubc.ca

Département de Mathématiques, Université de Genève, 2-4 rue du Lièvre, 1211 Genève, Switzerland

E-mail address: hugo.duminil@unige.ch

EURANDOM, Technische Universiteit Eindhoven, PO Box 513, 5600 MB Eindhoven, The Netherlands

E-mail address: j.a.goodman@tue.nl

Department of Mathematics, University of British Columbia, 1984 Mathematics Road, Vancouver, BC, Canada V6T 1Z2

E-mail address: slade@math.ubc.ca

Selected Published Titles in This Series

15 **David Ellwood, Charles Newman, Vladas Sidoravicius, and Wendelin Werner, Editors,** Probability and Statistical Physics in Two and More Dimensions, 2012

14 **David A. Ellwood and Emma Previato, Editors,** Grassmannians, Moduli Spaces and Vector Bundles, 2011

13 **James Arthur, James W. Cogdell, Steve Gelbart, David Goldberg, Dinakar Ramakrishnan, and Jiu-Kang Yu, Editors,** On Certain L-Functions, 2011

12 **Alan Carey, David Ellwood, Sylvie Paycha, and Steven Rosenberg, Editors,** Motives, Quantum Field Theory, and Pseudodifferential Operators, 2010

11 **Etienne Blanchard, David Ellwood, Masoud Khalkhali, Matilde Marcolli, Henri Moscovici, and Sorin Popa, Editors,** Quanta of Maths, 2010

10 **Manfred Leopold Einsiedler, David Alexandre Ellwood, Alex Eskin, Dmitry Kleinbock, Elon Lindenstrauss, Gregory Margulis, Stefano Marmi, and Jean-Christophe Yoccoz, Editors,** Homogeneous Flows, Moduli Spaces and Arithmetic, 2010

9 **Reza Akhtar, Patrick Brosnan, and Roy Joshua, Editors,** The Geometry of Algebraic Cycles, 2010

8 **Henri Darmon, David Alexandre Ellwood, Brendan Hassett, and Yuri Tschinkel, Editors,** Arithmetic Geometry, 2009

7 **William Duke and Yuri Tschinkel, Editors,** Analytic Number Theory, 2007

6 **Nigel Higson and John Roe, Editors,** Surveys in Noncommutative Geometry, 2006

5 **David A. Ellwood, Peter S. Ozsváth, András I. Stipsicz, and Zoltán Szabó, Editors,** Floer Homology, Gauge Theory, and Low-Dimensional Topology, 2006

4 **James Arthur, David Ellwood, and Robert Kottwitz, Editors,** Harmonic Analysis, the Trace Formula, and Shimura Varieties, 2005

3 **Michael Douglas, Jerome Gauntlett, and Mark Gross, Editors,** Strings and Geometry, 2004

2 **David Hoffman, Editor,** Global Theory of Minimal Surfaces, 2005

1 **Atish Dabholkar, Sunil Mukhi, and Spenta R. Wadia, Editors,** Strings 2001, 2002